STRUCTURAL ADHESIVES

DIRECTORY AND DATABOOK

Compiled by Bob Hussey and Jo Wilson
RJ Technical Consultants
Charente Maritime
France

CHAPMAN & HALL
London · Weinheim · New York · Tokyo · Melbourne · Madras

Published by Chapman & Hall, 2–6 Boundary Row, London SE1 8HN, UK

Chapman & Hall, 2–6 Boundary Row, London SE1 8HN, UK

Chapman & Hall GmbH, Pappelallee 3, 69469 Weinheim, Germany

Chapman & Hall USA, 115 Fifth Avenue, New York, NY 10003, USA

Chapman & Hall Japan, ITP-Japan, Kyowa Building, 3F, 2-2-1 Hirakawacho, Chiyoda-ku, Tokyo 102, Japan

Chapman & Hall Australia, 102 Dodds Street, South Melbourne, Victoria 3205, Australia

Chapman & Hall India, R. Seshadri, 32 Second Main Road, CIT East, Madras 600 035, India

First edition 1996

Printed in Great Britain by St Edmundsbury Press, Bury St Edmunds, Suffolk

ISBN 0 412 71470 1

A Catalogue record for this book is available from the British Library

∞ Printed on acid-free text paper, manufactured in accordance with ANSI/NISO Z39.48-1992 (Permanence of Paper).

PREFACE

The overall aim of this book is to aid the adhesive selection process. It provides concise, relevant technical data and details of sources for a wide range of synthetic polymer-based adhesives used for structural bonding in many industry sectors.

The increasing use of adhesive bonding for significant load-bearing applications reflects its advantages over traditional joining methods; particularly in the area of rapid, low-cost, automated production processes. With recent developments in materials technology and processing there are now many materials that can only be successfully joined using adhesives. There is also an increasing tendency for new adhesives to be developed in co-operation with materials producers or major industrial users.

For the design of bonded joints, basic information is required on available adhesives and their characteristics. This provides an indication of the capabilities of potential adhesives and also serves as a basis of comparison between commercially available products. However, finding and acquiring this information can be a difficult and very time consuming process. This book will greatly simplify this process for the reader.

The number and variety of commercially available adhesives are now very large and continue to increase. Adhesive manufacturers are active in developing new formulations to meet a widening demand, and in modifying current formulations to meet new environmental legislation. This is reflected in the reduction or removal of known hazardous chemicals, solvents and asbestos fillers and a general trend towards water-based adhesives.

Users of adhesives must determine the suitability of particular products for each intended use. Most manufacturers will provide assistance with this. The review section of this book emphasises that all aspects of the use of an adhesive should be fully evaluated before finalising a design.

THE STRUCTURE OF THIS BOOK

The book is divided into four main parts:

- ❑ **Part One - Review**: Provides a broad introduction to the subject of adhesives and adhesive bonding:
- ❑ **Part Two - Directory**: Contains comprehensive listings of manufacturers, their product types, tradenames and contact details for all those included in this edition.
- ❑ **Part Three - Data**: Consists of nine main sections, organised alphabetically by base-chemical constituent; often used by manufacturers to class their products and one basis for users to select an adhesive. A further section contains miscellaneous structural adhesives. Each section contains:
    - **Comparison Tables** - At-a-glance comparison of adhesive properties and characteristics.
    - **Data sheets** - Individual data sheets for each of the products.
- ❑ **Part Four - Appendices:**
    - A - Test Specifications and Standards.
    - B - Terminology: Definitions.
    - C - Glossary: Multilingual key words.
    - D - Conversion Factors.
- ❑ **Index:**
    - Product codes (giving manufacturer & type).
    - Manufacturers and adhesive types.

Note

The characteristics and adhesive property data in this book have been compiled from technical data sheets obtained from manufacturers, to whom we extend our gratitude. Great care has been taken to ensure that the information is accurately represented here, but no responsibility can be taken for errors or omissions.

For this first edition we have tried to include the largest possible number of manufacturers, suppliers and products. Inevitably some will have been missed, so any comments on alterations or additions would be very useful. The final page of the book shows how to contact us.

Acknowledgements

We would like to thank Roger Davies for acting as our leg-man in difficult circumstances, and Dave Bashford for bearing an endless stream of enquiries with fortitude. Thanks are due to many others who helped by replying promptly to our queries.

Thank you.

CONTENTS

Part 1: REVIEW

INTRODUCTION

The adhesive bonding of materials has been found in all of the ancient civilisations. The adhesives used were naturally occurring substances; gum arabic, bitumen, egg, starch pastes, protein from cheese or blood and gelatine from animal parts; hoof, horn and bones. These natural adhesives remained dominant until the twentieth century, but, although providing adequate strength, they did suffer from environmental degradation and had severe shortcomings when bonding synthetic materials and metals.

The demand for adhesives with improved durability, together with advances in polymer chemistry, saw the first range of synthetic adhesives becoming commercially available in the 1950's. Many of these offered a considerable improvement over natural adhesives, but still had some drawbacks - such as unstable curing and inconsistent bond strength. Great improvements were obtained by chemical modifications to the adhesive formulations. At the same time adhesion studies showed the great importance of surface preparation in achieving consistent bond strengths.

In recent decades the number and variety of commercially available adhesives has increased rapidly, and adhesive bonding has become a common joining method in most industry sectors. This is partly a result of the use of new materials and production processes, and partly a response to the economic and engineering benefits to be gained from replacing traditional mechanical joining methods, e.g. increased production rates, reduced machining operations, reduced corrosion. To ensure that the required bond performance (strength and durability) is achieved, design and testing procedures have evolved. The design of bonded joints is normally evaluated by a combination of analysis and testing. These are often specific procedures generated within an industry sector. Test standards and specifications for adhesive characteristics are now numerous.

Many current production materials can only be successfully joined using adhesives. The emergence of fibre-reinforced plastics as structural materials has meant that adhesive bonding has become an established production and assembly technique. Here, drilled holes for mechanical fasteners may affect the inherent strength of the material whereas bonded joints do not. The structural uses of adhesives are now very diverse, from the bonding of shoe uppers to soles to the assembly of aerospace structures and all points in-between.

This book concentrates on those synthetic adhesives which are required to have structural characteristics, i.e. they perform a significant load-bearing function within their given application. Classed by base-chemical constituent (a common method used by manufacturers to describe their products), and with comprehensive cross-referencing, each section of this book provides information and property data for comparison purposes. Details of manufacturers world-wide and their product ranges are listed in the directory.

BASIC CLASSES OF ADHESIVE

The term 'Structural Adhesive' covers many basic chemical classes of adhesives. Also, within chemical classes there may be both structural and non-structural adhesives, e.g. not all polyurethane adhesives are structural; some are sealants having some adhesion characteristics. As such, information is restricted to those types of adhesives used, or promoted for use, in structural applications. In some cases, the dividing line between sealant and adhesive becomes 'blurred'. For example; silicone-based materials may be used as adhesives or as sealants. In many applications, the ability of an adhesive to also 'seal' a joint against the environment is an important advantage, avoiding leakage or crevices where corrosion could occur.

What follows is a broad overview of the characteristics of the chemical classes of structural adhesives described in this book. Technical terms have been kept to a minimum such that it serves as an introductory guide rather than an in-depth description of the complexities of adhesive formulation.

Acrylic Adhesives

Acrylic-based polymers were first produced in Germany in 1901. Subsequent work has produced a range of acrylic-based adhesives with the following general features:

- Adhesion to a variety of substrates.
- Water resistance.
- Durability.
- Flexibility (with modified formulations).
- Good low temperature properties.
- Excellent optical properties.
- Low toxicity.

The main forms and uses of acrylic adhesives are:

- Water-borne emulsions and solvent solutions - used in non-structural applications, such as pressure-sensitive tapes and labels. Information on these types, of which there are many, is not given in this book. Those specifically for wood bonding are included.
- 100% solid reactive adhesives - used in structural bonding applications in the engineering industries for the bonding of metals and non-porous plastics.

Engineering acrylic adhesives are two-part systems which, when mixed or activated, polymerise (cure) to form an impact-resistant plastic layer which is well adhered to the surfaces of the adherends. The activator for the curing process may be a chemical (either mixed with the adhesive or applied to the adherend surfaces prior to the adhesive), or, for some varieties, UV or electron-beam radiation. The toughened or modified engineering adhesives have been widely used since the 1980's in vehicle construction, wood to metal bonding, aerospace applications, panels and computer equipment.

Anaerobic Adhesives

These adhesives were first produced in the 1940's. Their main feature being that they cure rapidly in the absence of oxygen to produce strong bond, i.e. the conditions found between adherends and in crevices. Other features are:

- Room temperature curing.
- Chemical resistance to a variety of substances, e.g. oils and solvents.
- Low shrinkage.
- Contain no solvents.
- Excess adhesive can be easily removed from adjacent surfaces.

Early anaerobic adhesives suffered from unstable curing, mainly as a result of the problems of maintaining the oxygen-level in the liquid adhesive prior to its use. This was subsequently solved by modification of formulations. Current versions no longer suffer from this problem, or the associated need for highly-controlled storage.

One of the most important applications for anaerobics is the thread-locking of mechanical fasteners which might loosen with vibration. A range of strengths are available to allow disassembly and use on a wide range of thread diameters. Higher strength versions may be used for the permanent fixing of rotating components, e.g. gears and bearings - avoiding the need for machined keyways and preventing mechanical fretting and crevice corrosion problems. Anaerobics are also widely employed in the assembly of flat load-bearing panels where ease of use and rapid curing has benefits over the more traditional classes of adhesives. Other formulations may be used for the filling and sealing of surface imperfections in manufactured parts and for gasket jointing. Adhesive primers, heat, metal ion catalysts and UV radiation can all be used to accelerate curing.

Cyanoacrylate Adhesives

Although the first alkyl cyanoacrylates were synthesised in 1947, they only appeared as adhesives in 1958. The first generation adhesives had some limitations because of brittleness (leading to poor tolerance to impact, peel, heat and thermal cycling). The cure was very sensitive to the nature of the surface to be bonded (porous and acidic surfaces were difficult). Available moisture also greatly affected curing. The adhesives showed poor gap filling, 'fogging' of nearby shiny or transparent materials and had a pungent odour. Many of these problems have now been solved by modifying the formulations. Toughened cyanoacrylates with rubbery additions have much improved mechanical properties. The addition of surface activating agents has reduced the cure sensitivity. Gelled varieties are now available which improve the handling and gap filling characteristics. Their main features are:

- One-part adhesive.
- 'Instant' bonding - curing achieved under ambient conditions (no additional heating required). Cure is activated by atmospheric moisture.
- Bond a wide variety of adherends.

Their use for joining small parts is now widely accepted in many industries, particularly in applications where an 'instant' bond is needed, avoiding the need for clamping, heating, etc.

Cyanoacrylates are now widely available for domestic use. When first introduced, their low viscosity, together with the unusual phenomenon of an 'instant' adhesive and good adherence to skin, resulted in some users finding themselves stuck to the parts. With public experience and new formulations, this problem has diminished.

Epoxy Adhesives

Epoxy-based adhesives were first introduced in the 1950's. Since then they have become widely used, particularly in the aerospace, vehicle and boat building industries, for applications requiring high strength and endurance.
The main features of epoxy-based adhesives are:

- Good adhesion to many different substrates.
- A range of mechanical properties: depending on formulation and curing.
- A wide range of cure characteristics.
- Low shrinkage.
- No by-products evolved during cure (except with phenolic modified types).

In addition to the various curing agents used, elastomeric modifiers have been added to improve peel strengths and moisture resistance. Newer varieties have been formulated to be 'tougher', i.e. having improved impact resistance and higher fracture surface energies, whilst retaining stiffness and hot strength. Epoxy-based adhesives are used extensively for bonding metals and fibre-reinforced plastic materials (FRP). Many FRP materials use epoxy-based resins as the matrix phase, showing an obvious compatibility with these adhesives.

Epoxy adhesives may have a variety of cure temperatures, from room temperature (RT) up to 175°C, and are available as pastes (one or two-part) or films. Two-part, RT curing varieties are widely available for domestic use.

Although epoxy adhesives represent a relatively small proportion of the total adhesives market, there are perhaps more varieties of formulations available from more sources than for any other adhesive class. For many years, they have dominated the high-performance adhesives market, but the desire for increased service temperatures has led to the development of new adhesives based on imide polymers (Polyimides and Bismaleimides).

Imide-Based Adhesives

This term is used within this book to group together Bismaleimide and Polyimide adhesives. Both types evolved from the development of resins for fibre-reinforced composites. BMI and PI composites have higher temperature resistance than those based on epoxy resins. Therefore an adhesive compatible with both the polymer matrix and the higher service temperatures was required. There are currently few commercial products available. The majority of these are designed for aerospace composite bonding applications, although metal bonding is also possible.

Bismaleimides

The main features of bismaleimide adhesives are:

- ❑ High-temperature capability (200°C to 230°C).
- ❑ Excellent electrical properties, e.g. high-energy radome applications.
- ❑ No volatiles evolved during cure (pure addition reaction), which simplifies processing and reduces porosity.
- ❑ Poor peel resistance, owing to their stiffness.

Modified BMI adhesives are now available with improved peel resistance.

Polyimides

The main features of polyimide adhesives are:

- ❑ High-temperature capability (up to 300°C).
- ❑ Excellent electrical properties, e.g. radome applications.
- ❑ Evolution of volatile materials during cure, so extraction or high processing pressures are needed.

Polyimide adhesives are mainly available as films, although some formulations are beginning to be available as pastes.

There are two main classes of polymerisation reaction (curing mechanism) for these adhesives and their processing characteristics are different:

- ❑ **Condensation**: These were the first type to be introduced and can withstand temperatures of 260°C to ~315°C. Volatiles (often water molecules) are evolved during the cure, so processing is normally carried out under vacuum to remove these from the curing polymer. The cure temperature is typically ~177°C followed by a post-cure at ~290°C.
- ❑ **Addition**: The first stage of curing, carried out at ~200°C, is a 'condensation' type. At the end of this the adhesive becomes a thermoplastic. A higher pressure and temperature stage (~290°C) is then used to consolidate the bondline and finalise the cure. These adhesives have a slightly lower oxidation resistance than the 'condensation' types, which limits their service temperature to ~260°C.

Phenolic Adhesives

Phenolic resins are synthesised from phenol (or phenol substitutes) and formaldehyde. They first appeared in the early 1900's, since when Bakelite has been a widely recognised name. Phenolic-based formulations are used as adhesives for structural applications, notably for wood products. Their main features are:

- ❑ High mechanical strength.
- ❑ Good solvent and water resistance.
- ❑ High thermal stability.
- ❑ Flame retardant.

The major uses of phenolic adhesives are:

- ❑ Wood, plywood, chip- and block-board manufacture, laminated beams and the joints between them.
- ❑ Abrasives - bonded abrasives (grinding wheels), coated abrasives (sand paper, discs and belts).
- ❑ Foundries - binders for moulding sands.
- ❑ Friction composites - binders for brake linings, clutches and automatic transmissions.
- ❑ Fibre-reinforced composite manufacture - with all of the major types of reinforcing materials. Used widely in the electronics industry, structural panelling (often flame retardant applications) and decorative laminates for furniture.

Phenolics are also used as additives in other synthetic adhesives, e.g. epoxies and 'contact' neoprenes, where they lend some level of the above properties to the adhesive.

Polyurethane Adhesives

This class contains both polyurethane- and isocyanate-based adhesives. The development of polyurethane adhesives is credited to the research of German scientists during World War II. Their main features are:

- ❑ Good adhesion to a variety of substrates.
- ❑ Good chemical resistance to oils, solvents, etc.
- ❑ Tough, flexible bonds.

- Available in many forms, e.g. liquids, dispersions, films, powders, etc.

Some of these products are described as 'hot-melt' or 'melt-applied', which may be misleading. They are provided as solids which must be heated to a molten state to be applied, however, the final properties are obtained by a subsequent room-temperature moisture curing mechanism rather than just solidification (true thermoplastics). Within this book, such products are classed in the polyurethane section, not as thermoplastic hot-melts.

The industrial uses of polyurethane adhesives are many and diverse. The major market sectors for structural applications are in the manufacture of furniture and in the construction, footwear and vehicle industries.

Silicone Adhesives

Silicones are synthetic polymeric materials which may be formulated to have a very wide range of properties. All have a basic polysiloxane structure, giving them a combination of organic and inorganic chemical properties. These properties have made them of interest for adhesive and sealing applications. Their main features are:

- A wide range of possible viscosities.
- Good flexibility.
- Very good resistance to high and low temperatures (-115°C to +265°C).
- Resistance to UV and IR radiation.
- Resistance to oxidation.

Depending on the formulation, curing is by:

- Exposure to atmospheric moisture (one-part systems), or
- Addition of cross-linking agents (two-part systems).

Some have an acetoxy cure mechanism, releasing acetic acid during cure which may cause corrosion problems. Others are described as 'neutral' curing. Most are cured at RT for several days (RTV - Room Temperature Vulcanisation), although higher temperatures may be used to reduce cure times.

Silicone adhesives/sealants have good resistance to water and weathering (UV and chemical attack). This, coupled with their high flexibility and good gap-filling makes them suitable for many applications in the construction, marine and vehicle assembly industries.

Thermoplastic Hot-Melt Adhesives

All thermoplastic adhesives require heating in order to melt the polymer, hence the name 'hot-melt'. Unlike the majority of structural adhesives, there is no chemical curing reaction. Initially based on natural substances, synthetic hot-melt adhesives first became available in the 1950's. A great many polymers exist as thermoplastics, providing a wide range of potential hot-melts. Each adhesive formulation contains a cocktail of compounds to modify the properties. These include tackifiers to promote adhesion and waxes to reduce viscosity and cost.

The general features and advantages of hot-melts, compared with conventional adhesives, include:

- Rapid setting times (increased production rates).
- A wide number of formulations available with different properties, e.g. melting temperature and viscosity. Most manufacturers will tailor properties to meet particular applications. Grades may also be blended by the user.
- Solvent free.
- Available in many solid forms, e.g. pellets, powders and granules, sheets and films, filaments, rods and billets.
- Ease of use (automated dispensing equipment or hand-held glue guns).
- No special storage conditions needed, i.e. long-term stability at ambient temperatures.

Within the 'hot-melt' class, the main types and base materials can be grouped as:

- **Low-performance** - based on Polyolefin or ethylene-based materials, e.g. EVA. These are widely used in the packaging industries. EVA is well-established for paper-based adherends, e.g. packaging, book-binding and textiles, with some, mainly semi-structural, uses in furniture manufacture.
- **High-performance** - based on Polyesters or Polyamides. These are a development of the high-performance hot-melts for synthetic fibres, e.g. Nylon and Polyester. The fibre polymers melt at too high a temperature and set too rapidly to be used directly as adhesives. This has led to the development of other forms (co-polymers) to be used as adhesives. These have good adhesion to a variety of surfaces, including wood, textiles, plastics and metals.

Some polyurethane adhesives are applied in a molten form but then undergo a moisture or reaction cure. This type (often called hot-, warm-melt or melt-applied) are not included in the thermoplastic hot-melt section. (See: Polyurethane)

Miscellaneous Adhesives

The 'Miscellaneous' section in this book gives information on other classes of adhesives used for structural bonding and products of unknown chemical base. Some of these are recent developments with a limited number of commercially available products, e.g. Cyanate Esters. The materials included have a very wide range of characteristics and applications. A brief description of some of them is given here.

Cyanate ester

These are a recent development for bonding high-performance, fibre-reinforced composites with a cyanate ester resin matrix. They are intended for aerospace applications requiring improved high temperature performance and moisture resistance.

Poly vinyl acetate

First available commercially in the 1930's, PVA became a widely accepted substitute for the cheaper hide-based glues in the 1940's. Easier to use and adaptable to automation,

PVA first found uses in packaging and paper converters. This was followed by the introduction of the 'white' glues for wood products. Other applications include textile bonding, building and construction industries.

Note: The term PVAC is used within this book to cover PVA and other co-polymer forms.

The main advantages of PVAC include:

- ❑ Adhesion to a wide variety of surfaces - cellulose-containing (paper and wood) and formulations for bonding plastic films and metal foils to paper and wood.
- ❑ High cohesive strength and toughness.
- ❑ A range of viscosities - applied by roller, spray or extrusion.
- ❑ Rapid setting - in emulsion types, the loss of a small amount of the water initiates rapid setting.
- ❑ A wide range of possible formulations - often including modifying agents such as plasticisers, tackifiers, wetting agents, setting agents, biocides, extenders and additions to alter the viscosity or rheology.
- ❑ A range of formulations showing differing solvent and water resistance.
- ❑ Good mechanical stability and machining characteristics.
- ❑ Resistance to degradation by oxygen, ozone and UV radiation.
- ❑ Low toxicity - some approved for use in food-packaging.

Other adhesives included in this group are:

- ❑ **Poly vinyl acetals** - formulated from the chemical components of PVA and polyvinyl alcohol (PVOH - also an adhesive in it's own right).
- ❑ **Poly vinyl butyrals** - which have good adhesion to glass. These are widely used in laminated 'safety' glass products, e.g. windscreens and architectural shaped-glass.

Plastisols

These are based on dispersions of plasticised PVC and cure on heating. The curing mechanism involves the absorption of the liquid carrier to cause gelling. A strong solid-phase is then produced on cooling. These adhesives are generally thought of as semi-structural, e.g. low resistance to peel and cleavage, creep under prolonged static loads, but they may show reasonable resistance to cyclic loading.

Plastisols are used in automotive assembly for lower performance applications, e.g. the bonding of panel reinforcements. An advantage here is that they are capable of bonding oily surfaces; some absorb the oil film during cure, reducing the required surface preparation. Also, the adhesive may be formulated such that it cures during the paint-stoving process.

Ceramic-based

There are many formulations of inorganic adhesives. Often they are one- or two-part paste or cement-like compounds for specific applications. The general areas of use are:

- ❑ Very high temperatures, i.e. 400 to ~1600°C. (Exceeding the capabilities of any polymer-based materials).
- ❑ Very aggressive chemical environments, e.g. the construction of acid reservoirs.

The composition of the ceramic phase largely dictates the temperature and chemical resistance of the adhesive.

Typical applications include the bonding of:

- ❑ Thermal insulation materials, to themselves or to metals, glass or wood.
- ❑ Refractory materials, e.g. kiln and furnace assembly.
- ❑ Foundry moulds and other liquid metal handling components.
- ❑ Teeth (dental cement) and bone.

ADHESIVES USED IN VARIOUS INDUSTRIES

Table 1 summarises the main industries using adhesives in structural applications and the classes of adhesives which are most frequently used.

Table 1 Adhesives Use by Industry Sector

	Acrylic	Anaerobic	Cyanoacrylate	Epoxy	Imide	Polyurethane	Phenolic	PVAC	Silicone	Hot-melt
Construction	●			●		●	●	●	●	●
Textile	●					●		●		●
Packaging	●		●			●		●		●
Wood Products				●		●	●	●		
Automotive	●	●	●			●			●	●
Electrical	●	●	●	●					●	●
Furniture	●			●		●		●		●
Aerospace				●	●		●		●	

GUIDE TO ADHESIVE SELECTION

Each bonding application has a set of requirements, mechanical, environmental and economic, which define the characteristics of the desired joint. These must be met by a combination of design, materials and process selection. A major element in this is the selection of an appropriate adhesive. In order to produce a successful adhesive bond, i.e. one which adequately fulfils all of the requirements, all of the contributing factors must be fully considered during the design development stage. The design of bonded joints is fundamentally different from that for other joining methods. Efficient and economic bonds can only be achieved by considering this fact at all stages. It is rarely acceptable to use adhesive bonds as a direct replacement for another joining method in an existing design.

Adhesive selection is ultimately a trade-off between:

- Adequate performance, and
- Cost and ease of use.

Adhesive manufacturers have worked hard to dispel the commonly-held belief that there is a single adhesive which can be used to bond any type of material. The transfer of loads through a bonded joint depends on complex interactions between the way in which loads are applied (load injection) and the mechanical properties of both adherends and the adhesive. This is further complicated by the effects of the temperature and chemical environments found in service. The selection of an adhesive (particularly for applications operating close to the limits of adhesive performance) must balance these, often conflicting, demands. This, for example, is one of the major reasons for the very large range of commercially available epoxy-based adhesives with slightly differing properties.

The failure of a bond is often blamed on the adhesive. In practice, poor bonding more usually results from:

- Inadequate design.
- The selection process which may have missed or underestimated one of the requirements.
- Poor surface preparation.
- Defects arising during the bonding process.

To assist in the selection process, manufacturers provide comprehensive guidance on the use of their products - including:

- Appropriate adherends.
- Surface preparation techniques.
- Curing conditions.
- Handling and safety aspects of the adhesives.

It is always advisable to contact the manufacturers to obtain their advice and full technical data.

Computer-based 'Adhesive Selector Guides' are now becoming available. Such programs aim to assist the adhesive selection process, taking into account the joint design and requirements. The programs often take the form of an on-screen questionnaire describing the application and any particular requirements. This information is then used to search a product database and list adhesives which meet the requirements. Such expert systems may also provide other functions relating to design and bonding.

Some examples of currently available computer programs are:

- **PAL** (Permabond Adhesive Locator) covering both Permabond products and other manufacturers. Available from Permabond.
- **CAT** (from TWI - The Welding Institute, UK) based on PAL, (with the co-operation of Permabond) and developed further to be an expert system; including other manufacturer's products, surface treatments and a stress analysis package. The program has been evaluated by various organisations.
- **QUALSIAT** (from TWI - The Welding Institute, UK) developed to meet the continuous monitoring and compliance with documented procedures requirements of an ISO 9000 Special Process (within which adhesive bonding is classed). Expanded, during development, to include design factors in addition to production engineering items.
- **JOINEXCELL** (from AJI - The American Joining Institute). An adhesive selection expert system for use with composite materials. The system consists of a job-planner subsystem which creates a joint schedule, complete joint description, adhesive selection, surface preparation, etc. Also a recorder for Quality Assurance criteria. JOINADSELECT is part of the expert system and identifies appropriate adhesives from 15 major classes of engineering adhesives, given information about adherends and adhesives. Databases for composite adherend properties are included.

Factors to be Considered

This section lists some common factors involved in adhesive selection.

Materials to be bonded

- These are generally termed adherends. A bonded joint may consist of one or more types of adherend material. Where the adherends are materials of a very different nature, e.g. metal to plastic, or metal to glass, the adhesive must also act as a transition layer. For example, where a large difference in thermal expansion coefficient exists between adherends, more flexible adhesives are normally used to allow differential expansion without risking failure.
- Manufacturers will give advice on which materials can be bonded by particular adhesives. In Table 2 some broad indications by adhesive class are given for common adherend combinations.
- The type of adherend is only one of the influences governing the type of adhesive which should be used. The final choice is always a trade-off between a large number of, often conflicting, design factors.

Table 2 - Examples of Adhesives Used to Bond Different Materials

	Textiles	Composites	Glass & Ceramics	Rubber	Plastics	Wood	Metals
Metals	☆✠	●■□◆	●■□	■□✧	●■□Λ	●□Λ💧	●○■□☆Λ✧
Wood	●💧✠	●■□💧	■□💧	●Λ✧	□☆Λ💧	□Λ💧	
Plastics	●💧	●□💧✧	●□☆💧	■□☆Λ	●■□Λ		
Rubber	●☆💧	□☆✧	●■□☆💧✧	☆✧			
Glass & Ceramics	●☆💧✠	■□✧	●■□				
Composites	●☆💧	□◆★Λ					
Textiles	💧✠						

Key:
- ● Acrylic
- ○ Anaerobic
- ■ Cyanoacrylate
- □ Epoxy
- ◆ Imide (polyimide & bismaleimide)
- ★ Polyester
- ☆ Polyurethane
- Λ Phenolic
- 💧 Polyvinylacetate
- ✧ Silicone
- ✠ Thermoplastic Hot Melt

Joint design

❑ **Adherends**

- Mechanical properties and manufacturing method.
- Surfaces of components may have degraded properties compared with the bulk material. These need to be removed/modified prior to bonding.

❑ **Surface Preparation**

- Many methods available to promote adhesion and reliable joint strengths.
- Depends on the type(s) of adherends.

❑ **Configuration**

- Arrangement of adherends in the joint region (See: Types of Joints).

❑ **Loading**

- Strength (tensile and shear).
- Stiffness (tensile and shear).
- Peel strength.
- Impact resistance.
- Environmental factors (vibration, changes in temperature, chemical resistance).

❑ **Durability**

- Required life for given operating environment.

❑ **Testing**

- Many specifications and standards (industry, national, international or military) exist for adhesive properties and bonded joints. (See also: Appendix A - Test Specifications).

❑ **Inspection**

- Adherends and materials used in bonding process.
- Finished bonded structures. Current non-destructive methods cannot assess joint strength. Numerous methods exist to examine bondlines for voids and defects.

Adhesive characteristics

❑ **Storage/Shelf life**

- Some require cold/freezer storage. Others can be stored under ambient conditions.

❑ **One-part**

- Will cure or set without additional chemicals (hardeners or catalysts).

❑ **Two-part**

- Normally require metering and thorough mixing before applying. Some have an activator which is applied to one surface, the adhesive to the other.

❑ **Physical state**

- Thermosetting adhesives available as pastes, liquids or films.
- Thermoplastics 'hot melts' available in many forms, e.g. filaments, granules, powders, sheet or films.

❑ **Viscosity**

- Wide range available. Some adhesives are formulated to flow easily, others are thixotropic or filled to be non-sag. (See also: Viscosity Guide, Page 9).

❑ **Gap filling**

- Can be related to viscosity and thixotropy.
- Low shrinkage to prevents voidage or cracking during cure of thicker layers.

❑ **Bondline thickness**

- Some adhesives contain spherical fillers or thin textiles to provide a controlled bondline thickness.

❑ **Use Temperature**

- Change in adhesive properties with temperature.
- All adhesives have a glass transition temperature (Tg) at which the mechanical properties of the adhesive change suddenly.

❑ **Moisture sensitivity**

- Some adhesives are sensitive to moisture. Working environments should control humidity to prevent irregular curing or reduced final properties.
- Other adhesives require moisture to initiate curing.
- Hot-melts are largely unaffected by moisture.

❑ **Packaging**

- Available quantities to minimise waste.
- May also provide the mixing or application equipment (e.g. cartridges).

Cost

This is often calculated as 'in-place' cost per unit area of cured/set bondline.

❑ **Adhesive**

- Generally low compared with other joining methods.
- Specialised adhesives, produced in low volumes, may be expensive.
- Adhesive costs are often quoted by weight for liquids, pastes and solid forms or by area for sheets and films.

❑ **Equipment**

- Disposable adhesive placement tools - often as part of packaging.
- Multi-function equipment for placement and clamping is used for high-volume production and high production rates.
- Heating and other curing equipment (ovens, lamps, UV sources). For some heat-cured adhesives, their cure temperature can be chosen to match another stage in the manufacturing process, e.g. adhesives which cure during the paint-stoving operation of automobile manufacture.

❑ **Clean-up**

- Time and materials (solvents, etc.) needed to clean equipment and assembled parts.

❑ **Waste**

- Loss from excess applied adhesive during assembly.
- Excess mixed or open adhesive.

❑ **Labour**

- Time to prepare, place and assemble parts to fully cured/set.

Application method

❑ **Manual**

- Many adhesives are applied and items assembled by hand.
- Gun-type applicators for thermosetting and hot-melt adhesives, cartridge dispensers.
- Clamping jigs for assembly.

❑ **Machine**

- Automated adhesive dispensing systems - usually for large batches.

Health and safety requirements

❑ **Surface preparation**

- May involve strong oxidising agents, corrosive, toxic or carcinogenic agents.
- Solvents.
- Dust.
- Extraction and ventilation required.

❑ **Adhesive**

- Organic compounds in adhesives, hardeners and catalysts.
- Skin and eye contact should avoided.
- Vapours and odours.
- Extraction and ventilation required.

Cure factors

❑ **Pot life**

- Applies to liquids and pastes which start to cure on mixing or at ambient temperature (1- and 2-part adhesives). The time before the adhesive gels and becomes unworkable.

❑ **Working life or Out-time**

- Applies to one-part thermosetting adhesive films which, after warming to RT (working temperature) will start to cure slowly and become less tacky. Time allowable for placing and assembling parts (cf. pot life for pastes and liquids).

❑ **Open time**

- Time available for applying an adhesive and assembling the parts before being cured.

❑ **Handling Strength/Time**

- Time for the assembly to obtain a sufficiently advanced cure or set that it can be handled without risk of damaging the joint.
- Clamping or assembly equipment will be tied-up for at least this time hence limiting production rates or increasing equipment costs.

❑ **Vapours and Odour**

- Extraction (sometimes vacuum) required.
- If adhesives which release vapours during cure are used (aqueous emulsions, solvent solutions and condensation curing adhesives), the joint design should allow venting.

❑ **Flash point**

- Some adhesive components or chemicals used in bonding release explosive vapours at relatively low temperatures.
- Storage and handling must limit the temperature and avoid sources of ignition.

❑ **Softening (melt) temperature**

- Processing temperature for 'hot-melt' adhesives.

❑ **Cure conditions**

- Temperature.
- Pressure.
- Time.
- Other factors - moisture curing, UV curing, etc.

❑ **Plant environment**

- Limit contaminants (dust, oil, grease) on adherends.
- Humidity.
- Temperature.

Appearance

- ❑ Adhesive colour (if visible after bonding).
- ❑ No discolouration of adjacent surfaces.
- ❑ Transparency (optical applications).
- ❑ Some adhesives are not suitable for painting.

Service environment

- ❑ **Humidity**
  - Some adhesives absorb moisture which can reduce mechanical properties and long-term resistance.
  - Absorbed moisture may alter the way in which the properties of some adhesives vary with temperature.
- ❑ **Temperature**
  - High temperature - often linked to the Tg and the temperature at which an adhesive cures or softens. Properties reduce as this temperature is approached or exceeded.
  - Low temperatures - most adhesives become brittle. Some are specially formulated for use at very low (cryogenic) temperatures.
  - Repeated changes in temperature (thermal cycling) can produce thermally induced cyclic loads from different expansion/contraction of the adherends.
- ❑ **Chemical resistance**
  - Water.
  - Industrial solvents.
  - Oils.
  - Grease.
- ❑ **Others factors**
  - UV.
  - Vacuum.
  - Radiation.

Viscosity Guide

Table 4 shows the viscosities of some common liquids as an aid to visualising the stated properties of adhesives.

Table 4 Typical Viscosities of some Liquids

LIQUID	VISCOSITY Centipoise	Pa.s	
Water	1	0.001	(Very Low)
Machine Oil	100	0.10	
#30 Motor Oil	250	0.25	(Low)
Castor Oil	1000	1.0	
Karo Syrup	3500	3.5	(Medium)
#40 Motor Oil	4500	4.5	
Ordinary Paint	10000	10	
Chocolate Syrup	25000	25	(High)
Molasses	100000	100	

BASIC DESIGN OF BONDED JOINTS

Research studies have been conducted in all industry sectors on the design of bonded joints and the influence of this on joint strength and durability. Many have found their way into published texts for engineers. Lack of space prevents any more than a brief summary of such a large topic.

Types of Joints

The arrangement of the adherends and the direction of loading in the joint region has a very strong influence on the effective strength of the joint - see Figure 1.

Figure 1 Types of Joints - Load Transfer

Tensile

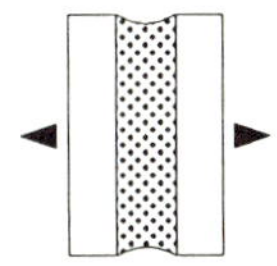

Strength relies on the adhesion to the substrate - which is heavily influenced by the surface preparation of the adherends. Thick or rigid adherends are required to avoid distortion under load, leading to peel stresses on the joint.

Tensile Shear

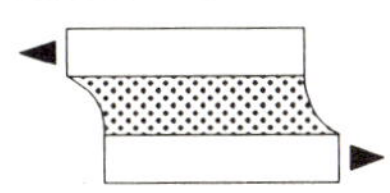

Whole of bond area is used to resist the applied load. This is the most common general type of adhesive joint, and the most efficient. Practical joint designs must avoid the tendency of adherends to distort and peel.

Cleavage (Thick sections)

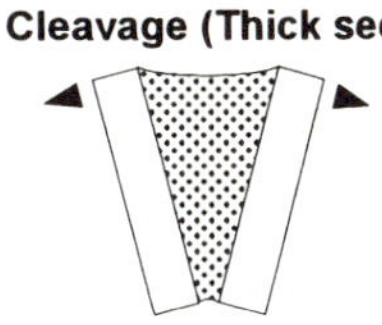

Inefficient because the bond area is not fully used for load distribution. Peel-type loading at the opening edge causes very high stress concentrations in this area.

Peel (Thin sections)

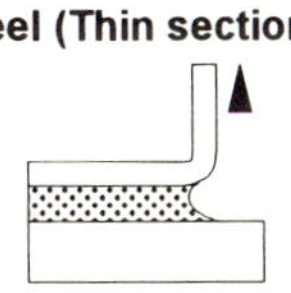

Very inefficient because all of the load is borne by a narrow strip along the opening edge of the bond area.

Torsion

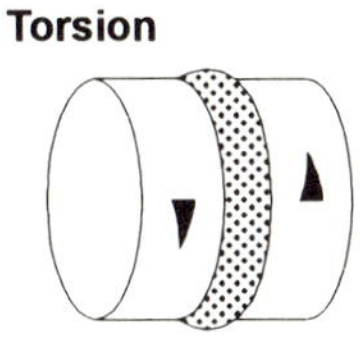

Torsional shear may be an acceptable loading mode, but adhesive butt joints should be generally avoided because the strength relies heavily on the properties of the adhesive, the bond area is often restricted and joints tend to be prone to off-axis loads.

Load transfer by tensile-shear helps to maximise the joint strength. In the simplest and most widely used form, two adherends are overlapped and bonded (single lap joint) - see Figure 2.

Figure 2 Single Lap Joint

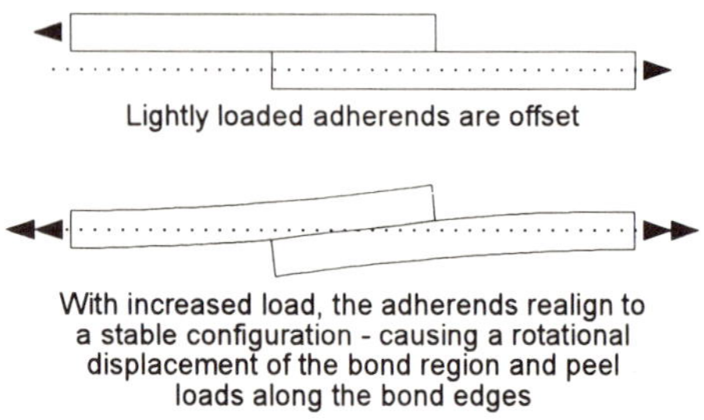

This type of joint, when loaded, can suffer from a peeling effect at the free edges of the adherends which reduces the strength. Methods of designing out or limiting peel loads are many and varied - some taper the adherends near the overlap edges, or use more complicated arrangements of adherends in the joint region, see Figure 3.

Figure 3 Joint Types for Reducing Peel Stresses

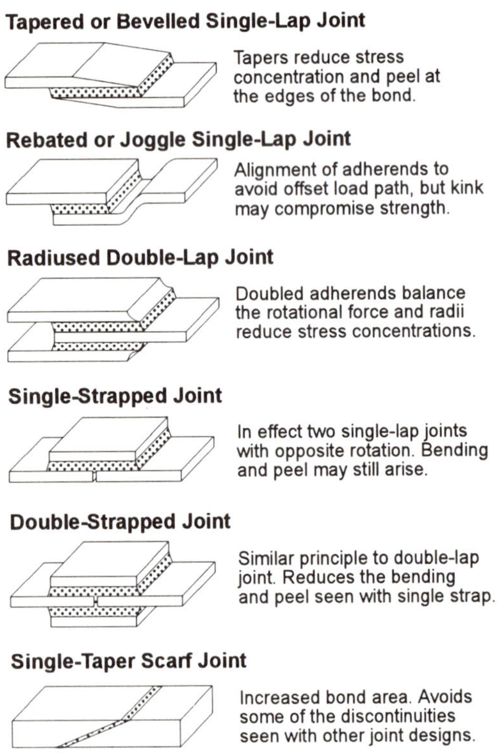

Such complications of the joint design are generally more expensive to produce, harder to manufacture and difficult to inspect. For the majority of industrial bonding applications, where strength and environmental requirements are fairly easily achievable, complex joint arrangements are usually unnecessary.

Surface Preparation

All adherends require some form of pre-bonding preparation in order to obtain a good quality bond with optimum mechanical performance and durability. The objective of surface preparation is to:

- Remove contaminants, and
- Create a chemically and physically compatible surface for the adhesive.

The extent and success of surface preparation may well be a major factor in determining the strength and durability of the finished structural bond.

Surface preparation methods normally consist of a combination of operations, each of which needs careful implementation and control. Some methods are very complex, and involve the use of several baths containing chemical reagents or solvents.

There are many methods recommended by adhesive manufacturers, government specifications, and industry standards. Some of these are currently under review because of growing ecological concern and legislation relating to the impact of industrial chemicals on the environment. Many surface preparation methods involve the use of aggressive chemicals which, if not stored, handled and disposed of properly, can be hazardous to operatives and the environment. Extreme care must be taken in the use of all chemicals. Health and safety notices and regulations must be strictly adhered to.

The subsequent handling of prepared adherend surfaces must avoid contamination. Gloves are normally recommended, as grease from the skin can be sufficient to reduce adhesion and hence reducing strength and/or durability. Pick-up of atmospheric moisture by the prepared surfaces can also affect the cured properties of the adhesive.

Degreasing

Metals are normally degreased using chlorinated organic solvents, e.g. trichloroethane, either as a vapour or liquid. Vapour degreasing baths are commonly used, where a rising vapour condenses on the material, running back into the bath and carrying contaminants with it. When this type of equipment is not available, the solvent can be wiped across the metal surface using clean, lint-free, cotton cloths. Cloths must be changed frequently to avoid build up and transfer of contaminants.

Plastics may be degreased using detergent solutions or polar solvents, such as acetone, methyl ethyl ketone (MEK) or methyl alcohol. Some solvents can attack plastics by absorbtion, solvent crazing or dissolution of the surface. Tests should be made to check that a solvent does not have an adverse affect on the plastic.

Mechanical treatments

Roughening of a smooth surface can improve bonding by a combination of cleaning and mechanical keying. Some adherends may have lower surface properties compared with their bulk as a result of their manufacturing method, e.g. oxidation of metals or the resin-rich surface layers of fibre composites. Mechanical treatments can be used to remove these external layers.

Mechanical surface treatments include:

- ❑ Abrasion - disc or sheet abrasives, e.g. emery paper, or grit blasting with fine abrasive particles. Whilst grit blasting is effective for thicker section metals, thin metals may require less aggressive methods to avoid distortion or damage.
- ❑ Vapour honing - where a high-velocity liquid or steam jet is used to carry the abrasive phase.
- ❑ Wire brushing - providing that there is no risk that broken bits of wire brush can become imbedded in the adherend.

All mechanical treatments should be followed by the removal of debris and degreasing to remove any abrasion residues.

Chemical treatments

There are a large number of standardised chemical and electrolytic pretreatments for the preparation of metal surfaces for bonding. These have been shown to greatly increase bond strengths and durability. Chemical treatments can etch the surface and convert naturally occurring oxide layers to an adherent form. Often described as anodising, bonding pretreatments vary from those anodising processes used for surface protection and aesthetic purposes. The composition of the chemical baths used varies for the type of metal being prepared. Careful monitoring is required as these treatments are generally sensitive to slight variations in the process parameters.

Primers

Primers protect prepared surfaces from the environment and may also be designed to promote adhesion. These are recommended by manufacturers for some adhesives. They are generally in the form of a liquid which is applied to the prepared surface and dried before applying the adhesive. Some contain a weak solution of adhesive or 'bond-promoting' substances. In some special forms of adhesive the primer also acts as a cure catalyst.

Bond Performance

In efficient or critical adhesive-bonding applications, it is necessary to demonstrate that a bonded joint will behave as required in service. Bond performance (the characteristics under the envisaged stress and environmental conditions) is assessed by a combination of analysis and testing. A large amount of effort has been devoted to the development of mathematical models to describe the performance of bonded joints and provide quantitative values for strength, stiffness, etc. It must be said that, despite this, it is still very difficult to produce realistic predictions of joint characteristics for all but the simplest designs and materials. It continues to be a topic of investigation.

Many books and conference papers have been published on all aspects of analysis and testing. A broad summary of the parameters determining bond performance is given here; keeping the mathematics to a minimum. Sources of more comprehensive information on the subject are given in the 'Further Reading' section of this book.

The performance of bonded joints is influenced by many interrelated factors, including:

- ❑ Adherend materials (same or dissimilar).
- ❑ Adhesive properties (brittle, toughened, etc.)
- ❑ Bondline (adhesive) thickness.
- ❑ Adhesion of the adhesive to the adherends (affected by surface preparation or contamination).
- ❑ Geometrical arrangement of the joint, and the overlap between the adherends.
- ❑ Cure history (may produce residual stresses in the joint).
- ❑ Magnitude and direction of loading.
- ❑ Environmental factors (UV, temperature, humidity, chemicals, etc.)

These factors may change in magnitude or importance during the life of the bonded joint. For example, residual stresses from curing may be partly relieved once operating stresses and temperatures have been experienced; the strength and elasticity of an adhesive may change in service as it absorbs moisture from the environment. In common with most areas of engineering, many of the more uncertain factors are coped with by including a generous factor of safety in the design.

Analysis

The combined effect of all of these factors is to create stresses and strains which are distributed through the joint region. The stress distribution is not (and should not be) even throughout the bond, i.e. in well-dimensioned joints, there is a central region of low stress and strain, necessary to resist creep. In very general terms, the ends or edges of the bond area, or the apparent 'ends' of steps (within complicated joints), carry higher stresses and strains. Because of this, the effect of defects on the strength of a bond depends on their position within the bond. In areas of low stress, the immediate effect is less marked than if the defect is located in an area of high stress. The presence and propagation of defects, which may be small areas of non bonding, or a crack produced by shrinkage or local overload, are important in studies that aim to predict the overall joint performance.

Various analytical techniques have been applied to bonded assemblies to predict the bond performance. In the simplest cases, the adhesive is assumed to behave elastically, i.e. all of the deformation occurring during loading is recovered when the load is removed. In practice adhesives have an elastic-plastic response which complicates the analysis. More brittle adhesives may have locally plastic behaviour near the crack front; toughened adhesives rely on the development of a plastic zone ahead of the crack front, so blunting the propagating defect. This is the mechanism which produces high resistance to fracture in toughened adhesives.

Calculations used for initial design may make assumptions about adhesive behaviour to allow an estimate of strength for comparison purposes. Other techniques, often involving computer modelling using finite-element stress analysis, may have provision to include non-elastic behaviour. These are most commonly used for very complex joint arrangements, e.g. those having one or more steps and tapers, or where the adherends are high-performance, fibre-reinforced composites. (See: Analytical Design Tools for some examples of commercially available data and computer programs).

Testing

Theoretical analysis is accompanied by testing. From a practical point of view, test results are used to confirm the calculations. In scientific studies, it also assists in understanding adhesion and failure mechanisms and in modifying mathematical predictions. The test methods used are based on fracture mechanics principles which have evolved steadily for bonded joints from those first applied to brittle solids by Griffith. These theories generally define three modes of loading (denoted I for tensile opening, II for in-plane shear and III for anti-plane shear). In practice, a single mode of loading in a bonded joint is difficult to achieve; mixed modes are much more frequent in real applications.

One of the basic concepts for assessing a bond is that of an energy balance. A joint will fail when the elastic energy released by the growing crack exceeds the energy required to cause a debond. The parameter G (energy release rate) is often used to indicate this, with a subscript to denote to loading mode, e.g. G_I - Energy Release Rate in tensile opening. In testing, for the (elastic) energy balance to remain valid, the test piece materials themselves must not deform permanently.

Adhesive bond failures are classified generally as either 'adhesive' or 'cohesive'.

- **Adhesive**: failure occurs at the contact surface between the adhesive and the adherend.
- **Cohesive**: failure occurs through the adhesive or through the adherend.

An adhesive failure implies that the strength of adhesion to an adherend is less than the material strength of the adhesive itself or of the weakest adherend ($G_a < G_c$). In cohesive failures the strength of adhesion is greater than the material strength of either the adhesive or adherend ($G_a > G_c$).

The energy balance assumes elastic behaviour, i.e. none of the energy in the system is lost by plastic (permanent) deformation. Many of the discussions on fracture mechanics in published texts use this basis. Where this assumption is not true, i.e. where there is a significant amount of plastic behaviour in or around the crack front, additional fracture parameters may be included, but this greatly complicates the mathematical analysis.

The test methods used for structural adhesives in fracture mechanics studies are often those based on either cleavage or torsion loading. Cleavage tests will produce a pure Mode I loading. Other test methods may be used to induce a mixture of Modes I and II.

Methods for characterising mechanical properties of adhesives are based on determining the stress at which failure occurs, usually an average stress. The properties of adhesives quoted by manufacturers are often obtained by such types of tests. They are also used in 'quality control' coupons alongside production to provide a fairly simple means of showing that the process conditions established during the design phase remain within acceptable limits. The use of these values for assessing bond performance for design purposes has, in some cases, been questioned.

Appendix A lists national, international and industry standards and specifications. This includes numerous test methods for adhesives and bonded joints.

HEALTH AND SAFETY ASPECTS OF ADHESIVE BONDING

Adhesive bonding and the associated processes are largely based on organic chemicals which are either known to have, or are suspected of having, an adverse affect on operator health or the environment. These may be present in the adhesives systems themselves or be required for surface preparation. Regulations covering the storage, handling and disposal of hazardous chemicals must be obeyed at all times.

Safety notices are provided by manufacturers. These normally state the hazardous components of the adhesive system and what action should be taken if they come into contact with eyes and skin, or are ingested. Many adhesives, and most solvents, produce hazardous vapours, so processing must be carried out with adequate ventilation. More recent developments show a marked trend towards low-vapour, water-based formulations to reduce these problems. As well as the chemical hazards, dust and debris generated during any abrasion process must be controlled. Regulations governing work-place exposure levels for the various hazards, along with Health and Safety boards and company recommendations on work-place practices, must be observed.

Adhesives are widely available and used in the home as well as the work-place. Although these tend to be less hazardous formulations, care must be taken to follow correctly the instructions provided on the containers and any accompanying notices. Storage in cool, dry conditions is normally recommended.

In addition to precautions taken against the degradation of adhesive products, it must now be recommended that all such substances should be locked away from children because of the recognised problems of solvent-abuse.

FURTHER READING

Books

Adams, R.D. and Wake, W.C.
Structural Adhesive Joints in Engineering
Chapman and Hall, London (1984)

Baker, A.A. et al (Editors)
Bonded Repairs for Aircraft Structures
Martinus Nijhoff Publishers (1987)
ISBN 90-247-3606-4

Department of Defense, USA.
Adhesive Bonded Aerospace Structure Repair
MIL-HDBK-337 (1982)
National Standards Association

Ellis, B (Editor)
Chemistry and Technology of Epoxy Resins
Blackie Chapman and Hall (1993)

European Space Agency
Adhesive Bonding Handbook for advanced structural materials
ESA-PSS-03-210 (Issue 2, March 1995)
ESA Publications Division
Noordwijk, NL.
ISSN 0379-4059

Hartshorn, S. A. (Editor)
Structural Adhesives - chemistry and technology
Plenum Press (1986)

Kinloch, A.J.
Adhesion and Adhesives - science and technology
Chapman and Hall, London (1987)
ISBN 0-412-27440X

Kinloch, A.J.
Durability of Structural Adhesives
Chapman and Hall, London (1983)
ISBN 0-85334-2-214-8

Kinloch, A.J.
Structural Adhesives
Chapman and Hall, London (1986)
ISBN 1-85166-002-X

Matthews, F.L. (Editor)
Joining Fibre-reinforced Plastics
Chapman and Hall, London (1987)
ISBN 1-85166-019-4

Mays, G. C. and Hutchinson, A. R.
Adhesives in Civil Engineering
Cambridge University Press (1992)

Mittal, K. L. (Editor)
Adhesive Joints - formation, characteristics and testing
Plenum Press (1984)

Pizzi, A. (Editor)
Wood Adhesives - chemistry and technology
Marcel Dekker Inc., New York
(Vol.1 - 1983. Vol. 2 -1989)

Pizzi, A. et al (Editors)
Handbook of Adhesive Technology
Marcel Dekker Inc., New York (1994)

Sadek, M. M.
Industrial Applications of Adhesive Bonding
Chapman and Hall, London (1987)
ISBN 1-85166-038-0

Skeist, I. (Editor)
Handbook of Adhesives,
Chapman and Hall, London (3rd Edition, 1990)
ISBN 0-442-28013-0

Stellbrink, K.
Preliminary Design of Composite Joints
DLR-Mitteilung 92-05, Stuttgart (1992)
ISSN 0939-298X

Waterman, N.A. (Editor)
Materials Selector
Chapman and Hall, London (1992)
ISBN 1-85166-605-2

2nd Edition, in press (1996)
ISBN 0-412-615800

Journals

Adhaesion (in German)
Bertelsmann Fachzeitschrifen GmbH
Munich, Germany.

Adhesion and Adhesives (in Japanese)
High Polymer Publishing Association
Kyoto, Japan.

Adhesives Age
Communications Channels Inc.
Atlanta, USA.

Assemblages Adhésifs (in French)
EDIREP, Paris, France.

International Journal of Adhesion and Adhesives
Elsevier Science Ltd., Oxford, UK

Journal of Adhesion
Gordon Breach Science Publishers Ltd.,
London, UK.

Journal of the Adhesion Society of Japan (in Japanese)
The Adhesion Society of Japan
Osaka, Japan.

Journal of Adhesion Science and Technology
VSP, Zeist, The Netherlands.

ANALYTICAL DESIGN TOOLS

ESDU details are given in the 'Useful Names and Addresses' directory.

ESDU 78042 - Bonded Joints 1
Shear stresses in the adhesive in bonded single-step, double-lap joints loaded in tension.

ESDU 79016 - Bonded Joints 2
Inelastic shear stresses and strains in the adhesive bonding lap joints loaded in tension or shear.

ESDU 80011 - Bonded Joints 3
Elastic stresses in the adhesive in single-step, double-lap bonded joints.

ESDU 80039 - Bonded Joints 4
Elastic adhesive stresses in multi-step lap joints loaded in tension.

ESDU 81022 - Bonded Joints 5
Guide to the use of data items in the design of bonded joints.

ESDU 86025 - Bonded Joints
Considerations in design against acoustic fatigue.

ESDU 92041 - Bonded Single-lap Joints:

I - Discussion of behaviour of adhesive stress distribution variation of adherend material and thickness, and of lap length.

II - Peel and shear stress distribution in adhesive under tension elastic analysis, including bending effects.

III - Shear strain/stress distribution in adhesive and stress distribution in adherends under tension or shear elastic or inelastic analysis, neglecting bending effects.

IV - Types of failure.

Part 2: DIRECTORY

MANUFACTURERS' PRODUCT RANGES

This cross-reference table indicates the classes of adhesives for each manufacturer or supplier, giving the page numbers for the corresponding product data sheets.

Manufacturer	Acrylic	An-aerobic	Cyano-acrylate	Epoxy	Imide	Phenolic	Poly-urethane	Silicone	Hot-melt	Misc.
3E	-	-	-	157	-	-	252	-	-	-
3M	69	-	122	157	-	-	252	-	325	-
American Tech. Int.	-	-	-	-	-	-	-	-	325	-
Apollo	-	-	-	-	-	-	254	-	-	-
Ashland Chemicals	-	-	-	-	-	-	256	-	-	-
Borden	-	-	-	-	-	-	-	-	325	-
Bostik	69	-	123	165	-	-	260	302	326	-
Wm. Canning	-	95	123	165	-	-	262	303	-	-
Casco Nobel	-	-	-	-	-	-	262	-	-	-
CECA	69	96	124	166	-	-	264	303	-	-
CFPI	-	-	-	-	-	-	264	-	-	-
Chemetall	-	-	-	166	-	-	-	-	-	-
Chryso	-	-	-	-	-	-	265	-	-	-
Ciba Polymers	-	-	-	167	237	243	266	-	-	-
Conap	-	-	-	-	-	-	266	-	-	-
Cytec	-	-	-	174	237	243	-	-	-	358
Degussa	70	-	125	178	-	-	267	-	-	-
Delo	73	97	125	178	-	-	-	304	-	-
Devcon	73	-	-	180	-	-	-	-	-	-
Dexter Hysol	-	-	-	181	238	-	-	-	326	-
Dow Corning	-	-	-	-	-	-	-	305	-	-
Dunlop	74	98	126	189	-	-	267	306	-	-
Dyckerhoff	-	-	-	191	-	-	268	306	-	-
Dymax	74	-	-	-	-	-	-	-	-	-
Electro-Seal	-	-	-	-	-	-	-	-	327	-
EMFI	-	-	-	-	-	-	268	307	-	358
EMS	-	-	-	-	-	-	-	-	331	-
Epirez	-	-	-	191	-	-	-	-	-	-
Evacor	-	-	-	-	-	-	-	-	333	-
Forbo Helmitin	83	-	128	193	-	-	270	308	339	359
Fortafix	-	-	-	-	-	-	-	-	-	361
Ed. Geistlich Söhne AG	-	-	-	-	-	-	274	-	-	364
GE Silicones	-	-	-	-	-	-	-	308	-	-
BF Goodrich	-	-	-	-	-	243	-	-	-	-
Grace Speciality Polymers	-	-	-	193	239	-	-	-	-	-
Hardman	-	-	-	198	-	-	275	-	-	-
(Henkel)	-	104	140	226	-	-	290	-	-	375
W. W. Henry	83	-	-	202	-	-	-	-	-	365
Hitachi-Kasei Polymer Co.	-	-	-	203	-	-	276	-	-	-

..ACTURERS & SUPPLIERS - ADDRESSES

3E

United States of America

3E Corporate Center
850 Glen Avenue
P.O. Box 392
Moorestown
NJ 08057-0392 - USA
Tel: +1 609 866 7600
Fax: +1 609 866 7603
Vice President: Neil R. Shearer

Andek Chemical Corp.
(Details as above)

Plastics & Resins Inc.
(Details as above)

3M

Australia

3M Australia Pty. Ltd.
Cnr. Blackburn & Ferntree Gully Rds.
Mount Waverly
Victoria 3149
Australia
Tel: +61 3 542 5222
Fax: +61 3 543 7438

3M Australia Pty. Ltd.
P.O. Box 99
Pymble
New South Wales 2073
Australia
Tel: +61 2 498 9333
Fax: +61 2 499 2818
Contact: Mr D. Powell

Belgium

3M (Europe) SA
106 Boulevard de la Woluwe
B-1200 Brussels, Belgium
Tel: +32 2 761 2211
Fax: +32 2 762 7978

3M Belgium NV/SA
Hermeslaan 7
B-1831 Deigem, Belgium
Tel: +32 722 5111
Fax: +32 762 7978

Canada

3M
1840 Oxford St. East
P.O. Box 5757, Terminal A
London, Ontario
Canada
N6A 4T1
Tel: +1 519 451 2500

Denmark

3M A/S
Fabriksparken 15
DK-2600 Glostrup
Denmark
Tel: +45 43 48 01 00
Fax: +45 43 96 85 96

France

3M France
Boulevard de l'Oise
F-95006 Cergy Pontoise Cedex
France
Tel: +33 1 30 31 61 61
Fax: +33 1 30 31 61 95

Germany

3M Deutschland GmbH
Carl-Schurz Str. 1
D-41453 Neuss, Germany
Tel: +49 2131 140
Fax: +49 2131 142649

Ireland

3M (Ireland) Ltd.
Adelphi Centre, Dun Laoghaire
Co. Dublin, Eire
Tel: +353 1 280 3555
Fax: +353 1 512 064

Italy

3M Italia SpA
P.O. Box 10411-10412
I-20110 Milano, Italy

Japan

Sumitomo 3M Ltd.
2-33-1 Tamagawadai
Setagayaku, Tokyo 158, Japan
Tel: +81 3 3709 8241
Fax: +81 3 3709 8453
General Manager: Reiji Shimura

Netherlands

3M Nederland BV
Industrieweg 24
NL-2382 NW Zoeterwoude
The Netherlands
Tel: +31 71 5450 450
Fax: +31 71 5450 212

Spain

3M España S.A.
J. Ignacio Luca de Tena
E-28027 Madrid, Spain
Tel: +34 1 321 6000
Fax: +34 1 321 6002

Switzerland

3M Schweiz AG
Eggstrasse 93
CH-8803 Rueschilikon
Switzerland
Tel: +41 724 9090
Fax: +41 724 0242

United Kingdom

3M (UK) plc.
P.O. Box 1
3M House, Market Place
Bracknell
RG12 1JU - England
Tel: +44 1334 858000
Aerospace Business: Sid Palmer

United States of America

Alaska:
3M
11151 Calaska Circle
Anchorage, AK 99515 - USA
Tel: +1 907 522 5200

California:
3M
6023 Garfield Ave.
P.O. Box 54019
Los Angeles, CA 90054 - USA
Tel: +1 213 726 6300
Fax: +1 213 727 2136

Georgia:
3M
2860 Bankers Industrial Dr.
Atlanta (Chamblee)
GA 30360-2764 - USA
Tel: +1 404 447 7000
Fax: +1 404 447 7161

Hawaii:
3M
4443 Malaii Street
Honolulu
Hawaii 96820 - USA
Tel: +1 808 422 2721

Mailing Address:
3M
P.O. Box 30048, Honolulu
Hawaii 96818 - USA

Michigan:
3M
P.O. Box 358
Southfield
MI 48037-0358 - USA
Tel: +1 313 827 2450
Fax: +1 313 827 6936

Minnesota:
3M Aerospace
Industrial Tape Div.
3M Centre Building 220-7E-01
St Paul, MN 55144 - USA
Tel: +1 612 733 9105
Fax: +1 612 733 1771
Marketing Comm. & Sales: Ann McGuiggan

New York:
3M
16 Henderson Drive
P.O. Box 2076
West Caldwell
NJ 07007-2078 - USA
Tel: +1 201 575 2000
Fax: +1 201 575 2108

Puerto Rico
3M
P.O. Box 100
Carolina
PR 00986-0100, Puerto Rico
Tel: +1 809 750 3000

Texas:
3M
2121 Santa Anna Ave.
P.O. Box 28158
Dallas
TX 75228-0158 - USA
Tel: +1 214 324 8100
Fax: +1 214 324 8225

Adhesives Australasia Ltd.

Agent: Henkel (New Zealand)

Aerospheres

Germany

Aerospheres
Forster Busch Str. 10
D-3510 Hann-Munden 21, Germany
Tel: +49 49 5545 781
Fax: +49 49 55451448

France

Aerospheres France Sarl
Prologue 1, La Pyreneenne voie No. 1
B.P. 27 - 37
F-31312 Labege Innople Cedex
France
Tel: +33 61 39 26 98
Fax: +33 61 39 26 96
Manager: David Normand

Italy

Aerospheres (Italy)
I-35031 Abano Terms (PD)
Italy
Tel: +39 49 866 8322
Fax: +39 49 866 8320

United Kingdom

Aerospheres (UK) Ltd.
(UK & European Head Office)
Trafalgar House, Grenville Place
London, NW7 3SA, England
Tel: +44 181 201 0924
Fax: +44 181 959 8199

United States of America

Aerospheres Inc.
(US Head Office)
216 Little Falls Road, Cedar Grove
NJ-07009 - USA
Tel: +1 201 239 0023
Fax: +1 201 239 5962

Amcor Trading (NZ) Ltd.

Agent: Witco (New Zealand)

American Cyanamid Co.

Now: Cytec

American Technologies Int.

United States of America

American Technologies International
3241, Brushwood Court
Clearwater, FL34621-2303 - USA
Tel: +1 813 785 9638
Fax: +1 813 785 9638
President: Dr. Sarfraz A. Saddiqui

Anchemco Ltd.

Agent: Teroson (India)

Andek Chemical Corp.

See: 3E (USA)

Apollo Chemicals Ltd.

United Kingdom

Apollo Chemicals Ltd.
Sandy Way
Amington Industrial Estate
Tamworth, Staffordshire
B77 4DS, England
Tel: +44 1827 54281
Fax: +44 1827 53030
Director: R.F. Saunders

Arania SA

Agent: Panacol-Elosol (Spain)

Arfeeti International Ltd.

Agent: Witco (Pakistan)

Argapura

Agent Teroson (Indonesia)

Arly Felthaus

Agent: Le Joint Français (Denmark, Scandinavia)

Ashland Chemical Co.

France

Ashland Chemical Co.
Div. d'Ashland Oil Inc.
64, rue Ernest Renan
F-92022 Nanterre Cedex, France
Tel: +33 1 41 37 92 92
Fax: +33 1 41 37 92 99
Sales: Jean-Claude Pierre

Germany

Ashland Suedchemie Kernfest
Reisholzstrasse 16
D-4010 Hilden, Germany
Tel: +49 211 711030
Fax: +49 211 7110335

Japan

Japan Synthetic Rubber Co.
2-11-24 Tsukiji
Chuo-Ku
Tokyo 104, Japan
Tel: +81 355 666620
Fax: +81 355 656636

United States of America

Michigan:
Ashland Chemical Co. Specialty Polymers & Adhesives Division
Wilshire Plaza North
900 Wilshire Drive, Suite #110
Troy, 48084 - USA
Tel: +1 313 244 9120
Fax: +1 313 244 8756

Ohio:
Ashland Chemical Co. Specialty Polymers & Adhesives Division
1745 Cottage Street
Ashland, 44805 - USA
Tel: +1 419 289 9588
Fax: +1 419 281 5991

Baagiel

Agent: Henkel (Saudi-Arabia)

Banite Inc.

See: Loctite (NY - USA)

... Polar SpA

Agent: Chemetall (Italy)

BASF Narmco

Now part of Cytec

M.G. Becker GmbH

See: Kleiberit Chimie Sarl. (France) & Klebchemie M.G. Becker GmbH & Co. KG (Germany)

Benecom

Agent: Le Joint Français (The Netherlands)

Bonder

Agent: Chemetall (Finland, Iran, Italy, Turkey, UK)

Borden

United Kingdom

Borden (UK) Ltd.
Rownhams Road, North Baddesley
Southampton, SO52 9ZB, England
Tel: +44 1703 732131
Fax: +44 1703 740294
Sales Manager: Vince Jones

Bostik

Austria

Bostik GmbH
Papierfabrikstrasse 1
A-4600 Wels, Austria
Tel: +43 724 253 010
(Also: Agent - Le Joint Français)

Denmark

Bostik A/S
Marielundvej 48 A
DK-2730 Herlev, Denmark
Tel: +45 42 84 15 00
Fax: +45 42 84 04 19
Contact: B. Rolf Larsen

France

Bostik S.A.
B.P. 46
F-77252 Brie Compte Robert Cedex
France
Tel: +33 1 64 42 12 12
Fax: +33 1 64 06 61 43
Contact: Eliane Dolhem

Bostik
Coubert
B.P. 46
F-77252 Brie Compte Robert Cedex
France
Tel: +33 1 64 42 12 12
Fax: +33 1 64 06 61 43

Germany

Bostik Gmbh
An Der Bundesstr. Nr 16
D-33829 Borgholzhausen, Germany
Tel: +49 5425 801 0
Fax: +49 5425 801 40

Ireland

Bostik Industrial Chemicals
Unit 9, Cookstown Ent. Pk.
Tallaght, Dublin 24, Eire
Tel: +353 1 512 322
Fax: +353 1 512 064

Portugal

Bostik Colas E Vedantes Lda.
Estra. Do Devisio P/A Odivelas, Adrião
P-2675 Odivelas, Portugal
Tel: +351 1 937 1972
Fax: +351 937 5810

Spain

Bostik SA
Berenguer De Palou, 64
E-08027 Barcelona, Spain
Tel: +34 (93) 340 3412
Fax: +34 (93) 349 4851

Sweden

Bostik AB
Box 903
S-25109 Helsingborg, Sweden
Tel: +46 42 19 50 00

Switzerland

Bostik AG
Industriestraße 35
CH-8112 Otelfingen, Switzerland
Tel: +41 1 844 5422
(Also: Agent - Le Joint Français)

United Kingdom

Bostik Ltd.
Ulverscroft Road, Leicester
LE4 6BW, England
Tel: +44 116 2510015
Fax: +44 116 2531943
Telex: 34625

United States of America

Emhart Fastening Systems Group
Bostik Division
Middleton, Massachusetts - USA

BTP plc.

See: Mydrin

CACI

Agent: CFPI (France)

Wm. Canning Ltd.

Wm. Canning Ltd.
Marston Bentley Division
Cale Lane, New Springs
Wigan WN2 1JR, England
Tel: +44 1942824242
Fax: +44 1942826653

Wm. Canning Ltd.
Aberdeen Office
c/o Scottish Express International
Howe Moss Drive
Kirkhill Industrial Estate
Dyce, Aberdeen AB2 OGL, Scotland
Tel: +44 1224 770199
Fax: +44 1224 723584

Casco Nobel

Denmark

Casco Nobel A/S
Praestemosevej 2-4
DK-3480 Fredensborg, Denmark
Tel: +45 42 28 10 66
Fax: +45 42 28 29 90
Contact for Denmark: Ole Noerland
Contact for export: Anders Edgren

Finland

Casco Nobel Oy
PB 33
SF-00741 Helsinki, Finland
Tel: +358 0 873 911
Fax: + 358 0 857 2268
Contact: Rolf Stolt

France

Casco Nobel Industrie SnC
34, Avenue Léon Jouhaux
B.P. 109
F-92164 Antony Cedex
Paris, France
Tel: +33 1 46 11 51 15
Fax: +33 1 46 66 49 22
Contact: Pierre Vachal

Germany

Casco-Nordsjo GmbH
Stobbenberger Straße 88
D-4300 Essen, Germany

Norway

Casco Nobel
Floisbonnveien 6, Sofiemyr
P.O. Box 565
N-1411 Kolbotn, Norway
Tel: +47 66 81 95 00
Fax: +47 66 80 28 13
Contact: Odd Arild Hansen

Singapore

Casco Nobel (SEA) Pte Ltd.
72, Hillview Avenue
01-02 Luwai Building
Singapore 2366
Tel: + 65 762 20 88
Fax: + 65 762 66 76
Contact: Bengt Carlsson

Sweden

Casco Nobel AB
La Cours Gata 4
Box 5016
S-250 05 Helsingborg, Sweden
Tel: +46 42 24 40 60
Fax: +46 42 24 40 64
Contact: Jens Elmegren

United Kingdom

Casco Nobel
Sadolin House
Meadow Lane, St. Ives
Cambridgeshire
PE17 4UY England
Tel: +44 1480 496414
Fax: +44 1480 496801
Contact: Steve Anslow

CECA

Belgium

CECA Belgium SA
Rue de Stalle, 63
Boîte 1
B-1180 Bruxelles
Tel: +32 2 370 2056, 2062 & 2064
Fax: +32 2 332 2901
Telex: 26022

France

CECA SA
12, place de l'Iris
Cedex 54 - F-92062
Paris La Défense 2, France
Tel: +33 1 47 96 90 90
Fax: +33 1 47 96 90 94
Telex: CECA 611444F

Germany

CECA Klebstoff GmbH
Otto Scheugenpflug Str. 8
D-63073 Offenbach, Germany
Tel: +49 69 89001 0
Fax: +49 69 89001 34

Italy

CECA Italiana SpA
Via G.Galilei 51/53
I-20096 Pioltello MI 046, Italy
Tel: +39 2 92 16 23 20
Fax: +39 2 92 69 507
Contact: Claudio Bonomo

Netherlands

CECA Nederlan BV
Molenvlietstraat 1
Postbus 22
NL-5140 AA Waalwijk
The Netherlands
Tel: +31 416 671100
Fax: +31 416 333329

Cetema

Agent: Chemetall (Netherlands, UK)

CFPI

Austria

ÖAG GmbH
Pfongauer Strasse 17
A-5202 Neumarkt Am Wallersee,
Austria
Tel: +43 62 16 6639 / 6689
Fax: +43 62 16 6689 85
Telex: 633242 OEAG

Egypt

Framchem
El Hoda Tower
16 El Falah Street
Mohendessin, Guiza, Egypt
Tel: +20 2 344 0385 / 347 6692
Fax: +20 2 345 0251
Telex: 92831 FRAM UN

France

CACI
Rue de la Bauve, Z.I. Nord
F-77100 Meaux, France
Tel: +33 1 64 36 54 54
Fax: +33 1 64 33 65 19
Telex: 694160

CFPI
(Corporate offices)
28, boulevard Camélinat, B.P. 75
F-92233 Gennevilliers Cedex, France
Tel: +33 1 40 85 50 50
Fax: +33 1 47 92 25 45
Telex: 610626 CFPI

SEAC
28, boulevard Camélinat, B.P. 75
F-92233 Gennevilliers Cedex, France
Tel: +33 1 40 85 54 00
Fax: +33 1 40 85 50 35
Telex: 620860 F LABCHIM

Germany

Gerhard Ruff GmbH
Postfach 1765
Oberbrühlstrasse 16 - 18
D-87700 Hemming
Germany
Tel: +49 83 31 83 600
Fax: +49 83 31 83 60 50
Telex: 54679

Spain

ETISA
Avenida Meridiana, 133, 2
Barcelona
E-08026, Spain
Tel: +34 3 245 87 24
Fax: +34 3 245 87 20
Telex: 59323 ETISA E

United States of America

South Carolina:
Lobeco Products Inc.
P.O. Drawer 630
Lobeco, SC 29931 - USA
Tel: +1 803 846 8171
Fax: +1 803 846 4777

Chemetall

Belgium

Parker Benelux NV
Mechelsesteenweg 592
B-1930 Zaventern
Belgium
Tel: +32 27 59 36 80
Fax: +32 27 59 55 72

Brazil

Diadema Industrias Quimicas Ltda.
Av. Fagundes de Oliveria 190
C-09950 Diadema SP, Brazil
Tel: +55 11 745 41 33
Fax: +55 11 745 44 38

Inbra Industrias Quimicas Ltda.
Av Fagundes de Oliveria 190
C-09950 Diadema SP, Brazil
Tel: +55 11 745 41 33
Fax: +55 11 745 44 38

Denmark

Chemetall GmbH Filial Danmark
Hørkaer 3 POB 770
DK-2730 Herlev, Denmark
Tel: +45 44 925 700
Fax: +45 44 925 901

Finland

Oy Bonder AB
Työpajakatu 10 A 8
SF-00580 Helsinki, Finland
Tel: +3580 730366
Fax: +3580 730411

France

Société Continental Parker SA
51 rue Pierre
F-92111 Clichy Cedex, France
Tel: +33 1 47 15 38 00
Fax: +33 1 47 30 01 92
+33 1 47 37 46 60

Germany

Chemetall GmbH
Reuterweg 14
D-60271 Frankfurt, Germany
Tel: +49 69 159 0
Fax: +49 69 159 3018
Telex: 4170090 cm-d
Contact: Dr. P. Hönel

Metasco chemisch-technische
Produkte GmbH
Aastraße 1
D-65195 Wiesbaden, Germany
Tel: +49 11 45 08-381
Fax: +49 11 45 08-380

Synthomer Chemie GmbH
Reuterweg 14
D-60271 Frankfurt a.M. Germany
Tel: +49 69 159-0
Fax: +49 69 159-29 60

Hungary

Hungarobond Kft.
Francia út 43
H-1143 Budapest, Hungary
Tel: +36 2 52 51 46
Fax: +36 2 52 51 36

Iran

Bonder Iran SSK
Av. Jahan-E-Koodak 43
Vanak Square, Teheran, Iran
Tel: +98 21 68 21 27
Telex: 212428 db ir: Attn. Bonder Iran

Italy

Barattini Polar SpA
Via Priv. G. Bugatti 5
I-20144 Mailand, Italy
Tel: +39 2 837 04 51
Fax: +39 2 89 40 45 29

Bonder SpA
Via della Tecnica 5-7
I-20034 Guissano-Birone (Milano)
Italy
Tel: +39 362 860811
Fax: +39 362 224913

Japan

PMG Corporation
15-1 chome, Nihinbashi
Chuo-ku, Parker Building
Tokyo 103, Japan
Tel: +81 3 278 45 87
Telex: 5333 nipak

Netherlands

Cetema BV
Usselstraat 41
NL-5340 AA Oss, The Netherlands
Tel: +31 412 681888
Fax: +31 412 631479

Poland

Gardobond Sp.zo.o
Przedawska 5
PL-03-879 Warsawa, Poland
Tel: +48 22 19 63 72
Telex: 816360

Portugal

Parker Lusitania Lda.
Rua das Vagens Ap. 23
P-2726 Mem-Martins, Portugal
Tel: +35 11 921 27 00/1/2
Fax: +35 11 922 10 27

South Africa

Chemrite (Pty) Ltd.
3 Molecule Street
POB 12370
Chloorkop 1624
Transvaal, South Africa
Tel: +27 11 976 1005
Fax: +27 11 976 3229

Reef Chemical Corp (pty) Ltd.
5 Sellers Road
POB 232
Boksburg East 1460, South Africa
Tel: +27 11 914 2500
Fax: +27 11 914 4792

Spain

Ozark Iberica SA
Calle Igorreko Indistrialdea E-7
E-48140 Igorre (Viscaya), Spain
Tel: +34 4 631 53 93
Fax: +34 4 631 55 87

Sociedad Continental Parker Hispania
SA
Calle Madrazo 14-16
E-08006 Barcelona 6, Spain
Tel: +34 3 415 22 55
Fax; +34 3 2 18 17 88

Sweden

AB Ytteknik
Dragets Ind.Område
S-74651 Bälsta, Sweden
Tel: +46 171 5 40 40
Fax: +46 171 5 75 33

Switzerland

CM-Celfa Mebrantrenntechnik
Bahnhofstraße
CH-6423 Seewen/Schwyz
Switzerland
Tel: +41 43 24 74 81
Fax: +41 43 24 74 50

CM Oberflächentechnik AG
Durisolstraße 12
CH-5612 Villmergen, Switzerland
Tel: +41 57 21 11 95-98
Fax: +41 57 22 36 58

Turkey

Bonder Kintya ve Sanayi AS
Kore Sehitleri Cad No 28
TR-80200 Zincirlikuyu-Istanbul,
Turkey
Tel: +9212 275 02 78
Fax: +9212 275 92 87

United Kingdom

Bonder UK (Branch of Cetema Ltd.)
35 Rickfords Hill
Aylesbury, Buckinghamshire
HP20 2RU, England
Tel: +44 1296 399233
Fax: +44 1296 3999211

Cetema Ltd.
2 Haig Court, Haig Road
Parkgate Industrial Estate
Knutsford, Cheshire
WA16 8XX, England
Tel: +44 1565 750058
Fax: +441565 750056

United States of America

New Jersey:
CM Chemical Products
50 Valley Road, Berkeley Heights
NJ 07922 - USA
Tel: +1 908 665 8593
Fax +1 908 665 8543

Oaklite Products Inc.
(Details as above)

Chemid Kenya Ltd.

Agent: Witco (Kenya)

Chempro

Agent: Witco (Egypt)

Chemrite (Pty) Ltd.

Agent: Chemetall (South Africa)

Chempro

Agent: Witco (Egypt, South Africa)

Chemserve Metal Sciences

Agent: Teroson (South Africa)

Chika Ltd.

Agent: Witco (India)

CHRYSO

France

CHRYSO (Mastics et Resines)
7 - 9, Rue de l'Europe Z.I.
F-45300 Sermaises du Loiret
France
Tel: +33 38 34 58 00
Fax: +33 38 39 01 16
Mastics & Resins:
Tel: +33 1 38 34 58 05/06
Lab. Director: Francois Mariotti

Ciba-Geigy

Australia

Ciba-Geigy Australia Ltd.
P.0. Box 332
Thomastown
Victoria 3074 Australia
Tel: +61 3 282 0600 Melbourne
Fax: +61 3 465 9070
Telex: 30435

Ciba Australia Ltd.
140 Bungaree Road, Pendle Hill
NSW 2145, Australia
Tel: +61 2 688 0444
Fax: +61 2 688 0555

Austria

Ciba-Geigy GmbH
Postfach 78
A-1231 Wien, Austria
Tel: +43 1 80111
Fax: +43 1 80111 421
Telex: 131923

Belgium

S.A. Ciba-Geigy N.V.
Noordkustlaan 18
B-1702 Groot-Bijgaarden, Belgium
Tel: +32 2 467 0611
Fax: +32 2 467 0774
Telex: 21559 cibgy bru

Brazil

Ciba-Geigy Quimica SA
Calxa postal 21.468
04698-São Paulo-SP
Brazil
Tel: +56 11 240 1011
Fax: +56 11 241 8175
Telex: 11 56777 ciba br

Canada

Ciba-Geigy Canada Ltd.
6860 Century Ave.
Mississauga
Ontario L5N 2W5
Canada
Tel: +1 416 821 4420
Fax: +1 416 821 9815
Telex: 06 217752

Denmark

Ciba-Geigy A/S
Lyngbyvej 172
DK-2100 København Ø, Denmark
Tel: +45 31 29 14 22
Fax: +45 31 29 14 21
Telex: 19282

Finland

Ciba Geigy OY
P.O. Box 16
SF-00241 Helsinki, Finland
Tel: +358 156911
Fax: +358 142976
Telex: 121463 cigisf

France

S.A. Ciba-Geigy
B.P. 308
F-92506 Rueil-Malmaison Cedex
France
Tel: +33 1 47 52 30 00
Fax: +33 1 47 52 31 50
Telex: 203012

Ciba
2-4 Rue Lionel Terray
B.P. 308
F-92506 Rueil Malmaison, France
Tel: +33 1 47 52 30 00
Fax: +33 1 47 52 33 93

Germany

Ciba-Geigy GmbH
Postfach 1160/1180
D-7867 Wehr/Baden, Germany
Tel: +49 7762 820
Fax: +49 7762 3727
Telex: 792702

Greece

Ciba-Geigy Hellas S.A.
B.P. 8282
GR-10010 Athenes
Greece
Tel: +30 1 666 6612
Fax: +30 1 666 7246
Telex: 215536

Hong Kong

Ciba-Geigy (Hong Kong) Ltd.
P.O. Box 285
Kwai Chung, N.T. Hong Kong
Tel: +852 424 9321
Fax: +852 487 1428
Telex: 39815 cigy hx

Italy

Ciba-Geigy S.p.A
Casella Postale 88
I-21047 Saronno VA, Italy
Tel: +39 2 96541
Fax: +39 2 96701084
Telex: 332407 cigy i

Japan

Ciba-Geigy (Japan) Ltd.
P.O. Box 1
665 Takarazuka, Japan
Tel: +81 797 74 2472
Fax: +81 797 74 2515
Telex: 63150 (cigytz j)
0 5645 684 (cigytz j)

Mexico

Ciba-Geigy Mexicana SA de CV
Apartado Postal 21-993
04021 México D.F.
Tel: +52 5 549 3000
Fax: +52 5 549 2511
Telex: 17 71 511

Netherlands

Ciba-Geigy BV
P.B. 211, NL-6800 4-7 Arnhem
The Netherlands
Tel: +31 26 365 6200
Fax: +31 26 364 0567
Telex: 45054

New Zealand

Ciba-Geigy New Zealand Ltd.
Private Bag
Avondale, Auckland 7, New Zealand
Tel: +64 9 883 149
Fax: +64 9 886 565
Telex: nz 61130 ciyak

Norway

Ciba-Geigy A/S
P.O. Box 6077
Etterstad, N-0601 Oslo 6, Norway
Tel: +47 2 194600
Fax: +47 2 686317
Telex: 78391 cigy n

Portugal

Ciba-Geigy Portuguese Lda.
Rua de Manuel Pinto de Azevedo 7
P-4100 Porto, Portugal
Tel: +351 2 6172266
Fax: +351 2 6172260
Telex: 22427 cigy p

Singapore

Ciba-Geigy S.E. Asia (Pte.) Ltd.
Jurong Town
P.O. Box 264, Singapore 9161
Tel: +65 2653622
Fax: +65 6638393
Telex: 51214

Spain

Ciba-Geigy S.A.
Apartado 744
E-08080 Barcelona, Spain
Tel: +34 3 40 40300
Fax: +34 3 40 40376
Telex: 54726 cigy

Switzerland

Ciba-Geigy AG
Postfach CH-4002, Basel, Switzerland
Tel: +41 61 69 71111
Fax: +41 61 69 64354
Telex: 962 355

Sweden

Ciba-Geigy AB
Box 605
S-42126 Västra Frölunda, Sweden
Tel: +46 31 892500
Fax: +46 31 452847
Telex: 21636 cigivfas

United Kingdom

Ciba Polymers
Duxford, Cambridge
CB2 4QA, England
Tel: +44 1223 832121
Fax: +44 1223 838404
Telex: 81101 cigypl g
Intl. Marketing: Max H. Rollin

United States of America

California:
Ciba-Geigy Corp.
Furane Aerospace Products
5121 San Femando Road West
Los Angeles, CA 90039 - USA
Tel: +1 818 247 6210
Fax: +1 818 247 6616

Michigan:
Ciba-Geigy Corp.
Formulated Systems Group
4917 Dawn Avenue, East Lansing
MI 48823 - USA
Tel: +1 517 351 5900
Fax: +1 517 351 9003

Collanti Parini Srl

Agent: Panacol-Elosol (Italy)

CM

Agent: Chemetall (Switzerland, USA)

Conap Inc.

(part of Cytec group)
Conap Inc.
1405, Buffalo Street
Olean, NY 14760 - USA
Tel: +1 716 372 9650
Fax: +1 716 372 1594

Coninter Ltda.

Agent: Witco (Chile)

Coprin

Agent: Henkel (Morocco)

CTA

Agent: Le Joint Français (Spain)

Cytec
(Formerly: American Cyanamid)

United Kingdom:

Cytec Aerospace Ltd.
Abenbury Way
Wrexham Industrial Estate, Wrexham
Clwyd, LL13 9UF Wales
Tel: +44 1978 661971
Fax: +44 1978 660220
Telex: 61348

United States of America

California:
Cytec
21444 Golden Triangle Road
Saugus, CA 91350 USA
Tel: +1 805 259 1415

Cytec Engineered Materials Inc.
1191 Hawk Circle
Anaheim, CA 92807 - USA
Tel: +1 714 632 8444

Cytec Engineered Materials Inc.
1440 N. Kraemer Blvd.
Anaheim, CA 92807 - USA
Tel: +1 714 630 9400

Connecticut:
Cytec Engineered Materials Inc.
South Cherry Street
Wallingford, CT 06492 - USA
Tel: +1 203 269 4481

Maryland:
Cytec Engineered Materials Inc.
1300 Revolution Street
Havre de Grace
MD 21078 USA
Tel: +1 301 939 1910
Fax: +1 301 939 0930

Cytec
Old Post Road
Havre de Grace
MD 21078 - USA
Tel: +1 301 939 1910

New Jersey:
Cytec
(Corporate Headquarters)
One Cyanamid Plaza
Wayne, NJ 07470 - USA
Tel: +1 201 831 2000

New York:
Cytec Engineered Materials Inc.
1405 Buffalo Street
Olean, NY 14760-1139 - USA
Tel: +1 716 372 9650

Daito Sangyo Co. Ltd.

Agent: Witco (Japan)

Degussa

Germany

Degussa AG
GB Industrie und Feinchemikalen
Monomere/Polymere
IC-MP-HK
Postfach 1345
D-63403 Hanau, Germany
Tel: +49 6181 59 0
Fax: +49 6181 59 2283
Product Manager: Dr. Hans Puff

DELO

Germany

DELO Industrieklebstoff GmbH & Co.
Ohmstr. 3
D-86899 Landsberg, Germany
Tel: +49 81 91 32040
Fax: +49 81 91 320444

DELO GmbH & Co. KG
Lohenstr. 11+13
D-82166 München, Germany
Tel: +49 89 85854 0
Fax: +49 89 85854 44

Devek On (International)

Agent: Panacol-Elosol (Israel)

Dexter Hysol

Europe

Hysol Aerospace Products
Knorrstrasse 83A
8 München 40, Germany
Tel: +49 89 351 8750
Fax: +49 89 354 1287
Marketing Manager: Tom Batot
Frazier

France

Structil
18, rue Lavoisier
BP 10
F-91710 Vert le Petit, France
Tel: +33 1 69 90 89 89
Fax: +33 1 64 93 44 57
Technical Contact: Mr. Vignando

Germany

Dexter
IM Knollen 3
D-69231 Rauemberg, Germany
Tel: +49 11 49 6222 61273
Fax: +49 11 49 1726 21734
Manager (Europe): Rainer Stricker

Singapore

Dexter Distributor Programs
6 Tuas Avenue
8, Singapore 2263
Tel: +65 11 65 862 0228
Fax: +65 11 65 861 8850
Business Manager (Far East): Henry Foo

United Kingdom

(Hot Melts Only):
Power Adhesives Ltd.
Unit 5, New England Estate
Gascoigne Road, Barking, Essex
IG11 7LN, England
Tel: +44 181 507 8433
Fax: +44 181 507 8407
Contact: Steve Sweeney

United States of America

Dexter Automotive Materials Division
Distributor Programs
One Dexter Drive
Seabrook, NH 03874 - USA
Tel: +1 663 474 5541 ext. 225
Fax: +1 603 474 7620
Marketing Dir.: Peter Ferris
Sales Manager: Jeff DeGrechie
Marketing: Dr. Jeffrey S. Payne

Dexter Adhesives & Structural Materials Div.
(Details as above)
Marketing: Tammy Campbell

Dexter Hysol
2850 Willow Pass Road
P.O. Box 312
Pittsburg, CA 94565-0031 - USA
Tel: +1 510 458 8259
Fax: +1 510 458 8030
Program Manager: Sid Quick

Hysol Aerospace Products
(Headquarters)
2850 Willow Pass Road
P.O. Box 312
Pittsburg, CA 94565-0031 - USA
Tel: +1 510 458 8000
Fax: +1 510 458 8030

Great Plains Area:
Hysol Aerospace Products
251, S. Hillside
Wichita, KS 67211 - USA
Tel: +1 316 685 2149
Fax: +1 316 685 4216
Technical Rep.: David L. Smizer

Mid Atlantic, Great Lakes & Eastern Canada:
Hysol Aerospace Products
16 Union Street
Medford, NJ 08055 - USA
Tel: +1 609 654 1651
Fax: +1 609 654 1151
NE Area Manager: Robert A. Golick

Northeast Area:
Hysol Aerospace Products
P.O. Box 216
Stafford Springs
CT 06076 - USA
Tel: +1 203 684 4525
Fax: +1 203 684 3846
Technical Rep.: Mike Leedie

Northwest U.S. & Western Canada:
Hysol Aerospace Products
Suite 300, 23830 Pacific Highway S.
Kent, WA 98032 - USA
Tel: +1 206 878 6607
Fax: +1 206 878 6606
NW Area Manager: John Montgomery
Technical Rep.: Larry Stelmack

Southern Area:
Hysol Aerospace Products
P.O. Box 554
Chico, TX 76431 - USA
Tel: +1 817 644 5155
Fax: +1 817 644 5155
S Area Manager: Robert L. Hammer

Southeastern Area:
Hysol Aerospace Products
1573 Halisport Lake Drive
Kennesaw, GA 30144 - USA
Tel: +1 404 427 1567
Fax: +1 404 427 1567
Snr. Technical Rep.: Joe Leonardo

Southwest Area:
Hysol Aerospace Products
Third Floor, 1915 Orangewood Ave.
Orange, CA 92668 - USA
Tel: +1 714 939 5540
Fax: +1 714 939 5543
SW Area Manager: John S. Hutcheon
Snr. Technical Rep.: John Meyn

Diadema Industrias Quimicas Ltda.

Agent: Chemetall (Brazil)

Dow Corning

Australia

Dow Corning Australia Pty Ltd.
(Area HQ)
4 Ray Road, Epping
NSW 2121, Australia
Tel: +61 612 8698555
Telex: 22108

Austria

Dow Corning GmbH
Ghegastraße 3
A-1030 Wien, Austria
Tel: +43 222 782501
Fax: +43 222 782179

Belgium

(Scandinavia & Benelux)
Dow Corning Europe
Rue Général de Gaulle 62
B-1310 La Hulpe
Belgium
Tel: +32 2 6552111
Fax: +32 2 6552001
Telex: 22530

(Technical Service):
Dow Corning SA.
Parc Indusriel
B-6198 Seneffe
Belgium
Tel: +02 064 272211
Fax: +02 64 272401
Telex: 84657028

Brazil

Dow Corning do Brazil Ltda.
Caixa Postal 4102-01000
São Paulo, Brazil
Tel: +55 5511/531 5199
Telex: 3911122843

France

Dow Corning France SA
Immeuble Britannia
20 blvd. Eugène Deruelle
F-69432 Lyon Cedex 3
France
Tel: +33 78 60 51 48
Fax: +33 78 62 78 98
Telex: 300537 F

Dow Corning Construction SA
191 rue du 1er Mai
F-92000 Nanterre
France
Tel: +33 1 47 80 71 65
Fax: +33 1 47 60 94 03
Telex: 610758

Germany

Dow Corning GmbH
Rheingaustraße 53
D-65201 Wiesbaden
Germany
Tel: +49 611 237 1
Fax: +49 611 25954
Telex: 04186177

(Technical Service):
Tel: +49 611 237 1
Fax: +49 611 25954
Telex: 22530

Italy

Dow Corning SA
Palazzo Canova-Centro Direzionale
I-20090 Milano 2 Segrate, Italy
Tel: +39 2 2641 4041
Fax: +39 2 2641 1661
Telex: 330279 dowsil I

Japan

Dow Corning Kabushiki Kaisha
Tokyo Office
15-1 Nishi Shimbashi 1-chome
Minato-Ku, Tokyo 105, Japan
Tel: +81 813 503 6051
Telex: +81 78102225183

Spain

Dow Corning Iberica S.A.
Avda. Diagonal 613, 5°
E-08028 Barcelona, Spain
Tel: +34 3 321 6991
Fax: +34 3 322 9952
Telex: 50425

Dow Corning Kkrafft SA
Barrio Buruntza s/n
E-20120 Andoain (Guipúzcoa)
Spain
Tel: +34 943 590490
Fax: +34 943 593483

United Kingdom

Dow Corning Hansil Ltd.
19 Wintersells Road
Byfleet, Surrey
KT14 7LH, England
Tel: +44 19323 51911
Telex: 928206

(Technical Service):
Tel: +44 19323 51911
Fax: +44 19323 40624
Telex: 928206

United States of America

Dow Corning USA
910 Auburn Court
Fremont, CA 94538 - USA
Tel: +1 415 490 9302 (Sales)

Dow Corning Center
Midland, MI 48686-0994 - USA
Tel: +1 517 496 4000
Telex: 227450

Dunlop

Germany

Dunlop GmbH
Birkenhainer Str. 77
D-63450 Hanau, Germany
Tel: +49 6181 361 0
Fax: +49 6181 361 396

United Kingdom

Dunlop Adhesives
Chester Road
Birmingham B35 7AL, England
Tel: +44 121 373 8101
Fax: +44 121 384 2826
Specialist Adhesives: Joan Howard
(Also Agent for McGhan NuSil)

Dyckerhoff

Austria

Dyckerhoff Austria Ausbauprodukte
GesmbH
Andritzer Reichsstraße 13
A-8045 Graz, Austria
Tel: +43 316 6721210
Fax: +43 316 672154
Telex: 311836 DyA

Germany

Dyckerhoff Sopro GmbH
Postfach 2247
D-65012 Wiesbaden, Germany

Germany - North:
Dyckerhoff Sopro GmbH
Lienener Straße 89
D-49525 Lengerich
Tel: +49 54 81 31314
Fax: +49 54 81 31414
Telex: 941310 dyle
Technical Advice:
Tel: +49 54 81 31310
Fax: +49 54 81 31414

Germany - East:
Dyckerhoff Sopro GmbH
Am Wüsteberg 2
D-01723 Kesselsdorf, Germany
Tel: +49 35204 40407
Fax: +49 35204 40405

Germany - South:
Dyckerhoff Sopro GmbH
Biebricher Straße 72
D-65203 Wiesbaden
Tel: +49 611 676 1841
Fax: +49 611 676 1830
Telex: 4186832 dy
Contact: Helmut Schäfer
Technical Advice:
Tel: +49 611 676 1912 / 1913
Fax: +49 611 676 1920

Dymax Corporation

France

ES
(Équipments Scientifiques S.A.)
127 rue de Buzenval
BP 26
F-92380 Garches, France
Tel: +33 1 47 95 99 00
Fax: +33 1 47 01 16 22

Germany

Dymax Europe GmbH
Reuterweg 14
D-60323 Frankfurt, Germany
Tel: +49 69 159 3068
Fax: +49 69 159 3830
Technical Manager: Dr. T Oehmichen

United States of America

Dymax Corporation
51, Greenwoods Road
Torrington
CT 06790 - USA
Tel: +1 203 482 1010
Fax: +1 203 496 0608
Marketing Communications Manager:
Martine M. Boutte

Eleco Produits SA

Agent: Panacol-Elosol (France)

Electro-Seal Corp.

United States of America

Electro-Seal Corp.
55, Wanaque Ave
Pompton Lakes
NJ 07442 - USA
Tel: +1 201 835 4100
Fax: +1 201 835 5986
Sales Manager: Steve Stein

Emerson & Cuming

See: Grace

EMFI

EMFI s.a.
B.P. 30
14, rue de la Ferme Clauss
F-67501 Haguenau Cedex, France
Tel: +33 88 90 60 00
Fax: +33 88 73 48 38
Telex: 870933

Emhart Fastening Systems Group

See: Bostik (USA)

EMS

Ecuador & Latin America

EMS de Latinoamerica SA
Av. Amazonas 3429
Apartado postal 6444
Quito, Ecuador
Tel: +593 2 24 78 90
Fax: +593 2 43 93 45
Telex: 22666

France

EMS-France SA
204, avenue Maréchal Juin
B.P. 52
F-92105 Boulogne Billancourt Cédex, France
Tel: +33 1 41 10 06 10
Fax: +33 1 48 25 46 07
Telex: 633430
Contact: Alain Pinay

Germany

EMS-Chemie (Deutschland) GmbH
Wendelinstrasse 1
D-50933 Köln, Germany
Tel: +49 221 9497040
Fax +49 221 4971549
Telex: 8882030

Japan

EMS-Japan Corp.
Kikuchi Building
2-12 Nihonbashi Hongoku-cho
3-chome, Chuo-ku
Tokyo 103, Japan
Tel: +81 3 3242 3854
Fox: +81 3 3241 8501

Switzerland

EMS-Chemie AG
CH-7013 Domat/Ems
Switzerland
Tel: 081 / 36 61 11
Fax: 081 / 36 74 02
Telex: 851 421

Taiwan

EMS-Far Eastern Ltd.
36, Kwang Fu South Road
Hsin Chu Industrial Park
Fu Kou Hsiong
Hsin Chu Hsien
Taiwan, R.O.C.
Tel: +886 35 985 335
Fax: +886 35 985 345

United Kingdom

EMS-Grilon (UK) Ltd.
Astonfields Industrial Estate
Drummond Road, Stafford
ST16 3HJ - England
Tel: +44 1785 59121
Fax: +44 1785 213068
Telex: 36254

United States of America

EMS-American Grilon Inc.
Industrial Park and Corporate Way
P.O. Box 1717
Sumter
South Carolina 29151-1948 - USA
Tel: +1 803 481 3172
Fax: +1 803 481 3820
Telex: 805077

EMS Acquisition Corp.
Summers Laboratories Div.
PO Box 162
Ft. Washington
PA 19034 - USA
Tel: +1 215 646 1477
Fax: +1 215 646 8931

Emulsio Danmark

Agent: Swift Adhesives

Endüstri Teknik Ltd.

Agent: Henkel (Turkey)

Entra Ibérica SA

Agent: Scott Bader (Spain)

Enzur S.A.

Agent: Witco (Uruguay)

Epirez

Australia

New South Wales:
(Head Office):
Epirez Construction Products
2 Seville Street
Villawood
NSW 2163 - Australia
Tel: +61 2 726 8899
Fax: +61 2 726 4410
Branch Manager: Max Simmons

North Queensland:
Epirez Construction Products
Unit 2, 197 Ingham Road
Garbutt
QLD 4814 - Australia
Tel: +61 77 25 3088
Fax: +61 77 25 1701
Branch Manager: Dennis Ashman

South Australia:
Epirez Construction Products
Unit 2, 55 Ashwin Parade
Torrensville
SA 5031 - Australia
Tel: +61 8 234 0171
Fax: +61 8 43 3374
Branch Manager: Phil McGowan

South Queensland:
Epirez Construction Products
Unit 1, 2 Darnick Street
Underwood
QLD 4119 - Australia
Tel: +61 7 841 1233
Fax: +61 7 841 1381
Branch Manager: Stoward Bolton

Victoria:
Epirez Construction Products
99B Highbury Road
Burwood, VIC 3125 - Australia
Tel: +61 3 808 7244
Fax: +61 3 808 5212
Branch Manager: Daryl Bartels

Western Australia:
Epirez Construction Products
106 Howe Street
Osborne Park
WA 6017 - Australia
Tel: +61 9 242 2833
Fax: +61 9 242 5598
Branch Manager: Rick Owens

ES

(Équipments Scientifiques S.A.)
Agent: Dymax (France)

ETISA

Agent: CFPI (Spain)

Eurobond Adhesives Ltd.

Agent: Panacol-Elosol (UK)

Evode / Evacor Resins Ltd.

United Kingdom

Evacor Resins Ltd.
Common Road
Stafford, ST16 3EH, England
Tel: +44 1785 57755
Fax: +44 1785 55093
Telex: 36474
Contact: Alf Stewart

Forbo-Helmitin

Austria

Forbo Industrieprodukte GesmbH
Eisenstraße 55
A-4400 Steyr, St. Ulrich, Austria
Tel: +43 7252 454400
Fax: +43 7252 47936

Canada

Helmitin (Canada) Inc.
99, Shorncliffe Road
Toronto, Ontario, Canada M8Z 5K7
Tel: +1 416 239 6487
Fax: +1 416 239 3105

France

Helmitin S.A. Colles Industrielles
Rue du Général de Gaulle
F-67250 Surbourg, France
Tel: +33 88 80 50 40
Fax: +33 88 80 54 14

Germany

Forbo-Helmitin GmbH
Zweibrücker Straße 185
D-66954 Pirmasens, Germany
Tel: +49 63 31 528 0
Fax: +49 63 31 528 201
Telex: 452230 helmi d
Contact: Mr. Gautsche

(Mailing Address):
Forbo-Helmitin GmbH
Postfach 19 61
D-66927 Pirmasens, Germany

Germany - East:
Forbo-Erfurt GmbH
Mittelhäuser Straße 65
D-99087 Erfurt, Germany
Tel: +49 3 617 3041 0
Fax: +49 3 617 3041 90
Telex: 3 41 263

(Mailing Address):
Forbo-Erfurt GmbH
Postfach 6 60
D-99013 Erfurt, Germany

Netherlands

Forbo Genderen B.V.
Hoofdstraat 46
NL-4265 HL Genderen
The Netherlands
Tel: +31 416 358200
Fax: +31 416 351408

Switzerland

Forbo CTU AG
CH-5012 Schönenwerd, Switzerland
Tel: +41 64 40 14 22
Fax: +41 64 40 20 84
Telex: 981504

Fortafix Ltd.

United Kingdom

Fortafix Ltd.
First Drove, Fengate
Peterborough, PE1 5BJ, England
Tel: +44 1733 66136
Fax: +44 1733 315393
Managing Director: J. Waddicor

Framchem

Agent: CFPI (Egypt)

France G.I.E.

Agent: Loctite (France)

Furane Products

See: Ciba-Geigy, USA

M. Gal Robert

See: Le Joint Français (France)

Garco AB

Agent: Panacol-Elosol (Sweden)

Gardobond

Agent: Chemetall (Poland)

GE Silicones

Australia

GE Silicones
775 Hammond Road
Dandenong
Victoria 3175, Australia
Tel:+61 3 791 6771

Canada

General Electric Canada Inc.
2300 Meadowvale Blvd.
Mississauga, Ontario
Canada L5N 5P9
Customer Service:
Tel: +1 800 668 464415
Technical Assistance:
Tel: +1 800 255 8886

Hong Kong

General Electric Plastics
3 Exchange Square
Room 7001, 10th Floor
Central, Hong Kong
Tel: +852 843 6700

Mexico

GE Silicones Mexico
Horacio 1855-104
Col. Polanco
C.P. 11510, Mexico, D.F.
Tel: +52 95 580 2722 /
+52 518 233 3734

Netherlands

General Electric Plastics B.V.
Plasticslaan 1
Box 117
NL-4600 AC Bergen op Zoom
The Netherlands
Tel:+31 164 232911
Fax:+31 164 292940
Telex: 78421

Singapore

GE Plastics Pacific Ltd.
GE Pacific PTE Ltd.
150 Beach Road
#24-00 Gateway West
Singapore 0718
Tel: +65 291 9700

South Korea

GE Silicones Korea
c/o Dongyang Silicones Co
Oriental Chem. Bldg.
50, Sokong-Dong, Chung-Ku
Seoul, South Korea, 100-070
Tel: +82 2 774 2641

Sweden

International General Electric AB
Box 1203
S-16428 Kista, Sweden

Taiwan

General Electric Tech. Service Co. Inc.
Taiwan Branch
7F, Formosa Plastics Building 201
Tun Hua North Road
Taipei, Taiwan
Tel: +886 2 514 9921

United Kingdom

GE Silicones
Old Hall Road
Sale
M33 2HG, England
Tel: +44 161 905 5020
Technical Sales: Nigel O'Sullivan

United States of America

New York:
General Electric Co.
260 Hudson River Road
Waterford
NY 12188 - USA
Customer Service:
Tel: +1 518 237 3330
Technical Assistance:
Tel: +1 518 233 3330

General Electric Co.
Business Information Centre
1, Winner Circle, AER
Albany
NY 12205 - USA
Tel: +1 518 438 6500
Fax: +1 518 438 2527

Geistlich Söhne AG

Switzerland

Ed. Geistlich Söhne AG
Endstringerstrasse 5
CH-8952 Schilieren
Switzerland
Tel: +41 1 733 88 33
Fax: +41 1 730 80 02
Contact: P. Böhlen

Grace

Benelux

Grace NV
Nijverheidsstraat 7
B-2260 Westerlo, Belgium
Tel: +32 14 57 56 11
Fax: + 32 14 58 55 30
Telex: 31500 eccoe b
Contact: Julien Souffriau

France, Spain & Portugal

Grace SA
Grace Specialty Polymers
Zone Industrielle St Denis
B.P. 9
F-28231 Epernon Cedex, France
Tel: +33 37 28 97 90
Fax: +33 37 28 97 95
Contact: Lionel Vincendeau

Germany, Austria & Switzerland

(Grace NV - Germany is due to move address in Oct 95)
Grace NV
Friedrich-Ebert-Anloge 11
D-66450 Hanau, Germany
Tel: +49 6181 31002
Fax: +49 6181 31007
Contact: Heinz J. Rautenberg

Italy

Grace Italiana S.p.A.
Grace Specialty Polymers
Via Trento, 7
I-20017 Passirana di Rho (MI), Italy
Tel: +39 2 93321
Fax: +39 2 9300927
Contact: Emilio ALLIEVI

Scandinavia

Grace AB
Grace Specialty Polymers
Box 622
S-251 06 Helsingborg, Sweden
Tel: +46 42 37 12 11
Fax : +46 42 37 13 86
Contact: Harri Riski

United Kingdom

Emerson & Cuming UK Ltd.
Parkway, Globe Park
Marlow, Buckinghamshire
SL7 1YB - England
Tel: +44 1628 488885
Fax: +44 1628 488890
Contact: Peter Brent

Gulf International

Agent: Henkel (United Arab Emirates)

Helmitin

See: Forbo Helmitin

Henkel

See also: Teroson

Algeria

Henkel Algeria
Algiers, Algeria

Argentina

Henkel Argentina SA
P-0- Box 3496
1000 Correo Central
Buenos Aires, Argentina
Tel: +54 1 204 2056
Fax: +54 1 205 3360

Austria

Henkel Austria GmbH
Erdbergstr. 29
A-1031 Wien 3, Austria
Tel: +43 1 7110 4510
Fax: +43 1 7110 4612

Belgium

Henkel Belgium NV
66, Avenue du Port (Havenlaan 66)
B-1210 Brüssel, Belgium

Brazil

Henkel S.A. Industrias Quimicas
Avenida das Nacoes
Unidas, 10989
CEP 04578, São Paulo 1, Brazil
Tel: +55 1 1 829 3602
Fax: +55 1 1 829 7270

Chile

Marson Chilena Ltda.
Valparaiso, Chile

Columbia

Pegatex/Molytec Ltda.
Bogotà, Columbia

Denmark

Henkel Barnängen A/S
Tåstrupgårdsvej 8
DK-2630 Tåstrup, Denmark
Tel: +45 43 99 70 22
Fax: +45 43 99 70 23
Contact: Iben Pockenauer

P. Højstrup ApS
Odense, Denmark

Egypt

MAS & Co.
Giza, Egypt

Finland

Oy Soumen Henkel Liimat AB
Helsinki, Finland

France

Henkel France S.A.
150 rue Gallieni
F-92100 Boulogne-Billancourt
France
Tel: +33 1 46 84 90 00
Fax: +33 1 46 84 91 90

Germany

Henkel KGaA
Henkelstr. 67
D-40191 Düsseldorf, Germany
Tel: +49 211 797 0
Fax: +49 211 798 2358

Sichel-Werke GmbH
Sichelstraße 1
D-3000 Hannover 91
Germany
Tel: +49 511 21 40 235
Fax: +49 511 21 40 230
Telex 922129

Guatemala

Quimica Henkel Centroamericana SA
Cuidad de Guatemala, Guatemala

Hong Kong

Henkel Chemicals (H-K) Ltd.
Hong Kong

India

Specifit Chemicals
Bombay, India

Indonesia

PT Zeta Aneka Kimia
Jakarta-Selatan, Indonesia

Iran

Iran Henkel AG
P.O. Box 15875-3117
Teheran, Iran

Italy

Henkel Chimica SpA
Fraz Zingonia
Via Madrid 21
I-24049 Verdellino BG 041
Italy
Tel: +39 35 88 71 111
Fax: +39 35 48 08 93
Telex: 300175
Contact: Dr. Assuero Mancini

Henkel Chem-Plast SpA
Milano, Italy

Jamaica

Henkel Chemicals (Caribbean) Ltd.
Kingston, Jamaica

Japan

Henkel Hakusui Corp.
Nissho Iwai Building
5-8, Imabashi 2-chome
Chuo-ku, Osaka 541, Japan
Tel: +81 6 231 9277
Fax: +81 6 223 1576
Contact: Masamori SekiKatsumi Morli

Kenya

Henkel Kenya Ltd.
Nairobi, Kenya

Korea

Henkel Korea Ltd.
13th Floor
Kumbo Electric Bldg.
418 Mapo-dong, Mapo-Ku
121-050 Seoul, Korea
Tel: +82 2 706 9701
Fax: +82 2 706 6091

Sudo International Trading Inc.
Seoul, Korea

Mexico

Quimica Henkel S.A de C.V,
Mexico City, Mexico

Morocco

Coprin
Casablanca, Morocco

Namibia

S. Thalmeier
Windhoek, Namibia

Netherlands

Henkel Nederland BV
Postbus 2100
NL-3430 CM Nieuwegenein
The Netherlands
Tel: +31 30 607 3911
Fax: +31 30 605 5203

New Zealand

Henkel New Zealand Ltd.
P.O. Box 58 493
Greenmount, East Tamaki
Auckland, New Zealand
Tel: +64 9 2743164
Fax: +64 9 2743172

Adhesives Australasia Ltd.
Auckland, New Zealand

Nigeria

Henkel Chemicals (Nigeria) Ltd.
Lagos, Nigeria

Norway

Industri Agentur ANS
Lier, Norway

Peru

Kimtec SRL
Lima, Peru

Philippines

Henkel Philippines Inc.
Makati, Philippines

Portugal

Henkel Iberia SA Suc Portugal
Estr. Nacional 10 km 127
P-2615 Alverca, Portugal
Tel: +351 1 9578160
Fax: +351 1 9578206

Henkel Portuguese
Lissabon, Portugal

Saudi-Arabia

Baagiel
Damrnan, Saudi-Arabia

Singapore

Malaysia/Singapur Yeochem (M) Sdn Bhd
Singapore

South Africa

Henkel South Africa (Pty.) Ltd.
Johannesburg, South Africa

Spain

Henkel Ibérica SA
Apartado 133
E-08025 Barcelona, Spain
Tel: +34 3 290 4361
Fax: +34 3 290 4741

Sweden

Henkel Kemi AB
P.0. Box 339
S-43124 Mölndal, Sweden

Tribotec AB
Mölnlycke, Sweden

Switzerland

Henkel & Cie. AG
Hardstraße 55
CH-4133 Pratteln
Switzerland
Tel: +41 61 825 01 11
Fax: +41 61 825 04 46
Telex: 968053

Taiwan

Henkel Taiwan Ltd.
9F, 315, Tun Hua S.Rd. Sec. 2
Taipei 106, Taiwan
Tel: + 886 2 7389832
Fax: +886 2 7372608

Tanzania

Henkel Chemicals East Africa Ltd.
Dar-es-Salaam
Tanzania

Thailand

Henkel Thai Ltd.
71/1 Phyathai Rd.
Bangkok 1044
Thailand
Tel: +66 2 2500820
Fax: +66 2 2532971

Turkey

Türk Henkel A.S.
P.K. 622
Sisli/Istanbul
Turkey
Tel: +90 212 2723393
Fax: +90 212 2113477

Endüstri Teknik Ltd.
Sti. Ankara, Turkey

United Arab Emirates

Gulf International
Dubai, U.A.E.

United Kingdom

Henkel Industrial Adhesives
Henkel House
292-308, Southbury Road
Enfield
EN1 1TS, England
Tel: +44 181 804 3343
Fax: +44 181 443 4321
Product Marketing Manager: J.L. Pipe

United States of America

Henkel Adhesives Corp.
Kankakee IL, USA

Venezuela

Henkel Venezolana SA
Caracas, Venezuela

Zimbabwe

Henkel Chemicals PTV Ltd.
Harare, Zimbabwe

W.W. Henry

United States of America

California:
The W.W. Henry Company
(Headquarters)
5608 Soto Street
Huntington Park
CA 90255 - USA
Tel: +1 213 583 4961
Fax: +1 213 582 8451

W.W. Henry Co.
(West Coast Sales / Maywood Factory)
5920 Alamo Street, Maywood
CA 90270 - USA
Tel: +1 213 583 4961
Fax: +1 213 771 6188

W.W. Henry Co.
(Orange Factory)
210 W. Taft Avenue
Orange
CA 92667 - USA
Tel: +1 714 282 1240
Fax: +1 714 974 6133

Georgia:
W.W. Henry Co.
(Dallas Factory)
350 S. Industrial Park Drive
Dallas
GA 30132 - USA
Tel: +1 404 445 5377
Fax: +1 404 445 9213

Illinois:
W.W. Henry Co.
(Chicago Sales Office)
3638 South Kedzie Avenue
Chicago
IL 60632 - USA
Tel: +1 312 254 3510
Fax: +1 312 254 7624

W.W. Henry Co.
(Bourbonnais Factory)
150 Mooney Drive
Bourbonnais
IL 60914 - USA
Tel: +1 815 933 8059
Fax: +1 815 933 4326

New Jersey:
W.W. Henry Co.
(East Coast Sales & S. River Factory)
Foot of Whitehead Avenue
South River, NJ 08882 - USA
Tel: +1 908 254 2424
Fax: +1 908 254 8683

Texas:
W.W. Henry Co.
(Southwest Sales & Arlington Factory)
1101 Avenue G East
Arlington, TX 76011 - USA
Tel: +1 817 649 8070
Fax: +1 817 633 4408

Hermann Ter Hell & Co. GmbH

Agent: Scott Bader (Germany)

Hirsch Montage-Technik

Agent: Panacol-Elosol (Austria)

Hitachi Kasei Polymer Co. Ltd.

Japan

Hitachi Kasei Polymer Co Ltd.
Shikoku Bldg. 13-7, 1-chome
Uchi-Kanda, Chi yoda-ku
Tokyo 101, Japan
Tel: +81 3 3294 6421
Fax: +81 3 3293 0898
Telex: 0222 6191
Contact: K. Tomomura

Hülss & Capelli

Agent: Scott Bader (Italy)

Hungarobond Kft.

Agent: Chemetall (Hungary)

Huntsman Chemical Co.

Australia

Huntsman Chemical Co. Australia Ltd.
Somerville Road, West Footscray
3012 Victoria, Australia
Tel: +61 3 316 3333
Fax: +61 3 314 2170
Technology Manager - Phenolic Resins: Dr. M.J. Delaney

United States of America

Huntsman Corporation
2000 Eagle Gate Tower
Salt Lake City
Utah 84111-1098, USA
Tel: +1 801 532 5200
Fax: +1 801 536 1581

Hutchinson

Agent: Le Joint Français (Germany, Switzerland)

Hysol Aerospace Products

See: Dexter Hysol

P. Højstrup ApS

Agent: Henkel (Denmark)

Inbra Industrias Quimicas Ltda.

Agent: Chemetall (Brazil)

Industri Agentur ANS

Agent: Henkel (Norway)

Industrias Quimicas Löwenberg SA (Quilosa)

Industrias Quimicas Löwenberg SA
(QUILOSA)
Av. San Pablo 22
E-28820 Coslada
Madrid, Spain
Tel: +34 91 671 2100
Fax: +34 91 673 3330

(Mailing address):
Industrias Quimicas Löwenberg SA
(QUILOSA)
Apartado 2
E-28820 Coslada, Madrid, Spain

Intec Ltd.

Agent: Witco (Greece)

ITW Adhesive Systems

United Kingdom

ITW Adhesive Systems
European Division
Suite D, Best House
Grange Business Park
Enderby Road, Whetstone
Leicestershire
LE8 3JL, England
Tel: +44 116 2789878
Fax: +44 116 2789800
Business Unit Manager: Ian Snowdon

Japan Synthetic Rubber Co.

See: Ashland Chemical Co. (USA)

Jebsen & Co. Ltd.

Agent: Witco (Hong Kong)

Jebsen & Jessen Ind. Serv. Pte. Ltd.

Agent: Witco (Singapore)

Kemira

Netherlands

Kemira Polymers BV
P.O. Box 83
Mijlweg 65
NL-3300AB Dordrecht
The Netherlands
Tel: +31 78 652 7527
Fax: +31 78 651 0886
Contact: Hans Cordes

United Kingdom

Kemira Coatings
Head Office:
Rookwood Way, Haverhill
Suffolk
CB9 8PQ, England
Tel: +44 1440 703611
Fax: +44 1440 61091
Telex: 817442

Kemira Coatings
5, Broadlands Crescent
Heywood Business Park
Pilsworth Rd. Heywood
Lancashire, OL10 2RG, England
Tel: +44 161 764 6017
Fax: +44 161 764 6102

Kemira Coatings
Argall Ave. Leyton
London, E10 7QA, England
Tel: +44 181 556 8833
Fax: +44 181 558 8352

Kemira Coatings
Kelvin Way, West Bromwich
West Midlands
B70 7JZ, England
Tel: +44 121 525 5665
Fax: +44 121 553 2787
Telex: 335398

Kemira Polymers
Station Road, Birch Vale
Stockport, Cheshire
SK12 5BR, England
Tel: +44 1663 74l6518
Fax: +44 1663 746605
Telex: 669258

Kemira Polymers
P.O. Box 56, Bury
Lancashire
BL9 9WW, England
Tel: +44 161 763 3702
Fax: +44 161 763 3729

Kernback Consulting GmbH

Agent: Teroson (Poland)

Kimtec SRL

Agent: Henkel (Peru)

Klebchemie (M.G. Becker GmbH)

France

Kleiberit Chimie Sarl.
CD 63
F-67116 Reichstett, France
Tel: +33 89 75 73 75
Fax: +33 89 75 72 76

Germany

Klebchemie M.G. Becker GmbH & Co. KG
Postfach 1204
D-76353 Weingarten
Germany
Tel: +49 7244 620
Fax: +49 7244 2961
Telex: 7826693 klch

Kömmerling

Germany

Kömmerling Chemische Fabrik KG
Zweibrücker Straße 200
D-66954 Pirmasens
Germany
Tel: +49 6331 56 0
Fax: +49 6331 56 2193

France

Kömmerling France Sarl
34, Route de Haguenau
B.P. 14
F-67340 Ingwiller, France
Tel: +33 88 71 52 00
Fax: +33 88 89 51 39
Agent: Gerard Heim

Le Joint Français

Austria

(Automobile & Glazing)
Bostik GmbH
Papierfabrikstrasse 1
A-4600 Wels, Austria
Tel: +43 724 253 010

(Aerospace - PRC Adhesives)
Sedelmayer
Angeligasse 29
A-1100 Vienne, Austria
Tel: +43 222 604 34 59
Fax: +43 222 604 83 23
Telex: 133717

Denmark

Lowener
Postbox 330 - Smedeland 2
DK-2600 Glostrup, Denmark
Tel: +45 42 45 33 45
Fax: +45 43 43 04 91
Telex: 33307

France

Le Joint Français
84-116 Rue Salvador Allende
F-95870 Bezons, France
Tel: +33 1 34 23 34 23
Fax: +33 1 34 23 34 99
Telex: 609 292
Technical Contact: Mr Parra

(Aerospace - PRC Adhesives)
M. Gal Robert
Résidence Pomègues - Bât. K
33 bd Baptistin Cayol
F-13008 Marseille, France
Tel: +33 91 73 47 84
Fax: +33 91 73 22 67

Région Ouest:
(Aerospace - PRC Adhesives)
Ets. Soredis
BP 73
F-44471 Carquefou Cedex, France
Tel: +33 40 30 18 39
Fax: +33 40 30 31 81

Germany

Deutsche Hutchinson GmbH
Internehehmensbereich TDS
D-61440 Oberursel /Taunus
Germany
Automobile & Glazing:
Tel: +49 6171 503 438
Aerospace (PRC Adhesives):
Tel: +49 6171 503 150

(Aerospace - PRC Adhesives)
TDS
Eschenheimer Anlage 1
Postfach 10.21.30
D-6000 Frankfurt am Main
Germany
Tel: +49 69 405 805 0
Fax: +49 69 405 805 40
Telex: 413015 / 412904

Italy

(Automobile & Glazing)
Retel Chemie S.r.l.
Via Rivera, 124
I-10040 Almese (TO), Italy
Tel: +39 11 9359264

(Aerospace - PRC Adhesives)
Turco Italiana
Via Artgianale, 29
I-25010 Montirone (Brescia), Italy
Tel: +39 30 267 444
Fax: +39 30 267 713 7
Telex: 305352

Netherlands

(Aerospace - PRC Adhesives)
Secoa bv
Bunsenstraat 21
NL-3316 GC Dordrecht
The Netherlands
Tel: +31 78 651 4400
Fax: +31 78 651 5191

(Aerospace - PRC Adhesives)
Benecom
PO Box 2533
NL-2940 AA Lekkerkerk
The Netherlands
Tel: +31 180 661399
Fax: +31 180 661022
Telex: 24695

Scandinavia

(Automobile & Glazing)
Arly Felthaus
Holtumvej 45
DK-7100 Vejle, Denmark
Tel: +45 75 85 14 07

Spain

(Automobile & Glazing):
LJF Iberica
Berenguer de Palou 64
E-08027 Barcelona
Spain
Tel: +34 3 340 3412

(Aerospace - PRC Adhesives):
A. Perez Rodriguez
Calle de Circe, 9
URB Pinar del Plantio
E-28220 Majadahonda (Madrid)
Spain
Tel: +34 1 638 4056

CTA
Calle de Circe, 9
URB Pinar del Plantio
E-28220 Majadahonda (Madrid),
Spain
Tel: +34 1 638 4056
Fax: +34 1 639 2314

Switzerland

(Automobile & Glazing)
Bostik AG
Industriestraße 35
CH-8112 Otelfingen, Switzerland
Tel: +41 1 844 5422

(Aerospace - PRC Adhesives)
Hutchinson (Suisse) SA
Spinnereistraße 12
CH-8135 Langnau a/A
Switzerland
Tel: +41 1 713 06 72
Fax : +41 1 713 04 74
Telex: 826595

United Kingdom

(Automobile & Glazing)
LJF UK
Centurion Way
Meridian Business Park
Leicester LE3 2WH, UK
Tel: +44 116 289 1888

LePage's Ltd.

Canada

LePage's Ltd.
50 West Drive, Brampton
Ontario, L6T 2J4
Canada
Tel: +1 416 459 1140
Fax: +1 416 453 8671
V.P. (Technical): Julio Fernandez

LJF

Agent: Le Joint Français (Spain, UK)

Lobeco Products Inc.

Agent: CFPI (USA: South Carolina)

Loctite

Argentina

Loctite Argentina S.A.
Agustin de Elia 856
C.P. 1704 - Ramos Mejia
Buenos Aires, Argentina
Tel: +54 1 656 5174
Fax: +54 1 656 5175
General Manager: Graciela Fernandez

Australia

Loctite Australia Pty. Ltd.
3, Endeavour Road
Caringbah, NSW 2229, Australia
Tel: +61 2 525 8366
Fax: +61 2 525 5643
Managing Director: Gary Van Ruth

Mailing Address:
Loctite Australia Pty. Ltd.
P.O. Box 2622
Tarren Point, NSW 2229, Australia

Loctite (Australia) Pty.
Private Bag 1, Ferntree Gully
Victoria 3156, Australia
Tel: +61 3 763 0555
Fax: +61 3 763 1534

Austria

Loctite Europa GesmbH
Akaziengasse 34
A-1234 Wien, Austria
Tel: +43 1 694521 / 22
Fax: +43 1 694526
Telex: 133412
Manager - Ind./AAM: Andreas Frey
Manager - Consumer: Josef Reuter

Loctite European Group (Central & Eastern):
Manager: Rainer M. Hirtel

Mailing Address:
Loctite Europa GesmbH
P.O. Box 29, A-1234 Wien, Austria

Benelux

Loctite Belgium
Mechelsesteenweg 313
B-2550 Kontich, Belgium
Tel: +32 3 457 78 33
Fax: +32 3 457 79 60
Manager - Industry: Jules Beernaert
Manager - Consumer/AAM: Guy Blauwbioeme

Brazil & Latin America

Loctite Brasil Ltda.
Av. Prof. Vernon Krieble, 91
06690-111 Itapevi, São Paulo
Brasil
Tel: +55 11 426 4011
Fax: +55 11 426 4820
Telex: 1171956 LOQLBR
Manager - Ind./AAM: Joseph Standish
Manager - Consumer: Melvyn Fox
Manager (Latin America): Flavio Lacerda

Bulgaria

Loctite ROEES Handelsges.m.b.H.
Representation Office Bulgaria
BG-1401 Sofia, PF. 122
Bulgaria
Tel: +359 2 586132
Fax: +359 2 590192
Managing Dir.: Rainer M. Härtel
Manager - Ind./Consumer/AAM: Juri Delin

Canada

Loctite Canada Inc.
270, Britannia Rd. East
Mississauga, Ontario
Canada L4Z 1S6
Tel: +1 905 890 6511
Fax: +1 905 890 6522
Director of Operations: Doug McLean
General Sales Manager (Canada/Ohio/Indiana/Michigan): Gary Steele

Chile

Loctite Corp. Chile Ltda.
Av. Pedro de Valdivia 6052
Macul, Santiago, Chile
Tel: +56 2 2380055
Fax: +56 2 2382588
General Manager: Gaston Vaidivieso

China

Loctite (China) Co. Ltd.

(Manufacturing Facility):
Yantai Development Zone
Shandong
China 264006
Tel: +86 535 371999
Fax: +86 535 371999

(Beijing Office):
Apartment No. 626
Jinjiang Asia Hotel Beijing
8 Xinzhong Xi Jie, Gongti Bei Lu,
Beijing
China 100027
Tel: +86 1 5007788 Ext. 626
Fax: +86 1 5007307
General Manager: Baosheng Xu

Colombia

Loctite Colombia S.A.
Calle 13 No. 68-D-62
Bogotá, Colombia
Tel: +57 1 292 1525
Fax: +57 1 292 4153
Telex: 396 41465 LOCCO CO.
General Manager: Armando Yanet

Costa Rica

Loctite de Costa Rica S.A.
P.O. Box 7 - 1007 Centro Colón
San José, Costa Rica
Central America
Tel: +506 232 2033
Fax: +506 232 6859
General Manager: Lellis Vieira

Croatia

Loctite ROEES Handeisges.m.b.H.
Predstavnistvo u Republici Hrvatskoj
Gorjanska 17/11
4100 Zagreb, Croatia
Tel: +385 41 /569 163
Fax: +385 41 569 163
Manager-Ind./Cons./AAM: Igor Pureta

Czech Republic

Loctite CZ S.r.o.
Vaclavkova ul 10
CZ-16000 Praha 6
The Czech Republic
Tel: +42 2 32 25 54
Fax: +42 2 32 53 16
Manager-Industry: Petr Smejcký
Manager-Consumer/AAM: Otomar Soldat

Denmark

A/S Plastic Padding
Fjeldhammervej 4
DK-2610 Rødovre, Denmark
Tel: +45 36708855
Fax: +45 36721316
Manager-Consumer/AAM: Per Hansen

Estonia and Baltic States

Loctite ROEES Handelsges.m.b.H.
Representation Office Estonia & Baltic States
Riia 132
EE-2400 Tartu, Estonia
Tel: +372 34 792 75
Fax: +372 34 792 75
Manager - Ind./Cons./AAM: Sari Odrats

Finland

Loctite Finland Oy
Räissitie 6
SF-01510 Vantaa, Finland
Tel: +358 8 254110
Fax: +358 8 275411
Managing Director: Frank Düsing

France

Loctite France, S.A.
10, Av. Eugene Gazeau
BP 100
F-60304 Senlis Cedex
France
Tel: +33 44 21 66 00
Fax: +33 44 60 92 57
Telex: 842 140049 CHAMET
Manager - Industry: Denis Morat
Manager - Consumer/AAM: Arthur Bonemeyer

France G.I.E.
27 Avenue Etienne Audibert
F-60300 Senlis Cedex, France
Tel: +33 44 21 68 00
Fax: +33 44 53 72 15

Germany

Loctite Deutschland GmbH
Arabellastrasse 17
D-81925 München
Germany
Tel: +49 89 92680
Fax: +49 89 910 1978
Telex: 841 523266 LTD D
Manager - Industry: Fritz Löhr
Manager - Consumer: Peter Helse

Mailing Address:
Loctite Deutschland GmbH
P.O. Box 810580
D-81905 München, Germany

Loctite Europa E.E.I.G.
European Group
(details as above)
President: Heinrich Grün
Fax: +49 89 9268168
V.P. Industry - Europe: Attilio Gatti
Fax: +49 89 9268182

Loctite Europa E.E.I.G.
New Business Development
Engineering Center
Gutenbergstrasse 3
D-85748 Garching-Hochbrück
Germany
Tel: +49 89 32 080041
Fax: +49 89 32 07774
Manager: Wolfgang Fleischmann

Hong Kong

Loctite (Asia) Ltd. &
Loctite (Pacific Region Headquarters)
Room 2014 Sun Hung Kai Centre
30, Harbour Road
Wanchai, Hong Kong
Tel: +852 2827 1223
Fax: +852 2827 0525
President: G. Martin Wiley
Manager - Consumer/AAM: Tom Muldowney
Manager, Chinas: John Chapman

Hungary

Loctite Hungary Ltd.
Tarogato ut 2-4
H-1021 Budapest, Hungary
Tel: +36 1 176 28 99
Fax: +36 1 176 27 37
Manager - Industry: Koppany Juhasz
Manager - Consumer/AAM: György Babos

India

Loctite India Pvt. Ltd.
38, Marol Co-op Industrial Estate
M.V. Road, Saki Naka
Andheri (East)
Bombay 400 059, India
Tel: +91 22 836 2775 / 4031
Fax: +91 22 836 8888
Managing Director: N. Viswanath

Ireland

Loctite (Ireland) Ltd.
Manufacturing Facility
Tallaght Business Park
Whitestown, Tallaght
Dublin 24
Ireland
Tel: +353 1 4519466
Fax: +353 1 4519494
Telex: 852 90798 LOIL EI
Vice President - Worldwide Manufacturing: Peter Dowling

New Business Development:
Tel: +353 1 4510433
Fax: +353 1 4519073
Telex: 852 31068 LOIL EI
Vice President - New Business Development: Bernard Bolger

Plastic Padding
Sevenoaks, Knockrobin
Wicklow
Ireland
Tel: +353 404 69675
Fax: +353 404 69141
Manager: Mat Canavan

Italy

Loctite Italia S.p.A.
Via Talete 56
I-20047 Brugherio (MI), Italy
Tel: +39 39 882001
Fax: +39 39 884672
Telex: 843 310049 LOCTIT I
Manager - Ind.: Cinzia Glangiacomi
Manager - Consumer: Piero Farah
Manager - AAM: Claudio Striuli

Loctite Europa E.E.I.G.
European Group
(details as above)
Fax: +39 39 2872659
Vice President - Consumer/AAM - Europe: Luigi Vanini

Loctite FAS S.p.A.
Via Vigevano, 27
I-28065 Cerano (No)
Italy
Tel: +39 321 721827
Fax: +39 321 726755
General Manager: Pat Owens

Japan

Loctite (Japan) Corp.
15-13 Fukuura 1 - Chome
Kanazawa-Ku
Yokohama 236, Japan
Tel: +81 45 784 2500
Fax: +81 45 785 0747
President: Leonard J. Parker

Pacific New Business Development
1-15-13, Fukuura, Kanazawa-Ku
Yokohama 236
Japan
Tel: +81 45 784 2151
Fax: +81 45 784 2093
Director: Masahiro (Yo) Yoneno

Nagaoka Distribution Center
Kumoide Industrial Park
2647-11 Seiryo-Machi
Nagaoka, Niigata 940-21
Japan
Tel: +81 258 46 6666
Fax: +81 258 46 8456

Korea

Loctite Korea Inc.
4th Fl. Chunasan Bldg
262-1, Hannam-Dong
Yongsan-Ku, Seoul
Rep. of Korea
Tel: +82 2 790 5676 / 5671
Fax: +82 2 790 5677
Gen. Manager: Donald C. Bargelski

Malaysia

Loctite (Malaysia) Sdn. Bhd.
5, Jalan 2/118 B
Desa Tun Razak Industrial Park
56000 Cheras
Kuala Lumpur
Malaysia
Tel: +60 3 972 5091
Fax: +60 3 972 5073
Managing Director: Stephen J. Ellis
Country Manager: Stanley Yap

Mexico

Loctite Co. de Mexico S.A. de C.V.
Bosques de Radiatas No. 18
3° y 4° pisos
Bosques de Las Lomas
C.P. 11700
Mexico, D.F.
Tel: +52 5 259 00 82
Fax: +52 5 259 10 08
Operational Manager: Luis Riquelme

Norway

Loctite Norway
Sandakerveien 102
N-0483 Oslo, Norway
Tel: +47 22 151490
Fax: +47 22 153765
Managing Director: Kris Finrud

Poland

Loctite Polska Sp.z o.o.
ul. Felinskiego 37
PL-01-569 Warszawa, Poland
Tel: +48 22 39 43 32
Fax: +48 22 39 49 90
Manager - Ind.: Dariusz Zychowicz
Manager - Consumer/AAM: Andrzej Upinski

Puerto Rico

Loctite Puerto Rico Inc.
9 V. Quilinchini Avenue
Sabana Grande, PR 00637
Puerto Rico
Tel: +1809 873 6500
Fax: +1809 873 2360
General Manager, Puerto Rico: James R. Pollfka

Mailing Address:
Loctite Puerto Rico Inc.
P.O. Box 669
Sabana Grande, PR 00637
Puerto Rico

Romania

Loctite ROEES Handelsges.m.b.H.
Representation Office Romania
bloc 20, etaj 2, Ap. 12
Str. Aurel Vlaicu nr. 147
RO-72114 Bucuresti 2, Romania
Tel: +40 1 211 1041
Fax: +40 1 211 1041
Manager - Industry/Consumer/AAM: Cristian Georgescu

Russia

Loctite ROEES Handeisges.m.b.H.
Representation Office Russia
EVAG Moskau Ul. Gubkina 14
Apt. 73
GUS-1 17312 Moscow
Russia
Tel: +7 95 9382846
Fax: +7 95 9382805
Managing Director: Rainer M. Härtel

Singapore

Loctite (Singapore) Pte. Ltd.
401 Commonwealth Drive
#04-01 / 02 Haw Par Technocentre
Singapore 0314
Tel: +65 473 1822
Fax: +65 473 1322
Managing Director: Stephen J. Ellis

Slovakia

Loctite SK s.r.o.
Prokopova 31
SK-851 01 Bratislava, Slovakia
Tel: +42 7 838 678
Fax: +42 7 831 669
Manager - Industry/Consumer/AAM: Josef Pavlik

Slovenia

Loctite d.o.o.
Kajuhova 17
SLO-61 000 Ljubljana
Slovenia
Tel: +386 61 4443384
Fax: +386 61 4448988
Manager - Industry/Consumer/AAM: Gorazd Kalcic

South Africa

Loctite (SA) (Pty) Ltd.
9 Yaron Avenue
Lea Glen, Florida, Tvl. 1709
Republic of South Africa
Tel: +27 11 674 1930 /1 /2 /3
Fax: +27 11 672 5803
Managing Director: Richard Burl

Mailing Address:
Loctite (SA) (Pty) Ltd.
P.O. Box 543
Florida 1710, Transvaal
Republic of South Africa

Spain

Loctite España, S.A.
Poligono Industrial Alparrache
Parcela 56
E-28600 Navalcarnero (Madrid), Spain
Tel: +34 1 811 2512
Fax: +34 1 811 2409
Manager - Industry: Jose Morueco
Manager - AAM: Benito Gomez
Manager - Consumer: Willem de Cock Buning

Sweden

Plastic Padding AB
P.O. Box 8823
S-40271 Gothenburg, Sweden
Tel: +46 31 505400
Fax: +46 31 515154
Vice President - Nordic Group: Lars E. Berntson
Managing Director: Bengt Wiberg

Taiwan

Loctite (Taiwan) Co. Ltd.
2 Fl., 87, Lane 155
Tun Hwa North Road
Taipei, Taiwan R.O.C.
Tel: +886 2 7191325
Fax: +886 2 5140583
Deputy General Manager: Paul Chyi

Thailand

Loctite (Thailand) Ltd.
4th Floor, TVO Building
149 Ratchadapisek (Thapra-Taksin) Rd.
Bukkhalo, Thonburi
Bangkok 10600, Thailand
Tel: +66 2 447 8022
Fax: +66 2 477 9757
General Manager: Mike Peplinski

Turkey

Loctite Yapistiricilar A.S.
Büyükdere Cad.
Kircicegi Sok. B2 Blok D.8
TR-80620 Levent-Istanbul, Turkey
Tel: +90 212 269 5937
Fax: +90 212 270 3419
General Manager: Nejat Yuceturk

United Kingdom

Loctite Holdings Ltd.
Watchmead, Welwyn Garden City
Hertfordshire AL7 1JB, England
Tel: +44 1707 821000
Fax: +44 1707 821200
Telex: 851 261277 LOCTUK
Manager - Nordic Group: S. Merkel
Manager - Industry: Paul Barnard
Manager - Consumer: Steve Morran

United States of America

California:
Loctite Corp.
West Coast Technical Center
3344 North Torrey Pines Court
Suite 100
La Jolla, CA 92037 - USA
Tel: +1 619 597 0844
Fax: +1 619 597 2041
Regional Manager: John DaVirro

Connecticut:
Loctite Corp.
World Headquarters
10 Columbus Boulevard
Hartford Square North
Hartford, CT 06106 - USA
Tel: +1 203 520 5000
Fax: +1 203 520 5073

Loctite North America
1001 Trout Brook Crossing
Rocky Hill, CT 06067 - USA
Tel: +1 203 571 5100
Fax: +1 203 571 5465

Kansas:
Loctite Corp.
Automotive & Consumer Group
Manufacturing Facility
3255 Harvester Road
Kansas City, KS 66115 - USA
Tel: +1 913 621 1400
Fax: +1 913 621 1470
Plant Manager: Dave Adams

Michigan:
Loctite Corp.
One Northfield Plaza
5600 Crooks Road, Suite 105
Troy, MI 48098 - USA
Tel: +1 810 828 8000
Fax: +1 810 828 8009
Manager, Midwest & Automotive Technical Center: Allen Forgach

New Hampshire:
Loctite Luminescent Systems Inc.
Etna Road, Lebanon, NH 03766 - USA
Tel: +1 603 448 3444
Fax: +1 603 448 3452
President: Jeff Piccolomini

New York:
Banite Inc.
47 East Market Street
Buffalo, NY 14204 - USA
Tel: +1 716 853 4700
Fax: +1 716 853 4708
Manufacturing Director: Michael Flora

Loctite VSI Inc.
405 Jordan Road
Troy, NY 12180 - USA
Tel: +1 518 285 6300
Fax: +1 518 285 6346
Operations Manager: William Gelinas

Ohio:
Loctite Corp.
Automotive & Consumer Group
4450 Cranwood Parkway
Cleveland, Ohio 44128 - USA
Tel: +1 216 475 3600
Fax: +1 216 475 7616
Director of Manufacturing: Mike Flora

Loctite Corp.
Solon Distribution Center
6875 Parkland Boulevard
Solon, Ohio 44139 - USA
Tel: +1 216 475 3600
Fax: +1 216 498 1048
Director of Distribution: John Falotico

Venezuela

Loctite de Venezuela C.A.
Centro Comercial 'Los Samanes'
Piso No. 1, OFC. 102/1 01 -
Urbanización Los Samanes
Las Minas de Baruta
Caracas 1080 - Venezuela
Tel: +58 2 943 31 55
Fax: +58 2 93 88 58
Telex: 23312 REGODOY
General Manager: Jaime Lopez

Mailing Address:
Loctite de Venezuela C.A.
Apartdo 206
Caracas 1010-A Venezuela

Yugoslavia

Loctite ROEES Handeisges.m.b.H.
Representation Office Yugoslavia
Albanske spomice 15
YU-11000 Beograd
Yugoslavia
Tel: +381 11 765266 (ALTA)
Fax: +381 11 766505 (ALTA)
Manager - Industry/Consumer/AAM: Mirce Zaharijevic

Lowener

Agent: Le Joint Français (Denmark)

Lugato Chemie

Germany

Lugato Chemie Dr. Büchtemann GmbH & Co
Helbingstr. 60-62
D-22047 Hamburg (Wandsbek)
Germany
Tel: +49 40 69407 0
Fax: +49 40 69407 110
Contact: Dr. M. Meyn

Manila Triton Ind. Services Inc.

Agent: Witco (Philippines)

Marson Chilena Ltda.

Agent: Henkel (Chile)

MAS & Co.

Agent: Henkel (Egypt)

Harald Mathisen AS

Agent: Witco (Norway)

McGhan - NuSil

Germany

Dunlop GmbH
Birkenhainer Str. 77
D-63450 Hanau, Germany
Tel: +49 6181 361 0
Fax: +49 6181 361 396

United Kingdom

Dunlop Adhesives
Chester Road
Birmingham, B35 7AL
England
Tel: +44 121 373 8101
Fax: +44 121 384 2826
Specialist Adhesives: Joan Howard

United States of America

McGhan NuSil Corp.
1040 Cindy Lane
Carpinteria, CA 93013 - USA
Tel: +1 805 684 8780
Fax: +1 805 684 2365

NuSil Technology
1040 Cindy Lane
Carpinteria, CA 93013 - USA
Tel: +1 805 684 8780
Fax: +1 805 684 2365

Metasco chemisch-technische

Agent: Chemetall (Germany)

Monsanto

Belgium

Monsanto Europe S.A./N.V.
Avenue de Tervuren 270-272
(Tervurenlaan 270-272)
Letter Box No. 1
B-1150 Brussels, Belgium
Tel: +32 2 776 41 11
Fax: +32 2 776 40 40
Telex: 62927 MESA B
Contact: D. Puttaert

Brasil

Monsanto do Brasil SA
Rua Paes Leme 524
Edificio Passarelli
05424 Pinheiros
São Paulo-SP-Brazil
Tel: +55 11 815 0211
Fax: +55 11 211 9922

Canada

Monsanto Canada Ltd.
Box 787, Streetsville Postal Sta.
2330 Argentina Road
Mississauga, Ontario
Canada, L5M 2G4
Tel: +1 416 826 9222
Fax: +1 416 826 3119

Monsanto Canada Ltd.
425 St. Patrick St.
West Lasalle, Quebec
Canada, H8N 2H3
Tel: +1 514 366 4850
Fax: +1 514 366-9933

Europe

Monsanto Europe S.A.
270-272 avenue de Tervuren
B-1150 Bruxelles, Belgium
Tel: +32 2 776 41 11
Fax: +32 2 776 40 40
Market Manager Specialty Resins:
Mr. Johannes Courtier

Hong Kong

Monsanto Far East Ltd.
1304-1308 Great Eagle Ctr
23 Harbour Road, Hong Kong
Tel: +852 5 740738
Fax: +852 5 838 2026

Japan

Monsanto Japan Ltd.
Rm. 520 Kokusai Building
1-1 Marunouchi 3-Chome
Chiyoda Ku, Tokyo 100
Japan
Tel: +81 3 287 1251
Fax: +81 3 287 1250

Singapore

Monsanto Singapore Co. PTE. Ltd.
101 Thomson Road
#19-00 & #20-01/3
United Square
Singapore 1130
Tel: +65 2502000
Fax: +65 2533723

Taiwan

Monsanto Far East Ltd.
Taiwan Branch,
11th Floor, 477 Tun Hwa South Road
Taipei, Taiwan
Tel: +886 2 776 6038
Fax: +886 2 721 4493

Thailand

Monsanto Thailand Ltd.
P.O. Box 2316
Bangkok, Thailand 10500
Tel: +66 2 234 4295 / 8
Fax: +66 2 236 7847

United States of America

California:
Monsanto
24012 Calle de la Plata, Suite 250
Laguna Hills, Los Angeles
CA 92654 - USA
Tel: +1 714 855 7700
Fax: +1 714-855-7734

Illinois:
Monsanto
500 Park Blvd. Suite 595
Itasca, Chicago
IL 60143 - USA
Tel: +1 708 250 4400
Fax: +1 708 250 4432

Michigan:
Monsanto
2401 Walton Blvd.
Auburn Hills, Detroit
MI 48326 - USA
Tel: +1 313 377 6200
Fax: +1 313 377 0526

Missouri:
Monsanto Co.
800 N. Lindbergh Blvd.
St. Louis, MO 63167 - USA
Tel: +1 314 694 1000
Fax: +1 314 694 5147
Manager Adhesives & Binders: James W. Gidcumb

Ohio:
Monsanto
Whitehall Building
8041 Hosbrook Road
Cincinnati, OH 45236 - USA
Tel: +1 513 792 2800
Fax: +1 513 792 2829

Oy Movetec AB

Agent: Panacol-Elosol (Finland)

Mydrin

Australia

Mydrin Australia
Div. of Polymer Coatings & Adhesives (BTP) Pty Ltd.
19-25 Anne Street
St Marys, NSW 2760 - Australia
Tel: +61 2 623 5555
Fax: +61 2 673 3180

Denmark

Mydrin Scandinavia
Smedevangen 5
DK-3540 Lynge, Denmark
Tel: +45 4218 8566
Fax: +45 4218 8036

France

Mydrin TRL SA
Z.I. Le Lac
B. P. 314
F-07003 Privas Cedex, France
Commercial Manager: A. Madebene
Tel: +33 75.64.42.66
Export Manager: M. Bondon
Tel: +33 75.64.42.20
Domestic Manager: J.P. Fournier
Tel: +33 75.64.42.66

Mydrin Lambiotte
7 Route de Nolay
F-58700 Premery, France
Tel: +33 86 68 12 87
Fax: +33 86 37 91 30
Telex: 800605 F

TRL S.A .
Z.I. Le Lac
F-07000 Privas, France
Tel: +33 75 64 42 66
Fax: +33 75 65 81 10
Telex: 345 072 F
Commercial Director: A. Madebene

Germany

Mydrin ags Industrieklebstoffe GmbH
Industriegebiet Ost
Feldstraße 37-45
D-63179 Obertshausen, Germany
Tel: +49 6104 70040
Fax: +49 6104 70054

Mydrin GmbH
Postfach 1620
D-47515 Kleve, Germany
Tel: +49 2 821 8020
Fax: +49 2 821 802181
Telex: 811819 mykl d

India

Mydrin Shoe Components (India)
Division of RPG-BTP (India) Ltd.
4th Floor, Spencer Plaza
769 Anna Salai
Madras 600 002 - India
Tel: +91 44 852 3172
Fax: +91 44 852 3242

Italy

Mydrin Srl
Viale Marelli 165 (5th Floor)
I-20099 Sesto San Giovanni (Milano)
Italy
Tel: +39 2 2626 0099
Fax: +39 2 2622 2900

Singapore

Mydrin SE Asia Pte. Ltd.
Block 134
Jurong East Street 13
#04-309J Singapore 2260
Tel: +65 5644702
Fax: +65 5641730

United Kingdom

Mydrin Ltd.
Albion Road, Carlton, Barnsley
South Yorkshire S71 3PL, England
Tel: +44 1226 723661
Fax: +44 1226 728298
Telex: 547445

United States of America

California
Mydrin Inc.
West Coast Division
5901 Telegraph Road, Commerce
CA 90040 - USA
Tel: +1 213 724 6161
Fax: +1 213 728 8738

Georgia
Mydrin Inc.
East Coast Division,
129 Nance Road, N.E.
P.O. Box 128, Calhoun
GA 30703-0128 - USA
Tel: +1 706 625 0720
Fax: +1 706 625 1404

National Casein International

Canada

National Casein Inc.
450 Tapscott Road Unit 6
Scarborough
Ontario
Canada M1B 1Y4
Tel: +1 416 298 6888

United States of America

California:
National Casein of California
3435, West Macarthur Blvd.
Santa Ana
CA 92704-6805 - USA
Tel: +1 714 979 8400
Fax: +1 714 979 2416
Intl. Sales Manager: Tracy Scott

Illinois:
National Casein Co.
601 W 80th Street
Chicago
IL 60620-2502 - USA
Tel: +1 312 846 7300
Fax: +1 312 487 5709

New Jersey:
Nationa Casein Co. of New Jersey
Broad street, Riverton
NJ 08077-0226 - USA
Tel: +1 609 829 1880

Texas:
National Casein Co.
13039 Gaut Avenue
Tyler
TX 75708-3315 - USA
Tel: +1 903 877 3443

NuSil Technology

See: McGhan NuSil (USA)

ÖAG GmbH

Agent: CFPI (Austria)

Oaklite Products Inc.

Agent: Chemetall (USA)

Ozark Iberica SA

Agent: Chemetall (Spain)

Paksoy Kimyevi Maddeler Sari.

Agent: Witco (Turkey)

Panacol-Elosol

Austria

Hirsch Montage-Technik
Allegasse 36
A-2120 Wolkersdorf
Austria
Tel: +43-22 452671

Belgium

Qualitape N.V.
Cargovil, Havendoklaan 9
B-1804 Vilvoorde
Belgium
Tel: +32 22 55 06 30

Denmark

Stokvis Danmark A/S
Teknisk Tapr & Lim
Frydensbervey 43
P.O.Box 43
DK-3660 Stenløse
Denmark
Tel: +45 42 17 16 22

Finland

Oy Movetec AB
Titinnityntic 4
SF-02230 Espoo, Finland
Tel: +35 80 804 13 11

(Mailing Address):
Oy Movetec AB
P.O.Box 85
SF-02211 Espoo, Finland

France

Eleco Produits SA
126 Bd. Victor Hugo
F-92110 Clichy, France
Tel: +33 1 42 70 34 02

Germany

Panacol-Elosol GmbH
Obere Zeil 6-8
D-61440 Oberursel
Germany
Tel: +49 6171 6202 0
Fax:+49 6171 6202 90

Israel

Devek On (International)
I.T.M. Ltd.
12, Bograshove Street
IL-63342 Tel-Aviv
Israel
Tel: +972 3 6285348

Italy

Collanti Parini Srl
Via Pucher, 5
I-20127 Milano, Italy
Tel: +39 2 26 14 90 40

Netherlands

Stokvis Tapes & Lijmen Srl
Touwslagerstraat 5
P.O.Box 4190
NL-2980 GD Ridderkerk
The Netherlands
Tel: +31 180 441400
Fax: +31 180 416141

Spain

Arania SA
Sepulvedo 80-2-4a
E-08015 Barcelona, Spain
Tel: +34 3 426 52 52

Sweden

Garco AB
Trädgärtgatan
P.O.Box 103
S-26522 Astrop 1, Sweden

Switzerland

Panacol AG/Elosol AG
Dufourstraße 101
CH-8024 Zürich, Switzerland
Tel: +41 1 383 22 44

United Kingdom

Eurobond Adhesives Ltd.
Unit A4 Smeed Dean Centre
Eurolink Industrial Estate
Sittingboume, Kent MEI0 3RN
England
Tel: +44 1795 427888

Parker

Agent: Chemetall (Belgium, France, Portugal, Spain)

Pegatex/Molytec Ltda.

Agent: Henkel (Columbia)

Permabond

France

Permabond
Rue Grange-Morin
B.P. 438, F-69655 Villefranche-sur-Saone Cedex
France
Tel: +33 74 62 70 80
Fax: +33 74 65 12 69
Contact: Patricia Mailloche

Germany

Permabond
Im Altenschemel 55
D-67435 Neustadt, Germany
Tel: +49 6327 382231
Fax: +49 6327 382220
Contact: Jeff Wigham

United Kingdom

Permabond
Woodside Road, Eastleigh
Hampshire SO50 4EX, England
Tel: +44 1703 629628
Fax: +44 1703 629629
UK Sales & Marketing Manager, Industrial: Russell Cradduck
European Distribution Manager: Rod Milne

United States of America

Permabond Intl.
480, S. Dean St.
Englewood, NJ 07631-9990 - USA
Tel: +1 201 567 9494
Fax: +1 201 567 3747

Philippine Industrial Sealants & Coatings Cooperation

Agent: Teroson (Philippines)

Plasquim C.A.

Agent: Witco (Venezuela)

Plastic Padding

See: Loctite (Denmark, Ireland, Sweden)

Plastics & Resins Inc.

See 3E (USA)

PMG Corporation

Agent: Chemetall (Japan)

Power Adhesives Ltd.

United Kingdom

Power Adhesives Ltd.
Unit 5, New England Estate
Gascoigne Road, Barking, Essex
IG11 7LN, England
Tel: +44 181 507 8433
Fax: +44 181 507 8407
Contact: Steve Sweeney

Also: Agent for Dexter Hysol Hot-melt adhesives.

Qualitape N.V.

Agent: Panacol-Elosol (Belgium)

Quimiventas Ltda.

Agent: Witco (Columbia)

Quilosa

See: Industrias Quimicas Löwenberg SA.

Rahn AG

Agent: Witco (Switzerland)

Reef Chemical Corp. (pty) Ltd.

Agent: Chemetall (South Africa)

Oy Resin Trading AB

Agent: Witco (Finland)

Reichold Group

See: Swift Adhesives

Retel Chemie S.r.l

Agent: Le Joint Français (Italy)

Ricon Resins Inc.

United States of America

Ricon Resins Inc.
569 24 1/4 Rd.
Grand Junction
CO 81505 - USA
Tel: +1 303 245 8148
Fax: +1 303 245 4348
Contact: Todd Speece

Rinaldi

Italy

S.L.C. Rinaldi Srl.
Via dell'Industria 13
I-35020 Brugine (PD)
Italy
Tel: +39 49 5806900
Fax: +39 49 5806588
Customer Technical Service:
D. Malagutti

A. Perez Rodriguez

Agent: Le Joint Français (Spain)

Gerhard Ruff GmbH

Agent: CFPI (Germany)

SABA

Germany

SABA Klebstoff &
Abdichtungssysteme GmbH
Morchfeldstr. 27
D-68199 Mannheim
Germany
Tel: +49 621 844780
Fax: +49 621 8620602

Netherlands

SABA Dinxperlo BV
Industriestraat 3
Postbus 3
NL-7090 AA Dinxperlo
The Netherlands
Tel: +31 315 658999
Fax: +31 315 653207
Telex: 45370

SAIM

Agent: Teroson (France)

Samaro

France

Samaro
Les Ambassadeurs
7, rue Louis Guerin
BP 1272
F-69607 Villeurbanne Cedex
France
Tel: +33 72 43 50 73
Fax: +33 78 94 17 78
Contact: Jean-Pierre Decost

Santos Mouta Ltda.

Agent: Witco (Portugal)

Scholten Lijmen BV

Netherlands

Scholten Lijmen BV
Postbus 322
NL-9640 Veendam
The Netherlands
Tel: +31 598 668015
Fax: +31 598 662550
Telex: 53571
Business Manager: T. Mulder

Scott Bader

Belgium

SOMAPI SA
Vredebaan 27/28 B1
B-2510 Mortsel, Belgium
Tel: +32 3 440 19 07
Fax: +32 3 440 77 42

France

Scott Bader SA
65 Rue Sully
F-80044 Amiens Cédex 01, France
Tel: +33 22 44 42 00
Fax: +33 22 44 06 24

Strand Glass SA
ZI du Vert Galant
7 Mail Joliot Curie
F-95310 St Ouen l'Autômne
France
Tel: +33 1 30 37 51 85
Fax: +33 1 30 37 49 49

Strand Bretagne
22 rue Mancel
F-56100 Lorient, France
Tel: +33 97 37 65 30
Fax: +33 97 83 92 73

Strand Sud-Ouest
ZA Alfred Daney
37 rue Francis Garnier
F-33300 Bordeaux Nord
France
Tel: +33 56 29 11 85
Fax: +33 56 39 47 66

Strand Sud-Est
ZI 19 rue Jaques Monod
F-69680 Chassieu, France
Tel: +33 78 90 87 34
Fax: +33 78 40 10 93

Germany

Hermann Ter Hell & Co. GmbH
Kattrepelsbrücke 1
D-2000 Hamburg 1, Germany
Tel: +49 40 300 5010
Fax: +49 40 33 5050

Italy

Hüls & Capelli
Via A Cechov 48, Centro Bonola
I-20152 Milano, Italy
Tel: +39 2 334 00770
Fax: +39 2 334 00774

Netherlands

J & W Wegman
Computerweg 5
Postbus 1648
NL-3600BP Maarssen
The Netherlands
Tel: +31 346 572864
Fax: +31 346 572842
Telex: 40346

Spain

Entra Ibérica SA
Puentecesures 1B Piso 6ºB
E-28029 Madrid, Spain
Tel: +34 1 314 36 55
Fax: +34 1 314 36 76

United Kingdom

Scott Bader Co. Ltd.
Williams Way, Wollaston
Wellingborough, Northamptonshire
NN9 7RL, England
Tel: +44 1933 663100
Fax: +44 1933 663474

Scott Bader - Agents & Licensees
Australia
Brazil
Canada
Canary Isles
Columbia
Cuba
Cyprus
Egypt
Finland
Greece
Honk Kong
Irak
Kenya
Koweit
Malta
New Zealand
Nigeria
Portugal
Singapore
South Africa
Sri Lanka
Sweden
Switzerland
Trinidad
Turkey
USA
Zimbabwe

SEA

France

SEA
(Société d'Exploitation des Adhésifs)
14, Avenue de Ile-de-France
BP 204
F-27202 Vernon Cedex
France
Tel: +33 32 51 70 72
Fax: +33 32 51 82 21
Telex: Seaver 180 264 F

SEAC

Agent: CFPI (France)

Secoa bv.

Agent: Le Joint Français (The Netherlands)

Sedelmayer

Agent: Le Joint Français (Austria)

Sichel-Werke GmbH

See: Henkel (Germany)

Sicomin

France

Sicomin
F-13161
Chateauneuf-les-Martigues Cedex
France
Tel: +33 42 81 51 51
Fax: +33 42 81 29 29
Telex: 402 511 France
Technical Sales: Mr. Desplats

Sicomin (Composite)
9 bis Rue Mejou
F-29120 Pont l'Abbe
France
Tel: +33 98 87 30 93
Fax: +33 98 87 33 00

Sika

France

Sika Division Industrie
84, Blvd. Edouard Vailliant
B.P. 104
F-93351 Le Bourget Cedex
France
Tel: +33 1 49 92 80 33
Fax: +33 1 49 92 80 97
Marketing Manager: Patrice Buisson

Japan

Sika Ltd.
Morl Building No. 29
2-1 Shinbashi 4-chome
Minato-Ku, Tokyo 105
Japan

Sweden

Svenka Sika AB
Box 6009
S-17506 Järfälla, Sweden
Tel: +46 8 36 26 40

Switzerland

Sika AG
Tüffenwies 16-22
Postfach
CH-8048 Zürich
Switzerland

United Kingdom

Sika Inertol
Watchmead, Welwyn Garden City
Hertfordshire AL7 1BQ
England
Tel: +44 1707 329241
Fax: +44 1707 375593
Development Manager: T.F. Watkins

United States of America

Sika Corp.
201 Polito Avenue
Lyndhurst, NJ 07071 - USA

SOMAPI SA

Agent: Scott Bader (Belgium)

Ets. Soredis

See: Le Joint Français (France)

Specifit Chemicals

Agent: Henkel (India)

Stokvis

Stokvis Danmark A/S
Agent: Panacol-Elosol (Demark)

Stokvis Tapes & Lijmen Srl
Agent: Panacol-Elosol (Netherlands)

Strand Glass

Agent: Scott Bader (France)

Structil

France

Structil
18, rue Lavoisier
BP 10
F-91710 Vert le Petit, France
Tel: +33 1 69 90 89 89
Fax: +33 1 64 93 44 57
Technical Contact: Mr. Vignando
(Also Agent for Dexter Hysol)

Structural Adhesives Ltd.

United Kingdom

Structural Adhesives Ltd.
Bushby Brook Works
16, Spencer Street
Leicester, LE5 3NW
England
Tel: +44 116 2460766
Fax: +44 116 2765753
Managing Director: K.W. Harrison

Sudo International Trading Inc.

Agent: Henkel (Korea)

Sumitomo 3M Ltd.

See 3M (Japan)

Sunkyong Industries

Korea

Sunkyong Industries
Head Office:
60-1, 3-ga Chungmu-ro, Chung-gu
Seoul, South Korea 100-705
Tel: +82 2 273 3131
Fax: +82 2 273 3920

Sunkyong Industries
600 Jungja-dong, Jangan-gu
Suwon, Kyounggi-do, South Korea
Tel: +82 331 40 8700
Fax: +82 331 42 8251

Sunnex Inc.

See: Sunstar Engineering Inc. (USA)

Sunstar Engineering Inc.

Japan

Sunstar Engineering Inc (Head Office)
7-1 Aketa-cho Takatsuki
Osaka 569, Japan
Tel: +81 726 81 3827
Fax: +81 726 81 7632
General Manager of Automotive
Marketing: Hirohide Tomoyasu

Sunstar Engineering Inc.
R&D Centre
5-30-1 Kamihamuro Takatsuki
Osaka 569, Japan
Tel: +81 726 93 3667
Fax: +81 726 95 3622
Manager of R&D Division: Dr.Yasuhiro Okuri

Thailand

Sunstar Chemical Thailand Co. Ltd.
632 Bangpoo Industrial Estate
Soi 6 Sukhumvit Road
Samutprakarn 10280, Thailand
Tel: + 66 2 324 0652
Fax: +66 2 324 0252

United States of America

Sunnex Inc.
403 Sette Drive
Paramus, NJ 07652 - USA
Tel: +1 201 261 8188
Fax: +1 201 261 8313

Swift Adhesives

Austria

Swift Austria GmbH
Percostrasse 20
A-1220 Vienne, Austria
Tel: +43 1250 32
Fax: +43 1250 3249

Belgium

Swift Kleefstoffen Bvba
Industriepark 'Den Hoek'
Vosveld 24
B-2110 Wijnegem, Belgium
Tel: +32 3326 2410
Fax: +32 3326 3932

Canada

Swift Adhesives Ltée
108 Hymus Street, Pointe Claire
Quebec, Canada H9R 1E4
Tel: +1 514 697 5920
Fax: +1 514 697 8803

Czech Republic

Swift Adhésifs
Czecho-Slovakia Development
Valdstejnska ul. 20
CZ-118 00 Prague 1
Czech Republic
Tel: +42 2536 275
Fax: +42 22451 1170

Denmark

Swift Adhesives Denmark
Strandvejen 203
DK-2900 Hellerup, Denmark
Tel: +45 3162 6355
Fax: +45 3162 5318
Telex: 22477

Emulsio Danmark
Strandvejen 203,1
DK-2900 Hellerup, Denmark
Tel: +45 3162 6355
Fax: +45 3162 5318
Contact: Michael Steincke

Finland

Swift Adhesives Finland
Mankimiehentie 10
SF-02780 Espoo, Finland
Tel: +358 0810 491
Fax: +358 0810 309
Telex: 124407

France

Swift Adhésifs S.A.
Allée Robert Schuman
BP 1308
F-41013 Blois Cedex
France
Tel: +33 54 78 49 81
Fax: +33 54 74 67 93
Telex: 750123

Germany

Swift Klebstoff GmbH
Am Freiheitsplatz 14
D-63450 Hanau, Germany
Tel: +49 6181 258341
Fax: +49 6181 23041
Telex: 4184044

Greece

Swift Adhesives S.A.I.C.
2 Bouhoulinas Street
Kallithea 1767S, GR-Athenes, Greece
Tel: +30 1952 2981
Fax: +30 1958 1070
Telex: 221419

Hungary

Swift Adhésifs / Framarket
Varosmajor u. 12 - 14 III em312
H-1122 Budapest, Hungary
Tel: +36 1201 4855
Fax: +36 1155 3909

Netherlands

Swift Adhesives BV
Energiestraat 2 - 4
P.O.B. 20, NL-2670 AA Naaldwijk 10
The Netherlands
Tel: +31 1740 26961
Fax: +31 1740 26776

Spain

Swift Adhésifs S.A.
Carretera Nacional 11 km 593
Poligono Ind. Torre Bovera
Nave E Derecha
E-08740 San Andres de la Barca
Barcelona, Spain
Tel: +34 36 821675
Fax: +34 36 534204

Sweden

Swift Adhesives Sweden
Box 600, Tomtebogatan 2
S-551 18 Jönköping, Sweden
Tel: +46 36 302750
Fax: +46 36 302755

Switzerland

Swift Adhesifs AG
Spis 7
CH-6072 Sachseln, Switzerland
Tel: +41 60 00 55
Fax: +41 64 00 77

United Kingdom

Swift Adhesives Ltd.
3, St. George's Place, Church Street
Twickenham, Middlesex
TW1 3NE, UK
Tel: +44 181891 3201 / 2
Fax: +44 181892 9450
Telex: 932495

United States of America

North Carolina:
Swift Adhesives Inc.
Reichold Chemicals Building
2400 Ellis Road, P.O. Box 13582
Durham, NC 27703-5543 - USA
Tel: +1 919 990 7796
Fax: +1 919 990 7711

Swiss Lloyd Ltd.

Agent: Witco (Sri Lanka)

Synthomer Chemie GmbH

Agent: Chemetall (Germany)

TACC (Transworld Adhesive & Chemical Corp.)

United States of America

Transworld Adhesive & Chemical Corp.
Air Station Industrial Park
Rockland, Massachusetts
MA 02370 - USA
Tel: +1 617 878 7015
Fax: +1 617 871 6727

TDS

Agent: Le Joint Français (Germany)

Teroson (Henkel)

See also: Henkel

Argentina

Henkel Argentina SA
P-0- Box 3496
1000 Correo Central
Buenos Aires, Argentina
Tel: +54 1 204 2056
Fax: +54 1 205 3360
Contact: P. Arkenberg

Australia

Henkel Australia Pty. Ltd.
135-141 Canterbury Road
Kilsyth
Victoria 3137
Australia
Tel: +61 3 7287200
Fax: +61 3 7285877
Contact: B. Turnbull

Austria

Henkel Austria GmbH
Erdbergstr. 29
A-1031 Wien 3
Austria
Tel: +43 1 7110 4510
Fax: +43 1 7110 4612
Contact: P. Ruiner

Belgium

Henkel Belgium NV
66, Avenue du Port
Havenlaan 66
B-1210 Brüssel
Belgium

Brazil

Henkel S.A. Industrias Quimicas
Avenida das Nacoes
Unidas, 10989
CEP 04578
São Paulo 1, Brazil
Tel: +55 1 1 829 3602
Fax: +55 1 1 829 7270
Contact: H. Munoz-Kampff

China

Shanghai Henkel Teroson Adhesives & Coatings Ltd.
Guoyuan, Yancang, Nanhui County
Shanghai 201324
Peoples Republic of China
Tel: +86 21 8097097
Fax: +86 21 8097654
Contact: G. Gibson

Czech Republic

(Henkel Teroson)
Lesnicka 3
CZ-15000 Praha 5, Czech Republic

Denmark

Henkel Teroson
P Broeste A/S
Lundtoftegaardsvej 95
P.O. Box 238
DK-2800 Lyngby, Denmark

France

Teroson
150 Rue Gallinéni
F-92100 Bilan Billancourt, France
Tel: +33 1 46 84 90 00

S.A.I.M.
82 Avenue du 85 ème R.I.
F-58200 Cosne-sur-Loire, France
Tel: +33 86 28 05 67
Fax: +33 86 28 55 89
Contact: P. Maréchal

Germany

Henkel KGaA
D-40191 Düsseldorf, Germany

Teroson GmbH
P.O. Box 10 56 20
D-69046 Heidelberg, Germany
Tel: +49 6221 704253
Fax: +49 6221 704745
Contact: E. Buse
Other Countries: D.R. Oudot

Teroson GmbH (Henkel)
Hans-Bunte-Str. 4
D-69123 Heidelberg, Germany
Tel: +49 6221 704 346
Fax: +49 6221 704 415
Manager - Bonding & Sealing: Dr. Michael Hirthammer

Hungary

Teroson Gmbh
Tolnai Str 40 IV 10
H-8000 Szekesfhervar, Hungary

India

Anchemco Ltd.
77, Indl Dev. Colony
Mehrauli Road, Gurgaon 122001, India

Indonesia

Argapura
Jl. Kedoya Agave Raya
Blok A1 No.2 Tomang Tol
Raya, Jakarta 11440, Indonesia

Iran

Henkel Industrie AG
P.O. Box 15875-3117
Teheran, Iran

Italy

Henkel Chimica SpA
Via Barela 6
I-20157 Milano, Italy
Tel: +39 2357 92270
Fax: +39 2320 0724
Contact: F. Ferrazzi

Japan

Teroson Overseas Tech. Service Lab.
Teroson Kabushiki Kaisha
13-2. Kumikae
Yokosuka-cho
Toyohasni-shi. Aichi-ken, Japan
Tel: +81 532 548808
Fax: +81 532 548805
Contact: T.Sato

Henkel Hakusui Corp
Nissho Iwai Building
5-8, Imabashi 2-chome
Chuo-ku, Osaka 541, Japan

Korea

Henkel Korea Ltd.
13th Floor
Kumbo Electric Bidg.
418 Mapo-dong, Mapo-Ku
121-050 Seoul, Korea
Tel: +82 2 706 9701
Fax: +82 2 706 6091
Contact: S. Loutfy

Mexico

Henkel Mexicana S.A. de C.V.
Calz. de la Viga s/n Fracc
Los Laureles, 55090
Ecatepec de morelos, Mexico
Tel: +52 5 729 9800
Fax: +52 5 729 9803
Contact: G. Didiake / H. Galonska

Netherlands

Henkel Nederland BV.
Postbus 2100
NL-3430 CM Nieuwegein
The Netherlands

New Zealand

Henkel New Zealand Ltd.
P.O. Box 58 493
Greenmount, East Tamaki, Auckland
New Zealand
Tel: +64 9 2743164
Fax: +64 9 2743172
Contact: A. Copeland

Philippines

Philippine Industrial Sealants & Coatings Cooperation
237 Cuenca St.
Ayala Alabang Village
Muntinlupa, Philippines

Poland

Kernback Consulting GmbH
UL Nawigatorow 13
PL-04628 Warzawa
Poland

Portugal

Henkel Ibérica S.A. Portugal
Apartado 237
Estrada Nacional 10, JM 127
P-2615 Alverea, Portugal

South Africa

Chemserve Metal Sciences
PO Box 12055
Chloorkop
1624 Transvaal
South Africa
Tel: +27 11 9762162
Fax: +27 11 9762320
Contact: N. Root

Spain

Henkel Ibérica S.A.
Apartado 133
E-08025 Barcelona, Spain
Tel: +34 3 290 4361
Fax: +34 3 290 4741
Contact: R. Bacardit

Sweden

Henkel Kemi AB
P.0. Box 339
S-43124 Mölndal
Sweden

Switzerland

Henkel & Cie AG
Hardstraße 55
CH-4133 Pratteln
Switzerland

Taiwan

Henkel Taiwan Ltd.
9F, 315, Tun Hua S.Rd. Sec. 2
Taipei 106, Taiwan
Tel: + 886 2 7389832
Fax: +886 2 7372608
Contact: C. Lindsay

Thailand

Henkel Thai Ltd.
71/1 Phyathai Rd.
Bangkok 1044
Thailand
Tel: +66 2 2500820
Fax: +66 2 2532971
Contact: H. Pilz

Turkey

Türk Henkel A.S.
P.K. 622
Sisli/Istanbul, Turkey
Tel: +90 212 2723393
Fax: +90 212 2113477
Contact: Y. Aktalay

United Kingdom

Teroson UK
Cromwell Road, St Neots
Cambridgeshire
PE19 2JW England
Tel: +44 1480 404505
Fax: +44 1480 406111
Contact: M. Hull

United States of America

Teroson Inc. (North America)
55, Hayden Avenue
Lexington, Mass 02173 - USA
Tel: + 1 617 674 2930
Fax: +1 617 861 6471
Contact: J. Marotta

Yugoslavia

Vialkem
Kochstr 33/2
A-9201 Krumpendorf
Austria

S. Thalmeier

Agent: Henkel (Namibia)

ThreeBond

Australia

ThreeBond Tec (S) Pte Ltd.
Australia Branch
8 Jarrah Drive
Braeside
Victoria 3195
Australia
Tel: +61 3 587 4911
Fax: +61 3 587 4922

Brazil

ThreeBond do Brasil
Industria e Comercio Ltda.
Av Prestes Máia 315
Vila Marina, Diadema
São Paulo
Brazil
Tel: +55 11 445 4011
Fax: +55 11 445 1308

France

ThreeBond Europe SA.
Avenue des Béthunes
Parc d'Activité des Béthunes
F-95310 Saint Ouen l'Autômne
France
Tel: +33 1 30 37 02 23
Fax: +33 1 34 64 77 02

ThreeBond Europe SA.
BP 9105
F-95073 Cergy Pontoise Cédex
France
Tel: +33 1 34 32 39 60
Fax: +33 1 34 32 39 61
Telex: 607598

Germany

ThreeBond Co. Ltd.
Düsseldorf Representative Office
Alexanderstraße 31
D-40210 Düsseldorf
Germany
Tel: +49 211 322558
Fax: +49 211 323309

Hong Kong

ThreeBond Co Ltd.
Hong Kong Liason Office
11th. Floor, Universal House
229-230 Gloucester Road
Causeway
Hong Kong
Tel: +852 8363068
+852 8363745
Fax: +852 8931727

Italy

ThreeBond Co. Ltd.
Milan Liason Office
Via Senato 7
I-20121 Milano
Italy
Tel: +39 2 7602 4222
Fax: +39 2 7602 4333

Japan

ThreeBond Co. Ltd.
(World HQ)
1456 Hazama-cho Hachioji-shi
Tokyo193
Japan
Tel: +81 426 61 1333
Fax: +81 426 67 6594

Korea

ThreeBond Co Ltd.
Seoul Liason Office
Victoria B/D 709
705-1 Yoksam-dong
Kangnam, Seoul
South Korea
Tel: +82 2 527 3263
Fax: +82 2 527 3262

Mexico

ThreeBond Co. Ltd.
Mexico Liason Office
Norte Defensores de la Republica
CP 07780 Mexico DF
Tel: +52 5 567 4792
Fax: +52 5 567 3181

Singapore

ThreeBond Tec (S) Pte Ltd.
No 5 Joo Koon Cresent
Jurong
Singapore 2262
Tel: +65 264 0022
Fax: +65 863 2324

Thailand

ThreeBond Tec (S) Pte Ltd.
Thai Representative Office
40 Sukhumvit Soi 42 Road
Bankok, Thailand
Tel: +66 2 381 2432
Fax: +66 2 381 1690

United Kingdom

ThreeBond Co. Ltd.
UK Division
Milton Keynes
England

United States of America

California:
ThreeBond of America Inc.
20815 Higgins Court
Torrance
CA 90501 - USA
Tel: +1 310 320 3342
Fax: +1 310 618 9507

Georgia:
ThreeBond USA Inc.
Atlanta Branch
2 Ravinia Drive, Suite 1760
Atlanta, GA 30328 - USA
Tel: +1 404 3955380
Fax: +1 404 395 5381

Michigan:
ThreeBond USA Inc.
Detroit Branch
2000 Town Center, Suite 1480
Southfield, MI 48075 - USA
Tel: +1 313 353 2225
Fax: +1 313 353 2633

New York:
ThreeBond International Inc.
101 East 52nd Street, 28th Floor
NY 10022 - USA
Tel: +1 212 750 0611
Fax: +1 212 750 0612

Ohio:
ThreeBond USA Inc.
6184 Schumacher Park Drive
West Chester, OH 45069 - USA
Tel: +1 513 779 7300
Fax: +1 513 779 7375

Tivoli

France

Tivoli France
29 Av. Allobroges
F-26100 Romans sur Isere, France
Tel: +33 75 02 55 88
Fax: +33 75 02 58 27
Contact: Mr. Ivier

Germany

Tivoli Werke AG
Reichsbahnstr. 99
D-22525 Hamburg, Germany
Tel: +49 40 540 02 9
Fax: +49 40 540 02 250
Contact: Heike Bunk

Transworld Adhesive & Chem. Corp.

See: TACC

Tremco

United Kingdom

England:
Tremco
86-88 Bestobell Road, Slough
Berkshlre SLI 4SZ, England
Tel: +44 1753 691696
Fax: +44 1753 822640
Tretobond Ltd. Customer Services:
Tel: +44 1753 692944

Scotland:
Tretobond Scottish Office
8 Flakefield, College Milton North
East Kilbride G74 1PF, Scotland
Tel: +44 13552 44224
+44 13552 35162

Tretobond

See: Tremco (Scotland)

Tribotec AB

Agent: Henkel (Sweden)

Triton (Thailand) Ltd.

Agent: Witco (Thailand)

TRL S.A .

See: Mydrin (France)

Turco Italiana

Agent: Le Joint Français (Italy)

UHU

France

UHU France
2 bis, Rue de Sébastopol
F-92404 Courbevoie
France
Tel: +33 1 46 67 59 20
Fax: +33 1 43 34 13 27
Marketing: Sandrine Bono

Germany

UHU GmbH
Hermannstr. 7
D-77813 Bühl, Germany
Tel: +49 7223 284 0
Fax: +49 7223 284 245

Union Carbide

See: McGhan NuSil & NuSil Technology (Silicones)

United Chemicals Int.

Agent: Witco (Indonesia)

Vialkem

Agent: Teroson (Yugoslavia - via Austria)

Wacker-Chemie

Australia

Wacker Chemicals Australia Pty. Ltd.
Office: 566 St. Kilda Road
Melbourne
Victoria 3004
Australia
Tel: +61 3 5251600
Fax: +61 3 5212743
Telex: 30367 hoechst aa

(Mailing Address):
Wacker Chemicals Australia Pty. Ltd.
P.O. Box 4337
Melbourne
Victoria 3001
Australia

Austria

Wacker-Chemie Ges.m.b.H.
Kirchengasse 39
A-1070 Wien, Austria
Tel: +43 2 22 934514/15
Fax: +43 2 22 93451433

Belgium

N.V. Wacker-Chemie (Belgium) S.A.
Chaussée de La Hulpe
Terhulpsesteenweg 177b.7
B-1170 Bruxelles, Belgium
Tel: +32 2 6738135
Fax: +32 2 6602553
Telex: 21241 wacker b

Brazil

Wacker-Quimica do Brasil Ltda.
Av. Adolfo Pinheiro 2056-9º andar
Santo Amaro
04734-003 São Paulo-SP, Brazil
Tel: +55 11 5488133
Fax: +55 11 2469975
Telex: 11-56946 wack br

Czech Republic

Wacker-Chemie GmbH
Obchodní zastupitelstvi
U nem. pojištovny 4
CZ-11 000 Praha 1, Czech Republic
Tel: +42 2 24 81 11 77
Fax: +42 2 2 31 22 62
Telex: 121075

Denmark

Wacker-Chemie Denmark A/S
Hovedvejen 91
DK-2600 Glostrop, Denmark
Tel: +45 43 430300
Fax: +45 43 430316

Finland

Wacker-Chemie Finland Oy
Vattuniemenranta 2
SF-00210 Helsinki, Finland
Tel: +358 9 0 692 5155
Fax: +358 9 0 692 5145

France

Wacker-Chimie S.A.
'Le Nobel'
99, Cours Gambetta
F-69446 Lyon Cedex 03, France
Tel: +33 72 61 03 00
Fax: +33 78 95 27 45

Germany

Wacker-Chemie Gmbh
Headquarters, Division S
Hanns-Seidel-Platz 4
D-81737 München, Germany
Tel: +49 89 627901
Fax: +49 89 62791771
Telex: 529121-56

Greece

Wacker-Chemle Hellas GmbH
3, Elef. Venizelou Ave.
GR-17671 Kallithea -Athens
Greece
Tel: +30 1 923 32 84
Fax: +30 1 922 16 96
Telex: 223256 wack gr

Hungary

Wacker-Chemie GmbH
Hegyalja u. 100 A
H-1112 Budapest XI, Hungary
Tel: +36 1 186 82 24
Fax: +36 1 186 82 24
Telex: 222076 wbu h

Ireland

Wacker Chemicals Ltd.
Dublin Branch Office
4 Dartmouth Place
IRL-Dublin 6, Eire
Tel:+353 1 96 11 77
+353 1 96 19 09
Fax: +353 1 97 31 56
Telex: 90492 wack ei

Italy

Wacker-Chemie Italia SpA
Via L. Einaudi 6
I-20068 Peschiera Borromeo Mi
Italy
Tel: + 39 2 51 691.1
Fax: + 39 2 51 691.499
Telex: 314060 wacker i

Japan

Wacker Chemicals East Asia Ltd.
TOHMA Nishi-Waseda Bldg.
2-14-1, Nishi Waseda
Shinjuku-ku, Tokyo 169
Japan
Tel: +81 3 52 723121
Fax: +81 3 52 723140

Mexico

Wacker Mexicana S.A. de G,V.
Apartado Postal 99-070
Periférico Sur 3343/PH
San Jerónimo
10200 México D.F.
Tel: +52 5 957599
Fax: +52 6 838434
Telex: 1772920 wack me

Netherlands

Wacker-Chemie Nederland B.V.
Postbus 249
NL-1560 AE Krommenie
Office: Heiligeweg 166
NL-1561 DM Krommenie
The Netherlands
Tel: +31 75 628 8511
Fax: +31 75 621 5061
Telex 19079 wack ni

Norway

Wilh. Willumsen AS
P.B.6
Kveldsrosvingen 2
N-1321 Stabekk
Norway
Tel: +47 67 59 06 10
Fax: +47 67 59 19 12

Poland

Wacker-Chemie GmbH
Oddzial w Warszawie
u. Stawki 2/P.9
Intraco I
PL-00-193 Warszawa
Poland
Tel: +48 2 635 33 31
Fax: +48 2 635 65 17
Telex 815036 wchopl

Portugal

Wacker Quimira Portuguesa Lda.
Rua Braankamp 9, 5º-Esqº
P-1200 Lisboa
Portugal
Tel: +351 1 3554655
Fax: +351 1 3554785

Singapore

Wacker-Chemicals (S. Asia) Pte. Ltd.
545 Orchard Road #16-07A
Far East Shopping Centre
Singapore 0923
Tel: +65 7 33 28 11
Fax: +65 7 32 38 67
Telex rs 37460 wac sin

South Africa

Wacker Chemical (S. Africa) (Pty) Ltd.
Petrob House, 343 Surrey Ave
Ferndale
Randburg 2194
Transvaal
South Africa
Tel: +27 11 789 2780
Fax: +27 11 789 2787
Telex: 4-27773 sa

(Mailing Address):
Wacker Chemical (S. Africa) (Pty) Ltd.
P.O. Box 4330
Randburg 2125
Transvaal
South Africa

Spain

Wacker Quimica Ibérica S.A.
Córcega, 303-2º 3a
E-08008 Barcelona
Spain
Tel: +34 93 217 5900
Fax: +34 93 217 5766
Telex: 97801 wqsa e

Sweden

Wacker-Kemi AB
Gustavslundsvägen 151f
S-16136 Bromma, Sweden
Tel: +46 8 80 60 70
Fax+ 46 8 25 83 30
Telex: 15898 wackem s

(Mailing Address):
Wacker-Kemi AB
P.O. Box 12228
S-10226 Stockholm, Sweden

Switzerland

Wacker-Chemie (Schweiz) AG
Postfach, Sonnhalde 15
CH-4410 Liestal, Switzerland
Tel: +41 61 9211031
Fax: +41 61 9211147
Telex: 966042 wack ch

Taiwan

Wacker Chemicals East Asia Ltd.
Taiwan Branch
11th Floor, No. 54, Sec. 4
Min-Shen East Road, Taiwan
Tel: +886 2 7182226
Fax: +886 2 5451416

(Mailing Address):
Wacker Chemicals East Asia Ltd.
P.O. Box 87 97, Taipei, Taiwan

Turkey

Wacker-Chemie GmbH
Türkiye Irtibat Bürosu
Yeni Sülün Sokak No- 58
TR-80620 Levent-Istanbul
Turkey
Tel: +90 212 2809900
Fax: +90 212 2784470

United Kingdom

Wacker Chemicals Ltd.
The Clock Tower
Mount Felix, Bridge Street
Walton-on-Thames
Surrey. KT12 1AS, England
Tel: +44 1932 246111
Fax: +44 1932 240141
Telex: 28391 wacker g
Contact: Roger Ellaby

United States of America

Wacker Silicones Corp.
3301 Sutton Road, Adrian
MI 49221-9397 - USA
Tel: +1 517 2 648500
Fax: +1 517 2 648246

J & W Wegman

Agent: Scott Bader (Netherlands)

Wilh. Willumsen AS

Agent: Wacker-Chemie (Norway)

Witco

Austria

Witco Handelsges.mbH
Zieglergasse 5
A-1072 Wien, Austria
Tel: +43 1 522 6761
Fax:+43 1 522 6750
Telex 112448 witco a

Australia

Witco Australia (Pty.) Ltd.
53 Restwell Street
Bankstown
NSW 2200 - Australia
Tel: +61 2 709 4699
Fax: +61 2 709 3637

Brasil

Witco do Brasil Ltda.
Rua Américo Brasiliense, 1490
5º andar-conj. 57 e 58
04715-002 São Paulo-SP, Brasil
Tel: +55 11 5245409
+55 11 5240892
Fax: +55 11 5249984

Canada

Witco Canada Inc.
2 Lansing Square, Suite 1200
Willowdale, Ontario
Canada M2J 4Z4
Tel: +1 416 497 9991
Fax: +1 416 497 7110

Chile

Coninter Ltda.
San Gerardo 770
Casilia 819, Santiago, Chile
Tel: +56 2 6213859
Fax: +56 2 6211759
Telex 340778

Columbia

Quimiventas Ltda.
Calle 10A No. 68C-20
P.O.Box 16864
Santafé de Bogóta, Columbia
Tel: +57 1 2627061
Fax: +57 1 2900560
Telex 41390

Egypt

Chempro
33 Safia Zaghloul Street
PO Box 722, Alexandria, Egypt
Tel: +20 3 4836635
Fax: +20 3 4821710
Telex 54338

Finland

Oy Resin Trading AB
Unioninkatu 45a 10b
SF-00170 Helsinki 17
Finland
Tel: +358 0 1354031
Fax: +358 0 1354076
Telex 125317

France

Witco S.A. Chimie des Polyméres &
Résines Synthétiques
20, Rue de la Ville l'Evéque
F-75008 Paris, France
Tel: +33 1 44 51 05 05
Fax: +33 1 44 51 05 00

Germany

Witco GmbH
Polymerchemikalien & Kunstharze
Ernst-Schering-Str. 14, Postfach 1620
D-59180 Bergkamen
Germany
Tel: +49 2307 65-7
Fax: +49 2307 652732

Witco GmbH
Informations- kontor Nordische Länder
Rundforbivej 2B
D-2950 Vedbaek
Germany
Tel: +45 45 66 05 55
Fax: +45 45 66 05 56

Greece

Intec Ltd.
1, Menandrou Str.
GR-14561 Kifissia
Greece
Tel: +30 1 8081255
Fax: +30 1 8084963
Telex 219931

Hong Kong

Jebsen & Co. Ltd. (Chemical Division)
28/F Caroline Centre
2-38 Yun Ping Road
Causeway Bay
G.P.O. Box 97
Hong Kong
Tel: +852 922 62222
+852 922 38777
Fax: +852 882636
+852 8821949
Telex 72331 & 73769

India

Chika Ltd.
P.O. Box 3927
Bombay 400004
India
Tel: +91 22 351456
Telex 1175562
1175534

Indonesia

Pt. United Chemicals Int.
Aneka
Jl. Kesehatan Raya 21
Jakarta 10160, Indonesia
Tel: +62 21 3803805
Fax: +62 21 3803837
Telex 45310

Italy

Witco Italiana srl
Viale Lunigiana 40
I-20125 Milano, Italy
Tel: +39 2 67 01 111
Fax: +39 2 66 98 22 09

Japan

Daito Sangyo Co. Ltd.
1- 1-10 Kitahama
Chuo-Ku, Osaka 541, Japan
Tel: +81 6 202 4123
Fax: +81 6 202 1108
Telex 2655032 bmarin-j

Kenya

Chemid Kenya Ltd.
P.O. Box 47585, Nairobi, Kenya
Tel: +254 2 55291 / 2
Telex 98724233

Malaysia

Witco Asia Office
c/o Jebsen & Jessen
Industrial Services (M) Sdn.Bhd.
No.5, Jalan Bersatu (13/4)
46200 Petaling Jaya,Selangor
Darul Ehsan
Malaysia
Tel: +60 3 7562134
Fax: +60 3 7565279

(Mailing Address):
Witco Asia Office
P.O. Box 463
46750 Petaling Jaya
Selangor Darul Ehsan, Malaysia

Netherlands

Witco Polymers & Resins BV
Postbus 163
NL-1380 AD Weesp
The Netherlands
Tel: +31 294 430401
Fax: +31 294 431329

New Zealand

Amcor Trading (NZ) Ltd.
Level 1, 407 Great South Rd
Private Bag 17908
Greenlane
Auckland 1130
New Zealand
Tel: +64 9 579 6767
Fax: +64 9 579 6766

Norway

Harald Mathisen AS
Nedre Vollgt. 1
Postboks 561 -Sentrum
N-0105 Oslo 1
Norway
Tel: +47 424630
Fax: +47 422857
Telex 71718

Pakistan

Arfeeti International Ltd.
P.O. Box 10 402
Karachi 4
Pakistan
Tel: +92 21 52036168
Fax: +92 21 220008
Telex 2712

Philippines

Manila Triton Ind. Services Inc.
3rd Floor, SGS Fouridation Building
1335 G, Araneta Ave.
Quezon City
Philippines 1105
Tel: +63 742 5210
+63 742 5207
Fax:: +63 632 7323361
Telex 29569

Portugal

Santos Mouta Ltda.
Rua de Instituto 18-3
P.O. Box
P-2622, Lisboa 2
Portugal
Tel: +351 1 66 31 69
Telex 18819

Singapore

Jebsen & Jessen Ind. Serv. Pte. Ltd.
10 Neythal Road
Singapore 2262
Tel: +65 2 659800
Fax: +65 2 659148
Telex 21487

South Africa

Chempro (Pty) Ltd.
249 Dyson Road
Private Bag 14786
Wadeville 1422
South Africa
Tel: +27 11 824 2020
Fax: +27 11 824 2820

South Korea

Young Shin Ind. Co.
C.P.O. Box 1358
Seoul
South Korea
Tel: +82 2 5495191
Fax: +82 2 5471902
Telex 29702

Spain

Witco Espana S.L.
Polimeros & Resinas
Dr. Esquerdo 125
E-28007 Madrid
Spain
Tel: +34 1 573 1005
Fax: +34 1 504 1398

Sri Lanka

Swiss Lloyd Ltd.
RM 2-205 BMICH
Bauddhaloka Mawatha
Colombo 7
Sri Lanka
Tel: +94 1 91604
+94 1 96421 / 4 Ext 69
Fax: +94 1 599632
+94 1 500544
Telex 22217 & 22094

Switzerland

Rahn AG
Dörflistr. 120
CH-8050 Zürich
Switzerland
Tel: +41 1 3121512
Fax: +41 1 3122150 / 52
Telex 822030

Thailand

Triton (Thailand) Ltd.
P.O. Box 175
Bangkok 10330
Thailand
Tel: +66 2 258 3958 65
Fax: +66 2 2531 705
Telex 84041

Turkey

Paksoy Kimyevi Maddeler Sari. ve
Tic AS
G.O.S.B
700 Sokak
TR-41480 Gebze, Turkey
Tel: +90 262 751 1351/10
Fax:+90 262 751 0258

United Kingdom

Witco Corp.UK Ltd.
Polymer Chemicals & Resins
Paragon Works - Baxenden
Nr. Accrington
Lancashire BB5 2SL
England
Tel: +44 1254 398616
Fax: +44 1254 398586

United States of America

Witco Corp.
One American Lane
Greenwich, CT 06831 - USA
Tel: +1 203 552 2000
Fax: +1 203 552 2860

Uruguay

Enzur S.A.
Azara 3787
Montivideo
Uruguay
Tel: +598 2 580505
Fax: +598 2 580504
Telex 22161

Venezuela

Plasquim C.A.
Avenida Tamanaco
Edificio Impres
Oficina 701/703/705
Caracas
Venezuela
Tel: +58 2 9513155
Fax: +58 2 9513390
Telex 28717

Wulff

Germany

Wulff GmbH & Co
Wersener Str. 3
D-49504 Lotte
Germany
Tel: +49 5404 8810
Fax: +49 5404 88119

Yeochem (M) Sdn Bhd

Agent: Henkel (Malaysia/Singapore)

Young Shin Ind. Co.

Agent: Witco (South Korea)

Ytteknik

Agent: Chemetall (Sweden)

Zeta Aneka Kimia

Agent: Henkel (Indonesia)

MANUFACTURERS & SUPPLIERS - COUNTRIES

ALGERIA

Henkel Algeria

ARGENTINA

Henkel Argentina SA (Teroson / Henkel)
Loctite Argentina SA

AUSTRALIA

New South Wales:
3M Australia Pty. Ltd.
Ciba Australia Ltd.
Dow Corning Australia Pty Ltd.
Epirez Construction Products
Loctite Australia Pty. Ltd.
Mydrin Australia
Witco Australia (Pty.) Ltd.

North Queensland:
Epirez Construction Products

South Australia:
Epirez Construction Products

South Queensland:
Epirez Construction Products

Victoria:
3M Australia Pty. Ltd.
Ciba-Geigy Australia Ltd.
Epirez Construction Products
GE Silicones
Henkel Australia Pty. Ltd. (Teroson / Henkel)
Huntsman Chemical Co. Australia Ltd.
Loctite (Australia) Pty.
ThreeBond Tec (S) Pte Ltd
Wacker Chemicals Australia Pty. Ltd.

Western Australia:
Epirez Construction Products

AUSTRIA

Bostik GmbH (Le Joint Français)
Ciba-Geigy GmbH
Dow Corning GmbH
Dyckerhoff Austria Ausbauprodukte GesmbH
Forbo Industrieprodukte GesmbH (See: Forbo-Helmitin)
Henkel Austria GmbH (Teroson / Henkel)
Hirsch Montage-Technik (Agent: Panacol-Elosol)
Loctite Europa GesmbH
ÖAG GmbH (See: CFPI)
Sedelmayer (Le Joint Français)
Swift Austria GmbH
Wacker-Chemie Ges.m.b.H.
Witco Handelsges.mbH

BELGIUM

3M (Europe) SA
3M Belgium NV/SA
CECA Belgium SA
S.A. Ciba-Geigy N.V.
Dow Corning Europe
Grace NV
Henkel Belgium NV (Teroson / Henkel)
Loctite Belgium
Monsanto Europe S.A./N.V.
Parker Benelux NV (Agent: Chemetall)
Qualitape N.V. (Agent: Panacol-Elosol)
SOMAPI SA (Agent: Scott Bader)
Swift Kleefstoffen Bvba
N.V. Wacker-Chemie (Belgium) S.A.

BRAZIL

Ciba-Geigy Quimica SA
Diadema Industrias Quimicas Ltda. (Agent: Chemetall)
Dow Corning do Brazil Ltda.
Henkel S.A. Industrias Quimicas (Teroson / Henkel)
Inbra Industrias Quimicas Ltda. (Agent: Chemetall)
Loctite Brasil Ltda.
Monsanto do Brasil SA
ThreeBond do Brasil
Wacker-Quimica do Brasil Ltda. (Wacker-Chemie)
Witco do Brasil Ltda.

BULGARIA

Loctite ROEES Handelsges.m.b.H.

CANADA

3M
Ciba-Geigy Canada Ltd.
Helmitin (Canada) Inc. (See: Forbo-Helmitin)
General Electric Canada Inc. (GE Silicones)
LePage's Ltd.
Loctite Canada Inc.
Monsanto Canada Ltd.
National Casein Inc.
Swift Adhesives Ltée
Witco Canada Inc.

CHILE

Coninter Ltda. (Agent: Witco)
Loctite Corp. Chile Ltda.
Marson Chilena Ltda. (See: Henkel)

CHINA

Loctite (China) Co. Ltd.
Shanghai Henkel Teroson Adhesives & Coatings Ltd.

COLUMBIA

Loctite Colombia S.A.
Pegatex/Molytec Ltda. (See: Henkel)
Quimiventas Ltda. (Agent: Witco)

COSTA RICA

Loctite de Costa Rica S.A.

CROATIA

Loctite ROEES Handeisges.m.b.H.

CZECH REPUBLIC

Loctite CZ S.r.o.
Henkel Teroson
Swift Adhésifs
Wacker-Chemie GmbH

DENMARK

3M A/S
Arly Felthaus (Le Joint Français)
Bostik A/S
Casco Nobel A/S
Chemetall GmbH Filial Danmark
Ciba-Geigy A/S
Emulsio Danmark (Agent: Swift Adhesives)
Henkel Barnängen A/S (Teroson / Henkel)
P. Højstrup ApS (See: Henkel)
Lowener (Le Joint Français)
Mydrin Scandinavia
A/S Plastic Padding (See: Loctite)
Stokvis Danmark A/S (Agent: Panacol-Elosol)
Swift Adhesives Denmark
Wacker-Chemie Denmark A/S

ECUADOR

EMS de Latinoamerica SA

EGYPT

Chempro (Agent: Witco)
Framchem (See: CFPI)
MAS & Co. (See: Henkel)

ESTONIA

Loctite ROEES Handelsges.m.b.H.

FINLAND

Oy Bonder AB (Agent: Chemetall)
Casco Nobel Oy
Ciba Geigy OY
Loctite Finland Oy
Oy Movetec AB (Agent: Panacol-Elosol)
Oy Resin Trading AB (Agent: Witco)
Oy Soumen Henkel Liimat AB
Swift Adhesives Finland
Wacker-Chemie Finland Oy

FRANCE

3M France
Aerospheres France
Ashland Chemical Co.
Bostik
CACI (See: CFPI)
Casco Nobel Industrie SnC
CECA SA
CFPI
CHRYSO (Mastics et Resines)
S.A. Ciba-Geigy
Dow Corning Construction SA
Eleco Produits SA (Agent: Panacol-Elosol)
EMFI
EMS-France SA
ES (Agent: Dymax)
Framchem (See: CFPI)
France G.I.E. (See: Loctite)
M. Gal Robert (Le Joint Français)
Grace SA
Helmitin S.A. Colles Industrielles (See: Forbo-Helmitin)
Henkel France S.A.
Kleiberit Chimie Sarl.
Kömmerling France Sarl
Le Joint Français
Loctite France, S.A.
Mydrin Lambiotte
Mydrin TRL SA
Permabond
S.A.I.M. (Agent: Teroson / Henkel)
Samaro
Scott Bader SA
SEA
SEAC (See: CFPI)
Sicomin
Sicomin (Composite)
Sika Division Industrie
Société Continental Parker SA (Agent: Chemetall)
Ets. Soredis (Le Joint Français)
Strand Bretagne (Agent: Scott Bader)
Strand Glass SA (Agent: Scott Bader)
Strand Sud-Ouest (Agent: Scott Bader)
Strand Sud-Est (Agent: Scott Bader)
Structil (Also Agent: Dexter Hysol)
Swift Adhésifs S.A.
Teroson (Teroson / Henkel)
ThreeBond Europe SA
Tivoli France
TRL S.A . (See: Mydrin)
UHU France
Wacker-Chimie S.A.
Witco S.A. Chimie des Polyméres & Résines Synthétiques

GERMANY

3M Deutschland GmbH
Aerospheres
Ashland Suedchemie Kernfest (Ashland Chemical Co.)
M.G. Becker GmbH (Klebchemie)
Bostik Gmbh
Casco-Nordsjo GmbH (Casco Nobel)
CECA Klebstoff GmbH
Chemetall GmbH
Ciba-Geigy GmbH
Degussa AG
DELO Industrieklebstoff GmbH & Co.
DELO GmbH & Co. KG
Dexter
Dow Corning GmbH
Dunlop GmbH (Also Agent: McGhan - NuSil)
Dyckerhoff Sopro GmbH
Dymax Europe GmbH
EMS-Chemie (Deutschland) GmbH
Forbo-Erfurt GmbH (See: Forbo-Helmitin)
Forbo-Helmitin GmbH
Grace NV
Henkel KGaA (Teroson / Henkel)
Hermann Ter Hell & Co. GmbH (Agent: Scott Bader)
Hutchinson (Deutsche) GmbH (Le Joint Français)
Hysol Aerospace Products
Klebchemie (M.G. Becker GmbH)
Kömmerling Chemische Fabrik KG
Loctite Deutschland GmbH
Loctite Europa E.E.I.G.
Lugato Chemie Dr. Büchtemann GmbH & Co
Metasco chemisch-technische Produkte GmbH (Agent: Chemetall)
Mydrin ags Industrieklebstoffe GmbH
Mydrin GmbH
Panacol-Elosol GmbH
Permabond
Gerhard Ruff GmbH (See: CFPI)
SABA Klebstoff & Abdichtungssysteme GmbH
Sichel-Werke GmbH (See: Henkel)
Swift Klebstoff GmbH

Germany (continued)

Synthomer Chemie GmbH (Agent: Chemetall)
TDS (Le Joint Français)
Teroson GmbH (Teroson / Henkel)
ThreeBond Co. Ltd
Tivoli Werke AG
UHU GmbH
Wacker-Chemie Gmbh
Witco GmbH
Wulff GmbH & Co

GREECE

Ciba-Geigy Hellas S.A.
Intec Ltd. (Agent: Witco)
Swift Adhesives S.A.I.C.
Wacker-Chemie Hellas GmbH

GUATEMALA

Quimica Henkel Centroamericana SA

HONG KONG

Ciba-Geigy (Hong Kong) Ltd.
General Electric Plastics (GE Silicones)
Henkel Chemicals (H-K) Ltd.
Jebsen & Co. Ltd. (Agent: Witco)
Loctite (Asia) Ltd.
Monsanto Far East Ltd.
ThreeBond Co Ltd.

HUNGARY

Hungarobond Kft. (Agent: Chemetall)
Loctite Hungary Ltd.
Teroson Gmbh (Teroson / Henkel)
Swift Adhésifs / Framarket
Wacker-Chemie GmbH

INDIA

Anchemco Ltd. (Agent: Teroson / Henkel)
Chika Ltd. (Agent: Witco)
Loctite India Pvt. Ltd.
Mydrin Shoe Components (India)
Specifit Chemicals (See: Henkel)

INDONESIA

Argapura (Agent: Teroson / Henkel)
Pt. United Chemicals Int. (Agent: Witco)
Pt Zeta Aneka Kimia (See: Henkel)

IRAN

Bonder Iran SSK (Agent: Chemetall)
Henkel Industrie AG (Teroson / Henkel)
Iran Henkel AG

IRELAND

3M (Ireland) Ltd.
Bostik Industrial Chemicals
Loctite (Ireland) Ltd.
Plastic Padding (See: Loctite)
Wacker Chemicals Ltd.

ISRAEL

Devek On (International) (Agent: Panacol-Elosol)

ITALY

3M Italia SpA
Aerospheres
Barattini Polar SpA (Agent: Chemetall)
Bonder SpA (Agent: Chemetall)
CECA Italiana SpA
Ciba-Geigy S.p.A
Collanti Parini Srl (Agent: Panacol-Elosol)
Dow Corning SA
Grace Italiana S.p.A.
Henkel Chem-Plast SpA
Henkel Chimica SpA (Teroson / Henkel)
Hüls & Capelli (Agent: Scott Bader)
Loctite Italia S.p.A.
Loctite Europa E.E.I.G.
Loctite FAS S.p.A.
Mydrin Srl
Retel Chemie Srl. (Le Joint Français)
S.L.C. Rinaldi Srl.
ThreeBond Co. Ltd.
Turco Italiana (Le Joint Français)
Wacker-Chemie Italia SpA
Witco Italiana srl

JAMAICA

Henkel Chemicals (Caribbean) Ltd.

JAPAN

Ciba-Geigy (Japan) Ltd.
Daito Sangyo Co. Ltd. (Agent: Witco)
Dow Corning Kabushiki Kaisha
EMS-Japan Corp.
Japan Synthetic Rubber Co. (See: Ashland Chemical Co.)
Henkel Hakusui Corp (Teroson / Henkel)
Hitachi Kasei Polymer Co. Ltd.
Loctite (Japan) Corp.
Monsanto Japan Ltd.
PMG Corporation (Agent: Chemetall)
Sika Ltd
Sumitomo 3M Ltd.
Sunstar Engineering Inc.
Teroson Overseas Tech. Service Lab. (Teroson / Henkel)
ThreeBond Co. Ltd.
Wacker Chemicals East Asia Ltd.

KENYA

Chemid Kenya Ltd. (Agent: Witco)
Henkel Kenya Ltd.

KOREA (SOUTH)

GE Silicones Korea
Henkel Korea Ltd (Teroson / Henkel)
Loctite Korea Inc.
Sudo International Trading Inc.
Sunkyong Industries
ThreeBond Co Ltd.
Young Shin Ind. Co. (Agent: Witco)

MALAYSIA

Loctite (Malaysia) Sdn. Bhd.
Witco Asia Office

MEXICO

Ciba-Geigy Mexicana SA de CV
GE Silicones Mexico
Henkel Mexicana S.A. de C.V. (Teroson / Henkel)
Loctite Co. de Mexico S.A. de C.V.
Quimica Henkel S.A de C.V,
ThreeBond Co. Ltd.
Wacker Mexicana S.A. de G,V. (See: Wacker-Chemie)

MOROCCO

Coprin (See: Henkel)

NAMIBIA

S. Thalmeier (See: Henkel)

THE NETHERLANDS

3M Nederland BV
Benecom (Le Joint Français)
CECA Nederlan BV
Cetema (Agent: Chemetall)
Ciba-Geigy BV
Forbo Genderen B.V. (See: Forbo-Helmitin)
General Electric Plastics B.V. (GE Silicones)
Henkel Nederland BV. (Teroson / Henkel)
Kemira Polymers BV
SABA Dinxperlo BV
Scholten Limjen BV
Secoa bv (Le Joint Français)
Stokvis Tapes & Lijmen Srl (Agent: Panacol-Elosol)
Swift Adhesives BV
Wacker-Chemie Nederland B.V.
J & W Wegman (Agent: Scott Bader)
Witco Polymers & Resins BV

NEW ZEALAND

Adhesives Australasia Ltd. (Agent: Henkel)
Amcor Trading (NZ) Ltd. (Agent: Witco)
Ciba-Geigy New Zealand Ltd.
Henkel New Zealand Ltd. (Teroson / Henkel)

NIGERIA

Henkel Chemicals (Nigeria) Ltd.

NORWAY

Casco Nobel
Ciba-Geigy A/S
Industri Agentur ANS (See: Henkel)
Loctite Norway
Harald Mathisen AS (Agent: Witco)
Wilh. Willumsen AS (Agent: Wacker-Chemie)

PAKISTAN

Arfeeti International Ltd (Agent: Witco)

PERU

Kimtec SRL (See: Henkel)

PHILIPPINES

Henkel Philippines Inc.
Manila Triton Ind. Services Inc. (Agent: Witco)
Philippine Industrial Sealants & Coatings Cooperation
(Agent: Teroson / Henkel)

POLAND

Gardobond Sp.zo.o (Agent: Chemetall)
Loctite Polska Sp.z o.o.
Kernback Consulting GmbH (Agent: Teroson / Henkel)
Wacker-Chemie GmbH

PORTUGAL

Bostik Colas E Vedantes Lda.
Ciba-Geigy Portuguese Lda.
Henkel Ibérica S.A. Portugal (Teroson / Henkel)
Henkel Portuguese
Parker Lusitania Lda (Agent: Chemetall)
Santos Mouta Ltda. (Agent: Witco)
Wacker Quimira Portuguesa Lda. (Wacker-Chemie)

PUERTO RICO

3M
Loctite Puerto Rico Inc.

ROMANIA

Loctite ROEES Handelsges.m.b.H.

RUSSIA

Loctite ROEES Handeisges.m.b.H.

SAUDI-ARABIA

Baagiel (Agent: Henkel)

SINGAPORE

Casco Nobel (SEA) Pte Ltd
Ciba-Geigy S.E. Asia (Pte.) Ltd.
Dexter Distributor Programs
GE Plastics Pacific Ltd. (GE Silicones)
Jebsen & Jessen Ind. Serv. Pte. Ltd. (Agent: Witco)
Loctite (Singapore) Pte. Ltd.
Malaysia/Singapur Yeochem (M) Sdn Bhd (See: Henkel)
Monsanto Singapore Co. PTE. Ltd.
Mydrin SE Asia Pte. Ltd.
ThreeBond Tec (S) Pte Ltd
Wacker-Chemicals (S. Asia) Pte. Ltd.

SLOVAKIA

Loctite SK s.r.o.

SLOVENIA

Loctite d.o.o.

SOUTH AFRICA

Chempro (Pty) Ltd. (Agent: Witco)
Chemrite (Pty) Ltd (Agent: Chemetall)
Chemserve Metal Sciences (Agent: Teroson / Henkel)
Henkel South Africa (Pty.) Ltd.
Loctite (SA) (Pty) Ltd.
Reef Chemical Corp (pty) Ltd. (Agent: Chemetall)
Wacker Chemical (S. Africa) (Pty) Ltd.

SPAIN

3M España S.A.
Arania SA (Agent: Panacol-Elosol)
Bostik SA
Ciba-Geigy S.A.
CTA (Le Joint Français)
Dow Corning Kkrafft SA
Entra Ibérica SA (Agent: Scott Bader)
ETISA (See: CFPI)

Spain (continued)

Henkel Ibérica S.A. (Teroson / Henkel)
Industrias Quimicas Löwenberg SA (QUILOSA)
LJF Iberica (Le Joint Français)
Loctite España, S.A.
Ozark Iberica SA (Agent: Chemetall)
A. Perez Rodriguez (Le Joint Français)
Sociedad Continental Parker Hisp. SA (Agent: Chemetall)
Swift Adhésifs S.A.
Wacker Quimica Ibérica S.A. (Wacker-Chemie)
Witco Espana S.L.

SRI LANKA

Swiss Lloyd Ltd. (Agent: Witco)

SWEDEN

Bostik AB
Casco Nobel AB
Ciba-Geigy AB
International General Electric AB (GE Silicones)
Garco AB (Agent: Panacol-Elosol)
Grace AB
Henkel Kemi AB (Teroson / Henkel)
Plastic Padding AB (See: Loctite)
Svenka Sika AB
Swift Adhesives Sweden
Tribotec AB (See: Henkel)
Wacker-Kemi AB (Wacker-Chemie)
AB Ytteknik (Agent: Chemetall)

SWITZERLAND

3M Schweiz AG
Bostik AG (Le Joint Français)
Ciba-Geigy AG
CM-Celfa Mebrantrenntechnik (Agent: Chemetall)
CM Oberflächentechnik AG (Agent: Chemetall)
EMS-Chemie AG
Forbo CTU AG (See: Forbo-Helmitin)
Geistlich Söhne AG
Henkel & Cie AG (Teroson / Henkel)
Hutchinson (Suisse) SA (Le Joint Français)
Panacol AG/Elosol AG
Rahn AG (Agent: Witco)
Sika AG
Swift Adhesifs AG
Wacker-Chemie (Schweiz) AG

TAIWAN

EMS-Far Eastern Ltd.
General Electric Tech. Service Co. Inc. (GE Silicones)
Henkel Taiwan Ltd. (Teroson / Henkel)
Loctite (Taiwan) Co. Ltd
Monsanto Far East Ltd.
Triton (Thailand) Ltd. (Agent: Witco)
Wacker Chemicals East Asia Ltd.

TANZANIA

Henkel Chemicals East Africa Ltd.

THAILAND

Henkel Thai Ltd. (Teroson / Henkel)
Loctite (Thailand) Ltd.
Monsanto Thailand Ltd.
Sunstar Chemical Thailand Co. Ltd.
ThreeBond Tec (S) Pte Ltd.

TURKEY

Bonder Kintya ve Sanayi AS (Agent: Chemetall)
Endüstri Teknik Ltd. (See: Henkel)
Loctite Yapistiricilar A.S.
Paksoy Kimyevi Maddeler Sari. ve (Agent: Witco)
Türk Henkel A.S. (Teroson / Henkel)
Wacker-Chemie GmbH

UNITED ARAB EMIRATES

Gulf International (See: Henkel)

UNITED KINGDOM

3M (UK) plc.
Aerospheres (UK) Ltd.
Apollo Chemicals Ltd.
Bonder UK - Branch of Cetema Ltd. (Agent: Chemetall)
Borden (UK) Ltd.
Bostik Ltd.
BTP plc. (See: Mydrin)
Wm. Canning Ltd.
Casco Nobel
Cetema Ltd. (Agent: Chemetall)
Ciba Polymers
Cytec Aerospace Ltd.
Dow Corning Hansil Ltd.
Dunlop Adhesives (Also Agent: McGhan - NuSil)
Emerson & Cuming UK Ltd. (See: Grace)
EMS-Grilon (UK) Ltd.
Eurobond Adhesives Ltd. (Agent: Panacol-Elosol)
Evacor Resins Ltd.
Fortafix Ltd.
GE Silicones
Henkel Industrial Adhesives
ITW Adhesive Systems
Kemira Coatings
Kemira Polymers
LJF UK (Le Joint Français)
Loctite Holdings Ltd.
Mydrin Ltd.
Permabond
Power Adhesives Ltd.
Scott Bader Co. Ltd.
Sika Inertol
Structural Adhesives Ltd.
Swift Adhesives Ltd.
Teroson UK (Teroson / Henkel)
ThreeBond Co. Ltd.
Tremco
Tretobond (See: Tremco)
Wacker Chemicals Ltd.
Witco Corp.UK Ltd.

UNITED STATES OF AMERICA

Alaska:
3M

Califomia:
3M
Ciba-Geigy Corp. / Furane Products
Cytec
Cytec Engineered Materials Inc.
Dexter Hysol
Dow Corning USA
W.W. Henry Co.
Hysol Aerospace Products
Loctite Corp.
McGhan - NuSil

United States of America - California (continued)

Monsanto
Mydrin Inc.
National Casein of California
NuSil Technology (McGhan - NuSil)
ThreeBond of America Inc.

Colorado:
Ricon Resins Inc.

Connecticut:
Cytec Engineered Materials Inc.
Dymax Corporation
Hysol Aerospace Products
Loctite Corp.
Loctite North America
Witco Corp.

Florida:
American Technologies International

Georgia:
3M
W.W. Henry Co.
Hysol Aerospace Products
Mydrin Inc.
ThreeBond USA Inc.

Hawaii:
3M

Illinois:
Aerospheres
Henkel Adhesives Corp.
W.W. Henry Co.
Monsanto
National Casein Co.

Kansas:
Hysol Aerospace Products
Loctite Corp.

Maryland:
Cytec Engineered Materials Inc.
Cytec

Massachusetts:
Emhart Fastening Systems Group (See: Bostik)
TACC (Transworld Adhesive & Chemical Corp.)
Teroson Inc. - North America (Teroson / Henkel)

Michigan:
3M
Ashland Chemical Co.
Ciba-Geigy Corp.
Dow Corning Center
Loctite Corp.
Monsanto
ThreeBond USA Inc.
Wacker Silicones Corp. (Wacker-Chemie)

Minnesota:
3M Aerospace

Missouri:
Monsanto Co.

New Hampshire:
Dexter Automotive Materials Division
Dexter Adhesives & Structural Materials Div.
Loctite Luminescent Systems Inc.

New Jersey:
3E Corp.
Aerospheres
Andek Chemical Corp. (See: 3E Corp.)
CM Chemical Products (Agent: Chemetall)
Cytec
Electro-Seal Corp.
W.W. Henry Co.
Hysol Aerospace Products
Oaklite Products Inc. (Agent: Chemetall)
Plastics & Resins Inc. (See: 3E Corp.)
Nationa Casein Co. of New Jersey
Permabond Intl.
Sika Corp.
Sunnex Inc. (See: Sunstar Engineering Inc.)

New York:
3M
Banite Inc. (See: Loctite)
Conap Inc. (See: Cytec)
Cytec Engineered Materials Inc.
General Electric Co. (GE Silicones)
Loctite VSI Inc.
ThreeBond International Inc.

North Carolina:
Swift Adhesives Inc.

Ohio:
Ashland Chemical Co.
Loctite Corp.
Monsanto
ThreeBond USA Inc.

Pennsylvania:
EMS Acquisition Corp.

South Carolina:
EMS-American Grilon Inc.
Lobeco Products Inc. (See: CFPI)

Texas:
3M
W.W. Henry Co.
Hysol Aerospace Products
National Casein Co.

Washington:
Hysol Aerospace Products

URUGUAY

Enzur S.A. (Agent: Witco)

VENEZUELA

Henkel Venezolana SA
Loctite de Venezuela C.A.
Plasquim C.A. (Agent: Witco)

YUGOSLAVIA

Loctite ROEES Handeisges.m.b.H.
Vialkem (Agent: Teroson / Henkel)

ZIMBABWE

Henkel Chemicals PTV Ltd.

USEFUL NAMES & ADDRESSES

Adhesive & Sealant Manufacturers Associations

AUSTRALIA

ASMAA - Adhesives & Sealants Manufacturers Association of Australia Inc.
Private Bag No. 938
North Sydney
New South Wales 2059
Australia

ASMAA
Chamber of Manufacturers of NSW (West Sydney Office)
P.O. Box 2538
North Parramatta
New South Wales 2151
Australia
Tel: +61 02 630 4988
Fax: +61 02 630 4641

AUSTRIA

Vereingung Österreichischer Klebstofferzeuger
c/o Wegscheider Farben GmbH & Co. KG
P.O. Box 327
A-4021 Linz /00
Austria
Dr. H. Bukowiecki - President
Tel: +43 732 85080
Fax: +43 732 86836

BELGIUM

DETIC
Comité Professionel BELCAM
Square Marie Louise 49
B-1040 Bruxelles
Belgium
Robert Castelein - President
A. Vercruysse - Secretary
Tel: +32 2 2389711
Fax: +32 2 2308288

CANADA

Adhesives & Sealants Manufacturers Association of Canada,
208, Brimorton Drive
Scarborough
Ontario M1H 2C6
Canada
John Martin - Secretary Treasurer
Tel: +1 416 431 1330
Fax: +1 416 431 5223

DENMARK

Brancheforeningen for Lim og Fugemasser
(Danish Adhesives & Sealants Manufacturers)
Norre Volgade 48
DK-1358 Kobenhavn K
Denmark
Preben Andersen - President
Vibeke Plambeck - Secretary
Tel: +45 33 159222
Fax: +45 33 939322

EUROPE

FEICA - Association of European Adhesive Manufacturers
P.O. Box 230169
D-40087 Düsseldorf
Germany

Ivo-Beucker-Strasse 43
D-40237 Düsseldorf
Germany

Jürgen Wolter - Permanent Secretary
Tel: +49 211 67931 30
Fax: +49 211 67931 88

FINLAND

Kemianteollisuus RY
(Chemical Industry Federation)
P.O. Box 4
Eteläranta 10
FIN-00131 Helsinki
Finland
Errki Solja - President
Arto Jokinen - Secretary
Tel: +358 0 172841
Fax: +358 0 630225

FRANCE

Syndicat Français des Colles et Adhésifs
c/o FIPEC
42 Avenue Marceau
F-75008 Paris
France
Daniel Brulé - President
Jean-Noël Simonot - Secretary
Tel: +33 1 47233612
Fax: +33 1 47209030

GERMANY

IVK - Industrieverband Klebstoffe e.V.
Ivo-Beucker-Strasse 43
D-40237 Düsseldorf
Germany

P.O. Box 230169
D-40087 Düsseldorf
Germany

Tel: +49 211 67931 30
Fax: +49 211 67931 88

ITALY

AVISA - Associazione Vernici Inchiostri Sigillanti e Adesivi
Via Accademia 33
I-20131 Milano
Italy
Dr. Assuero Mancini - President
Dr. Marco Surdi - Secretary
Tel: +39 2 26 810 1
Fax: +39 2 26 810 350

JAPAN

Japan Adhesive Industry Association
Fukushima Bldg
1-15-1-Chome, Uchikanda
Chiyoda-Ku, Tokyo, Japan
Hisashi Okazaki - Managing Director
Tel: +81 03 3291 3303
Fax: +81 03 3291 3305

Japan Sealant Industries Association
Kyoda Bldg
2-2-17 Soto-Kanda
Chiyoda-Ku, Tokyo, Japan

THE NETHERLANDS

Vereniging Nederlandse Lijmindustrie
P.O. Box 418
NL-2260 AK Leidschendam
The Netherlands
Robert J. Baruch - President
Jaap W. Adrian - Secretary
Tel: +31 70 3177243
Fax: +31 70 3177412

NORWAY

Norsk Limforening
c/o Mr. Gunnar Skjöld
Casco Nordsjo A/S
P.O. Box 65
N-1410 Kolbotn, Norway
Gunnar Skjöld - President
Tel: +47 66 819400
Fax: +47 66 803435

PORTUGAL

Associação da Indústrias e Comércio de Colas e Similares
Avenida Guerra Junqueiro, 8, 2° -E
P-1000 Lisboa, Portugal
José Belmiro R. Ferreira - President
A.M. Rodrigues Gonçalves - Secretary
Tel: +351 1 8494502
Fax: +351 1 8494502

SPAIN

ASEFCA - Asociación Española de Fabricants de Colas y Adhesivos
Zurbano 8
E-28010 Madrid
Spain
García Martín - President
J.M. Larraz - Secretary
Tel: +34 1 4190243
Fax: +34 1 3085450

SWEDEN

Sveriges Limleverantörers Förening
(The Swedish Adhesive Suppliers Association)
Klarabergviadukten 90
S-106 13 Stockholm
Sweden
Jan Drejevik - President
Christina Mattsson - Secretary
Tel: +46 8 698 81 20
Fax: +46 8 698 81 29

SWITZERLAND

Fachverband Klebstoffindustrie Schweiz
Löwenstrasse 42
CH-8001 Zurich
Switzerland
Marcel Widmer - President
Dr. Andreas Coradi - Secretary
Tel: +41 1 2213655
Fax: +41 1 2213678

UNITED KINGDOM

BASA - British Adhesives & Sealants Association
33, Fellowes Way
Stevenage
Hertfordshire
SG2 8BW
UK
Neil MacDonald - President
David Williams - Secretary
Tel: +44 1438 358514
Fax: +44 1438 742565

UNITED STATES OF AMERICA

AMA - Adhesives Manufacturers Association
401, North Michigan Avenue
Chicago
Illinois 60611
USA
J. Dollard Carey - Secretary
Tel: +1 312 644 6610
Fax:

ASC - Adhesive & Sealant Council
1627 K St. Ste. 1000
Washington
DC 20006-1707
USA
Kerry L. Lake - Executive Vice President
Tel:
Fax: +1 202 452-1501

Other Trade & Research Organisations

EUROPE

The Adhesion Science Group
City University, Northampton Square
London EC1V 0HB
United Kingdom

AFTPV, Section Adhésif
5 rue Etex
F-75018 Paris, France
Tel: +33 1 42 63 45 91
Fax: +33 1 42 63 31 50

BRE (Building Research Establishment)
Bucknalls Lane, Garston, Watford
Herts. WD2 7JR
United Kingdom
Tel: +44 1923 664040
Fax: +44 1923 664010
Telex: 923220 BRSBRE G

CAT (Centre for Adhesive Technology)
TWI (The Welding Institute)
Abington Hall, Abington
Cambridge CB1 6AL
United Kingdom
Tel: +44 1223 891162
Fax: +44 1223 892588
Telex: 81183 WELDEX G
Clare Rushforth - Secretary
Provides technical consultancy to industry on all aspects of adhesive bonding technology, including expert adhesive selection and quality assurance computer software.

Chambre Syndicate des Société d'Etude et de Conseil
3, rue Léon Bonnat
F-75016 Paris, France
Tel: +33 1 45 24 43 53
Fax: +33 1 42 88 26 84
French engineering industries.

Comité des Constructeurs Français d'Automobiles
2, rue Presbourg
F-75008 Paris, France
Tel: +33 1 49 52 51 00
Fax: +33 1 47 23 74 73
Association of French car constructers.

Cranfield Institute of Technology
School of Industrial and Manufacturing Sciences
Cranfield, Bedford MK43 0AL
United Kingdom
Tel: +44 1234 750111
Fax: +44 1234 750875

EFNARC
(European Federation of National Associations of Specialist Repair Contractors and Material Suppliers to the Construction Industry)
First Floor, 241 High Street
Aldershot, Hants GU11 1TJ
United Kingdom
Tel: +44 1252 342072
Fax: +44 1252 333901

ESDU International plc
27 Corsham Street
London, N1 6UA
United Kingdom
Tel: +44 171 490 5151
Fax: +44 171 490 2701
Produce design data and computer programs for engineers working in a range of engineering fields - including bonded joint analysis.

European Resin Manufacturers Association
Queensway House
2, Queensway
Redhill, Surrey RH1 1QS
United Kingdom
Tel: +44 1737 768611

Fédération des Chambres Syndicates des Minerais et Metaux Non Ferreux
30, Avenue de Messine
F-75008 Paris, France
Tel: +33 1 45 63 02 66
Fax: +33 1 45 63 61 54
Federation of French non-ferrous metal industries.

Fédération Français de l'Acier
Immeuble Elysées La Defense, Cedex 35
F-75008 Paris, France
Tel: +33 1 45 63 02 66
Fax: +33 1 45 63 61 54
Federation of French ferrous metal industries.

Fédération des Industries Electriques et Electroniques
11 - 17 rue Hamelin
F-75116 Paris, France
Tel: +33 1 45 05 70 70
Fax: +33 1 45 53 03 93
Federation of French electrical & electronic industries.

Fédération Nationale des Travaux Publics
3, rue de Berri
F-75008 Paris, France
Tel: +33 1 44 13 31 44
Fax: +33 1 44 56 04 47
Federation of French civil engineering industries.

Fédération Nationale du Bâtiment
9, rue de la Pérouse
F-75016 Paris, France
Tel: +33 1 40 69 51 00
Fax: +33 1 45 53 58 77
Federation of French civil building and construction industries.

FERFA
(The Trade Federation of Specialist Contractors and Material Suppliers to the Construction Industry)
First Floor, 241 High Street
Aldershot, Hants GU11 1TJ
United Kingdom
Tel: +44 1252 342072
Fax: +44 1252 333901
Association of companies engaged in the manufacture and application of epoxy, polyester and resin-based materials for the building and civil engineering industries.

FIRA (Furniture Industry Research Association)
Maxwell Road, Stevenage, Herts. SG1 2EW
United Kingdom
Tel: +44 1438 313433
Fax: +44 1438 727607
Telex: 827653 FIRA G
Testing and advice on adhesives in the furniture industry.

Glued Laminated Timber Association
Chiltern House, Stocking Lane
Hughenden Valley, High Wycombe
Bucks. HP14 4ND - United Kingdom
Tel: +44 1494 565180
Fax: +44 1494 565487
Telex: 83292

Groupement des Industries Françaises Aéronautiques et Spatiales
4, rue Galilée
F-75782 Paris Cedex 16, France
Tel: +33 1 44 43 17 00
Fax: +33 1 40 70 91 41
French aerospace industries.

The Institute of Materials
1, Carlton House Terrace
London SW1Y 5DB
United Kingdom
Tel: +44 171 976 1338
Fax:: +44 171 839 2078
H. Turkdogan - Marketing Manager
Professional association of materials scientists. Publish monthly journal and organise meetings, seminars, etc.

The Institute of Packaging
Sysonby Lodge, Nottingham Road
Melton Mowbray
Leics. LE11 3TU
United Kingdom
Tel: +44 1664 500055
Fax: +44 1664 64164

RAPRA Technology Ltd. (Rubber and Plastics Research Association)
Shawbury, Shrewsbury
Shropshire SY4 4NR
United Kingdom
Tel: +44 1939 250383
Fax: +44 1939 251118
Telex: 35134
Mr. Brian Evans - Head of Enquiry Bureau
All aspects of rubbers, plastics, adhesives and composites.

SAMPE (Society for the
Advancement of Material & Process Engineering)
European Chapter
Unterloostraße 12
CH-8461 Oelingen
Switzerland
Tel: +41 52 43 31 43
Fax: +41 52 43 26 25
Heidi Müller -Membership Administration
Non-profit making association of European-based materials and process engineers. Organises conferences to promote information exchange between members.

SATRA - Footwear Technology Centre
SATRA House, Rockingham Road
Kettering
Northants. NN16 9JH
United Kingdom
Tel: +44 1536 410000
Fax: +44 1536 410626
Advice on adhesives for footwear and allied industries.

Solvents Industry Association
57, Mymms Drive
Brookmans Park, Hatfield
Herts. AL9 7AE
United Kingdom
Tel: +44 1707 55461
Fax: +44 1707 55461

TRADA (Timber Research and Development Association)
Chiltern House, Stocking Lane
Hughenden Valley, High Wycombe
Bucks. HP14 4ND
United Kingdom
Tel: +44 1494 563091
Fax: +44 1494 565487
Telex: 83292 TRADA G

Union des Industries Chimiques
Tour Aurore
18 Place des Reflets Cedex 5
F-92080 Paris la Defense
France
Tel: +33 1 47 20 56 03
French chemical industries.

Union des Industries Métallurgiques et Minières
56, Avenue de Wagram
F-75017 Paris
France
Tel: +33 1 40 54 20 20
Fax: +33 1 47 66 22 74
French metals industries.

Union des Industries Textiles
37, rue de Neuilly
F-92110 Clichy
France
Tel: +33 1 47 56 31 00
Fax: +33 1 47 30 25 28
French textile industries.

UNITED STATES OF AMERICA

The Adhesion Society
Ross Yates, Kendall Co.
17 Hartwell Avenue
Lexington
MA 02173-3195 - USA

The Adhesives Council
Society of Manufacturing Engineers
One SME Drive
PO Box 930
Dearborn
MI 48121 - USA

AJI - American Joining Institute
10628 Dutchtown Rd
Knoxville
TN-37932 - USA
Tel: +1 615 675 2150
Fax: +1 615 675 6081
Robert B. Allen - JOINEXCELL Project Manager
Activities include development of computer-based expert system for adhesive bonding technology.

Epoxy Resin Formulators - Society of the Plastics Industry
1275 K Street NW
Suite 400
Washington DC 20005 - USA

SAMPE
(Society for the Advancement of Material & Process Engineering)
1161 Parkview Drive
Covina
CA-91724-3748, USA
Tel: +1 818 331-0616
Fax: +1 818 332-8929
Non-profit making association of American-based materials and process engineers. Organises conferences to promote information exchange between members. Publishes bimonthly SAMPE Journal and quarterly Journal of Advanced Materials.

Standards Organisations

ASTM Standards

(American Society for Testing Materials)

ASTM
1916 Race Street
Philadelphia
PA-19103-1187
USA

American Technical Publishers Ltd.
68A Wilbury Way
Hitchin
Herts.
SG4 0SX
UK
Tel: +44 1462 31525

BS Standards

(British Standards)

BSI - British Standards Institute
Linford Wood
Milton Keynes
MK14 6LE
UK
Tel: +44 1908 220022
Fax: +44 1908 320856
Telex: 825777 BSIMK G

BSI - British Standards Institute
2 Park Street
London
W1A 2BS
UK
Tel: +44 171 629 9000
Fax: +44 171 629 0506

DIN Standards

(Deutsche Institut für Normung)

Beuth Verlag
Burggrafenstraße 6
D-1000 Berlin 30
Germany

EN Standards

(Euro Norme)

Information is available from:
- National standards institutions, and
- National CEN/CENELEC organisations.

ISO Standards

(International Standards Organisation)

ISO Central Secretariat
1 rue de Varembe
Case Postale 56
CH-1211 Geneve 20
Switzerland
Tel: +41 22 749 01 11
Fax: +41 22 734 10 79
Email: (Internet) central@isocs.iso.ch

NF Standards

(Norme Française)

AFNOR - Association Française de Normalisation
Tour Europe
F-92049 Paris la Defense
France
Tel: +33 1 42 91 55 55
Fax: +33 1 42 91 56 56
Telex: AFNOR 611 974 F
Minitel: 3616 AFNOR
Minitel: 3616 CATAFNOR - *search & order AFNOR catalogue items*

Part 3: DATA

PRESENTATION OF DATA

The data provided here is divided into 10 sections, reflecting the chemical classes of the adhesives. For each of these the information is presented in two distinct forms:

- ❑ **Comparison Tables** - To provide a review of all of the included adhesives, at a glance. These give the manufacturer and product name/code with a set of basic characteristics, common for the majority of the adhesives. The actual properties and characteristics quoted for each class differ to reflect their relative importance (and availability).
  - Mechanical properties appear as a standard set of symbols:
    Higher: ● Moderate: ◕ Lower: ○
    These codings were determined by examining all of the particular property data for all of the adhesives included in that class.
  - For other properties, H = High, M = Medium, L = Low, ✖ = No and ✓ = Yes are shown in place of numeric data where these were stated in the manufacturers product data.
  - Where adhesives within a class have very different formulations, a comparison of properties is not directly valid. Here, typical uses have been substituted for numeric data to provide a basis of comparison. This applies to the *Phenolic Adhesives* and *Miscellaneous* sections.
  - Individual comparison tables are given for epoxy one-part, two-part and film adhesives to aid comparing like-with-like.
  - The final column in each table gives the page number of the corresponding data sheet for that product.

- ❑ **Data Sheets** - Each data sheet contains:
  - Manufacturer and product code
  - Manufacturer's comments
  - Typical uses for the particular product.
  - A date - either that given in the manufacturer's product information, or (if shown in brackets) the date that the information was received.
  - Curing or processing information - in some cases given as a range of temperatures, times and conditions.
  - A set of physical properties
  - A set of mechanical properties.
  - Test Method, where available, shows the initials of the test-standards organisations, i.e. National - NF: French; DIN: German; BS: UK; ASTM: USA, etc.; International (ISO, EN, etc.); Industry standards, e.g. Airbus, General Motors. Specification numbers for particular properties, if given, are stated in the 'Notes' of each data sheet.
  - Notes showing any other relevant information.

  Again, the actual properties included may vary between the classes of adhesives. This reflects the differing intended uses and the type of data required for these. For example, anaerobic adhesives commonly quote torque properties relating to thread-locking; thermoplastics concentrate on physical characteristics at various temperatures. Few manufacturers quote a full set of property data for each adhesive product. All numerical data given in this book has been converted, where necessary, to a standard set of units (based on S.I. units), e.g. strength values given by manufacturers in pounds per square inch (psi) have been converted to megapascal (MPa). Appendix D is a table of conversion factors relevant to adhesive properties and uses.

KEY

The terms stated in the key are an alphabetical compilation of all of those used within the data sections, i.e. comparison tables and data sheets. Many are common to all adhesive classes, whereas some may have a slightly different interpretation for particular cases (as stated in the descriptions).

τ (RT): Lap shear strength at room temperature, in MPa.

τ (°C): Lap shear strength at temperature stated, in MPa. Often upper service temperature.

τ (compressive): Compressive lap shear strength, in MPa.

τ (wet): Lap shear strength at room temperature after exposure to water/moisture, in MPa.

σ_c : Compressive strength, in MPa.

σ_t : Tensile strength, in MPa.

Adhesive: Manufacturers full product name or code.

Application Temp.: *Thermoplastic Hot-Melts & some polyurethanes* - Temperature, or range, at which the adhesive is applied to the adherend, °C

Assembly Time: *Thermoplastic Hot-Melts* - Indication of time available to assemble, in minutes or seconds. Some are stated (by the manufacturer) as F= fast or M = medium (Comparison Table).

Base: *Imide-based*: Formulation and cure type. BMI = bismaleimide; PI = Polyimide; a = addition cure; c = condensation. *Thermoplastic Hot-Melts*: Principal polymer constituent.

Break Torque: *Anaerobics* - The initial torque (N.m) required to break the adhesive bond locking a threaded fastener.

Colour: Mixed colour for two-part adhesives, colour after cure. Those shown as, e.g. white→clear, have a colour change on curing. *Thermoplastics* - may state a colour numerically to the Gardner Scale. *Silicones*: Various - several colour available, but product code may differ for coloured variants.

Company: Manufacturers name.

CTE (°C): Coefficient of thermal expansion (for a given temperature or range, if stated), in $°C^{-1}$ or K^{-1}.

Cure °C or **(°C)** RT- room temperature. At temperature or range stated, in °C.

Cure Rate: *Silicones* - Depth of cure from the surface, in mm per 24 hrs.

Cure Shrinkage: *Silicones* - % after cure, often linear value.

Cure Time (sec): *Cyanoacrylates* -Time for initial cure to produce useful strength. Final strength occurs after approximately 24 hours.

Cure: Method and conditions. RT = Room temperature; UV = Ultraviolet radiation exposure; Vis. = Visible light exposure. *Acrylics*: Aerobic = Exposure to the atmosphere. *Anaerobics*: Cat. = Catalytic curing (usually by contact with metallic surfaces or primers). *Polyurethanes*: Temperature, in °C. Heat - unstated temperature; RF - Radio Frequency.

Description: *Miscellaneous* - Chemical base, if given, and general type (one-part, two-part or film).

E: Tensile modulus, in MPa (if not stated otherwise).

El (%): Elongation to break during tensile test, in %.

Elongation: to break (unless otherwise stated), in %.

Flash Point: The temperature at which the vapour above the liquid adhesive forms a combustible (explosive) mixture with air, in °C.

Form: *Thermoplastic Hot-Melts* - Available solid product forms, e.g. granules, rods, etc.

G: Shear modulus, in MPa.

Gap Filling: *Anaerobics & Cyanoacrylates* - Maximum gap bridgeable by the liquid adhesive, in mm. Some shown (by the manufacturer) as YES = ✓ or NO = ✖.

H (Shore): Hardness to stated Shore A or D scale. (Data Sheets).

Hs: Hardness to stated Shore A or D scale, e.g. 75D. (Comparison Tables).

LSS: Lap Shear Strength. Normally determined by an overlap test configuration. *Phenolics*: d = 'disc shear'.

Manufacturers' Comments: Gives main features of the adhesive product, including any industry approvals.

Paste: One or two-part adhesives, of any viscosity, that are not films.

Peel ( ): Peel strength for stated test sample configuration, e.g. T peel, floating roller, bell, sandwich, etc. Units vary for different test pieces.

Peel: Peel Strength.

Pg. No.: Page number of the data sheet for the adhesive.

Pot life (mins) or **Pot**: Pot life, in minutes, for mixed two-part adhesives. For one-part paste and film adhesives, 'Pot Life' = work life (time available to use an adhesive exposed to working conditions). *Phenolics*: Time before adhesive gels to an unusable working state. *Silicones*: Pot life or working life (before skin forms).

Pot/Work life: Minutes at stated temperature (°C). Volume of mix (if stated).

Prevail Torque: *Anaerobics* - The torque (N.m) required during the first turn to unscrew the fastener after 'breaking' (See: Break Torque).

Refractive Index: *Cyanoacrylates* - Of interest in the bonding of clear plastics and glass. See data sheet 'Notes' for other classes (if stated).

Resistivity: Vol. - volumetric, in Ω.cm; **Surf**. - surface, in Ω.

Service (°C): Service temperature range; or maximum continuous service temperature, in °C.

SG: Specific gravity (density relative to that of water).

Shelf (yrs): Shelf life of the stored adhesive in years.

Shelf life: Time/temperature (if stated).

Skin Time: *Silicones* - Time for the applied adhesive to form a surface skin, in mins. Indication of working or assembly time available.

Softening Temp or **Soft. Temp.:** *Thermoplastic Hot-Melts:* Temperature, or range, when the adhesive softens, in °C

Tear strength: Resistance to tearing. Units vary for different test configurations. Often $N.mm^{-1}$.

Temp. Range: *Thermoplastic Hot-Melt:* Temperature range for the application or processing of the adhesive, in °C.

Test Methods: Gives the initials of the principal national standards organisations used for the testing of the physical and mechanical properties.

Tg or Tg (°C): Glass transition temperature, in °C.

Tg dry (°C): *Film adhesives (Epoxy)*: Glass transition temperature, in °C, for product in a dry condition.

Thermal Cond.: Thermal conductivity, in $W.m^{-1}.°C^{-1}$.

TM: Tensile Modulus

TS: Tensile Strength.

$T_{Service}$: Service temperature range, if stated, or maximum continuous use temperature, in °C. An 'intermittent use' temperature (i.e. brief high temperature pulses) is often given in the 'Notes'.

Type: 1 = one-part; 2 = two-part; F - film. *Acrylics*: m = two parts mixed; a = activator applied to adherend; am = use mixed or with activator. *Cyanoacrylates*: Principal chemical constituent. *Polyurethanes*: m = moisture curing; * = require melting before applying. *Imide-based*: foam = foaming (during cure). *Silicones*: a = acetic acid; n = neutral (non-corrosive); m = moisture curing.

Typical Uses: Data sheets - examples of uses for the particular adhesive, as stated by the manufacturer. Also given in *Phenolics* and *Miscellaneous* Comparison Tables.

Useable: *Polyurethanes* - Indicates the available time, in minutes, to make the assembly.

Various Carriers: *Epoxy film*:- Indicates availability of products with other support types and materials. YES = ✓ or NO = ×.

Various Wts.: *Epoxy film* - Indicates availability of products of other areal weights. YES = ✓ or NO = ×.

Visc (Pa.s): Viscosity (Pascal.seconds). Some manufacturers state only L = low, M = medium, H = high (includes thixotropic for *polyurethanes*).

Viscosity: in Pa.s (Pascal.seconds). Normally for usable form, e.g. two parts mixed. Some state each part separately. Many are thixotropic. Some manufacturers quote only L = low, M = medium, H= high.

Viscosity (°C): *Thermoplastic Hot-Melts* - Viscosity, in Pa.s (Pascal.seconds), at stated temperature. Temperatures given are generally related to processing temperatures, where available they are the upper and lower temperatures of the stated usage range.

ACRYLIC ADHESIVES

NOTES

There are several different cure mechanisms within the acrylic adhesive family.

- One-part solvent drying.
- One-part aqueous emulsion (cure initiated by slight drop in water content on exposure to atmosphere).
- Two-part mixed.
- Two-part (no mix) using one of the parts as an activator or primer applied to one of the adherends.
- Exposure to light (visible or UV).
- Some have a multi-cure by heat, activator or UV.

The various cure types allow adhesive selection to be made on the basis of the production processes.

Main Uses

In general, the one-part dispersions are used for bonding high-duty, industrial flooring. The others are used to bond structural materials (metals, wood, plastics, ceramics and glass) in various industry sectors (engineering, automotive, construction, etc.).

COMPARISON TABLE - ACRYLIC ADHESIVES

Company	Adhesive	Type	Colour	LSS	TS	Peel	Hs	El. (%)	SG	Visc (Pa.s)	Service (°C)	Pot life (mins)	Cure	Shelf (yrs)	Pg. No.
3M	DP-805	2m	Lt. Yellow	●	-	●	-	-	-	-	-	3.5	RT	-	69
Bostik	Hyperbond M890	2a	Beige	●	-	-	-	-	1	20	-	2.5	RT	1	69
	Hyperbond M895	2	Opaque	●	-	-	-	-	1.02	-	-40/+150	15	RT	1	69
CECA	804	1	Off-white	-	-	-	-	-	1.38	36	-	30	RT	1	69
	Sader Adhésif	1	White	-	-	-	-	-	1.19	2.2	-	-	RT	1	70
	Sader Fix	1	White	-	-	-	-	-	1.35	39	-	45	RT	1	70
	Sader Tac	1	White	-	-	-	-	-	1.25	40	-	35	RT	1	70
Degussa	Agomet F300	2a	-	◐	-	○	-	-	1.2	20	-	Low	RT	0.5	70
	Agomet F305	2a	-	●	-	-	-	-	1	4	-	Low	RT	0.5	71
	Agomet F310	2am	-	●	-	●	-	-	1.0	22	-40/+130	10	RT	0.5	71
	Agomet F330	2am	-	●	-	●	-	-	1.01	18	-40/+130	6	RT	0.75	71
	Agomet F347 HI	2am	-	◐	-	-	-	-	1.1	70	-40/+90	9	RT	0.5	71
	Agomet U1	2m	-	◐	-	◐	-	-	1.0	0.25	-	20	RT	1	72
	Agovit	2m	Transparent	○	-	-	-	-	1.0	3.2	-25/+100	20	RT	1	72
	Agovit 1074	2m	Transparent	○	-	-	-	-	1.0	0.4	-	18	RT	-	72
	Agovit 1093 AB	2m	Transparent	○	-	-	-	-	1.0	1.5	-	15	RT	-	72
	Agovit 1900	2m	Transparent	○	-	-	-	-	1.0	0.5	-	20	RT	-	73
Delo	Maflex 241	2a	-	●	-	-	-	-	-	30	-55/+100	-	RT	0.5	73
	Maflex 245	2a	-	◐	-	-	-	-	-	50	-55/+100	-	RT	0.25	73
Devcon ITW	MVP Type 11	2m	Yellow	●	-	-	-	-	1.2	35	<93	5	RT	-	73
Dunlop	Megabond SAS 300	2a	Translucent	○	-	○	-	-	1.1	50	<100	-	RT	0.5	74
	Megabond SAS 350	2a	Translucent	●	-	-	-	-	1.05	6	<100	-	RT	0.5	74
	Megabond SAS 700	1	Transparent	◐	○	-	-	-	1.05	6	<100	-	UV	0.5	74
Dymax	1-20270M	1	-	-	○	-	60D	160	-	1.4	-	-	UV/vis aerobic	-	74
	1-20271M	1	-	-	◐	-	70D	9	-	4	-	-	UV/vis aerobic	-	75
	3-20256	1	-	-	●	-	80D	6	-	-	-	-	UV/vis aerobic	-	75
	3-20261	1	-	-	○	-	60D	160	-	-	-	-	UV/vis aerobic	-	75
	4-20260	1	-	-	●	-	80D	15	-	-	-	-	UV/vis aerobic	-	75
	6-20265	1	-	-	◐	-	60D	30	-	-	-	-	UV/vis aerobic	-	76
	828	2a	Amber	◐	-	-	-	-	-	0.65	-18/+177	-	RT	1	76
	840	2a	Amber	◐	-	-	-	-	-	0.5	-18/+177	-	RT	1	76
	845	2a	Amber	●	-	-	-	-	-	3.5	-18/+177	-	RT	1	76
	846	2a	Translucent	●	-	-	-	-	-	3.5	-18/+177	-	RT	1	77
	847	2a	Amber	◐	-	-	-	-	-	0.5	-18/+177	-	RT	1	77
	Light-Weld 304	1	Transparent	-	-	-	-	-	-	-	-	-	UV	-	77
	Light-Weld 401	1	Trans. yellow	-	-	-	75D	-	-	0.08	-	-	UV	-	77
	Light-Weld 415	1	Trans. yellow	-	-	-	85D	-	-	5	-	-	UV	-	78
	Light-Weld 416A	1	Trans. Straw	-	○	-	40D	225	-	8	-	-	UV	-	78
	Light-Weld 425	1	Transparent	-	-	-	80D	-	-	4	-	-	UV	-	78
	Light-Weld 481A	1	Trans. Straw	-	●	-	80D	10	-	7.5	-	-	UV	-	78
	Light-Weld 484A	1	Trans. Straw	-	●	-	80D	13	-	0.08	-	-	UV	-	79
	Light-Weld 486A	1	Trans. Straw	-	●	-	85D	5	-	0.14	-	-	UV	-	79
	Light-Weld 488	1	Trans. yellow	-	-	-	50D	-	-	0.5	-	-	UV	-	79
	MD - 128M	1	Transl. white	-	◐	-	75D	30	-	0.6	-	-	UV + Heat	-	79
	MD - 136M	1	Transp. Straw	-	●	-	85D	10	-	0.1	-	-	UV aerobic	-	80
	MD - 138M	1	Transparent	-	○	-	40D	300	-	0.35	-	-	UV aerobic	-	80
	MD - 190M	1	Transparent	-	●	-	80D	10	-	0.08	-	-	UV aerobic	-	80
	MD - 191M	1	-	-	◐	-	60D	190	-	0.5	-	-	UV aerobic	-	80
	MD - 195M	1	-	-	●	-	80D	10	-	6.5	-	-	UV aerobic	-	81
	MD - 197M	1	-	-	○	-	60D	160	-	1.4	-	-	UV aerobic	-	81

Continued...

Company	Adhesive	Type	Colour	LSS	TS	Peel	Hs	El. (%)	SG	Visc (Pa.s)	Service (°C)	Pot life (mins)	Cure	Shelf (yrs)	Pg. No.
Dymax	Multi-Cure 427	1	Transp. yellow	-	-	-	80D	-	-	4	-	-	UV/Heat	-	81
(Continued)	Multi-Cure 428	1	Transp. yellow	-	-	-	60D	-	-	0.15	-	-	UV/Heat	-	81
	Multi-Cure 602	2a	Amber	◑	-	-	70D	20	-	4	-18/+177	0.3	RT/UV/Heat	-	82
	Multi-Cure 625	2a	Amber	◑	-	-	40D	25	-	5	-7/+177	0.5	RT/UV/Heat	-	82
	Multi-Cure 628	2a	Amber	◑	-	-	75D	20	-	0.7	-7/+177	0.5	RT/UV/Heat	-	82
	Multi-Cure 652	2a	Amber	○	-	-	30D	80	-	3	-29/+93	0.5	RT/UV/Heat	-	82
Forbo Helmitin	Helmitin 34040	2m	White	-	-	-	-	-	1.03	16	-	3 days	RT + 80-100	0.5	83
W W Henry	660	1	White	-	-	-	-	-	-	-	-	-	RT	-	83
Loctite	Adhesive 3295	2m	Green	◑	-	◑	-	-	1.03	25	<80	4	RT	1	83
	Multi-Bond 329	2a	-	-	-	-	-	-	-	-	<80	15	RT	-	83
	Multi-Bond 330	2a	-	-	-	◑	-	-	-	-	<80	15	RT	-	84
	Structural Adhesive 326	2a	-	-	-	-	-	-	-	-	<130	-	RT +120	-	84
	Structural Adhesive 327	2m	-	-	-	-	-	-	-	-	<130	-	RT	-	84
	Structural Adhesive 341	2m	-	-	-	-	-	-	-	-	<80	-	RT +<200	-	84
Permabond	Flexon 245	1	Off-white	●	-	◑	-	-	1.0	80	-60/+120	-	RT	1	85
	Flexon 246	1	Off-white	●	-	●	-	-	1.0	30	-60/+120	-	RT	1	85
Plexus ITW	MA 300	2m	Lt. Yellow	●	○	-	78D	-	-	28	<93	4	RT	-	85
	MA 310	2m	Lt. Yellow	◑	○	-	75D	-	-	45	<93	10	RT	-	85
	MA 320	2m	Off-white	○	-	-	72D	-	-	123	<93	5	RT	-	86
	MA 550	2m	Off-white	○	○	-	65D	-	-	216	<93	35	RT	-	86
Quilosa	Aklesil L-71	1	Beige	-	-	-	-	-	-	30	-	45	RT	1	86
	Aklesil L-80	1	Beige	-	-	-	-	-	-	27	-	30	RT	1	86
SEA	A 406	1	White	-	-	-	-	-	1.25	-	-	60	RT	1	87
	Stykisol	1	White	-	-	-	-	-	1.9	-	-	15	RT	1	87
Structural	S-511A/S-10A	2am	Off-white	●	-	-	-	-	1.0	20	-	7	RT	1	87
Adhesives	S-515A/S-34A	2am	Lt. Brown	◑	-	-	-	-	1.0	65	-	7	RT	1	87
	S-545A/S-47A	2am	Off-white/grey	◑	-	-	-	-	1.25	-	-	20	RT	1	88
	S-570A/S-70A	2am	Purple	◑	-	-	-	-	1.0	4.5	-	2.5	RT	1	88
	S-600A/S-12A	2am	Lt. Brown	◑	-	●	-	-	1.05	4.5	-40/+150	8	RT	1	88
Sunstar	Tufflock 6201	1	Off-white	●	-	-	-	-	-	23	-	-	RT	-	88
Engineering	Tufflock 6202	1	Off-white	◑	-	-	-	-	-	15	-	-	RT	-	89
	Tufflock 6203	1	Off-white	●	-	-	-	-	-	23	-	-	RT	-	89
	Tufflock 6506	1	Transp./Brown	○	-	-	-	-	-	23	-	-	RT	-	89
UHU	"1"	1	Transparent	-	○	-	-	-	0.84	0.8	-20/+80	-	RT	3	89
	Hard plastic adhesive	1	Transparent	-	-	-	-	-	0.85	0.8	-30/+80	-	RT	-	90
	Plus Acrylit	2m	Off-white	-	○	-	-	-	1.00	20	-30/+90	8	RT	-	90
Wulff	AC 1	1	-	-	-	-	-	-	1.42	23	<40	20	RT	-	90
(Naturaian)	ES 14	1	-	-	-	-	-	-	1.18	22	<40	20	RT	-	90
	Supra Coll	1	-	-	-	-	-	-	1.15	26	<40	20	RT	-	91
	Supra Tex	1	-	-	-	-	-	-	1.19	38	<40	20	RT	-	91
	Tep-Coll	1	-	-	-	-	-	-	1.26	22	<40	20	RT	-	91

See Key to Comparison Tables - Pages 63 to 65

3M

3M	DP-805		(March 95)
Acrylic - Two part			
Manufacturers Comments: Duo-pak, self-metering and mixing cartridge. High peel and shear strength. Bonds to slightly oily metal. Good elevated temperature performance.			
Typical Use: General engineering assemblies.			
Cure: RT.			
Test Methods: Not stated			
Physical		**Mechanical**	
Colour	Pale Yellow	τ (RT)	48.3 MPa
Viscosity	A:195; B:75 Pa.s	τ (82°C)	30.3 MPa
S.G.	-	τ (wet)	-
Shelf life	-	Peel (T)	6.13 N mm^{-1}
Pot/Work life	3 - 4 mins	σ_f	-
$T_{Service}$	-	E	-
CTE (30°C)	105 × 10^{-6} °C^{-1}	Elongation	-
Thermal cond.	-	H (Shore)	-
Notes: Volume Resistivity: 1.1 × 10^{14} Ω.cm.			

Bostik	Hyperbond M895		May 93
Methacrylic - Two part, viscous paste			
Manufacturers Comments: 100% solids, solvent-free. High performance for general industrial assembly. Non-slump. Mechanical and environmental resistance.			
Typical Use: General engineering assemblies: Metals, some plastics, wood, rubber.			
Cure: RT (>15°C)/25 mins, low pressure (handling).			
Test Methods: ASTM, BSI			
Physical		**Mechanical**	
Colour	Opaque	τ (RT)	28.0 MPa
Viscosity	-	τ (°C)	-
S.G.	~1.02	τ (wet)	-
Shelf life	1 year (<25°C)	Peel (FR)	13.50 N mm^{-1}
Pot/Work life	15 mins	σ_f	-
$T_{Service}$	-40 to +150°C	E	-
CTE (°C)	-	Elongation	-
Thermal cond.	-	H (Shore)	-
Notes: τ ASTM D-1002AI, acid etched. Peel - Floating Roller BS5350 ptC9. Al. acid etched. Gap filling: 0.75 mm max. Flash point >11°C (adhesive).			

Bostik

Bostik	Hyperbond M890		July 92
Methacrylic - Two part, viscous liquid			
Manufacturers Comments: High performance for many industrial uses. No mixing (activator applied to one surface). Fast curing. Mechanical and environmental resistance.			
Typical Use: General engineering assemblies: Steels, Al-alloys, Cu-alloys, some plastics, wood, hard rubbers.			
Cure: RT (>15°C)/ 3-5 mins, low pressure (handling); ~24 hrs (full cure)			
Test Methods: Not stated			
Physical		**Mechanical**	
Colour	Beige	τ (RT)	27.9 MPa
Viscosity	~20 Pa.s	τ (°C)	-
S.G.	~1	τ (wet)	24.7 MPa
Shelf life	1 year (5-15°C)	Peel ( )	-
Pot/Work life	2-3 mins/RT	σ_f	-
$T_{Service}$	-	E	-
CTE (°C)	-	Elongation	-
Thermal cond.	-	H (Shore)	-
Notes: τ Al-alloy, acid etched. τ(wet) 1 month, water, 23°C. Flash point >10°C (adhesive).			

CECA

CECA	CECA 804		Jan 93
Acrylic - One part, emulsion			
Manufacturers Comments: ~76% solids, <5% solvent. High strength, long work-life flooring adhesive. Resists wheel-chair traffic.			
Typical Use: Flooring - PVC, rubber-backed carpets, vinyl.			
Cure: RT (10-25°C)/24 hrs; 48-72 hrs (full cure).			
Test Methods: NF			
Physical		**Mechanical**	
Colour	Off-white	τ (RT)	-
Viscosity	36 Pa.s	τ (°C)	-
S.G.	1.38	τ (wet)	-
Shelf life	1 year (10-30°C)	Peel ( )	>0.80 N mm^{-1}
Pot/Work life	~30 mins	σ_f	-
$T_{Service}$	-	E	-
CTE (°C)	-	Elongation	-
Thermal cond.	-	H (Shore)	-
Notes: Peel - NF T 76.126. Flash point 53°C.			

CECA	Sader Adhésif		Sept 91
Acrylic - One part, emulsion			
Manufacturers Comments: ~62% solids, <5% solvent. Heavy-duty flooring adhesive. Sprayable. Retains bitumen- and PVC-backed floor tiles.			
Typical Use: Flooring - PVC-, bitumen-backed tiles, untreated wood products and metals.			
Cure: RT (12-30°C)/24 hrs; 48-72 hrs (full cure).			
Test Methods: NF			
Physical		**Mechanical**	
Colour	White	τ (RT)	-
Viscosity	2.2 Pa.s	τ (°C)	-
S.G.	1.19	τ (wet)	-
Shelf life	1 year (10-30°C)	Peel ( )	-
Pot/Work life	*	σ_f	-
$T_{Service}$	-	*E*	-
CTE (°C)	-	Elongation	-
Thermal cond.	-	H (Shore)	-
Notes: * Requires 4 hrs to become 'tacky' before floor covering is applied. Flash point 60°C; pH 7.			

CECA	Sader Tac		(July 95)
Acrylic - One part, water-based			
Manufacturers Comments: 75% solids, <5% solvent. General purpose, high performance flooring installation. Porous and non-porous surfaces. Good resistance to wheel-chairs. Specification: DTU 53-2.			
Typical Use: Flooring - linoleum, rubber, PVC.			
Cure: RT (10-25°C)/24 hrs; 48-72 hrs (full cure).			
Test Methods: NF			
Physical		**Mechanical**	
Colour	White	τ (RT)	-
Viscosity	40 Pa.s	τ (°C)	-
S.G.	1.25	τ (wet)	-
Shelf life	1 year (10-30°C)	Peel (-)	>1.20 N mm^{-1}
Pot/Work life	35 - 40 mins	σ_f	-
$T_{Service}$	-	*E*	-
CTE (°C)	-	Elongation	-
Thermal cond.	-	H (Shore)	-
Notes: Peel - NF T 76.126. Flash point 65°C; pH 6.1.			

CECA	Sader Fix		Jan 94
Acrylic - One part, emulsion			
Manufacturers Comments: ~75% solids, <5% solvent. General purpose flooring adhesive.			
Typical Use: Flooring - PVC, rubber-backed carpets, vinyl.			
Cure: RT (12-30°C)/24 hrs; 48-72 hrs (full cure).			
Test Methods: NF			
Physical		**Mechanical**	
Colour	White	τ (RT)	-
Viscosity	39 Pa.s	τ (°C)	-
S.G.	1.35	τ (wet)	-
Shelf life	1 year (10-30°C)	Peel (-)	>1.00 N mm^{-1}
Pot/Work life	35-55 mins	σ_f	-
$T_{Service}$	-	*E*	-
CTE (°C)	-	Elongation	-
Thermal cond.	-	H (Shore)	-
Notes: Peel - NF T 76.126. Flash point 66°C; pH 6. Similar products: SADERFLEX, SADER ROLL.			

Degussa

Degussa	Agomet F 300		Feb 95
Acrylic - Two part, with surface activator			
Manufacturers Comments: Multi-purpose, fast curing. For high production rate assembly processes, e.g. loud-speaker assembly.			
Typical Use: Metals, plastics, wood-based products.			
Cure: RT/2-4 mins; 2 hrs (full strength).			
Test Methods: DIN			
Physical		**Mechanical**	
Colour	-	τ (RT)	23 MPa
Viscosity	20 Pa.s	τ (80°C)	~9 MPa
S.G.	~1.2	τ (wet)	-
Shelf life	6 months	Peel (-)	~2 N mm^{-1}
Pot/Work life	Short	σ_f	-
$T_{Service}$	-	*E*	-
CTE (°C)	-	Elongation	-
Thermal cond.	-	H (Shore)	-
Notes: τ DIN 53283 aluminium: Rigid-PVC 10 MPa (substrate failure). Peel (angle) DIN 53282, Al F13.3.			

Degussa	Agomet F 305		Feb 95
Acrylic - Two part, with surface activator			
Manufacturers Comments: Multi-purpose, low viscosity, fast curing. For high production rate assembly processes, e.g. loud-speaker assembly.			
Typical Use: Metals, plastics, wood-based products.			
Cure: RT/2-4 mins; 2 hrs (full strength).			
Test Methods: DIN			
Physical		**Mechanical**	
Colour	-	τ (RT)	25 MPa
Viscosity	4 Pa.s	τ (°C)	-
S.G.	~1.0	τ (wet)	-
Shelf life	6 months	Peel ( )	-
Pot/Work life	Short	σ_f	-
$T_{Service}$	-	*E*	-
CTE (°C)	-	Elongation	-
Thermal cond.	-	H (Shore)	-
Notes: τ DIN 53283 aluminium.			

Degussa	Agomet F 330		Feb 95
Methylmethacrylate - Two part, with surface activator (no mix), or - Two part, mixed			
Manufacturers Comments: Can be used with an activator, or mixed with one of two hardeners. Good shear and peel strength. Chemical and temperature resistance. Mainly used for metal assemblies.			
Typical Use: Metals, plastics.			
Cure: RT/15 mins (handling strength); 24 hrs (full strength).			
Test Methods: DIN			
Physical		**Mechanical**	
Colour	-	τ (RT)	33 MPa
Viscosity	15-20 Pa.s	τ (120°C)	31.6 MPa
S.G.	~1.01	τ (wet)	-
Shelf life	9 months	Peel (-)	6 N mm^{-1}
Pot/Work life	6 (4) mins	σ_f	-
$T_{Service}$	-40 to +130°C	*E*	-
CTE (°C)	-	Elongation	-
Thermal cond.	-	H (Shore)	-
Notes: Pot-life for: Harter paste (Harter D). τ DIN 53283 aluminium. Peel DIN 53282.			

Degussa	Agomet F 310		(July 95)
Methylmethacrylate - Two part, with surface activator (no mix), or - Two part, mixed			
Manufacturers Comments: Family of multi-purpose general assembly adhesives (Agomet F310 to F 317) with various viscosities. Can be used with an activator, or mixed with one of two hardeners.			
Typical Use: Metals, plastics, wood-based products.			
Cure: RT/15-120 mins (activator); 2 hrs (full strength).			
Test Methods: DIN			
Physical		**Mechanical**	
Colour	-	τ (RT)	30 MPa
Viscosity	22* Pa.s	τ (80°C)	~19 MPa
S.G.	~1.0	τ (wet)	-
Shelf life	6 months	Peel (-)	6 N mm^{-1}
Pot/Work life	10 (4) mins	σ_f	-
$T_{Service}$	-40 to +130°C	*E*	-
CTE (°C)	-	Elongation	-
Thermal cond.	-	H (Shore)	-
Notes: * same for F311 and F312. F313-200 Pa.s; F315-2.5 Pa.s; F317-3 Pa.s. Pot-life for: Harter paste (Harter D). τ DIN 53283 aluminium. Peel DIN 53282. Flash point +10°C. Ignition temperature: 430°C.			

Degussa	Agomet F 347 HI		Dec 94
Methylmethacrylate - Two part, with surface activator (no mix), or - Two part, mixed			
Manufacturers Comments: Can be used with an activator, or mixed with one of two hardeners. Good shear and peel strength at low temperatures. Chemical and temperature resistance.			
Typical Use: Metals, plastics.			
Cure: RT/15-20 mins (handling strength); 24 hrs (full strength).			
Test Methods: DIN			
Physical		**Mechanical**	
Colour	-	τ (RT)	18 MPa
Viscosity	70 Pa.s	τ (°C)	-
S.G.	~1.1	τ (wet)	-
Shelf life	6 months	Peel ( )	-
Pot/Work life	8-10 mins	σ_f	-
$T_{Service}$	-40 to +90°C	*E*	-
CTE (°C)	-	Elongation	-
Thermal cond.	-	H (Shore)	-
Notes: τ DIN 53283 aluminium. Impact: +90°C/6.5 kJ/m^2, RT/9.0 kJ/m^2, -20°C /10.5 kJ/m^2,-40°C /15 kJ/m^2.			

Degussa	Agomet U1		Dec 94
Methylmethacrylate - Two part			
Manufacturers Comments: Family of adhesives (U1, U3, U4) various viscosities. Good shear and peel strength. Good temperature resistance. Chemical and temperature resistance.			
Typical Use: Metals, porcelain, ceramics, marble, glass, wood, some plastics (not suitable for fluoro-elastomers, polyolefin, PTFE).			
Cure: RT/15-20 mins (handling strength); 24 hrs (full strength).			
Test Methods: DIN			
Physical		**Mechanical**	
Colour	-	τ (RT)	~20 MPa
Viscosity	0.25 Pa.s*	τ (°C)	-
S.G.	~1.0	τ (wet)	-
Shelf life	12 months	Peel (-)	~3 N mm^{-1}
Pot/Work life	20 mins	σ_t	-
$T_{Service}$	-	E	-
CTE (°C)	-	Elongation	-
Thermal cond.	-	H (Shore)	-
Notes: *U3:16 Pa.s; U4:40 Pa.s. τ DIN 53283 aluminium. Peel - DIN53282			

Degussa	Agovit 1074		May 95
Methylmethacrylate - Two part			
Manufacturers Comments: Transparent adhesive for bonding acrylic glass. Fast curing, high optical quality, flexible.			
Typical Use: Acrylic glass- extruded.			
Cure: RT/2 hrs.			
Test Methods: DIN			
Physical		**Mechanical**	
Colour	Transparent	τ (RT)	>5 MPa
Viscosity	~0.4 Pa.s	τ (°C)	-
S.G.	~1.0	τ (wet)	-
Shelf life	-	Peel ( )	-
Pot/Work life	18 mins	σ_t	-
$T_{Service}$	-	E	-
CTE (°C)	-	Elongation	-
Thermal cond.	-	H (Shore)	-
Notes: τ DIN 53283 (substrate failure).			

Degussa	Agovit		May 94
Methylmethacrylate - Two part			
Manufacturers Comments: Family of adhesives various viscosities for acrylic glass (manual or machine). High optical value. Fast cure.			
Typical Use: Acrylic glass, e.g. Paraglas®, Decarglas®.			
Cure: RT/40 mins (handling strength).			
Test Methods: DIN			
Physical		**Mechanical**	
Colour	Transparent	τ (RT)	5 MPa
Viscosity	3.2 Pa.s*	τ (°C)	-
S.G.	~1.0	τ (wet)	-
Shelf life	12 months/-5+23°C	Peel ( )	-
Pot/Work life	20 mins	σ_t	-
$T_{Service}$	-25 to +100°C	E	-
CTE (°C)	-	Elongation	-
Thermal cond.	-	H (Shore)	-
Notes: * Agovit 1900 0.5Pa.s; Agovit 1074 0.4 Pa.s. τ DIN 53283 (substrate failure). Flash point 10°C.			

Degussa	Agovit 1093 AB		Feb 94
Methylmethacrylate - Two part			
Manufacturers Comments: Transparent adhesive for bonding acrylic glass.			
Typical Use: Acrylic glass- cast, extruded or injection moulded.			
Cure: RT/40 mins (handling strength).			
Test Methods: DIN			
Physical		**Mechanical**	
Colour	Transparent	τ (RT)	4 MPa
Viscosity	~1.5 Pa.s	τ (°C)	-
S.G.	~1.0	τ (wet)	-
Shelf life	-	Peel ( )	-
Pot/Work life	15 mins	σ_t	-
$T_{Service}$	-	E	-
CTE (°C)	-	Elongation	-
Thermal cond.	-	H (Shore)	-
Notes: τ DIN 53283 (substrate failure).			

Degussa	Agovit 1900		May 95
Methylmethacrylate - Two part			
Manufacturers Comments: Transparent adhesive for bonding acrylic glass.			
Typical Use: Acrylic glass: cast, extruded, injection moulded.			
Cure: RT/2 hrs.			
Test Methods: DIN			
Physical		**Mechanical**	
Colour	Transparent	τ (RT)	6 MPa
Viscosity	~0.5 Pa.s	τ (°C)	-
S.G.	~1.0	τ (wet)	-
Shelf life	-	Peel ( )	-
Pot/Work life	20 mins	σ_t	-
$T_{Service}$	-	E	-
CTE (°C)	-	Elongation	-
Thermal cond.	-	H (Shore)	-
Notes: τ DIN 53283 (substrate failure).			

Delo	Maflex 245		(July 95)
Acrylic - Two part, with surface activator			
Manufacturers Comments: No mix assembly adhesive. Two activators available, No.5 faster curing than No.1. High shear strength, peel and impact resistance. Very good chemical resistance (not acetone). Low odour. Higher flash point than Maflex 241.			
Typical Use: Metals, most plastics, glass, wood, textiles, stoneware.			
Cure: RT 3-5 mins (firm to touch); 20-30 (full cure) (Optimum layer thickness 0.25mm).			
Test Methods: DIN			
Physical		**Mechanical**	
Colour	-	τ (RT)	19 (19) MPa
Viscosity	50 Pa.s (resin)	τ (°C)	-
S.G.	-	τ (wet)	-
Shelf life	3 months	Peel (-)	400 N mm^{-1}
Pot/Work life	-	σ_t	-
$T_{Service}$	-55 to +100°C	E	1100-1400 MPa
CTE (°C)	75-85 $\times 10^{-6}$ K^{-1}	Elongation	-
Thermal cond.	0.2 W mK^{-1}	H (Shore)	-
Notes: Resistivity DIN 53482: 10^{15} Ωcm . Dielectric Index DIN 53482 (10Hz/RT) 5.2. τ DIN 53283 data Al/Al - Activator No.1 (Activator No. 5). steel/steel: 13 MPa (14 MPa). Impact strength: 49J.			

Delo

Delo	Maflex 241		(July 95)
Acrylic - Two part, with surface activator			
Manufacturers Comments: No mix assembly adhesive. Two activators available, No.5 faster curing than No.1. High shear strength, peel and impact resistance. Very good chemical resistance (not acetone).			
Typical Use: Metals, most plastics, glass, wood, textiles, stoneware.			
Cure: RT 1-5 mins (firm to touch); 20-30 (full cure) (Optimum layer thickness 0.25mm).			
Test Methods: DIN			
Physical		**Mechanical**	
Colour	-	τ (RT)	32 (26) MPa
Viscosity	20 - 40 Pa.s (resin)	τ (°C)	-
S.G.	-	τ (wet)	-
Shelf life	6 months	Peel (-)	600 N mm^{-1}
Pot/Work life	-	σ_t	-
$T_{Service}$	-55 to +100°C	E	1100-1400 MPa
CTE (°C)	75-85 $\times 10^{-6}$ K^{-1}	Elongation	-
Thermal cond.	0.2 W mK^{-1}	H (Shore)	-
Notes: Resistivity DIN 53482: 10^{15} Ωcm. Dielectric Index DIN 53482 (10Hz/RT) 5.2. τ DIN 53283 data Al/Al - Activator No.1 (Activator No. 5). PVC/PVC: 7 MPa (11 MPa). Impact strength: 49J.			

Devcon (ITW)

Devcon	MVP Type 11		(Feb 95)
Acrylic - Two part methacrylate			
Manufacturers Comments: High viscosity, high strength.			
Typical Use: Plastics, metals, wood, ceramics.			
Cure: Useful strength: 16 - 24hr/RT, Full cure: Several days.			
Test Methods: ASTM			
Physical		**Mechanical**	
Colour	Yellow	τ (RT)	31.7 MPa
Viscosity	35.2 Pa.s	τ (°C)	-
S.G.	1.2	τ (wet)	-
Shelf life	-	Peel ( )	-
Pot/Work life	3 - 7 min.	σ_t	-
$T_{Service}$	<93°C	E	-
CTE (°C)	-	Elongation	-
Thermal cond.	-	H (Shore)	-
Notes: Breakdown potential: 490 V/0.25mm.			

Dunlop

Dunlop	Megabond SAS 300		April 94
Toughened methacrylate - Two part, with surface activator (no mix)			
Manufacturers Comments: Rapid RT cure. High shear, peel and impact strength. Non-flammable, low odour, low toxicity. Tolerates oily surfaces. Assembly of vehicle panels - replacing welds and rivets. Office equipment, sports goods. Machine tools.			
Typical Use: Metals, glass, wood, GRP, most plastics (not polyethylene or PTFE).			
Cure: RT 2-4 mins (handling strength); 24 hrs (full cure).			
Test Methods: Not stated			
Physical		**Mechanical**	
Colour	Translucent	τ (RT)	13.8 MPa
Viscosity	44 - 54 Pa.s (resin)	τ (°C)	-
S.G.	~1.1	τ (wet)	-
Shelf life	6 months/10-25°C	Peel (T)	2.45 - 2.80Nmm^{-1}
Pot/Work life	-	σ_f	-
$T_{Service}$	+100°C*	E	-
CTE (°C)	-	Elongation	-
Thermal cond.	-	H (Shore)	-
Notes: *+150°C, intermittently. τ (Al/Al, etched) substrate failure. Oily Steel: 11 MPa; Steel (clean/abraded): 16.5 MPa; Oak: 10 MPa (substrate fail). Gap filling: 0.5mm.			

Dunlop	Megabond SAS 700		April 91
Acrylic - One part, UV Curing			
Manufacturers Comments: Fast RT cure. High optical quality. Transparent bondline. Easy to use, repositionable before UV exposure.			
Typical Use: Glass display units. Lenses, optical fibres, jewellery.			
Cure: RT 10 - 15 seconds (UV Exposure).			
Test Methods: Not stated			
Physical		**Mechanical**	
Colour	Clear*	τ (RT)	17 MPa
Viscosity	3.5 - 7.5 Pa.s (resin)	τ (°C)	-
S.G.	~1.05	τ (wet)	-
Shelf life	6 months	Peel ( )	-
Pot/Work life	-	σ_f	4 - 7 MPa
$T_{Service}$	+100°C*	E	-
CTE (°C)	-	Elongation	-
Thermal cond.	-	H (Shore)	-
Notes: * Light transmittance >95%. σ- glass/metal bond. τ glass/glass bond. Gap filling: 0.1mm.			

Dunlop	Megabond SAS 350		May 92
Elastomeric methacrylate - Two part, with surface activator (no mix)			
Manufacturers Comments: RT cure. Excellent shear and peel strength. Tolerates oily surfaces. High heat and chemical resistance. General assembly (metals). Vehicle body panels, loudspeakers, electronics.			
Typical Use: Metals, glass, GRP, most plastics.			
Cure: RT 20-25 mins (handling strength); 24 hrs (full cure)			
Test Methods: Not stated			
Physical		**Mechanical**	
Colour	Translucent	τ (RT)	25 MPa
Viscosity (resin)	3.5 - 7.5 Pa.s	τ (°C)	-
S.G.	~1.05	τ (wet)	-
Shelf life	6 months/10-25°C	Peel ( )	-
Pot/Work life	-	σ_f	-
$T_{Service}$	+100°C*	E	-
CTE (°C)	-	Elongation	-
Thermal cond.	-	H (Shore)	-
Notes: *+150°C, intermittently. τ (Al/Al). Steel: 28 MPa. Gap filling: 0.5mm.			

Dymax

Dymax	1-20270M		(July 95)
UV/visible light curing aerobic acrylic - One part,			
Manufacturers Comments: 100% solids, 100% solvent-free, structural-strength. Fast cure, flexible, moisture resistant. For assembly of medical devices and equipment. Bonds a variety of materials, resists sterilisation (ETO, gamma, electron beam). Meets: USP Class VI.			
Typical Use: Plastic/plastic reservoirs: blood oxygenators, heat exchangers and pressure transducer. Polycarbonate, St. steel & PVC.			
Cure: UV/ visible light.			
Test Methods: Not stated			
Physical		**Mechanical**	
Colour	-	τ (RT)	-
Viscosity	1.4 Pa.s standard*	τ (°C)	-
S.G.	-	τ (wet)	-
Shelf life	-	Peel ( )	-
Pot/Work life	-	σ_f	13.8 MPa
$T_{Service}$	-	E	-
CTE (°C)	-	Elongation	160%
Thermal cond.	-	H (Shore D)	60
Notes: * other viscosity grades: T (thick) - 7 Pa.s; G(gel) - 25 Pa.s.			

See Key to Data Sheets - Pages 63 to 65

Dymax	**1-20271M**		(July 95)
UV/visible light curing aerobic acrylic - One part,			
Manufacturers Comments: 100% solids, 100% solvent-free, structural-strength. Fast cure, impact resistant. For assembly of medical devices and equipment. Bonds a variety of materials, resists sterilisation (ETO, gamma, electron beam). Meets: USP Class VI.			
Typical Use: Medical cannulae assembly: metals, glass, ceramic, some plastics.			
Cure: UV/ visible light.			
Test Methods: Not stated			
Physical		**Mechanical**	
Colour	-	τ (RT)	-
Viscosity	4.0 Pa.s standard*	τ (°C)	-
S.G.	-	τ (wet)	-
Shelf life	-	Peel ( )	-
Pot/Work life	-	σ_t	22.8 MPa
$T_{Service}$	-	*E*	-
CTE (°C)	-	Elongation	9%
Thermal cond.	-	H (Shore D)	70
Notes: * other viscosity grades: T (thick) - 7 Pa.s; VT(very thick) - 13 Pa.s. G(gel): 25 Pa.s.			

Dymax	**3-20261**		(July 95)
UV/visible light curing aerobic acrylic - One part,			
Manufacturers Comments: Solvent-free, structural-strength. Fast cure, moisture resistance. Engineering assembly. Ultra Light-Weld series.			
Typical Use: Plastic appliance housings, display cases. Lenses. Polycarbonate.			
Cure: UV/ visible light.			
Test Methods: Not stated			
Physical		**Mechanical**	
Colour	-	τ (RT)	-
Viscosity	-	τ (°C)	-
S.G.	-	τ (wet)	-
Shelf life	-	Peel ( )	-
Pot/Work life	-	σ_t	13.8 MPa
$T_{Service}$	-	*E*	-
CTE (°C)	-	Elongation	160%
Thermal cond.	-	H (Shore D)	60
Notes:			

Dymax	**3-20256**		(July 95)
UV/visible light curing aerobic acrylic - One part,			
Manufacturers Comments: Solvent-free, structural-strength. Fast cure, surface wetting. Engineering assemblies. Ultra Light-Weld series.			
Typical Use: Assembly of flexible PVC. Plastics.			
Cure: UV/ visible light.			
Test Methods: Not stated			
Physical		**Mechanical**	
Colour	-	τ (RT)	-
Viscosity	-	τ (°C)	-
S.G.	-	τ (wet)	-
Shelf life	-	Peel ( )	-
Pot/Work life	-	σ_t	48.3 MPa
$T_{Service}$	-	*E*	-
CTE (°C)	-	Elongation	6%
Thermal cond.	-	H (Shore D)	80
Notes:			

Dymax	**4-20260**		(July 95)
UV/visible light curing aerobic acrylic - One part,			
Manufacturers Comments: Solvent-free, structural-strength. Fast cure. Transparent. Dish-washer safe. Engineering assembly. Ultra Light-Weld series.			
Typical Use: Glass, ceramics.			
Cure: UV/ visible light.			
Test Methods: Not stated			
Physical		**Mechanical**	
Colour	-	τ (RT)	-
Viscosity	-	τ (°C)	-
S.G.	-	τ (wet)	-
Shelf life	-	Peel ( )	-
Pot/Work life	-	σ_t	34.5 MPa
$T_{Service}$	-	*E*	-
CTE (°C)	-	Elongation	15%
Thermal cond.	-	H (Shore D)	80
Notes:			

Dymax	6-20265		(July 95)
UV/visible light curing aerobic acrylic - One part,			
Manufacturers Comments: Solvent-free, structural-strength. Fast cure. General purpose engineering assembly. Ultra Light-Weld series.			
Typical Use: Metals, glass, some plastics.			
Cure: UV/ visible light. Thermal cure.			
Test Methods: Not stated			
Physical		**Mechanical**	
Colour	-	τ (RT)	-
Viscosity	-	τ (°C)	-
S.G.	-	τ (wet)	-
Shelf life	-	Peel ( )	-
Pot/Work life	-	σ_t	22.8 MPa
$T_{Service}$	-	*E*	-
CTE (°C)	-	Elongation	30%
Thermal cond.	-	H (Shore D)	60
Notes:			

Dymax	840		(Oct 95)
Two-part (no mix), (meth)acrylate-urethane			
Manufacturers Comments: Fast cure. Activator, no-mix. For rapid bonding of variety of materials. Tough, resilient bonds with environmental resistance (solvents and temperature).			
Typical Use: Assembly adhesive: Metal (inc. plated and painted steel, stainless steel), ceramics, glass, some plastics. Honing abrasives to brackets, cutlery, magnet bonding, DC motor assembly.			
Cure: RT/ ~20 secs (handling), 4 hrs (full cure).			
Test Methods: ASTM			
Physical		**Mechanical**	
Colour	Straw/amber	τ (RT)	17.2 MPa
Viscosity	0.5 Pa.s (standard)*	τ (°C)	-
S.G.	-	τ (wet)	12.0 MPa (70%)
Shelf life	1 year/<32°C	Peel ( )	-
Pot/Work life	-	σ_t	-
$T_{Service}$	-18 to +177°C	*E*	-
CTE (°C)	-	Elongation	-
Thermal cond	-	H (Shore)	-
Notes: * other viscosity grades: T(thick) - 5 Pa.s; VT(very thick) - 45 Pa.s; G(gel) - thixotropic, non-flow. τ data: ASTM D-1002. τ (wet): After 100hrs water immersion at 25°C (% strength retention). Gap filling: 0.5 mm max. Flash point >93.3°C.			

Dymax	828		(Oct 95)
Two-part (no mix), (meth)acrylate-urethane			
Manufacturers Comments: Fast cure. Activator, no-mix. For rapid bonding of variety of materials. Tough, hard bonds with high environmental resistance (solvents and temperature).			
Typical Use: Assembly adhesive: Metal (inc. copper and brass), glass, ceramics, plastics. Electro-mechanical devices, brass hand-rails.			
Cure: RT/ ~30 secs (handling), 24 hrs (full cure).			
Test Methods: ASTM			
Physical		**Mechanical**	
Colour	Straw/amber	τ (RT)	20.7 MPa
Viscosity	0.65 Pa.s (standard)*	τ (°C)	-
S.G.	-	τ (wet)	14.5 MPa (70%)
Shelf life	1 year/<32°C	Peel ( )	-
Pot/Work life	-	σ_t	-
$T_{Service}$	-18 to +177°C	*E*	-
CTE (°C)	-	Elongation	-
Thermal cond	-	H (Shore)	-
Notes: * other viscosity grades: T(thick) - 3.25 Pa.s; VT(very thick) - 9 Pa.s; G(gel) - thixotropic, non-flow. τ data: ASTM D-1002. τ (wet): After 100hrs water immersion at 25°C (% strength retention). Gap filling: 0.25 mm max. Flash point >93.3°C.			

Dymax	845		(Oct 95)
Two-part (no mix), (meth)acrylate-urethane			
Manufacturers Comments: Fast cure. Activator, no-mix. For rapid bonding of variety of materials. Tough, high strength bonds (steel) with environmental resistance (solvents and temperature). Best gap filling.			
Typical Use: Assembly adhesive: Metals (iron, steel, aluminium), glass, ceramics. Architectural steel panels, stainless steel table, wear strips on machinery, rare earth magnets.			
Cure: RT/ ~40 secs (handling), 24 hrs (full cure).			
Test Methods: ASTM			
Physical		**Mechanical**	
Colour	Straw/amber	τ (RT)	24.1 MPa
Viscosity	3.5 Pa.s (standard)*	τ (°C)	-
S.G.	-	τ (wet)	14.5 MPa (60%)
Shelf life	1 year/<32°C	Peel ( )	-
Pot/Work life	-	σ_t	-
$T_{Service}$	-18 to +177°C	*E*	-
CTE (°C)	-	Elongation	-
Thermal cond	-	H (Shore)	-
Notes: * other viscosity grades: T(thick) - 7 Pa.s; VT(very thick) - 45 Pa.s; G(gel) - thixotropic, non-flow. τ data: ASTM D-1002. τ (wet): After 100hrs water immersion at 25°C (% strength retention). Gap filling: 1.0 mm max. Flash point >93.3°C.			

Dymax	846		(Oct 95)
Two-part (no mix), (meth)acrylate-urethane			
Manufacturers Comments: Fast cure. Choice of activator, no-mix. High strength, general purpose assembly adhesive for rapid bonding of variety of materials. Good thermal shock resistance.			
Typical Use: Assembly adhesive: Metals (inc. plated), glass, ceramics, plastics (filled nylon, thermoplastics, epoxy-based).			
Cure: RT/ ~15 secs (handling), 24 hrs (full cure).			
Test Methods: ASTM			
Physical		**Mechanical**	
Colour	Translucent/straw	τ (RT)	24.1 MPa
Viscosity	3.5 Pa.s	τ (°C)	-
S.G.	-	τ (wet)	16.9 MPa (70%)
Shelf life	1 year/<32°C	Peel ( )	-
Pot/Work life	-	σ_t	-
$T_{Service}$	-18 to +177°C	*E*	-
CTE (°C)	-	Elongation	-
Thermal cond	-	H (Shore)	-
Notes: τ data: ASTM D-1002. τ (wet): After 8 hrs steam exposure (% strength retention). Gap filling: 0.5 mm max. Flash point >93.3°C.			

Dymax	Light-Weld 304		(Oct 95)
One-part, (meth)acrylate-urethane, UV cure			
Manufacturers Comments: UV curing. Optically transparent adhesive for fibre-optic applications (close-tolerance parts).			
Typical Use: Fibre optic assembly: fibre-to-metal, fibre-to-fibre.			
Cure: UV.			
Test Methods: -			
Physical		**Mechanical**	
Colour	Transparent	τ (RT)	-
Viscosity	-	τ (°C)	-
S.G.	-	τ (wet)	-
Shelf life	-	Peel ( )	-
Pot/Work life	-	σ_t	-
$T_{Service}$	-	*E*	-
CTE (°C)	-	Elongation	-
Thermal cond	-	H (Shore)	-
Notes: For looser tolerance parts, use Light-Weld 305.			

Dymax	847		(Oct 95)
Two-part (no mix), (meth)acrylate-urethane			
Manufacturers Comments: Fastest cure. Activator, no-mix. For rapid bonding of variety of materials. Tough, resilient bonds with excellent impact, vibration, thermal shock resistance.			
Typical Use: Assembly adhesive: Metals (iron, steel, aluminium), glass, ceramics, epoxy-based plastics. DC motors (steel-ferrites), gauge wear surfaces (carbide-tool steel), speaker magnets, mirror brackets to windscreens.			
Cure: RT/ ~15 secs (handling), 4 hrs (full cure).			
Test Methods: ASTM			
Physical		**Mechanical**	
Colour	Straw/amber	τ (RT)	13.8 MPa
Viscosity	0.5 Pa.s (standard)*	τ (°C)	-
S.G.	-	τ (wet)	6.9 MPa (50%)
Shelf life	1 year/<32°C	Peel ( )	-
Pot/Work life	-	σ_t	-
$T_{Service}$	-18 to +177°C	*E*	-
CTE (°C)	-	Elongation	-
Thermal cond	-	H (Shore)	-
Notes: * other viscosity grades: T(thick) - 7 Pa.s; VT(very thick) - 45 Pa.s; G(gel) - thixotropic, non-flow. τ data: ASTM D-1002. τ (wet): After 100hrs water immersion at 25°C (% strength retention). Gap filling: 0.25 mm max. Flash point >93.3°C.			

Dymax	Light-Weld 401		(Oct 95)
One-part, (meth)acrylate-urethane, UV cure			
Manufacturers Comments: UV curing. General purpose, high clarity for assembly of glass. Provides 'invisible' bondlines.			
Typical Use: Glass-to-glass, glass-to-metal. Syringe bodies. Repair of glass seals.			
Cure: UV.			
Test Methods: ASTM			
Physical		**Mechanical**	
Colour	Clear/Lt. Yellow	τ (RT)	-
Viscosity	0.06-0.1 Pa.s	τ (°C)	-
S.G.	-	τ (wet)	-
Shelf life	-	Peel ( )	-
Pot/Work life	-	σ_t	-
$T_{Service}$	-	*E*	-
CTE (°C)	-	Elongation	-
Thermal cond	-	H (Shore D)	75
Notes: Max. cure depth: 6.35 mm.			

Dymax	Light-Weld 415		(Oct 95)
One-part, (meth)acrylate-urethane, UV cure			
Manufacturers Comments: UV curing. Provides 'invisible' bondlines. Glass assembly for small cross-section joints with filet. Will cure outside joint (hard, clear coating). Deep section cures.			
Typical Use: Glass-bonding/coating.			
Cure: UV.			
Test Methods: ASTM			
Physical		**Mechanical**	
Colour	Clear/Lt. Yellow	τ (RT)	-
Viscosity	5 Pa.s	τ (°C)	-
S.G.	-	τ (wet)	-
Shelf life	-	Peel ( )	-
Pot/Work life	-	σ_t	-
$T_{Service}$	-	E	-
CTE (°C)	-	Elongation	-
Thermal cond	-	H (Shore D)	85
Notes: Max. cure depth: 12.7 mm			

Dymax	Light-Weld 425		(Oct 95)
One-part, (meth)acrylate-urethane, UV cure			
Manufacturers Comments: UV curing. Provides 'invisible' bondlines. Glass assembly with environmental/aesthetic requirements. High clarity, dishwasher safe. High strength and durability.			
Typical Use: Glass-bonding. Display case assembly.			
Cure: UV.			
Test Methods: ASTM			
Physical		**Mechanical**	
Colour	Clear	τ (RT)	-
Viscosity	4 Pa.s	τ (°C)	-
S.G.	-	τ (wet)	-
Shelf life	-	Peel ( )	-
Pot/Work life	-	σ_t	-
$T_{Service}$	-	E	-
CTE (°C)	-	Elongation	-
Thermal cond	-	H (Shore D)	80
Notes: Max. cure depth: 6.35 mm. Similar use: Light-Weld 426 lower viscosity (1 Pa.s).			

Dymax	Light-Weld 416A		(Oct 95)
One-part, (meth)acrylate-urethane, UV cure			
Manufacturers Comments: UV curing. Provides 'invisible' bondlines in transparent plastic materials. General purpose bonding and laminating. Flexible, strain-relief for dissimilar materials.			
Typical Use: Plastics (inc. PVC, polycarbonate, ABS).			
Cure: UV (30-50 mW/cm²)/1-2 secs.			
Test Methods: -			
Physical		**Mechanical**	
Colour	Clear/Straw	τ (RT)	-
Viscosity	8 Pa.s (standard)	τ (°C)	-
S.G.	-	τ (wet)	-
Shelf life	-	Peel ( )	-
Pot/Work life	-	σ_t	17.2 MPa (film)
$T_{Service}$	-	E	-
CTE (°C)	-	Elongation	225%
Thermal cond	-	H (Shore D)	40
Notes: Replaces Dymax product 183. Nominal Cure depth: 6.35mm.			

Dymax	Light-Weld 481A		(Oct 95)
One-part, (meth)acrylate-urethane, UV cure			
Manufacturers Comments: UV curing. Provides 'invisible' bondlines in transparent plastic materials. Rigid adhesive with strength > most plastics.			
Typical Use: Plastics (inc. assembly of polycarbonate, PVC).			
Cure: UV (30-50 mW/cm²)/1-2 secs.			
Test Methods: -			
Physical		**Mechanical**	
Colour	Clear/Straw	τ (RT)	-
Viscosity	7.5 Pa.s (standard)*	τ (°C)	-
S.G.	-	τ (wet)	-
Shelf life	-	Peel ( )	-
Pot/Work life	-	σ_t	44.9 MPa (film)
$T_{Service}$	-	E	-
CTE (°C)	-	Elongation	10%
Thermal cond	-	H (Shore D)	80
Notes: * G (gel) version available. Nominal cure depth: 9.53 mm. Replaces Dymax product 181.			

Dymax	Light-Weld 484A		(Oct 95)
One-part, (meth)acrylate-urethane, UV cure			
Manufacturers Comments: UV curing. Provides 'invisible' bondlines in transparent plastic materials. Hard, moisture resistant bonds. Deep section curing.			
Typical Use: Plastics (inc. polycarbonate, PVC).			
Cure: UV (30-50 mW/cm²)/~1 secs.			
Test Methods: -			
Physical		**Mechanical**	
Colour	Clear/Straw	τ (RT)	-
Viscosity	0.08 Pa.s (standard)*	τ (°C)	-
S.G.	-	τ (wet)	-
Shelf life	-	Peel ( )	-
Pot/Work life	-	σ_f	41.4 MPa (film)
$T_{Service}$	-	*E*	-
CTE (°C)	-	Elongation	12%
Thermal cond	-	H (Shore D)	80
Notes: * G (gel) version available. Nominal cure depth: 12.7 mm.			

Dymax	Light-Weld 488		(Oct 95)
One-part, (meth)acrylate-urethane, UV cure			
Manufacturers Comments: UV curing. Provides 'invisible' bondlines in transparent materials. Flexible, moisture resistant adhesive-sealant.			
Typical Use: Plastics, glass and metals.			
Cure: UV or heat.			
Test Methods: ASTM			
Physical		**Mechanical**	
Colour	Clear/Lt. Yellow	τ (RT)	-
Viscosity	0.5 Pa.s	τ (°C)	-
S.G.	-	τ (wet)	-
Shelf life	-	Peel ( )	-
Pot/Work life	-	σ_f	-
$T_{Service}$	-	*E*	-
CTE (°C)	-	Elongation	-
Thermal cond	-	H (Shore D)	50
Notes: Max. cure depth: 3.17 mm.			

Dymax	Light-Weld 486A		(Oct 95)
One-part, (meth)acrylate-urethane, UV cure			
Manufacturers Comments: UV curing. Provides 'invisible' bondlines in transparent plastic materials. High strength bonds in other engineering materials.			
Typical Use: Plastics (inc. styrene-SAN, ABS), metals, glass, ceramic.			
Cure: UV (30-50 mW/cm²)/1-3 secs.			
Test Methods: -			
Physical		**Mechanical**	
Colour	Clear/Straw	τ (RT)	-
Viscosity	0.14 Pa.s (standard)*	τ (°C)	-
S.G.	-	τ (wet)	-
Shelf life	-	Peel ( )	-
Pot/Work life	-	σ_f	41.4 MPa (film)
$T_{Service}$	-	*E*	-
CTE (°C)	-	Elongation	5%
Thermal cond	-	H (Shore D)	85
Notes: * other viscosity grades available: T(thick) - 6.7 Pa.s, G (gel). Nominal cure depth: 9.53 mm. Replaces Dymax product 186.			

Dymax	MD - 128M		(July 95)
UV + heat curing acrylic - One part			
Manufacturers Comments: 100% solids, 100% solvent-free, structural-strength. Fast cure. For assembly of medical devices and equipment. Bonds a variety of materials, resists sterilisation (ETO, gamma, electron beam). Meets: USP Class VI.			
Typical Use: Cannulae (tube/needle) to hub or reservoir. Metal, glass, ceramics, some plastics (inc. ABS, nylon, polystyrene).			
Cure: UV + heat cure. Cure rate: <5.			
Test Methods: Not stated			
Physical		**Mechanical**	
Colour	Translucent/white	τ (RT)	-
Viscosity	0.6 Pa.s (standard)*	τ (°C)	-
S.G.	-	τ (wet)	-
Shelf life	-	Peel ( )	-
Pot/Work life	-	σ_f	26.2 MPa
$T_{Service}$	-	*E*	-
CTE (°C)	-	Elongation	30%
Thermal cond.	-	H (Shore D)	75
Notes: * other viscosity grades: T(thick) - 3.25 Pa.s; VT(very thick) - 11 Pa.s; G(gel): 25 Pa.s. Max. depth cure: 3.17mm.			

Dymax **MD - 136M** (July 95)

UV curing aerobic acrylic - One part

Manufacturers Comments: 100% solids, 100% solvent-free, structural-strength. Fast cure. For assembly of medical devices and equipment. Bonds a variety of materials, resists sterilisation (ETO, gamma, electron beam). Meets: USP Class VI.

Typical Use: Cannulae (tube/needle) to hub or reservoir. Metal, glass, ceramics, some plastics (inc. ABS, nylon, polystyrene).

Cure: UV. Cure rate <3.

Test Methods: Not stated

Physical		Mechanical	
Colour	Clear straw	τ (RT)	-
Viscosity	0.1 Pa.s (standard)*	τ (°C)	-
S.G.	-	τ (wet)	-
Shelf life	-	Peel ( )	-
Pot/Work life	-	σ_t	41.1 MPa
$T_{Service}$	-	E	-
CTE (°C)	-	Elongation	10%
Thermal cond.	-	H (Shore D)	85

Notes: * other viscosity grades: T(thick) - 76.5 Pa.s; G(gel) - 25 Pa.s. Max. depth cure: 6.35 mm.

Dymax **MD - 190M** (July 95)

UV curing aerobic acrylic - One part

Manufacturers Comments: 100% solids, 100% solvent-free, structural-strength. Fast cure, wicking. For assembly of medical devices and equipment. Bonds a variety of materials, resists sterilisation (ETO, gamma, electron beam). Meets: USP Class VI.

Typical Use: Respiratory, oxygen and anaesthesia masks. IV tube sets. Some plastics, inc. rigid plastics to flexible PVC.

Cure: UV Cure rate <2.

Test Methods: Not stated

Physical		Mechanical	
Colour	Clear	τ (RT)	-
Viscosity	0.08 Pa.s	τ (°C)	-
S.G.	-	τ (wet)	-
Shelf life	-	Peel ( )	-
Pot/Work life	-	σ_t	42.7 MPa
$T_{Service}$	-	E	-
CTE (°C)	-	Elongation	10%
Thermal cond.	-	H (Shore D)	80

Notes: Max. depth cure: 6.35 mm.

Dymax **MD - 138M** (July 95)

UV curing aerobic acrylic - One part

Manufacturers Comments: 100% solids, 100% solvent-free, structural-strength. Fast cure, flexible. For assembly of medical devices and equipment. Bonds a variety of materials, resists sterilisation (ETO, gamma, electron beam). Meets: USP Class VI.

Typical Use: Respiratory, oxygen and anaesthesia masks. Catheters. Some plastics, inc. rigid plastics to flexible PVC, some metals.

Cure: UV Cure rate:<3.

Test Methods: Not stated

Physical		Mechanical	
Colour	Clear	τ (RT)	-
Viscosity	0.35 Pa.s standard*	τ (°C)	-
S.G.	-	τ (wet)	-
Shelf life	-	Peel ( )	-
Pot/Work life	-	σ_t	9.7 MPa
$T_{Service}$	-	E	-
CTE (°C)	-	Elongation	300%
Thermal cond.	-	H (Shore D)	40

Notes: * other viscosity grades: G (gel) - 25 Pa.s. Max. depth cure: 6.35 mm.

Dymax **MD - 191M** (July 95)

UV curing aerobic acrylic - One part

Manufacturers Comments: 100% solids, 100% solvent-free, structural-strength. Fast cure, flexible. For assembly of medical devices and equipment. Bonds a variety of materials, resists sterilisation (ETO, gamma, electron beam). Meets: USP Class VI.

Typical Use: Respiratory, oxygen and anaesthesia masks. IV tube sets. Catheters. Some plastics, inc. rigid plastics to flexible PVC.

Cure: UV.

Test Methods: Not stated

Physical		Mechanical	
Colour	-	τ (RT)	-
Viscosity	0.5 Pa.s	τ (°C)	-
S.G.	-	τ (wet)	-
Shelf life	-	Peel ( )	-
Pot/Work life	-	σ_t	20.7 MPa
$T_{Service}$	-	E	-
CTE (°C)	-	Elongation	190%
Thermal cond.	-	H (Shore D)	60

Notes:

Dymax | **MD - 195M** | (July 95)

UV curing aerobic acrylic - One part

Manufacturers Comments: 100% solids, 100% solvent-free, structural-strength. Fast cure, flexible. For assembly of medical devices and equipment. Bonds a variety of materials, resists sterilisation (ETO, gamma, electron beam). Meets: USP Class VI.

Typical Use: Plastic/plastic reservoirs: blood oxygenators, heat exchangers and pressure transducers. Some plastics - polycarbonate/PVC.

Cure: UV.

Test Methods: Not stated

Physical		Mechanical	
Colour	-	τ (RT)	-
Viscosity	6.5 Pa.s standard*	τ (°C)	-
S.G.	-	τ (wet)	-
Shelf life	-	Peel ( )	-
Pot/Work life	-	σ_f	41.4 MPa
$T_{Service}$	-	E	-
CTE (°C)	-	Elongation	10%
Thermal cond.	-	H (Shore D)	80

Notes: * other viscosity grades: G(gel) - 25 Pa.s.

Dymax | **Multi-Cure 427** | (Oct 95)

One-part, (meth)acrylate-urethane, UV or heat cure

Manufacturers Comments: UV or heat curing. Provides 'invisible' bondlines. High strength, high clarity for structural glass-to-metal bonds. Hard, glossy surface when exposed to UV.

Typical Use: Glass-to-metal bonding.

Cure: UV or heat (177°C/10 mins).

Test Methods: ASTM

Physical		Mechanical	
Colour	Clear/Lt. Yellow	τ (RT)	-
Viscosity	4 Pa.s	τ (°C)	-
S.G.	-	τ (wet)	-
Shelf life	-	Peel ( )	-
Pot/Work life	-	σ_f	-
$T_{Service}$	-	E	-
CTE (°C)	-	Elongation	-
Thermal cond	-	H (Shore D)	80

Notes: Max. cure depth: 3.17 mm.

Dymax | **MD - 197M** | (July 95)

UV curing aerobic acrylic - One part

Manufacturers Comments: 100% solids, 100% solvent-free, structural-strength. Fast cure, flexible, tough. For assembly of medical devices and equipment. Bonds a variety of materials, resists sterilisation (ETO, gamma, electron beam). Meets: USP Class VI.

Typical Use: Plastic/plastic reservoirs: blood oxygenators, heat exchangers and pressure transducer (polycarbonate/PVC). IV tube sets.

Cure: UV.

Test Methods: Not stated

Physical		Mechanical	
Colour	-	τ (RT)	-
Viscosity	1.4 Pa.s standard*	τ (°C)	-
S.G.	-	τ (wet)	-
Shelf life	-	Peel ( )	-
Pot/Work life	-	σ_f	13.8 MPa
$T_{Service}$	-	E	-
CTE (°C)	-	Elongation	160%
Thermal cond.	-	H (Shore D)	60

Notes: * other viscosity grades: T (thick) - 7 Pa.s; G(gel) - 25 Pa.s.

Dymax | **Multi-Cure 428** | (Oct 95)

One-part, (meth)acrylate-urethane, UV or heat cure

Manufacturers Comments: UV or heat curing. Provides 'invisible' bondlines. Low viscosity, high clarity for laminating glass or metal-glass bonds.

Typical Use: Laminating glass (lenses). Glass-to-metal bonding.

Cure: UV or heat.

Test Methods: ASTM

Physical		Mechanical	
Colour	Clear/Lt. Yellow	τ (RT)	-
Viscosity	0.15 Pa.s	τ (°C)	-
S.G.	-	τ (wet)	-
Shelf life	-	Peel ( )	-
Pot/Work life	-	σ_f	-
$T_{Service}$	-	E	-
CTE (°C)	-	Elongation	-
Thermal cond	-	H (Shore D)	60

Notes: Max. cure depth: 3.17 mm.

Dymax	Multi-Cure 602		(Oct 95)
Two-part, (meth)acrylate-urethane Activator (no-mix),UV- or heat cure			
Manufacturers Comments: Rapid setting. Choice of curing method (activator, UV or heat) to suit production methods. High strength bonds (some metals). Good heat resistance, not for humid environments.			
Typical Use: Assembly adhesive: Steel, ceramics, some plastics. Heavy gauge wire-tacking, speaker housing textiles, screw-locking.			
Cure: Activator: ~20 secs (handling), 4 hrs (full cure). UV: <10 secs (full cure). Heat: 93°C/1hr; 149°C/5 mins.			
Test Methods: ASTM			
Physical		**Mechanical**	
Colour	Straw/amber	τ (RT, steel)	20.7 MPa
Viscosity	4 Pa.s (standard)*	τ (°C)	-
S.G.	-	τ (wet)	-
Shelf life	-	Peel ( )	-
Pot/Work life	20-45 secs (tack-free)	σ_f	-
$T_{Service}$	-18 to +177°C	E	-
CTE (°C)	-	Elongation	20%
Thermal cond	-	H (Shore D)	70
Notes: * other viscosity grades: T(thick) - 7 Pa.s; VT(very thick) - 13 Pa.s; G(gel) - thixotropic. τ data (ASTM D-1002): steel, activator cured. Gap filling: 0.5mm max. (activator), 3.2 mm max. (UV). Water absorption (after 8hrs/boiling water): 13%. Dielectric strength: 18.9 kV/mm.			

Dymax	Multi-Cure 628		(Oct 95)
Two-part, (meth)acrylate-urethane Activator (no-mix),UV- or heat cure			
Manufacturers Comments: Rapid setting. Choice of curing method (activator, UV or heat) to suit production methods. High strength, rigid bonds for a variety of materials.			
Typical Use: Assembly adhesive: Metals (inc. brass and copper), glass, ceramics, some plastics. Electrical uses, e.g. Bridge-bonding transformer parts, formed-in-place insulators/isolators.			
Cure: Activator: ~30 secs (handling), 24 hrs (full cure). UV: <15 secs (full cure). Heat: 93°C/1hr; 149°C/5 mins.			
Test Methods: ASTM			
Physical		**Mechanical**	
Colour	Straw/amber	τ (RT, steel)	20.7 MPa
Viscosity	0.7 Pa.s (standard)*	τ (°C)	-
S.G.	-	τ (wet)	-
Shelf life	-	Peel ( )	-
Pot/Work life	30-45 secs (tack-free)	σ_f	-
$T_{Service}$	-7 to +177°C	E	-
CTE (°C)	-	Elongation	20%
Thermal cond	-	H (Shore D)	75
Notes: * other viscosity grades: T(thick) - 3 Pa.s; VT(very thick) - 11 Pa.s; G(gel) - thixotropic. τ data (ASTM D-1002): steel, activator cured. Gap filling: 0.25mm max. (activator), 3.2 mm max. (UV). Water absorption (after 8hrs/boiling water): 7%. Dielectric strength: 63 kV/mm.			

Dymax	Multi-Cure 625		(Oct 95)
Two-part, (meth)acrylate-urethane Activator (no-mix),UV- or heat cure			
Manufacturers Comments: Rapid setting. Choice of curing method (activator, UV or heat) to suit production methods. General purpose grade for variety of materials. Tough, moisture resistant bonds for materials with different expansion characteristics.			
Typical Use: Assembly adhesive: Metals, glass, ceramics, some plastics. Metal brackets to glass, glass-metal furniture assembly.			
Cure: Activator: ~30 secs (handling), 24 hrs (full cure). UV: <15 secs (full cure). Heat: 93°C/1hr; 149°C/5 mins.			
Test Methods: ASTM			
Physical		**Mechanical**	
Colour	Straw/amber	τ (RT, steel)	15.9 MPa
Viscosity	5 Pa.s (standard)*	τ (°C)	-
S.G.	-	τ (wet)	-
Shelf life	-	Peel ( )	-
Pot/Work life	30-45 secs (tack-free)	σ_f	-
$T_{Service}$	-7 to +177°C	E	-
CTE (°C)	-	Elongation	25%
Thermal cond	-	H (Shore D)	40
Notes: * other viscosity grades: T(thick) - 25 Pa.s; VT(very thick) - 40 Pa.s; G(gel) - thixotropic. τ data (ASTM D-1002): steel, activator cured. Gap filling: 0.5mm max. (activator), 3.2 mm max. (UV). Water absorption (after 8hrs/boiling water): 5%. Dielectric strength: 13.1 kV/mm.			

Dymax	Multi-Cure 652		(Oct 95)
Two-part, (meth)acrylate-urethane Activator (no-mix),UV- or heat cure			
Manufacturers Comments: Rapid setting. Choice of curing method (activator, UV or heat) to suit production methods. Flexible adhesive-sealant suited for cold temperature use. Mainly used as sealant (rigid materials), adhesive for textiles.			
Typical Use: Adhesive-sealant: Textiles. Head lamp seals, strain-relief in coil windings.			
Cure: Activator: ~30 secs (handling), 24 hrs (full cure). UV: <15 secs (full cure). Heat: 93°C/1hr; 149°C/5 mins.			
Test Methods: ASTM			
Physical		**Mechanical**	
Colour	Straw/amber	τ (RT, steel)	8.3 MPa
Viscosity	3 Pa.s (standard)*	τ (°C)	-
S.G.	-	τ (wet)	-
Shelf life	-	Peel ( )	-
Pot/Work life	30-45 secs (tack-free)	σ_f	-
$T_{Service}$	-29 to +93°C	E	-
CTE (°C)	-	Elongation	80%
Thermal cond	-	H (Shore D)	30
Notes: * other viscosity grades on special request. τ data (ASTM D-1002): steel, activator cured. Gap filling: 0.25mm max. (activator), 3.2 mm max. (UV). Water absorption (after 8hrs/boiling water): 6.9%.			

Forbo Helmitin

Forbo Helmitin	**Helmitin 34040**		May 92
Acrylic dispersion - Two part			
Manufacturers Comments: Cross-linking, heat-sealing. For inner roof linings.			
Typical Use: PVC and PUR foam, foils to expanded Styrofoam panels.			
Cure: RT/8-10 mins (drying); Open time: <24hrs. Heat-seal: 80-100°C.			
Test Methods: Not Stated			
Physical		**Mechanical**	
Colour	White	τ (RT)	-
Viscosity	~16 Pa.s	τ (°C)	-
S.G.	~1.03	τ (wet)	-
Shelf life	6 months/RT	Peel ( )	-
Pot/Work life	3 days	σ_f	-
$T_{Service}$	-	E	-
CTE (°C)	-	Elongation	-
Thermal cond.	-	H (Shore)	-
Notes: Hardener (4.6% addition): 49926.			

Loctite

Loctite	**Adhesive 3295**		(Feb 95)
Acrylic - Two part (mix), toughened methacrylate			
Manufacturers Comments: Fast RT cure. Strong tough bond, withstands bending. Cures tack-free. Withstands paint-bake process. Exposed adhesive can be painted.			
Typical Use: General purpose: metals, wood, most plastics (inc. GRP), glass. Metal mesh to frames, stiffeners to metal sheet, hinges to glass doors.			
Cure: RT/5-10 mins (metals), 10-20 mins (plastics). 24 hrs (full cure).			
Test Methods: ASTM/DIN			
Physical		**Mechanical**	
Colour	Green	τ (RT, Al-alloy)	17.5 MPa
Viscosity	25 Pa.s	τ (RT, steel)	25.0 MPa
S.G.	1.03	τ (wet)	~1.5 MPa (6%)
Shelf life	1 yr./<21°C	Peel (T)	4 N mm^{-1}
Pot/Work life	4 mins (open)	σ_f	-
$T_{Service}$	~80°C	E	-
CTE (°C)	10^{-4} K^{-1}	Elongation	-
Thermal cond.	0.1 W mK^{-1}	H (Shore)	-
Notes: Gap filling:<3mm. Flash point 120°C (flammable). Specific heat: 0.3 kJ/kg/K. τ ASTM D1002/DIN 53283: 22°C/48 hr cure. τ(wet): Steel, after 1000 hrs in water (% strength retention). T-peel: ASTM D1876/DIN 53282, Al-alloy. CTE: ASTM D696. Thermal Conductivity: ASTM C177.			

W. W. Henry

W.W. Henry	**660**		Feb 92
Acrylic-latex - one part			
Manufacturers Comments: Multi-purpose flooring adhesive.			
Typical Use: Flooring: Vinyl, wood parquet, rubber-backed <30x30cm.			
Cure: RT(>18°C)/~5 days.			
Test Methods: Not Stated			
Physical		**Mechanical**	
Colour	White→Clear	τ (RT)	-
Viscosity	-	τ (°C)	-
S.G.	-	τ (wet)	-
Shelf life	-	Peel ( )	-
Pot/Work life	-	σ_f	-
$T_{Service}$	-	E	-
CTE (°C)	-	Elongation	-
Thermal cond.	-	H (Shore)	-
Notes: Volatiles content: 50-60%.			

Loctite	**Multi-Bond 329**		(Feb 95)
Acrylic - Two part (no mix), toughened			
Manufacturers Comments: RT cure. Bonding oily or wax coated 'as received' parts. No mix (activator + adhesive) system. Fast handling strength in factory conditions. Can be paint-stoved. Choice of activator for materials and assembly procedures.			
Typical Use: General purpose: metals (ferrous and non-ferrous), ferrites, GRP, concrete, masonry, laminates.			
Cure: RT/1-2 mins (handling); few hours (high strength).			
Test Methods: -			
Physical		**Mechanical**	
Colour	-	τ (RT)	-
Viscosity	-	τ (°C)	-
S.G.	-	τ (wet)	-
Shelf life	-	Peel ( )	-
Pot/Work life	15 mins (open)	σ_f	-
$T_{Service}$	80°C	E	-
CTE (°C)	-	Elongation	-
Thermal cond.	-	H (Shore)	-
Notes: Gap filling: 1 mm max.			

Loctite	Multi-Bond 330		(Feb 95)
Acrylic - Two part (no mix), toughened			
Manufacturers Comments: RT cure. Bonding oily or wax coated 'as received' parts. No mix (activator + adhesive) system. Fast handling strength in factory conditions. Can be paint-stoved. Low odour, non-flammable for use in confined spaces.			
Typical Use: General purpose: metals (ferrous and non-ferrous), ferrites, GRP, concrete, masonry, laminates.			
Cure: RT. Handling: 1-2 mins (metals), 2-4 mins (plastics, wood). 50% strength: ~10 mins (0.25 mm thick bondline).			
Test Methods: -			
Physical		**Mechanical**	
Colour	-	τ (RT)	-
Viscosity	-	τ (°C)	-
S.G.	-	τ (wet)	~90% retention
Shelf life	-	Peel (T)	2.5 - 3.5 N mm^{-1}
Pot/Work life	15 mins (open)	σ_f	-
$T_{Service}$	80°C	E	-
CTE (°C)	-	Elongation	-
Thermal cond.	-	H (Shore)	-
Notes: Gap filling: 1mm max. τ (wet): grit-blasted mild steel, after 21 days 90%RH, 40°C.			

Loctite	Structural Adhesive 327		(Feb 95)
Acrylic - Two part (mix), modified methacrylate			
Manufacturers Comments: Fast RT cure for general purpose assembly. Non-flammable, non-volatile.			
Typical Use: Electric motor assembly, jewellery. Passivated metals to metals and ferrites.			
Cure: RT/ 40 secs (steel, 0.05mm bondline). 10 mins (50% strength).			
Test Methods: -			
Physical		**Mechanical**	
Colour	-	τ (RT)	-
Viscosity	-	τ (°C)	-
S.G.	-	τ (wet)	-
Shelf life	-	Peel ( )	-
Pot/Work life	-	σ_f	-
$T_{Service}$	130°C max.	E	-
CTE (°C)	-	Elongation	-
Thermal cond.	-	H (Shore)	-
Notes: Gap filling: 1 mm max.			

Loctite	Structural Adhesive 326		(Feb 95)
Acrylic - Two part (no mix), modified methacrylate			
Manufacturers Comments: RT cure for general purpose assembly. Activator for fast fixing. Also heat-cured to obtain full strength.			
Typical Use: Metals (degreased steel, stainless, Zn-plated). Glass.			
Cure: RT/ 30 secs with activator. 120°C/15 mins (full cure).			
Test Methods: -			
Physical		**Mechanical**	
Colour	-	τ (RT)	-
Viscosity	-	τ (°C)	-
S.G.	-	τ (wet)	-
Shelf life	-	Peel ( )	-
Pot/Work life	-	σ_f	-
$T_{Service}$	130°C max.	E	-
CTE (°C)	-	Elongation	-
Thermal cond.	-	H (Shore)	-
Notes: Gap filling: 0.25 max.			

Loctite	Structural Adhesive 341		(Feb 95)
Acrylic - Two part (mix), toughened			
Manufacturers Comments: RT cure. Bonding oily or wax coated 'as received' parts (absorbs rust-protection oils in bondline). Fast cure accelerated by spot-heating. Part-cured adhesive can be painted. Can be paint-stoved (completes cure).			
Typical Use: General purpose: metals (ferrous and non-ferrous), ferrites, GRP, concrete, masonry, laminates.			
Cure: RT. Handling: 5 mins (0.05 mm bondline, steel), time reduced by spot-heating. Full cure (paint-stoving 185-200°C): 30 mins.			
Test Methods: -			
Physical		**Mechanical**	
Colour	-	τ (RT)	-
Viscosity	-	τ (°C)	-
S.G.	-	τ (wet)	-
Shelf life	-	Peel ( )	-
Pot/Work life	-	σ_f	-
$T_{Service}$	80°C	E	-
CTE (°C)	-	Elongation	-
Thermal cond.	-	H (Shore)	-
Notes: Gap filling: 0.5 to 3 mm max.			

Permabond

Permabond		Flexon 245	Mar 94
Acrylic - One part, rubber modified methacrylate			
Manufacturers Comments: Fast RT cure. Very good mechanical properties. Flexible. May be used with either Permabond No. 1 or No. 5 initiators to increase cure speed.			
Typical Use: General purpose - high performance.			
Cure: 24 hrs/RT.			
Test Methods: DIN			
Physical		**Mechanical**	
Colour	Off-white	τ (RT, steel)	25 MPa
Viscosity	80 Pa.s	τ (80°C)	~13 MPa
S.G.	1.0	τ (wet)	-
Shelf life	1 yr./RT	Peel (-)	4.8 N mm^{-1}
Pot/Work life	-	σ_t	-
$T_{Service}$	-60°C to +120°C	E	-
CTE (°C)	-	Elongation	-
Thermal cond.	-	H (Shore)	-
Notes: Impact strength: >50 J. Gap filling: <0.75mm.			

Permabond		Flexon 246	Mar 94
Acrylic - One part, rubber modified methacrylate			
Manufacturers Comments: Fast RT cure. Very good mechanical properties. Flexible. May be used with either Permabond No. 1 or No. 5 initiators to increase cure speed.			
Typical Use: General purpose - very high performance.			
Cure: 24 hrs/RT.			
Test Methods: DIN			
Physical		**Mechanical**	
Colour	Off-white	τ (RT, steel)	35 MPa
Viscosity	30 Pa.s	τ (100°C)	~18 MPa
S.G.	1.0	τ (wet)	-
Shelf life	1 yr./RT	Peel (-)	6 N mm^{-1}
Pot/Work life	-	σ_t	-
$T_{Service}$	-60°C to +120°C	E	-
CTE (°C)	-	Elongation	-
Thermal cond.	-	H (Shore)	-
Notes: Impact strength: >50 J. Gap filling: <0.75mm.			

Plexus (ITW)

Plexus (ITW)		MA 300	(Feb 95)
Acrylic - Two part methacrylate			
Manufacturers Comments: Good mechanical properties. Flexible. No solvents. Tolerant of poor surface preparation.			
Typical Use: Construction. PVC, ABS, Acrylics, GRP, Styrene, metals, Urethanes, wood.			
Cure: RT, Initial cure: 7 - 10 mins, Final cure: 24hrs.			
Test Methods: DIN			
Physical		**Mechanical**	
Colour	Pale Yellow	τ (RT, steel)	25 MPa
Viscosity	27.5 Pa.s	τ (°C)	-
S.G.	-	τ (wet)	-
Shelf life	-	Peel ( )	-
Pot/Work life	3 - 5 mins	σ_t	6 - 7 MPa
$T_{Service}$	+93°C	E	-
CTE (°C)	-	Elongation	-
Thermal cond.	-	H (Shore D)	78
Notes:			

Plexus (ITW)		MA 310	(Feb 95)
Acrylic - Two part methacrylate			
Manufacturers Comments: Good mechanical properties. Flexible. No solvents. Tolerant of poor surface preparation.			
Typical Use: Construction. Polyamides, Polyesters, Polyolefins, ABS, Acrylics, GRP, Styrene, metals, wood, Epoxy/Kevlar.			
Cure: RT, Initial cure: 30 - 35 mins, Final cure: 24hrs.			
Test Methods: DIN			
Physical		**Mechanical**	
Colour	Pale Yellow	τ (RT, steel)	14 MPa
Viscosity	45 Pa.s	τ (°C)	-
S.G.	-	τ (wet)	-
Shelf life	-	Peel ( )	-
Pot/Work life	10 - 11 mins	σ_t	2.6 - 3.5 MPa
$T_{Service}$	+93°C	E	-
CTE (°C)	-	Elongation	-
Thermal cond.	-	H (Shore D)	75
Notes: MA340 is a grey pigment filled version of MA310 with improved flexibility and gap filling properties.			

Quilosa
(Industrias Quimicas Löwenberg SA)

Plexus (ITW)	**MA 320**		(Feb 95)
Acrylic - Two part methacrylate			
Manufacturers Comments: Good mechanical properties. Flexible. No solvents. Tolerant of poor surface preparation.			
Typical Use: Construction. PVC, ABS, metals, wood, Epoxy/Kevlar.			
Cure: RT, Initial cure: 20 mins.			
Test Methods: DIN			
Physical		**Mechanical**	
Colour	Off-white	τ (RT, steel)	10.2 MPa
Viscosity	123 Pa.s	τ (°C)	-
S.G.	-	τ (wet)	-
Shelf life	-	Peel ( )	-
Pot/Work life	4 - 5 mins	σ_t	-
$T_{Service}$	+93°C	E	-
CTE (°C)	-	Elongation	-
Thermal cond.	-	H (Shore D)	72
Notes:			

Quilosa	**Aklesil L-71**		Feb 92
Acrylic - One part modified - aqueous dispersion			
Manufacturers Comments: Single-side application.			
Typical Use: Laminating and bonding of PVC, etc.			
Cure: Fast initial cure, full cure: 24hrs.			
Test Methods: -			
Physical		**Mechanical**	
Colour	Beige	τ (RT, steel)	-
Viscosity	25 - 35 Pa.s	τ (°C)	-
S.G.	-	τ (wet)	-
Shelf life	1 yr./RT	Peel ( )	-
Pot/Work life	45 mins	σ_t	-
$T_{Service}$	-	E	-
CTE (°C)	-	Elongation	-
Thermal cond.	-	H (Shore)	-
Notes: Coverage: ~0.35 kg.m^{-2}.			

Plexus (ITW)	**MA 550**		(Feb 95)
Acrylic - Two part methacrylate			
Manufacturers Comments: Good mechanical properties. Flexible. No solvents. Tolerant of poor surface preparation.			
Typical Use: Construction. GRP, PMMA, ABS, PVC, metals.			
Cure: RT, Initial cure: 60 mins.			
Test Methods: DIN			
Physical		**Mechanical**	
Colour	Off-white	τ (RT, steel)	10.2 MPa
Viscosity	216 Pa.s	τ (°C)	-
S.G.	-	τ (wet)	-
Shelf life	-	Peel ( )	-
Pot/Work life	35 - 40 mins	σ_t	2.6 - 3.5 MPa
$T_{Service}$	+93°C	E	-
CTE (°C)	-	Elongation	-
Thermal cond.	-	H (Shore D)	65
Notes:			

Quilosa	**Aklesil L-80**		Feb 92
Acrylic - One part modified - aqueous dispersion			
Manufacturers Comments: Single-side application. Resistant to alkaline water.			
Typical Use: Laminating and bonding of PVC, etc.			
Cure: Full cure: Several days to 1 month.			
Test Methods: -			
Physical		**Mechanical**	
Colour	Beige	τ (RT, steel)	-
Viscosity	27 Pa.s	τ (°C)	-
S.G.	-	τ (wet)	-
Shelf life	1 yr./RT	Peel ( )	-
Pot/Work life	~30 mins (open)	σ_t	-
$T_{Service}$	-	E	-
CTE (°C)	-	Elongation	-
Thermal cond.	-	H (Shore)	-
Notes: Coverage: ~0.3 kg.m^{-2}.			

SEA
(Société d'Exploitation des Adhésifs)

SEA	A 406		Jan 93
Acrylic - One part, aqueous emulsion.			
Manufacturers Comments: 72% solids.			
Typical Use: Construction industry (flooring).			
Cure: Initial: 15 mins/RT. Full cure: 48 hr.			
Test Methods: Not stated.			
Physical		**Mechanical**	
Colour	White	τ (RT)	-
Viscosity	(Paste)	τ (°C)	-
S.G.	1.25	τ (wet)	-
Shelf life	1 yr./RT	Peel ( )	-
Pot/Work life	60 mins/RT (open)	σ_f	-
$T_{Service}$	-	E	-
CTE (°C)	-	Elongation	-
Thermal cond.	-	H (Shore)	-
Notes: -			

SEA	Stykisol		Jun 91
Acrylic - One part, aqueous emulsion.			
Manufacturers Comments: 91% solids.			
Typical Use: Construction industry.			
Cure: 24 - 48 hrs/RT.			
Test Methods: Not stated.			
Physical		**Mechanical**	
Colour	White	τ (RT)	-
Viscosity	(Paste)	τ (°C)	-
S.G.	1.90	τ (wet)	-
Shelf life	1 yr./RT	Peel ( )	-
Pot/Work life	15 mins/RT (open)	σ_f	-
$T_{Service}$	-	E	-
CTE (°C)	-	Elongation	-
Thermal cond.	-	H (Shore)	-
Notes: -			

Structural Adhesives Ltd.

Structural Adhesives Ltd	S - 511A / S - 10A		Sept 93
Acrylic - Two part, reactive.			
Manufacturers Comments: Fast RT cure, flexible. Low flash point (7°C), both components are highly inflammable.			
Typical Use: Metallic and plastic panels. Developed for SMC bonding.			
Cure: RT.			
Test Methods: Not stated.			
Physical		**Mechanical**	
Colour	Off-white	τ (RT, steel)	24.3 MPa
Viscosity	20 Pa.s	τ (RT, SMC)	7.6 MPa
S.G.	~1.0	τ (wet)	-
Shelf life	1 yr./RT	Peel ( )	-
Pot/Work life	7 mins/RT	σ_f	-
$T_{Service}$	-	E	-
CTE (°C)	-	Elongation	-
Thermal cond.	-	H (Shore)	-
Notes: Structural Adhesives Ltd. have a wide range of two-part acrylic adhesives/activators, all of which can be used (in various combinations) either as mixed two-part adhesives or with each component applied to a different mating surface. Properties are for pre-mixed adhesive.			

Structural Adhesives Ltd	S - 515A / S - 34A		Feb 92
Acrylic - Two part, reactive, thixotropic.			
Manufacturers Comments: RT cure, flexible. Low flash point (11°C), both components are highly inflammable.			
Typical Use: Plastic bonding.			
Cure: RT.			
Test Methods: Not stated.			
Physical		**Mechanical**	
Colour	Pale brown	τ (RT, steel)	19.5 MPa
Viscosity	65 Pa.s	τ (RT, Al)	9.5 MPa
S.G.	~1.0	τ (120°C, steel)	5.8 MPa
Shelf life	1 yr./RT	Peel ( )	-
Pot/Work life	7 mins/RT	σ_f	-
$T_{Service}$	-	E	-
CTE (°C)	-	Elongation	-
Thermal cond.	-	H (Shore)	-
Notes: Structural Adhesives Ltd. have a wide range of two-part acrylic adhesives/activators, all of which can be used (in various combinations) either as mixed two-part adhesives or with each component applied to a different mating surface. Properties are for pre-mixed adhesive.			

See Key to Data Sheets - Pages 63 to 65

Structural Adhesives Ltd	S - 545A / S - 47A		May 93
Acrylic - Two part, reactive, thixotropic.			
Manufacturers Comments: RT cure, flexible. Flash point (50°C), both components are highly inflammable. Low odour.			
Typical Use: Metal bonding. Panel stiffeners.			
Cure: RT.			
Test Methods: Not stated.			
Physical		**Mechanical**	
Colour	Off-white or grey	τ (RT, steel)	15.9 MPa
Viscosity	-	τ (°C)	-
S.G.	~1.25	τ (wet)	-
Shelf life	1 yr./RT	Peel ( )	-
Pot/Work life	20 mins/RT	σ_t	-
$T_{Service}$	-	*E*	-
CTE (°C)	-	Elongation	-
Thermal cond.	-	H (Shore)	-
Notes: Structural Adhesives Ltd. have a wide range of two-part acrylic adhesives/activators, all of which can be used (in various combinations) either as mixed two-part adhesives or with each component applied to a different mating surface. Properties are for pre-mixed adhesive.			

Structural Adhesives Ltd	S - 600A / S - 12A		May 93
Acrylic - Two part, reactive, thixotropic.			
Manufacturers Comments: RT cure. Flash point (95°C). Low odour.			
Typical Use: Structural bonding of steel, aluminium, painted or primed metals and some rigid plastics.			
Cure: RT.			
Test Methods: Not stated.			
Physical		**Mechanical**	
Colour	Pale brown	τ (RT, steel)	11 - 18 MPa
Viscosity	~4.5 Pa.s	τ (RT, Al)	10 - 21 MPa
S.G.	~1.05	τ (wet)	-
Shelf life	1 yr./<RT	Peel (-)	9.1 N mm^{-1}
Pot/Work life	8 mins/RT	σ_t	-
$T_{Service}$	-40°C to +150°C	*E*	-
CTE (°C)	-	Elongation	-
Thermal cond.	-	H (Shore)	-
Notes: Structural Adhesives Ltd. have a wide range of two-part acrylic adhesives/activators, all of which can be used (in various combinations) either as mixed two-part adhesives or with each component applied to a different mating surface. Properties are for pre-mixed adhesive.			

Structural Adhesives Ltd	S - 570A / S - 70A		May 93
Acrylic - Two part, reactive, thixotropic.			
Manufacturers Comments: Fast RT cure. Low flash point (11°C), both components are highly inflammable.			
Typical Use: Metal bonding. Panel stiffeners.			
Cure: RT.			
Test Methods: Not stated.			
Physical		**Mechanical**	
Colour	Purple	τ (RT, steel)	20.7 MPa
Viscosity	~4.5 Pa.s	τ (RT, Al)	21.1 MPa
S.G.	~1.0	τ (210°C, steel)	3.8 MPa
Shelf life	1 yr./<RT	Peel ( )	-
Pot/Work life	2.5 mins/RT	σ_t	-
$T_{Service}$	-	*E*	-
CTE (°C)	-	Elongation	-
Thermal cond.	-	H (Shore)	-
Notes: Structural Adhesives Ltd. have a wide range of two-part acrylic adhesives/activators, all of which can be used (in various combinations) either as mixed two-part adhesives or with each component applied to a different mating surface. Properties are for pre-mixed adhesive.			

Sunstar Engineering

Sunstar Engineering	Tufflock 6201		(Mar 95)
Acrylic - One part, liquid			
Manufacturers Comments:			
Typical Use: Automotive, construction, electronics, etc. Metals, thermoplastics, etc.			
Cure: 10 - 20 mins/RT.			
Test Methods: Not stated.			
Physical		**Mechanical**	
Colour	Off-white	τ (RT, steel)	24.5 MPa
Viscosity	15 - 30 Pa.s	τ (°C)	-
S.G.	-	τ (wet)	-
Shelf life	-	Peel ( )	-
Pot/Work life	-	σ_t	-
$T_{Service}$	-	*E*	-
CTE (°C)	-	Elongation	-
Thermal cond.	-	H (Shore)	-
Notes:			

Sunstar Engineering	Tufflock 6202		(Mar 95)
Acrylic - One part, liquid			
Manufacturers Comments:			
Typical Use: Automotive, construction, electronics, etc. Metals, thermoplastics, etc.			
Cure: 3 - 6 mins/RT.			
Test Methods: Not stated.			
Physical		**Mechanical**	
Colour	Off-white	τ (RT, steel)	20 MPa
Viscosity	10 - 20 Pa.s	τ (°C)	-
S.G.	-	τ (wet)	-
Shelf life	-	Peel ( )	-
Pot/Work life	-	σ_f	-
$T_{Service}$	-	E	-
CTE (°C)	-	Elongation	-
Thermal cond.	-	H (Shore)	-
Notes:			

Sunstar Engineering	Tufflock 6506		(Mar 95)
Acrylic - One part, liquid			
Manufacturers Comments:			
Typical Use: Automotive, construction, electronics. Plastics, FRP, etc.			
Cure: 10 - 20 mins/RT.			
Test Methods: Not stated.			
Physical		**Mechanical**	
Colour	Brown (transp.)	τ (RT, ABS)	6.2 MPa
Viscosity	15 - 30 Pa.s	τ (°C)	-
S.G.	-	τ (wet)	-
Shelf life	-	Peel ( )	-
Pot/Work life	-	σ_f	-
$T_{Service}$	-	E	-
CTE (°C)	-	Elongation	-
Thermal cond.	-	H (Shore)	-
Notes:			

Sunstar Engineering	Tufflock 6203		(Mar 95)
Acrylic - One part, liquid			
Manufacturers Comments:			
Typical Use: Automotive, construction, electronics, etc. Metals, thermoplastics, etc.			
Cure: 50 - 100 mins/RT.			
Test Methods: Not stated.			
Physical		**Mechanical**	
Colour	Off-white	τ (RT, steel)	24.5 MPa
Viscosity	15 - 30 Pa.s	τ (°C)	-
S.G.	-	τ (wet)	-
Shelf life	-	Peel ( )	-
Pot/Work life	-	σ_f	-
$T_{Service}$	-	E	-
CTE (°C)	-	Elongation	-
Thermal cond.	-	H (Shore)	-
Notes:			

UHU

UHU	'1'		Jan 94
Acrylic - One part			
Manufacturers Comments: Retains strength (~5.5 MPa) between -20°C and +80°C.			
Typical Use: Bonds and repairs most hard or soft domestic materials.			
Cure: Pre-coat both surfaces. Initial cure: 5 mins, Full cure: 12 hr.			
Test Methods: Not stated.			
Physical		**Mechanical**	
Colour	Transparent	τ (RT)	-
Viscosity	0.8 Pa.s	τ (95°C)	-
S.G.	0.84	τ (wet)	-
Shelf life	>3 years.	Peel ( )	-
Pot/Work life	-	σ_f	5.5 MPa
$T_{Service}$	-20 to +80°C	E	-
CTE (°C)	-	Elongation	-
Thermal cond.	-	H (Shore)	-
Notes:			

UHU	Hard Plastic Adhesive		April 94
Acrylic - One part			
Manufacturers Comments: Three variants: Standard, Liquid and Super Liquid. Retains properties between -30°C and +80°C. Resistant to water, oils and most domestic fluids.			
Typical Use: Bonds and repairs polystyrene, and most hard or opaque plastics (domestic).			
Cure: 1 hour/RT.			
Test Methods: Not stated.			
Physical		**Mechanical**	
Colour	Transparent	τ (RT)	-
Viscosity	0.8 Pa.s	τ (°C)	-
S.G.	0.85 (0.88, 0.88)	τ (wet)	-
Shelf life	-	Peel ( )	-
Pot/Work life	-	σ_t	-
$T_{Service}$	-30 to +80°C	*E*	-
CTE (°C)	-	Elongation	-
Thermal cond.	-	H (Shore)	-
Notes: Liquid and Super Liquid variants both have stated viscosities of 0.05 Pa.s.			

Wulff

Wulff (Naturalan)	AC 1		(Feb 95)
Acrylic - One part, liquid			
Manufacturers Comments: Natural resin dispersion adhesive, solvent free. Water-based acrylic dispersion. Flexible with good initial tack.			
Typical Use: Bonds PVC sheets and tiles, various flooring materials. Castor wheel and radiant floor heating resistant.			
Cure: Full cure: 72 hr.			
Test Methods: DIN			
Physical		**Mechanical**	
Colour	-	τ (RT)	-
Viscosity	23 Pa.s	τ (°C)	-
S.G.	1.42	τ (wet)	-
Shelf life	-	Peel ( )	-
Pot/Work life	20 min (RT, 75%RH)	σ_t	-
$T_{Service}$	~40°C	*E*	-
CTE (°C)	-	Elongation	-
Thermal cond	-	H (Shore)	-
Notes: Conforms to DIN 53 269 (tensile shear and peel strength) and DIN 68 131 (castor wheel use). 81% dry substance. Coverage 350 - 450 g.m^{-2}.			

UHU	Plus Acrylit		June 94
Acrylic - 2 part kit (liquid + powder)			
Manufacturers Comments: RT cure. Quick-setting, high strength. Retains strength (~10 MPa) between -30°C and +90°C.			
Typical Use: Bonds most plastics (no solvent deformation), wood, stone, glass, concrete (Domestic).			
Cure: Initial cure: 15 mins, Full cure: 1 hr.			
Test Methods: Not stated.			
Physical		**Mechanical**	
Colour	Off-white, opaque	τ (RT)	-
Viscosity	20 Pa.s	τ (°C)	-
S.G.	1.00	τ (wet)	-
Shelf life	-	Peel ( )	-
Pot/Work life	7 - 10 min.	σ_t	10 MPa
$T_{Service}$	-30 to +90°C	*E*	-
CTE (°C)	-	Elongation	-
Thermal cond.	-	H (Shore)	-
Notes:			

Wulff (Naturalan)	ES 14		(Feb 95)
Acrylic - One part, liquid			
Manufacturers Comments: One side dispersion adhesive, solvent free. Water-based acrylic dispersion. Flexible.			
Typical Use: Bonds PVC sheets and tiles, various flooring materials. Castor wheel and radiant floor heating resistant.			
Cure: Full cure: 72 hr.			
Test Methods: DIN			
Physical		**Mechanical**	
Colour	-	τ (RT)	-
Viscosity	22 Pa.s	τ (°C)	-
S.G.	1.18	τ (wet)	-
Shelf life	-	Peel ( )	-
Pot/Work life	20 min (RT, 75%RH)	σ_t	-
$T_{Service}$	~40°C	*E*	-
CTE (°C)	-	Elongation	-
Thermal cond	-	H (Shore)	-
Notes: Conforms to DIN 53 269 (tensile shear and peel strength) and DIN 68 131 (castor wheel use). 78% dry substance. Coverage 350 - 450 g.m^{-2}.			

Wulff (Naturalan)	**Supra Coll**		(Feb 95)
Acrylic - One part, liquid			
Manufacturers Comments: One side special dispersion adhesive, solvent free. Water-based acrylic dispersion. Flexible with very high initial tack.			
Typical Use: Bonds PVC sheets and tiles, various flooring materials. Castor wheel and radiant floor heating resistant.			
Cure: Full cure: 72 hr.			
Test Methods: DIN			
Physical		**Mechanical**	
Colour	-	τ (RT)	-
Viscosity	26 Pa.s	τ (°C)	-
S.G.	1.15	τ (wet)	-
Shelf life	-	Peel ( )	-
Pot/Work life	20 min(20°C, 75%RH)	σ_t	-
$T_{Service}$	~40°C	E	-
CTE (°C)	-	Elongation	-
Thermal cond	-	H (Shore)	-
Notes: Conforms to DIN 53 269 (tensile shear and peel strength) and DIN 68 131 (castor wheel use). 74% dry substance. Coverage 350 - 400 g.m^{-2}.			

Wulff (Naturalan)	**Tep-Coll**		(Feb 95)
Acrylic - One part, liquid			
Manufacturers Comments: Natural resin dispersion adhesive, solvent free. Water-based acrylic dispersion. Flexible with high initial tack.			
Typical Use: Bonds PVC sheets and tiles, various flooring materials. Castor wheel and radiant floor heating resistant.			
Cure: Full cure: 72 hr.			
Test Methods: DIN			
Physical		**Mechanical**	
Colour	-	τ (RT)	-
Viscosity	22 Pa.s	τ (°C)	-
S.G.	1.26	τ (wet)	-
Shelf life	-	Peel ( )	-
Pot/Work life	20 min (RT, 75%RH)	σ_t	-
$T_{Service}$	~40°C	E	-
CTE (°C)	-	Elongation	-
Thermal cond	-	H (Shore)	-
Notes: Conforms to DIN 53 269 (tensile shear and peel strength) and DIN 68 131 (castor wheel use). 82% dry substance. Coverage 350 - 450 g.m^{-2}.			

Wulff (Naturalan)	**Supra Tex**		(Feb 95)
Acrylic - One part, liquid			
Manufacturers Comments: One side special dispersion adhesive, solvent free. Water-based acrylic dispersion. Flexible with very high initial tack.			
Typical Use: Bonds PVC sheets and tiles, various flooring materials. Castor wheel and radiant floor heating resistant.			
Cure: Full cure: 72 hr.			
Test Methods: DIN			
Physical		**Mechanical**	
Colour	-	τ (RT)	-
Viscosity	38 Pa.s	τ (°C)	-
S.G.	1.19	τ (wet)	-
Shelf life	-	Peel ( )	-
Pot/Work life	20 min (20°C,75%RH)	σ_t	-
$T_{Service}$	~40°C	E	-
CTE (°C)	-	Elongation	-
Thermal cond	-	H (Shore)	-
Notes: Conforms to DIN 53 269 (tensile shear and peel strength) and DIN 68 131 (castor wheel use). 75% dry substance. Coverage 350 - 400 g.m^{-2}.			

ANAEROBIC ADHESIVES

NOTES

These adhesives cure at room temperature in the absence of oxygen (an-aerobic), making them ideal for bonding and sealing of joints where oxygen access is restricted, i.e. large bond areas, threads and crevices. The adhesive will only begin to cure when the joint surfaces are closed, excluding the atmosphere. An unusual advantage of this is that any excess adhesive outside the joint will remain liquid and can be wiped away.

Common variations on this basic cure mechanism are:

- Catalytic (Cat/RT) - using ions from a metallic surface as a catalyst to accelerate cure. Some metal surfaces catalyse better than others, e.g. copper - good; galvanised (zinc) - bad. Non-metallic surfaces and poor catalysts may require a primer.
- Visible Light (Vis) or UV (UV) Exposure - cures the externally exposed surface, so excluding oxygen from the bulk of the adhesive and initiating anaerobic curing.

A variety of adhesive formulations are available from most manufacturers to provide a range of strengths. This is important because some of the standard uses of anaerobic adhesives allow for subsequent disassembly, e.g. Threadlocking to prevent loosening in service. In this, the strength required depends on the component dimensions, e.g. low strength products for large diameter threads/ components (otherwise impossible to dismantle) and high strength for small threads/components (small bond areas). Most products are coloured to indicate their presence and strength class. The properties currently available tend to relate to their uses for thread- and component-locking. With a greater interest for structural bonding and laminating, the amount of published bond strength data (e.g. lap shear, peel, modulus, etc.) may increase in the future.

Main uses

The majority of these products are used in general engineering for threadlocking and fixing of gears, bearings, keys, etc. to rotating machinery. The adhesive also seals against corrosion and prevents fretting of mechanical fits. In some applications, the permanent bonding of parts (bearings, gears, etc.) can replace expensive, high-tolerance machining of interference fits and keyways. Other uses include electronic assembly, e.g. component placement, tacking of wires and windings (cures well with copper), magnets, etc. Small mechanical component assembly and placement, e.g. computer disk drives. There are also many forms of anaerobic resin coating; both optical and protective.

COMPARISON TABLE - ANAEROBIC ADHESIVES

Company	Adhesive	Colour	Break Torque	Prevail Torque	LSS	SG	Visc. (Pa.s)	Service (°C)	Flash Point (°C)	Gap Filling (mm max)	Cure	Pg. No.
W. Canning	Hylomar Grade 520	Green	●	○	-	-	0.015	-	-	-	RT	95
(Hylomar)	Hylomar Grade 540 'Bearing Fit'	Yellow	◑	○	-	-	0.4	-	-	-	RT	95
	Hylomar Grade 550 'Screw Lock'	Purple	○	○	-	-	1	-	-	-	RT	95
	Hylomar Grade 580 'Nut Lock'	Blue	◑	○	-	-	1.5	-	-	-	RT	95
	Hylomar Grade 600 'Retaining'	Green	●	-	-	-	0.65	-	-	-	RT	96
	Hylomar Grade 640 'Stud Lock'	Green	●	-	-	-	0.65	-	-	-	RT	96
	Hylomar Grade 690 'High Strength Retainer'	Green	●	-	-	-	0.65	-	-	-	RT	96
CECA	Topfix NR116	-	H	-	-	-	-	-	-	-	RT	96
	Topfix NR203	-	M	-	-	-	-	-	-	-	RT	97
	Topfix NR210	-	M	-	-	-	-	-	-	-	RT	97
Delo	ML-5302	Green	H	-	-	-	0.015	-60/+150	>100	0.05	RT	97
	ML-5328	Green	H	-	-	-	0.3	-60/+150	>100	0.13	RT	97
	ML-5349	Green	H	-	-	-	1.0	-60/+150	>100	0.25	RT	98
	ML-5549	Clear Brown	H	-	-	-	0.6	-40/+150	>100	0.15	Vis/ RT	98
	ML-5849	Clear Brown	H	-	-	-	1.0	-40/+150	>100	0.15	Vis/ RT	98
Dunlop	SAS100 Screwlock	Red	○	○	-	-	0.15	-40/+150	100	0.1	RT	98
	SAS110 Nutlock	Blue	○	○	-	-	0.15	-40/+150	100	0.1	RT	99
	SAS120 Studlock	Green	◑	◑	-	-	0.5	-40/+150	100	0.15	RT	99
	SAS123 Studlock	Green	◑	-	◑	-	0.5	-55/+150	>100	0.15	RT	99
	SAS130 Bearing Fit	Purple	◑	◑	-	-	0.4	-55/+150	100	0.2	RT	99
	SAS 140 Pipe Seal	Brown	○	○	-	-	0.4	-55/+150	100	0.2	RT	100
	SAS 150 Pipe Seal	Caramel	○	○	-	-	0.5	-55/+150	100	0.2	RT	100
Loctite	243	-	◑	○	-	-	2	<150	>100	-	RT	100
	262	-	◑	◑	-	-	1.7	<150	>100	-	RT	100
	290	-	○	●	-	-	0.1	<150	>100	-	RT	101
	648	-	H	-	●	-	0.5	<175	>100	0.075	RT	101
	661	-	H	-	●	-	0.5	<175	>100	0.075	UV	101
	662	-	H	-	●	-	2.5	<175	>100	0.125	UV	101
	Bearing Fit 641	-	-	-	◑	-	0.5	<150	>100	0.075	RT	102
	High Strength Retainer 638	-	H	-	●	-	2.5	<150	>100	0.125	RT	102
	Hydraulic Sealant 542	-	○	○	-	-	0.5	<150	>100	-	RT	102
	Nutlock 242	-	○	○	-	-	1.2	<150	>100	-	RT	102
	Pipe Sealant 577	-	○	◑	-	-	24	<150	>100	-	RT	103
	Retainer 601	-	H	-	●	-	0.125	<150	>100	0.075	RT	103
	Retainer 603	-	H	-	●	-	0.125	<150	>100	0.075	RT	103
	Screwlock 222	Purple	○	○	-	-	1.2	<150	>100	-	RT	103
	Studlock 270	-	●	●	-	-	0.5	<150	>100	-	RT	104
Omnitechnic	Omnifit 15 M	Yellow	○	○	○	-	0.7	-60/+80	>100	0.01	Cat/ RT	104
(Henkel)	Omnifit 50 H	Blue	○	○	◑	-	-	-60/+150	>100	0.01	Cat/ RT	104
	Omnifit 50 M	Blue	○	○	◑	-	0.7	-60/+150	>100	0.01	Cat/ RT	104
	Omnifit 80 M	Red	◑	◑	◑	-	1.1	-60/+150	>100	0.01	Cat/ RT	105
	Omnifit 100 H	Red	◑	◑	◑	-	-	-60/+150	>100	0.01	Cat/ RT	105
	Omnifit 100 M	Red	◑	◑	◑	-	0.7	-60/+150	>100	0.01	Cat/ RT	105
	Omnifit 150 M	Green	●	●	●	-	1.1	-60/+150	>100	0.01	Cat/ RT	105
	Omnifit 200 H	Green	◑	◑	●	-	-	-60/+150	>100	0.01	Cat/ RT	106
	Omnifit 200 LL	Green	○	○	●	-	0.03	-60/+150	>100	0.01	Cat/ RT	106
	Omnifit 200 M	Green	●	●	●	-	0.7	-60/+150	>100	0.01	Cat/ RT	106
	Omnifit 220 M	Violet	●	●	●	-	0.4	-60/+220	>100	0.01	Cat/ RT	106
	Omnifit 222 M	Violet	●	●	●	-	0.7	-60/+220	>100	0.01	Cat/ RT	107
	Omnifit 230 L	Violet	●	●	●	-	0.15	-60/+200	>100	0.01	Cat/ RT	107
	Omnifit 230 M	Violet	●	●	●	-	0.7	-60/+200	>100	0.01	Cat/ RT	107
	Omnifit 280 M	Violet	●	●	●	-	0.7	-60/+200	>100	0.01	Cat/ RT	107

Continued...

See Key to Comparison Tables - Pages 63 to 65

Company	Adhesive	Colour	Break Torque	Prevail Torque	LSS	SG	Visc. (Pa.s)	Service (°C)	Flash Point (°C)	Gap Filling (mm max)	Cure	Pg. No.
Panacol	Penloc A	Transparent	◑	◑	◑	1.11	0.01	-55/+150	100	0.05	Cat/ RT	108
Elosol	Penloc NL	Blue	○	○	○	1.11	6	-55/+150	100	0.2	Cat/ RT	108
	Penloc NT	Blue	◑	◑	◑	1.11	0.15	-55/+150	100	0.1	Cat/ RT	108
	Penloc NTT	Blue	◑	◑	◑	1.11	20	-55/+150	100	0.35	Cat/ RT	108
	Penloc R2	Red	●	●	●	1.11	0.4	-55/+200	100	0.1	Cat/ RT	109
	Penloc RL	Green	●	●	●	1.11	0.5	-55/+200	100	0.15	Cat/ RT	109
	Penloc RT	Red	●	●	●	1.11	0.1	-55/+200	100	0.1	Cat/ RT	109
	Penloc SEET	Red	●	●	●	1.11	15	-55/+150	100	0.3	Cat/ RT	109
	Penloc SET	Blue	◑	◑	●	1.11	1.5	-55/+150	100	0.2	Cat/ RT	110
	Penloc ST	Green	◑	◑	◑	1.11	0.4	-55/+150	100	0.1	Cat/ RT	110
	Penloc ULT	Brown	○	○	◑	1.11	0.03	-55/+150	100	0.05	Cat/ RT	110
	Penloc UT	Yellow	○	○	○	1.11	0.125	-55/+130	100	0.1	Cat/ RT	110
	Penloc VF	Red	●	●	●	1.11	2.5	-55/+120	100	0.2	Cat/ RT	111
	Penloc VFF	Green	●	●	●	1.11	3	-55/+120	100	0.25	Cat/ RT	111
	Penloc ZT	Yellow	○	○	◑	1.11	0.2	-55/+150	100	0.1	Cat/ RT	111
	Penloc ZTT	Violet	○	○	◑	1.11	3	-55/+150	100	0.25	Cat/ RT	111
	Vitralit 4280	-	-	-	◑	1.25	0.5	-40/+150	94	-	UV	112
	Vitralit 4285	-	-	-	◑	1.34	4	-40/+150	96	-	UV	112
Permabond	A134	Green	◑	-	●	1.0	70	-55/+150	-	0.5	RT	112
	F202	Brown	◑	-	●	1.0	135	-55/+100	-	0.25	RT	112
	Pipe Sealant A131	White	○	-	○	1.0	40	-55/+150	-	0.5	RT	113
	Retainer A118	Green	◑	-	●	1.0	0.5	-55/+150	-	0.12	RT	113
	Threadlock A130	Blue	○	-	◑	1.0	0.6	-55/+150	-	0.12	RT	113
ThreeBond	1303	Green	●	●	-	-	0.15	-80/+150	-	0.15	RT	113
	1305	Green	●	●	-	-	0.6	-80/+150	-	0.3	RT	114
	1322	Red	◑	●	-	-	0.15	-80/+150	-	0.15	RT	114
	1324	Red	◑	●	-	-	0.6	-80/+150	-	0.3	RT	114
	1327	Red	◑	◑	-	-	2.5	-80/+150	-	0.3	RT	114
	1342	Blue	○	○	-	-	0.15	-80/+150	-	0.15	RT	115
	1344	Blue	○	○	-	-	0.6	-80/+150	-	0.3	RT	115
	1360	Red	◑	◑	-	-	0.8	-80/+200	-	0.15	RT	115
	1361B	Green	○	●	-	-	0.015	-80/+150	-	0.1	RT	115
	1373B	Green	●	●	-	-	0.125	-80/+150	-	0.15	RT	116
	1374	Red	◑	◑	-	-	0.75	-80/+150	-	0.3	RT	116
	1375B	Green	●	●	-	-	0.85	-80/+175	-	0.15	RT	116
	1377B	Green	●	●	-	-	2	-80/+150	-	0.3	RT	116
	1379B	Green	●	●	-	-	7.5	-80/+150	-	0.3	RT	117
	3062	Cl. Yellow	-	-	-	1.05	8	-	-	-	UV	117
	3062B	Cl. Green	-	-	-	1.10	4	-	-	-	UV	117
	3062D	Cl. Blue	-	-	-	1.10	0.15	-	-	-	UV	117

See Key to Comparison Tables - Pages 63 to 65

William Canning (Hylomar)

Wm. Canning	**Hylomar Grade 520**		(Feb 95)
Anaerobic			
Manufacturers Comments: High strength.			
Typical Use: Fixing and locking of components.			
Cure: RT/5 mins.			
Test Methods: BSI			
Physical		**Mechanical**	
Colour	Green	τ (RT)	-
Viscosity	0.015 Pa.s	τ (compressive)	-
S.G.	-	σ_t	-
Shelf life	-	σ_c	-
$T_{Service}$	-	Break Torque	28 N.m
CTE	-	Prevail Torque	14 N.m
Flash Point	-	Elongation	-
Gap Filling	-	H (Shore)	-
Notes: Cure times and torque values: BS 5350 Pt. G1 1987.			

Wm. Canning	**Hylomar Grade 550 'Screw Lock'**		(Feb 95)
Anaerobic			
Manufacturers Comments: Low strength.			
Typical Use: Thread locking of items that require dismantling.			
Cure: RT/5 mins.			
Test Methods: BSI			
Physical		**Mechanical**	
Colour	Purple	τ (RT)	-
Viscosity	1.0 Pa.s	τ (compressive)	-
S.G.	-	σ_t	-
Shelf life	-	σ_c	-
$T_{Service}$	-	Break Torque	4 N.m
CTE	-	Prevail Torque	2 N.m
Flash Point	-	Elongation	-
Gap Filling	-	H (Shore)	-
Notes: Cure times and torque values: BS 5350 Pt. G1 1987.			

Wm. Canning	**Hylomar Grade 540 'Bearing Fit'**		(Feb 95)
Anaerobic			
Manufacturers Comments: Medium strength.			
Typical Use: Retaining bearings, liners, bushes, keys and splines that require dismantling.			
Cure: RT/6 mins.			
Test Methods: BSI			
Physical		**Mechanical**	
Colour	Yellow	τ (RT)	-
Viscosity	0.4 Pa.s	τ (compressive)	-
S.G.	-	σ_t	-
Shelf life	-	σ_c	-
$T_{Service}$	-	Break Torque	18 N.m
CTE	-	Prevail Torque	11 N.m
Flash Point	-	Elongation	-
Gap Filling	-	H (Shore)	-
Notes: Cure times and torque values: BS 5350 Pt. G1 1987.			

Wm. Canning	**Hylomar Grade 580 'Nut Lock'**		(Feb 95)
Anaerobic			
Manufacturers Comments: Medium strength.			
Typical Use: Thread locking of items that require dismantling.			
Cure: RT/4 mins.			
Test Methods: BSI			
Physical		**Mechanical**	
Colour	Blue	τ (RT)	-
Viscosity	1.5 Pa.s	τ (compressive)	-
S.G.	-	σ_t	-
Shelf life	-	σ_c	-
$T_{Service}$	-	Break Torque	16 N.m
CTE	-	Prevail Torque	13 N.m
Flash Point	-	Elongation	-
Gap Filling	-	H (Shore)	-
Notes: Cure times and torque values: BS 5350 Pt. G1 1987.			

Wm. Canning	Hylomar Grade 600		(Feb 95)
'Retaining' Anaerobic			
Manufacturers Comments: High strength.			
Typical Use: Fixing of close fitting items, not requiring dismantling.			
Cure: RT/4 mins.			
Test Methods: BSI			
Physical		**Mechanical**	
Colour	Green	τ (RT)	-
Viscosity	0.125 Pa.s	τ (compressive)	-
S.G.	-	σ_t	-
Shelf life	-	σ_c	-
$T_{Service}$	-	Break Torque	27 N.m
CTE	-	Prevail Torque	(thread failure)
Flash Point	-	Elongation	-
Gap Filling	-	H (Shore)	-
Notes: Cure times and torque values: BS 5350 Pt. G1 1987.			

Wm. Canning	Hylomar Grade 690		(Feb 95)
'High Strength Retainer' Anaerobic			
Manufacturers Comments: High strength, fast curing. Allows larger machine tolerances.			
Typical Use: Thread locking and sealing of threads and cylindrical parts, not requiring dismantling.			
Cure: RT/4 mins.			
Test Methods: BSI			
Physical		**Mechanical**	
Colour	Green	τ (RT)	-
Viscosity	0.65 Pa.s	τ (compressive)	-
S.G.	-	σ_t	-
Shelf life	-	σ_c	-
$T_{Service}$	-	Break Torque	26 N.m
CTE	-	Prevail Torque	(thread failure)
Flash Point	-	Elongation	-
Gap Filling	-	H (Shore)	-
Notes: Cure times and torque values: BS 5350 Pt. G1 1987.			

Wm. Canning	Hylomar Grade 640		(Feb 95)
'Stud Lock' Anaerobic			
Manufacturers Comments: High strength.			
Typical Use: Thread locking of studs, nuts, bolts and threaded fasteners, not requiring dismantling.			
Cure: RT/4 mins.			
Test Methods: BSI			
Physical		**Mechanical**	
Colour	Green	τ (RT)	-
Viscosity	0.65 Pa.s	τ (compressive)	-
S.G.	-	σ_t	-
Shelf life	-	σ_c	-
$T_{Service}$	-	Break Torque	26 N.m
CTE	-	Prevail Torque	(thread failure)
Flash Point	-	Elongation	-
Gap Filling	-	H (Shore)	-
Notes: Cure times and torque values: BS 5350 Pt. G1 1987.			

CECA

CECA	Topfix NR116		(Jul 95)
Anaerobic, liquid			
Manufacturers Comments: High strength.			
Typical Use: Mounting bearing assemblies onto shafts.			
Cure: RT.			
Test Methods:			
Physical		**Mechanical**	
Colour	-	τ (RT)	-
Viscosity	-	τ (compressive)	-
S.G.	-	σ_t	-
Shelf life	-	σ_c	-
$T_{Service}$	-	Break Torque	High
CTE (°C)	-	Prevail Torque	-
Flash Point	-	Elongation	-
Gap Filling	-	H (Shore)	-
Notes: Other version: NR199 - high strength. Cannot be disassembled.			

Delo

CECA	Topfix NR203		(Jul 95)
Anaerobic, thixotropic paste			
Manufacturers Comments: Medium strength. Sealing pneumatic and hydraulic fittings. Resistant to oils and pressure.			
Typical Use: Thread sealing, pipe fittings.			
Cure: RT.			
Test Methods:			
Physical		**Mechanical**	
Colour	-	τ (RT)	-
Viscosity	-	τ (compressive)	-
S.G.	-	σ_f	-
Shelf life	-	σ_c	-
$T_{Service}$	-	Break Torque	Medium
CTE (°C)	-	Prevail Torque	-
Flash Point	-	Elongation	-
Gap Filling	-	H (Shore)	-
Notes: Other versions: NR603 - medium strength, thread-locking and sealing, contains PTFE. Thermal resistance; NR650 - sealing surfaces. Flexible joint.			

Delo	ML-5302		(July 95)
Methylacrylester-based Anaerobic			
Manufacturers Comments: One part, solvent free. High strength, low viscosity. Can be used with an accelerator/activator to shorten cure. Good chemical resistance.			
Typical Use: Mounting rotational symmetrical joints (bearings, bushes); Thread-locking/sealing. Sealing porous weld seams.			
Cure: RT/5-10 mins (firm to touch); 24 hrs (full strength).			
Test Methods: DIN, ISO			
Physical		**Mechanical**	
Colour	Green	τ (RT)	-
Viscosity	~0.015 Pa.s	τ (compressive)	25 MPa
S.G.	-	σ_f	-
Shelf life	1 year/+5°C	σ_c	-
$T_{Service}$	-60 to +150°C	Break Torque	High strength
CTE (°C)	85 × 10^{-6} K^{-1}	Prevail Torque	-
Flash Point	>100°C (liquid)	Elongation	-
Gap Filling	0.05mm max.	H (Shore)	-
Notes: Thread: ≤ M10. Optimum clearance (D-d): 0.01 mm. τ DIN 54 452. Off-torque ISO 10964/M10/8.8 screw: M_L (at M_A 46 N.m): 50 N.m, (without M_A) ~32 N.m.			

CECA	Topfix NR210		(Jul 95)
Anaerobic, thixotropic paste			
Manufacturers Comments: Medium strength. Can be disassembled			
Typical Use: Thread locking mechanical fasteners.			
Cure: RT.			
Test Methods: BSI			
Physical		**Mechanical**	
Colour	-	τ (RT)	-
Viscosity	-	τ (compressive)	-
S.G.	-	σ_f	-
Shelf life	-	σ_c	-
$T_{Service}$	-	Break Torque	Medium
CTE (°C)	-	Prevail Torque	-
Flash Point	-	Elongation	-
Gap Filling	-	H (Shore)	-
Notes: Other versions: NR213 - medium strength, thread-locking and sealing; NR216 - high strength (cannot be disassembled), thread-locking.			

Delo	ML-5328		(July 95)
Methylacrylester-based Anaerobic			
Manufacturers Comments: One part, solvent free. High strength, higher viscosity. Can be used with an accelerator/activator to shorten cure. Very good chemical resistance.			
Typical Use: Mounting rotational symmetrical joints (bearings, bushes); Thread-locking/sealing.			
Cure: RT/5-10 mins (firm to touch); 24 hrs (full strength).			
Test Methods: DIN, ISO			
Physical		**Mechanical**	
Colour	Green	τ (RT)	-
Viscosity	0.3 Pa.s	τ (compressive)	27 MPa
S.G.	-	σ_f	-
Shelf life	1 year/+5°C	σ_c	-
$T_{Service}$	-60 to +150°C	Break Torque	High strength
CTE (°C)	85 × 10^{-6} K^{-1}	Prevail Torque	-
Flash Point	>100°C (liquid)	Elongation	-
Gap Filling	0.13mm max.	H (Shore)	-
Notes: Thread: ≤ M10. Optimum clearance (D-d): 0.04 mm. τ DIN 54 452. Off-torque ISO 10964/M10/8.8 screw: M_L (at M_A 46 N.m): 60 N.m, (without M_A) 30 N.m.			

Delo	ML-5349		(July 95)
Methylacrylester-based Anaerobic			
Manufacturers Comments: One part, solvent free. High strength, medium, thixotropic viscosity. Can be used with an accelerator/activator to shorten cure. Very good chemical resistance.			
Typical Use: Mounting rotationally symmetrical joints (bearings, bushes); Thread-locking/sealing. Sealing pipe fittings.			
Cure: RT/5-20 mins (firm to touch); 24 hrs (full strength).			
Test Methods: DIN, ISO			
Physical		**Mechanical**	
Colour	Green	τ (RT)	-
Viscosity	1.0 Pa.s thixotropic	τ (compressive)	20 MPa
S.G.	-	σ_t	-
Shelf life	1 year/+5°C	σ_c	-
$T_{Service}$	-60 to +150°C	Break Torque	High strength
CTE (°C)	85×10^{-6} K^{-1}	Prevail Torque	-
Flash Point	>100°C (liquid)	Elongation	-
Gap Filling	0.25mm max.	H (Shore)	-
Notes: Thread: All. Optimum clearance (D-d): 0.08 mm. τ DIN 54 452. Off-torque ISO 10964/M10/8.8 screw: M_L (at M_A 46 N.m): 60 N.m, (without M_A) ~22 N.m.			

Delo	ML-5849		(July 95)
Dual-curing dimethylacrylate-based Anaerobic			
Manufacturers Comments: Visible light (400-500nm wavelength) and/or anaerobic cure. Fast cure of margins to fix joint, with anaerobic cure in shadow zones. One part, solvent free. High strength, medium viscosity. Very good chemical resistance. KTW recommendation of German Association of Gas and Water Specialists for gas-burning installations (DVGW). Drinking water systems +90°C.			
Typical Use: Mounting rotationally symmetrical joints (bearings, bushes); Thread-locking/sealing. Potable water system assembly.			
Cure: Visible light RT/8 secs (fix time). Anaerobic: RT/5-10 mins (firm to touch); 24 hrs (full strength).			
Test Methods: DIN, ISO			
Physical		**Mechanical**	
Colour	Clear/brown tint	τ (RT)	-
Viscosity	1.0 Pa.s thixotropic	τ (compressive)	~20 MPa
S.G.	-	σ_t	-
Shelf life	1 year/+5°C	σ_c	-
$T_{Service}$	-40 to +150°C	Break Torque	High strength
CTE (°C)	85×10^{-6} K^{-1}	Prevail Torque	-
Flash Point	>100°C (liquid)	Elongation	-
Gap Filling	0.15mm max.	H (Shore)	-
Notes: Thread: ≤M18/≤R1. Optimum clearance (D-d): 0.05 mm. τ (compressive) DIN 54 452. Off-torque ISO 10964/M10/8.8 screw: M_L (at M_A 46 N.m): 48 N.m, (without M_A) ~37 N.m.			

Delo	ML-5549		(July 95)
Dual-curing dimethylacrylate-based Anaerobic			
Manufacturers Comments: Visible light (400-500nm wavelength) and/or anaerobic cure. Fast cure of margins to fix joint, with anaerobic cure in shadow zones. One part, solvent free. High strength, medium viscosity. Very good chemical resistance.			
Typical Use: Mounting rotationally symmetrical joints (bearings, bushes); Thread-locking/sealing.			
Cure: Visible light RT/5 secs (fix time). Anaerobic: RT/5-10 mins (firm to touch); 24 hrs (full strength).			
Test Methods: DIN, ISO			
Physical		**Mechanical**	
Colour	Clear/brown tint	τ (RT)	-
Viscosity	~0.6 Pa.s	τ (compressive)	~23 MPa
S.G.	-	σ_t	-
Shelf life	1 year/+5°C	σ_c	-
$T_{Service}$	-40 to +150°C	Break Torque	High strength
CTE (°C)	85×10^{-6} K^{-1}	Prevail Torque	-
Flash Point	>100°C (liquid)	Elongation	-
Gap Filling	0.15mm max.	H (Shore)	-
Notes: Thread: ≤M10. Optimum clearance (D-d): 0.05 mm. τ (compressive) DIN 54 452. Off-torque ISO 10964/M10/8.8 screw: M_L (at M_A 46 N.m): 55 N.m, (without M_A) ~37 N.m.			

Dunlop

Dunlop	SAS 100 Screwlock		March 91
Anaerobic - One part			
Manufacturers Comments: Medium viscosity, solvent-free, low strength. For thread-locking/sealing small screws and threads which need to be disassembled. Some chemical resistance. Thin oil film tolerance.			
Typical Use: Thread-locking/sealing (small threads). Can be disassembled. Most metals, but Stainless steels, cadmium, aluminium, zinc may require activator.			
Cure: RT/ 0.5 to 3 hrs (set time); ~24 hrs (practical cure).			
Test Methods: Not stated			
Physical		**Mechanical**	
Colour	Red	τ (RT)	-
Viscosity	0.15 Pa.s	τ (compressive)	-
S.G.	-	σ_t	-
Shelf life	1 year	σ_c	-
$T_{Service}$	-40 to +150°C*	Break Torque	5 - 10 N.m
CTE (°C)	-	Prevail Torque	5 - 10 N.m
Flash Point	100°C	Elongation	-
Gap Filling	0.1mm max.	H (Shore)	-
Notes: Service +150°C, intermittently. Break: Bright Mild steel 220 M07/M8 Nuts and bolts. Prevail: can be affected by engagement length and thread form.			

Dunlop	SAS 110 Nutlock		March 91
Anaerobic - One part			
Manufacturers Comments: Medium viscosity, solvent-free, medium strength. For thread-locking/sealing small screws and threads which need to be disassembled. Some chemical resistance. Thin oil film tolerance.			
Typical Use: Thread-locking/sealing (small threads). Can be disassembled. Most metals, but Stainless steels, cadmium, aluminium, zinc may require activator.			
Cure: RT/ 0.5 to 3 hrs (set time); ~24 hrs (practical cure).			
Test Methods: Not stated			
Physical		**Mechanical**	
Colour	Blue	τ (RT)	-
Viscosity	0.15 Pa.s	τ (compressive)	-
S.G.	-	σ_t	-
Shelf life	1 year	σ_c	-
$T_{Service}$	-40 to +150°C*	Break Torque	10 - 15 N.m
CTE (°C)	-	Prevail Torque	10 - 15 N.m
Flash Point	100°C	Elongation	-
Gap Filling	0.1mm max.	H (Shore)	-
Notes: Service +150°C, intermittently. Break: Bright Mild steel 220 M07/M8 Nuts and bolts. Prevail: can be affected by engagement length and thread form.			

Dunlop	SAS 123 Studlock		Feb 90
Anaerobic - One part			
Manufacturers Comments: High viscosity, solvent-free, high strength. For thread-locking/sealing studs and other cylindrical assemblies. Permanent. Some chemical resistance. Thin oil film tolerance. Meets: UK MoD DTD 5633 Grade 3.			
Typical Use: Studs, bolts, keyways, splines and press fits. Most metals, inc. cadmium, zinc. Some plastics - may require activator.			
Cure: RT/ 24 hrs (full cure).			
Test Methods: Not stated			
Physical		**Mechanical**	
Colour	Green	τ (RT) static	> 17 MPa
Viscosity	0.5 Pa.s	τ (heat ageing)	95% (70%)
S.G.	-	σ_t	-
Shelf life	1 year	σ_c	-
$T_{Service}$	-55 to +150°C	Break Torque	>19 N.m*
CTE (°C)	-	Prevail Torque	-
Flash Point	>100°C	Elongation	-
Gap Filling	0.15mm max.	H (Shore)	-
Notes: Maximum torque. Min. strength retention after ageing: 95% (1000hrs/100°C); 70% (1000hrs/150°C). 70% (boiling water, 1 week). τ (heat ageing) data - min. strength retention after: 1000hrs/100°C (1000hrs/150°C).			

Dunlop	SAS 120 Studlock		April 91
Anaerobic - One part			
Manufacturers Comments: High viscosity, solvent-free, high strength. For thread-locking/sealing studs and other cylindrical assemblies. Permanent. Some chemical resistance. Thin oil film tolerance.			
Typical Use: Studs, bolts, keyways, splines and press fits. Most metals, but Stainless steels, cadmium, aluminium, zinc may require activator.			
Cure: RT/ 0.5 to 3 hrs (set time); ~24 hrs (practical cure).			
Test Methods: Not stated			
Physical		**Mechanical**	
Colour	Green	τ (RT)	-
Viscosity	0.5 Pa.s	τ (compressive)	-
S.G.	-	σ_t	-
Shelf life	1 year	σ_c	-
$T_{Service}$	-40 to +150°C*	Break Torque	20 - 25 N.m
CTE (°C)	-	Prevail Torque	20 - 25 N.m
Flash Point	100°C	Elongation	-
Gap Filling	0.15mm max.	H (Shore)	-
Notes: Service +150°C, intermittently. Break: Bright Mild steel 220 M07/M8 Nuts and bolts. Prevail: can be affected by engagement length and thread form.			

Dunlop	SAS 130 Bearing Fit		April 91
Anaerobic - One part			
Manufacturers Comments: Thixotropic, solvent-free, medium strength. For retention and mounting of mechanical parts. Some chemical resistance. Thin oil film tolerance.			
Typical Use: Retention of bearings, bushes, keys and splines. Mounting oil seals and wear liners. Most metals, cadmium, zinc, stainless steel, aluminium may need activator.			
Cure: RT/ 0.5 to 3 hrs (set time); 24 hrs (full cure).			
Test Methods: Not stated			
Physical		**Mechanical**	
Colour	Purple	τ (RT)	-
Viscosity	0.3 - 0.5 Pa.s	τ (compressive)	-
S.G.	-	σ_t	-
Shelf life	1 year	σ_c	-
$T_{Service}$	-55 to +150°C*	Break Torque	15 - 20 N.m
CTE (°C)	-	Prevail Torque	15 - 20 N.m
Flash Point	100°C	Elongation	-
Gap Filling	0.2mm max.	H (Shore)	-
Notes: Service +150°C, intermittently. Break: Bright Mild steel 220 M07/M8 Nuts and bolts. Prevail: can be affected by engagement length and thread form.			

Loctite

Dunlop	SAS 140 Pipe Seal		April 91
Anaerobic - One part			
Manufacturers Comments: Thixotropic, solvent-free, medium strength. For locking and sealing small hydraulic and pneumatic pipes (fine threads). Some chemical resistance. Thin oil film tolerance.			
Typical Use: Most suitable for pipes. Not for flammable gases or drinking water. Most metals, cadmium, zinc, stainless steel, aluminium may need activator.			
Cure: RT/ 0.5 - 3 hrs (set time); 24 hrs (full cure).			
Test Methods: Not stated			
Physical		**Mechanical**	
Colour	Brown	τ (RT)	-
Viscosity	0.3 - 0.5 Pa.s	τ (compressive)	-
S.G.	-	σ_t	-
Shelf life	1 year	σ_c	-
$T_{Service}$	-55 to +150°C*	Break Torque	10 - 15 N.m
CTE (°C)	-	Prevail Torque	10 - 15 N.m
Flash Point	100°C	Elongation	-
Gap Filling	0.20 mm max.	H (Shore)	-
Notes: Service +150°C, intermittently. Break: Bright Mild steel 220 M07/M8 Nuts and bolts. Prevail: can be affected by engagement length and thread form.			

Loctite	243		(Feb 95)
One part anaerobic			
Manufacturers Comments: Medium strength torque grade. Oil tolerant. Higher viscosity.			
Typical Use: Metals (inc. passive - stainless steel, plated).			
Cure: RT/10 mins (fixturing) on M10 steel bolts.			
Test Methods: MIL			
Physical		**Mechanical**	
Colour	-	τ (RT)	-
Viscosity	1.45-2.65 Pa.s	τ (compressive)	-
S.G.	-	σ_t	-
Shelf life	1 yr/5-25°C	σ_c	-
$T_{Service}$	150°C (continuous)	Break Torque	15-25 N.m
CTE (°C)		Prevail Torque	4-10 N.m
Flash Point	>100°C	Elongation	-
Gap Filling	-	H (Shore)	-
Notes: Break and Prevail Torque: MIL-S-46163.			

Dunlop	SAS 150 Pipe Seal		April 91
Anaerobic - One part			
Manufacturers Comments: Thixotropic, solvent-free, low strength. Contains PTFE. For sealing small hydraulic and pneumatic pipes (fine threads). Some chemical resistance. Thin oil film tolerance.			
Typical Use: Most suitable for 0.25" (8 mm) pipes. Not for flammable gases or drinking water. Most metals, cadmium, zinc, stainless steel, aluminium may need activator.			
Cure: RT/ 0.5 - 3 hrs (set time); 24 hrs (full cure).			
Test Methods: Not stated			
Physical		**Mechanical**	
Colour	Caramel	τ (RT)	-
Viscosity	0.35 - 0.70 Pa.s	τ (compressive)	-
S.G.	-	σ_t	-
Shelf life	1 year	σ_c	-
$T_{Service}$	-55 to +150°C*	Break Torque	5 - 10 N.m
CTE (°C)	-	Prevail Torque	5 - 10 N.m
Flash Point	100°C	Elongation	-
Gap Filling	0.20 mm max.	H (Shore)	-
Notes: Service +150°C, intermittently. Break: Bright Mild steel 220 M07/M8 Nuts and bolts. Prevail: can be affected by engagement length and thread form. Similar product: SAS 160 - instant gasket.			

Loctite	262		(Feb 95)
One part anaerobic			
Manufacturers Comments: High strength torque grade.			
Typical Use: Metals (inc. passive - stainless steel, plated).			
Cure: RT/15 mins (fixturing) on M10 steel bolts.			
Test Methods: MIL			
Physical		**Mechanical**	
Colour	-	τ (RT)	-
Viscosity	1.3-2.4 Pa.s	τ (compressive)	-
S.G.	-	σ_t	-
Shelf life	1 yr/5-25°C	σ_c	-
$T_{Service}$	150°C (continuous)	Break Torque	12-25 N.m
CTE (°C)		Prevail Torque	10-30 N.m
Flash Point	>100°C	Elongation	-
Gap Filling	-	H (Shore)	-
Notes: Break and Prevail Torque: MIL-S-46163.			

Loctite	290		(Feb 95)
One part anaerobic			
Manufacturers Comments: Penetrating grade (low viscosity).			
Typical Use: Metals. Threadlocking.			
Cure: RT/15 mins (fixturing) on M10 steel bolts.			
Test Methods: MIL			
Physical		**Mechanical**	
Colour	-	τ (RT)	-
Viscosity	0.01 - 0.015 Pa.s	τ (compressive)	-
S.G.	-	σ_t	-
Shelf life	1 yr/5-25°C	σ_c	-
$T_{Service}$	150°C (continuous)	Break Torque	5-16 N.m
CTE (°C)		Prevail Torque	20-50 N.m
Flash Point	>100°C	Elongation	-
Gap Filling	-	H (Shore)	-
Notes: Break and Prevail Torque: MIL-S-46163			

Loctite	661		(Feb 95)
One part anaerobic, UV (surface) cure			
Manufacturers Comments: UV surface cure version of 648.			
Typical Use:			
Cure: UV/15-20 secs (fixturing time).			
Test Methods: see notes			
Physical		**Mechanical**	
Colour	-	τ (RT)	18-35 MPa
Viscosity	0.4-0.6 Pa.s	τ (compressive)	-
S.G.	-	σ_t	-
Shelf life	1 yr/5-25°C	σ_c	-
$T_{Service}$	175°C (continuous)	Break Torque	-
CTE (°C)		Prevail Torque	-
Flash Point	>100°C	Elongation	-
Gap Filling	0.15 mm*	H (Shore)	-
Notes: *Gap filling: value on diameter. τ : MIL-R-46086.			

Loctite	648		(Feb 95)
One part anaerobic.			
Manufacturers Comments: High strength, medium viscosity retaining adhesive. Temperature resistance. Solvent resistance.			
Typical Use: Retention of oil seals, bearings on shafts and in housings, bushes and journals in soft materials.			
Cure: RT/5 mins (fixturing time).			
Test Methods: see notes			
Physical		**Mechanical**	
Colour	-	τ (RT)	18-35 MPa
Viscosity	0.4-0.6 Pa.s	τ (compressive)	-
S.G.	-	σ_t	-
Shelf life	1 yr/5-25°C	σ_c	-
$T_{Service}$	175°C (continuous)	Break Torque	-
CTE (°C)		Prevail Torque	-
Flash Point	>100°C	Elongation	-
Gap Filling	0.15 mm*	H (Shore)	-
Notes: *Gap filling: value on diameter. τ : MIL-R-46086.			

Loctite	662		(Feb 95)
One part anaerobic, UV (surface) cure			
Manufacturers Comments: UV surface cure version of Loctite High Strength Retainer 638.			
Typical Use:			
Cure: UV/15-20 secs (fixturing time).			
Test Methods: see notes			
Physical		**Mechanical**	
Colour	-	τ (RT)	22-40 MPa
Viscosity	2-3 Pa.s	τ (compressive)	-
S.G.	-	σ_t	-
Shelf life	1 yr/5-25°C	σ_c	-
$T_{Service}$	175°C (continuous)	Break Torque	-
CTE (°C)		Prevail Torque	-
Flash Point	>100°C	Elongation	-
Gap Filling	0.25 mm*	H (Shore)	-
Notes: *Gap filling: value on diameter. τ : MIL-R-46086.			

Loctite	Bearing Fit 641		(Feb 95)
One part anaerobic.			
Manufacturers Comments: Recommended for cylindrical parts that require disassembly.			
Typical Use: Retention of oil seals, bearings on shafts and in housings, bushes and journals in soft materials.			
Cure: RT/20 mins (fixturing time).			
Test Methods: see notes			
Physical		**Mechanical**	
Colour	-	τ (RT)	7-16 MPa
Viscosity	0.35 - 0.65 Pa.s	τ (compressive)	-
S.G.	-	σ_t	-
Shelf life	1 yr/5-25°C	σ_c	-
$T_{Service}$	150°C	Break Torque	-
CTE (°C)		Prevail Torque	-
Flash Point	>100°C	Elongation	-
Gap Filling	0.15 mm*	H (Shore)	-
Notes: *Gap filling: value on diameter. τ : MIL-R-46086.			

Loctite	Hydraulic Sealant 542		(Feb 95)
One part anaerobic			
Manufacturers Comments: Medium strength, medium viscosity, sealant. Resistant to all common hydraulic fluids, inc. fire-resistant synthetics. Thixotropic. Can be dismantled.			
Typical Use: Hydraulic and pneumatic connections, <20 mm, fine threads.			
Cure: RT/15 mins (fixturing) on M10 steel bolts.			
Test Methods: MIL			
Physical		**Mechanical**	
Colour	-	τ (RT)	-
Viscosity	0.35 - 0.65 Pa.s	τ (compressive)	-
S.G.	-	σ_t	-
Shelf life	1 yr/5-25°C	σ_c	-
$T_{Service}$	150°C (continuous)	Break Torque	8-20 N.m
CTE (°C)		Prevail Torque	8-15 N.m
Flash Point	>100°C	Elongation	-
Gap Filling	-	H (Shore)	-
Notes: Break and Prevail Torque: MIL-S-46163. Other sealant grades: 567 - low strength, very high viscosity for coarse threads (St. steel) 572 - low strength, high viscosity for pipe seals.			

Loctite	High Strength Retainer 638		(Feb 95)
One part anaerobic.			
Manufacturers Comments: Maximum strength, high-viscosity retaining adhesive.			
Typical Use: Cylindrical parts under dynamic loads, structural tubing, splines.			
Cure: RT/5 mins (fixturing time).			
Test Methods: see notes			
Physical		**Mechanical**	
Colour	-	τ (RT)	22-40 MPa
Viscosity	2 - 3 Pa.s	τ (compressive)	-
S.G.	-	σ_t	-
Shelf life	1 yr/5-25°C	σ_c	-
$T_{Service}$	150°C	Break Torque	-
CTE (°C)		Prevail Torque	-
Flash Point	>100°C	Elongation	-
Gap Filling	0.25 mm*	H (Shore)	-
Notes: *Gap filling: value on diameter. τ : MIL-R-46086.			

Loctite	Nutlock 242		(Feb 95)
One part anaerobic,			
Manufacturers Comments: Medium strength torque grade. Thixotropic. Used on 'as received' fasteners. Can be disassembled.			
Typical Use: Metals (inc. passive - stainless steel, plated).			
Cure: RT/10 mins (fixturing) on M10 steel bolts.			
Test Methods: MIL			
Physical		**Mechanical**	
Colour	-	τ (RT)	-
Viscosity	0.84 - 1.56 Pa.s	τ (compressive)	-
S.G.	-	σ_t	-
Shelf life	1 yr/5-25°C	σ_c	-
$T_{Service}$	150°C (continuous)	Break Torque	8-18 N.m
CTE (°C)		Prevail Torque	2-10 N.m
Flash Point	>100°C	Elongation	-
Gap Filling	-	H (Shore)	-
Notes: Break and Prevail Torque: MIL-S-46163.			

Loctite	Pipe Sealant 577		(Feb 95)
One part anaerobic			
Manufacturers Comments: High viscosity, medium strength for pipes and fittings (up to R3"), coarse threaded components. Cure resists low temperatures. Resists high pressures (200 bar). Lead-free. Approved for contact with potable water (by Water Research Centre).			
Typical Use: Copper, iron, steel, galvanised pipes and fittings.			
Cure: RT/30 mins (fixturing) on M10 steel bolts.			
Test Methods: MIL			
Physical		**Mechanical**	
Colour	-	τ (RT)	-
Viscosity	17 - 31 Pa.s	τ (compressive)	-
S.G.	-	σ_t	-
Shelf life	1 yr/5-25°C	σ_c	-
$T_{Service}$	150°C (continuous)	Break Torque	6-15 N.m
CTE (°C)		Prevail Torque	9-25 N.m
Flash Point	>100°C	Elongation	-
Gap Filling	-	H (Shore)	-
Notes: Break and Prevail Torque: MIL-S-46163.			

Loctite	Retainer 603		(Feb 95)
One part anaerobic.			
Manufacturers Comments: Oil tolerant high strength retainer for 'as received' parts.			
Typical Use: Oily metals. Oil-impregnated sintered bushes.			
Cure: RT/10 mins (fixturing time).			
Test Methods: see notes			
Physical		**Mechanical**	
Colour	-	τ (RT)	20-32 MPa
Viscosity	0.1 - 0.15 Pa.s	τ (compressive)	-
S.G.	-	σ_t	-
Shelf life	1 yr/5-25°C	σ_c	-
$T_{Service}$	150°C	Break Torque	-
CTE (°C)		Prevail Torque	-
Flash Point	>100°C	Elongation	-
Gap Filling	0.15 mm*	H (Shore)	-
Notes: *Gap filling: value on diameter. τ : MIL-R-46086.			

Loctite	Retainer 601		(Feb 95)
One part anaerobic.			
Manufacturers Comments: High strength retaining of close-fitting parts. High shear strength. For slip-fit and press-fit parts.			
Typical Use: Retention of shafts and rotors (electric motors), gears, pulleys, sleeves, bushes, oil seals in housings, pins, keys, dowels.			
Cure: RT/15 mins (fixturing time).			
Test Methods: see notes			
Physical		**Mechanical**	
Colour	-	τ (RT)	18-30 MPa
Viscosity	0.1 - 0.15 Pa.s	τ (compressive)	-
S.G.	-	σ_t	-
Shelf life	1 yr/5-25°C	σ_c	-
$T_{Service}$	150°C	Break Torque	-
CTE (°C)		Prevail Torque	-
Flash Point	>100°C	Elongation	-
Gap Filling	0.15 mm*	H (Shore)	-
Notes: *Gap filling: value on diameter. τ : MIL-R-46086.			

Loctite	Screwlock 222		(Feb 95)
One part anaerobic,			
Manufacturers Comments: Lower strength, threadlocking grade for fasteners that may require subsequent adjustment. Thixotropic. Used on 'as received' fasteners.			
Typical Use: Low-strength, non-ferrous metals (brass, aluminium). Locking precision screws.			
Cure: RT/15 mins (fixturing) on M10 steel bolts.			
Test Methods: MIL			
Physical		**Mechanical**	
Colour	Purple	τ (RT)	-
Viscosity	0.84 - 1.56 Pa.s	τ (compressive)	-
S.G.	-	σ_t	-
Shelf life	1 yr/5-25°C	σ_c	-
$T_{Service}$	150°C (continuous)	Break Torque	3-11 N.m
CTE (°C)		Prevail Torque	1-5 N.m
Flash Point	>100°C	Elongation	-
Gap Filling	-	H (Shore)	-
Notes: Break and Prevail Torque: MIL-S-46163			

Loctite	Studlock 270		(Feb 95)
One part anaerobic			
Manufacturers Comments: High strength threadlocking grade. For studs, bolts and parts that do not require disassembly.			
Typical Use: Metals. Stud in castings.			
Cure: RT/10 mins (fixturing) on M10 steel bolts.			
Test Methods: MIL			
Physical		**Mechanical**	
Colour	-	τ (RT)	-
Viscosity	0.4 - 0.6 Pa.s	τ (compressive)	-
S.G.	-	σ_f	-
Shelf life	1 yr/5-25°C	σ_c	-
$T_{Service}$	150°C (continuous)	Break Torque	16-36 N.m
CTE (°C)		Prevail Torque	22-50 N.m
Flash Point	>100°C	Elongation	-
Gap Filling	-	H (Shore)	-
Notes: Break and Prevail Torque: MIL-S-46163			

Omnitechnic (Henkel)	Omnifit 50 H		(Feb 95)
Catalytic (metal) curing, one part anaerobic.			
Manufacturers Comments: Rapid cure. Thixotropic, low strength.			
Typical Use: Thread locking (< 40mm) for removable components. Fixing of small components.			
Cure: Initial cure (70% strength): <3hrs/RT, Final cure: 6hrs/RT.			
Test Methods: DIN			
Physical		**Mechanical**	
Colour	Blue	τ (RT)	12.5 MPa
Viscosity	-	τ (compressive)	-
S.G.	-	σ_f	-
Shelf life	1 yr./cool, dark	σ_c	
$T_{Service}$	-60 to +150°C	Break Torque	13 N.m
CTE (°C)	81×10^{-6} K^{-1}	Prevail Torque	< 13 N.m
Flash Point	>100°C	Elongation	-
Gap Filling	0.01 mm max.	H (Shore)	-
Notes: Break torque: DIN 993-8, 8/934-8, Shear; DIN 54452, Resistivity (DIN 53482): 2.3×10^{14} ohm-cm. Dielectric constant (DIN 53483): 4.5.			

Omnitechnic (Henkel)

Omnitechnic (Henkel)	Omnifit 15 M		(Feb 95)
Catalytic (metal) curing, one part anaerobic.			
Manufacturers Comments: Rapid cure. Medium viscosity, very low strength.			
Typical Use: Thread locking (large diameters) for removable components.			
Cure: Initial cure (70% strength): <3hrs/RT, Final cure: 8hrs/RT.			
Test Methods: DIN			
Physical		**Mechanical**	
Colour	Yellow	τ (RT)	4.5 MPa
Viscosity	0.7 Pa.s	τ (compressive)	-
S.G.	-	σ_f	-
Shelf life	1 yr./cool, dark	σ_c	
$T_{Service}$	-60 to +80°C	Break Torque	5 N.m
CTE (°C)	81×10^{-6} K^{-1}	Prevail Torque	< 5 N.m
Flash Point	>100°C	Elongation	-
Gap Filling	0.01 mm max.	H (Shore)	-
Notes: Break torque: DIN 993-8, 8/934-8, Shear; DIN 54452, Resistivity (DIN 53482): 2.3×10^{14} ohm-cm. Dielectric constant (DIN 53483): 4.5.			

Omnitechnic (Henkel)	Omnifit 50 M		(Feb 95)
Catalytic (metal) curing, one part anaerobic.			
Manufacturers Comments: Rapid cure. Medium viscosity, low strength.			
Typical Use: Thread locking (< 40mm) for removable components. Fixing of small components.			
Cure: Initial cure (70% strength): <3hrs/RT, Final cure: 6hrs/RT.			
Test Methods: DIN			
Physical		**Mechanical**	
Colour	Blue	τ (RT)	11 MPa
Viscosity	0.7 Pa.s	τ (compressive)	-
S.G.	-	σ_f	-
Shelf life	1 yr./cool, dark	σ_c	
$T_{Service}$	-60 to +150°C	Break Torque	12 N.m
CTE (°C)	81×10^{-6} K^{-1}	Prevail Torque	< 12 N.m
Flash Point	>100°C	Elongation	-
Gap Filling	0.01 mm max.	H (Shore)	-
Notes: Break torque: DIN 993-8, 8/934-8, Shear; DIN 54452, Resistivity (DIN 53482): 2.3×10^{14} ohm-cm. Dielectric constant (DIN 53483): 4.5.			

Omnitechnic (Henkel) Omnifit 80 M (Feb 95)

Catalytic (metal) curing, one part anaerobic.

Manufacturers Comments: Medium viscosity, medium strength.

Typical Use: Thread locking (< 40mm) for removable components. Fixing small components. Sealing threaded pipework joints (< 40mm).

Cure: Initial cure (70% strength): <6hrs/RT, Final cure: 24hrs/RT.

Test Methods: DIN

Physical		Mechanical	
Colour	Red	τ (RT)	20 MPa
Viscosity	1.1 Pa.s	τ (compressive)	-
S.G.	-	σ_t	-
Shelf life	1 yr./cool, dark	σ_c	
$T_{Service}$	-60 to +150°C	Break Torque	20 N.m
CTE (°C)	81×10^{-6} K^{-1}	Prevail Torque	< 20 N.m
Flash Point	>100°C	Elongation	-
Gap Filling	0.01 mm max.	H (Shore)	-

Notes: Break torque: DIN 993-8, 8/934-8, Shear; DIN 54452, Resistivity (DIN 53482): 2.3 x 10^{14} ohm-cm. Dielectric constant (DIN 53483): 4.5.

Omnitechnic (Henkel) Omnifit 100 M (Feb 95)

Catalytic (metal) curing, one part anaerobic.

Manufacturers Comments: Rapid cure. Medium viscosity, medium strength.

Typical Use: Thread locking (< 40mm) for removable components. Fixing small components. Sealing threaded pipework joints (< 40mm).

Cure: Initial cure (70% strength): <3hrs/RT, Final cure: 6hrs/RT.

Test Methods: DIN

Physical		Mechanical	
Colour	Red	τ (RT)	20 MPa
Viscosity	0.7 Pa.s	τ (compressive)	-
S.G.	-	σ_t	-
Shelf life	1 yr./cool, dark	σ_c	
$T_{Service}$	-60 to +150°C	Break Torque	17 N.m
CTE (°C)	81×10^{-6} K^{-1}	Prevail Torque	< 17 N.m
Flash Point	>100°C	Elongation	-
Gap Filling	0.01 mm max.	H (Shore)	-

Notes: Break torque: DIN 993-8, 8/934-8, Shear; DIN 54452, Resistivity (DIN 53482): 2.3 x 10^{14} ohm-cm. Dielectric constant (DIN 53483): 4.5.

Omnitechnic (Henkel) Omnifit 100 H (Feb 95)

Catalytic (metal) curing, one part anaerobic.

Manufacturers Comments: Rapid cure. Thixotropic, medium strength.

Typical Use: Thread locking (< 40mm) for removable components. Fixing small components. Sealing threaded pipework joints (< 40mm).

Cure: Initial cure (70% strength): <3hrs/RT, Final cure: 6hrs/RT.

Test Methods: DIN

Physical		Mechanical	
Colour	Red	τ (RT)	20 MPa
Viscosity	-	τ (compressive)	-
S.G.	-	σ_t	-
Shelf life	1 yr./cool, dark	σ_c	
$T_{Service}$	-60 to +150°C	Break Torque	18 N.m
CTE (°C)	81×10^{-6} K^{-1}	Prevail Torque	< 18 N.m
Flash Point	>100°C	Elongation	-
Gap Filling	0.01 mm max.	H (Shore)	-

Notes: Break torque: DIN 993-8, 8/934-8, Shear; DIN 54452, Resistivity (DIN 53482): 2.3 x 10^{14} ohm-cm. Dielectric constant (DIN 53483): 4.5.

Omnitechnic (Henkel) Omnifit 150 M (Feb 95)

Catalytic (metal) curing, one part anaerobic.

Manufacturers Comments: Medium viscosity. High strength.

Typical Use: Permanent mounting of rotating components (pinions, flanges, etc.). Fixing of mechanical keys. Thread locking.

Cure: Initial cure (70% strength): <6hrs/RT, Final cure: 24hrs/RT.

Test Methods: DIN

Physical		Mechanical	
Colour	Green	τ (RT)	31 MPa
Viscosity	1.1 Pa.s	τ (compressive)	-
S.G.	-	σ_t	-
Shelf life	1 yr./cool, dark	σ_c	
$T_{Service}$	-60 to +150°C	Break Torque	28 N.m
CTE (°C)	81×10^{-6} K^{-1}	Prevail Torque	< 28 N.m
Flash Point	>100°C	Elongation	-
Gap Filling	0.01 mm max.	H (Shore)	-

Notes: Break torque: DIN 993-8, 8/934-8, Shear; DIN 54452, Resistivity (DIN 53482): 2.3 x 10^{14} ohm-cm. Dielectric constant (DIN 53483): 4.5.

Omnitechnic (Henkel) **Omnifit 200 H** (Feb 95)			
Catalytic (metal) curing, one part anaerobic.			
Manufacturers Comments: Rapid cure. Thixotropic. High strength.			
Typical Use: Permanent mounting of rotating components (pinions, flanges, etc.). Fixing of mechanical keys. Thread locking.			
Cure: Initial cure (70% strength): <3hrs/RT, Final cure: 8hrs/RT.			
Test Methods: DIN			
Physical		**Mechanical**	
Colour	Green	τ (RT)	31 MPa
Viscosity	-	τ (compressive)	-
S.G.	-	σ_t	-
Shelf life	1 yr./cool, dark	σ_c	
$T_{Service}$	-60 to +150°C	Break Torque	22 N.m
CTE (°C)	81 × 10^{-6} K^{-1}	Prevail Torque	< 22 N.m
Flash Point	>100°C	Elongation	-
Gap Filling	0.01 mm max.	H (Shore)	-
Notes: Break torque: DIN 993-8, 8/934-8, Shear; DIN 54452, Resistivity (DIN 53482): 2.3 x 10^{14} ohm-cm. Dielectric constant (DIN 53483): 4.5.			

Omnitechnic (Henkel) **Omnifit 200 M** (Feb 95)			
Catalytic (metal) curing, one part anaerobic.			
Manufacturers Comments: Rapid cure. Medium viscosity. High strength.			
Typical Use: Permanent mounting of rotating components (pinions, flanges, etc.). Fixing of mechanical keys. Thread locking.			
Cure: Initial cure (70% strength): <3hrs/RT, Final cure: 6hrs/RT.			
Test Methods: DIN			
Physical		**Mechanical**	
Colour	Green	τ (RT)	34 MPa
Viscosity	0.7 Pa.s	τ (compressive)	-
S.G.	-	σ_t	-
Shelf life	1 yr./cool, dark	σ_c	
$T_{Service}$	-60 to +150°C	Break Torque	28 N.m
CTE (°C)	81 × 10^{-6} K^{-1}	Prevail Torque	< 28 N.m
Flash Point	>100°C	Elongation	-
Gap Filling	0.01 mm max.	H (Shore)	-
Notes: Break torque: DIN 993-8, 8/934-8, Shear; DIN 54452, Resistivity (DIN 53482): 2.3 x 10^{14} ohm-cm. Dielectric constant (DIN 53483): 4.5.			

Omnitechnic (Henkel) **Omnifit 200 LL** (Feb 95)			
Catalytic (metal) curing, one part anaerobic.			
Manufacturers Comments: Rapid cure. Very low viscosity. High strength.			
Typical Use: Permanent mounting of rotating components (pinions, flanges, etc.). Fixing of mechanical keys. Thread locking.			
Cure: Initial cure (70% strength): <3hrs/RT, Final cure: 6hrs/RT.			
Test Methods: DIN			
Physical		**Mechanical**	
Colour	Green	τ (RT)	30 MPa
Viscosity	0.03 Pa.s	τ (compressive)	-
S.G.	-	σ_t	-
Shelf life	1 yr./cool, dark	σ_c	
$T_{Service}$	-60 to +150°C	Break Torque	10 N.m
CTE (°C)	81 × 10^{-6} K^{-1}	Prevail Torque	< 10 N.m
Flash Point	>100°C	Elongation	-
Gap Filling	0.01 mm max.	H (Shore)	-
Notes: Break torque: DIN 993-8, 8/934-8, Shear; DIN 54452, Resistivity (DIN 53482): 2.3 x 10^{14} ohm-cm. Dielectric constant (DIN 53483): 4.5.			

Omnitechnic (Henkel) **Omnifit 220 M** (Feb 95)			
Catalytic (metal) curing, one part anaerobic.			
Manufacturers Comments: Rapid cure. Medium viscosity. Very high strength. Good temperature resistance.			
Typical Use: Permanent mounting of rotating components (pinions, flanges, etc.). Fixing of mechanical keys. Thread locking. Higher temperature applications.			
Cure: Initial cure (70% strength): <3hrs/RT, Final cure: 8hrs/RT.			
Test Methods: DIN			
Physical		**Mechanical**	
Colour	Violet	τ (RT)	28 MPa
Viscosity	0.4 Pa.s	τ (compressive)	-
S.G.	-	σ_t	-
Shelf life	1 yr./cool, dark	σ_c	
$T_{Service}$	-60 to +220°C	Break Torque	32 N.m
CTE (°C)	81 × 10^{-6} K^{-1}	Prevail Torque	≥ 32 N.m
Flash Point	>100°C	Elongation	-
Gap Filling	0.01 mm max.	H (Shore)	-
Notes: Break torque: DIN 993-8, 8/934-8, Shear; DIN 54452, Resistivity (DIN 53482): 2.3 x 10^{14} ohm-cm. Dielectric constant (DIN 53483): 4.5.			

Omnitechnic (Henkel)	**Omnifit 222 M**		(Feb 95)
Catalytic (metal) curing, one part anaerobic.			
Manufacturers Comments: Rapid cure. Medium viscosity. Very high strength. Good temperature resistance.			
Typical Use: Permanent mounting of rotating components (pinions, flanges, etc.). Fixing of mechanical keys. Thread locking. Higher temperature applications.			
Cure: Initial cure (70% strength): <3hrs/RT, Final cure: 8hrs/RT.			
Test Methods: DIN			
Physical		**Mechanical**	
Colour	Violet	τ (RT)	30 MPa
Viscosity	0.7 Pa.s	τ (compressive)	-
S.G.	-	σ_f	-
Shelf life	1 yr./cool, dark	σ_c	
$T_{Service}$	-60 to +220°C	Break Torque	35 N.m
CTE (°C)	81×10^{-6} K^{-1}	Prevail Torque	$\geq$ 35 N.m
Flash Point	>100°C	Elongation	-
Gap Filling	0.01 mm max.	H (Shore)	-
Notes: Break torque: DIN 993-8, 8/934-8, Shear; DIN 54452, Resistivity (DIN 53482): 2.3 x 10^{14} ohm-cm. Dielectric constant (DIN 53483): 4.5.			

Omnitechnic (Henkel)	**Omnifit 230 M**		(Feb 95)
Catalytic (metal) curing, one part anaerobic.			
Manufacturers Comments: Rapid cure. Medium viscosity. Very high strength. Good temperature resistance.			
Typical Use: Permanent mounting of rotating components (pinions, flanges, etc.). Fixing of mechanical keys. Thread locking. Higher temperature applications.			
Cure: Initial cure (70% strength): <3hrs/RT, Final cure: 8hrs/RT.			
Test Methods: DIN			
Physical		**Mechanical**	
Colour	Violet	τ (RT)	40 MPa
Viscosity	0.7 Pa.s	τ (compressive)	-
S.G.	-	σ_f	-
Shelf life	1 yr./cool, dark	σ_c	
$T_{Service}$	-60 to +200°C	Break Torque	47 N.m
CTE (°C)	81×10^{-6} K^{-1}	Prevail Torque	$\geq$ 47 N.m
Flash Point	>100°C	Elongation	-
Gap Filling	0.01 mm max.	H (Shore)	-
Notes: Break torque: DIN 993-8, 8/934-8, Shear; DIN 54452, Resistivity (DIN 53482): 2.3 x 10^{14} ohm-cm. Dielectric constant (DIN 53483): 4.5.			

Omnitechnic (Henkel)	**Omnifit 230 L**		(Feb 95)
Catalytic (metal) curing, one part anaerobic.			
Manufacturers Comments: Rapid cure. Low viscosity. Very high strength. Good temperature resistance.			
Typical Use: Permanent mounting of rotating components (pinions, flanges, etc.). Fixing of mechanical keys. Thread locking. Higher temperature applications.			
Cure: Initial cure (70% strength): <3hrs/RT, Final cure: 8hrs/RT.			
Test Methods: DIN			
Physical		**Mechanical**	
Colour	Violet	τ (RT)	36 MPa
Viscosity	0.15 Pa.s	τ (compressive)	-
S.G.	-	σ_f	-
Shelf life	1 yr./cool, dark	σ_c	
$T_{Service}$	-60 to +200°C	Break Torque	43 N.m
CTE (°C)	81×10^{-6} K^{-1}	Prevail Torque	$\geq$ 43 N.m
Flash Point	>100°C	Elongation	-
Gap Filling	0.01 mm max.	H (Shore)	-
Notes: Break torque: DIN 993-8, 8/934-8, Shear; DIN 54452, Resistivity (DIN 53482): 2.3 x 10^{14} ohm-cm. Dielectric constant (DIN 53483): 4.5.			

Omnitechnic (Henkel)	**Omnifit 280 M**		(Feb 95)
Catalytic (metal) curing, one part anaerobic.			
Manufacturers Comments: Slow cure. Medium viscosity. Very high strength. Good temperature resistance.			
Typical Use: Permanent mounting of rotating components (pinions, flanges, etc.). Fixing of mechanical keys. Thread locking. Higher temperature applications.			
Cure: Initial cure (70% strength): <12hrs/RT, Final cure: 72hrs/RT.			
Test Methods: DIN			
Physical		**Mechanical**	
Colour	Violet	τ (RT)	40 MPa
Viscosity	0.7 Pa.s	τ (compressive)	-
S.G.	-	σ_f	-
Shelf life	1 yr./cool, dark	σ_c	
$T_{Service}$	-60 to +200°C	Break Torque	45 N.m
CTE (°C)	81×10^{-6} K^{-1}	Prevail Torque	$\geq$ 45 N.m
Flash Point	>100°C	Elongation	-
Gap Filling	0.01 mm max.	H (Shore)	-
Notes: Break torque: DIN 993-8, 8/934-8, Shear; DIN 54452, Resistivity (DIN 53482): 2.3 x 10^{14} ohm-cm. Dielectric constant (DIN 53483): 4.5			

See Key to Data Sheets - Pages 63 to 65

Panacol Elosol

Panacol Elosol	Penloc A		Jun 94
Catalytic (metal) curing, one part anaerobic.			
Manufacturers Comments: RT cure. Low viscosity, medium strength. Good temperature and chemical resistance. Passive surfaces (e.g. glass or plastics) require an activating primer.			
Typical Use: Threadlocking and small component fixing.			
Cure: Handling strength: 5 - 20min/RT, Final cure: 6 - 24hrs/RT.			
Test Methods: DIN			
Physical		**Mechanical**	
Colour	Transparent	τ (RT)	15.5 MPa
Viscosity	0.01 Pa.s	τ (compressive)	-
S.G.	1.05 - 1.18	σ_t	-
Shelf life	1 yr./RT	σ_c	-
$T_{Service}$	-55 to +150°C	Break Torque	17.0 N.m
CTE (°C)	-	Prevail Torque	17.5 N.m
Flash Point	100°C	Elongation	-
Gap Filling	0.05 mm max.	H (Shore)	-
Notes: Shear: DIN 54452. Density: DIN 51757. Flash point: DIN 55213. Breakdown potential: 8.5 kV.mm^{-1}.			

Panacol Elosol	Penloc NT		Jun 94
Catalytic (metal) curing, one part anaerobic.			
Manufacturers Comments: RT cure. Medium viscosity, medium strength. Good temperature and chemical resistance. Passive surfaces (e.g. glass or plastics) require an activating primer.			
Typical Use: Threadlocking and small component fixing.			
Cure: Handling strength: 5 - 20min/RT, Final cure: 6 - 24hrs/RT.			
Test Methods: DIN			
Physical		**Mechanical**	
Colour	Blue	τ (RT)	17.5 MPa
Viscosity	0.15 Pa.s	τ (compressive)	-
S.G.	1.05 - 1.18	σ_t	-
Shelf life	1 yr./RT	σ_c	-
$T_{Service}$	-55 to +150°C	Break Torque	18.5 N.m
CTE (°C)	-	Prevail Torque	18.5 N.m
Flash Point	100°C	Elongation	-
Gap Filling	0.1 mm max.	H (Shore)	-
Notes: Shear: DIN 54452. Density: DIN 51757. Flash point: DIN 55213. Breakdown potential: 8.5 kV.mm^{-1}.			

Panacol Elosol	Penloc NL		Jun 94
Catalytic (metal) curing, one part anaerobic.			
Manufacturers Comments: RT cure. High viscosity, low strength. Good temperature and chemical resistance. Passive surfaces (e.g. glass or plastics) require an activating primer.			
Typical Use: Threadlocking and small component fixing.			
Cure: Handling strength: 5 - 20min/RT, Final cure: 6 - 24hrs/RT.			
Test Methods: DIN			
Physical		**Mechanical**	
Colour	Blue	τ (RT)	10 MPa
Viscosity	6 Pa.s	τ (compressive)	-
S.G.	1.05 - 1.18	σ_t	-
Shelf life	1 yr./RT	σ_c	-
$T_{Service}$	-55 to +150°C	Break Torque	12.5 N.m
CTE (°C)	-	Prevail Torque	12.0 N.m
Flash Point	100°C	Elongation	-
Gap Filling	0.2 mm max.	H (Shore)	-
Notes: Shear: DIN 54452. Density: DIN 51757. Flash point: DIN 55213. Breakdown potential: 8.5 kV.mm^{-1}.			

Panacol Elosol	Penloc NTT		Jun 94
Catalytic (metal) curing, one part anaerobic.			
Manufacturers Comments: RT cure. High viscosity, medium strength. Good temperature and chemical resistance. Passive surfaces (e.g. glass or plastics) require an activating primer.			
Typical Use: Threadlocking and small component fixing.			
Cure: Handling strength: 5 - 20min/RT, Final cure: 6 - 24hrs/RT.			
Test Methods: DIN			
Physical		**Mechanical**	
Colour	Blue	τ (RT)	20 MPa
Viscosity	20 Pa.s	τ (compressive)	-
S.G.	1.05 - 1.18	σ_t	-
Shelf life	1 yr./RT	σ_c	-
$T_{Service}$	-55 to +150°C	Break Torque	22.5 N.m
CTE (°C)	-	Prevail Torque	22.0 N.m
Flash Point	100°C	Elongation	-
Gap Filling	0.35 mm max.	H (Shore)	-
Notes: Shear: DIN 54452. Density: DIN 51757. Flash point: DIN 55213. Breakdown potential: 8.5 kV.mm^{-1}.			

See Key to Data Sheets - Pages 63 to 65

Panacol Elosol	Penloc R2		Jun 94
Catalytic (metal) curing, one part anaerobic.			
Manufacturers Comments: RT cure. Medium viscosity, high strength. Good temperature and chemical resistance. Passive surfaces (e.g. glass or plastics) require an activating primer.			
Typical Use: Threadlocking and small component fixing.			
Cure: Handling strength: 5 - 20min/RT, Final cure: 6 - 24hrs/RT.			
Test Methods: DIN			
Physical		**Mechanical**	
Colour	Red	τ (RT)	28.5 MPa
Viscosity	0.4 Pa.s	τ (compressive)	-
S.G.	1.05 - 1.18	σ_t	-
Shelf life	1 yr./RT	σ_c	-
$T_{Service}$	-55 to +200°C	Break Torque	32.0 N.m
CTE (°C)	-	Prevail Torque	35.5 N.m
Flash Point	100°C	Elongation	-
Gap Filling	0.1 mm max.	H (Shore)	-
Notes: Shear: DIN 54452. Density: DIN 51757. Flash point: DIN 55213. Breakdown potential: 8.5 kV.mm^{-1}.			

Panacol Elosol	Penloc RT		Jun 94
Catalytic (metal) curing, one part anaerobic.			
Manufacturers Comments: RT cure. Low viscosity, high strength. Good temperature and chemical resistance. Passive surfaces (e.g. glass or plastics) require an activating primer.			
Typical Use: Component fixing. Shafts, pinions, keys, etc.			
Cure: Handling strength: 5 - 20min/RT, Final cure: 6 - 24hrs/RT.			
Test Methods: DIN			
Physical		**Mechanical**	
Colour	Red	τ (RT)	28.5 MPa
Viscosity	0.1 Pa.s	τ (compressive)	-
S.G.	1.05 - 1.18	σ_t	-
Shelf life	1 yr./RT	σ_c	-
$T_{Service}$	-55 to +200°C	Break Torque	30.5 N.m
CTE (°C)	-	Prevail Torque	33.0 N.m
Flash Point	100°C	Elongation	-
Gap Filling	0.1 mm max.	H (Shore)	-
Notes: Shear: DIN 54452. Density: DIN 51757. Flash point: DIN 55213. Breakdown potential: 8.5 kV.mm^{-1}.			

Panacol Elosol	Penloc RL		Jun 94
Catalytic (metal) curing, one part anaerobic.			
Manufacturers Comments: RT cure. Medium viscosity, high strength. Good temperature and chemical resistance. Passive surfaces (e.g. glass or plastics) require an activating primer.			
Typical Use: Component fixing. Shafts, pinions, keys, etc.			
Cure: Handling strength: 5 - 20min/RT, Final cure: 6 - 24hrs/RT.			
Test Methods: DIN			
Physical		**Mechanical**	
Colour	Green	τ (RT)	28.5 MPa
Viscosity	0.5 Pa.s	τ (compressive)	-
S.G.	1.05 - 1.18	σ_t	-
Shelf life	1 yr./RT	σ_c	-
$T_{Service}$	-55 to +200°C	Break Torque	32.0 N.m
CTE (°C)	-	Prevail Torque	35.0 N.m
Flash Point	100°C	Elongation	-
Gap Filling	0.15 mm max.	H (Shore)	-
Notes: Shear: DIN 54452. Density: DIN 51757. Flash point: DIN 55213. Breakdown potential: 8.5 kV.mm^{-1}.			

Panacol Elosol	Penloc SEET		Jun 94
Catalytic (metal) curing, one part anaerobic.			
Manufacturers Comments: RT cure. High viscosity, high strength. Good temperature and chemical resistance. Passive surfaces (e.g. glass or plastics) require an activating primer.			
Typical Use: Threadlocking and small component fixing.			
Cure: Handling strength: 5 - 20min/RT, Final cure: 6 - 24hrs/RT.			
Test Methods: DIN			
Physical		**Mechanical**	
Colour	Red	τ (RT)	25.5 MPa
Viscosity	15 Pa.s	τ (compressive)	-
S.G.	1.05 - 1.18	σ_t	-
Shelf life	1 yr./RT	σ_c	-
$T_{Service}$	-55 to +150°C	Break Torque	27.0 N.m
CTE (°C)	-	Prevail Torque	29.5 N.m
Flash Point	100°C	Elongation	-
Gap Filling	0.3 mm max.	H (Shore)	-
Notes: Shear: DIN 54452. Density: DIN 51757. Flash point: DIN 55213. Breakdown potential: 8.5 kV.mm^{-1}.			

See Key to Data Sheets - Pages 63 to 65

Panacol Elosol	Penloc SET		Jun 94
Catalytic (metal) curing, one part anaerobic.			
Manufacturers Comments: RT cure. High viscosity, medium strength. Good temperature and chemical resistance. Passive surfaces (e.g. glass or plastics) require an activating primer.			
Typical Use: Component fixing. Shafts, pinions, keys, etc.			
Cure: Handling strength: 5 - 20min/RT, Final cure: 6 - 24hrs/RT.			
Test Methods: DIN			
Physical		**Mechanical**	
Colour	Blue	τ (RT)	22.5 MPa
Viscosity	1.5 Pa.s	τ (compressive)	-
S.G.	1.05 - 1.18	σ_t	-
Shelf life	1 yr./RT	σ_c	-
$T_{Service}$	-55 to +150°C	Break Torque	22.0 N.m
CTE (°C)	-	Prevail Torque	22.0 N.m
Flash Point	100°C	Elongation	-
Gap Filling	0.2 mm max.	H (Shore)	-
Notes: Shear: DIN 54452. Density: DIN 51757. Flash point: DIN 55213. Breakdown potential: 8.5 kV.mm^{-1}.			

Panacol Elosol	Penloc ULT		Jun 94
Catalytic (metal) curing, one part anaerobic.			
Manufacturers Comments: RT cure. Low viscosity, low strength. Good temperature and chemical resistance. Passive surfaces (e.g. glass or plastics) require an activating primer.			
Typical Use: Threadlocking and small component fixing.			
Cure: Handling strength: 5 - 20min/RT, Final cure: 6 - 24hrs/RT.			
Test Methods: DIN			
Physical		**Mechanical**	
Colour	Brown	τ (RT)	12.5 MPa
Viscosity	0.03 Pa.s	τ (compressive)	-
S.G.	1.05 - 1.18	σ_t	-
Shelf life	1 yr./RT	σ_c	-
$T_{Service}$	-55 to +150°C	Break Torque	11.5 N.m
CTE (°C)	-	Prevail Torque	11.5 N.m
Flash Point	100°C	Elongation	-
Gap Filling	0.05 mm max.	H (Shore)	-
Notes: Shear: DIN 54452. Density: DIN 51757. Flash point: DIN 55213. Breakdown potential: 8.5 kV.mm^{-1}.			

Panacol Elosol	Penloc ST		Jun 94
Catalytic (metal) curing, one part anaerobic.			
Manufacturers Comments: RT cure. Medium viscosity, medium strength. Good temperature and chemical resistance. Passive surfaces (e.g. glass or plastics) require an activating primer.			
Typical Use: Component fixing. Shafts, pinions, keys, etc.			
Cure: Handling strength: 5 - 20min/RT, Final cure: 6 - 24hrs/RT.			
Test Methods: DIN			
Physical		**Mechanical**	
Colour	Green	τ (RT)	18.5 MPa
Viscosity	0.4 Pa.s	τ (compressive)	-
S.G.	1.05 - 1.18	σ_t	-
Shelf life	1 yr./RT	σ_c	-
$T_{Service}$	-55 to +150°C	Break Torque	19.5 N.m
CTE (°C)	-	Prevail Torque	20.0 N.m
Flash Point	100°C	Elongation	-
Gap Filling	0.1 mm max.	H (Shore)	-
Notes: Shear: DIN 54452. Density: DIN 51757. Flash point: DIN 55213. Breakdown potential: 8.5 kV.mm^{-1}.			

Panacol Elosol	Penloc UT		Jun 94
Catalytic (metal) curing, one part anaerobic.			
Manufacturers Comments: RT cure. Medium viscosity, low strength. Good temperature and chemical resistance. Passive surfaces (e.g. glass or plastics) require an activating primer.			
Typical Use: Threadlocking and small component fixing.			
Cure: Handling strength: 5 - 20min/RT, Final cure: 6 - 24hrs/RT.			
Test Methods: DIN			
Physical		**Mechanical**	
Colour	Yellow	τ (RT)	6 MPa
Viscosity	0.125 Pa.s	τ (compressive)	-
S.G.	1.05 - 1.18	σ_t	-
Shelf life	1 yr./RT	σ_c	-
$T_{Service}$	-55 to +130°C	Break Torque	5.0 N.m
CTE (°C)	-	Prevail Torque	4.5 N.m
Flash Point	100°C	Elongation	-
Gap Filling	0.1 mm max.	H (Shore)	-
Notes: Shear: DIN 54452. Density: DIN 51757. Flash point: DIN 55213. Breakdown potential: 8.5 kV.mm^{-1}.			

Panacol Elosol	Penloc VF		Jun 94
Catalytic (metal) curing, one part anaerobic.			
Manufacturers Comments: RT cure. High viscosity, high strength. Good temperature and chemical resistance. Passive surfaces (e.g. glass or plastics) require an activating primer.			
Typical Use: Threadlocking and small component fixing.			
Cure: Handling strength: 5 - 20min/RT, Final cure: 6 - 24hrs/RT.			
Test Methods: DIN			
Physical		**Mechanical**	
Colour	Red	τ (RT)	28 MPa
Viscosity	2.5 Pa.s	τ (compressive)	-
S.G.	1.05 - 1.18	σ_t	-
Shelf life	1 yr./RT	σ_c	-
$T_{Service}$	-55 to +120°C	Break Torque	30.5 N.m
CTE (°C)	-	Prevail Torque	33.0 N.m
Flash Point	100°C	Elongation	-
Gap Filling	0.2 mm max.	H (Shore)	-
Notes: Shear: DIN 54452. Density: DIN 51757. Flash point: DIN 55213. Breakdown potential: 8.5 kV.mm^{-1}.			

Panacol Elosol	Penloc ZT		Jun 94
Catalytic (metal) curing, one part anaerobic.			
Manufacturers Comments: RT cure. Medium viscosity, low strength. Good temperature and chemical resistance. Passive surfaces (e.g. glass or plastics) require an activating primer.			
Typical Use: Component fixing. Shafts, pinions, keys, etc.			
Cure: Handling strength: 5 - 20min/RT, Final cure: 6 - 24hrs/RT.			
Test Methods: DIN			
Physical		**Mechanical**	
Colour	Yellow	τ (RT)	11.5 MPa
Viscosity	0.2 Pa.s	τ (compressive)	-
S.G.	1.05 - 1.18	σ_t	-
Shelf life	1 yr./RT	σ_c	-
$T_{Service}$	-55 to +150°C	Break Torque	12.5 N.m
CTE (°C)	-	Prevail Torque	12.0 N.m
Flash Point	100°C	Elongation	-
Gap Filling	0.1 mm max.	H (Shore)	-
Notes: Shear: DIN 54452. Density: DIN 51757. Flash point: DIN 55213. Breakdown potential: 8.5 kV.mm^{-1}.			

Panacol Elosol	Penloc VFF		Jun 94
Catalytic (metal) curing, one part anaerobic.			
Manufacturers Comments: RT cure. High viscosity, high strength. Good temperature and chemical resistance. Passive surfaces (e.g. glass or plastics) require an activating primer.			
Typical Use: Component fixing. Shafts, pinions, keys, etc.			
Cure: Handling strength: 5 - 20min/RT, Final cure: 6 - 24hrs/RT.			
Test Methods: DIN			
Physical		**Mechanical**	
Colour	Green	τ (RT)	30.5 MPa
Viscosity	3 Pa.s	τ (compressive)	-
S.G.	1.05 - 1.18	σ_t	-
Shelf life	1 yr./RT	σ_c	-
$T_{Service}$	-55 to +120°C	Break Torque	30.5 N.m
CTE (°C)	-	Prevail Torque	37.5 N.m
Flash Point	100°C	Elongation	-
Gap Filling	0.25 mm max.	H (Shore)	-
Notes: Shear: DIN 54452. Density: DIN 51757. Flash point: DIN 55213. Breakdown potential: 8.5 kV.mm^{-1}.			

Panacol Elosol	Penloc ZTT		Jun 94
Catalytic (metal) curing, one part anaerobic.			
Manufacturers Comments: RT cure. High viscosity, low strength. Good temperature and chemical resistance. Passive surfaces (e.g. glass or plastics) require an activating primer.			
Typical Use: Component fixing. Shafts, pinions, keys, etc.			
Cure: Handling strength: 5 - 20min/RT, Final cure: 6 - 24hrs/RT.			
Test Methods: DIN			
Physical		**Mechanical**	
Colour	Violet	τ (RT)	13 MPa
Viscosity	3 Pa.s	τ (compressive)	-
S.G.	1.05 - 1.18	σ_t	-
Shelf life	1 yr./RT	σ_c	-
$T_{Service}$	-55 to +150°C	Break Torque	13.5 N.m
CTE (°C)	-	Prevail Torque	13.0 N.m
Flash Point	100°C	Elongation	-
Gap Filling	0.25 mm max.	H (Shore)	-
Notes: Shear: DIN 54452. Density: DIN 51757. Flash point: DIN 55213. Breakdown potential: 8.5 kV.mm^{-1}.			

Permabond

Panacol Elosol	Vitralit 4280		Jun 94
UV curing anaerobic.			
Manufacturers Comments: Combination curing product. Solvent-free, 100% solids.			
Typical Use: Metals, plastics.			
Cure: UV-A, 60 mW.cm^2: 10 sec. Anaerobic cure: 10 - 15 mins. Final strength: 12 hrs.			
Test Methods: DIN/ASTM			
Physical		**Mechanical**	
Colour	-	τ (RT)	15.8 MPa
Viscosity	0.5 Pa.s	τ (compressive)	-
S.G.	1.25	σ_t	-
Shelf life	6 months/RT	σ_c	-
$T_{Service}$	-40 to +150°C	Break Torque	-
CTE	75.5 x 10^{-6} K^{-1}	Prevail Torque	-
Flash Point	94°C	Elongation	-
Gap Filling	-	H (Shore D)	83
Notes: Shear: ASTM D 1002. Density: DIN 51757. Flash point: DIN 51755. Shrinkage: 3.4%. Tensile Modulus (DIN 53457): 0.845 GPa. Moisture Absorption (ASTM 570): 0.6%. Thermal Conductivity (DIN 52612): 0.16 $Wm^{-1}K^{-1}$. Dielectric Constant (DIN 53483): 4.5. Breakdown potential (DIN 53481): 22.3 $kV.mm^{-1}$.			

Permabond	A134		Mar 94
Threadlocking, fixing and sealing anaerobic			
Manufacturers Comments: High strength.			
Typical Use: Locking and sealing of metal parts where larger gaps exist, i.e. worn or damaged components.			
Cure: Initial cure: 1 hr/RT, Final cure: 24hr/RT.			
Test Methods: Not stated.			
Physical		**Mechanical**	
Colour	Green	τ (RT)	21 MPa
Viscosity	70 Pa.s	τ (compressive)	-
S.G.	1.0	σ_t	-
Shelf life	1 yr./RT	σ_c	-
$T_{Service}$	-55°C to +150°C	Break Torque	25 N.m
CTE (°C)	-	Prevail Torque	-
Flash Point	-	Elongation	-
Gap Filling	0.5 mm.	H (Shore)	-
Notes:			

Panacol Elosol	Vitralit 4285		Jun 94
UV curing anaerobic.			
Manufacturers Comments: Combination curing product. Solvent-free, 100% solids.			
Typical Use: Metals, plastics.			
Cure: UV-A, 60 mW.cm^2: 10 sec. Anaerobic cure: 10 - 15 mins. Final strength: 12 hrs.			
Test Methods: DIN/ASTM			
Physical		**Mechanical**	
Colour	-	τ (RT)	10.2 MPa
Viscosity	4 Pa.s	τ (compressive)	-
S.G.	1.34	σ_t	-
Shelf life	6 months/RT	σ_c	-
$T_{Service}$	-40 to +150°C	Break Torque	-
CTE	65.5 x 10^{-6} K^{-1}	Prevail Torque	-
Flash Point	96°C	Elongation	-
Gap Filling	-	H (Shore D)	85
Notes: Shear: ASTM D 1002. Density: DIN 51757. Flash point: DIN 51755. Shrinkage: 2.52%. Tensile Modulus (DIN 53457): 0.975 GPa. Moisture Absorption (ASTM 570): 0.6%. Thermal Conductivity (DIN 52612): 0.19 $Wm^{-1}K^{-1}$. Dielectric Constant (DIN 53483): 3.7. Breakdown potential (DIN 53481): 27.5 $kV.mm^{-1}$.			

Permabond	F202		Mar 94
Toughened (rubber modified), thixotropic anaerobic			
Manufacturers Comments: Threadlocking, fixing and sealing, high strength. High resistance to impact and peel stresses.			
Typical Use: Permanent locking and sealing of metal parts where larger gaps exist, i.e. worn or damaged components. Useful for dissimilar substrates where thermal stresses may occur.			
Cure: Initial cure: 2 hr/RT, Final cure: 24hr/RT.			
Test Methods: Not stated.			
Physical		**Mechanical**	
Colour	Brown	τ (RT)	30 MPa
Viscosity	135 Pa.s	τ (compressive)	-
S.G.	1.0	σ_t	-
Shelf life	1 yr./RT	σ_c	-
$T_{Service}$	-55°C to +100°C	Break Torque	18 N.m
CTE (°C)	-	Prevail Torque	-
Flash Point	-	Elongation	-
Gap Filling	0.25 mm.	H (Shore)	-
Notes:			

Permabond	Pipe Sealant A131		Mar 94
Threadlocking and sealing, thixotropic, anaerobic			
Manufacturers Comments: Moderate/low strength.			
Typical Use: Locking and sealing of threaded pipe connections.			
Cure: Initial cure: 2 hr/RT, Final cure: 24hr/RT.			
Test Methods: Not stated.			
Physical		**Mechanical**	
Colour	White	τ (RT)	8 MPa
Viscosity	40 Pa.s	τ (compressive)	-
S.G.	1.0	σ_t	-
Shelf life	1 yr./RT	σ_c	-
$T_{Service}$	-55°C to +150°C	Break Torque	6 N.m
CTE (°C)	-	Prevail Torque	-
Flash Point	-	Elongation	-
Gap Filling	0.5 mm.	H (Shore)	-
Notes:			

Permabond	Threadlock A130		Mar 94
Threadlocking anaerobic			
Manufacturers Comments: Moderate strength.			
Typical Use: Locking of threads and components which may require disassembly.			
Cure: Initial cure: 1 hr/RT, Final cure: 24hr/RT.			
Test Methods: Not stated.			
Physical		**Mechanical**	
Colour	Blue	τ (RT)	12 MPa
Viscosity	0.6 Pa.s	τ (compressive)	-
S.G.	1.0	σ_t	-
Shelf life	1 yr./RT	σ_c	-
$T_{Service}$	-55°C to +150°C	Break Torque	10 N.m
CTE (°C)	-	Prevail Torque	-
Flash Point	-	Elongation	-
Gap Filling	0.12 mm.	H (Shore)	-
Notes:			

Permabond	Retainer A118		Mar 94
Threadlocking and bonding anaerobic			
Manufacturers Comments: High strength.			
Typical Use: Permanent locking and sealing. Retaining bearings, permanent threadlocking, fixing pulleys, pinions and studs.			
Cure: Initial cure: 1 hr/RT, Final cure: 24hr/RT.			
Test Methods: Not stated.			
Physical		**Mechanical**	
Colour	Green	τ (RT)	21 MPa
Viscosity	0.5 Pa.s	τ (compressive)	-
S.G.	1.0	σ_t	-
Shelf life	1 yr./RT	σ_c	-
$T_{Service}$	-55°C to +150°C	Break Torque	25 N.m
CTE (°C)	-	Prevail Torque	-
Flash Point	-	Elongation	-
Gap Filling	0.12 mm.	H (Shore)	-
Notes:			

ThreeBond

ThreeBond	1303		(Aug 95)
Anaerobic			
Manufacturers Comments: High strength and quick-setting at RT. Good sealing. Very low shrinkage with good resistance to oils, chemicals, weathering and vibration.			
Typical Use: Thread locking. Permanent fixing and sealing of stud bolts and screws.			
Cure: Initial cure: 1 - 2hr/RT, Full cure: 24hr/RT.			
Test Methods: Not stated.			
Physical		**Mechanical**	
Colour	Green	τ (RT)	-
Viscosity	0.15 Pa.s	τ (compressive)	-
S.G.	-	σ_t	-
Shelf life	-	σ_c	-
$T_{Service}$	-80°C to +150°C	Break Torque	25 - 35 N.m
CTE (°C)	-	Prevail Torque	25 - 35 N.m
Flash Point	-	Elongation	-
Gap Filling	0.15mm max.	H (Shore)	-
Notes: Optimum gap 0.005 - 0.01 mm. Fit strength (6mm ∅ x 15mm): 25 - 35 MPa, (12.7mm ∅ x 11mm): 20 - 25 MPa.			

ThreeBond	1305		(Aug 95)
Anaerobic			
Manufacturers Comments: High strength and quick-setting at RT. Good sealing. Very low shrinkage with good resistance to oils, chemicals, weathering and vibration.			
Typical Use: Thread locking. Permanent fixing and sealing of stud bolts and screws.			
Cure: Initial cure: 1 - 2hr/RT, Full cure: 24hr/RT.			
Test Methods: Not stated.			
Physical		**Mechanical**	
Colour	Green	τ (RT)	-
Viscosity	0.6 Pa.s	τ (compressive)	-
S.G.	-	σ_t	-
Shelf life	-	σ_c	-
$T_{Service}$	-80°C to +150°C	Break Torque	25 - 35 N.m
CTE (°C)	-	Prevail Torque	25 - 35 N.m
Flash Point	-	Elongation	-
Gap Filling	0.3mm max.	H (Shore)	-
Notes: Optimum gap 0.01 - 0.02 mm. Fit strength (6mm ∅ x 15mm): 25 - 35 MPa, (12.7mm ∅ x 11mm): 20 - 25 MPa.			

ThreeBond	1324		(Aug 95)
Anaerobic			
Manufacturers Comments: Medium strength and quick-setting at RT. Good sealing. Very low shrinkage with good resistance to oils, chemicals, weathering and vibration.			
Typical Use: Thread locking. Medium strength fixing and sealing of removable bolts and nuts.			
Cure: Initial cure: 1 - 2hr/RT, Full cure: 24hr/RT.			
Test Methods: Not stated.			
Physical		**Mechanical**	
Colour	Red	τ (RT)	-
Viscosity	0.6 Pa.s	τ (compressive)	-
S.G.	-	σ_t	-
Shelf life	-	σ_c	-
$T_{Service}$	-80°C to +150°C	Break Torque	10 - 20 N.m
CTE (°C)	-	Prevail Torque	25 - 35 N.m
Flash Point	-	Elongation	-
Gap Filling	0.3mm max.	H (Shore)	-
Notes: Optimum gap 0.01 - 0.02 mm. Fit strength (6mm ∅ x 15mm): 20 - 28 MPa, (12.7mm ∅ x 11mm): 16 - 21 MPa.			

ThreeBond	1322		(Aug 95)
Anaerobic			
Manufacturers Comments: Medium strength and quick-setting at RT. Good sealing. Very low shrinkage with good resistance to oils, chemicals, weathering and vibration.			
Typical Use: Thread locking. Medium strength fixing and sealing of removable bolts and nuts.			
Cure: Initial cure: 1 - 2hr/RT, Full cure: 24hr/RT.			
Test Methods: Not stated.			
Physical		**Mechanical**	
Colour	Red	τ (RT)	-
Viscosity	0.15 Pa.s	τ (compressive)	-
S.G.	-	σ_t	-
Shelf life	-	σ_c	-
$T_{Service}$	-80°C to +150°C	Break Torque	10 - 20 N.m
CTE (°C)	-	Prevail Torque	25 - 35 N.m
Flash Point	-	Elongation	-
Gap Filling	0.15mm max.	H (Shore)	-
Notes: Optimum gap 0.005 - 0.01 mm. Fit strength (6mm ∅ x 15mm): 16 - 21 MPa, (12.7mm ∅ x 11mm): 14 - 19 MPa.			

ThreeBond	1327		(Aug 95)
Anaerobic			
Manufacturers Comments: Medium strength and quick-setting at RT. Good sealing. Very low shrinkage with good resistance to oils, chemicals, weathering and vibration.			
Typical Use: Thread locking. Medium strength fixing and sealing of removable bolts and nuts.			
Cure: Initial cure: 1 - 2hr/RT, Full cure: 24hr/RT.			
Test Methods: Not stated.			
Physical		**Mechanical**	
Colour	Red	τ (RT)	-
Viscosity	2.5 Pa.s	τ (compressive)	-
S.G.	-	σ_t	-
Shelf life	-	σ_c	-
$T_{Service}$	-80°C to +150°C	Break Torque	20 - 30 N.m
CTE (°C)	-	Prevail Torque	20 - 30 N.m
Flash Point	-	Elongation	-
Gap Filling	0.3mm max.	H (Shore)	-
Notes: Optimum gap 0.02 - 0.05 mm. Fit strength (6mm ∅ x 15mm): 22 - 30 MPa, (12.7mm ∅ x 11mm): 18 - 23 MPa.			

ThreeBond	1342		(Aug 95)
Anaerobic			
Manufacturers Comments: Low strength and quick-setting at RT. Good sealing. Very low shrinkage with good resistance to oils, chemicals, weathering and vibration.			
Typical Use: Thread locking. Fixing and sealing of screws requiring frequent removal.			
Cure: Initial cure: 1 - 2hr/RT, Full cure: 24hr/RT.			
Test Methods: Not stated.			
Physical		**Mechanical**	
Colour	Blue	τ (RT)	-
Viscosity	0.15 Pa.s	τ (compressive)	-
S.G.	-	σ_t	-
Shelf life	-	σ_c	-
$T_{Service}$	-80°C to +150°C	Break Torque	5 - 10 N.m
CTE (°C)	-	Prevail Torque	5 - 7 N.m
Flash Point	-	Elongation	-
Gap Filling	0.15mm max.	H (Shore)	-
Notes: Optimum gap 0.005 - 0.01 mm. Fit strength (6mm Ø x 15mm): 10 - 16 MPa, (12.7mm Ø x 11mm): 2 - 6 MPa.			

ThreeBond	1360		(Aug 95)
Anaerobic			
Manufacturers Comments: Low strength, thermo-resistive and quick-setting at RT. Good sealing. Very low shrinkage with good resistance to oils, chemicals, weathering and vibration.			
Typical Use: Thread locking. Fixing of screws requiring thermal resistance.			
Cure: Initial cure: 6 - 12hr/RT, Full cure: 48hr/RT.			
Test Methods: Not stated.			
Physical		**Mechanical**	
Colour	Red	τ (RT)	-
Viscosity	0.8 Pa.s	τ (compressive)	-
S.G.	-	σ_t	-
Shelf life	-	σ_c	-
$T_{Service}$	-80°C to +200°C	Break Torque	20 - 25 N.m
CTE (°C)	-	Prevail Torque	20 - 25 N.m
Flash Point	-	Elongation	-
Gap Filling	0.15mm max..	H (Shore)	-
Notes: Optimum gap 0.01 - 0.02 mm. Fit strength (6mm Ø x 15mm): 20 - 28 MPa, (12.7mm Ø x 11mm): 16 - 21 MPa.			

ThreeBond	1344		(Aug 95)
Anaerobic			
Manufacturers Comments: Low strength and quick-setting at RT. Good sealing. Very low shrinkage with good resistance to oils, chemicals, weathering and vibration.			
Typical Use: Thread locking. Fixing and sealing of screws requiring frequent removal.			
Cure: Initial cure: 1 - 2hr/RT, Full cure: 24hr/RT.			
Test Methods: Not stated.			
Physical		**Mechanical**	
Colour	Blue	τ (RT)	-
Viscosity	0.6 Pa.s	τ (compressive)	-
S.G.	-	σ_t	-
Shelf life	-	σ_c	-
$T_{Service}$	-80°C to +150°C	Break Torque	5 - 10 N.m
CTE (°C)	-	Prevail Torque	5 - 7 N.m
Flash Point	-	Elongation	-
Gap Filling	0.3mm max.	H (Shore)	-
Notes: Optimum gap 0.01 - 0.02 mm. Fit strength (6mm Ø x 15mm): 10 - 16 MPa, (12.7mm Ø x 11mm): 5 - 9 MPa.			

ThreeBond	1361B		(Aug 95)
Anaerobic			
Manufacturers Comments: Post impregnating, low viscosity. Quick-setting at RT. Good sealing. Very low shrinkage with good resistance to oils, chemicals, weathering and vibration.			
Typical Use: Thread locking. Fixing of screws and sealing of pin-holes by post-impregnation.			
Cure: Initial cure: 0.5 - 1hr/RT, Full cure: 24hr/RT.			
Test Methods: Not stated.			
Physical		**Mechanical**	
Colour	Green	τ (RT)	-
Viscosity	0.015 Pa.s	τ (compressive)	-
S.G.	-	σ_t	-
Shelf life	-	σ_c	-
$T_{Service}$	-80°C to +150°C	Break Torque	7 - 12 N.m
CTE (°C)	-	Prevail Torque	25 - 35 N.m
Flash Point	-	Elongation	-
Gap Filling	0.1mm max.	H (Shore)	-
Notes: Optimum gap 0.005 - 0.01 mm. Fit strength (6mm Ø x 15mm): 15 - 20 MPa, (12.7mm Ø x 11mm): 7 - 14 MPa.			

ThreeBond	1373B		(Aug 95)
Anaerobic			
Manufacturers Comments: Quick-setting at RT. Good sealing. Very low shrinkage with good resistance to oils, chemicals, weathering and vibration.			
Typical Use: 'Fit' locking. Sealing and fixing of shafts and components.			
Cure: Initial cure: 0.5 - 1hr/RT, Full cure: 6hr/RT.			
Test Methods: Not stated.			
Physical		**Mechanical**	
Colour	Green	τ (RT)	-
Viscosity	0.125 Pa.s	τ (compressive)	-
S.G.	-	σ_t	-
Shelf life	-	σ_c	-
$T_{Service}$	-80°C to +150°C	Break Torque	30 - 40 N.m
CTE (°C)	-	Prevail Torque	30 - 40 N.m
Flash Point	-	Elongation	-
Gap Filling	0.15mm max.	H (Shore)	-
Notes: Optimum gap 0.005 - 0.01 mm. Fit strength (6mm Ø x 15mm): 30 - 40 MPa, (12.7mm Ø x 11mm): 25 - 30 MPa.			

ThreeBond	1375B		(Aug 95)
Anaerobic			
Manufacturers Comments: Quick-setting at RT. Good sealing. Very low shrinkage with good resistance to oils, chemicals, weathering and vibration.			
Typical Use: 'Fit' locking. Sealing and fixing of shafts and components requiring thermal resistance.			
Cure: Initial cure: 1 - 2hr/RT, Full cure: 12hr/RT.			
Test Methods: Not stated.			
Physical		**Mechanical**	
Colour	Green	τ (RT)	-
Viscosity	0.85 Pa.s	τ (compressive)	-
S.G.	-	σ_t	-
Shelf life	-	σ_c	-
$T_{Service}$	-80°C to +175°C	Break Torque	30 - 40 N.m
CTE (°C)	-	Prevail Torque	30 - 40 N.m
Flash Point	-	Elongation	-
Gap Filling	0.15mm max.	H (Shore)	-
Notes: Optimum gap 0.005 - 0.03 mm. Fit strength (6mm Ø x 15mm): 30 - 40 MPa, (12.7mm Ø x 11mm): 23 - 28 MPa.			

ThreeBond	1374		(Aug 95)
Anaerobic			
Manufacturers Comments: Lubricating and quick-setting at RT. Good sealing. Very low shrinkage with good resistance to oils, chemicals, weathering and vibration.			
Typical Use: Thread locking. Fixing of high tensile bolts with stabilised axial force.			
Cure: Initial cure: 1 - 2hr/RT, Full cure: 24hr/RT.			
Test Methods: Not stated.			
Physical		**Mechanical**	
Colour	Red	τ (RT)	-
Viscosity	0.75 Pa.s	τ (compressive)	-
S.G.	-	σ_t	-
Shelf life	-	σ_c	-
$T_{Service}$	-80°C to +150°C	Break Torque	10 - 20 N.m
CTE (°C)	-	Prevail Torque	15 - 20 N.m
Flash Point	-	Elongation	-
Gap Filling	0.3mm max.	H (Shore)	-
Notes: Optimum gap 0.01 - 0.02 mm. Fit strength (6mm Ø x 15mm): 19 - 27 MPa, (12.7mm Ø x 11mm): 15 - 20 MPa.			

ThreeBond	1377B		(Aug 95)
Anaerobic			
Manufacturers Comments: Quick-setting at RT. Good sealing. Very low shrinkage with good resistance to oils, chemicals, weathering and vibration.			
Typical Use: 'Fit' locking. Sealing and fixing of medium to large components.			
Cure: Initial cure: 1 - 2hr/RT, Full cure: 12hr/RT.			
Test Methods: Not stated.			
Physical		**Mechanical**	
Colour	Green	τ (RT)	-
Viscosity	2 Pa.s	τ (compressive)	-
S.G.	-	σ_t	-
Shelf life	-	σ_c	-
$T_{Service}$	-80°C to +150°C	Break Torque	30 - 40 N.m
CTE (°C)	-	Prevail Torque	30 - 40 N.m
Flash Point	-	Elongation	-
Gap Filling	0.3mm max.	H (Shore)	-
Notes: Optimum gap 0.01 - 0.05 mm. Fit strength (6mm Ø x 15mm): 30 - 40 MPa, (12.7mm Ø x 11mm): 23 - 28 MPa.			

ThreeBond	1379B	(Aug 95)
Anaerobic		

Manufacturers Comments: Quick-setting at RT. Good sealing. Very low shrinkage with good resistance to oils, chemicals, weathering and vibration.

Typical Use: 'Fit' locking. Sealing and fixing of large components.

Cure: Initial cure: 1 - 2hr/RT, Full cure: 12hr/RT.

Test Methods: Not stated.

Physical		Mechanical	
Colour	Green	τ (RT)	-
Viscosity	7.5 Pa.s	τ (compressive)	-
S.G.	-	σ_t	-
Shelf life	-	σ_c	-
$T_{Service}$	-80°C to +150°C	Break Torque	30 - 40 N.m
CTE (°C)	-	Prevail Torque	30 - 40 N.m
Flash Point	-	Elongation	-
Gap Filling	0.3mm max.	H (Shore)	-

Notes: Optimum gap 0.01 - 0.05 mm. Fit strength (6mm Ø x 15mm): 30 - 40 MPa, (12.7mm Ø x 11mm): 23 - 28 MPa.

ThreeBond	3062B	(Aug 95)
UV curing Anaerobic		

Manufacturers Comments: Initial (surface) cure by UV irradiation - sealed resin then undergoes anaerobic cure. Good moisture and low temperature resistance. Excellent heat and impact resistance.

Typical Use: Metal and plastic bonding.

Cure: 4kW mercury lamp (3.5 J.cm^{-2}): 30 sec, then anaerobic cure.

Test Methods:

Physical		Mechanical	
Colour	Clear green	τ (RT)	-
Viscosity	4 Pa.s	τ (compressive)	-
S.G.	1.10	σ_t	21.6 MPa
Shelf life	-	σ_c	-
$T_{Service}$	-	Break Torque	-
CTE (°C)	-	Prevail Torque	-
Flash Point	-	Elongation	200%
Gap Filling	-	H (Shore)	95-A, 70-D

Notes: Tensile modulus (Young's modulus): 98 MPa. Surface resistivity: 4.0 x 10^{14} ohm.cm. Replaces grades: 3013C, 3060C and 3060D. Recommends 3095B primer.

ThreeBond	3062	(Aug 95)
UV curing Anaerobic		

Manufacturers Comments: Initial (surface) cure by UV irradiation - sealed resin then undergoes anaerobic cure. Flexible with good moisture and low temperature resistance.

Typical Use: Metal and plastic bonding.

Cure: 4kW mercury lamp (3.5 J.cm^{-2}): 30 sec, then anaerobic cure.

Test Methods:

Physical		Mechanical	
Colour	Clear yellow	τ (RT)	-
Viscosity	8 Pa.s	τ (compressive)	-
S.G.	1.05	σ_t	24.5 MPa
Shelf life	-	σ_c	-
$T_{Service}$	-	Break Torque	-
CTE (°C)	-	Prevail Torque	-
Flash Point	-	Elongation	200%
Gap Filling	-	H (Shore)	95-A, 70-D

Notes: Tensile modulus (Young's modulus): 156 MPa. Surface resistivity: 4.2 x 10^{14} ohm.cm. Dielectric constant (1 MHz): 4.4. Replaces grades: 3014, 3031D, 3037, 3060B and 3060E. Recommends 3095B primer.

ThreeBond	3062D	(Aug 95)
UV curing Anaerobic		

Manufacturers Comments: Low viscosity. Initial (surface) cure by UV irradiation - sealed resin then undergoes anaerobic cure. Good moisture and low temperature resistance.

Typical Use: Metal and plastic bonding.

Cure: 4kW mercury lamp (3.5 J.cm^{-2}): 30 sec, then anaerobic cure.

Test Methods:

Physical		Mechanical	
Colour	Clear blue	τ (RT)	-
Viscosity	0.15 Pa.s	τ (compressive)	-
S.G.	1.10	σ_t	19.6 MPa
Shelf life	-	σ_c	-
$T_{Service}$	-	Break Torque	-
CTE (°C)	-	Prevail Torque	-
Flash Point	-	Elongation	10%
Gap Filling	-	H (Shore)	95-A, 80-D

Notes: Tensile modulus (Young's modulus): 735 MPa. Surface resistivity: 2.6 x 10^{14} ohm.cm. Dielectric constant (1 MHz): 4.0. Replaces grades: 3066, 3067 and 3067B. Recommends 3095B primer.

CYANOACRYLATE ADHESIVES

NOTES

The majority of these adhesives are based on either methyl- or ethyl-cyanoacrylate. However there are a number based on other alkyl or ester groups. All are room temperature curing in the presence of atmospheric moisture. Bonding is 'instant', i.e. several seconds before parts can be handled, with full cure several hours later. The cure speed depends mainly on the adherend materials. Activators (primers) can be used for conditions where moisture-curing alone is of limited success, e.g. large, non-porous bond areas or difficult to bond materials. Most are colourless, although a few filled versions exist (black or white). Both liquid (very low viscosity) and gel versions are available. Gap-filling ability is generally low, unless specifically stated otherwise. Few mechanical properties are given compared with other structural adhesive classes.

Main uses

Numerous industrial and domestic applications requiring 'Instant bonding' for metals, plastics, rubbers, wood, glass and ceramics. Widely used for the precision assembly of small mechanical and electrical components, i.e. small bond areas. Special grades are available for bonding difficult materials (porous, polyethylene, some elastomers, etc.). Gel products with improved gap filling properties are also available.

COMPARISON TABLE - CYANOACRYLATE ADHESIVES

Company	Adhesive	Cyanoacrylate Type	LSS	TS	SG	Viscosity (Pa.s)	Gap Fill (mm)	Refractive Index	Flash Point (°C)	Service Temp. (°C)	Cure Time (sec)	Pg. No.
3M	Pronto CA-4	Ethyl	○	-	-	0.12	×	-	-	-	4 - 40	122
	Pronto CA-7	Methyl	◐	-	-	0.04	×	-	-	-	1 - 30	122
	Pronto CA-8	Ethyl	◐	-	-	0.12	×	-	-	-	5 - 40	122
	Pronto CA-40	Ethyl	○	-	-	0.01	×	-	-	-	1 - 30	122
	Pronto CA-50	Ethyl	○	-	-	65	✓	-	-	-	60 - 120	123
	Pronto CA-100	Ethyl	●	-	-	4.5	✓	-	-	-	20 - 70	123
Bostik	7431	Ethyl	◐	-	1.1	0.002	×	-	82	-20/+80	5 - 10	123
	7432	Ethyl	◐	-	1.1	0.1	×	-	82	-20/+80	5 - 10	123
	7433	Ethyl	◐	-	1.1	0.3	✓	-	82	-20/+80	5 - 10	123
	7434	Ethyl	◐	-	1.1	1.5	✓	-	82	-20/+80	5 - 10	123
	7435	Ethyl	◐	-	1.1	5	✓	-	82	-20/+80	5 - 10	123
W. Canning	Superglue - Gel	Ethyl	-	-	1.05	-	0.2	-	-	-	-	123
(Hylomar)	Superglue - High Viscosity	Ethyl	-	-	1.05	1.5	0.1	-	-	-	-	124
	Superglue - Low Viscosity	Ethyl	-	-	1.05	0.003	0.05	-	-	-	-	124
	Superglue - Medium Viscosity	Ethyl	-	-	1.05	0.1	0.1	-	-	-	-	124
CECA	CN4	-	-	-	-	M	-	-	-	-	-	124
	Topfix ACE50	-	-	-	-	M	-	-	-	-	-	124
	Topfix ACE88	-	-	-	-	L	-	-	-	-	-	124
	W100	-	-	-	-	H	-	-	-	-	-	124
	W2	-	-	-	-	H	-	-	-	-	-	124
	Z72X	-	-	-	-	L	-	-	-	-	-	124
Degussa	Agomet C 550	-	○	-	1.1	0.003	-	-	-	-30/+90	3 - 120	125
	Agomet C 890	-	●	-	1.1	0.003	×	-	-	-30/+90	10 - 180	125
Delo	CA 2106 (CA 16)	Methyl Ester	●	-	1.09	0.035	0.03	-	84	-30/+100	10 - 20	125
	CA 2153 (CA 140)	Methyl Ester	●	-	1.11	0.2	0.1	-	87	-30/+100	10 - 20	125
	CA 2902 (CA 11)	Ethyl Ester	●	-	1.05	0.004	0.02	-	84	-30/+100	10 - 20	126
	CA 2905 (CA 15)	Ethyl Ester	●	-	1.06	0.025	0.03	-	84	-30/+100	10 - 20	126
	CA 2994 (CA 1400)	Ethyl Ester	●	-	1.12	20	0.3	-	84	-30/+100	20 - 40	126
Dunlop	Maxibond SAS 200	Ethyl	◐	-	1.24	~0.005	×	1.49	80	-40/+80	5 - 30	126
	Maxibond SAS 205	Ethyl	◐	-	1.24	0.1	×	1.49	80	-40/+80	10 - 30	127
	Maxibond SAS 210	Ethyl	◐	-	-	0.5	-	-	83	-60/+80	20 - 50	127
	Maxibond SAS 211	Ethyl	◐	-	-	0.5	-	-	83	-60/+80	20 - 50	127
	Maxibond SAS 220	Ethyl	◐	-	-	1.8	-	1.49	83	-60/+80	30 - 60	127
	Maxibond SAS 231	Ethyl	-	-	-	~0.001	-	1.49	80	-60/+80	15 - 40	128
	Maxibond SAS 232	Modified Ethyl	-	-	1.0	2.75	-	1.49	83	-60/+130	10 - 50	128
Forbo	Helmitin 121	-	-	-	1.05	0.04	-	-	-	-80/+80	4 - 6	128
Helmitin	Helmitin 122	-	-	-	1.05	0.12	-	-	-	-80/+80	4 - 6	128
LePage's	Accu-flo Super Glue Auto. Pen #373	-	◐	-	1.05	0.12	×	-	-	<70	10 - 300	129
	Super Glue #377	-	●	-	1.1	0.12	×	-	-	<80	10 - 300	129
Loctite	Jet Black Structural Adhesive 380	-	-	-	-	-	0.25	-	-	<105	60	129
	Prism 401	Ethyl	●	-	-	0.11	0.1	-	-	-50/+80	2 - 10	129
	Prism 406	Ethyl	●	-	-	0.002	0.1	-	-	-50/+80	1 - 5	130
	Prism 454	Ethyl	●	-	-	0.002	0.2	-	-	-50/+80	5 - 30	130
	Prism 460	-	◐	-	-	0.04	0.1	-	-	-50/+80	20 - 50	130
	Prism 480	Modified Ethyl	●	-	-	0.3	0.1	-	-	-50/+100	-	130
Panacol	Cyanolit 101	Methyl	◐	◐	~1.1	0.002	-	1.487	83	<100	10 - 60	131
Elosol	Cyanolit 102	Methyl	◐	◐	~1.1	0.1	-	1.487	83	<100	20 - 80	131
	Cyanolit 103	Methyl	◐	◐	~1.1	1.5	0.2	1.487	83	<100	30 - 120	131
	Cyanolit 141	Methyl	◐	●	~1.1	0.04	-	1.487	83	<100	10 - 60	131
	Cyanolit 200	Ethyl	○	●	~1.1	0.002	-	1.487	83	<100	3 - 12	132

Continued...

See Key to Comparison Tables - Pages 63 to 65

Company	Adhesive	Cyanoacrylate Type	LSS	TS	SG	Viscosity (Pa.s)	Gap Fill (mm)	Refractive Index	Flash Point (°C)	Service Temp. (°C)	Cure Time (sec)	Pg. No.
Panacol	Cyanolit 201	Ethyl	○	●	~1.1	0.002	-	1.487	83	<100	7 - 20	132
Elosol	Cyanolit 202	Ethyl	○	●	~1.1	0.1	-	1.487	83	<100	13 - 40	132
(continued)	Cyanolit 203	Ethyl	○	●	~1.1	1.5	0.2	1.487	83	<100	30 - 75	132
	Cyanolit 208	Ethyl	○	●	~1.1	0.04	-	1.487	83	<100	3 - 12	133
	Cyanolit 241	Ethyl	○	●	~1.1	0.04	-	1.487	83	<100	8 - 22	133
	Cyanolit 302	Alkyl Ester	●	-	~1.1	0.01	-	1.487	83	<150	-	133
	Cyanolit 303	Alkyl Ester	●	-	~1.1	0.006	-	1.487	83	<150	-	133
	Cyanolit 401	-	-	-	~1.1	0.002	-	1.487	83	<150	-	134
	Cyanolit 403 T	-	◑	-	~1.1	1.0	0.2	1.487	83	-60/+100	40 - 120	134
	Cyanolit 403 TB	(filled)	●	-	~1.1	H	-	-	83	-60/+100	80 - 200	134
	Cyanolit 432 TW	(filled)	●	-	~1.1	M	-	-	83	-60/+100	60 - 180	134
	Cyanolit 503	Ethyl	-	○	~1.1	0.002	-	1.487	83	-	5 - 7	135
	Cyanolit 504	Ethyl	-	○	~1.1	0.01	-	1.487	83	-	12 - 19	135
	Cyanolit 601	-	○	-	~1.1	0.002	-	1.487	83	-	45 - 55	135
	Cyanolit 602	-	○	-	~1.1	0.1	-	1.487	83	-	70 - 80	135
	Cyanolit 631	-	○	-	~1.1	0.03	-	1.487	83	-	50 - 55	136
	Cyanolit 732 F	-	○	●	~1.1	0.3	-	1.487	83	-	10 - 40	136
	Cyanolit 802	-	○	◑	~1.1	0.12	-	1.487	83	-	18 - 30	136
	Cyanolit 811	-	○	◑	~1.1	0.012	-	1.487	83	-	10 - 18	136
	Cyanolit Cristal 212 F	Ethyl	○	○	1.05	0.1	0.1	1.480	83	-80/+80	2 - 15	137
	Cyanolit Cristal 221 F	Ethyl	◑	○	1.05	0.02	0.05	1.438	83	-80/+100	2 - 10	137
	Cyanolit Cristal 351	Alkyl Ester	◑	○	1.06	0.04	0.05	1.458	83	-80/+150	20 - 50	137
	Cyanolit Cristal 402	Methyl	◑	○	1.1	0.1	0.1	1.492	81	-80/+100	10 - 40	137
	Cyanolit Cristal 403 TB	Ethyl (filled)	●	◑	1.07	1	0.2	-	83	-80/+110	20 - 180	138
	Cyanolit Cristal 432 TW	Ethyl (filled)	●	◑	1.07	0.3	0.2	-	83	-80/+120	5 - 240	138
	Cyanolit Cristal 631 F	Isopropyl	○	○	1.05	0.3	0.05	1.434	83	-80/+100	5 - 20	138
	Cyanolit Cristal 811 F	Alkyl	◑	◑	1.07	0.01	0.05	1.434	83	-80/+80	15 - 60	138
	Cyanolit Cristal GEL	Ethyl	○	◑	1.07	10	1	1.480	83	-80/+80	5 - 120	139
	Cyanolit GEL	Ethyl	○	●	~1.1	-	3	1.487	83	-	5 - 120	139
Permabond	Plus C791	Ethyl	-	◑	1.05	0.04	0.075	-	-	-30/+80	1 - 7	139
	Plus C792	Ethyl	-	◑	1.05	0.075	0.125	-	-	-30/+85	1 - 7	139
	Plus C910	Methyl	-	◑	1.09	0.085	0.125	-	-	-30/+95	15 - 40	140
	Plus C2010	Ethyl	-	◑	1.05	-	0.375	-	-	-30/+85	3 - 15	140
Sichel	Sicomet 40	Ethyl Ester	●	●	1.01	0.01	-	1.49	87	-30/+110	1 - 25	140
(Henkel)	Sicomet 50	Ethyl	●	●	1.06	0.04	-	1.49	87	-30/+110	1 - 30	140
	Sicomet 63	Ethyl	●	●	1.06	0.08	-	1.49	87	-30/+110	2 - 35	141
	Sicomet 65	Ethyl	●	-	1.12	20	-	1.49	87	-30/+110	15 - 150	141
	Sicomet 77	Ethyl	●	●	1.08	1.1	-	1.49	87	-30/+110	5 - 50	141
	Sicomet 85	Methyl	●	●	1.09	0.035	-	1.49	87	-30/+110	2 - 60	141
	Sicomet 99	Methyl	●	●	1.11	2	0.2	1.49	87	-30/+110	3 - 80	142
	Sicomet 5132	Ethyl	●	-	1.08	-	-	1.49	87	-30/+110	15 - 95	142
	Sicomet 5139	Ethyl	●	-	1.09	2	-	-	87	-30/+110	15 - 95	142
	Sicomet 6021	Butyl	◑	-	1.01	0.2	-	-	87	-30/+110	3 - 60	142
	Sicomet 6085	Butyl	◑	-	1.01	0.02	-	-	87	-30/+110	3 - 60	143
	Sicomet 7000	Methyl	●	●	1.09	0.12	-	1.49	87	-30/+110	2 - 60	143
	Sicomet 8200	Ethyl Ester	●	●	1.06	0.04	-	1.49	87	-30/+110	2 - 75	143
	Sicomet 8300	Ethyl Ester	●	●	1.05	0.03	-	1.49	87	-30/+110	1 - 30	143
	Sicomet 8400	Ethyl	●	●	1.06	0.21	-	1.49	87	-30/+110	2 - 45	144
	Sicomet 9000	Alkoxyethyl	◑	-	1.06	0.015	-	-	87	-30/+110	7 - 50	144
	Sicomet 9010	Alkoxyethyl	◑	-	1.06	0.16	-	-	87	-30/+110	6 - 50	144
	Sicomet Thixotrope	Ethyl	●	●	1.08	-	-	1.49	87	-30/+110	15 - 95	144

Continued...

See Key to Comparison Tables - Pages 63 to 65

Company	Adhesive	Cyanoacrylate Type	LSS	TS	SG	Viscosity (Pa.s)	Gap Fill (mm)	Refractive Index	Flash Point (°C)	Service Temp. (°C)	Cure Time (sec)	Pg. No.
ThreeBond	TB1701	Methyl	○	-	1.1	0.005	×	-	-	-	<40	145
	TB1702/N	Methyl	○	-	1.1	0.025	×	-	-	-	<40	145
	TB1703	Methyl	○	-	1.1	0.1	×	-	-	-	<30	145
	TB1741	Ethyl	○	-	1.05	0.005	×	-	-	-	<15	145
	TB1742D	Ethyl	○	-	1.05	0.025	×	-	-	-	<20	146
	TB1743	Ethyl	○	-	1.05	0.1	×	-	-	-	<30	146
	TB1747	Ethyl	○	-	1.05	1.6	✓	-	-	-	<60	146
	TB1781	Ethyl	◕	-	1.05	0.003	×	-	-	-	<15	146
	TB1782	Ethyl	◕	-	1.05	0.08	×	-	-	-	<20	147
	TB1783	Ethyl	◕	-	1.05	0.9	✓	-	-	-	<30	147
	TB1785	Ethyl	○	-	1.05	0.85	×	-	-	-	<30	147
	TB1786	Ethyl	○	-	1.05	0.15	×	-	-	-	<30	147
	TP127H	Ethyl	●	-	1.05	12 - 42	✓	-	-	-	<20	148
UHU	Flash	-	-	-	1.07	0.07	×	-	82	-	<60	148
	Flash Gel	-	-	-	1.03	0.07	×	-	82	-	<180	148

3M

3M	Pronto CA-4		Liquid
(March 95)			
Manufacturers Comments: Ethyl cyanoacrylate. Multi-purpose, fast setting. Acid adherends may need surface activator. Meets: USA MIL-A-46050C, Type II, Class 2.			
Typical Use: Rubbers and plastics.			
Cure: RT 5-40 seconds (handling strength).			
Test Methods: Not Stated			
Physical		**Mechanical**	
Colour	-	τ (RT, Al)	10.3 MPa
Viscosity	<0.12 Pa.s	τ (°C)	
SG	-	τ (wet)	
Shelf life	-	Peel (T)	~0.35 N mm^{-1}
$T_{Service}$	-	σ_f	
CTE (°C)	-	E	
Flash Point	-	Elongation	
Gap Filling	No*	H (Shore)	
Notes: τ(nitrile rubber) 0.2MPa - substrate failure. T-peel, RT. *CA-5: Higher viscosity, slower set version for gap-filling. Meets: MIL-A-46050 Type II, Class 3.			

3M	Pronto CA-8		Liquid
(March 95)			
Manufacturers Comments: Ethyl cyanoacrylate. Fast setting, low viscosity. Acid adherends may need surface activator. Meets: USA MIL-A-46050C, Type II, Class 2.			
Typical Use: Metals, rubbers and plastics.			
Cure: RT 5 - 40 seconds (handling strength).			
Test Methods: Not Stated			
Physical		**Mechanical**	
Colour	-	τ (RT, Al)	14.7 MPa
Viscosity	<0.12 Pa.s	τ (°C)	
SG	-	τ (wet)	
Shelf life	-	Peel (T)	~0.71 N mm^{-1}
$T_{Service}$	-	σ_f	
CTE (°C)	-	E	
Flash Point	-	Elongation	
Gap Filling	No*	H (Shore)	
Notes: τ(nitrile rubber) 0.2MPa - substrate failure. * CA-9 high viscosity, slow-set version for gap-filling. Meets MIL-A-46050C Type II, Class 1.			

3M	Pronto CA-7		Liquid
(March 95)			
Manufacturers Comments: Methyl cyanoacrylate. Very fast setting, low viscosity. Acid adherends may need surface activator. Meets: USA MIL-A-46050C, Type I, Class 1.			
Typical Use: Metals, rubbers and plastics.			
Cure: RT 1 - 30 seconds (handling strength).			
Test Methods: Not Stated			
Physical		**Mechanical**	
Colour	-	τ (RT, Al)	16.5 MPa
Viscosity	<0.04 Pa.s	τ (°C)	
SG	-	τ (wet)	
Shelf life	-	Peel (T)	~0.71 N mm^{-1}
$T_{Service}$	-	σ_f	
CTE (°C)	-	E	
Flash Point	-	Elongation	
Gap Filling	No	H (Shore)	
Notes: τ(nitrile rubber) 0.2MPa - substrate failure.			

3M	Pronto CA-40		Liquid
(March 95)			
Manufacturers Comments: Ethyl cyanoacrylate. Fast setting, low viscosity. Acid adherends may need surface activator. Meets: USA MIL-A-46050C, Type II, Class 1.			
Typical Use: Most substrates, inc. flexible vinyl EPDM rubber.			
Cure: RT 1 - 30 seconds (handling strength).			
Test Methods: Not Stated			
Physical		**Mechanical**	
Colour	-	τ (RT, Al)	9.8 MPa
Viscosity	<0.01 Pa.s	τ (°C)	
SG	-	τ (wet)	
Shelf life	-	Peel (T)	~0.35 N mm^{-1}
$T_{Service}$	-	σ_f	
CTE (°C)	-	E	
Flash Point	-	Elongation	
Gap Filling	No*	H (Shore)	
Notes: τ(nitrile rubber) 0.2MPa - substrate failure. * CA-40H higher viscosity, slow set version for gap-filling. Meets MIL-A-46050C Type II, Class 3.			

3M — Pronto CA-50 — Gel

(March 95)

Manufacturers Comments: Ethyl cyanoacrylate. High viscosity, non-sag. Less sensitive to surface acidity.

Typical Use: Metals, plastics and rubbers.

Cure: RT 1 - 2 minutes (handling strength).

Test Methods: Not Stated

Physical		Mechanical	
Colour	-	τ (RT, Al)	6.3 MPa
Viscosity	45 - 85 Pa.s	τ (°C)	
SG	-	τ (wet)	
Shelf life	-	Peel (T)	~0.35 N mm^{-1}
$T_{Service}$	-	σ_f	
CTE (°C)	-	E	
Flash Point	-	Elongation	
Gap Filling	Yes	H (Shore)	

Notes: τ(nitrile rubber) 0.7MPa - substrate failure.

Bostik

Bostik — 7431 — Liquid

(July 92)

Manufacturers Comments: Ethyl cyanoacrylate. 100% solids, solvent-free. Numerous applications needing 'instant' bonding for wide number of materials (not polyethylene or Teflon). Refractive index - similar to glass. Good water and chemical resistance.

Typical Use: Precision work (clock, jewellery, electronics) - metals, plastics, rubbers, ceramic, glass, wood.

Cure: RT(20°C)/5-10 secs (Al); <5secs (glass); Full cure: 12-24 hrs. Parts often used after 30 mins.

Test Methods: Not Stated

Physical		Mechanical	
Colour	Translucent	τ (RT, Al)	16.7 MPa
Viscosity	~0.002 Pa.s	τ (°C)	-
SG	~1.1	τ (wet)	-
Shelf life	1 year/ 5-14°C	Peel ( )	-
$T_{Service}$	-20 to +80°C	σ_f	-
CTE (°C)	-	E	-
Flash Point	82°C	Elongation	-
Gap Filling	*	H (Shore)	-

Notes: Cure: DIN 53281/1. * Other Bostik products with similar uses:
7432 (0.1 Pa.s) - general purpose; 7433 (0.3 Pa.s) - vertical surfaces;
7434 (1.5 Pa.s) - general purpose, bead; 7435 (5 Pa.s) - high viscosity, bead.

3M — Pronto CA-100 — Liquid

(March 95)

Manufacturers Comments: Ethyl cyanoacrylate. High peel and impact resistance. Thermal shock resistance and improved temperature resistance. Acid substrates may require surface activator.
Meets: MIL-A-46050C Type II, Class 3.

Typical Use: Metals, plastics and rubbers.

Cure: RT 20 - 70 seconds (handling strength).

Test Methods: Not Stated

Physical		Mechanical	
Colour	-	τ (RT, Al)	20.0 MPa
Viscosity	<4.5 Pa.s	τ (°C)	
SG	-	τ (wet)	
Shelf life	-	Peel (T)	2.64 N mm^{-1}
$T_{Service}$	-	σ_f	
CTE (°C)	-	E	
Flash Point	-	Elongation	
Gap Filling	Yes	H (Shore)	

Notes: τ(nitrile rubber) 0.7MPa - substrate failure.

William Canning Ltd. (Hylomar)

W. Canning — Superglue - Gel — Gel

(Feb 95)

Manufacturers Comments: Ethyl cyanoacrylate. General purpose. High strength, rapid cure.

Typical Use: Metal, plastics, rubber, wood, leather, ceramics and cork.

Cure: RT.

Test Methods: Not Stated

Physical		Mechanical	
Colour	Transparent	τ (RT)	-
Viscosity	-	τ (°C)	-
SG	1.05	τ (wet)	-
Shelf life	-	Peel ( )	-
$T_{Service}$	-	σ_f	-
CTE (°C)	-	E	-
Flash Point	-	Elongation	-
Gap Filling	<0.2 mm.	H (Shore)	-

Notes: -

W. Canning — **Superglue - High Viscosity** — **Liquid**

(Feb 95)

Manufacturers Comments: Ethyl Cyanoacrylate. General purpose, high viscosity. High strength, rapid cure.

Typical Use: Metal, plastics, rubber, wood, leather, ceramics and cork.

Cure: RT.

Test Methods: Not Stated

Physical		**Mechanical**	
Colour	Transparent	τ (RT)	-
Viscosity	1.5 Pa.s	τ (°C)	-
SG	1.05	τ (wet)	-
Shelf life	-	Peel ( )	-
$T_{Service}$	-	σ_f	-
CTE (°C)	-	E	-
Flash Point	-	Elongation	-
Gap Filling	<0.1 mm.	H (Shore)	-

Notes: -

W. Canning — **Superglue - Medium Viscosity** — **Liquid**

(Feb 95)

Manufacturers Comments: Ethyl Cyanoacrylate. General purpose, medium viscosity. High strength, rapid cure.

Typical Use: Metal, plastics, rubber, wood, leather, ceramics & cork.

Cure: RT.

Test Methods: Not Stated

Physical		**Mechanical**	
Colour	Transparent	τ (RT)	-
Viscosity	0.1 Pa.s	τ (°C)	-
SG	1.05	τ (wet)	-
Shelf life	-	Peel ( )	-
$T_{Service}$	-	σ_f	-
CTE (°C)	-	E	-
Flash Point	-	Elongation	-
Gap Filling	<0.1 mm.	H (Shore)	-

Notes: -

W. Canning — **Superglue - Low Viscosity** — **Liquid**

(Feb 95)

Manufacturers Comments: Ethyl Cyanoacrylate. General purpose, low viscosity. High strength, rapid cure.

Typical Use: Metal, plastics, rubber, wood, leather, ceramics and cork.

Cure: RT.

Test Methods: Not Stated

Physical		**Mechanical**	
Colour	Transparent	τ (RT)	-
Viscosity	0.003 Pa.s	τ (°C)	-
SG	1.05	τ (wet)	-
Shelf life	-	Peel ( )	-
$T_{Service}$	-	σ_f	-
CTE (°C)	-	E	-
Flash Point	-	Elongation	-
Gap Filling	<0.05 mm.	H (Shore)	-

Notes: -

CECA

CECA — **Topfix ACE88** — **Liquid**

(July 95)

Manufacturers Comments: General purpose, low viscosity.

Typical Use: Plastics, rubbers and metals.

Cure: RT.

Test Methods: Not Stated

Physical		**Mechanical**	
Colour	-	τ (RT)	-
Viscosity	-	τ (°C)	-
SG	-	τ (wet)	-
Shelf life	-	Peel ( )	-
$T_{Service}$	-	σ_f	-
CTE (°C)	-	E	-
Flash Point	-	Elongation	-
Gap Filling	-	H (Shore)	-

Notes: * Other Topfix products: ACE E50 - medium viscosity, general purpose;
W100 & W2 - high viscosity, porous materials;
CN4 - medium viscosity, metals and alloys;
Z72X - low viscosity, rubbers and engineering plastics.

Degussa

Degussa	Agomet C 550	Liquid
	May 94	

Manufacturers Comments: Cyanoacrylate. Numerous applications needing 'instant' bonding for wide number of materials.

Typical Use: Metals, plastics, rubbers, ceramic, glass.

Cure: RT 1-2 mins (metals); 10 - 60 seconds (synthetic materials); 3-10 seconds (rubbers). Full strength: within 24 hrs.

Test Methods: DIN

Physical		Mechanical	
Colour	-	τ (RT, plastic)	4 - 6 MPa*
Viscosity	~0.003 Pa.s	τ (°C)	-
SG	~1.1	τ (wet)	-
Shelf life	6 months	Peel ( )	-
$T_{Service}$	-30 to +90°C	σ_f	-
CTE (°C)	-	E	-
Flash Point	-	Elongation	-
Gap Filling	-	H (Shore)	-

Notes: * substrate failure. τ PTFE, polyethylene, polyamide: 1.5-3.3 N/mm.

Degussa	Agomet C 890	Liquid
	May 94	

Manufacturers Comments: Cyanoacrylate. Numerous applications needing 'quick' bonding for wide number of materials. Slower curing, allowing time to reposition.

Typical Use: Metals, plastics, rubbers, ceramic, glass.

Cure: RT 1.5-2.5 mins (metals); 1 - 3 minutes (synthetic materials); 10-30 seconds (rubbers). Full strength: within 24 hrs.

Test Methods: DIN

Physical		Mechanical	
Colour	-	τ (RT, metal)	~26 MPa
Viscosity	~0.003 Pa.s	τ (°C)	-
SG	~1.1	τ (wet)	-
Shelf life	6 months	Peel ( )	-
$T_{Service}$	-30 to +90°C	σ_f	-
CTE (°C)	-	E	-
Flash Point	-	Elongation	-
Gap Filling	no	H (Shore)	-

Notes: τ (rubbers): ~10 N/mm.

Delo

Delo	CA 2106 (CA-16)	Liquid
	(July 95)	

Manufacturers Comments: Methyl ester cyanoacrylate. Solvent free. Fast cure, high strength (fully cured). Low viscosity.

Typical Use: Metals, plastics, rubbers and elastomers (synthetic rubbers).

Cure: RT(23°C/50%RH) 10-20 sec (firm to touch). Full strength: 20 hrs.

Test Methods: DIN

Physical		Mechanical	
Colour	-	τ (RT)	~24 MPa
Viscosity	0.035 Pa.s	τ (°C)	-
SG	~1.09	τ (wet)	-
Shelf life	6 months/RT	Peel ( )	-
$T_{Service}$	-30 to +100°C	σ_f	-
CTE (°C)	-	E	-
Flash Point	84°C	Elongation	-
Gap Filling	0.03 mm, max.	H (Shore)	-

Notes: <40%RH increases cure time. τ DIN 53283. Flash point DIN 53213.

Delo	CA 2153 (CA 140)	Liquid
	(July 95)	

Manufacturers Comments: Methyl ester cyanoacrylate. Solvent free. Fast cure, high strength (fully cured) also on uneven surfaces. Low viscosity.

Typical Use: Metals, plastics, rubbers and elastomers (synthetic rubbers).

Cure: RT(23°C/50%RH) 10-20 sec (firm to touch). Full strength: 20 hrs.

Test Methods: DIN

Physical		Mechanical	
Colour	-	τ (RT)	~22 MPa
Viscosity	0.2 Pa.s	τ (°C)	-
SG	~1.11	τ (wet)	-
Shelf life	6 months/RT	Peel ( )	-
$T_{Service}$	-30 to +100°C	σ_f	-
CTE (°C)	-	E	-
Flash Point	87°C	Elongation	-
Gap Filling	0.1 mm max.	H (Shore)	-

Notes: <40%RH increases cure time. τ DIN 53283. Flash point DIN 53213.

Delo	CA 2902 (CA 11)		Liquid
(July 95)			
Manufacturers Comments: Ethyl ester cyanoacrylate. Solvent free. Fast cure, high strength (fully cured). Low viscosity.			
Typical Use: Metals, plastics, rubbers and elastomers (synthetic rubbers).			
Cure: RT(23°C/50%RH) 10-20 sec (firm to touch). Full strength: 20 hrs.			
Test Methods: DIN			
Physical		**Mechanical**	
Colour	-	τ (RT)	~24 MPa
Viscosity	0.004 Pa.s	τ (°C)	-
SG	~1.05	τ (wet)	-
Shelf life	6 months/RT	Peel ( )	-
$T_{Service}$	-30 to +100°C	σ_f	-
CTE (°C)	-	*E*	-
Flash Point	84°C	Elongation	-
Gap Filling	0.02 mm, max.	H (Shore)	-
Notes: <40%RH increases cure time. τ DIN 53283. Flash point DIN 53213.			

Delo	CA 2994 (CA 1400)		Gel
(July 95)			
Manufacturers Comments: Ethyl ester cyanoacrylate. Solvent free. Slower cure, high strength (fully cured). High viscosity.			
Typical Use: Metals, plastics, rubbers and elastomers (synthetic rubbers).			
Cure: RT(23°C/50%RH) 20-40 sec (firm to touch). Full strength: 20 hrs.			
Test Methods: DIN			
Physical		**Mechanical**	
Colour	-	τ (RT)	~22 MPa
Viscosity	20 Pa.s	τ (°C)	-
SG	~1.12	τ (wet)	-
Shelf life	6 months/RT	Peel ( )	-
$T_{Service}$	-30 to +100°C	σ_f	-
CTE (°C)	-	*E*	-
Flash Point	84°C	Elongation	-
Gap Filling	0.30 mm, max.	H (Shore)	-
Notes: <40%RH increases cure time. τ DIN 53283. Flash point DIN 53213.			

Delo	CA 2905 (CA 15)		Liquid
(July 95)			
Manufacturers Comments: Ethyl ester cyanoacrylate. Solvent free. Fast cure, high strength (fully cured). Low viscosity.			
Typical Use: Metals, plastics, rubbers and elastomers (synthetic rubbers).			
Cure: RT(23°C/50%RH) 10-20 sec (firm to touch). Full strength: 20 hrs.			
Test Methods: DIN			
Physical		**Mechanical**	
Colour	-	τ (RT)	~22 MPa
Viscosity	0.025 Pa.s	τ (°C)	-
SG	~1.06	τ (wet)	-
Shelf life	6 months/RT	Peel ( )	-
$T_{Service}$	-30 to +100°C	σ_f	-
CTE (°C)	-	*E*	-
Flash Point	84°C	Elongation	-
Gap Filling	0.03 mm, max.	H (Shore)	-
Notes: <40%RH increases cure time. τ DIN 53283. Flash point DIN 53213.			

Dunlop

Dunlop	Maxibond SAS 200		Liquid
Jan 94			
Manufacturers Comments: Ethyl cyanoacrylate. Solvent free. Very fast cure. For close-fitting, non-porous substrates. Capillary action. Not for applications with water immersion.			
Typical Use: Metals, plastics, rubbers and elastomers (synthetic rubbers).			
Cure: RT/ 20-30 secs (Al); 5-10 sec (SBR); 10-15 secs (polycarbonate) Full cure: ~24-48 hrs.			
Test Methods: Not Stated			
Physical		**Mechanical**	
Colour	Clear	τ (RT, Al)	18.5 MPa
Viscosity	0.001-0.010 Pa.s	τ (°C)	-
SG	1.24	τ (wet)	0 MPa
Shelf life	6 months/<25°C	Peel ( )	-
$T_{Service}$	-40 to +80°C	σ_f	-
CTE (°C)	-	*E*	-
Flash Point	80°C	Elongation	-
Gap Filling	No	H (Shore)	-
Notes: τ (steel) 21.6 MPa; Rigid PVC 5.5 MPa. τ(wet): steel, after 1 year water soak. Softening Point 145°C; Refractive Index: 1.490. Dielectric Constant (10MHz): 3.5. Dielectric Loss (10MHz): 0.067. Soluble in acetone & some other solvents.			

See Key to Data Sheets - Pages 63 to 65

Dunlop	Maxibond SAS 205		Liquid
Jan 94			
Manufacturers Comments: Ethyl cyanoacrylate. Solvent free. Fast cure. General purpose, thin bondlines. Moisture resistance (not immersion). Meets: UK Def. Stan 80-146/1.			
Typical Use: Rubber-to-plastic, ceramics. O-ring sizing. Electronics.			
Cure: RT/ 20-30secs (Al); 10-15 sec (SBR); 20-30 sec (polycarbonate) Full cure: ~24-48 hrs.			
Test Methods: Not Stated			
Physical		**Mechanical**	
Colour	Clear	τ (RT, Al)	17.0 MPa
Viscosity	0.09-0.11 Pa.s	τ (°C)	-
SG	1.24	τ (wet)	0 (22) MPa
Shelf life	6 months/<25°C	Peel ( )	-
$T_{Service}$	-40 to +80°C	σ_f	-
CTE (°C)	-	E	-
Flash Point	80°C	Elongation	-
Gap Filling	No	H (Shore)	-
Notes: τ (steel) 19.5 MPa, Rigid PVC 5.4 MPa (substrate failure) τ(wet) data after 1 year water soak : steel (PVC). Softening Point 145°C; Refractive Index: 1.490. Dielectric Constant (10MHz): 3.5. Dielectric Loss (10MHz): 0.067. Soluble in acetone & some other solvents.			

Dunlop	Maxibond SAS 211		Liquid
Jan 94			
Manufacturers Comments: Ethyl cyanoacrylate. Solvent-free, fast cure. Medium viscosity, high strength. General purpose, thin bondlines. Moisture resistance (not immersion).			
Typical Use: Plastics, rubber, ceramics, wood. Double-glazing gaskets. Drive belts assembly and repair. O-ring sizing. Electronics.			
Cure: RT/ 20 secs (rubber); 50 sec (steel). Full cure: ~24-48 hrs.			
Test Methods: Not Stated			
Physical		**Mechanical**	
Colour	Clear	τ (RT, Fe)	15.0 MPa*
Viscosity	0.5 Pa.s	τ (°C)	-
SG	-	τ (wet)	-
Shelf life	6 months/<25°C	Peel ( )	-
$T_{Service}$	-60 to +80°C	σ_f	-
CTE (°C)	-	E	-
Flash Point	83°C	Elongation	-
Gap Filling	-	H (Shore)	-
Notes: *20°C/48 hrs/60%RH. Coupon size 1-2 mmx25mmx100mm.			

Dunlop	Maxibond SAS 210		Liquid
Jan 94			
Manufacturers Comments: Ethyl cyanoacrylate. Solvent free. Fast cure. Medium viscosity. General purpose, thin bondlines. Moisture resistance (not immersion).			
Typical Use: Plastics, rubber, ceramics, wood. Double-glazing gaskets. Drive belts assembly and repair. O-ring sizing. Electronics.			
Cure: RT/ 20 secs (rubber); 50 sec (steel). Full cure: ~24-48 hrs.			
Test Methods: Not Stated			
Physical		**Mechanical**	
Colour	Clear	τ (RT, Fe)	15.0 MPa*
Viscosity	0.5 Pa.s	τ (°C)	-
SG	-	τ (wet)	-
Shelf life	6 months/<25°C	Peel ( )	-
$T_{Service}$	-60 to +80°C	σ_f	-
CTE (°C)	-	E	-
Flash Point	83°C	Elongation	-
Gap Filling	-	H (Shore)	-
Notes: *20°C/48 hrs/60%RH. Coupon size 25mmx100mm.			

Dunlop	Maxibond SAS 220		Liquid
Jan 94			
Manufacturers Comments: Ethyl cyanoacrylate. Solvent-free, slower cure. High viscosity, high strength. Gap-filling. Moisture resistance (not immersion).			
Typical Use: Wood joint assembly. Magnets, ceramics - stone setting. Electronics.			
Cure: RT/ 30 secs (rubber); 60 sec (steel). Full cure: ~24-48 hrs.			
Test Methods: Not Stated			
Physical		**Mechanical**	
Colour	Clear	τ (RT, Fe)	15.0 MPa*
Viscosity	1.8 Pa.s	τ (°C)	-
SG	-	τ (wet)	-
Shelf life	6 months/<25°C	Peel ( )	-
$T_{Service}$	-60 to +80°C	σ_f	-
CTE (°C)	-	E	-
Flash Point	83°C	Elongation	-
Gap Filling	-	H (Shore)	-
Notes: *20°C/48 hrs/60%RH. Coupon size 1-2 mmx25mmx100mm. Refractive Index: 1.490.			

Forbo Helmitin

Dunlop	Maxibond SAS 231		Liquid
Oct 93			
Manufacturers Comments: Ethyl cyanoacrylate. Solvent-free, fast cure. Low viscosity, high strength for a wide variety of woods (also metal, plastics, rubber, ceramics). Moisture resistance (not immersion).			
Typical Use: Joint assembly in wood (window-frames).			
Cure: RT/ 20-30 secs (oak); 25-40 sec (mahogany); 15-20 (beech). Full cure: ~24-48 hrs.			
Test Methods: Not Stated			
Physical		**Mechanical**	
Colour	Clear	τ (RT, wood)	*
Viscosity	0.001-0.002 Pa.s	τ (°C)	-
SG	-	τ (wet)	-
Shelf life	6 months/<25°C	Peel ( )	-
$T_{Service}$	-60 to +80°C	σ_f	-
CTE (°C)	-	*E*	-
Flash Point	80°C	Elongation	-
Gap Filling	-	H (Shore)	-
Notes: * Substrate failure when used as wood adhesive. Refractive Index: 1.490. Softening point: 80°C.			

Forbo Helmitin	Helmitin 121		Liquid
April 92			
Manufacturers Comments: Cyanoacrylate. Solvent-free, fast cure. Good adhesion to a variety of materials. Good resistance to moisture, petrol, oil, grease and most solvents.			
Typical Use: Rubbers to metal, glass and various plastics.			
Cure: RT/ 4-6 seconds. Full cure: ~12 hrs.			
Test Methods: Not Stated			
Physical		**Mechanical**	
Colour	Transparent	τ (RT)	-
Viscosity	~0.04 Pa.s	τ (°C)	-
SG	~1.05	τ (wet)	-
Shelf life	6 months/RT	Peel ( )	-
$T_{Service}$	-80 to +80°C	σ_f	-
CTE (°C)	-	*E*	-
Flash Point	-	Elongation	-
Gap Filling	-	H (Shore)	-
Notes:			

Dunlop	Maxibond SAS 232		Liquid
March 94			
Manufacturers Comments: Modified ethyl cyanoacrylate. Solvent-free, fast cure. Medium viscosity, high strength, good impact resistance for a variety of materials. Moisture resistance (not immersion). Solvent resistance.			
Typical Use: Metal, plastics, rubber, wood, ceramics. Double-glazing. Drive belt assembly. Electronics. O-ring sizing.			
Cure: RT/ 10-20 sec (rubber); 40-50 sec (steel).Full cure: ~24-48 hrs.			
Test Methods: Not Stated			
Physical		**Mechanical**	
Colour	Black	τ (RT)	-
Viscosity	2.75 Pa.s	τ (°C)	-
SG	1.0	τ (wet)	-
Shelf life	6 months/<25°C	Peel ( )	-
$T_{Service}$	-60 to +130°C	σ_f	-
CTE (°C)	-	*E*	-
Flash Point	83°C	Elongation	-
Gap Filling	-	H (Shore)	-
Notes:			

Forbo Helmitin	Helmitin 122		Liquid
April 92			
Manufacturers Comments: Cyanoacrylate. Solvent-free, fast cure. Good adhesion to a variety of materials. Good resistance to moisture, petrol, oil, grease and most solvents.			
Typical Use: Rubbers to metal, glass and various plastics.			
Cure: RT/ 4-6 seconds. Full cure: ~12 hrs.			
Test Methods: Not Stated			
Physical		**Mechanical**	
Colour	Transparent	τ (RT)	-
Viscosity	~0.12 Pa.s	τ (°C)	-
SG	~1.05	τ (wet)	-
Shelf life	6 months/RT	Peel ( )	-
$T_{Service}$	-80 to +80°C	σ_f	-
CTE (°C)	-	*E*	-
Flash Point	-	Elongation	-
Gap Filling	-	H (Shore)	-
Notes:			

LePage's

LePage's	Accu-Flo Super Glue Auto. Pen #373		Liquid
(April 95)			
Manufacturers Comments: Alpha-cyanoacrylate. 100% solids. Solvent-free, no shrinkage. Moisture resistance (not immersion), most chemicals (not acetone, nitromethane) and low temperatures (not heat). Low shock resistance.			
Typical Use: Precision bonds. Metals, ceramics, plastics.			
Cure: RT/ 10-30 secs. (rubber, glass, ceramic). 1-2 mins. (metals). Handling: 3-5 mins. Full cure: 3-12 hrs.			
Test Methods: Not Stated			
Physical		**Mechanical**	
Colour	Transparent	τ (RT, steel)	17.4 MPa
Viscosity	~0.12 Pa.s	τ (°C)	-
SG	~1.05	τ (wet)	-
Shelf life	-	Peel ( )	-
$T_{Service}$	70°C (max.)	σ_t	-
CTE (°C)	-	E	-
Flash Point	-	Elongation	-
Gap Filling	No	H (Shore)	-
Notes: Volume resistivity: 8.6 x 10^{16} Ω.cm (30°C).			

LePage's	Super Glue #377		Liquid
(April 95)			
Manufacturers Comments: Alpha-cyanoacrylate. 100% solids. Solvent-free, no shrinkage. Moisture resistance (not immersion), most chemicals (not acetone, nitromethane) and low temperatures (not heat). Low shock resistance.			
Typical Use: Precision bonds. Metals, ceramics, plastics.			
Cure: RT/ 10-60 secs (varies for material). Handling: 3-5 mins. Full cure: 3-12 hrs.			
Test Methods: Not Stated			
Physical		**Mechanical**	
Colour	Transparent	τ (RT, steel)	24.0 MPa
Viscosity	~0.12 Pa.s	τ (°C)	-
SG	1.1	τ (wet)	-
Shelf life	-	Peel ( )	-
$T_{Service}$	80°C (max.)	σ_t	-
CTE (°C)	-	E	-
Flash Point	-	Elongation	-
Gap Filling	No	H (Shore)	-
Notes: Dielectric strength: 9800 V/mm.			

Loctite

Loctite	Jet Black Structural Adhesive 380		Liquid
(Feb 1995)			
Manufacturers Comments: Rubber toughened cyanoacrylate. Tough bonds. High peel and impact strength. Excellent resistance to thermal cycling, humidity, solvents, coolants and fuels.			
Typical Use: Precision bonds. Plastics, rubber, some metals.			
Cure: RT/ ~60 secs (varies for material and bondline thickness), e.g. 60% strength: 1 hr (abraded aluminium, 0.05 mm bondline).			
Test Methods: Not Stated			
Physical		**Mechanical**	
Colour	Black	τ (RT)	-
Viscosity	-	τ (°C)	-
SG	-	τ (wet)	-
Shelf life	-	Peel ( )	-
$T_{Service}$	105°C (max.)	σ_t	-
CTE (°C)	-	E	-
Flash Point	-	Elongation	-
Gap Filling	0.25 mm	H (Shore)	-
Notes:			

Loctite	Prism 401		Liquid
(Feb 1995)			
Manufacturers Comments: Ethyl cyanoacrylate. General purpose, particularly suitable for porous materials. Fast cure even with low humidity (20%RH) conditions.			
Typical Use: General assembly: plated metals, composites, wood, cork, foam, leather, textiles.			
Cure: RT/5-10 secs (metals); 5 secs (ceramic, leather); 2 secs (paper, textiles).			
Test Methods: ASTM			
Physical		**Mechanical**	
Colour	-	τ (RT, steel)	18-26 MPa
Viscosity	0.09-0.125 Pa.s	τ (°C)	-
SG	-	τ (wet)	-
Shelf life	6 months/<25°C	Peel ( )	-
$T_{Service}$	-50 to +80°C	σ_t	-
CTE (°C)	-	E	-
Flash Point	-	Elongation	-
Gap Filling	0.1 mm (max.)	H (Shore)	-
Notes: τ : ASTM D1002, grit-blasted mild steel.			

Loctite	Prism 406	Liquid
	(Feb 1995)	

Manufacturers Comments: Ethyl cyanoacrylate. Fast cure for 'difficult' materials to bond (with Prism 757 primer for polyolefin plastics).

Typical Use: Rubber and plastics: Polyethylene, polypropylene, polyolefin.

Cure: RT/1-5 secs (nitrile rubber); 5 secs (ceramic, leather); 2 secs (paper, textiles).

Test Methods: ASTM

Physical		Mechanical	
Colour	-	τ (RT, steel)	18 - 26 MPa
Viscosity	0.001-0.003 Pa.s	τ (°C)	-
SG	-	τ (wet)	-
Shelf life	6 months/<25°C	Peel ( )	-
$T_{Service}$	-50 to +80°C	σ_t	-
CTE (°C)	-	*E*	-
Flash Point	-	Elongation	-
Gap Filling	0.1 mm (max.)	H (Shore)	-

Notes: τ data (ASTM D1002): 9 MPa - nitrile rubber; 7 MPa - polypropylene/steel (with primer).

Loctite	Prism 460	Liquid
	(Feb 1995)	

Manufacturers Comments: High molecular weight cyanoacrylate. No odour, for use in confined spaces. Low bloom grade for dark shiny components. Electrical and electronic assemblies near optical parts.

Typical Use: General assembly: plated metals, composites, wood, cork, foam, leather, textiles.

Cure: RT/~20 secs (metals); 50 secs (PVC); 40 secs (ABS, polycarbonate).

Test Methods: ASTM

Physical		Mechanical	
Colour	-	τ (RT, steel)	14 - 22 MPa
Viscosity	0.025-0.06 Pa.s	τ (°C)	-
SG	-	τ (wet)	-
Shelf life	6 months/<25°C	Peel ( )	-
$T_{Service}$	-50 to +80°C	σ_t	-
CTE (°C)	-	*E*	-
Flash Point	-	Elongation	-
Gap Filling	0.1 mm (max.)	H (Shore)	-

Notes: τ ASTM D1002: grit-blasted, mild steel.

Loctite	Prism 454	Gel
	(Feb 1995)	

Manufacturers Comments: Ethyl cyanoacrylate. General purpose, particularly suitable for porous materials. Gel version of Prism 401. Non-drip, no run off on overhead/vertical surfaces.

Typical Use: General assembly: plated metals, composites, wood, cork, foam, leather, textiles.

Cure: RT/~20 secs (metals); 5 secs (textiles, PVC, neoprene); 10 secs (ceramic, leather); 30 secs (ABS, pine).

Test Methods: ASTM

Physical		Mechanical	
Colour	-	τ (RT, steel)	18 - 26 MPa
Viscosity	0.001-0.003 Pa.s	τ (°C)	-
SG	-	τ (wet)	-
Shelf life	6 months/<25°C	Peel ( )	-
$T_{Service}$	-50 to +80°C	σ_t	-
CTE (°C)	-	*E*	-
Flash Point	-	Elongation	-
Gap Filling	0.2 mm (max.)	H (Shore)	-

Notes: τ ASTM D1002: grit-blasted, mild steel.

Loctite	Prism 480	Liquid
	(Feb 1995)	

Manufacturers Comments: Modified ethyl cyanoacrylate. Toughened (rubber) adhesive for fast metal bonding. High peel strength, shock resistance.

Typical Use: Metals.

Cure: RT.

Test Methods: ASTM

Physical		Mechanical	
Colour	Black	τ (RT, steel)	22 - 30 MPa
Viscosity	0.1-0.5 Pa.s	τ (°C)	-
SG	-	τ (wet)	-
Shelf life	6 months/<25°C	Peel (T)	4 N mm^{-1}
$T_{Service}$	-50 to +100°C	σ_t	-
CTE (°C)	-	*E*	-
Flash Point	-	Elongation	-
Gap Filling	0.1 mm (max.)	H (Shore)	-

Notes: τ ASTM D1002: grit-blasted, mild steel. T-peel: ASTM D1876. BLACKTAK 410 is a high viscosity (2.7-5.4 Pa.s) version of PRISM 480.

Panacol Elosol

Panacol Elosol		Cyanolit 101	Liquid
June 94			
Manufacturers Comments: Methyl cyanoacrylate. Very low viscosity. Fast cure.			
Typical Use: Metals and other materials.			
Cure: 10 - 60 sec.			
Test Methods: DIN / ASTM			
Physical		**Mechanical**	
Colour	Transparent	τ (RT, Al)	17.4 MPa
Viscosity	0.002 Pa.s	τ (RT, steel)	28.7 MPa
SG	1.01 - 1.18	τ (RT, copper)	21.8 MPa
Shelf life	6 months/RT	Peel ( )	-
$T_{Service}$	~100°C	σ_t (RT, steel)	25.3 MPa
CTE (°C)	126×10^{-6} K^{-1}	*E*	-
Flash Point	83°C	Elongation	-
Gap Filling	-	H (Shore)	-
Notes: Softening temperature: 175°C. Refractive index: 1.487. Thermal conductivity (DIN 52612): 0.25 W.m^{-1}K^{-1}. Dielectric constant (10 MHz, DIN 53483): 3.54. Flash point: DIN 51755. Tensile: ASTM D1002.			

Panacol Elosol		Cyanolit 103	Liquid
June 94			
Manufacturers Comments: Methyl cyanoacrylate. High viscosity. Slow cure.			
Typical Use: Metals and other materials.			
Cure: 30 - 120 sec.			
Test Methods: DIN / ASTM			
Physical		**Mechanical**	
Colour	Transparent	τ (RT, Al)	18.7 MPa
Viscosity	1.5 Pa.s	τ (RT, steel)	29.5 MPa
SG	1.01 - 1.18	τ (RT, copper)	23.1 MPa
Shelf life	6 months/RT	Peel ( )	-
$T_{Service}$	~100°C	σ_t (RT, steel)	20.2 MPa
CTE (°C)	126×10^{-6} K^{-1}	*E*	-
Flash Point	83°C	Elongation	-
Gap Filling	< 0.2 mm.	H (Shore)	-
Notes: Softening temperature: 175°C. Refractive index: 1.487. Thermal conductivity (DIN 52612): 0.25 W.m^{-1}K^{-1}. Dielectric constant (10 MHz, DIN 53483): 3.54. Flash point: DIN 51755. Tensile: ASTM D1002.			

Panacol Elosol		Cyanolit 102	Liquid
June 94			
Manufacturers Comments: Methyl cyanoacrylate. Medium viscosity.			
Typical Use: Metals and other materials.			
Cure: 20 - 80 sec.			
Test Methods: DIN / ASTM			
Physical		**Mechanical**	
Colour	Transparent	τ (RT, Al)	15.2 MPa
Viscosity	0.1 Pa.s	τ (RT, steel)	26.3 MPa
SG	1.01 - 1.18	τ (RT, copper)	19.5 MPa
Shelf life	6 months/RT	Peel ( )	-
$T_{Service}$	~100°C	σ_t (RT, steel)	23.2 MPa
CTE (°C)	126×10^{-6} K^{-1}	*E*	-
Flash Point	83°C	Elongation	-
Gap Filling	-	H (Shore)	-
Notes: Softening temperature: 175°C. Refractive index: 1.487. Thermal conductivity (DIN 52612): 0.25 W.m^{-1}K^{-1}. Dielectric constant (10 MHz, DIN 53483): 3.54. Flash point: DIN 51755. Tensile: ASTM D1002.			

Panacol Elosol		Cyanolit 141	Liquid
June 94			
Manufacturers Comments: Methyl cyanoacrylate. Low viscosity. Fast cure.			
Typical Use: Metals and other materials.			
Cure: 10 - 60 sec.			
Test Methods: DIN / ASTM			
Physical		**Mechanical**	
Colour	Transparent	τ (RT, Al)	17.9 MPa
Viscosity	0.04 Pa.s	τ (RT, steel)	29.5 MPa
SG	1.01 - 1.18	τ (RT, copper)	22.4 MPa
Shelf life	6 months/RT	Peel ( )	-
$T_{Service}$	~100°C	σ_t (RT, steel)	27.4 MPa
CTE (°C)	126×10^{-6} K^{-1}	*E*	-
Flash Point	83°C	Elongation	-
Gap Filling	-	H (Shore)	-
Notes: Softening temperature: 175°C. Refractive index: 1.487. Thermal conductivity (DIN 52612): 0.25 W.m^{-1}K^{-1}. Dielectric constant (10 MHz, DIN 53483): 3.54. Flash point: DIN 51755. Tensile: ASTM D1002.			

Panacol Elosol	Cyanolit 200		Liquid
June 94			
Manufacturers Comments: Ethyl cyanoacrylate. Very low viscosity. Fast cure.			
Typical Use: Elastomers, plastics and other materials.			
Cure: 3 - 12 sec.			
Test Methods: DIN / ASTM			
Physical		**Mechanical**	
Colour	Transparent	τ (RT, PVC)	5.1 MPa
Viscosity	0.002 Pa.s	τ (RT, ABS)	5.3 MPa
SG	1.01 - 1.18	τ (RT, Nylon)	4.0 MPa
Shelf life	6 months/RT	Peel ( )	-
$T_{Service}$	~100°C	σ_t (RT, PVC)	30.2 MPa
CTE (°C)	126 x 10^{-6} K^{-1}	*E*	-
Flash Point	83°C	Elongation	-
Gap Filling	-	H (Shore)	-
Notes: Softening temperature: 175°C. Refractive index: 1.487. Thermal conductivity (DIN 52612): 0.25 $W.m^{-1}K^{-1}$. Dielectric constant (10 MHz, DIN 53483): 3.54. Flash point: DIN 51755. Tensile: ASTM D1002.			

Panacol Elosol	Cyanolit 202		Liquid
June 94			
Manufacturers Comments: Ethyl cyanoacrylate. Medium viscosity.			
Typical Use: Plastics and other materials.			
Cure: 13 - 40 sec.			
Test Methods: DIN / ASTM			
Physical		**Mechanical**	
Colour	Transparent	τ (RT, PVC)	5.1 MPa
Viscosity	0.1 Pa.s	τ (RT, ABS)	5.4 MPa
SG	1.01 - 1.18	τ (RT, Nylon)	4.7 MPa
Shelf life	6 months/RT	Peel ( )	-
$T_{Service}$	~100°C	σ_t (RT, PVC)	33.2 MPa
CTE (°C)	126 x 10^{-6} K^{-1}	*E*	-
Flash Point	83°C	Elongation	-
Gap Filling	-	H (Shore)	-
Notes: Softening temperature: 175°C. Refractive index: 1.487. Thermal conductivity (DIN 52612): 0.25 $W.m^{-1}K^{-1}$. Dielectric constant (10 MHz, DIN 53483): 3.54. Flash point: DIN 51755. Tensile: ASTM D1002.			

Panacol Elosol	Cyanolit 201		Liquid
June 94			
Manufacturers Comments: Ethyl cyanoacrylate. Very low viscosity. Fast cure.			
Typical Use: Plastics and other materials.			
Cure: 7 - 20 sec.			
Test Methods: DIN / ASTM			
Physical		**Mechanical**	
Colour	Transparent	τ (RT, PVC)	5.5 MPa
Viscosity	0.002 Pa.s	τ (RT, ABS)	5.4 MPa
SG	1.01 - 1.18	τ (RT, Nylon)	4.3 MPa
Shelf life	6 months/RT	Peel ( )	-
$T_{Service}$	~100°C	σ_t (RT, PVC)	35.5 MPa
CTE (°C)	126 x 10^{-6} K^{-1}	*E*	-
Flash Point	83°C	Elongation	-
Gap Filling	-	H (Shore)	-
Notes: Softening temperature: 175°C. Refractive index: 1.487. Thermal conductivity (DIN 52612): 0.25 $W.m^{-1}K^{-1}$. Dielectric constant (10 MHz, DIN 53483): 3.54. Flash point: DIN 51755. Tensile: ASTM D1002.			

Panacol Elosol	Cyanolit 203		Liquid
June 94			
Manufacturers Comments: Ethyl cyanoacrylate. High viscosity. Slow cure.			
Typical Use: Plastics and other materials.			
Cure: 30 - 75 sec.			
Test Methods: DIN / ASTM			
Physical		**Mechanical**	
Colour	Transparent	τ (RT, PVC)	5.4 MPa
Viscosity	1.5 Pa.s	τ (RT, ABS)	5.5 MPa
SG	1.01 - 1.18	τ (RT, Nylon)	4.5 MPa
Shelf life	6 months/RT	Peel ( )	-
$T_{Service}$	~100°C	σ_t (RT, PVC)	29.8 MPa
CTE (°C)	126 x 10^{-6} K^{-1}	*E*	-
Flash Point	83°C	Elongation	-
Gap Filling	< 0.2 mm.	H (Shore)	-
Notes: Softening temperature: 175°C. Refractive index: 1.487. Thermal conductivity (DIN 52612): 0.25 $W.m^{-1}K^{-1}$. Dielectric constant (10 MHz, DIN 53483): 3.54. Flash point: DIN 51755. Tensile: ASTM D1002.			

Panacol Elosol	Cyanolit 208		Liquid
June 94			
Manufacturers Comments: Ethyl cyanoacrylate. Low viscosity. Fast cure.			
Typical Use: Elastomers, plastics and other materials.			
Cure: 3 - 12 sec.			
Test Methods: DIN / ASTM			
Physical		**Mechanical**	
Colour	Transparent	τ (RT, PVC)	5.2 MPa
Viscosity	0.04 Pa.s	τ (RT, ABS)	5.4 MPa
SG	1.01 - 1.18	τ (RT, Nylon)	3.9 MPa
Shelf life	6 months/RT	Peel ( )	-
$T_{Service}$	~100°C	σ_f (RT, PVC)	32.4 MPa
CTE (°C)	$126 \times 10^{-6} K^{-1}$	E	-
Flash Point	83°C	Elongation	-
Gap Filling	-	H (Shore)	-

Notes: Softening temperature: 175°C. Refractive index: 1.487. Thermal conductivity (DIN 52612): 0.25 $W.m^{-1}K^{-1}$. Dielectric constant (10 MHz, DIN 53483): 3.54. Flash point: DIN 51755. Tensile: ASTM D1002.

Panacol Elosol	Cyanolit 302		Liquid
June 94			
Manufacturers Comments: Alkylester cyanoacrylate. Very low viscosity. Fast cure. Good higher temperature properties.			
Typical Use: Elastomers, plastics and other materials.			
Cure: -			
Test Methods: DIN / ASTM			
Physical		**Mechanical**	
Colour	Transparent	τ (RT)	~19 MPa
Viscosity	0.01 Pa.s	τ (100°C)	~15 MPa
SG	1.01 - 1.18	τ (150°C)	~8 MPa
Shelf life	6 months/RT	Peel ( )	-
$T_{Service}$	> 150°C	σ_f	-
CTE (°C)	$126 \times 10^{-6} K^{-1}$	E	-
Flash Point	83°C	Elongation	-
Gap Filling	-	H (Shore)	-

Notes: Softening temperature: 175°C. Refractive index: 1.487. Thermal conductivity (DIN 52612): 0.25 $W.m^{-1}K^{-1}$. Dielectric constant (10 MHz, DIN 53483): 3.54. Flash point: DIN 51755. Tensile: ASTM D1002.

Panacol Elosol	Cyanolit 241		Liquid
June 94			
Manufacturers Comments: Ethyl cyanoacrylate. Low viscosity. Fast cure.			
Typical Use: Plastics, metals and other materials.			
Cure: 8 - 22 sec.			
Test Methods: DIN / ASTM			
Physical		**Mechanical**	
Colour	Transparent	τ (RT, PVC)	5.2 MPa
Viscosity	0.04 Pa.s	τ (RT, ABS)	5.6 MPa
SG	1.01 - 1.18	τ (RT, Nylon)	4.2 MPa
Shelf life	6 months/RT	Peel ( )	-
$T_{Service}$	~100°C	σ_f (RT, PVC)	35.7 MPa
CTE (°C)	$126 \times 10^{-6} K^{-1}$	E	-
Flash Point	83°C	Elongation	-
Gap Filling	-	H (Shore)	-

Notes: Softening temperature: 175°C. Refractive index: 1.487. Thermal conductivity (DIN 52612): 0.25 $W.m^{-1}K^{-1}$. Dielectric constant (10 MHz, DIN 53483): 3.54. Flash point: DIN 51755. Tensile: ASTM D1002.

Panacol Elosol	Cyanolit 303		Liquid
June 94			
Manufacturers Comments: Alkylester cyanoacrylate. Very low viscosity. Fast cure. Good higher temperature properties.			
Typical Use: Elastomers, plastics and other materials.			
Cure: -			
Test Methods: DIN / ASTM			
Physical		**Mechanical**	
Colour	Transparent	τ (RT)	~19 MPa
Viscosity	0.006 Pa.s	τ (100°C)	~15 MPa
SG	1.01 - 1.18	τ (150°C)	~8 MPa
Shelf life	6 months/RT	Peel ( )	-
$T_{Service}$	> 150°C	σ_f	-
CTE (°C)	$126 \times 10^{-6} K^{-1}$	E	-
Flash Point	83°C	Elongation	-
Gap Filling	-	H (Shore)	-

Notes: Softening temperature: 175°C. Refractive index: 1.487. Thermal conductivity (DIN 52612): 0.25 $W.m^{-1}K^{-1}$. Dielectric constant (10 MHz, DIN 53483): 3.54. Flash point: DIN 51755. Tensile: ASTM D1002.

Panacol Elosol	Cyanolit 401		Liquid
June 94			
Manufacturers Comments: Cyanoacrylate. Very low viscosity. Fast cure.			
Typical Use: -			
Cure: -			
Test Methods: DIN / ASTM			
Physical		**Mechanical**	
Colour	Transparent	τ (RT)	-
Viscosity	0.002 Pa.s	τ (°C)	-
SG	1.01 - 1.18	τ (wet)	-
Shelf life	6 months/RT	Peel ( )	-
$T_{Service}$	> 150°C	σ_t	-
CTE (°C)	126 x 10^{-6} K^{-1}	E	-
Flash Point	83°C	Elongation	-
Gap Filling	-	H (Shore)	-
Notes: Softening temperature: 175°C. Refractive index: 1.487. Thermal conductivity (DIN 52612): 0.25 $W.m^{-1}K^{-1}$. Dielectric constant (10 MHz, DIN 53483): 3.54. Flash point: DIN 51755. Tensile: ASTM D1002.			

Panacol Elosol	Cyanolit 403 TB		Liquid
June 94			
Manufacturers Comments: Flexible filled cyanoacrylate. High viscosity. Slow cure. Meets the requirements of (USA) MIL-A-46050B.			
Typical Use: -			
Cure: 80 - 200 sec.			
Test Methods: DIN / ASTM			
Physical		**Mechanical**	
Colour	Black	τ (RT, Al)	23.5 MPa
Viscosity	-	τ (RT, steel)	30.4 MPa
SG	1.01 - 1.18	τ (wet)	-
Shelf life	6 months/RT	Peel ( )	-
$T_{Service}$	-60°C to +100°C	σ_t	
CTE (°C)	126 x 10^{-6} K^{-1}	E	-
Flash Point	83°C	Elongation	-
Gap Filling	-	H (Shore)	-
Notes: Softening temperature: 175°C. Thermal conductivity (DIN 52612): 0.25 $W.m^{-1}K^{-1}$. Dielectric constant (10 MHz, DIN 53483): 3.54. Flash point: DIN 51755. Tensile: ASTM D1002. Impact strength (Al): 0.034 $J.mm^{-2}$.			

Panacol Elosol	Cyanolit 403 T		Liquid
June 94			
Manufacturers Comments: Flexible cyanoacrylate. High viscosity. Slow cure.			
Typical Use: -			
Cure: 40 - 120 sec.			
Test Methods: DIN / ASTM			
Physical		**Mechanical**	
Colour	Transparent	τ (RT, Al)	16.3 MPa
Viscosity	1.0 Pa.s	τ (RT, steel)	26.4 MPa
SG	1.01 - 1.18	τ (wet)	-
Shelf life	6 months/RT	Peel ( )	-
$T_{Service}$	-60°C to +100°C	σ_t	
CTE (°C)	126 x 10^{-6} K^{-1}	E	-
Flash Point	83°C	Elongation	-
Gap Filling	< 0.2 mm.	H (Shore)	-
Notes: Softening temperature: 175°C. Refractive index: 1.487. Thermal conductivity (DIN 52612): 0.25 $W.m^{-1}K^{-1}$. Dielectric constant (10 MHz, DIN 53483): 3.54. Flash point: DIN 51755. Tensile: ASTM D1002. Impact strength (Al): 0.03 $J.mm^{-2}$.			

Panacol Elosol	Cyanolit 432 TW		Liquid
June 94			
Manufacturers Comments: Flexible filled cyanoacrylate. Medium viscosity. Slow cure.			
Typical Use: -			
Cure: 60 - 180 sec.			
Test Methods: DIN / ASTM			
Physical		**Mechanical**	
Colour	White	τ (RT, Al)	22.4 MPa
Viscosity	-	τ (RT, steel)	28.9 MPa
SG	1.01 - 1.18	τ (wet)	-
Shelf life	6 months/RT	Peel (-)	5.5 N mm^{-1}
$T_{Service}$	-60°C to +100°C	σ_t	
CTE (°C)	126 x 10^{-6} K^{-1}	E	-
Flash Point	83°C	Elongation	-
Gap Filling	-	H (Shore)	-
Notes: Softening temperature: 175°C. Thermal conductivity (DIN 52612): 0.25 $W.m^{-1}K^{-1}$. Dielectric constant (10 MHz, DIN 53483): 3.54. Flash point: DIN 51755. Tensile: ASTM D1002. Impact strength (Al): 0.031 $J.mm^{-2}$.			

Panacol Elosol	Cyanolit 503		Liquid
June 94			
Manufacturers Comments: Ethyl cyanoacrylate. Very low viscosity. Fast cure.			
Typical Use: Elastomers and plastics.			
Cure: 5 - 7 sec.			
Test Methods: DIN / ASTM			
Physical		**Mechanical**	
Colour	Transparent	τ (RT)	-
Viscosity	0.002 Pa.s	τ (°C)	-
SG	1.01 - 1.18	τ (wet)	-
Shelf life	6 months/RT	Peel ( )	-
$T_{Service}$	-	σ_t	6.3 - 10.4 MPa
CTE (°C)	126 x 10^{-6} K^{-1}	*E*	-
Flash Point	83°C	Elongation	-
Gap Filling	-	H (Shore)	-

Notes: Softening temperature: 175°C. Refractive index: 1.487. Thermal conductivity (DIN 52612): 0.25 $W.m^{-1}K^{-1}$. Dielectric constant (10 MHz, DIN 53483): 3.54. Flash point: DIN 51755. Tensile: ASTM D1002.

Panacol Elosol	Cyanolit 601		Liquid
June 94			
Manufacturers Comments: Cyanoacrylate. Very low viscosity. Fast cure.			
Typical Use: Particularly suitable for galvanised, chromed or coated surfaces.			
Cure: 45 - 55 sec.			
Test Methods: DIN / ASTM			
Physical		**Mechanical**	
Colour	Transparent	τ (RT)	8.5 - 10.3 MPa
Viscosity	0.002 Pa.s	τ (°C)	-
SG	1.01 - 1.18	τ (wet)	-
Shelf life	6 months/RT	Peel ( )	-
$T_{Service}$	-	σ_t	-
CTE (°C)	126 x 10^{-6} K^{-1}	*E*	-
Flash Point	83°C	Elongation	-
Gap Filling	-	H (Shore)	-

Notes: Softening temperature: 175°C. Refractive index: 1.487. Thermal conductivity (DIN 52612): 0.25 $W.m^{-1}K^{-1}$. Dielectric constant (10 MHz, DIN 53483): 3.54. Flash point: DIN 51755. Tensile: ASTM D1002.

Panacol Elosol	Cyanolit 504		Liquid
June 94			
Manufacturers Comments: Ethyl cyanoacrylate. Low viscosity. Moderate cure.			
Typical Use: Elastomers and plastics.			
Cure: 12 - 19 sec.			
Test Methods: DIN / ASTM			
Physical		**Mechanical**	
Colour	Transparent	τ (RT)	-
Viscosity	0.01 Pa.s	τ (°C)	-
SG	1.01 - 1.18	τ (wet)	-
Shelf life	6 months/RT	Peel ( )	-
$T_{Service}$	-	σ_t	8.4 - 12.3 MPa
CTE (°C)	126 x 10^{-6} K^{-1}	*E*	-
Flash Point	83°C	Elongation	-
Gap Filling	-	H (Shore)	-

Notes: Softening temperature: 175°C. Refractive index: 1.487. Thermal conductivity (DIN 52612): 0.25 $W.m^{-1}K^{-1}$. Dielectric constant (10 MHz, DIN 53483): 3.54. Flash point: DIN 51755. Tensile: ASTM D1002.

Panacol Elosol	Cyanolit 602		Liquid
June 94			
Manufacturers Comments: Cyanoacrylate. Medium viscosity. Fast cure.			
Typical Use: Particularly suitable for galvanised, chromed or coated surfaces.			
Cure: 70 - 80 sec.			
Test Methods: DIN / ASTM			
Physical		**Mechanical**	
Colour	Transparent	τ (RT)	8.9 - 12.4 MPa
Viscosity	0.1 Pa.s	τ (°C)	-
SG	1.01 - 1.18	τ (wet)	-
Shelf life	6 months/RT	Peel ( )	-
$T_{Service}$	-	σ_t	-
CTE (°C)	126 x 10^{-6} K^{-1}	*E*	-
Flash Point	83°C	Elongation	-
Gap Filling	-	H (Shore)	-

Notes: Softening temperature: 175°C. Refractive index: 1.487. Thermal conductivity (DIN 52612): 0.25 $W.m^{-1}K^{-1}$. Dielectric constant (10 MHz, DIN 53483): 3.54. Flash point: DIN 51755. Tensile: ASTM D1002.

Panacol Elosol	Cyanolit 631		Liquid
June 94			
Manufacturers Comments: Cyanoacrylate. Low viscosity. Fast cure.			
Typical Use: Particularly suitable for galvanised, chromed or coated surfaces.			
Cure: 50 - 55 sec.			
Test Methods: DIN / ASTM			
Physical		**Mechanical**	
Colour	Transparent	τ (RT)	8.6 - 10.6 MPa
Viscosity	0.03 Pa.s	τ (°C)	-
SG	1.01 - 1.18	τ (wet)	-
Shelf life	6 months/RT	Peel ( )	-
$T_{Service}$	-	σ_t	-
CTE (°C)	126 x 10^{-6} K^{-1}	E	-
Flash Point	83°C	Elongation	-
Gap Filling	-	H (Shore)	-
Notes: Softening temperature: 175°C. Refractive index: 1.487. Thermal conductivity (DIN 52612): 0.25 $W.m^{-1}K^{-1}$. Dielectric constant (10 MHz, DIN 53483): 3.54. Flash point: DIN 51755. Tensile: ASTM D1002.			

Panacol Elosol	Cyanolit 802		Liquid
June 94			
Manufacturers Comments: Cyanoacrylate. Medium viscosity. No skin irritation problems.			
Typical Use: DIY, optical and electronic.			
Cure: 18 - 30 sec.			
Test Methods: DIN / ASTM			
Physical		**Mechanical**	
Colour	Transparent	τ (RT, PVC)	10.2 MPa
Viscosity	0.12 Pa.s	τ (RT, ABS)	9.8 MPa
SG	1.01 - 1.18	τ (wet)	-
Shelf life	6 months/RT	Peel ( )	-
$T_{Service}$	-	σ_t (RT, PVC)	23.4 MPa
CTE (°C)	126 x 10^{-6} K^{-1}	E	-
Flash Point	83°C	Elongation	-
Gap Filling	-	H (Shore)	-
Notes: Softening temperature: 175°C. Refractive index: 1.487. Thermal conductivity (DIN 52612): 0.25 $W.m^{-1}K^{-1}$. Dielectric constant (10 MHz, DIN 53483): 3.54. Flash point: DIN 51755. Tensile: ASTM D1002.			

Panacol Elosol	Cyanolit 732 F		Liquid
June 94			
Manufacturers Comments: Cyanoacrylate. Medium viscosity.			
Typical Use: Particularly suitable for galvanised, chromed or coated surfaces.			
Cure: 10 - 40 sec.			
Test Methods: DIN / ASTM			
Physical		**Mechanical**	
Colour	Transparent	τ (RT, PVC)	10.7 MPa
Viscosity	0.3 Pa.s	τ (RT, ABS)	5.2 MPa
SG	1.01 - 1.18	τ (RT, wood)	15.2 MPa
Shelf life	6 months/RT	Peel ( )	-
$T_{Service}$	-	σ_t (RT, PVC)	35.5 MPa
CTE (°C)	126 x 10^{-6} K^{-1}	E	-
Flash Point	83°C	Elongation	-
Gap Filling	-	H (Shore)	-
Notes: Softening temperature: 175°C. Refractive index: 1.487. Thermal conductivity (DIN 52612): 0.25 $W.m^{-1}K^{-1}$. Dielectric constant (10 MHz, DIN 53483): 3.54. Flash point: DIN 51755. Tensile: ASTM D1002.			

Panacol Elosol	Cyanolit 811		Liquid
June 94			
Manufacturers Comments: Cyanoacrylate. Low viscosity. No skin irritation problems.			
Typical Use: DIY, optical and electronic.			
Cure: 10 - 18 sec.			
Test Methods: DIN / ASTM			
Physical		**Mechanical**	
Colour	Transparent	τ (RT, PVC)	10.2 MPa
Viscosity	0.012 Pa.s	τ (RT, ABS)	9.8 MPa
SG	1.01 - 1.18	τ (wet)	-
Shelf life	6 months/RT	Peel ( )	-
$T_{Service}$	-	σ_t (RT, PVC)	25.2 MPa
CTE (°C)	126 x 10^{-6} K^{-1}	E	-
Flash Point	83°C	Elongation	-
Gap Filling	-	H (Shore)	-
Notes: Softening temperature: 175°C. Refractive index: 1.487. Thermal conductivity (DIN 52612): 0.25 $W.m^{-1}K^{-1}$. Dielectric constant (10 MHz, DIN 53483): 3.54. Flash point: DIN 51755. Tensile: ASTM D1002.			

Panacol Elosol	Cyanolit Cristal 212 F		Liquid
(July 95)			
Manufacturers Comments: Ethyl cyanoacrylate. Low viscosity.			
Typical Use: Absorbent and porous substrates.			
Cure: 2 - 15 sec.			
Test Methods: Not Stated.			
Physical		**Mechanical**	
Colour	Transparent	τ (RT)	6 - 15 MPa
Viscosity	0.1 Pa.s	τ (°C)	-
SG	1.05	τ (wet)	-
Shelf life	-	Peel ( )	-
$T_{Service}$	-80°C to +80°C	σ_f (RT)	2 - 30 MPa
CTE (°C)	-	*E*	-
Flash Point	83°C	Elongation	-
Gap Filling	< 0.1 mm.	H (Shore)	-
Notes: Refractive index: 1.480.			

Panacol Elosol	Cyanolit Cristal 351		Liquid
(July 95)			
Manufacturers Comments: Alkyl cyanoacrylate. Low viscosity.			
Typical Use: Higher temperature applications.			
Cure: 20 - 50 sec.			
Test Methods: Not Stated.			
Physical		**Mechanical**	
Colour	Transparent	τ (RT)	10 - 18 MPa
Viscosity	0.04 Pa.s	τ (°C)	-
SG	1.06	τ (wet)	-
Shelf life	-	Peel ( )	-
$T_{Service}$	-80°C to +150°C	σ_f (RT)	8 - 25 MPa
CTE (°C)	-	*E*	-
Flash Point	83°C	Elongation	-
Gap Filling	< 0.05 mm.	H (Shore)	-
Notes: Refractive index: 1.458.			

Panacol Elosol	Cyanolit Cristal 221 F		Liquid
(July 95)			
Manufacturers Comments: Ethyl cyanoacrylate. Low viscosity.			
Typical Use: Plastics and elastomers.			
Cure: 2 - 10 sec.			
Test Methods: Not Stated.			
Physical		**Mechanical**	
Colour	Transparent	τ (RT)	3 - 28 MPa
Viscosity	0.02 Pa.s	τ (°C)	-
SG	1.05	τ (wet)	-
Shelf life	-	Peel ( )	-
$T_{Service}$	-80°C to +100°C	σ_f (RT)	2 - 19 MPa
CTE (°C)	-	*E*	-
Flash Point	83°C	Elongation	-
Gap Filling	< 0.05 mm.	H (Shore)	-
Notes: Refractive index: 1.438.			

Panacol Elosol	Cyanolit Cristal 402		Liquid
(July 95)			
Manufacturers Comments: Methyl cyanoacrylate. Low viscosity.			
Typical Use: Metals and other materials.			
Cure: 10 - 40 sec.			
Test Methods: Not Stated.			
Physical		**Mechanical**	
Colour	Transparent	τ (RT)	12 - 25 MPa
Viscosity	0.1 Pa.s	τ (°C)	-
SG	1.1	τ (wet)	-
Shelf life	-	Peel ( )	-
$T_{Service}$	-80°C to +100°C	σ_f (RT)	10 - 27 MPa
CTE (°C)	-	*E*	-
Flash Point	81°C	Elongation	-
Gap Filling	< 0.1 mm.	H (Shore)	-
Notes: Refractive index: 1.492.			

Panacol Elosol	Cyanolit Cristal 403 TB		Liquid
(July 95)			
Manufacturers Comments: Filled ethyl cyanoacrylate. Very good resistance to thermal shock, vibration and impact. Meets the requirements of (USA) MIL-A-46050B.			
Typical Use: -			
Cure: 20 - 180 sec.			
Test Methods: Not Stated.			
Physical		**Mechanical**	
Colour	Black	τ (RT)	15 - 30 MPa
Viscosity	1 Pa.s	τ (°C)	-
SG	1.07	τ (wet)	-
Shelf life	-	Peel ( )	-
$T_{Service}$	-80°C to +110°C	σ_t (RT)	12 - 33 MPa
CTE (°C)	-	*E*	-
Flash Point	83°C	Elongation	-
Gap Filling	< 0.2 mm.	H (Shore)	-
Notes:			

Panacol Elosol	Cyanolit Cristal 631 PF		Liquid
(July 95)			
Manufacturers Comments: Isopropyl cyanoacrylate. Low viscosity.			
Typical Use: Particularly suitable for galvanised or chrome-plated metals.			
Cure: 5 - 20 sec.			
Test Methods: Not Stated.			
Physical		**Mechanical**	
Colour	Transparent	τ (RT)	8 - 10 MPa
Viscosity	0.03 Pa.s	τ (°C)	-
SG	1.05	τ (wet)	-
Shelf life	-	Peel ( )	-
$T_{Service}$	-80°C to +100°C	σ_t (RT)	4 - 22 MPa
CTE (°C)	-	*E*	-
Flash Point	83°C	Elongation	-
Gap Filling	< 0.05 mm.	H (Shore)	-
Notes: Refractive index: 1.434.			

Panacol Elosol	Cyanolit Cristal 432 TW		Liquid
(July 95)			
Manufacturers Comments: Filled ethyl cyanoacrylate. Medium viscosity. Very good peel strength.			
Typical Use: -			
Cure: 5 - 240 sec.			
Test Methods: Not Stated.			
Physical		**Mechanical**	
Colour	White	τ (RT)	22 - 28 MPa
Viscosity	0.3 Pa.s	τ (°C)	-
SG	1.07	τ (wet)	-
Shelf life	-	Peel ( )	-
$T_{Service}$	-80°C to +120°C	σ_t (RT)	10 - 37 MPa
CTE (°C)	-	*E*	-
Flash Point	83°C	Elongation	-
Gap Filling	< 0.2 mm.	H (Shore)	-
Notes:			

Panacol Elosol	Cyanolit Cristal 811 F		Liquid
(July 95)			
Manufacturers Comments: Alkyl cyanoacrylate. Low viscosity. No skin irritation problems.			
Typical Use: -			
Cure: 15 - 60 sec.			
Test Methods: Not Stated.			
Physical		**Mechanical**	
Colour	Transparent	τ (RT)	8 - 17 MPa
Viscosity	0.01 Pa.s	τ (°C)	-
SG	1.07	τ (wet)	-
Shelf life	-	Peel ( )	-
$T_{Service}$	-80°C to +80°C	σ_t (RT)	8 - 35 MPa
CTE (°C)	-	*E*	-
Flash Point	83°C	Elongation	-
Gap Filling	< 0.05 mm.	H (Shore)	-
Notes: Refractive index: 1.434.			

Permabond

Panacol Elosol	Cyanolit Cristal GEL		Gel
(July 95)			
Manufacturers Comments: Ethyl cyanoacrylate.			
Typical Use: Highly porous substrates and large gaps.			
Cure: 5 - 120 sec.			
Test Methods: Not Stated.			
Physical		**Mechanical**	
Colour	Transparent	τ (RT)	5 - 10 MPa
Viscosity	10 Pa.s	τ (°C)	-
SG	1.07	τ (wet)	-
Shelf life	-	Peel ( )	-
$T_{Service}$	-80°C to +80°C	σ_t (RT)	8 - 35 MPa
CTE (°C)	-	*E*	-
Flash Point	83°C	Elongation	-
Gap Filling	< 1 mm.	H (Shore)	-
Notes: Refractive index: 1.480.			

Permabond	Plus C791		Liquid
March 94			
Manufacturers Comments: Ethyl cyanoacrylate. Low viscosity. Ultra-fast cure.			
Typical Use: Plastics, rubbers and other materials.			
Cure: RT: <1 to 7 sec. Full strength: 24 hrs.			
Test Methods: Not Stated			
Physical		**Mechanical**	
Colour	Transparent	τ (RT)	-
Viscosity	0.04 Pa.s	τ (°C)	-
SG	1.05	τ (wet)	-
Shelf life	9 mnth/0°C - 5°C	Peel ( )	-
$T_{Service}$	-30°C to +80°C	σ_t	20 MPa
CTE (°C)	-	*E*	-
Flash Point	-	Elongation	-
Gap Filling	<0.075mm	H (Shore)	-
Notes:			

Panacol Elosol	Cyanolit GEL		Liquid
June 94			
Manufacturers Comments: Ethyl cyanoacrylate. High viscosity, thixotropic. No skin irritation problems.			
Typical Use: Plastics, rubbers and porous substrates.			
Cure: 5 sec (ABS) - 120 sec (FRP).			
Test Methods: DIN / ASTM			
Physical		**Mechanical**	
Colour	Transparent	τ (RT, PVC)	10.0 MPa
Viscosity	-	τ (RT, ABS)	5.0 MPa
SG	1.01 - 1.18	τ (RT, Nylon)	3.1 MPa
Shelf life	6 months/RT	Peel ( )	-
$T_{Service}$	-	σ_t (RT, PVC)	30.0 MPa
CTE (°C)	126×10^{-6} K^{-1}	*E*	-
Flash Point	83°C	Elongation	-
Gap Filling	< 3 mm.	H (Shore)	-
Notes: Softening temperature: 175°C. Refractive index: 1.487. Thermal conductivity (DIN 52612): 0.25 $W.m^{-1}K^{-1}$. Dielectric constant (10 MHz, DIN 53483): 3.54. Flash point: DIN 51755. Tensile: ASTM D1002.			

Permabond	Plus C792		Liquid
March 94			
Manufacturers Comments: Ethyl cyanoacrylate. Low viscosity. Ultra-fast cure.			
Typical Use: Plastics, rubbers and other materials.			
Cure: RT: 1 to 7 sec. Full strength: 24 hrs.			
Test Methods: Not Stated			
Physical		**Mechanical**	
Colour	Transparent	τ (RT)	-
Viscosity	0.075 Pa.s	τ (°C)	-
SG	1.05	τ (wet)	-
Shelf life	9 mnth/0°C - 5°C	Peel ( )	-
$T_{Service}$	-30°C to +85°C	σ_t	20 MPa
CTE (°C)	-	*E*	-
Flash Point	-	Elongation	-
Gap Filling	<0.125mm	H (Shore)	-
Notes:			

Sichel Werke (Henkel)

Permabond	Plus C910		Liquid
March 94			
Manufacturers Comments: Methyl cyanoacrylate. Low viscosity. Fast cure.			
Typical Use: Plastics, rubbers, metals and other materials.			
Cure: RT: 15 to 40 sec. Full strength: 24 hrs.			
Test Methods: Not Stated			
Physical		**Mechanical**	
Colour	Transparent	τ (RT)	-
Viscosity	0.085 Pa.s	τ (°C)	-
SG	1.09	τ (wet)	-
Shelf life	1 yr/0°C - 5°C	Peel ( )	-
$T_{Service}$	-30°C to +95°C	σ_f	25 MPa
CTE (°C)	-	E	-
Flash Point	-	Elongation	-
Gap Filling	<0.125mm	H (Shore)	-
Notes:			

Sichel (Henkel)	Sicomet 40		Liquid
December 91			
Manufacturers Comments: Ester and Ethyl cyanoacrylate. Low viscosity.			
Typical Use: Plastics and porous materials.			
Cure: RT: 25 secs. (Al), 3 - 5 secs. (Plastic), 1 - 2 secs. (Elastomer).			
Test Methods: DIN			
Physical		**Mechanical**	
Colour	Transparent	τ (RT, Al)	20 - 26 MPa
Viscosity	0.01 Pa.s	τ (100°C, Al)	~18 MPa
SG	1.01	τ (RT, 90%RH)	10 - 14 MPa
Shelf life	-	Peel ( )	-
$T_{Service}$	-30°C to +110°C	σ_f	~28 MPa
CTE (°C)	-	E	-
Flash Point	87°C	Elongation	-
Gap Filling	-	H (Shore)	-
Notes: Volume resistivity (DIN 53482): 5.37 x 10^{14} ohm.cm. Thermal conductivity: 0.2 $W.m^{-1}K^{-1}$. Dielectric constant (1 MHz, DIN 53483): 5.4. Refractive index: 1.49. Softening temperature: 165°C. Breakdown potential: 11.6 $kV.mm^{-1}$.			

Permabond	Plus C2010		Gel
March 94			
Manufacturers Comments: Ethyl cyanoacrylate. Thixotropic. Fast cure.			
Typical Use: Plastics, rubbers, metals and other materials. Suitable for porous materials.			
Cure: RT: 3 to 15 sec. Full strength: 24 hrs.			
Test Methods: Not Stated			
Physical		**Mechanical**	
Colour	Transparent	τ (RT)	-
Viscosity	(thixotropic)	τ (°C)	-
SG	1.05	τ (wet)	-
Shelf life	9 mnth/0°C - 5°C	Peel ( )	-
$T_{Service}$	-30°C to +85°C	σ_f	20 MPa
CTE (°C)	-	E	-
Flash Point	-	Elongation	-
Gap Filling	<0.375mm	H (Shore)	-
Notes:			

Sichel (Henkel)	Sicomet 50		Liquid
December 91			
Manufacturers Comments: Ethyl cyanoacrylate. Low viscosity. Meets the requirements of (USA) MIL-A-46050 C.			
Typical Use: Plastics and other materials.			
Cure: RT: 30 secs. (Al), 3 - 6 secs. (Plastic), 1 - 2 secs. (Elastomer).			
Test Methods: DIN			
Physical		**Mechanical**	
Colour	Transparent	τ (RT, Al)	20 - 26 MPa
Viscosity	0.04 Pa.s	τ (100°C, Al)	~18 MPa
SG	1.06	τ (RT, 90%RH)	10 - 14 MPa
Shelf life	-	Peel ( )	-
$T_{Service}$	-30°C to +110°C	σ_f	~28 MPa
CTE (°C)	-	E	-
Flash Point	87°C	Elongation	-
Gap Filling	-	H (Shore)	-
Notes: Volume resistivity (DIN 53482): 5.37 x 10^{14} ohm.cm. Thermal conductivity: 0.2 $W.m^{-1}K^{-1}$. Dielectric constant (1 MHz, DIN 53483): 5.4. Refractive index: 1.49. Softening temperature: 165°C. Breakdown potential: 11.6 $kV.mm^{-1}$.			

Sichel (Henkel)	Sicomet 63		Liquid
December 91			
Manufacturers Comments: Ethyl cyanoacrylate.			
Typical Use: Plastics and other materials.			
Cure: RT: 35 secs. (Al), 4 - 7 secs. (Plastic), 2 - 3 secs. (Elastomer).			
Test Methods: DIN			
Physical		**Mechanical**	
Colour	Transparent	τ (RT, Al)	20 - 26 MPa
Viscosity	0.08 Pa.s	τ (100°C, Al)	~18 MPa
SG	1.06	τ (RT, 90%RH)	10 - 14 MPa
Shelf life	-	Peel ( )	-
$T_{Service}$	-30°C to +110°C	σ_f	~28 MPa
CTE (°C)	-	*E*	-
Flash Point	87°C	Elongation	-
Gap Filling	-	H (Shore)	-
Notes: Volume resistivity (DIN 53482): 5.37 x 10^{14} ohm.cm. Thermal conductivity: 0.2 $W.m^{-1}K^{-1}$. Dielectric constant (1 MHz, DIN 53483): 5.4. Refractive index: 1.49. Softening temperature: 165°C. Breakdown potential: 11.6 $kV.mm^{-1}$.			

Sichel (Henkel)	Sicomet 77		Liquid
December 91			
Manufacturers Comments: Ethyl cyanoacrylate. High viscosity. Meets the requirements of (USA) MIL-A-46050 C.			
Typical Use: General purpose.			
Cure: RT: 50 secs. (Al), 5 - 10 secs. (Plastic), 5 - 8 secs. (Elastomer).			
Test Methods: DIN			
Physical		**Mechanical**	
Colour	Transparent	τ (RT, Al)	20 - 26 MPa
Viscosity	1.1 Pa.s	τ (100°C, Al)	~18 MPa
SG	1.08	τ (RT, 90%RH)	10 - 14 MPa
Shelf life	-	Peel ( )	-
$T_{Service}$	-30°C to +110°C	σ_f	~28 MPa
CTE (°C)	-	*E*	-
Flash Point	87°C	Elongation	-
Gap Filling	-	H (Shore)	-
Notes: Volume resistivity (DIN 53482): 5.37 x 10^{14} ohm.cm. Thermal conductivity: 0.2 $W.m^{-1}K^{-1}$. Dielectric constant (1 MHz, DIN 53483): 5.4. Refractive index: 1.49. Softening temperature: 165°C. Breakdown potential: 11.6 $kV.mm^{-1}$.			

Sichel (Henkel)	Sicomet 65		Liquid
December 91			
Manufacturers Comments: Ethyl cyanoacrylate. High viscosity.			
Typical Use: Plastics and metals.			
Cure: RT: 150 secs. (Al), 50 secs. (Plastic), 15 - 20 secs. (Elastomer).			
Test Methods: DIN			
Physical		**Mechanical**	
Colour	Transparent	τ (RT, Al)	24.5 - 27.6 MPa
Viscosity	20 Pa.s	τ (°C)	-
SG	1.12	τ (wet)	-
Shelf life	-	Peel ( )	-
$T_{Service}$	-30°C to +110°C	σ_f	-
CTE (°C)	-	*E*	-
Flash Point	87°C	Elongation	-
Gap Filling	-	H (Shore)	-
Notes: Volume resistivity (DIN 53482): 5.37 x 10^{14} ohm.cm. Thermal conductivity: 0.2 $W.m^{-1}K^{-1}$. Dielectric constant (1 MHz, DIN 53483): 5.4. Refractive index: 1.49. Softening temperature: 165°C. Breakdown potential: 11.6 $kV.mm^{-1}$.			

Sichel (Henkel)	Sicomet 85		Liquid
December 91			
Manufacturers Comments: Methyl cyanoacrylate. Low viscosity. Meets the requirements of (USA) MIL-A-46050 C.			
Typical Use: Metals and other materials.			
Cure: RT: 60 secs. (Al), 15 - 17 secs. (Plastic), 2 - 5 secs. (Elastomer).			
Test Methods: DIN			
Physical		**Mechanical**	
Colour	Transparent	τ (RT, Al)	20 - 26 MPa
Viscosity	0.035 Pa.s	τ (100°C, Al)	~18 MPa
SG	1.09	τ (RT, 90%RH)	10 - 14 MPa
Shelf life	-	Peel ( )	-
$T_{Service}$	-30°C to +110°C	σ_f	~28 MPa
CTE (°C)	-	*E*	-
Flash Point	87°C	Elongation	-
Gap Filling	-	H (Shore)	-
Notes: Volume resistivity (DIN 53482): 5.37 x 10^{14} ohm.cm. Thermal conductivity: 0.2 $W.m^{-1}K^{-1}$. Dielectric constant (1 MHz, DIN 53483): 5.4. Refractive index: 1.49. Softening temperature: 165°C. Breakdown potential: 12.9 $kV.mm^{-1}$.			

Sichel (Henkel)	Sicomet 99		Liquid
December 91			
Manufacturers Comments: Methyl cyanoacrylate. High viscosity. Meets the requirements of (USA) MIL-A-46050 C.			
Typical Use: Metals and other materials.			
Cure: RT: 80 secs. (Al), 15 - 20 secs. (Plastic), 3 - 8 secs. (Elastomer).			
Test Methods: DIN			
Physical		**Mechanical**	
Colour	Transparent	τ (RT, Al)	20 - 26 MPa
Viscosity	2 Pa.s	τ (100°C, Al)	~18 MPa
SG	1.11	τ (RT, 90%RH)	10 - 14 MPa
Shelf life	-	Peel ( )	-
$T_{Service}$	-30°C to +110°C	σ_t	~28 MPa
CTE (°C)	-	*E*	-
Flash Point	87°C	Elongation	-
Gap Filling	<0.2 mm.	H (Shore)	
Notes: Volume resistivity (DIN 53482): 5.37 x 10^{14} ohm.cm. Thermal conductivity: 0.2 W.m^{-1}K^{-1}. Dielectric constant (1 MHz, DIN 53483): 5.4. Refractive index: 1.49. Softening temperature: 165°C. Breakdown potential: 12.9 kV.mm^{-1}.			

Sichel (Henkel)	Sicomet 5139		Liquid
December 91			
Manufacturers Comments: Ethyl cyanoacrylate. High viscosity.			
Typical Use: Metals and plastics.			
Cure: RT: 95 secs (Al), 15 - 25 secs (Plastic), 15 - 20 secs (Elastomer)			
Test Methods: DIN			
Physical		**Mechanical**	
Colour	Black	τ (RT, Al)	24.5 - 27.6 MPa
Viscosity	2 Pa.s	τ (°C)	-
SG	1.09	τ (wet)	-
Shelf life	-	Peel ( )	-
$T_{Service}$	-30°C to +110°C	σ_t	-
CTE (°C)	-	*E*	-
Flash Point	87°C	Elongation	-
Gap Filling	-	H (Shore)	-
Notes: Volume resistivity (DIN 53482): 5.37 x 10^{14} ohm.cm. Dielectric constant (1 MHz, DIN 53483): 5.4. Softening temperature: 165°C. Breakdown potential: 11.6 kV.mm^{-1}.			

Sichel (Henkel)	Sicomet 5132		Gel
December 91			
Manufacturers Comments: Ethyl cyanoacrylate. Gel formulation.			
Typical Use: Metals and plastics.			
Cure: RT: 95 secs (Al), 15 - 25 secs (Plastic), 15 - 20 secs (Elastomer)			
Test Methods: DIN			
Physical		**Mechanical**	
Colour	Transparent	τ (RT, Al)	24.5 - 27.6 MPa
Viscosity	(Gel)	τ (°C)	-
SG	1.08	τ (wet)	-
Shelf life	-	Peel ( )	-
$T_{Service}$	-30°C to +110°C	σ_t	-
CTE (°C)	-	*E*	-
Flash Point	87°C	Elongation	-
Gap Filling	-	H (Shore)	-
Notes: Volume resistivity (DIN 53482): 5.37 x 10^{14} ohm.cm. Dielectric constant (1 MHz, DIN 53483): 5.4. Refractive index: 1.49. Softening temperature: 165°C. Breakdown potential: 11.6 kV.mm^{-1}.			

Sichel (Henkel)	Sicomet 6021		Liquid
December 91			
Manufacturers Comments: Butyl cyanoacrylate. Medium viscosity.			
Typical Use: -			
Cure: RT: 60 secs (Al), 4 - 8 secs (Plastic), 3 - 5 secs (Elastomer).			
Test Methods: DIN			
Physical		**Mechanical**	
Colour	Transparent	τ (RT, Al)	18 - 20 MPa
Viscosity	0.2 Pa.s	τ (°C)	-
SG	1.01	τ (wet)	-
Shelf life	-	Peel ( )	-
$T_{Service}$	-30°C to +110°C	σ_t	-
CTE (°C)	-	*E*	-
Flash Point	87°C	Elongation	-
Gap Filling	-	H (Shore)	-
Notes: Volume resistivity (DIN 53482): 5.37 x 10^{14} ohm.cm. Dielectric constant (1 MHz, DIN 53483): 5.4. Softening temperature: 165°C. Breakdown potential: 11.6 kV.mm^{-1}.			

Sichel (Henkel) — Sicomet 6085 — Liquid

December 91

Manufacturers Comments: Butyl cyanoacrylate. Low viscosity.

Typical Use: -

Cure: RT: 60 secs (Al), 4 - 8 secs (Plastic), 3 - 5 secs (Elastomer).

Test Methods: DIN

Physical		Mechanical	
Colour	Transparent	τ (RT, Al)	18 - 20 MPa
Viscosity	0.02 Pa.s	τ (°C)	-
SG	1.01	τ (wet)	-
Shelf life	-	Peel ( )	-
$T_{Service}$	-30°C to +110°C	σ_f	-
CTE (°C)	-	*E*	-
Flash Point	87°C	Elongation	-
Gap Filling	-	H (Shore)	-

Notes: Volume resistivity (DIN 53482): 5.37 x 10^{14} ohm.cm. Dielectric constant (1 MHz, DIN 53483): 5.4. Softening temperature: 165°C. Breakdown potential: 11.6 kV.mm^{-1}.

Sichel (Henkel) — Sicomet 8200 — Liquid

December 91

Manufacturers Comments: Ester and Ethyl cyanoacrylate. Low viscosity.

Typical Use: Elastomers.

Cure: RT: 75 secs. (Al), 15 - 16 secs. (Plastic), 2 - 3 secs. (Elastomer).

Test Methods: DIN

Physical		Mechanical	
Colour	Transparent	τ (RT, Al)	20 - 26 MPa
Viscosity	0.04 Pa.s	τ (100°C, Al)	~18 MPa
SG	1.06	τ (RT, 90%RH)	10 - 14 MPa
Shelf life	-	Peel ( )	-
$T_{Service}$	-30°C to +110°C	σ_f	~28 MPa
CTE (°C)	-	*E*	-
Flash Point	87°C	Elongation	-
Gap Filling	-	H (Shore)	-

Notes: Volume resistivity (DIN 53482): 5.37 x 10^{14} ohm.cm. Thermal conductivity: 0.2 W.m^{-1}K^{-1}. Dielectric constant (1 MHz, DIN 53483): 5.4. Refractive index: 1.49. Softening temperature: 165°C. Breakdown potential: 11.6 kV.mm^{-1}.

Sichel (Henkel) — Sicomet 7000 — Liquid

December 91

Manufacturers Comments: Methyl cyanoacrylate. Medium viscosity.

Typical Use: Metals and other materials.

Cure: RT: 60 secs. (Al), 15 - 18 secs. (Plastic), 2 - 6 secs. (Elastomer).

Test Methods: DIN

Physical		Mechanical	
Colour	Transparent	τ (RT, Al)	20 - 26 MPa
Viscosity	0.12 Pa.s	τ (100°C, Al)	~18 MPa
SG	1.09	τ (RT, 90%RH)	10 - 14 MPa
Shelf life	-	Peel ( )	-
$T_{Service}$	-30°C to +110°C	σ_f	~28 MPa
CTE (°C)	-	*E*	-
Flash Point	87°C	Elongation	-
Gap Filling	-	H (Shore)	-

Notes: Volume resistivity (DIN 53482): 5.37 x 10^{14} ohm.cm. Thermal conductivity: 0.2 W.m^{-1}K^{-1}. Dielectric constant (1 MHz, DIN 53483): 5.4. Refractive index: 1.49. Softening temperature: 165°C. Breakdown potential: 12.9 kV.mm^{-1}.

Sichel (Henkel) — Sicomet 8300 — Liquid

December 91

Manufacturers Comments: Ester and Ethyl cyanoacrylate. Low viscosity.

Typical Use: Elastomers.

Cure: RT: 30 secs. (Al), 3 - 6 secs. (Plastic), 1 - 2 secs. (Elastomer).

Test Methods: DIN

Physical		Mechanical	
Colour	Transparent	τ (RT, Al)	20 - 26 MPa
Viscosity	0.03 Pa.s	τ (100°C, Al)	~18 MPa
SG	1.05	τ (RT, 90%RH)	10 - 14 MPa
Shelf life	-	Peel ( )	-
$T_{Service}$	-30°C to +110°C	σ_f	~28 MPa
CTE (°C)	-	*E*	-
Flash Point	87°C	Elongation	-
Gap Filling	-	H (Shore)	-

Notes: Volume resistivity (DIN 53482): 5.37 x 10^{14} ohm.cm. Thermal conductivity: 0.2 W.m^{-1}K^{-1}. Dielectric constant (1 MHz, DIN 53483): 5.4. Refractive index: 1.49. Softening temperature: 165°C. Breakdown potential: 11.6 kV.mm^{-1}.

Sichel (Henkel)	Sicomet 8400		Liquid
December 91			
Manufacturers Comments: Ethyl cyanoacrylate. Medium viscosity.			
Typical Use: Plastics and other materials.			
Cure: RT: 45 secs. (Al), 4 - 8 secs. (Plastic), 2 - 4 secs. (Elastomer).			
Test Methods: DIN			
Physical		**Mechanical**	
Colour	Transparent	τ (RT, Al)	20 - 26 MPa
Viscosity	0.21 Pa.s	τ (100°C, Al)	~18 MPa
SG	1.06	τ (RT, 90%RH)	10 - 14 MPa
Shelf life	-	Peel ( )	-
$T_{Service}$	-30°C to +110°C	σ_f	~28 MPa
CTE (°C)	-	*E*	-
Flash Point	87°C	Elongation	-
Gap Filling	-	H (Shore)	-

Notes: Volume resistivity (DIN 53482): 5.37 x 10^{14} ohm.cm. Thermal conductivity: 0.2 $W.m^{-1}K^{-1}$. Dielectric constant (1 MHz, DIN 53483): 5.4. Refractive index: 1.49. Softening temperature: 165°C. Breakdown potential: 11.6 $kV.mm^{-1}$.

Sichel (Henkel)	Sicomet 9010		Liquid
December 91			
Manufacturers Comments: Alkoxyethyl cyanoacrylate. Medium viscosity.			
Typical Use: -			
Cure: RT: 50 secs (Al), 15 - 20 secs (Plastic), 6 - 10 secs (Elastomer).			
Test Methods: DIN			
Physical		**Mechanical**	
Colour	Transparent	τ (RT, Al)	16 - 18 MPa
Viscosity	0.16 Pa.s	τ (°C)	-
SG	1.06	τ (wet)	-
Shelf life	-	Peel ( )	-
$T_{Service}$	-30°C to +110°C	σ_f	-
CTE (°C)	-	*E*	-
Flash Point	87°C	Elongation	-
Gap Filling	-	H (Shore)	-

Notes: Volume resistivity (DIN 53482): 5.37 x 10^{14} ohm.cm. Dielectric constant (1 MHz, DIN 53483): 5.4. Softening temperature: 165°C. Breakdown potential: 11.6 $kV.mm^{-1}$.

Sichel (Henkel)	Sicomet 9000		Liquid
December 91			
Manufacturers Comments: Alkoxyethyl cyanoacrylate. Low viscosity.			
Typical Use: -			
Cure: RT: 50 secs (Al), 15 - 20 secs (Plastic), 7 - 10 secs (Elastomer).			
Test Methods: DIN			
Physical		**Mechanical**	
Colour	Transparent	τ (RT, Al)	16 - 18 MPa
Viscosity	0.015 Pa.s	τ (°C)	-
SG	1.06	τ (wet)	-
Shelf life	-	Peel ( )	-
$T_{Service}$	-30°C to +110°C	σ_f	-
CTE (°C)	-	*E*	-
Flash Point	87°C	Elongation	-
Gap Filling	-	H (Shore)	-

Notes: Volume resistivity (DIN 53482): 5.37 x 10^{14} ohm.cm. Dielectric constant (1 MHz, DIN 53483): 5.4. Softening temperature: 165°C. Breakdown potential: 11.6 $kV.mm^{-1}$.

Sichel (Henkel)	Sicomet Thixotrope		Liquid
December 91			
Manufacturers Comments: Ethyl cyanoacrylate. Thixotropic.			
Typical Use: General purpose.			
Cure: RT: 95secs (Al), 15 - 25 secs. (Plastic), 15 - 20 secs (Elastomer)			
Test Methods: DIN			
Physical		**Mechanical**	
Colour	Transparent	τ (RT, Al)	20 - 26 MPa
Viscosity	-	τ (100°C, Al)	~18 MPa
SG	1.08	τ (RT, 90%RH)	10 - 14 MPa
Shelf life	-	Peel ( )	-
$T_{Service}$	-30°C to +110°C	σ_f	~28 MPa
CTE (°C)	-	*E*	-
Flash Point	87°C	Elongation	-
Gap Filling	-	H (Shore)	-

Notes: Volume resistivity (DIN 53482): 5.37 x 10^{14} ohm.cm. Thermal conductivity: 0.2 $W.m^{-1}K^{-1}$. Dielectric constant (1 MHz, DIN 53483): 5.4. Refractive index: 1.49. Softening temperature: 165°C. Breakdown potential: 11.6 $kV.mm^{-1}$.

ThreeBond

ThreeBond	TB1701		Liquid
(August 95)			
Manufacturers Comments: Methyl cyanoacrylate. Meets the requirements of (USA) MIL-A-46050C, Type 1, Class 1.			
Typical Use: Metals, wicking.			
Cure: <40 sec.			
Test Methods: Not Stated (USA)			
Physical		**Mechanical**	
Colour	-	τ (steel)	>11.7 MPa
Viscosity	<0.005 Pa.s	τ (°C)	-
SG	1.10	τ (wet)	-
Shelf life	-	Peel ( )	-
$T_{Service}$	-	σ_f	-
CTE (°C)	-	*E*	-
Flash Point	-	Elongation	-
Gap Filling	Low	H (Shore)	-
Notes:			

ThreeBond	TB1703		Liquid
(August 95)			
Manufacturers Comments: Methyl cyanoacrylate. Meets the requirements of (USA) MIL-A-46050C, Type 1, Class 2.			
Typical Use: Metals, general purpose.			
Cure: <30 sec.			
Test Methods: Not Stated (USA)			
Physical		**Mechanical**	
Colour	-	τ (steel)	>7.6 MPa
Viscosity	0.1 Pa.s	τ (°C)	-
SG	1.10	τ (wet)	-
Shelf life	-	Peel ( )	-
$T_{Service}$	-	σ_f	-
CTE (°C)	-	*E*	-
Flash Point	-	Elongation	-
Gap Filling	Low	H (Shore)	-
Notes:			

ThreeBond	TB1702/N		Liquid
(August 95)			
Manufacturers Comments: Methyl cyanoacrylate. Meets the requirements of (USA) MIL-A-46050C, Type 1, Class 1.			
Typical Use: Metals, general purpose.			
Cure: <40 sec.			
Test Methods: Not Stated (USA)			
Physical		**Mechanical**	
Colour	-	τ (steel)	>9.6 MPa
Viscosity	0.025 Pa.s	τ (°C)	-
SG	1.10	τ (wet)	-
Shelf life	-	Peel ( )	-
$T_{Service}$	-	σ_f	-
CTE (°C)	-	*E*	-
Flash Point	-	Elongation	-
Gap Filling	Low	H (Shore)	-
Notes:			

ThreeBond	TB1741		Liquid
(August 95)			
Manufacturers Comments: Ethyl cyanoacrylate. Meets the requirements of (USA) MIL-A-46050C, Type 2, Class 1.			
Typical Use: Plastics, rubber, wicking.			
Cure: <15 sec.			
Test Methods: Not Stated (USA)			
Physical		**Mechanical**	
Colour	-	τ (steel)	>7.6 MPa
Viscosity	<0.005 Pa.s	τ (°C)	-
SG	1.05	τ (wet)	-
Shelf life	-	Peel ( )	-
$T_{Service}$	-	σ_f	-
CTE (°C)	-	*E*	-
Flash Point	-	Elongation	-
Gap Filling	Low	H (Shore)	-
Notes:			

See Key to Data Sheets - Pages 63 to 65

ThreeBond	TB1742D		Liquid
(August 95)			
Manufacturers Comments: Ethyl cyanoacrylate. Meets the requirements of (USA) MIL-A-46050C, Type 2, Class 1.			
Typical Use: Plastics, rubber, general purpose.			
Cure: <20 sec.			
Test Methods: Not Stated (USA)			
Physical		**Mechanical**	
Colour	-	τ (steel)	>9.6 MPa
Viscosity	0.025 Pa.s	τ (°C)	-
SG	1.05	τ (wet)	-
Shelf life	-	Peel ( )	-
$T_{Service}$	-	σ_t	-
CTE (°C)	-	E	-
Flash Point	-	Elongation	-
Gap Filling	Low	H (Shore)	-
Notes:			

ThreeBond	TB1747		Liquid
(August 95)			
Manufacturers Comments: Ethyl cyanoacrylate. Meets the requirements of (USA) MIL-A-46050C, Type 2, Class 3.			
Typical Use: Plastics, rubber, general purpose.			
Cure: <60 sec.			
Test Methods: Not Stated (USA)			
Physical		**Mechanical**	
Colour	-	τ (steel)	>9.6 MPa
Viscosity	1.6 Pa.s	τ (°C)	-
SG	1.05	τ (wet)	-
Shelf life	-	Peel ( )	-
$T_{Service}$	-	σ_t	-
CTE (°C)	-	E	-
Flash Point	-	Elongation	-
Gap Filling	Good	H (Shore)	-
Notes:			

ThreeBond	TB1743		Liquid
(August 95)			
Manufacturers Comments: Ethyl cyanoacrylate. Meets the requirements of (USA) MIL-A-46050C, Type 2, Class 2.			
Typical Use: Plastics, rubber, general purpose.			
Cure: <30 sec.			
Test Methods: Not Stated (USA)			
Physical		**Mechanical**	
Colour	-	τ (steel)	>9.6 MPa
Viscosity	0.1 Pa.s	τ (°C)	-
SG	1.05	τ (wet)	-
Shelf life	-	Peel ( )	-
$T_{Service}$	-	σ_t	-
CTE (°C)	-	E	-
Flash Point	-	Elongation	-
Gap Filling	Low	H (Shore)	-
Notes:			

ThreeBond	TB1781		Liquid
(August 95)			
Manufacturers Comments: Ethyl cyanoacrylate. Meets the requirements of (USA) MIL-A-46050C, Type 2, Class 1.			
Typical Use: All substrates, wicking.			
Cure: <15 sec.			
Test Methods: Not Stated (USA)			
Physical		**Mechanical**	
Colour	-	τ (steel)	>13.8 MPa
Viscosity	0.003 Pa.s	τ (°C)	-
SG	1.05	τ (wet)	-
Shelf life	-	Peel ( )	-
$T_{Service}$	-	σ_t	-
CTE (°C)	-	E	-
Flash Point	-	Elongation	-
Gap Filling	Low	H (Shore)	-
Notes: -			

ThreeBond	TB1782		Liquid
(August 95)			
Manufacturers Comments: Ethyl cyanoacrylate. Meets the requirements of (USA) MIL-A-46050C, Type 2, Class 2.			
Typical Use: All substrates, general purpose.			
Cure: <20 sec.			
Test Methods: Not Stated (USA)			
Physical		**Mechanical**	
Colour	-	τ (steel)	>13.8 MPa
Viscosity	0.08 Pa.s	τ (°C)	-
SG	1.05	τ (wet)	-
Shelf life	-	Peel ( )	-
$T_{Service}$	-	σ_f	-
CTE (°C)	-	*E*	-
Flash Point	-	Elongation	-
Gap Filling	Low	H (Shore)	-
Notes:			

ThreeBond	TB1785		Liquid
(August 95)			
Manufacturers Comments: Ethyl cyanoacrylate. Meets the requirements of (USA) MIL-A-46050C, Type 2, Class 2.			
Typical Use: All substrates, general purpose.			
Cure: <30 sec.			
Test Methods: Not Stated (USA)			
Physical		**Mechanical**	
Colour	-	τ (steel)	>9.6 MPa
Viscosity	0.085 Pa.s	τ (°C)	-
SG	1.05	τ (wet)	-
Shelf life	-	Peel ( )	-
$T_{Service}$	-	σ_f	-
CTE (°C)	-	*E*	-
Flash Point	-	Elongation	-
Gap Filling	Low	H (Shore)	-
Notes: -			

ThreeBond	TB1783		Liquid
(August 95)			
Manufacturers Comments: Ethyl cyanoacrylate. Meets the requirements of (USA) MIL-A-46050C, Type 2, Class 3.			
Typical Use: All substrates, large gaps.			
Cure: <30 sec.			
Test Methods: Not Stated (USA)			
Physical		**Mechanical**	
Colour	-	τ (steel)	>13.8 MPa
Viscosity	0.9 Pa.s	τ (°C)	-
SG	1.05	τ (wet)	-
Shelf life	-	Peel ( )	-
$T_{Service}$	-	σ_f	-
CTE (°C)	-	*E*	-
Flash Point	-	Elongation	-
Gap Filling	Good	H (Shore)	-
Notes:			

ThreeBond	TB1786		Liquid
(August 95)			
Manufacturers Comments: Ethyl cyanoacrylate. Meets the requirements of (USA) MIL-A-46050C, Type 2, Class 2.			
Typical Use: All substrates, general purpose.			
Cure: <30 sec.			
Test Methods: Not Stated (USA)			
Physical		**Mechanical**	
Colour	-	τ (steel)	>9.6 MPa
Viscosity	0.15 Pa.s	τ (°C)	-
SG	1.05	τ (wet)	-
Shelf life	-	Peel ( )	-
$T_{Service}$	-	σ_f	-
CTE (°C)	-	*E*	-
Flash Point	-	Elongation	-
Gap Filling	Low	H (Shore)	-
Notes:			

ThreeBond	TP127H		Liquid
(August 95)			
Manufacturers Comments: Ethyl/Isopropyl cyanoacrylate.			
Typical Use: All substrates, large gaps.			
Cure: <20 sec.			
Test Methods: Not Stated (USA)			
Physical		**Mechanical**	
Colour	-	τ (steel)	>19.2 MPa
Viscosity	12 - 42 Pa.s	τ (°C)	-
SG	1.07	τ (wet)	-
Shelf life	-	Peel ( )	-
$T_{Service}$	-	σ_f	-
CTE (°C)	-	*E*	-
Flash Point	-	Elongation	-
Gap Filling	Very good	H (Shore)	-
Notes:			

UHU	Flash Gel		Gel
February 94			
Manufacturers Comments: Instant bonding and repair of domestic materials. Non-drip. Bonds are more flexible than those of the liquid version.			
Typical Use: Pottery, porcelain, metals, alloys, plastics and glass (Domestic). Not suitable for polyethylene or polypropylene.			
Cure: A few seconds to 3 minutes.			
Test Methods:			
Physical		**Mechanical**	
Colour	Transparent	τ (RT)	-
Viscosity	0.07 Pa.s	τ (°C)	-
SG	1.03	τ (wet)	-
Shelf life	-	Peel ( )	-
$T_{Service}$	-	σ_f	-
CTE (°C)	-	*E*	-
Flash Point	82°C	Elongation	-
Gap Filling	Low	H (Shore)	-
Notes:			

UHU

UHU	Flash		Liquid
February 94			
Manufacturers Comments: Instant bonding and repair of non-porous materials.			
Typical Use: Pottery, porcelain, metals, alloys, plastics and glass (Domestic). Not suitable for polyethylene or polypropylene.			
Cure: A few seconds to 1 minute.			
Test Methods:			
Physical		**Mechanical**	
Colour	Transparent	τ (RT)	-
Viscosity	0.07 Pa.s	τ (°C)	-
SG	1.07	τ (wet)	-
Shelf life	-	Peel ( )	-
$T_{Service}$	-	σ_f	-
CTE (°C)	-	*E*	-
Flash Point	82°C	Elongation	-
Gap Filling	Low	H (Shore)	-
Notes:			

EPOXY ADHESIVES

NOTES

There are numerous commercially available epoxy adhesives supplied as One-part Pastes, Two-part Pastes or Films.

- One-part pastes and films are generally heat curing, although a few one-part paste adhesives are formulated to cure on exposure to UV.
- Films may be supplied with or without the adhesive impregnated into a thin textile carrier or support (aids handling, placement and control of bondline thickness). Films are usually provided with a protective (contrast coloured) backing sheet to avoid contamination during handling and cutting. This must be removed completely during assembly as it will prevent adhesion.
- Two-part pastes are normally room temperature (RT) curing, although heating can be used to accelerate the cure. This may be essential to realise the full service temperature potential of the adhesive and provide an acceptable rate of cure. The paste viscosities range from thin liquids to putty. Many are thixotropic or non slumping to aid handling and placement in the joint.

Special grades exist which foam during cure to fill gaps, giving expansions between 150% and 500%. Some are filled with metal powders to be electrically conductive or receptive to induction heating. Others may be unfilled or filled with non-conductive phases (thickening agents). Traditionally asbestos was widely used as a filler material, but this is now being replaced where possible. Some have an addition of small glass micro-spheres which aid control of the bondline thickness; these are usually described as 'syntactic' or 'bondline-control'.

Epoxy adhesives are generally well represented by property data for a variety of test conditions and exposure to different environments. Separate comparison tables have been provided for ONE PART, TWO PART and FILM epoxy adhesives.

Main Uses

Epoxy adhesives are widely used in most industries for the structural bonding of most types of materials (metals, wood, plastics, composites, ceramics, glass, concrete, masonry, etc). Primers may be recommended for some materials. They are also used to repair structures; bonding patches and crack-filling. Some adhesives are specifically designed for aerospace applications or specialised mechanical and electrical assembly.

COMPARISON TABLES

One Part Epoxy Paste Adhesives

Company	Adhesive	Colour	LSS	TS	TM	Peel	Hs	El (%)	SG	Viscosity (Pa.s)	Service (°C)	Tg (°C)	Cure (°C)	Pg. No.
3M	1386	Cream	●	-	-	○	-	-	-	-	-	-	177	157
	1469	Cream	◕	-	-	○	-	-	-	-	-	-	177	157
	2086	Grey	●	-	-	●	-	-	-	-	-	-	177	159
	2214 Hi-Density	Grey	●	-	-	○	-	-	-	-	-	-	120	159
	2214 Hi-Flex	Grey	◕	-	-	○	-	-	-	-	-	-	120	159
	2214 Hi-Temp	Grey	○	-	-	○	-	-	-	-	-	-	120	160
	2214 Non-metallic	Cream	◕	-	-	○	-	-	-	-	-	-	120	160
	2214 Regular	Grey	●	-	-	○	-	-	-	-	-	-	120	160
	2290	Amber	●	-	-	○	-	-	-	-	-	-	171	161
	New 2214 Hi-Temp	Grey	◕	-	-	○	-	-	-	-	-	-	120	165
Chemetall	Naftotech 7140	Blue	○	●	-	◕	-	3	1.4	-	<60	87	130 - 180	166
Ciba Polymers	Epocast 1610	Off-white	-	-	-	-	-	-	0.50	-	-55/+80	-	120	170
	Epocast 1614	Off-white	○	-	-	-	-	-	0.75	-	<180	-	180	170
	Redux 204	Black	○	-	-	-	-	-	1.06	3700	-55/+90	-	120 - 170	171
Delo	Monopox 1196	Metallic	●	●	●	-	-	1.4	1.4	290	-55/+200	113	130 - 180	179
	Monopox 1197	Metallic	◕	◕	◕	-	-	1.4	1.4	580	-55/+200	112	130 - 180	180
	Monopox 1494	-	●	-	●	-	-	-	-	280	-55/+150	-	130 - 180	180
Dexter Hysol	EA 9346.2	Burgundy	●	○	◕	-	80D	3	1.2	1000	<150	120	120 - 175	184
	EA 9432 NA	Grey	◕	●	●	-	90D	1.2	1.38	225	<150	-	120 - 175	186
Dunlop	SAS 510	Grey	◕	-	-	◕	-	-	1.25	-	-	-	150 - 190	190
	SAS 515	Black	-	-	-	-	-	-	-	-	-	-	150 -190	190
Grace	Eccobond 68C	Metallic	○	-	-	-	-	-	-	-	-	-	80 - 160	194
	Eccobond 144A	Black	○	-	-	-	-	-	1.5	95	-62/+230	-	120 - 180	195
	Eccobond 927-10E	Lt. Yellow	○	-	-	-	80D	-	1.33	11	-	-	90 - 120	196
	Eccobond 928-69-10	Lt. Yellow	-	-	-	-	80D	-	1.21	38	-	-	85	196
	Eccobond 2332-17	Black	◕	-	-	-	75D	-	1.2	105	-	-	100 - 150	196
	Eccobond 50248 F	Dk. Yellow	○	-	-	-	85D	-	1.2	12	-	75	80 - 100	196
	Eccobond A 304-6	Lt. Blue	○	-	-	-	80D	-	1.5	22	-	-	120 - 180	197
	Eccobond A 316-34	Tan	-	-	-	-	85D	-	1.55	43	-	-	120 - 180	197
	Eccobond A 329-1	Beige	◕	-	-	-	85D	-	1.3	43	<155	-	100 - 180	197
	Eccobond A 401-12	Black	-	-	-	-	80D	-	1.75	45	-	-	120 - 180	197
Hardman	Monopoxy 16365	Blue	○	-	-	○	85D	-	1.38	80	<125	-	107	202
	Monopoxy 16551	Grey	●	-	-	○	90D	-	1.59	-	<135	-	121	202
	Monopoxy 18512	Grey	◕	-	-	○	85D	-	1.52	-	>150	-	121	202
Hitachi Kasei	Hi-Bon 3530	Cream	◕	-	○	●	70D	8	1.1	900	-	130	120	203
	Hi-Bon 3550	Cream	●	-	○	●	80D	8	1.15	130	-	97	150	203
Loctite	Structural Epoxy 344	-	-	-	-	-	-	-	-	-	<100	-	230	208
Panacol Elosol	Vitralit 1500LI	-	-	-	-	-	85D	-	1.45	8	<175	150	UV	208
	Vitralit 1505	-	-	-	-	-	86D	-	1.12	0.45	<175	150	UV	209
	Vitralit 1550	-	-	-	-	-	60D	-	1.46	6	<135	30	UV	209
	Vitralit 1555	-	-	-	-	-	76D	-	1.1	0.28	<150	35	UV	209
	Vitralit 1722	-	-	-	-	-	75D	-	1.14	5.5	<120	40	UV	209
	Vitralit 1731	Transp.	-	-	-	-	65D	-	1.12	0.65	<80	40	UV	210
	Vitralit 4102	-	-	-	○	-	83D	-	1.23	32	-50/+180	-	UV/heat 120	210
Structil	ST-1039	Dk. Grey	◕	-	-	-	-	-	-	300	-55/+200	-	150 - 200	214
	ST-1110	-	-	-	-	-	-	-	-	-	-	-	Heat	215
Structural Adhesives	S - 106 E	Grey	◕	-	-	●	-	-	1.3	-	-	-	140 - 200	216

Continued...

See Key to Comparison Tables - Pages 63 to 65

One-Part Epoxy Paste Adhesives - continued.

Company	Adhesive	Colour	LSS	TS	TM	Peel	Hs	El (%)	SG	Viscosity (Pa.s)	Service (°C)	Tg (°C)	Cure (°C)	Pg. No.
Sunstar Engineering	Penguin Cement 1063	Brown	◑	-	-	●	-	-	-	1100	-	-	170	220
	Penguin Cement 1064	Black	◑	-	-	●	-	-	-	700	-	-	170	220
	Penguin Cement 1065T	Brown	◑	-	-	●	-	-	-	425	-	-	170	220
	Penguin Cement 1069	Black	○	-	-	●	-	-	-	1700	-	-	170	221
	Penguin Cement 1069B	Black	○	-	-	●	-	-	-	1000	-	-	170	221
	Penguin Cement 1081L	Brown	◑	-	-	●	-	-	-	500	-	-	170	221
	Penguin Cement 1086	Brown	◑	-	-	●	-	-	-	500	-	-	170	221
TACC	EK 2241	Metallic	◑	●	-	-	84D	2	1.35	800	-50/+150	-	Ind/125 - 175	222
	ER 2241	Metallic	-	●	-	-	90D	-	1.44	-	-50/+150	-	125 - 175	223
	ER 3126	Black	◑	●	-	-	86D	-	1.57	130	-	-	120 - 175	223
	ER 7040	-	-	-	-	-	50D	-	-	-	-	-50	120 - 175	224
	ER 7045	Grey	-	-	-	-	-	-	1.64	-	-	-	120 - 175	224
	ER 7124	Green	◑	-	-	-	75D	-	1.33	350	-40/+155	-	125 - 150	224
	ER 7138	Lt. Grey	-	○	-	○	75D	-	1.25	-	-	-	100 - 160	225
	ER 7820	Beige	-	-	-	-	80D	-	1.49	60	-	-	100 - 175	225
	ER 7821	Beige	-	-	-	-	80D	-	1.49	20	-40/+200	-	100 - 160	225
	KR 2241	Metallic	◑	●	-	-	78D	-	1.4	800	-	-	Ind.	225
Teroson	Terokal 3919 S	Black	○	-	-	●	-	-	1.2	58	-40/+90	-	160 - 230	226
	Terokal 4520-34	Black	◑	◑	◑	-	64D	13	1.3	65	-40/+90	-	180	226
	Terokal 5026	Black	◑	-	-	-	68D	-	1.3	65	-40/+90	-	180	226
	Terokal 5027	Black	◑	-	-	-	75D	-	1.4	65	-40/+90	-	180	226
	Terokal 5811	Black	○	○	-	-	-	15	1.5	-	-40/+80	-	160 -230	227
	Terokal 6049	Yellow	-	-	-	-	-	-	1.67	-	-40/+90	-	180	227
	Terokal 8554	White	○	○	-	◑	-	45	1.5	-	-40/+80	-	170 - 220	228
	Terokal 8554-3	Yellow	○	-	-	-	-	-	1.5	-	-40/+80	-	170 - 220	228
	Terorehm 2000	Black	◑	-	-	-	-	-	1.35	-	-40/+90	-	180	228
Tivoli	Tivo 9870/2	Black	○	-	-	-	-	-	1.5	300	-	-	180	228
	Tivo 9870/3	Black	○	-	-	-	-	-	1.5	500	-	-	180	229

Two Part Epoxy Paste Adhesives

Company	Adhesive	Colour	LSS	TS	TM	Peel	Hs	El. (%)	SG	Visc. (Pa.s)	Service (°C)	Tg (°C)	Pot life (mins)	Cure (°C)	Pg. No.
3E	Polarjoint	Grey	-	◑	-	-	95A	-	1.3	-	+121	-	-	RT	157
3M	1648 B/A	Green	◑	-	-	○	-	-	-	275	-	-	-	RT	157
	1751 B/A	Grey	○	-	-	○	-	-	-	700	-	-	-	RT	158
	1751L B/A	Grey	○	-	-	○	-	-	-	250	-	-	-	RT	158
	1838 B/A	Green	◑	-	-	○	-	-	-	400	-	-	-	RT	158
	1838L B/A	Green	◑	-	-	○	-	-	-	10	-	-	-	RT	158
	2158 B/A	Grey	○	-	-	○	-	-	-	375	-	-	-	RT	159
	2216 B/A Clear	Translucent	◑	-	-	○	-	-	-	10	-	-	-	RT	160
	2216 B/A Grey	Grey	◑	-	-	○	-	-	-	80	-	-	-	RT	161
	3501 B/A	Grey	◑	-	-	○	-	-	-	500	-	-	-	RT	161
	DP-100 (Duo-Pak)	Transparent	○	-	-	○	-	-	-	13	-	-	-	RT	163
	DP-100 FR	White	○	-	-	○	-	-	-	50	-	-	-	RT	163
	DP-100 NS	Translucent	○	-	-	○	-	-	-	95	-	-	-	RT	163
	DP-110	Translucent	◑	-	-	○	-	-	-	50	-	-	-	RT	163
	DP-190	Grey	◑	-	-	○	-	-	-	80	-	-	-	RT	164
	DP-260	Translucent	◑	-	-	○	-	-	-	60	-	-	-	RT	164
	DP-420	Off-white	●	-	-	●	-	-	-	45	-	-	-	RT	164
	DP-460	Off-white	●	-	-	●	-	-	-	45	-	-	-	RT	164

Continued...

Two-Part Epoxy Paste Adhesives - continued.

Company	Adhesive	Colour	LSS	TS	TM	Peel	Hs	El. (%)	SG	Visc. (Pa.s)	Service (°C)	Tg (°C)	Pot life (mins)	Cure (°C)	Pg. No.
Bostik	2003	Grey	-	-	-	-	-	-	1.6	-	-30/+120	-	-	RT	165
	EP 500	Transp.	◑	-	-	-	-	-	-	-	<120	-	-	RT - 60	165
Wm. Canning	Epoxy Putty	Lt. Grey	-	-	-	-	-	-	-	-	-	-	-	RT	165
(Hylomar)	Marston Tough Bond	Translucent	-	-	-	-	-	-	-	-	-	-	-	RT	166
	Tilebond	-	-	-	-	-	-	-	-	-	<150	-	-	-	166
CECA	Epoglue	Pink	-	-	-	◑	-	-	1.73	40	-	-	-	RT	166
Ciba Polymers	Araldite 2010	Lt. yellow	◑	-	-	◑	-	-	1.2	80	-	32	-	RT - 60	167
	Araldite 2011	Lt. yellow	●	-	-	○	-	-	1.05	45	-	-	-	RT - 60	167
	Araldite 2012	Lt. yellow	◑	-	-	○	-	-	1.18	30	-	-	-	RT - 60	167
	Araldite 2013	Grey	◑	-	-	○	-	-	1.2	-	-	-	-	RT - 60	167
	Araldite 2014	Grey	◑	-	-	◑	-	-	1.6	90	-	-	-	RT - 60	168
	Araldite 2015	Beige	◑	-	-	○	-	-	1.4	-	-	-	-	RT - 60	168
	Araldite 2016	Green	◑	○	○	●	-	58	1.05	60	<40	32.5	-	RT - 60	168
	Araldite 2017	Yellow	◑	○	○	●	-	190	1.15	80	-	25	-	RT - 60	168
	Araldite 2019	Lt. yellow	◑	-	-	○	-	-	1.0	3.5	<50	57	-	RT - 60	169
	Araldite 2020	Transp.	◑	-	-	○	-	-	1.1	0.15	-	40	-	RT - 60	169
	AV 100/HV 100	-	◑	-	◑	-	-	-	-	-	<150	-	-	RT - 100	169
	AV 138M/HV 998	Beige	○	-	-	-	-	-	-	90	<150	-	-	RT - 60	169
	Epibond 1559 A/B	Lt. Grey	○	-	-	-	-	-	1.2	80	-55/+80	-	-	RT - 70	170
	Epocast 1617 A/B	Off-white	-	-	-	-	-	-	0.7	-	-55/+80	-	-	RT - 52	170
	Redux 252	Blue	-	-	-	-	-	-	0.63	-	-55/+140	-	-	RT - 70	171
	Redux 255 A/B	Green	-	-	-	-	-	-	0.65	-	-55/+140	-	-	RT - 70	171
	Redux 408 A/B	Grey/Green	◑	-	-	-	-	-	1.25	40	-55/+100	-	-	RT - 70	173
	Redux 420 A/B	Green	●	-	-	-	-	-	1.2	40	-	-	-	RT - 120	173
	Redux 501 A/B	Blue	-	-	-	-	-	-	1.17	2.5	-	-	-	RT - 70	173
Degussa	Agomet P54	-	◑	-	-	○	-	-	1.5	40	-30/+80	-	-	RT - 120	178
	Agomet P76	-	◑	-	-	○	-	-	1.6	100	-30/+80	-	-	RT - 120	178
	Agomet P79	-	◑	-	-	○	-	-	1.6	120	-30/+80	-	-	RT - 120	178
Delo	Delomet 02-Rapid	Cl. Yellow	◑	-	-	-	-	-	1.2	9.5	-30/+80	-	-	RT - 120	178
	Delomet 03-Rapid	Cl. Yellow	◑	-	-	-	-	-	1.15	11.5	-30/+80	-	-	RT - 120	179
	Delopox 333	Grey	-	○	-	-	-	-	2	-	-30/+180	-	-	RT	179
	Dupox	Grey	◑	◑	●	○	75D	1.4	1.3	100	-40/+100	112	-	RT	179
Devcon	2-Ton Clear Epoxy	Transp.	◑	-	-	-	-	-	1.1	15.5	<93	-	10	RT	180
	5 Minute Epoxy	Transp.	○	-	-	-	-	-	1.1	15.5	<93	-	3	RT	180
	5 Minute Epoxy Gel	White	◑	-	-	-	-	-	1.17	-	<93	-	3	RT	181
	Magic Bond	White	-	-	-	-	-	-	-	-	-	-	-	RT	181
	Special F Aluminium	Metallic	◑	-	-	-	-	-	1.72	-	<121	-	45	RT	181
Dexter Hysol	EA 901 NA/B-1	Red	◑	◑	●	-	90D	0.9	-	650	<105	-	50	RT - 150	181
	EA 934 NA	Grey	◑	◑	●	-	85D	1.2	1.36	80	<149	71	40	RT - 93	182
	EA 956	Amber	○	◑	◑	-	85D	2.35	1.15	20	<149	69	30	RT - 93	182
	EA 9309 NA	Grey	●	◑	◑	●	80D	10	1.10	~7	<71	56	40	RT - 93	182
	EA 9309.3 NA	Violet	●	◑	◑	●	80D	10	-	250	<80	80	35	RT - 80	182
	EA 9313	Pink	●	●	◑	●	76D	8	1.10	1.2	<49	49	60	RT - 80	183
	EA 9317	Grey	◑	-	-	-	90D	-	1.36	320	<175	-	35	RT - 80	183
	EA 9320 NA	Lt. Blue	●	◑	◑	◑	81D	9	1.36	170	<80	80	25	RT - 80	183
	EA 9321	Grey	●	●	●	○	84D	6	1.36	350	<121	110	40	RT - 80	183
	EA 9323	Amber	●	○	◑	●	86D	9	1.15	47	<121	97	30	RT - 80	184
	EA 9330	Cream	●	◑	◑	●	-	2.4	1.15	150	<82	57	60	RT - 93	184
	EA 9330.3	Grey	●	●	◑	●	81D	9	1.11	300	<80	55	60	RT - 95	184
	EA 9390	Violet	◑	●	●	-	89D	2.5	1.1	8	<175	175	120	95 - 150	185
	EA 9394	Grey	●	●	●	○	88D	1.66	1.33	160	<177	78	90	RT - 93	185
	EA 9395	Grey/blue	●	●	●	○	90D	2.6	1.27	200	<175	72	100	RT - 66	185

Continued...

Two-Part Epoxy Paste Adhesives - continued.

Company	Adhesive	Colour	LSS	TS	TM	Peel	Hs	El. (%)	SG	Visc. (Pa.s)	Service (°C)	Tg (°C)	Pot life (mins)	Cure (°C)	Pg. No.
Dexter Hysol	EA 9396	-	◑	-	-	-	-	-	1.14	3.5	<175	-	75	RT - 65	185
(cont.)	EA 9412	Pink	●	●	◑	●	76D	8	1.1	1.2	-	-	60	RT - 82	186
	EA 9430	Off-white	●	◑	◑	●	75D	6	1.12	12	-	-	50	RT - 82	186
	EA 9460	Grey	◑	◑	◑	◑	80D	3.5	1.33	2	<80	-	40	RT - 150	186
	EA 9462	Beige	◑	-	-	○	80D	-	1.17	70	-	-	55	RT - 93	187
	EPK 1C	Various	○	-	-	-	80D	-	1.55	500	-	98	20	RT - 82	188
	XEA 9359.3	Green	●	-	-	●	-	-	1.12	250	<95	-	40	RT - 105	189
	XEA 9361	Grey	◑	-	○	◑	-	>50	1.28	100	<60	-	120	RT - 93	189
Dunlop	SAS 400	Grey	-	-	-	-	80D	-	-	-	<120	-	30	RT	189
	SAS 520	Cl. Amber	◑	-	-	◑	-	-	1.25	-	-10/+70	-	20	RT	190
	SAS 530	-	○	-	-	-	-	-	1.25	11	0/+60	-	<10	RT	190
	SAS 535	-	○	-	-	-	-	-	1.25	-	-20/+60	-	6	RT	191
Dyckerhoff	DünnbettEpoxi #500	-	-	-	-	-	-	-	-	-	<80	-	90	RT	191
Epirez	8141	Grey	-	◑	-	-	-	-	-	-	-	-	60	RT	191
	8242	Grey	◑	-	-	-	-	-	-	-	<90	-	45	RT	191
	8641	Grey	◑	-	-	-	-	-	-	-	-10/+130	-	120	RT	192
	Epistop Standard	Brown	○	○	-	-	70D	-	-	-	-	-	30	RT	192
	Highway Marker 611	Lt. Grey	◑	-	-	-	-	-	-	-	-	-	3	RT	192
	Highway Marker 712	Lt. Grey	○	-	-	-	-	-	-	-	-	-	45	RT	192
	Highway Marker 8242	Lt. Grey	○	-	-	-	-	-	-	-	-	-	45	RT	193
Forbo Helmitin	Helmitin 670	Grey	-	-	-	-	-	-	1.46	90	-	-	60	RT	193
	Helmitin A + B	Yellow	-	-	-	-	-	-	1.08	20	-	-	90	RT	193
Grace	Eccobond 24	Transp.	○	-	-	-	-	-	1.72	0.8	<120	-	30	RT	193
	Eccobond 45	Black	○	-	-	-	70A	-	-	-	-55/+90	-	120	RT - 105	194
	Eccobond 56C	Metallic	○	-	-	-	85D	-	3.5	-	-60/+120	75	-	50 - 65	194
	Eccobond 72C	Metallic	○	-	-	-	-	-	-	-	-54/+150	-	-	RT - 90	194
	Eccobond 104	Black	○	-	-	-	90D	-	1.5	-	<230	-	720	120 - 200	195
	Eccobond 285	-	○	-	-	-	-	-	1.5	-	<180	-	30	RT - 190	195
	Eccobond 286	White	○	-	-	-	-	-	-	-	-	-	-	RT	195
	Eccolite 82 EP PB	White	○	-	-	-	-	-	0.7	-	-50/+100	-	20	RT	198
Hardman	Epoweld 3243	Ivory	◑	-	-	○	85D	-	1.17	15	<80	-	150	RT	198
	Epoweld 7897	Black	◑	-	-	○	85D	-	1.17	2.2	<60	-	86	100	198
	Epoweld 7929	Black	●	-	-	○	83D	-	1.09	4.5	<60	-	57	RT	198
	Epoweld 8173	Amber	◑	-	-	○	80D	-	1.16	40	<60	-	5	RT	199
	Epoweld 13230	Grey	◑	-	-	◑	65D	-	1.27	50	<80	-	90	RT	199
	Epoweld 16469	Grey	◑	-	-	○	90D	-	1.52	52	<70	-	12	RT	199
	Epoweld 16480	Blue	◑	-	-	○	85D	-	2.1	4.6	<70	-	198	RT	199
	Epoweld 16530	Black	◑	-	-	●	75D	-	1.17	16	<70	-	45	RT	200
	Epoweld 16535	Amber	◑	-	-	○	85D	-	1.11	10	<130	-	42	RT	200
	Epoweld 16562	Grey	◑	-	-	●	82D	-	1.38	126	<80	-	45	RT	200
	Epoweld 16580	Translucent	◑	-	-	○	83D	-	1.17	-	<70	-	5	RT	200
	Epoweld 19157	Black	●	-	-	○	80D	-	1.18	8	<90	-	74	80	201
	Epoweld 20018	Grey	●	-	-	○	95D	-	1.53	-	<176	-	32	RT	201
	Epoweld 20214	Grey	◑	-	-	◑	65D	-	1.30	42	<60	-	116	RT	201
	Epoweld 20215	Amber	◑	-	-	○	80D	-	1.09	408	<90	-	125	RT	201
W.W. Henry	191	Lt. Grey	-	-	-	-	-	-	-	-	-	-	30	RT	202
Hitachi-Kasei	Hi-Bon 3430	Violet	◑	-	○	●	60D	10	1.14	72	-	55	40	RT	203
Klebchemie	Kleiberit EP 554	-	○	-	-	-	-	-	1.5	-	-	-	30	RT - 100	203
(M.G. Becker)	Plastic-Mastic 528	Grey	-	-	-	-	-	-	1.93	-	-	-	40	RT	204
	Plastic-Mastic 531.1	Grey	-	-	-	-	-	-	1.66	-	-50/+100	-	75	RT - 100	204
	Supra-Epoxy 560.1	-	◑	-	-	-	-	-	1.1	30	-	-	90	RT - 130	204
	Suprapox 560.5	-	-	-	-	-	-	-	1.2	45	-	-	90	RT - 130	204

Continued...

See Key to Comparison Tables - Pages 63 to 65

Two-Part Epoxy Paste Adhesives - continued.

Company	Adhesive	Colour	LSS	TS	TM	Peel	Hs	El. (%)	SG	Visc. (Pa.s)	Service (°C)	Tg (°C)	Pot life (mins)	Cure (°C)	Pg. No.
Kömmerling	Corapox 733	Lt. Yellow	-	-	-	-	78D	-	1.1	28	-	-	120	RT	205
Le Joint	PR 1219	-	-	-	-	-	-	-	-	-	-	-	-	-	205
Français	PR 1826	-	-	-	-	-	-	-	-	-	<203	-	-	-	205
	PR 1828	-	-	-	-	-	-	-	-	-	<203	-	-	-	205
	PR 1829	-	-	-	-	-	-	-	-	-	-	-	-	-	206
	PR 1875	-	-	-	-	-	-	-	-	-	-	-	-	-	206
LePage's	5 Minute Epoxy	Transp.	●	-	-	-	-	-	1.1	16	<60	-	<5	RT	206
	Epoxy Putty	Brown	○	◑	-	-	-	-	1.8	-	<121	-	60	RT	206
	Epoxy (Regular)	Transl.	◑	-	-	-	-	-	1.2	12	<70	-	<120	RT	207
	Epoxy Steel #400	Transp.	◑	-	-	-	-	-	1.1	500	<60	-	<120	RT	207
	LEP 502 Epoxy Glue	Lt. Yellow	◑	-	-	-	-	-	-	900	<71	-	120	RT	207
	LEP 503 5 Minute	Lt. Yellow	●	-	-	-	-	-	-	900	<60	-	-	RT	207
	Permabond 1700	Beige	-	-	-	-	-	-	-	-	-	-	60	RT	208
Lugato	Kraftkleber	Grey	○	-	-	-	-	-	-	-	-50/+70	-	90	RT	208
Permabond	Double Bubble	Transp.	◑	-	-	-	-	-	-	-	-	-	-	RT	210
	E01	Transp.	◑	-	-	-	-	-	1.16	35	-55/+100	-	3	RT - 60	210
	E04	Grey	◑	-	-	-	-	-	1.34	-	-55/+70	-	8	RT - 60	211
	E18	Transp.	◑	-	-	-	-	-	1.16	35	-55/+100	-	3	RT - 60	211
	E32	Grey	●	-	-	-	-	-	1.22	-	-55/+80	-	120	RT - 60	211
	Fe Rapid	Dk. Metallic	◑	○	-	-	80D	-	2.6	-	-55/+120	-	4	RT	211
Ricon Resins	Ricotuff 1000	-		○	○	-	82D	6	-	168	-	125	(7 days)	85 - 130	212
	Ricotuff 1100	-	●	○	○	-	75D	6	-	40	-	119	(7 days)	85 - 130	212
	Ricotuff 1110	-	●	○	○	-	75D	6	-	34	-	119	60	RT - 75	212
SEA	Epox	Lt. Brown	-	-	-	-	-	-	1.65	100	-	-	120	RT	212
	SIKI 440	Grey	-	-	-	-	-	-	1.75	48	-	-	120	RT - 40	213
Sicomin	Aerobond 7303	Grey	●	-	-	-	-	-	1.11	48	-	65	150	RT - 130	213
Structil	ST-1013	-	-	-	-	-	-	-	-	-	-55/+80	-	35	RT	213
	ST-1015	-	-	-	-	-	-	-	-	-	-250/+150	-	40	RT	213
	ST-1016	-	-	-	-	-	-	-	-	-	-55/+150	-	30	RT	214
	ST-1021	-	-	-	-	-	-	-	-	-	-55/+120	-	40	RT	214
	ST-1050	-	◑	-	-	-	-	-	-	-	-55/+80	-	>360	RT - 60	215
	ST-1060	-	◑	-	-	-	-	85	-	-	-50/+60	-	120	RT - 60	215
	ST-1060 F	-	◑	-	-	-	-	85	-	-	-50/+60	-	120	RT - 60	215
Structural	S - 203 E	Grey	◑	-	-	-	-	-	1.5	400	-30/+120	-	330	RT	216
Adhesives Ltd.	S - 206 E	Grey	◑	-	-	-	-	-	1.0	90	-	-	180	RT - 180	216
	S - 241 E	Grey	◑	-	-	-	-	-	1.7	100	-	-	90	RT	216
	S - 255 E	Grey	○	-	-	-	-	-	1.45	100	-	-	90	RT	217
	S - 264 E	Various	○	-	-	-	-	-	1.34	200	-	-	60	RT	217
	S - 288 E	Various	○	-	-	-	-	-	1.55	-	-	-	60	RT	217
	S - 305 E	Off-white	○	-	-	-	-	-	1.07	-	-	-	60	RT	217
	S - 307 E	Khaki	○	-	-	-	-	-	-	-	-	-	60	RT	218
	S - 334 E	Grey	◑	-	-	-	-	-	1.55	50	-	-	180	RT	218
	S - 360 E	Grey	○	-	-	-	-	-	1.0	40	-	-	12.5	RT	218
	S - 361 E	Off-white	○	-	-	-	-	-	1.5	80	-	-	40	RT	218
	S - 384 E	Cream	○	-	-	-	-	-	1.03	65	-	-	120	RT	219
	S - 386 E	Lt. Pink	○	-	-	-	-	-	1.15	25	-	-	25	RT	219
	S - 2022 E	Grey	○	-	-	-	-	-	-	20	-	-	50	RT	219
	S - 2029 E	Off-white	○	-	-	-	-	-	1.5	20	-	-	30	RT	219
	S - 2045 E	Transp.	○	-	-	-	-	-	1.15	8.5	-	-	5	RT	220

Continued...

See Key to Comparison Tables - Pages 63 to 65

Two-Part Epoxy Paste Adhesives - continued.

Company	Adhesive	Colour	LSS	TS	TM	Peel	Hs	El. (%)	SG	Visc. (Pa.s)	Service (°C)	Tg (°C)	Pot life (mins)	Cure (°C)	Pg. No.
TACC	ER 1007	Grey	-	-	-	-	80D	-	-	-	-	-	240	100 - 160	222
	ER 1100	Transp.	◐	-	-	-	80D	-	1.2	19	-40/+105	-	5	RT	222
	ER 1218	Amber	-	◐	-	-	75D	-	1.2	-	-	-	35	RT - 60	222
	ER 1650	Grey	-	◐	-	-	75D	-	1.4	-	-60/+160	-	40	RT - 100	223
	ER 4765	Various	-	-	-	-	-	-	1.1	-	-	-	-	RT - 65	223
	ER 7093	Grey	-	-	-	-	-	-	-	-	-	-	-	RT	224
Teroson	Terokal 5045	Red/Brown	◐	-	○	-	70D	-	1.25	100	-40/+80	-	30	RT - 100	227
	Terokal 5050	Lt. Yellow	●	-	-	-	-	-	1.12	50	-40/+80	-	90	RT - 150	227
Tremco	ES 1122	Off-white	○	-	-	-	80D	-	-	140	-20/+80	-	60	RT - 80	229
UHU	Plus	Yellow	-	-	-	-	-	-	1.1	35	-	-	60	RT	229
	Ultra Rapid	Transp.	-	-	-	-	-	-	1.2	30	<100	-	2	RT	229
Witco	710 + 10	-	○	-	●	◐	-	-	~1.15	7.5	-	-	10	RT	230
(Eurepox #A	710 + 12	-	◐	-	●	◐	-	-	~1.15	1.5	-	-	12	RT	230
+ Euredur #B)	710 + 26	-	◐	-	●	◐	-	-	~1.15	0.9	-	-	35	RT - 80	230
	710 + 27	-	●	-	●	●	-	-	~1.15	0.5	-	-	70	RT - 40	230
	710 + 115	-	●	-	●	◐	-	-	~1.15	130	<80	-	160	RT - 150	231
	710 + 125	-	●	-	●	◐	-	-	~1.15	21	<90	-	100	RT - 150	231
	710 + 140	-	●	-	●	◐	-	-	~1.15	12	<130	-	100	RT - 60	231
	710 + 200	-	○	-	●	◐	-	-	~1.15	3.5	-	-	35	RT	231
	710 + 250	-	●	-	●	◐	-	-	~1.15	3	-	-	60	RT	232
	710 + 350	-	●	-	●	◐	-	-	~1.15	2.5	-	-	360	RT	232
	710 + 370	-	●	-	●	◐	-	-	~1.15	2.5	-	-	100	RT	232
	783 + 115	-	●	-	●	◐	-	-	~1.15	39	-	-	190	RT	232
	783 + 125	-	●	-	●	◐	-	-	~1.15	6	-	-	120	RT	233
	783 + 140	-	●	-	●	◐	-	-	~1.15	4	-	-	120	RT	233
	783 + 200	-	○	-	●	◐	-	-	~1.15	1.5	-	-	40	RT	233
	783 + 250	-	●	-	●	◐	-	-	~1.15	0.8	-		70	RT	233
	783 + 350	-	●	-	●	◐	-	-	~1.15	0.5	-	-	540	RT	234
	783 + 370	-	●	-	●	◐	-	-	~1.15	0.6	-	-	110	RT	234

Epoxy Film Adhesives

Company	Adhesive	Colour	LSS	TS	TM	Peel	El. (%)	Various Carriers	Various Wts.	Service (°C)	Tg dry (°C)	Cure (°C)	Pg. No.
3M	AF-126 (0.06)	Green	●	○	○	●	23.7	-	✓	-55/+121	-	121	161
	AF-126-2 (0.06)	Grey	◐	-	-	●	-	-	✓	-55/+121	-	121	162
	AF-143 (0.1)	Tan	○	-	-	◐	-	-	-	-	-	177	162
	AF-163-2M	Red	●	◐	○	●	-	✓	✓	-55/+121	108	177	162
	AF-191 (0.08)	Tan	◐	-	-	◐	-	✓	✓	-	226	177	162
Ciba	Redux 308	Black	●	-	-	○	-	-	-	<120	-	170	171
Polymers	Redux 312	Grey	●	-	-	◐	-	✓	✓	<100	-	120	172
	Redux 319	Grey	●	-	-	◐	-	✓	✓	<150	-	175	172
	Redux 322	Grey/Green	○	-	-	◐	-	✓	✓	<200	-	175	172
	Redux 330	Lt Green	●	●	●	●	6.5	-	✓	<150	-	175	172
	Redux 609	Blue	◐	-	-	○	-	-	-	<100	-	120	173
Cytec	FM 27	Tan	-	-	-	-	-	✗	✓	-55/+149	-	121 - 177	174
	FM 61	Lt. Brown	○	-	-	-	-	-	-	-55/+177	-	157 - 177	174
	FM 73 M	Yellow	●	-	-	●	-	✓	✓	-55/+82	95	120 - 177	174
	FM 87-1 M	Tan	◐	-	-	●	-	✓	✓	-55/+120	134	105 - 175	174
	FM 123-2M	Purple	◐	-	-	◐	-	✓	✓	<82	101	120	175
	FM 275 M	-	◐	-	-	◐	-	-	✓	<121	138	120 - 149	175

Continued...

Epoxy Film Adhesives - continued.

Company	Adhesive	Colour	LSS	TS	TM	Peel	El. (%)	Various Carriers	Various Wts.	Service (°C)	Tg dry (°C)	Cure (°C)	Pg. No.
Cytec	FM 300 M	Green	●	-	-	◑	-	✓	✓	-55/+150	144	175	175
(cont.)	FM 350 NA	Grey	◑	-	-	○	-	✓	-	-55/+177	174	177 - 185	175
	FM 355	Grey	○	-	-	○	-	✓	✓	<177	180	177 - 185	176
	FM 404 NA	Grey	-	-	-	-	-	×	-	<210	-	170	176
	FM 410-1	Green	-	-	-	-	-	×	✓	-55/+175	-	105 - 175	176
	FM 1000	White	●	-	-	●	-	-	✓	-250/+95	125	170	176
	FM 1020	Yellow	-	-	-	-	-	×	✓	-55/+175	-	120 - 175	177
	Metlbond 1113	Yellow	●	-	-	●	-	✓	✓	-55/+82	104	93 - 143	177
	Metlbond 1146 M	-	●	-	-	◑	-	✓	✓	<107	137	121 - 177	177
	Metlbond 1515-1	Brown	◑	-	-	○	-	✓	✓	-54/+149	160	149 - 177	177
Dexter	EA 9601	Beige	●	-	-	◑	-	✓	✓	<121	-	107 - 121	187
Hysol	EA 9628	-	●	●	◑	●	7.5	✓	×	<120	115	115 - 120	187
	EA 9657	-	●	-	-	●	-	✓	✓	<205	-	175	187
	EA 9689	-	○	-	-	○	-	✓	✓	<215	225	175	188
	EA 9814 NA	-	-	-	-	-	-	×	✓	-	-	120 - 175	188
	X 9680.1	-	○	-	-	-	-	×	✓	<150	155	120 - 175	188
	X 9690	-	●	-	-	●	-	✓	-	<120	115	120	189
Structil	ST-1035	Ivory	●	-	-	●	-	✓	✓	<120	-	95 - 200	214

3E (USA)

3E	Polarjoint		Paste
Epoxy - two part			
Manufacturers Comments: RT cure. 100% solids. Pourable. Forms a semi-hard, durable joint. Good mechanical damage and chemical resistance.			
Typical Use: Crack bonding in concrete and steel construction industry.			
Cure: RT.			
Test Methods: ASTM		(March 95)	
Physical		**Mechanical**	
Colour	Grey	τ (RT)	-
Viscosity	-	τ (°C)	-
S.G.	1.3	τ (wet)	-
Shelf life	1 yr./RT	Peel ( )	-
Pot/Work life	-	σ_t	38.6 MPa
Tg	-	σ_c	-
$T_{Service}$	121°C max.	*E*	-
CTE (°C)	-	*G*	-
Thermal cond.	-	Elongation	-
Resistivity: Vol.	-	Tear strength	-
Surf.	-	H (Shore A)	95
Notes: Mix ratio 4:1. Hardness - ASTM D 2240; S.G. - ASTM D792. Flash point - FTMS 141A(M4293): 400°C. Water absorption - ASTM D560-77: 0 Water vapour transmission (24 hrs)- ASTM E96: 5.2 g/m².			

3M	1469		Paste
Epoxy - one part			
Manufacturers Comments: 177°C cure. Superior performance at elevated temperatures. Meets: MMM-A-132 Type II Class 3 Group 4.			
Typical Use: Engineering assemblies.			
Cure: Optimum 177°C/ 2 hr/ 0.07MPa.			
Test Methods: Not stated		(March 95)	
Physical		**Mechanical**	
Colour	Cream	τ (RT)	25.5 MPa
Viscosity	-	τ (121°C)	24.8 MPa
S.G.	-	τ (wet)	39.8 MPa
Shelf life	-	Peel (T)	0.35 N mm^{-1}
Pot/Work life	-	σ_t	-
Tg	-	σ_c	-
$T_{Service}$	-	*E*	-
CTE (°C)	-	*G*	-
Thermal cond.	-	Elongation	-
Resistivity: Vol.	-	Tear strength	-
Surf.	-	H (Shore)	-
Notes: τ(wet) 30days, tap water, RT. T-Peel Al-alloy adherends, RT.			

3M

3M	1386		Paste
Epoxy - one part			
Manufacturers Comments: 177°C cure. High strength, impact resistant bonds on metal/metal. Meets: MMM-A-132 Type III.			
Typical Use: Engineering assemblies.			
Cure: Optimum 177°C/ 1 hr/ 0.07MPa.			
Test Methods: Not stated		(March 95)	
Physical		**Mechanical**	
Colour	Cream	τ (RT)	37.9 MPa
Viscosity	-	τ (121°C)	17.2 MPa
S.G.	-	τ (wet)	39.3 MPa
Shelf life	-	Peel (T)	1.75 N mm^{-1}
Pot/Work life	-	σ_t	-
Tg	-	σ_c	-
$T_{Service}$	-	*E*	-
CTE (°C)	-	*G*	-
Thermal cond.	-	Elongation	-
Resistivity: Vol.	-	Tear strength	-
Surf.	-	H (Shore)	-
Notes: τ(wet) 30days, tap water, RT. T-Peel Al-alloy adherends, RT.			

3M	1648 B/A		Paste
Epoxy - two part paste			
Manufacturers Comments: RT cure. Version of 1838B/A with superior environmental resistance and high temperature performance.			
Typical Use: Metals, woods, reinforced plastics, masonry.			
Cure: RT Handling strength: 8-12 hrs; Full cure: 7 days.			
Test Methods: Not stated		(March 95)	
Physical		**Mechanical**	
Colour	Green	τ (RT)	17.2 MPa
Viscosity	275 Pa.s	τ (121°C)	2.8 MPa
S.G.	-	τ (wet)	17.2 MPa
Shelf life	-	Peel (T)	0.70 N mm^{-1}
Pot/Work life	1hr/RT	σ_t	-
Tg	-	σ_c	-
$T_{Service}$	-	*E*	-
CTE (°C)	-	*G*	-
Thermal cond.	-	Elongation	-
Resistivity: Vol.	-	Tear strength	-
Surf.	-	H (Shore)	-
Notes: τ(wet) 30days, tap water. T-peel RT.			

See Key to Data Sheets - Pages 63 to 65

3M	1751 B/A		Paste
Epoxy - two part paste			
Manufacturers Comments: RT cure. Aluminium filled, mastic consistency. Good gap-filling.			
Typical Use: Metals. Good adhesion to steel.			
Cure: RT Handling strength: 8-12 hrs; Full cure: 7 days.			
Test Methods: Not stated		(March 95)	
Physical		**Mechanical**	
Colour	Grey	τ (RT)	13.8 MPa
Viscosity	700 Pa.s	τ (121°C)	2.1 MPa
S.G.	-	τ (wet)	13.8 MPa
Shelf life	-	Peel (T)	0.70 N mm^{-1}
Pot/Work life	45 mins/RT	σ_t	-
Tg	-	σ_c	-
$T_{Service}$	-	E	-
CTE (°C)	-	G	-
Thermal cond.	-	Elongation	-
Resistivity: Vol.	-	Tear strength	-
Surf.	-	H (Shore)	-
Notes: τ(wet) 30days, tap water. T-peel RT.			

3M	1838 B/A		Paste
Epoxy - two part paste			
Manufacturers Comments: RT cure. Excellent environmental resistance. Meets MIL-A-23941A.			
Typical Use: Metals, woods, reinforced plastics, masonry.			
Cure: RT Handling strength: 8-12 hrs; Full cure: 7 days.			
Test Methods: Not stated		(March 95)	
Physical		**Mechanical**	
Colour	Green*	τ (RT)	20.7 MPa
Viscosity	400 Pa.s	τ (121°C)	2.1 MPa
S.G.	-	τ (wet)	17.2 MPa
Shelf life	-	Peel (T)	0.70 N mm^{-1}
Pot/Work life	1 hr/RT	σ_t	-
Tg	-	σ_c	-
$T_{Service}$	-	E	-
CTE (°C)	-	G	-
Thermal cond.	-	Elongation	-
Resistivity: Vol.	-	Tear strength	-
Surf.	-	H (Shore)	-
Notes: τ(wet) 30days, tap water. T-peel RT * Tan colour version available with lower viscosity (250 Pa.s).			

3M	1751L B/A		Paste
Epoxy - two part paste			
Manufacturers Comments: RT cure. Aluminium filled. Lower viscosity version of 1751 B/A.			
Typical Use: Metals. Good adhesion to steel.			
Cure: RT Handling strength: 8-12 hrs; Full cure: 7 days.			
Test Methods: Not stated		(March 95)	
Physical		**Mechanical**	
Colour	Grey	τ (RT)	13.8 MPa
Viscosity	250 Pa.s	τ (121°C)	2.1 MPa
S.G.	-	τ (wet)	13.8 MPa
Shelf life	-	Peel (T)	0.70 N mm^{-1}
Pot/Work life	45 mins/RT	σ_t	-
Tg	-	σ_c	-
$T_{Service}$	-	E	-
CTE (°C)	-	G	-
Thermal cond.	-	Elongation	-
Resistivity: Vol.	-	Tear strength	-
Surf.	-	H (Shore)	-
Notes: τ(wet) 30days, tap water. T-peel RT.			

3M	1838L B/A		Paste
Epoxy - two part paste			
Manufacturers Comments: RT cure. Lower viscosity, translucent version of 1838 B/A.			
Typical Use: Metals, woods, reinforced plastics, masonry.			
Cure: RT Handling strength: 8-12 hrs; Full cure: 7 days.			
Test Methods: Not stated		(March 95)	
Physical		**Mechanical**	
Colour	Translucent	τ (RT)	20.7 MPa
Viscosity	10 Pa.s	τ (121°C)	1.4 MPa
S.G.	-	τ (wet)	24.1 MPa
Shelf life	-	Peel (T)	0.88 N mm^{-1}
Pot/Work life	1 hr/RT	σ_t	-
Tg	-	σ_c	-
$T_{Service}$	-	E	-
CTE (°C)	-	G	-
Thermal cond.	-	Elongation	-
Resistivity: Vol.	-	Tear strength	-
Surf.	-	H (Shore)	-
Notes: τ(wet) 30days, tap water. T-peel RT.			

3M	**2086**		**Paste**
Epoxy - one part			
Manufacturers Comments: 177°C cure. Similar to 1386, but filled for improved flow control.			
Typical Use: Engineering assemblies.			
Cure: Optimum 177°C/ 1 hr/ 0.07MPa.			
Test Methods: Not stated		(March 95)	
Physical		**Mechanical**	
Colour	Grey	τ (RT)	34.5 MPa
Viscosity	-	τ (121°C)	15.2 MPa
S.G.	-	τ (wet)	39.3 MPa
Shelf life	-	Peel (T)	8.76 N mm^{-1}
Pot/Work life	-	σ_t	-
Tg	-	σ_c	-
$T_{Service}$	-	E	-
CTE (°C)	-	G	-
Thermal cond.	-	Elongation	-
Resistivity: Vol.	-	Tear strength	-
Surf.	-	H (Shore)	-
Notes: τ(wet) 30days, tap water, RT. T-Peel steel adherends, RT.			

3M	**2214 Hi-Density**		**Paste**
Epoxy - one part paste			
Manufacturers Comments: 121°C cure. 100% solids. Non-sag, aluminium filled. Deaerated. Formulated for dense, void-free bondlines.			
Typical Use: Clean metals, glass and most plastics.			
Cure: Optimum 120°C/ 1 hr/ 0.07MPa.			
Test Methods: ASTM		Aug 84	
Physical		**Mechanical**	
Colour	Grey	τ (RT)	31.0 MPa
Viscosity	-	τ (121°C)	11.7 MPa
S.G.	-	τ (wet)	33.1 MPa
Shelf life	storage <3°C	Peel (T)	0.88 N mm^{-1}
Pot/Work life	-	σ_t	-
Tg	-	σ_c	-
$T_{Service}$	-	E	-
CTE (0-80°C)	49 × 10^{-6} °C^{-1}	G	-
Thermal cond.	0.40 W m.K^{-1} (24°C)	Elongation	-
Resistivity: Vol.	2.8 × 10^{13} Ω.cm	Tear strength	-
Surf.	9.8 × 10^{17} Ω	H (Shore)	-
Notes: τ Al-alloy adherends, FPL etch ASTM D1002-64. τ(wet) 30days, tap water. T-Peel Al-alloy adherends, FPL etch, RT. ASTM D-1867-61T Resistivity ASTM D-257 (22°C). Optimum bondline thickness: 0.05 - 0.13mm.			

3M	**2158 B/A**		**Paste**
Epoxy - two part paste			
Manufacturers Comments: RT cure. Equal mix ratio. Contrasting colours (adhesive and catalyst) for easy metering and mixing.			
Typical Use: General purpose assembly.			
Cure: RT Handling strength: 8-12 hrs; Full cure: 7 days.			
Test Methods: Not stated		(March 95)	
Physical		**Mechanical**	
Colour	Grey	τ (RT)	13.8 MPa
Viscosity	375 Pa.s	τ (121°C)	2.8 MPa
S.G.	-	τ (wet)	13.8 MPa
Shelf life	-	Peel (T)	0.70 N mm^{-1}
Pot/Work life	2 hr/RT	σ_t	-
Tg	-	σ_c	-
$T_{Service}$	-	E	-
CTE (°C)	-	G	-
Thermal cond.	-	Elongation	-
Resistivity: Vol.	-	Tear strength	-
Surf.	-	H (Shore)	-
Notes: τ(wet) 30days, tap water. T-peel RT.			

3M	**2214 Hi-Flex**		**Paste**
Epoxy - one part paste			
Manufacturers Comments: 121°C cure. 100% solids. Non-sag, aluminium filled. Deaerated. High strength , impact resistant bonds.			
Typical Use: Clean metals, glass and most plastics.			
Cure: Optimum 120°C/ 1 hr/ 0.07MPa.			
Test Methods: ASTM		Aug 84	
Physical		**Mechanical**	
Colour	Grey	τ (RT)	27.6 MPa
Viscosity	-	τ (121°C)	3.1 MPa
S.G.	-	τ (wet)	24.1 MPa
Shelf life	storage <3°C	Peel (T)	1.75 N mm^{-1}
Pot/Work life	-	σ_t	-
Tg	-	σ_c	-
$T_{Service}$	-	E	-
CTE (0-80°C)	80 × 10^{-6} °C^{-1}	G	-
Thermal cond.	0.33 W m.K^{-1} (25°C)	Elongation	-
Resistivity: Vol.	2.8 × 10^{13} Ω.cm	Tear strength	-
Surf.	1.2 × 10^{15} Ω	H (Shore)	-
Notes: τ Al-alloy adherends, FPL etch ASTM D1002-64. τ(wet) 30days, tap water. T-Peel Al-alloy adherends, FPL etch, RT. ASTM D-1867-61T Resistivity ASTM D-257 (22°C). Optimum bondline thickness: 0.05 - 0.13mm.			

See Key to Data Sheets - Pages 63 to 65

3M	2214 Hi-Temp		Paste
Epoxy - one part paste			
Manufacturers Comments: 121°C cure. 100% solids. Non-sag, aluminium filled. Deaerated. High strength >80°C.			
Typical Use: Clean metals, glass and most plastics.			
Cure: Optimum 120°C/ 1 hr/ 0.07MPa.			
Test Methods: ASTM		Aug 84	
Physical		**Mechanical**	
Colour	Grey	τ (RT)	13.8 MPa
Viscosity	-	τ (121°C)	17.2 MPa
S.G.	-	τ (wet)	13.8 MPa
Shelf life	storage <3°C	Peel (T)	0.35 N mm^{-1}
Pot/Work life	-	σ_t	-
Tg	-	σ_c	-
$T_{Service}$	-	E	-
CTE (0-80°C)	48 × 10^{-6} °C^{-1}	G	-
Thermal cond.	0.40 W m.K^{-1} (25°C)	Elongation	-
Resistivity: Vol.	9.4 × 10^{14} Ω.cm	Tear strength	-
Surf.	1.1 × 10^{17} Ω	H (Shore)	-

Notes: τ Al-alloy adherends, FPL etch ASTM D1002-64. τ(wet) 30days, tap water.
T-Peel Al-alloy adherends, FPL etch, RT. ASTM D-1867-61T
Resistivity ASTM D-257 (22°C).
Optimum bondline thickness: 0.05 - 0.13mm.

3M	2214 Regular		Paste
Epoxy - one part paste			
Manufacturers Comments: 121°C cure. 100% solids. Non-sag, aluminium filled.			
Typical Use: Clean metals, glass and most plastics.			
Cure: Optimum 120°C/ 1 hr/ 0.07MPa.			
Test Methods: ASTM		Aug 84	
Physical		**Mechanical**	
Colour	Grey	τ (RT)	31.0 MPa
Viscosity	-	τ (121°C)	10.3 MPa
S.G.	-	τ (wet)	33.1 MPa
Shelf life	storage <3°C	Peel (T)	0.88 N mm^{-1}
Pot/Work life	-	σ_t	-
Tg	-	σ_c	-
$T_{Service}$	-	E	-
CTE (0-80°C)	49 × 10^{-6} °C^{-1}	G	-
Thermal cond.	0.40 W m.K^{-1} (25°C)	Elongation	-
Resistivity: Vol.	2.8 × 10^{13} Ω.cm	Tear strength	-
Surf.	9.8 × 10^{12} Ω	H (Shore)	-

Notes: τ Al-alloy adherends, FPL etch ASTM D1002-64. τ(wet) 30days, tap water.
T-Peel Al-alloy adherends, FPL etch, RT. ASTM D-1867-61T
Resistivity ASTM D-257 (22°C).
Optimum bondline thickness: 0.05 - 0.13mm.

3M	2214 Non-Metallic		Paste
Epoxy - one part			
Manufacturers Comments: 121°C cure. 100% solids. Non-sag. Filled. with non-metallic material.			
Typical Use: Electrical uses requiring 'resistance'. Clean metals, glass and most plastics.			
Cure: Optimum 120°C/ 1 hr/ 0.07MPa.			
Test Methods: ASTM		Aug 84	
Physical		**Mechanical**	
Colour	Cream	τ (RT)	27.6 MPa
Viscosity	-	τ (121°C)	10.3 MPa
S.G.	-	τ (wet)	26.2 MPa
Shelf life	storage <3°C	Peel (T)	1.23 N mm^{-1}
Pot/Work life	-	σ_t	-
Tg	-	σ_c	-
$T_{Service}$	-	E	-
CTE (°C)	-	G	-
Thermal cond.	-	Elongation	-
Resistivity: Vol.	2.5 × 10^{14} Ω.cm	Tear strength	-
Surf.	-	H (Shore)	-

Notes: τ Al-alloy adherends, FPL etch ASTM D1002-64. τ(wet) 30days, tap water.
T-Peel Al-alloy adherends, FPL etch, RT. ASTM D-1867-61T
Resistivity ASTM D-257 (22°C).
Optimum bondline thickness: 0.05 - 0.13mm.

3M	2216 B/A Clear		Paste
Epoxy - two part paste			
Manufacturers Comments: RT cure. Translucent, low viscosity version of 2216B/A Grey. Properties differ slightly. Slower cure.			
Typical Use: Rubber, metal, wood, most plastics and masonry.			
Cure: RT Handling strength: 16-20 hrs; Full cure: 30 days.			
Test Methods: ASTM		June 84	
Physical		**Mechanical**	
Colour	Translucent	τ (RT)	8.3 MPa*
Viscosity	10 Pa.s	τ (81°C)	1.4 MPa
S.G.	-	τ (wet)	13.4 MPa
Shelf life	-	Peel (T)	0.44 N mm^{-1}
Pot/Work life	2 hr/RT	σ_t	-
Tg	-	σ_c	-
$T_{Service}$	-	E	-
CTE (60-150°C)	207 × 10^{-6} °C^{-1}	G	-
Thermal cond.	0.2 W m.K^{-1}	Elongation	-
Resistivity: Vol.	3.0 × 10^{12} Ω.cm	Tear strength	-
Surf.	-	H (Shore)	-

Notes: τ Al-alloy, FPL etch, 7 day cure ASTM D-1002-64
*17.2 MPa after 30 day cure. τ(wet) 30days, tap water.
T-peel Al-alloy, FPL etch, RT. ASTM D-1876-61T. Resistivity 23°C.

3M	**2216 B/A Grey**		**Paste**
Epoxy - two part paste			
Manufacturers Comments: RT cure. High shear and peel strengths.			
Typical Use: Rubber, metal, wood, most plastics and masonry.			
Cure: RT Handling strength: 8-12 hrs; Full cure: 7 days.			
Test Methods: ASTM		June 84	
Physical		**Mechanical**	
Colour	Grey	τ (RT)	17.2 MPa
Viscosity	80 Pa.s	τ (82°C)	2.8 MPa
S.G.	-	τ (wet)	20.3 MPa
Shelf life	-	Peel (T)	0.44 N mm^{-1}
Pot/Work life	90 mins/RT	σ_t	-
Tg	-	σ_c	-
$T_{Service}$	-	E	-
CTE (0-40°C)	102×10^{-6} °C^{-1}	G	-
Thermal cond.	0.39 W m.K^{-1}	Elongation	-
Resistivity: Vol.	1.9×10^{12} Ω.cm	Tear strength	-
Surf.	5.5×10^{16} Ω	H (Shore)	-

Notes: τ Al-alloy (FPL etch) 7day cure ASTM D-1002-64.τ(wet) 30days, tap water. T-peel Al-alloy, FPL etch, RT. ASTM D-1876-61T. Resistivity 23°C.

3M	**3501 B/A**		**Paste**
Epoxy - two part paste			
Manufacturers Comments: Rapid RT cure modified epoxy.			
Typical Use: Metal, wood, most plastics and masonry.			
Cure: RT Handling strength: 25-30 mins; Full cure: 2 days.			
Test Methods: Not stated		(March 95)	
Physical		**Mechanical**	
Colour	Grey	τ (RT)	16.5 MPa
Viscosity	500 Pa.s	τ (121°C)	1.4 MPa
S.G.	-	τ (wet)	13.1 MPa
Shelf life	-	Peel (T)	0.88 N mm^{-1}
Pot/Work life	5-7 mins/RT	σ_t	-
Tg	-	σ_c	-
$T_{Service}$	-	E	-
CTE (°C)	-	G	-
Thermal cond.	-	Elongation	-
Resistivity: Vol.	-	Tear strength	-
Surf.	-	H (Shore)	-

Notes: τ(wet) 30days, tap water. T-peel RT.

3M	**2290**		**Paste**
Epoxy - one part paste			
Manufacturers Comments: 21% solids. B-stageable.			
Typical Use: Laminating steel cores for motor stators and rotors. Magnetic tape head laminations.			
Cure: B-stage: 121°C/ 15 mins/no pressure Optimum 171°C/ 30 mins/0.35MPa.			
Test Methods: Not stated		(March 95)	
Physical		**Mechanical**	
Colour	Amber	τ (RT)	34.5 MPa
Viscosity	-	τ (121°C)	8.3 MPa
S.G.	-	τ (wet)	34.5 MPa
Shelf life	-	Peel (T)	1.75 N mm^{-1}
Pot/Work life	-	σ_t	-
Tg	-	σ_c	-
$T_{Service}$	-	E	-
CTE (°C)	-	G	-
Thermal cond.	-	Elongation	-
Resistivity: Vol.	-	Tear strength	-
Surf.	-	H (Shore)	-

Notes: τ(wet) 30days, tap water. T-Peel Al-alloy adherends, RT.

3M	**AF-126 (0.06)**		**Film**
Modified Epoxy film. Areal wt*. 0.29 kg/m²			
Manufacturers Comments: Non-volatile film. High shear and peel strengths. With primer, meets MIL-A-132 Type I Class 2 and MIL-A-25463 Type I Class 2.			
Typical Use: Aerospace. Metals and honeycombs.			
Cure: Optimum: 121°C/ 1 hr/0.3 MPa (metal); ~0.17 MPa (honeycomb)			
Test Methods: MMM-A-132		March 86	
Physical		**Mechanical**	
Colour	Green	τ (RT)	36.5 MPa
Viscosity	-	τ (121°C)	7.4 MPa
S.G.	-	τ (wet)	37.2 MPa
Shelf life	6 month/ -18°C	Peel (T)	7.53 N mm^{-1}
Pot/Work life	1hr /RT	σ_t	41.2 MPa
Tg	-	σ_c	-
$T_{Service}$	-55 to +121°C	E	703 MPa
CTE (°C)	-	G	-
Thermal cond.	-	Elongation	23.7%
Resistivity: Vol.	-	Tear strength	-
Surf.	-	H (Shore)	-

Notes: Other areal wts: 0.03(grey) 0.15 kg/m²; 0.08(blue) 0.39 kg/m². Mechanical data with EC-2320 primer. τ(wet) 30days, tap water. T-peel RT. Volatiles NASA SP-R-0022A: %TWL 1.97; %VCM 0.85.

See Key to Data Sheets - Pages 63 to 65

3M	AF-126-2 (0.06)		Film
Modified Epoxy film. Areal wt*. 0.29 kg/m²			
Manufacturers Comments: Non-volatile film. High shear and peel strengths. With primer, meets MIL-A-132 Type I Class 2 and MIL-A-25463 Type I Class 2.			
Typical Use: Aerospace. Metals and honeycombs.			
Cure: Optimum: 121°C/ 1 hr/0.3 MPa (metal); ~0.17 MPa (honeycomb)			
Test Methods: MMM-A-132		March 86	
Physical		**Mechanical**	
Colour	Grey	τ (RT)	35.9 MPa
Viscosity	-	τ (121°C)	7.3 MPa
S.G.	-	τ (wet)	34.5 MPa
Shelf life	6 months/ -18°C	Peel (T)	7.27 N mm^{-1}
Pot/Work life	1hr /RT	σ_f	-
Tg	-	σ_c	-
$T_{Service}$	-55 to +121°C	E	-
CTE (°C)	-	G	-
Thermal cond.	-	Elongation	-
Resistivity: Vol.	-	Tear strength	-
Surf.	-	H (Shore)	-
Notes: Other areal wts: 0.03(grey) 0.15 kg/m²; 0.08(blue) 0.39 kg/m². Mechanical data with EC-2320 primer. τ(wet) 30days, tap water. T-peel RT. Volatiles NASA SP-R-0022A: %TWL2.62; %VCM 1.34 .			

3M	AF-163-2M		Film
Supported (matt) Modified Epoxy film. Areal wt. 0.29 kg/m²			
Manufacturers Comments: Family of film adhesives (supported and non-supported). Various areal weights. High mechanical properties. X-ray opaque. Improved work-time. Limited self-extinguishing.			
Typical Use: Aerospace. Metals and honeycombs.			
Cure: 177°C/ 1 hr/0.3 MPa (metal).			
Test Methods: AC&S		March 86	
Physical		**Mechanical**	
Colour	Red	τ (RT)	39.4 MPa
Viscosity	-	τ (82°C)	24.8 MPa
S.G.	-	τ (wet)	40.4 MPa
Shelf life	Storage -18°C	Peel (T)	7.71 N mm^{-1}
Pot/Work life	-	σ_f	48.3 MPa
Tg	108(dry); 82(wet)°C	σ_c	-
$T_{Service}$	-55 to +121°C	E	1110 MPa
CTE (°C)	-	G	414 MPa
Thermal cond.	-	Elongation	-
Resistivity: Vol.	-	Tear strength	-
Surf.	-	H (Shore)	-
Notes: Film thickness: 2.4mm. σ, E, G (ASTM D-3039) non-supported adhesive. τ(wet) 30 days, tap water, 0.49kg/m².			

3M	AF-143 (0.1)		Film
Modified Epoxy film. Areal wt. 0.49 kg/m²			
Manufacturers Comments: Asbestos-free version of AF-143. High strength at 177°C. Supported, embossed liner for improved porosity control.			
Typical Use: Aerospace. Metals and honeycombs.			
Cure: 177°C/ 1 hr/0.3 MPa (metal).			
Test Methods: MMM-A-132		March 86	
Physical		**Mechanical**	
Colour	Tan	τ (RT)	22.4 MPa
Viscosity	-	τ (177°C)	15.9 MPa
S.G.	-	τ (wet)	34.5 MPa
Shelf life	-	Peel (Bell)	3.15 N mm^{-1}
Pot/Work life	-	σ_f	-
Tg	-	σ_c	-
$T_{Service}$	-	E	-
CTE (°C)	-	G	-
Thermal cond.	-	Elongation	-
Resistivity: Vol.	-	Tear strength	-
Surf.	-	H (Shore)	-
Notes: Film thickness: 0.41mm.Mechanical data with EC-3917 primer. Bell Peel Al. 2024-T3.			

3M	AF-191 (0.08)		Film
Modified Epoxy film. Areal wt.* 0.39 kg/m²			
Manufacturers Comments: Non-volatile film. High strength at 177°C. High strength and peel. Durable bonds. Pre-bond humidity resistance.			
Typical Use: Aerospace. Metals and honeycombs.			
Cure: Recommended: 177°C/ 1 hr/0.3 MPa (metal).			
Test Methods: MMM-A-132		April 87	
Physical		**Mechanical**	
Colour	Tan	τ (RT)	35.2 MPa
Viscosity	-	τ (177°C)	15.2 MPa
S.G.	-	τ (wet)	34.4 MPa
Shelf life	Storage -18°C	Peel (FR)	5.25 N mm^{-1}
Pot/Work life	-	σ_f	-
Tg	226(dry);190(wet)°C	σ_c	-
$T_{Service}$	-	E	-
CTE (°C)	-	G	-
Thermal cond.	-	Elongation	-
Resistivity: Vol.	-	Tear strength	-
Surf.	-	H (Shore)	-
Notes: *Areal wts. 0.035(0.17kg/m², supported); 0.05(0.24kg/m²,unsupported); 0.1 (0.49kg/m²,supported). Mechanical data with EC-3917 primer. τ(wet) 30 days,95-100%RH, 49°C. Peel - Floating Roller.			

3M	DP-100		Paste
Epoxy - two part paste			
Manufacturers Comments: Duo-pak. Rapid RT cure. Flowable and machinable.			
Typical Use: General assembly. Most adherends types.			
Cure: RT Handling strength: 15-20 mins; Full cure: 1 day.			
Test Methods: Not stated		(March 95)	
Physical		**Mechanical**	
Colour	Clear	τ (RT)	10.3 MPa
Viscosity	13 Pa.s	τ (121°C)	1.4 MPa
S.G.	-	τ (wet)	10.3 MPa
Shelf life	-	Peel (T)	0.35 N mm^{-1}
Pot/Work life	3-5 mins/RT	σ_t	-
Tg	-	σ_c	-
$T_{Service}$	-	*E*	-
CTE (50-100°C)	209 × 10^{-6} $°C^{-1}$	*G*	-
Thermal cond.	0.18 W m.K^{-1}	Elongation	-
Resistivity: Vol.	3.5 × 10^{12} Ω.cm	Tear strength	-
Surf.	-	H (Shore)	-
Notes: τ(wet) 30days, tap water. T-peel RT.			

3M	DP-100 NS		Paste
Epoxy - two part paste			
Manufacturers Comments: Duo-pak. RT cure. Non-sag/flow version of DP-100.			
Typical Use: General assembly- most materials.			
Cure: RT Handling strength: 25-30 mins; Full cure: 2 day.			
Test Methods: Not stated		(March 95)	
Physical		**Mechanical**	
Colour	Translucent	τ (RT)	10.3 MPa
Viscosity	95 Pa.s	τ (121°C)	1.4 MPa
S.G.	-	τ (wet)	10.3 MPa
Shelf life	-	Peel (T)	0.35 N mm^{-1}
Pot/Work life	5-7 mins/RT	σ_t	-
Tg	-	σ_c	-
$T_{Service}$	-	*E*	-
CTE (50-110°C)	149 × 10^{-6} $°C^{-1}$	*G*	-
Thermal cond.	0.18 W m.K^{-1}	Elongation	-
Resistivity: Vol.	2.2 × 10^{14} Ω.cm	Tear strength	-
Surf.	-	H (Shore)	-
Notes: τ(wet) 30days, tap water. T-peel RT.			

3M	DP-100 FR		Paste
Epoxy - two part paste			
Manufacturers Comments: Duo-pak. RT cure. Fire retardant, self-extinguishing version of DP-100. Passes FAA-14CFR 25.853 vertical burn. Meets UL 94 VO.			
Typical Use: General assembly- fire retardant applications.			
Cure: RT Handling strength: 25-30 mins; Full cure: 2 day.			
Test Methods: Not stated		(March 95)	
Physical		**Mechanical**	
Colour	White	τ (RT)	9.7 MPa
Viscosity	50 Pa.s	τ (121°C)	1.4 MPa
S.G.	-	τ (wet)	10.3 MPa
Shelf life	-	Peel (T)	0.35 N mm^{-1}
Pot/Work life	5-7 mins/RT	σ_t	-
Tg	-	σ_c	-
$T_{Service}$	-	*E*	-
CTE (80-100°C)	125 × 10^{-6} $°C^{-1}$	*G*	-
Thermal cond.	0.19 W m.K^{-1}	Elongation	-
Resistivity: Vol.	1.7 × 10^{14} Ω.cm	Tear strength	-
Surf.	-	H (Shore)	-
Notes: τ(wet) 30days, tap water. T-peel RT.			

3M	DP-110		Paste
Epoxy - two part paste			
Manufacturers Comments: Duo-pak. Fast RT cure. Flexible, translucent with controlled flow. Good peel strength.			
Typical Use: General assembly- most materials.			
Cure: RT Handling strength: 25-30 mins; Full cure: 2 day.			
Test Methods: Not stated		(March 95)	
Physical		**Mechanical**	
Colour	Translucent*	τ (RT)	17.2 MPa
Viscosity	50 Pa.s	τ (121°C)	1.0 MPa
S.G.	-	τ (wet)	15.9 MPa
Shelf life	-	Peel (T)	2.63 N mm^{-1}
Pot/Work life	8-10 mins/RT	σ_t	-
Tg	-	σ_c	-
$T_{Service}$	-	*E*	-
CTE (50-110°C)	165 × 10^{-6} $°C^{-1}$	*G*	-
Thermal cond.	0.18 W m.K^{-1}	Elongation	-
Resistivity: Vol.	4.5 × 10^{14} Ω.cm	Tear strength	-
Surf.	-	H (Shore)	-
Notes: τ(wet) 30days, tap water. T-peel RT. * Grey colour version also available.			

3M	DP-190		Paste
Epoxy - two part paste			
Manufacturers Comments: Duo-pak. RT cure. Long-working life, slow cure, flexible. Good peel strength.			
Typical Use: Plastics, rubber, metal and glass.			
Cure: RT Handling strength: 8-12 hrs; Full cure: 7 day.			
Test Methods: Not stated		(March 95)	
Physical		**Mechanical**	
Colour	Grey*	τ (RT)	15.2 MPa
Viscosity	80 Pa.s	τ (121°C)	2.1 MPa
S.G.	-	τ (wet)	17.2 MPa
Shelf life	-	Peel (T)	2.50 N mm^{-1}
Pot/Work life	90 mins/RT	σ_t	-
Tg	-	σ_c	-
$T_{Service}$	-	*E*	-
CTE (50-110°C)	177×10^{-6} °C^{-1}	*G*	-
Thermal cond.	0.34 W m.K^{-1}	Elongation	-
Resistivity: Vol.	5×10^{12} Ω.cm	Tear strength	-
Surf.	-	H (Shore)	-

Notes: τ(wet) 30days, tap water. T-peel RT.
* Translucent version also available.

3M	DP-420		Paste
Epoxy - two part paste			
Manufacturers Comments: Duo-pak. RT cure. Medium-working life. High shear peel strength.			
Typical Use: General assembly.			
Cure: RT Handling strength: 1-2 hrs; Full cure: 3 day.			
Test Methods: Not stated		(March 95)	
Physical		**Mechanical**	
Colour	Off-white	τ (RT)	31.0 MPa
Viscosity	45 Pa.s	τ (121°C)	2.1 MPa
S.G.	-	τ (wet)	31.0 MPa
Shelf life	-	Peel (T)	8.76 N mm^{-1}
Pot/Work life	20 mins/RT	σ_t	-
Tg	-	σ_c	-
$T_{Service}$	-	*E*	-
CTE (50-110°C)	147×10^{-6} °C^{-1}	*G*	-
Thermal cond.	0.18 W m.K^{-1}	Elongation	-
Resistivity: Vol.	1.3×10^{14} Ω.cm	Tear strength	-
Surf.	-	H (Shore)	-

Notes: τ(wet) 30days, tap water. T-peel RT.

3M	DP-260		Paste
Epoxy - two part paste			
Manufacturers Comments: Duo-pak. RT cure. Controlled flow. Good electrical properties. Good corrosion resistance.			
Typical Use: Electrical assemblies.			
Cure: RT Handling strength: 8-12 hrs; Full cure: 7 day.			
Test Methods: Not stated		(March 95)	
Physical		**Mechanical**	
Colour	Translucent	τ (RT)	19.3 MPa
Viscosity	60 Pa.s	τ (121°C)	1.4 MPa
S.G.	-	τ (wet)	13.8 MPa
Shelf life	-	Peel (T)	0.35 N mm^{-1}
Pot/Work life	1hr /RT	σ_t	-
Tg	-	σ_c	-
$T_{Service}$	-	*E*	-
CTE (°C)	-	*G*	-
Thermal cond.	-	Elongation	-
Resistivity: Vol.	-	Tear strength	-
Surf.	-	H (Shore)	-

Notes: τ(wet) 30days, tap water. T-peel RT.

3M	DP-460		Paste
Epoxy - two part paste			
Manufacturers Comments: Duo-pak. RT cure. High performance. Longer working life. Meets MIL-A-23941A.			
Typical Use: General assembly.			
Cure: RT Handling strength: 2-4 hrs; Full cure: 5 day.			
Test Methods: Not stated		(March 95)	
Physical		**Mechanical**	
Colour	Off-white	τ (RT)	31.0 MPa
Viscosity	45 Pa.s	τ (121°C)	2.1 MPa
S.G.	-	τ (wet)	31.0 MPa
Shelf life	-	Peel (T)	12.26 N mm^{-1}
Pot/Work life	1hr /RT	σ_t	-
Tg	-	σ_c	-
$T_{Service}$	-	*E*	-
CTE (50-110°C)	159×10^{-6} °C^{-1}	*G*	-
Thermal cond.	0.18 W m.K^{-1}	Elongation	-
Resistivity: Vol.	4.2×10^{14} Ω.cm	Tear strength	-
Surf.	-	H (Shore)	-

Notes: τ(wet) 30days, tap water. T-peel RT.

See Key to Data Sheets - Pages 63 to 65

3M	New 2214 Hi-Temp		Paste
Epoxy - one part paste			
Manufacturers Comments: 121°C cure. New formulation of 2214 with increased resistance to elevated temperatures and ethylene glycol. Low exotherm.			
Typical Use: Clean metals, glass and most plastics.			
Cure: Optimum 120°C/ 1 hr/ 0.07MPa.			
Test Methods: Not stated		(March 95)	
Physical		**Mechanical**	
Colour	Grey	τ (RT)	19.3 MPa
Viscosity	-	τ (121°C)	13.8 MPa
S.G.	-	τ (wet)	13.8 MPa
Shelf life	-	Peel (T)	0.35 N mm^{-1}
Pot/Work life	-	σ_t	-
Tg	-	σ_c	-
$T_{Service}$	-	E	-
CTE (°C)	-	G	-
Thermal cond.	-	Elongation	-
Resistivity: Vol.	-	Tear strength	-
Surf.	-	H (Shore)	-
Notes: τ(wet) 30days, tap water. T-Peel Al-alloy adherends, RT.			

Bostik	EP 500		Paste
Epoxy - two part paste			
Manufacturers Comments: RT cure. Wide number of uses. General engineering assemblies. Large bond areas.			
Typical Use: Metal, wood, marble, glass, masonry, plastics.			
Cure: 20°C/ 6 hrs; 60°C/40 mins. Full cure: 24 hrs.			
Test Methods: Not stated		July 92	
Physical		**Mechanical**	
Colour	Transparent	τ (RT)	15 MPa
Viscosity	-	τ (°C)	-
S.G.	-	τ (wet)	-
Shelf life	1 yr. (+5 to +15°C)	Peel ( )	-
Pot/Work life	1 hr/RT (for 1 kg)	σ_t	-
Tg	-	σ_c	-
$T_{Service}$	+120°C	E	-
CTE (°C)	-	G	-
Thermal cond.	-	Elongation	-
Resistivity: Vol.	-	Tear strength	-
Surf.	-	H (Shore)	-
Notes: Mix ratio 1:1. τ steel/steel bond.			

Bostik

Bostik	2003		Paste
Epoxy - two part paste			
Manufacturers Comments: RT cure. 100% solids. Many uses. Large bond areas. General engineering assembly, building structural panels and claddings. High strength, good ageing characteristics.			
Typical Use: Metal, wood, glass, masonry, plastics, flooring and tiles.			
Cure: RT Handling: 24 hrs; Full cure: 3 days.			
Test Methods: Not stated		July 89	
Physical		**Mechanical**	
Colour	Grey (mix)	τ (RT)	-
Viscosity	High	τ (°C)	-
S.G.	~1.6	τ (wet)	-
Shelf life	1 yr. (+5 to +25°C)	Peel ( )	-
Pot/Work life	2 hr/RT	σ_t	-
Tg	-	σ_c	-
$T_{Service}$	-30 to +120°C	E	-
CTE (°C)	-	G	-
Thermal cond.	-	Elongation	-
Resistivity: Vol.	-	Tear strength	-
Surf.	-	H (Shore)	-
Notes:			

William Canning (Hylomar)

W. Canning	Epoxy Putty		Paste
Epoxy - Two part.			
Manufacturers Comments: RT cure. Putty. Hard and tough. May be drilled, machined, etc.			
Typical Use: General purpose. Bonding and repairing of a wide range of materials.			
Cure: RT.			
Test Methods: -		(Feb 95)	
Physical		**Mechanical**	
Colour	Pale grey	τ (RT)	-
Viscosity	-	τ (°C)	-
S.G.	-	τ (wet)	-
Shelf life	-	Peel ( )	-
Pot/Work life	-	σ_t	-
Tg	-	σ_c	-
$T_{Service}$	-	E	-
CTE (°C)	-	G	-
Thermal cond.	-	Elongation	-
Resistivity: Vol.	-	Tear strength	-
Surf.	-	H (Shore)	-
Notes:			

See Key to Data Sheets - Pages 63 to 65

W. Canning	Marston Tough Bond	Paste
Epoxy - Two part.		

Manufacturers Comments: Rapid cure. Double syringe dispenser. Good gap filling. Bonds well to many materials.

Typical Use: General purpose.

Cure: RT/5 mins (initial cure).

Test Methods: -		(Feb 95)	
Physical		**Mechanical**	
Colour	Translucent	τ (RT)	-
Viscosity	-	τ (°C)	-
S.G.	-	τ (wet)	-
Shelf life	-	Peel ( )	-
Pot/Work life	-	σ_t	-
Tg	-	σ_c	-
$T_{Service}$	-	*E*	-
CTE (°C)	-	*G*	-
Thermal cond.	-	Elongation	-
Resistivity: Vol.	-	Tear strength	-
Surf.	-	H (Shore)	-

Notes: -

CECA

CECA	Epoglue	Paste
Epoxy- two part paste		

Manufacturers Comments: RT cure. Heavy-duty flooring adhesive, e.g. factory, sports halls and airports. Internal/external use. High peel and humidity resistance.

Typical Use: Flooring adhesive: rubber, wood, metal, concrete.

Cure: RT (12-30°C)/24 hrs; 72 hrs (full cure).

Test Methods: NF		Aug 95	
Physical		**Mechanical**	
Colour	Pink	τ (RT)	-
Viscosity	~40 Pa.s	τ (°C)	-
S.G.	1.73	τ (wet)	-
Shelf life	1 yrs (10-30°C)	Peel (-)	>3.00 N mm^{-1}
Pot/Work life	2hrs(work)*	σ_t	-
Tg	-	σ_c	-
$T_{Service}$	-	*E*	-
CTE (°C)	-	*G*	-
Thermal cond.	-	Elongation	-
Resistivity: Vol.	-	Tear strength	-
Surf.	-	H (Shore)	-

Notes: * 7.2 kg mix Pot Life: 30 mins. Mix ratio 1:0.2(wt). Flash point >55°C.

W. Canning	Tilebond	Paste
Epoxy - Two part.		

Manufacturers Comments: Tough, flexible, impact resistant epoxy for industrial applications involving impact, vibration and deflection.

Typical Use: Ceramic tiles in coal shale and other bulk solids transportation systems. Ceramic wear plates (sub-soiler machines). Lead radiation screening materials.

Cure: -

Test Methods: -		(Feb 95)	
Physical		**Mechanical**	
Colour	-	τ (RT)	-
Viscosity	-	τ (°C)	-
S.G.	-	τ (wet)	-
Shelf life	-	Peel ( )	-
Pot/Work life	-	σ_t	-
Tg	-	σ_c	-
$T_{Service}$	<150°C	*E*	-
CTE	-	*G*	-
Thermal cond.	-	Elongation	-
Resistivity: Vol.	-	Tear strength	-
Surf.	-	H (Shore)	-

Notes: -

Chemetall

Chemetall	Naftotech 7140	Paste
Epoxy- 1 part thixotropic paste		

Manufacturers Comments: Heat cure. Solvent-free. High strength, durable bonds.

Typical Use: Metals (steels and galvanised, aluminium). Some plastics.

Cure: 130°C/1 hr. or 180°C/20 mins.

Test Methods: DIN		April 93	
Physical		**Mechanical**	
Colour	Blue	τ (RT)	18 MPa
Viscosity	70 Pa.s (60°C)	τ (80°C)	14 MPa
S.G.	1.4	τ (wet)	-
Shelf life	3 months/20°C*	Peel (T)	5.00 N mm^{-1}
Pot/Work life	-	σ_t	56 MPa
Tg	87°C	σ_c	-
$T_{Service}$	40-60°C	*E*	-
CTE (°C)	-	*G*	-
Thermal cond.	-	Elongation	3%
Resistivity: Vol.	-	Tear strength	-
Surf.	-	H (Shore)	-

Notes: τ DIN53283, steel. T-peel DIN 53282, steel 23°C
* Storage: 5-15°C extends shelf-life.

Ciba Polymers

Ciba Polymers	Araldite 2010		Paste
Epoxy- two part toughened, thixotropic paste			
Manufacturers Comments: Fast RT cure. Multi-purpose. High shear and peel strengths, tough and resilient. Cartridge pack.			
Typical Use: metals, ceramics, rubbers, most rigid plastics.			
Cure: RT (23°C)/3 hrs; 60°C/20 mins.			
Test Methods: ISO		Nov 93	
Physical		**Mechanical**	
Colour	Lt. Yellow	τ (RT)	23 (29) MPa
Viscosity	~80 Pa.s	τ (80°C)	~3 (~3) MPa
S.G.	~1.2	τ (wet)	26 MPa
Shelf life	3 yrs./RT	Peel (FR)	7.80 N mm^{-1}
Pot/Work life	10 mins (in nozzle)	σ_t	-
Tg	32°C	σ_c	-
$T_{Service}$	-	E	-
CTE (°C)	-	G	-
Thermal cond.	-	Elongation	-
Resistivity: Vol.	-	Tear strength	-
Surf.	-	H (Shore)	-

Notes: Mix ratio 1:1. Peel: ISO 4587
τ data: ISO 4587 RT/7 day cure (23°C/24 hr + 80°C/3 mins cure) ;
Industrial Bulk No. XD4443/XD4444.

Ciba Polymers	Araldite 2012		Paste
Epoxy- two part viscous liquid			
Manufacturers Comments: Rapid RT cure. Multi-purpose. High shear and peel strengths. Cartridge & bulk packs.			
Typical Use: metals, ceramics, rubbers, most rigid plastics.			
Cure: RT (23°C)/100 mins; 60°C/25 mins.			
Test Methods: ISO, DIN		Sept 93	
Physical		**Mechanical**	
Colour	Lt. Yellow	τ (RT)	16 (38) MPa
Viscosity	~30 Pa.s	τ (80°C)	~6 MPa
S.G.	~1.18	τ (wet)	19 MPa
Shelf life	3 yrs/RT	Peel (FR)	5.50 N mm^{-1}
Pot/Work life	4 mins (100g/RT)	σ_t	-
Tg	-	σ_c	-
$T_{Service}$	-	E	-
CTE (°C)	-	G	-
Thermal cond.	-	Elongation	-
Resistivity: Vol.	-	Tear strength	-
Surf.	-	H (Shore)	-

Notes: Mix ratio 1:1 .Peel: ISO 4578; τ(wet) 30days, water, 23°C
τ data: DIN 53283 23°C/7 day cure (23°C/24 hr+80°C/30 mins)
Original Product Code: 2002. Industrial Bulk No. AW2104/HW2934

Ciba Polymers	Araldite 2011		Paste
Epoxy- two part viscous liquid			
Manufacturers Comments: RT cure. Multi-purpose. High shear and peel strengths, tough and resilient to dynamic loads. Cartridge & bulk packs.			
Typical Use: Metals, ceramics, rubbers, rigid plastics.			
Cure: RT (23°C)/10 hrs; 60°C/45 mins.			
Test Methods: ISO, DIN		Sept 93	
Physical		**Mechanical**	
Colour	Lt. Yellow	τ (RT)	26 MPa
Viscosity	~45 Pa.s	τ (80°C)	8 MPa
S.G.	~1.05	τ (wet)	19 MPa
Shelf life	3 yrs./RT	Peel (FR)	5.00 N mm^{-1}
Pot/Work life	100 mins (100g/RT)	σ_t	-
Tg	-	σ_c	-
$T_{Service}$	-	E	-
CTE (°C)	-	G	-
Thermal cond.	0.28 W m.K^{-1}	Elongation	-
Resistivity: Vol.	-	Tear strength	-
Surf.	-	H (Shore)	-

Notes: Mix ratio 100:80 (wt) τ DIN 53283; Peel: ISO 4578;
τ(wet) 30days, water, 23°C
Original Product Code: 2001. Industrial Bulk No. AW106/HV953U.

Ciba Polymers	Araldite 2013		Sept. 93
Epoxy- two part thixotropic paste			
Manufacturers Comments: RT cure. Metal bonding. Good environment and chemical resistance. Gap-filling, non-sag. Cartridge & bulk packs.			
Typical Use: Metals, ceramics, glass and rigid plastics.			
Cure: RT (23°C)/10 hrs ; 60°C/40 mins.			
Test Methods: ISO, DIN		Sept 93	
Physical		**Mechanical**	
Colour	Grey	τ (RT)	21 MPa
Viscosity	Thixotropic	τ (80°C)	4 MPa
S.G.	1.2	τ (wet)	19 MPa
Shelf life	3 yrs/RT	Peel (FR)	4.00 N mm^{-1}
Pot/Work life	50-80mins(100g/RT)	σ_t	-
Tg	-	σ_c	-
$T_{Service}$	-	E	-
CTE (°C)	-	G	-
Thermal cond.	-	Elongation	-
Resistivity: Vol.	-	Tear strength	-
Surf.	-	H (Shore)	-

Notes: Mix ratio 100:60 (wt). τ data: DIN 53283; Peel: ISO 4578;
τ(wet) 30days, water, 23°C
Industrial Bulk No. AV144-2/HV997.

Ciba Polymers	Araldite 2014		Paste
Epoxy- two part paste			
Manufacturers Comments: RT cure. Heat resistant to 120°C. Good water and chemical resistance. Gap-filling, non-sag. Cartridge and bulk pack.			
Typical Use: GRP structures in higher temperature/aggressive environments.			
Cure: RT (23°C)/5 hrs ; 60°C/25 mins.			
Test Methods: DIN, ISO		Sept 93	
Physical		**Mechanical**	
Colour	Grey	τ (RT)	16 MPa
Viscosity	~90 Pa.s	τ (80°C)	16 MPa
S.G.	1.6	τ (wet)	17 MPa
Shelf life	3 yrs/RT	Peel (-)	3.00 N mm^{-1}
Pot/Work life	40 mins(100g/RT)	σ_t	-
Tg	-	σ_c	-
$T_{Service}$	-	E	-
CTE (°C)	-	G	-
Thermal cond.	-	Elongation	-
Resistivity: Vol.	-	Tear strength	-
Surf.	-	H (Shore)	-
Notes: Mix ratio 100:60 (wt). τ: DIN 53283 23°C/7day cure; Peel: ISO 4578; τ(wet) 90days, water, 23°C Industrial Bulk No. AW139/XB5323. Nearest equivalent to AV138M/HV998.			

Ciba Polymers	Araldite 2016		Paste
Epoxy- two part viscous liquid			
Manufacturers Comments: RT Flexible, tough bonds. High peel strength, long pot-life. Cartridge pack.			
Typical Use: Metals, ceramics, glass, rubbers and rigid plastics.			
Cure: RT (23°C)/16 hrs ; 60°C/1 hr.			
Test Methods: ISO		Nov 93	
Physical		**Mechanical**	
Colour	Green	τ (RT)	16 MPa
Viscosity	50-70 Pa.s	τ (80°C)	~2 MPa
S.G.	~1.05	τ (wet)	11 MPa
Shelf life	3 yrs/RT	Peel (FR)	8.90 N mm^{-1}
Pot/Work life	3 hrs(100g/RT)	σ_t	16.5 MPa
Tg	30-35°C	σ_c	-
$T_{Service}$	<40°C	E	680 MPa
CTE (°C)	-	G	-
Thermal cond.	-	Elongation	58%
Resistivity: Vol.	-	Tear strength	-
Surf.	-	H (Shore)	-
Notes: Mix ratio 100:85 (wt); τ: ISO 4587; Peel: ISO 4578; τ(wet) 60days, water, 23°C Industrial Bulk No.XD4417/XD4418.			

Ciba Polymers	Araldite 2015		Paste
Epoxy- two part thixotropic paste			
Manufacturers Comments: RT cure. Toughened, high shear and peel strengths. Gap-filling, non-sag. Cartridge and bulk pack.			
Typical Use: SMC and GRP.			
Cure: RT (23°C)/10 hrs ; 60°C/35 mins.			
Test Methods: DIN, ISO		Sept 93	
Physical		**Mechanical**	
Colour	Beige	τ (RT)	21 MPa
Viscosity	Thixotropic	τ (80°C)	8 MPa
S.G.	1.4	τ (wet)	14 MPa
Shelf life	3 yrs/RT	Peel (FR)	4.50 N mm^{-1}
Pot/Work life	30-40mins(100g/RT)	σ_t	-
Tg	-	σ_c	-
$T_{Service}$	-	E	-
CTE (°C)	-	G	-
Thermal cond.	-	Elongation	-
Resistivity: Vol.	-	Tear strength	-
Surf.	-	H (Shore)	-
Notes: Mix ratio 1:1 (wt); τ: DIN 53283; Peel: ISO 4578; τ(wet) 60days, water, 23°C Industrial Bulk No.AV5308/HV5309. Nearest equivalent to 2005 (AV4076/HY4076)			

Ciba Polymers	Araldite 2017		Paste
Epoxy- two part liquid			
Manufacturers Comments: Rapid RT cure. High peel strength. Cartridge pack.			
Typical Use: Metals, ceramics, glass, rubbers and rigid plastics.			
Cure: RT (23°C)/4 hrs ; 60°C/30 mins.			
Test Methods: ISO		Nov 93	
Physical		**Mechanical**	
Colour	Yellow	τ (RT)	23 MPa
Viscosity	80 Pa.s	τ (80°C)	~4 MPa
S.G.	1.15	τ (wet)	9 MPa
Shelf life	3 yrs/RT	Peel (FR)	10.00 N mm^{-1}
Pot/Work life	5 mins(100g/RT)	σ_t	4.6 MPa
Tg	25°C	σ_c	-
$T_{Service}$	-	E	55 MPa
CTE (°C)	-	G	-
Thermal cond.	-	Elongation	190%
Resistivity: Vol.	-	Tear strength	-
Surf.	-	H (Shore)	-
Notes: Mix ratio 100:106 (wt); τ: ISO 4587; Peel: ISO 4578; τ(wet) 30days, water, 23°C Industrial Bulk No.XD4417/XD4420.			

Ciba Polymers	Araldite 2019		Paste
Epoxy- two part liquid			
Manufacturers Comments: RT cure. Multi-purpose. Low viscosity, solvent-free. High strength. Large bond areas.			
Typical Use: Metals, ceramics, glass, rubbers and rigid plastics.			
Cure: RT (23°C)/22 hrs ; 60°C/75 mins.			
Test Methods: ISO, DIN		Nov 93	
Physical		**Mechanical**	
Colour	Lt. Yellow/clear	τ (RT)	15 MPa
Viscosity	3-4 Pa.s	τ (80°C)	4 MPa
S.G.	~1.0	τ (wet)	9 MPa
Shelf life	3 yrs/RT	Peel (FR)	0.15 N mm^{-1}
Pot/Work life	90 mins(100g/RT)	σ_t	-
Tg	~57°C	σ_c	-
$T_{Service}$	+50°C	E	-
CTE (°C)	-	G	-
Thermal cond.	-	Elongation	-
Resistivity: Vol.	-	Tear strength	-
Surf.	-	H (Shore)	-
Notes: Mix ratio 100:40 (wt); τ:DIN 53282; Peel: ISO 4578; τ(wet) 90 days, water, 23°C Industrial Bulk No.AY103/HY991.			

Ciba Polymers	AV100/HV100		Paste
Epoxy- two part paste			
Manufacturers Comments: RT cure. Heavy and light engineering uses. Solvent-free. Meets MoD Spec DTD 900/6031(metals to CFRP). DQA/TS Schedule AFS18A.			
Typical Use: Metals, ceramics, glass, rubber, plastics and wood.			
Cure: RT (25°C)/24 hrs; 60°C/3 hrs; 100°C/20 mins.			
Test Methods: ASTM, BS		April 86	
Physical		**Mechanical**	
Colour	-	τ (RT)	15 (25) MPa
Viscosity	-	τ (-60°C)	12.5 (20.5) MPa
S.G.	-	τ (wet)	-
Shelf life	3 yrs (5-20°C)	Peel ( )	-
Pot/Work life	2hrs (250g/RT)	σ_t	-
Tg	-	σ_c	-
$T_{Service}$	+150°C max.	E	2700 MPa
CTE (°C)	67 × 10^{-6} °C^{-1}	G	-
Thermal cond.	-	Elongation	-
Resistivity: Vol.	-	Tear strength	-
Surf.	-	H (Shore)	-
Notes: Mix ratio 1:1. CTE - ASTM D-696-70; E - BS2782 Method 335A τ data: RT cure (100°C cure) Al-alloy BSL152 adherends, 3.2 cm² bond area.			

Ciba Polymers	Araldite 2020		Paste
Epoxy- two part liquid			
Manufacturers Comments: RT cure. Transparent for bonding glass and other clear materials. Refractive index similar to glass. Low viscosity.			
Typical Use: Glass bonding.			
Cure: RT (23°C)/25 hrs ; 60°C/2.5 hrs.			
Test Methods: ISO		Nov 93	
Physical		**Mechanical**	
Colour	Transparent	τ (RT)	16 MPa
Viscosity	0.15 Pa.s	τ (80°C)	3 MPa
S.G.	1.1	τ (wet)	13 MPa
Shelf life	2 yrs/RT	Peel (FR)	0.21 N mm^{-1}
Pot/Work life	40-50mins(100g/RT)	σ_t	-
Tg	~40°C	σ_c	-
$T_{Service}$	-	E	-
CTE (°C)	-	G	-
Thermal cond.	-	Elongation	-
Resistivity: Vol.	-	Tear strength	-
Surf.	-	H (Shore)	-
Notes: Mix ratio 100:30 (wt); Refractive Index: 1.553 τ:ISO 4587; Peel: ISO 4578; τ(wet) 90 days, water, 23°C Industrial Bulk No.XW396/XW397.			

Ciba Polymers	AV138M/HV998		Paste
Epoxy- two part paste			
Manufacturers Comments: RT cure. Heavy and light engineering uses. Temperature and chemical resistance. Tolerance to low temperatures during cure. Solvent-free.			
Typical Use: Metals.			
Cure: RT (23°C)/24 hrs; 5°C/48 hrs; 60°C/1hr.			
Test Methods: DIN, ISO		Feb 93	
Physical		**Mechanical**	
Colour	Beige	τ (RT)	13 MPa
Viscosity	80-100 Pa.s	τ (132°C)	16 MPa
S.G.	-	τ (wet)	15 MPa
Shelf life	-	Peel (CD)	1.8 N mm^{-1}
Pot/Work life	35 mins (100g/RT)	σ_t	-
Tg	-	σ_c	-
$T_{Service}$	+150°C max.	E	-
CTE (16-93)	67 × 10^{-6} K^{-1}	G	-
Thermal cond.	-	Elongation	-
Resistivity: Vol.	-	Tear strength	-
Surf.	-	H (Shore)	-
Notes: Mix ratio 100:40 (wt) τ(wet) DIN 53283, 90 days, water, 20°C. Peel- Drum ISO 4578.			

See Key to Data Sheets - Pages 63 to 65

Ciba Polymers	Epibond 1559 A/B		Paste
Epoxy - two part paste			
Manufacturers Comments: Fast RT cure. Tolerant to low curing temperatures. Repair system. Self-extinguishing. Cartridge pack.			
Typical Use: Metals, ceramics, glass, wood, most plastics.			
Cure: RT (25°C)/ 1 hr ; 70°C/30 mins.			
Test Methods: Not stated		Jan 91	
Physical		**Mechanical**	
Colour	Light Grey	τ (RT)	13.8 MPa
Viscosity	80 Pa.s	τ (°C)	-
S.G.	1.2	τ (wet)	4.1 MPa
Shelf life	1 yr./ 18 to 25°C	Peel ( )	-
Pot/Work life	6 mins (100g/RT)	σ_t	-
Tg	-	σ_c	-
$T_{Service}$	-55 to +80°C	E	-
CTE (°C)	-	G	-
Thermal cond.	-	Elongation	-
Resistivity: Vol.	-	Tear strength	-
Surf.	-	H (Shore)	-
Notes: τ Al-alloy adherends; τ(wet) 14 days, 90%RH, 25°C. Flammability BSS 7230; FAR 25.853B: Self-extinguishing 2 sec; burn length: 10.2 cm.			

Ciba Polymers	Epocast 1614		Paste
Epoxy - one part paste (supplied frozen)			
Manufacturers Comments: 180°C cure. Syntactic product. Cartridge pack. Long work-life and good strength to +180°C. Self-extinguishing. Meets BMS 5-28 Type 14.			
Typical Use: Fastener, potting and edge members for honeycomb panels.			
Cure: Thaw before use (<80°C). Stepped cure to 180°C/1 hr.			
Test Methods: Not stated		Jan 91	
Physical		**Mechanical**	
Colour	Off-white	τ (RT)	6.9 MPa
Viscosity	Soft paste	τ (°C)	-
S.G.	0.75	τ (wet)	-
Shelf life	90 days/ -40°C	Peel ( )	-
Pot/Work life	8 hrs (100g/RT)	σ_t	-
Tg	-	σ_c	90 MPa
$T_{Service}$	+180°C max.	E	
CTE (°C)	-	G	-
Thermal cond.	-	Elongation	-
Resistivity: Vol.	-	Tear strength	-
Surf.	-	H (Shore)	-
Notes: τ Al-alloy. Water absorption (distilled water,25°C, 24hrs) +0.2%. Self-extinguishing (BMS 5.28, horizontal) EPOCAST 1614ATF - antimony-free version.			

Ciba Polymers	Epocast 1610		Paste
Epoxy - one part paste			
Manufacturers Comments: 120°C cure. Low density, filler paste. Co-curing with composites or honeycomb potting. Long pot life. Self-extinguishing. Meets BMS 5-28 Type 10.			
Typical Use: Composite and honeycomb panels.			
Cure: 120°C/90 mins.			
Test Methods: Not stated		Jan 91	
Physical		**Mechanical**	
Colour	Off-white	τ (RT)	-
Viscosity	Non-flow paste	τ (°C)	-
S.G.	0.50	τ (wet)	-
Shelf life	6 months/ -18°C	Peel ( )	-
Pot/Work life	30 days (100g/RT)	σ_t	-
Tg	-	σ_c	14 MPa
$T_{Service}$	-55 to +80°C	E	
CTE (°C)	-	G	-
Thermal cond.	-	Elongation	-
Resistivity: Vol.	-	Tear strength	-
Surf.	-	H (Shore)	-
Notes: Self-extinguishing time: 1-3 secs. Water absorption 4% max.			

Ciba Polymers	Epocast 1617 A/B		Paste
Epoxy - two part paste			
Manufacturers Comments: RT cure. 100% solids syntactic compound. Flame resistant, high strength, low density. Meets BMS 5-28 Type 17.			
Typical Use: Edge seals for honeycomb panels.			
Cure: RT/ 24 hrs; 52°C/5 hrs.			
Test Methods: ASTM		Sept 89	
Physical		**Mechanical**	
Colour	Off-white	τ (RT)	-
Viscosity	Soft paste	τ (°C)	-
S.G.	0.7	τ (wet)	-
Shelf life	6 months/RT	Peel ( )	-
Pot/Work life	90 mins (100g/RT)	σ_t	-
Tg	-	σ_c	37.9 MPa
$T_{Service}$	-55 to +80°C	$E_{(compressive)}$	206.8 MPa
CTE (°C)	-	G	-
Thermal cond.	-	Elongation	-
Resistivity: Vol.	-	Tear strength	-
Surf.	-	H (Shore)	-
Notes: σ, E ASTM D-695. Flammability: self-extinguishing. Water absorption (distilled water,25°C, 24hrs) 2%.			

See Key to Data Sheets - Pages 63 to 65

Ciba Polymers	Redux 204		Paste
Epoxy - one part thixotropic paste			
Manufacturers Comments: Elevated temperature cure. For honeycomb cores and structural edge members. Foaming to gap-fill. Volatiles <1%. Extrudable.			
Typical Use: Honeycomb cores/panels.			
Cure: 120°C/1 hr; 150°C/30 mins; 170°C/20 mins.			
Test Methods: Not stated		Jan 91	
Physical		**Mechanical**	
Colour	Black	τ (RT)	12 MPa
Viscosity	~3700 Pa.s	τ (80°C)	10 MPa
S.G.	1.06	τ (wet)	-
Shelf life	1 yrs/ 0°C	Peel ( )	-
Pot/Work life	-	σ_t	-
Tg	-	σ_c	-
$T_{Service}$	-55 to +90°C *	E	-
CTE (°C)	-	G	-
Thermal cond.	-	Elongation	-
Resistivity: Vol.	-	Tear strength	-
Surf.	-	H (Shore)	-
Notes: * +120°C intermittent. τ double lap shear, 1.63 mm glueline.			

Ciba Polymers	Redux 255 A/B		Paste
Epoxy - two part paste			
Manufacturers Comments: RT cure. Flame retardant, low density. Edge filler in 'fire-retardant' honeycomb sandwich panels. Meets FAR 25.853B.			
Typical Use: Honeycomb panels.			
Cure: RT /16 hrs; 70°C/2 hrs.			
Test Methods: Not stated		Jan 91	
Physical		**Mechanical**	
Colour	Green	τ (RT)	-
Viscosity	-	τ (°C)	-
S.G.	0.65	τ (wet)	-
Shelf life	2 yrs/ 18 to 25°C	Peel ( )	-
Pot/Work life	1 hr (100g/RT)	σ_t	-
Tg	-	σ_c	30 MPa
$T_{Service}$	-55 to +140°C	E	-
CTE (°C)	-	G	-
Thermal cond.	-	Elongation	-
Resistivity: Vol.	-	Tear strength	-
Surf.	-	H (Shore)	-
Notes: Mix ratio 100:30(wt). Flammability FAR 25.853B: immediate self-extinguishing in vertical burn.			

Ciba Polymers	Redux 252 A/B		Paste
Epoxy - two part paste			
Manufacturers Comments: RT cure. Low density, low volatiles during cure, self extinguishing. Edge filler for honeycomb sandwich panels. Meets BMS 5-28J.			
Typical Use: Honeycomb panels.			
Cure: RT /16 hrs. Full cure: 3 days. 70°C/2 hrs.			
Test Methods: Not stated		April 91	
Physical		**Mechanical**	
Colour	Blue	τ (RT)	-
Viscosity	-	τ (°C)	-
S.G.	0.63	τ (wet)	-
Shelf life	2 yrs/ 18 to 25°C	Peel ( )	-
Pot/Work life	90 mins (100g/RT)	σ_t	-
Tg	-	σ_c	35 MPa
$T_{Service}$	-55 to +140°C	$E_{(compressive)}$	1400 MPa
CTE (°C)	-	G	-
Thermal cond.	-	Elongation	-
Resistivity: Vol.	-	Tear strength	-
Surf.	-	H (Shore)	-
Notes: Mix ratio 100:40(wt). Water absorption 1.3%wt after 24 hrs immersion.			

Ciba Polymers	Redux 308		Film
Modified epoxy film (unsupported)			
Manufacturers Comments: 170°C cure. High duty uses up to +120°C. Solvent-free, low volatiles (1%). Meets UK DTD 5577Type2 Class 2PH; MMM-A-132 Type I Class 2; MIL-A-25463(ASG) Type I Class1 & 2.			
Typical Use: Aerospace. Metal/metal and honeycomb panels.			
Cure: 170°C/1 hr /0.17-0.35 MPa (metals), lower for honeycombs.			
Test Methods: MMM-A-132		May 85	
Physical		**Mechanical**	
Colour	Black	τ (RT)	47.1 MPa
Viscosity	-	τ (100°C)	25 MPa
S.G.	-	τ (wet)	42.4 MPa
Shelf life	1 yr/-18°C	Peel (T)	2.73 N mm^{-1}
Pot/Work life	6 months/RT	σ_t	-
Tg	-	σ_c	-
$T_{Service}$	+120°C max.	E	-
CTE (°C)	-	G	-
Thermal cond.	-	Elongation	-
Resistivity: Vol.	-	Tear strength	-
Surf.	-	H (Shore)	-
Notes: Areal wt. 0.3 kg/m². τ(wet) 30 days, tap water, 24°C . T-peel RT. Redux 308A-NA (0.3 kg/m²) Asbestos-free & glass beads (bondline control)			

See Key to Data Sheets - Pages 63 to 65

Ciba Polymers	Redux 312		Film
Modified epoxy film (unsupported)*			
Manufacturers Comments: 120°C cure. High duty uses up to +100°C. Solvent-free, low volatiles. Meets MMM-A-132.			
Typical Use: Aerospace. Metal/metal and honeycomb cores.			
Cure: 120°C/30 mins/0.17-0.35 MPa (metals), lower for honeycombs.			
Test Methods: MMM-A-132		Aug 89	
Physical		**Mechanical**	
Colour	Grey	τ (RT)	44.8 MPa
Viscosity	-	τ (100°C)	30 MPa
S.G.	-	τ (wet)	44.5 MPa
Shelf life	1 yr/-18°C	Peel (-)	4.86 N mm^{-1}
Pot/Work life	3 months/RT	σ_t	-
Tg	-	σ_c	-
$T_{Service}$	+100°C max.	E	-
CTE (°C)	-	G	-
Thermal cond.	-	Elongation	-
Resistivity: Vol.	-	Tear strength	-
Surf.	-	H (Shore)	-

Notes: *Redux 312/5 supported version 0.3 kg/m². Light-weight versions: Redux 312L (0.15 kg/m²); Redux 312UL(0.10 kg/m².) Peel metal/metal, RT. τ(wet) 30 days, tap water, 24°C.

Ciba Polymers	Redux 322		Film
Modified epoxy film (supported)*			
Manufacturers Comments: 175°C cure. High duty applications to temperatures +200°C (long-term). Solvent free, low volatiles.			
Typical Use: Aerospace. metals and honeycomb skin to core bonds.			
Cure: 175°C/1 hr/ 0.14-0.70 MPa (metal); lower for honeycomb.			
Test Methods: DTD-5577		Sept 89	
Physical		**Mechanical**	
Colour	Grey/Green	τ (RT)	20 MPa
Viscosity	-	τ (220°C)	10 MPa
S.G.	-	τ (wet)	-
Shelf life	1 yr/-18°C	Peel (CD)	53 Nmm^{-1}
Pot/Work life	3 months/RT	σ_t	-
Tg	-	σ_c	-
$T_{Service}$	+200(c)+220(i) °C	E	-
CTE (°C)	-	G	-
Thermal cond.	-	Elongation	-
Resistivity: Vol.	-	Tear strength	-
Surf.	-	H (Shore)	-

Notes: T service: (c) continuous, (i) intermittent maximum
Areal wt: 0.38 kg/m², 0.35 kg/m². Unsupported 0.38 kg/m².
Peel-Climbing Drum to Concorde spec. CMO 37A.

Ciba Polymers	Redux 319		Film
Modified epoxy film (unsupported)*			
Manufacturers Comments: 175°C cure. High duty uses up to +150°C (long-term). High strength, peel and retained toughness. Meets: UK DTD 5577 Type 4 Class 4PH.			
Typical Use: Aerospace. Metal/metal, composites and honeycomb cores.			
Cure: 175°C/1 hr /0.14-0.70 MPa (metals), lower for honeycombs.			
Test Methods: DTD-5577		Mar 89	
Physical		**Mechanical**	
Colour	Grey	τ (RT)	42 MPa
Viscosity	-	τ (175°C)	17 MPa
S.G.	-	τ (wet)	42.4 MPa
Shelf life	1 yr/-18°C	Peel (Bell)	6.30 N mm^{-1}
Pot/Work life	1 month/RT	σ_t	-
Tg	-	σ_c	-
$T_{Service}$	+150°C max.	E	-
CTE (°C)	-	G	-
Thermal cond.	-	Elongation	-
Resistivity: Vol.	-	Tear strength	-
Surf.	-	H (Shore)	-

Notes: Areal wt: 0.38 kg/m². Peel 60° Bell, with Redux 119 primer, RT.
* Redux 319A supported version. Redux 319L areal wt. 0.18kg/m².

Ciba Polymers	Redux 330		Film
Modified epoxy film (supported)			
Manufacturers Comments: 175°C cure. Composite co-curing or secondary bonding, no metallic filler, X-ray opaque. High fracture toughness. Good drape and tack. Meets BMS 8-245.			
Typical Use: Aerospace. Advanced composites, metals and honeycomb.			
Cure: 175°C/1-2 hrs.			
Test Methods: BMS 8-245*		Oct 91	
Physical		**Mechanical**	
Colour	Light Green	τ (RT)	41 (30.4) MPa
Viscosity	~10 Pa.s (155°C)	τ (132°C)	20 (10.6) MPa
S.G.	1.28	τ (wet)	(18.6) MPa
Shelf life	6 months/-18°C	Peel (Bell)	7.00 N mm^{-1}
Pot/Work life	6 weeks/RT	σ_t	78 MPa
Tg	-	σ_c	335 MPa
$T_{Service}$	+150°C max.	E	3030 MPa
CTE (°C)	-	G	1100 MPa
Thermal cond.	-	Elongation	6.5%
Resistivity: Vol.	-	Tear strength	-
Surf.	-	H (Shore)	-

Notes: Areal wt: 0.39 kg/m², 0.25 kg/m². * Composite materials
τ data: metal/metal bonds (composite/composite bonds).
τ(wet) 14 days, 100%RH, 70°C. Peel- Bell
Redux 330SP - surface finishing ply (0.15 kg/m²).

See Key to Data Sheets - Pages 63 to 65

Ciba Polymers	Redux 408 A/B		Paste
Epoxy - two part paste			
Manufacturers Comments: RT cure. For use with rivets and weld-bonding. Structural repairs. Good strength retention at elevated temperatures.			
Typical Use: Metals, carbon fibre composites, honeycombs.			
Cure: RT /16 hrs + 70°C/4 hrs.			
Test Methods: Not stated		Jan 91	
Physical		**Mechanical**	
Colour	Grey/Green	τ (RT)	19 MPa
Viscosity	~40 Pa.s	τ (100°C)	17 MPa
S.G.	~1.25	τ (wet)	21 MPa
Shelf life	3 yrs/ 18 to 25°C	Peel ( )	-
Pot/Work life	90 mins (100g/RT)	σ_t	-
Tg	-	σ_c	-
$T_{Service}$	-55 to +100°C	E	-
CTE (°C)	-	G	-
Thermal cond.	-	Elongation	-
Resistivity: Vol.	-	Tear strength	-
Surf.	-	H (Shore)	-
Notes: Mix ratio 100:7.5(wt). τ(wet) 1000hrs, distilled water, 22°C. Corrosion inhibited version: Redux 408 SCF.			

Ciba Polymers	Redux 501 A/B		Paste
Epoxy - two part paste			
Manufacturers Comments: RT cure. Repair adhesive for composites, e.g. ATR aircraft repair kits. Good temperature resistance to 120°C.			
Typical Use: Composite laminate repairs.			
Cure: RT /2 days (50% strength), 7weeks (full cure) 45°C/16 hrs; 70°C/2 hrs.			
Test Methods: Not stated		Jan 91	
Physical		**Mechanical**	
Colour	Blue	τ (RT)	-
Viscosity	2.5 Pa.s	τ (°C)	-
S.G.	1.17	τ (wet)	-
Shelf life	3 yrs/ 18 to 25°C	Peel ( )	-
Pot/Work life	90 mins (100g/RT)	σ_t	-
Tg	-	σ_c	35 MPa
$T_{Service}$	-	E(compressive)	1400 MPa
CTE (°C)	-	G	-
Thermal cond.	-	Elongation	-
Resistivity: Vol.	-	Tear strength	-
Surf.	-	H (Shore)	-
Notes: Mix ratio 100:15(wt).			

Ciba Polymers	Redux 420 A/B		Paste
Epoxy- two part paste			
Manufacturers Comments: RT cure. High strength and toughness. Good moisture resistance.			
Typical Use: metals, wood, rubber, GRP and many plastics.			
Cure: RT (25°C)/24 hrs (handling); 1-2 weeks (full cure). 50°C/4 hrs; 120°C/1 hr.			
Test Methods: Not stated		May 92	
Physical		**Mechanical**	
Colour	Green	τ (RT)	33 (35) MPa
Viscosity	~40 Pa.s	τ (80°C)	4 (3) MPa
S.G.	~1.2	τ (wet)	25 MPa
Shelf life	3 yrs/RT	Peel (Bell)	6.00 N mm^{-1}
Pot/Work life	1 hr (100g/RT)	σ_t	-
Tg	-	σ_c	-
$T_{Service}$	-	E	-
CTE (°C)	-	G	-
Thermal cond.	-	Elongation	-
Resistivity: Vol.	-	Tear strength	-
Surf.	-	H (Shore)	-
Notes: Mix ratio 100:40. τ data:RT/7 day cure (50°C/4 hr cure) ; Peel 60° Bell - Al-alloy 2024-T3 clad, RT.			

Ciba Polymers	Redux 609		Film
Modified epoxy film (supported)			
Manufacturers Comments: 120°C cure. General engineering use. Low cost, high mechanical properties, chemical resistance, no solvents, low volatiles.			
Typical Use: Metals and honeycombs, reinforced plastics.			
Cure: 120°C/1 hr pressure to maintain contact.			
Test Methods: MMM-A-132		Jan 91	
Physical		**Mechanical**	
Colour	Blue	τ (RT)	33 MPa
Viscosity	-	τ (100°C)	12 MPa
S.G.	-	τ (wet)	-
Shelf life	1 yr/-18°C	Peel (-)	2.20 N mm^{-1}
Pot/Work life	1 month/RT	σ_t	-
Tg	-	σ_c	-
$T_{Service}$	+100°C	E	-
CTE (°C)	-	G	-
Thermal cond.	-	Elongation	-
Resistivity: Vol.	-	Tear strength	-
Surf.	-	H (Shore)	-
Notes: Peel-metal/metal, RT.			

Cytec

Cytec	FM 27		Film
Modified Epoxy Foaming Film (unsupported)			
Manufacturers Comments: 121 and 171°C cure. Low density, non-asbestos, non-metallic filler. High expansion ratio. Radar transparent.			
Typical Use: Aerospace. Core splicing (metal and composite).			
Cure: 121°C/ 1 hr; 177°C/ 45 mins.			
Test Methods: Not stated		Mar 90	
Physical		**Mechanical**	
Colour	Maroon→Tan	τ (RT) Tube	6.2 (6.2) MPa
Viscosity	-	τ (149°C) Tube	- (2.8) MPa
S.G.	~ 0.2 cured	τ (wet)	-
Shelf life	6 months/-18°C	Peel ( )	-
Pot/Work life	10 days/32°C	σ_t	-
Tg	-	σ_c	-
$T_{Service}$	-55 to +149°C	E	-
CTE (°C)	-	G	-
Thermal cond.	-	Elongation	-
Resistivity: Vol.	-	Tear strength	-
Surf.	-	H (Shore)	-

Notes: Thickness: 0.64 mm, 1.27 mm. Expansion ratio: 3.5 - 5 x thickness
τ data [45g/tube]: 121°C cure (171°C cure).

Cytec	FM 73M		Film
Toughened Epoxy Film (supported)* Areal wt: 0.3 kg/m²			
Manufacturers Comments: 120 or 170°C cure. General purpose aerospace use. High performance, high peel. Humidity resistance. Meets: MMM-132A Type I, Class 2; MIL-A-25463 Type I, Class2.			
Typical Use: Aerospace. Metals (primer for Al-alloys), composites and honeycombs.			
Cure: 120°C/ 1 hr/ 0.3 MPa; 177°C/10 mins 95°C/4 hrs.			
Test Methods: MMM-A-132A		- 89	
Physical		**Mechanical**	
Colour	Yellow	τ (RT)	46.1 MPa
Viscosity	-	τ (80°C)	27.9 MPa
S.G.	-	τ (wet)	34.4 MPa
Shelf life	1 yr/-18°C	Peel (FR)	73.00 N mm⁻¹
Pot/Work life	30 days/24°C)	σ_t	-
Tg	95°C	σ_c	-
$T_{Service}$	-55 to +82°C	E	-
CTE (°C)	-	G	-
Thermal cond.	-	Elongation	-
Resistivity: Vol.	-	Tear strength	-
Surf.	-	H (Shore)	-

Notes: *other supports & unsupported versions available.
Other areal wts. 0.15 kg/m²; 0.43 kg/m².
Outgassing (NASA): TWL 0.78%, CVM 0.00%;
Peel: ASTM D-1781-76(metal/metal); τ(wet) 30 days, 95-100%RH, 50°C
FM-94 higher temperature version (+107°C).

Cytec	FM 61		Film
Duplex Nitrile Epoxy Film (supported) Areal wt: 0.37 kg/m²			
Manufacturers Comments: 177°C cure. One side elastomeric, the other modified epoxy (colour-coded). Elevated temperature and high sandwich peel. Meets: MMM-A-132 Type I Class 3; MIL-A-25463A Type I.			
Typical Use: Metals and honeycombs (BR227 primer recommended).			
Cure: 157 - 177°C: 177°C/ 1hr/0.17-0.34 MPa.			
Test Methods: MMM-A-132		Apr 83	
Physical		**Mechanical**	
Colour	Pale + Brown	τ (RT)	24.1 MPa
Viscosity	-	τ (177°C)	6.9 MPa
S.G.	-	τ (wet)	25.5 MPa
Shelf life	6 months/<29°C	Peel (S/wich)	120 Nm m⁻¹
Pot/Work life		σ_t	-
Tg		σ_c	-
$T_{Service}$	-55 to +177°C	E	-
CTE (°C)	-	G	-
Thermal cond.	-	Elongation	-
Resistivity: Vol.	-	Tear strength	-
Surf.	-	H (Shore)	-

Notes: Thickness: 0.38mm. Mechanical properties with primer.
τ(wet) 30 days, tap water. Peel - Sandwich, with BR227A primer.

Cytec	FM 87-1M		Film
Modified Epoxy Film (supported)* Areal wt: 0.3 kg/m²			
Manufacturers Comments: 120°C+ cure. High temperature performance, toughness and moisture resistance. X-ray opaque. BR127 primer for Al-alloys. Meets: MMM-132A; MIL-A-25463B.			
Typical Use: Aerospace: metals & composites (inc. wet Nomex cores).			
Cure: 105°C to 175°C: e.g. 120°C/ 3 hrs/ 0.3 MPa			
Test Methods: Not stated		April 89	
Physical		**Mechanical**	
Colour	Tan	τ (RT)	31.0 MPa
Viscosity	-	τ (120°C)	20.0 MPa
S.G.	-	τ (wet)	-
Shelf life	1 yr/-18°C	Peel (CD)	290 Nm m⁻¹
Pot/Work life	15 days/<30°C)	σ_t	-
Tg	134°C	σ_c	-
$T_{Service}$	-55 to +120°C	E	-
CTE (°C)	-	G	-
Thermal cond.	-	Elongation	-
Resistivity: Vol.	-	Tear strength	-
Surf.	-	H (Shore)	-

Notes: *Thickness 0.25mm. Other supports & unsupported versions.
Other areal wt. 0.15 kg/m². τ area wide-lap shear. Peel- climbing drum.

Cytec	FM 123-2M		Film
Modified Epoxy Film (supported)* Areal wt: 0.3 kg/m²			
Manufacturers Comments: 120°C cure. Structural bonding aerospace use. Volatiles <2%. Meets: MMM-132 Type I, Class 2; MIL-A-25463 Type I, Class 2.			
Typical Use: Metals (BR127 primer for Al-alloys), composites and honeycombs.			
Cure: 120°C/ 1 hr/ 0.34 MPa ; 105°C/90 mins (test data).			
Test Methods: MMM-A-132		May 84	
Physical		**Mechanical**	
Colour	Purple	τ (RT)	35.5 MPa
Viscosity	-	τ (80°C)	22.8 MPa
S.G.	-	τ (wet)	35.6 MPa
Shelf life	6 months/-18°C	Peel (T)	5.80 N mm⁻¹
Pot/Work life	10 days/25°C)	σ_t	-
Tg	101°C	σ_c	-
$T_{Service}$	+ 82°C	E	-
CTE (°C)	-	G	-
Thermal cond.	-	Elongation	-
Resistivity: Vol.	-	Tear strength	-
Surf.	-	H (Shore)	-
Notes: *unsupported versions available. FM 123-5 supported (knit) carrier. Other areal wts. 0.15 kg/m²; 0.22 kg/m², 0.43 kg/m². τ(wet) 30 days, tap water, 24°C.			

Cytec	FM 300M		Film
Modified Epoxy Film (supported)* Areal wt: 0.4 kg/m²			
Manufacturers Comments: 175°C+ cure. Wide industry acceptance. Moisture and corrosion resistance. High shear strength. X-ray opaque. Meets MMM-A-132A Type II; MIL-A-25463 Rev B Type II Class 2.			
Typical Use: Aerospace. Metals and composites. BR127 primer for Al-alloys.			
Cure: 175°C/ 1hr/0.1-0.7 MPa.			
Test Methods: MMM-A-132		Jan 87	
Physical		**Mechanical**	
Colour	Green	τ (RT)	36.4 MPa
Viscosity	-	τ (149°C)	20.4 MPa
S.G.	-	τ (wet)	38.2 MPa
Shelf life	6 months/-18°C	Peel (FR)	5.08 N mm⁻¹
Pot/Work life	10 days/ 32°C	σ_t	-
Tg	144°C	σ_c	-
$T_{Service}$	-55 to +150°C	E	-
CTE (°C)	-	G	-
Thermal cond.	-	Elongation	-
Resistivity: Vol.	-	Tear strength	-
Surf.	-	H (Shore)	-
Notes: *supported & unsupported versions. Thickness 0.32 mm. Other areal wt. 0.15 kg/m², 0.4kg/m². τ 2024-T3, clad; Peel floating-roller . τ (wet) 30days, 65°C/95-100%RH,RT test.			

Cytec	FM 275-M		Film
Modified Epoxy Film (supported) Areal wt: 0.3 kg/m²			
Manufacturers Comments: 120°C+ cure. Aircraft repair bonding, without need for refrigerated storage. High strength and peel retention in humidity. Rapid cure.			
Typical Use: Aerospace repairs. Metals and composites. BR127 primer for Al-alloys.			
Cure: 120°C/ 2 hrs; 135°C/ 90 mins; 149°C/ 50mins. All with 0.3 MPa.			
Test Methods: Not stated		Feb 88	
Physical		**Mechanical**	
Colour	-	τ (RT)	33.6 MPa
Viscosity	-	τ (121°C)	21.0 MPa
S.G.	-	τ (wet)	-
Shelf life	1 yr/22°C	Peel (FR)	5.60 N mm⁻¹
Pot/Work life	-	σ_t	-
Tg	138°C	σ_c	-
$T_{Service}$	+121°C	E	-
CTE (°C)	-	G	-
Thermal cond.	-	Elongation	-
Resistivity: Vol.	-	Tear strength	-
Surf.	-	H (Shore)	-
Notes: *6 months/32°C. Thickness 0.30mm. Peel- floating roller Other repair adhesive with similar uses: FM235-M (0.4kg/m²).			

Cytec	FM 350 NA		Film
Modified Epoxy Film (supported)* Areal wt: 0.5 kg/m²			
Manufacturers Comments: 177°C+ cure. Non-asbestos, aluminium filled. Toughness at elevated temperatures (long-term exposure). X-ray opaque.			
Typical Use: Aerospace. Engine nacelles. Metals (sheets and honeycombs). BR152 primer for Al-alloys.			
Cure: 177-185°C: e.g. 177°C/ 1hr/0.2 MPa.			
Test Methods: Not stated		Feb 88	
Physical		**Mechanical**	
Colour	Grey	τ (RT)	28.9 MPa
Viscosity	-	τ (176°C)	21.4 MPa
S.G.	-	τ (wet)	26.2 MPa
Shelf life	6 months/-18°C	Peel (FR)	0.508 N mm⁻¹
Pot/Work life	5 days/ 32°C	σ_t	-
Tg	174°C	σ_c	-
$T_{Service}$	-55 to +177°C	E	-
CTE (°C)	-	G	-
Thermal cond.	-	Elongation	-
Resistivity: Vol.	-	Tear strength	-
Surf.	-	H (Shore)	-
Notes: *FM-350U-NA unsupported version. τ (wet) 7 days, 70°C/98%RH,RT test. Peel - Floating Roller .			

See Key to Data Sheets - Pages 63 to 65

Cytec — FM 355 — Film

Modified Epoxy Film (supported)* Areal wt: 0.4 kg/m²

Manufacturers Comments: 177°C+ cure. Non-asbestos, aluminium filled. Toughness at elevated temperatures (long-term). X-ray opaque.

Typical Use: Aerospace. Engine nacelles. Metals (sheets and honeycombs). BR152 primer for Al-alloys.

Cure: 177-185°C: e.g. 177°C/ 1hr/0.2 MPa.

Test Methods: Not stated		Feb 88	
Physical		**Mechanical**	
Colour	Grey	τ (RT)	21.4 MPa
Viscosity	-	τ (177°C)	20.7 MPa
S.G.	-	τ (wet)	26.2 MPa
Shelf life	6 months/-18°C	Peel (FR)	1.00 N mm^{-1}
Pot/Work life	9 days/ 24°C	σ_t	-
Tg	180°C	σ_c	-
$T_{Service}$	+177°C	*E*	-
CTE (°C)	-	*G*	-
Thermal cond.	-	Elongation	-
Resistivity: Vol.	-	Tear strength	-
Surf.	-	H (Shore)	-

Notes: Thickness 0.33mm. *other supports & unsupported versions available. Other areal wt. : 0.25 kg/m². τ Al 2024-T3 clad, FPL etch, no primer. Peel - Floating Roller .

Cytec — FM 410-1 — Film

Modified Epoxy Foaming Film (unsupported)

Manufacturers Comments: 121°C+cure. Non-asbestos, non-metallic filler. Radar transparent. Ambient pressure splicing.

Typical Use: Aerospace. Core splicing (metal and composite). Edge members and inserts to core.

Cure: 105°C/ 90 mins; 121°C/ 1 hr; 175°C/ 1 hr.

Test Methods: Not stated		Oct 92	
Physical		**Mechanical**	
Colour	Blue→Green	τ (RT) Tube	6.2 (7.6) MPa
Viscosity	-	τ (150°C) Tube	- (5.5) MPa
S.G.	0.24-0.56 cured	τ (wet)	-
Shelf life	6 months/-18°C	Peel ( )	-
Pot/Work life	10 days/32°C	σ_t	-
Tg	-	σ_c	-
$T_{Service}$	-55 to +175°C	*E*	-
CTE (°C)	-	*G*	-
Thermal cond.	-	Elongation	-
Resistivity: Vol.	-	Tear strength	-
Surf.	-	H (Shore)	-

Notes: Thickness: 0.64 mm, 1.27 mm, 2.54. Expansion ratio: 1.7 - 3.5 x thickness τ data [tube]: 120°C cure (175°C cure).

Cytec — FM 404NA — Film

Modified Epoxy Foaming Film (unsupported)

Manufacturers Comments: 170°C cure. Non-asbestos. For use with FM 400 adhesive. X-ray opaque.

Typical Use: Aerospace. Core splicing (metal and composite), edge-members and inserts to core.

Cure: 170°C/ 1 hr/ 0.1 - 0.3 MPa.

Test Methods: Not stated		Oct 86	
Physical		**Mechanical**	
Colour	Grey	τ (RT) Beam	4.5 MPa
Viscosity	-	τ (185°C) Beam	3.2 MPa
S.G.	~ 0.2- 0.4 cured	τ (wet)	-
Shelf life	6 months/-18°C	Peel ( )	-
Pot/Work life	5 days/<32°C	σ_t	-
Tg	-	σ_c	-
$T_{Service}$	+210°C	*E*	-
CTE (°C)	-	*G*	-
Thermal cond.	-	Elongation	-
Resistivity: Vol.	-	Tear strength	-
Surf.	-	H (Shore)	-

Notes: Thickness: - Areal wt. 0.98 kg/m². Expansion ratio: 4 5 x thickness τ Beam shear specimen, with 2 core splices.

Cytec — FM 1000 — Film

Modified Polyamide Epoxy Film (unsupported) Areal wt*: 0.3 kg/m²

Manufacturers Comments: 170°C cure. High elongation, high peel strength. Wide temperature use, inc. low temperatures. Often no primer used. 'Tack' primers available.

Typical Use: Metal helicopter blades. Metals, composites, structural plastic sandwich panels.

Cure: 170°C/ 1hr/0.3 MPa.

Test Methods: MMM-A-132		Feb 82	
Physical		**Mechanical**	
Colour	White	τ (RT)	48.9 MPa
Viscosity	-	τ (82°C)	25.3 MPa
S.G.	-	τ (wet)	37.9 MPa
Shelf life	4-6 months/<30°C	Peel (T)	10.51 N mm^{-1}
Pot/Work life	-	σ_t	-
Tg	125°C	σ_c	-
$T_{Service}$	-250 to +95°C	*E*	-
CTE (°C)	-	*G*	-
Thermal cond.	-	Elongation	-
Resistivity: Vol.	-	Tear strength	-
Surf.	-	H (Shore)	-

Notes: Thickness 0.25mm. *other areal wts Mechanical properties without primer. τ(wet) 30 days, tap water, 24°C. T-Peel .

Cytec — FM 1020 — Film

Modified Epoxy Foaming Film (unsupported)

Manufacturers Comments: 121°C+cure. High density, non-asbestos, no metallic filler. Radar transparent. Fuel cell, fuel barrier.

Typical Use: Aerospace. Core splicing. Edge members, honeycombs.

Cure: 120/ 1 hr; 150°C/ 2 hr; 175°C/ 1 hr.

Test Methods: Not stated		(Mar 95)	
Physical		**Mechanical**	
Colour	Dk. Red→Yellow	τ (RT) Tube	9.3 (9.0) MPa
Viscosity	-	τ (175°C) Tube	2.9 (5.4) MPa
S.G.	0.4 - 0.6 cured	τ (wet)	-
Shelf life	6 months/-18°C	Peel ( )	-
Pot/Work life	10 days/<30°C	σ_t	-
Tg	-	σ_c	-
$T_{Service}$	-55 to +175°C	E	-
CTE (°C)	-	G	-
Thermal cond.	-	Elongation	-
Resistivity: Vol.	-	Tear strength	-
Surf.	-	H (Shore)	-

Notes: Thickness: 0.64 mm to 2.54mm. Areal wt. 1.1 kg/m² (1.25mm thick).
Expansion ratio: 1.4 - 2 x thickness
τ data [tube]: 120°C cure (150°C cure).

Cytec — Metlbond 1146M — Film

Modified Epoxy Film (supported)* Areal wt: 0.3 kg/m²

Manufacturers Comments: 120°C+ cure. High toughness and hot/wet strength and environmental resistance. X-ray opaque.

Typical Use: Aerospace-metals and composites. BR127 primer for Al-alloys.

Cure: 121-177°C/ 90 mins/ 0.3 MPa.

Test Methods: ASTM		Feb 92	
Physical		**Mechanical**	
Colour	-	τ (RT)	45.5 MPa
Viscosity	400% flow	τ (121°C)	27.8 MPa
S.G.	-	τ (wet)	46.2 MPa
Shelf life	6 months/-18°C	Peel (T)	6.48 N mm^{-1}
Pot/Work life	30 days/ RT	σ_t	-
Tg °C	137(dry) 127(wet)	σ_c	-
$T_{Service}$ °C	+121(wet) 107(dry)	E	-
CTE (24-204°C)	62 × 10^{-6} °C^{-1}	G	-
Thermal cond.	-	Elongation	-
Resistivity: Vol.	-	Tear strength	-
Surf.	-	H (Shore)	-

Notes: ex-BASF product. *also knit support. Thickness 0.25 mm.
Other areal wts. 0.15 kg/m², 0.4kg/m². τ ASTM D-4027; T-peel ASTM1876
τ (wet) 30days, 71°C/95%RH,RT test. Loss tangent (10GHz), RT: 0.1;
Dielectric constant (10 GHz), RT: 0.96.

Cytec — Metlbond 1113 — Film

Modified Epoxy Film (supported)* Areal wt: 0.3 kg/m²

Manufacturers Comments: 120°C cure. High peel, environmentally tolerant. Long shop-life. Meets: MMM-132 Type I Class 2; MIL-A-25463 Type I Class 1 & 2; Bell BHT-299-947-121; Grumman APS-1127; Aerospatiale DCC37.1; McDonnelDouglas DMS1911; Piper PMSC1014; Sikorsky SS8612: Gulfstream GMS4301; Vought 207-8-415.

Typical Use: Helicopter blades, military and civil aircraft (metals, with primer for Al-alloys, and composites).

Cure: 93°C to 143°C, Standard: 127°C/ 1 hr/ 0.34 MPa.

Test Methods: MMM-A-132		May 87	
Physical		**Mechanical**	
Colour	Yellow	τ (RT)	43.2 MPa
Viscosity	-	τ (80°C)	34.2 MPa
S.G.	-	τ (wet)	41.0 MPa
Shelf life	> 1 yr/-18°C	Peel (T)	8.00 N mm^{-1}
Pot/Work life	45 days/24°C)	σ_t	-
Tg	104°C	σ_c	-
$T_{Service}$	-55 to + 82°C	E	-
CTE (°C)	-	G	-
Thermal cond.	-	Elongation	-
Resistivity: Vol.	-	Tear strength	-
Surf.	-	H (Shore)	-

Notes: Ex-BASF product. * Metlbond 1113-2 - unsupported version.
Other areal wts. 0.15 kg/m²; 0.43 kg/m². τ 2024-T3, primer 6726.
τ(wet) 30 days, tap water, 24°C.

Cytec — Metlbond 1515-1 — Film

Modified Epoxy Film (supported)* Areal wt: 0.24 kg/m²

Manufacturers Comments: 170°C cure. Wide aerospace acceptance. Co-cure and secondary bonding (composites). Meets: Boeing BMS8-245B [1515-1 (0.05) & 1515-3]; Northrop MS208[1515.2].

Typical Use: Metals, composites, honeycombs. Surfacing.

Cure: 149 - 177°C: e.g. 177°C/ 2hr/0.3 MPa.

Test Methods: BMS 8-245B		April 87	
Physical		**Mechanical**	
Colour	Brown	τ (RT)	29.0 MPa
Viscosity	-	τ (149°C)	19.3 MPa
S.G.	-	τ (wet)	-
Shelf life	1 yr/-18°C	Peel (Bell)	1.20 N mm^{-1}
Pot/Work life	10 days/RT	σ_t	-
Tg	160°C	σ_c	-
$T_{Service}$	-54 to +149°C	E	-
CTE (°C)	-	G	-
Thermal cond.	-	Elongation	-
Resistivity: Vol.	-	Tear strength	-
Surf.	-	H (Shore)	-

Notes: Ex-BASF product. *other supports & unsupported versions + areal wts.
Mechanical properties without primer.
τ 2024-T3, no primer. Peel - Bell 2024T3, no primer.

See Key to Data Sheets - Pages 63 to 65

Degussa

Degussa	Agomet P54		Paste
Epoxy - Two part, paste			
Manufacturers Comments: RT cure. Good shear strength. Good temperature and moisture resistance.			
Typical Use: Metals, some plastics, wood, glass, ceramics.			
Cure: RT/ 24 hrs; 50°C/ 4 hr; 120°C/ 15 mins			
Test Methods: DIN		July 94	
Physical		**Mechanical**	
Colour	-	τ (RT)	16 MPa
Viscosity	~40 Pa.s	τ (°C)	-
S.G.	~1.5	τ (wet)	-
Shelf life	1 year	Peel (-)	1.5 N mm^{-1}
Pot/Work life	30-40 mins	σ_t	-
Tg	-	σ_c	-
$T_{Service}$	-30 to +80°C	E	-
CTE (°C)	-	G	-
Thermal cond.	-	Elongation	-
Resistivity: Vol.	-	Tear strength	-
Surf.	-	H (Shore)	-
Notes: τ DIN 53282 (aluminium); Peel DIN 53282 (aluminium). Flash point >100°C			

Degussa	Agomet P79		Paste
Epoxy - Two part, paste			
Manufacturers Comments: RT cure. Good shear strength. Good temperature and moisture resistance.			
Typical Use: Metals, some plastics, wood, glass, ceramics.			
Cure: RT/ 24 hrs; 50°C/ 4 hr; 120°C/ 10 mins.			
Test Methods: DIN		Aug 94	
Physical		**Mechanical**	
Colour	-	τ (RT)	15 MPa
Viscosity	~120 Pa.s	τ (°C)	-
S.G.	~1.6	τ (wet)	-
Shelf life	1 year/RT	Peel (-)	1.5 N mm^{-1}
Pot/Work life	1 hr	σ_t	-
Tg	-	σ_c	-
$T_{Service}$	-30 to +80°C	E	-
CTE (°C)	-	G	-
Thermal cond.	-	Elongation	-
Resistivity: Vol.	-	Tear strength	-
Surf.	-	H (Shore)	-
Notes: τ DIN 53282 (aluminium); Peel DIN 53282 (aluminium). Flash point >100°C			

Degussa	Agomet P76		Paste
Epoxy - Two part, paste			
Manufacturers Comments: RT cure. High peel and shear strength. Good temperature and moisture resistance.			
Typical Use: Metals, some plastics, wood, glass, ceramics.			
Cure: RT/ 48 hrs; 50°C/ 6 hr; 120°C/ 15 mins			
Test Methods: DIN		July 94	
Physical		**Mechanical**	
Colour	-	τ (RT)	>19 MPa
Viscosity	~100 Pa.s	τ (°C)	-
S.G.	~1.6	τ (wet)	-
Shelf life	1 year/RT	Peel (-)	2.1 N mm^{-1}
Pot/Work life	2-3 hr	σ_t	-
Tg	-	σ_c	-
$T_{Service}$	-30 to +80°C	E	-
CTE (°C)	-	G	-
Thermal cond.	-	Elongation	-
Resistivity: Vol.	-	Tear strength	-
Surf.	-	H (Shore)	-
Notes: τ DIN 53282 (aluminium); Peel DIN 53282 (aluminium). Flash point >100°C			

Delo

Delo	Delomet 02-Rapid		Paste
Epoxy - Two part, liquid			
Manufacturers Comments: RT + heat cured. Assembly adhesive and fast, reliable repairs. Short pot-life. Good strength static and dynamic loads. (Preliminary data).			
Typical Use: Metals and non-metals.			
Cure: RT/ 24 hrs; 60°C/90 mins; 120°C/~25mins (full τ-strength).			
Test Methods: DIN		(July 95)	
Physical		**Mechanical**	
Colour	Clear Yellow	τ (RT)	18 MPa
Viscosity	~9.5 Pa.s	τ (°C)	-
S.G.	~1.2	τ (wet)	-
Shelf life	1 year/RT	Peel ( )	-
Pot/Work life	7 mins (3 g mix)	σ_t	-
Tg	-	σ_c	-
$T_{Service}$	-30 to +80°C	E	-
CTE (°C)	-	G	-
Thermal cond.	-	Elongation	-
Resistivity: Vol.	-	Tear strength	-
Surf.	-	H (Shore)	-
Notes: Mix ratio 1:1 (wt). τ DIN 53283 (after 24 hrs).			

Delo	Delomet 03-Rapid		Paste
Epoxy - Two part, liquid			
Manufacturers Comments: RT + heat cured. Assembly adhesive and fast, reliable repairs. Short pot-life, reaching 50% strength after 15 mins. Good strength static and dynamic loads. (Preliminary data).			
Typical Use: Metals and non-metals.			
Cure: RT/ 24 hrs; 120°C/~15mins (full τ-strength).			
Test Methods: DIN		(July 95)	
Physical		**Mechanical**	
Colour	Clear Yellow	τ (RT)	18.7 (10.0) MPa
Viscosity	~11.5 Pa.s	τ (°C)	-
S.G.	~1.15	τ (wet)	-
Shelf life	1 year/RT	Peel ( )	-
Pot/Work life	3.5 mins (3 g mix)	σ_t	-
Tg	-	σ_c	-
$T_{Service}$	-30 to +80°C	E	-
CTE (°C)	-	G	-
Thermal cond.	-	Elongation	-
Resistivity: Vol.	-	Tear strength	-
Surf.	-	H (Shore)	-
Notes: Mix ratio 1:1 (wt). τ data DIN 53283 al/al, 1.6mm: after 24 hrs (after 15 mins).			

Delo	Dupox 1895/1		Paste
Epoxy - Two part, paste			
Manufacturers Comments: RT cured. General assembly adhesive. Good strength static and dynamic loads. Chemical resistance.			
Typical Use: Metals and non-metals.			
Cure: RT/ 24 hrs.			
Test Methods: DIN		(July 95)	
Physical		**Mechanical**	
Colour	Grey	τ (RT)	21 (26) MPa
Viscosity	~100 Pa.s	τ (100°C)	5 MPa
S.G.	~1.3	τ (wet)	-
Shelf life	1 year/RT	Peel (FR)	4.0 N mm^{-1}
Pot/Work life	~1 hr (100g)	σ_t	38 MPa
Tg	112°C	σ_c	-
$T_{Service}$	-40 to +100°C	E	3200 MPa
CTE (°C)	-	G	-
Thermal cond.	-	Elongation	1.4% (tear)
Resistivity: Vol.	-	Tear strength	-
Surf.	-	H (Shore D)	75
Notes: τ data: DIN 53283 - al/al, 1.6mm (DIN54451 - al/al, 6mm); τ°C DIN 53286: Peel - Floating Roller, DIN 53289 (steel), σt DIN 53455; E DIN53457; Elong DIN 54455; Thermal conductivity: DIN 52612. Hardness DIN 53505. Water absorption DIN 53495, 24 hr, RT: 0.13%.			

Delo	Delopox 333		Paste
Epoxy - Two part, thick paste			
Manufacturers Comments: RT cure. Mineral filler. Repair/filler adhesive for non-corrosive repairs. Good chemical resistance.			
Typical Use: Crack repairs - engine blocks, castings, pipelines, machine beds, car body-work (weld seams), etc.			
Cure: RT/ 24 hrs.			
Test Methods: DIN		(July 95)	
Physical		**Mechanical**	
Colour	Black + Beige	τ (RT)	-
Viscosity	Paste	τ (°C)	-
S.G.	~2.0	τ (wet)	-
Shelf life	1 year/RT min.	Peel ( )	-
Pot/Work life	40 mins (200 g mix)	σ_t	25 MPa
Tg	-	σ_c	-
$T_{Service}$	-30 to +180°C	E	-
CTE (°C)	-	G	-
Thermal cond.	-	Elongation	-
Resistivity: Vol.	-	Tear strength	-
Surf.	-	H (Shore)	-
Notes: Mix ratio 1:1 (wt). Break bearing pressure DIN 53456 92.5 MPa; Pressure strength DIN 53454 73 MPa. Flash point >100°C. Shrinkage: 0.5 - 1% Products with similar uses: DELOPOX 334 (faster cure); DELOPOX - steel powder filled (2 hr/RT cure); DELOPOX 202 aluminium powder filled (12hr/RT cure).			

Delo	Monopox 1196 (VE 299)		Paste
Epoxy - One part, thixotropic paste			
Manufacturers Comments: Heat cured, aluminium-filled. General assembly adhesive. High strength, tough joints. Chemical and thermal resistance.			
Typical Use: Metals (inc. ferrites) and ceramics.			
Cure: 130°C/75 mins; 150°C/ 40 mins; 180°C/ 15 mins.			
Test Methods: DIN		(July 95)	
Physical		**Mechanical**	
Colour	Silver Grey	τ (RT)	36 (57) MPa
Viscosity	290 Pa.s	τ (150°C)	5 MPa
S.G.	~1.4	τ (wet)	-
Shelf life	6 months/+5°C	Peel (FR)	12.4 N mm^{-1}
Pot/Work life	-	σ_t	46 MPa
Tg	113°C	σ_c	-
$T_{Service}$	-55 to +200°C	E	4000 MPa
CTE (°C)	-	G	-
Thermal cond.	0.7 W m.K^{-1}	Elongation	1.4% (tear)
Resistivity: Vol.	-	Tear strength	-
Surf.	-	H (Shore)	-
Notes: τ data: DIN 53283 - al/al, 1.6mm (DIN54451 - al/al, 6mm); τ°C DIN 53286: Peel - Floating Roller, DIN 53289 (aluminium), σt DIN 53455; E DIN53457; Elong DIN 54455; Thermal conductivity: DIN 52612. Indentation hardness DIN 53456: 96 N/mm^2: Water absorption DIN 53495, 24 hr, RT: 0.21%.			

See Key to Data Sheets - Pages 63 to 65

Devcon (ITW)

Delo	Monopox 1197 (VE 314)		Paste
Epoxy - One part, thixotropic paste			
Manufacturers Comments: Heat cured, aluminium-filled. General assembly adhesive. High strength, tough joints. Chemical and thermal resistance.			
Typical Use: Metals (inc. ferrites) and ceramics.			
Cure: 130°C/75 mins; 150°C/ 40 mins; 180°C/ 15 mins.			
Test Methods: DIN		(July 95)	
Physical		**Mechanical**	
Colour	Silver Grey	τ (RT)	26 (55) MPa
Viscosity	580 Pa.s	τ (150°C)	5 MPa
S.G.	~1.4	τ (wet)	-
Shelf life	6 months/+5°C	Peel (FR)	11.6 N mm^{-1}
Pot/Work life	-	σ_f	40 MPa
Tg	112°C	σ_c	-
$T_{Service}$	-55 to +200°C	*E*	3300 MPa
CTE (°C)	-	*G*	-
Thermal cond.	0.7 W m.K^{-1}	Elongation	1.4% (tear)
Resistivity: Vol.	-	Tear strength	-
Surf.	-	H (Shore)	-
Notes: τ data: DIN 53283 - al/al, 1.6mm (DIN54451 - al/al, 6mm); τ°C DIN 53286: Peel - Floating Roller, DIN 53289 (aluminium), σt DIN 53455; *E* DIN53457; Elong DIN 54455; Thermal conductivity: DIN 52612. Indentation hardness DIN 53456: 95 N/mm²: Water absorption DIN 53495, 24 hr, RT: 0.21%.			

Devcon	2-Ton Clear Epoxy		Paste
Epoxy - Two part.			
Manufacturers Comments: RT cure. Good impact properties and resistant to water, oils and chemicals.			
Typical Use: General purpose. Bonding of metals, ceramics, wood, concrete and most plastics.			
Cure: Useful strength: 8 - 16 hr/RT, Full cure: several days.			
Test Methods: ASTM		(Feb 95)	
Physical		**Mechanical**	
Colour	Transparent	τ (RT)	15.5 MPa
Viscosity	15.5 Pa.s	τ (°C)	-
S.G.	1.1	τ (wet)	-
Shelf life	-	Peel ( )	-
Pot/Work life	8 - 12 min	σ_f	-
Tg	-	σ_c	-
$T_{Service}$	<93°C	*E*	-
CTE (°C)	-	*G*	-
Thermal cond.	-	Elongation	-
Resistivity: Vol.	-	Tear strength	-
Surf.	-	H (Shore)	-
Notes: Breakdown potential: 600 V/0.25mm.			

Delo	Monopox 1494 (ESP 106)		Paste
Epoxy - One part, paste			
Manufacturers Comments: Heat cured, mineral-filled. Electrical insulator.			
Typical Use: Metals (inc. ferrites) and ceramics.			
Cure: Heat cured. 130 to 180°C.			
Test Methods: DIN		(July 95)	
Physical		**Mechanical**	
Colour	-	τ (RT)	60 MPa
Viscosity	280 Pa.s	τ (°C)	-
S.G.	-	τ (wet)	-
Shelf life	-	Peel ( )	-
Pot/Work life	-	σ_f	-
Tg	-	σ_c	-
$T_{Service}$	-55 to +150°C	*E*	5900 MPa
CTE (°C)	-	*G*	-
Thermal cond.	-	Elongation	-
Resistivity: Vol.	-	Tear strength	-
Surf.	-	H (Shore)	-
Notes: τ: DIN 54451 - al/al, 6mm; *E* DIN53457.			

Devcon	5 Minute Epoxy		Paste
Epoxy - Two part.			
Manufacturers Comments: Fast RT cure. Good tensile and impact properties.			
Typical Use: General purpose. Bonding of metals, ceramics, wood, concrete and most plastics.			
Cure: Initial cure: 5/RT, Useful strength: 1hr/RT, Full cure: several days			
Test Methods: ASTM		(Feb 95)	
Physical		**Mechanical**	
Colour	Transparent	τ (RT)	9.7 MPa
Viscosity	15.5 Pa.s	τ (°C)	-
S.G.	1.1	τ (wet)	-
Shelf life	-	Peel ( )	-
Pot/Work life	3 - 6 min	σ_f	-
Tg	-	σ_c	-
$T_{Service}$	<93°C	*E*	-
CTE (°C)	-	*G*	-
Thermal cond.	-	Elongation	-
Resistivity: Vol.	-	Tear strength	-
Surf.	-	H (Shore)	-
Notes: Breakdown potential: 490 V/0.25mm.			

Devcon	5 Minute Epoxy Gel		Paste
Epoxy - Two part thixotropic			
Manufacturers Comments: Fast RT cure. Good shear properties. 100% solids.			
Typical Use: General purpose. Bonding of metals, ceramics, wood, concrete and most plastics.			
Cure: Initial cure: 5/RT, Useful strength: 1hr/RT, Full cure: several days			
Test Methods: ASTM		(Feb 95)	
Physical		**Mechanical**	
Colour	White	τ (RT)	17.2 MPa
Viscosity	(gel)	τ (°C)	-
S.G.	1.17	τ (wet)	-
Shelf life	-	Peel ( )	-
Pot/Work life	3 - 6 min	σ_t	-
Tg	-	σ_c	-
$T_{Service}$	<93°C	*E*	-
CTE (°C)	-	*G*	-
Thermal cond.	-	Elongation	-
Resistivity: Vol.	-	Tear strength	-
Surf.	-	H (Shore)	-
Notes: Breakdown potential: 440 V/0.25mm.			

Devcon	Special F Epoxy Aluminium		Paste
Epoxy - Two part, aluminium filled.			
Manufacturers Comments: RT cure. Good tensile and impact properties and resistant to chemicals.			
Typical Use: General purpose. Bonding of aluminium and other metals, ceramics, wood and concrete.			
Cure: Useful strength: 16 - 24 hr/RT, Full cure: several days.			
Test Methods: ASTM		(Feb 95)	
Physical		**Mechanical**	
Colour	Metallic (Al)	τ (RT)	17.2 MPa
Viscosity	(thick paste)	τ (°C)	-
S.G.	1.72	τ (wet)	-
Shelf life	-	Peel ( )	-
Pot/Work life	45 - 60 min	σ_t	-
Tg	-	σ_c	-
$T_{Service}$	<121°C	*E*	-
CTE (°C)	-	*G*	-
Thermal cond.	-	Elongation	-
Resistivity: Vol.	-	Tear strength	-
Surf.	-	H (Shore)	-
Notes:			

Devcon	Magic Bond		Paste
Epoxy - Two part putty.			
Manufacturers Comments: RT cure. Supplied as co-extruded product, the required amount is cut off and mixed by manipulation with the fingers. Adheres to damp surfaces and usable under water.			
Typical Use: General purpose, 'instant' domestic repairs. Metal, wood, concrete, GRP, ceramics.			
Cure: Initial cure: 10min/RT, Full cure: 24hr/RT.			
Test Methods: ASTM		(Feb 95)	
Physical		**Mechanical**	
Colour	White	τ (RT)	-
Viscosity	(putty)	τ (°C)	-
S.G.	-	τ (wet)	-
Shelf life	-	Peel ( )	-
Pot/Work life	-	σ_t	-
Tg	-	σ_c	-
$T_{Service}$	-	*E*	-
CTE (°C)	-	*G*	-
Thermal cond.	-	Elongation	-
Resistivity: Vol.	-	Tear strength	-
Surf.	-	H (Shore)	-
Notes:			

Dexter Hysol

Dexter Hysol	EA 901NA/B-1		Paste
Epoxy - 2 part, thixotropic			
Manufacturers Comments: RT cure. Good sag resistance and gap filling. Bonds to a variety of surfaces. Changes colour on mixing (Grey + amber → red).			
Typical Use: Suitable for bonding a variety of substrates: metal, rubber, plastics, wood and glass.			
Cure: 5-7 days /RT, accelerated cures (for small masses only) up to 150°C possible, e.g. 1hr/95°C will give complete cure.			
Test Methods: ASTM		October 92	
Physical		**Mechanical**	
Colour	Red	τ (RT)	19.3 MPa
Viscosity	650 Pa.s	τ (95°C)	17.2 MPa
S.G.	-	τ (wet)	21.4 MPa
Shelf life	1 yr/5°C	Peel ( )	-
Pot/Work life	50 min.	σ_t	29.7 MPa
Tg	-	σ_c	122.8 MPa
$T_{Service}$	105°C (95°C cure)	*E*	3740 MPa
CTE (°C)	-	*G*	-
Thermal cond.	-	Elongation	0.9 %
Resistivity: Vol.	2.70 Ω.cm	Tear strength	-
Surf.	1.30 Ω	H (Shore D)	90
Notes: Tensile lap shear strength and Service temperature - ASTM D 1002, Tensile properties - ASTM D 638. Electrical ASTM D 149/150: Dielectric constant @1kHz RT: 6.48. Dissipation factor @1kHz RT: 0.026.			

Dexter Hysol	EA934NA		Paste
Epoxy - 2 part, thixotropic			
Manufacturers Comments: RT cure. Good strength performance to 177°C. Qualified to MMM-A-132, Type 1, Class 3 with RT cure. Good gap filler. Potting material. Develops strength rapidly.			
Typical Use: Gap-filling, potting, fairing and shim applications.			
Cure: 5-7 days /RT, accelerated cures up to 1hr/93°C will give complete cure for small masses.			
Test Methods: ASTM		March 92	
Physical		**Mechanical**	
Colour	Grey	τ (RT)	21.4 MPa
Viscosity	80 Pa.s	τ (149°C)	7.6 MPa
S.G.	1.36	τ (wet)	24.1 MPa
Shelf life	1 yr/4°C	Peel ( )	-
Pot/Work life	40 min.	σ_t	40 MPa
Tg	71°C*	σ_c	65.5 MPa
$T_{Service}$	149°C	*E*	3800 MPa
CTE (°C)	-	*G*	-
Thermal cond.	-	Elongation	1.2 %
Resistivity: Vol.	-	Tear strength	-
Surf.	-	H (Shore D)	85
Notes: Tensile lap shear strength and Service temperature - ASTM D 1002, Tensile properties - ASTM D 638. * Tg: 129°C, for 93°C cure. Electrical ASTM D 149/150: Dielectric constant @1kHz RT: 7.24. Dissipation factor @1kHz RT: 0.028.			

Dexter Hysol	EA 9309 NA		Paste
Epoxy - two part			
Manufacturers Comments: RT cure. High peel strength. Bonds to many surfaces.			
Typical Use: Bonds: metal, wood, plastics and glass. Bonds are flexible and resist most fluids. Excellent peel strength to Aluminium.			
Cure: 3 - 5 days /RT, accelerated cures up to 1hr/93°C (for small masses), 1hr/66°C will give complete cure.			
Test Methods: ASTM		March 92	
Physical		**Mechanical**	
Colour	Grey	τ (RT)	34.5 MPa
Viscosity	A=8, B=0.04 Pa.s	τ (82°C)	4.1 MPa
S.G.	1.1	τ (wet)	-
Shelf life	1 yr./RT	Peel (-)	6.83 Nmm^{-1}
Pot/Work life	40 min.	σ_t	37.9 MPa
Tg	56°C (54°C wet)	σ_c	48.3 MPa
$T_{Service}$	71°C	*E*	2070 MPa
CTE (°C)	-	*G*	900 MPa
Thermal cond.	-	Elongation	10 %
Resistivity: Vol.	-	Tear strength	-
Surf.	-	H (Shore D)	80
Notes: Tensile lap shear strength and Service temperature - ASTM D 1002. Tensile properties - ASTM D 638. Compressive - ASTM D 638. Peel - ASTM D 1876. Electrical ASTM D 149/150: Dielectric constant @1kHz RT: 4.29. Dissipation factor @1kHz RT: 0.016.			

Dexter Hysol	EA956		Liquid
Epoxy - 2 part, low-viscosity			
Manufacturers Comments: RT cure. Good elevated temperature strength (<149°C). Low viscosity. Unfilled version of EA 934 NA.			
Typical Use: Repairs, e.g. laminating, injecting and coating.			
Cure: 5-7 days /RT, accelerated cures up to 1hr/93°C will give complete cure for small masses.			
Test Methods: ASTM		Sept 91	
Physical		**Mechanical**	
Colour	Amber	τ (RT)	12.3 MPa
Viscosity	A=35, B=2 Pa.s	τ (149°C)	6.9 MPa
S.G.	1.15	τ (wet)	-
Shelf life	1 yr/4°C (3mth/RT)	Peel ( )	-
Pot/Work life	>30 min.	σ_t	40 MPa
Tg	69°C*	σ_c	82.7 MPa
$T_{Service}$	149°C (1hr/93°C)	*E*	2600 MPa
CTE (°C)	-	*G*	-
Thermal cond.	0.205 $Wm^{-1}K^{-1}$	Elongation	2.35 %
Resistivity: Vol.	8.53 x 10^{14} Ω.cm	Tear strength	-
Surf.	2.43 x 10^{15} Ω	H (Shore D)	85
Notes: * Tg: 123°C for 1hr/93°C cure. Tensile lap shear strength and TService - ASTM D 1002: Tensile properties - ASTM D 638. Electrical ASTM D 149/150: Dielectric constant @1kHz RT: 3.59. Dissipation factor @1kHz RT: 0.017.			

Dexter Hysol	EA 9309.3 NA		Paste
Epoxy - two part, toughened			
Manufacturers Comments: Contains 0.13mm dia. glass beads for bondline thickness control. Tough, permanently flexible joints resistant to most fluids. High shear and peel strengths to aluminium.			
Typical Use: Metal, composite, thermoplastic, wood, and glass. Structural repair. Bonds are flexible and resist most fluids. Excellent peel strength to aluminium.			
Cure: 3 - 5 days /RT or 1hr/80°C will give normal cure.			
Test Methods: ASTM		Aug 92	
Physical		**Mechanical**	
Colour	(Pink, Blue)	τ (RT)	29.0 MPa
Viscosity	A=300, B=0.02 Pa.s	τ (80°C)	6.9 MPa
S.G.	-	τ (RT, wet)	32.4 MPa
Shelf life	1 yr./RT	Peel (-)	6.13 N mm^{-1}
Pot/Work life	35 min.	σ_t	31.0 MPa
Tg	80°C (60°C wet)	σ_c	51.7 MPa
$T_{Service}$	80°C	*E*	2230 MPa
CTE (°C)	-	*G*	860 MPa
Thermal cond.	0.189 $Wm^{-1}°C^{-1}$	Elongation	10 %
Resistivity: Vol.	1.36 x 10^{14} Ω.cm	Tear strength	-
Surf.	4.94 x 10^{14} Ω	H (Shore D)	80
Notes: Tensile lap shear strength and Service temperature - ASTM D 1002. Tensile properties - ASTM D 638. Compressive - ASTM D 638. Peel - ASTM D 1876. Electrical ASTM D 149/150: Dielectric constant @1kHz RT: 4.29. Dissipation factor @1kHz RT: 0.014.			

Dexter Hysol	EA 9313		Paste
Epoxy - two part, low viscosity			
Manufacturers Comments: RT cure. Low viscosity, two-part paste with good; flexibility, peel strength and low temperature properties.			
Typical Use: Useful for dissimilar adherends and may be injected into pre-assembled parts. Also used for laminating.			
Cure: 5 days /RT or 1hr/80°C (small masses) will give normal cure.			
Test Methods: ASTM		Jan 90	
Physical		**Mechanical**	
Colour	Pink	τ (RT)	27.58 MPa
Viscosity	1.2 Pa.s	τ (71°C)	4.14 MPa
S.G.	1.1	τ (wet)	-
Shelf life	1 yr./RT	Peel (-)	10.33 N mm^{-1}
Pot/Work life	60 min.	σ_f	43.44 MPa
Tg	49°C	σ_c	62.33 MPa
$T_{Service}$	49°C	E	2280 MPa
CTE (°C)	-	G	890 MPa
Thermal cond.	-	Elongation	8 %
Resistivity: Vol.	-	Tear strength	-
Surf.	-	H (Shore D)	76
Notes: Tensile lap shear strength and Service temperature - ASTM D 1002. Tensile properties - ASTM D 638. Compressive - ASTM D 638. Peel - ASTM D 1876. Electrical ASTM D 149/150: Dielectric constant @1kHz RT: 3.86. Dissipation factor @1kHz RT: 0.012.			

Dexter Hysol	EA 9320 NA		Paste
Epoxy - two part			
Manufacturers Comments: RT cure. Two-part paste with good elevated temperature performance and peel strength.			
Typical Use: Meets the requirements of MMM-A-132 Type 1 Class 2.			
Cure: 5 - 7 days /RT or 1hr/80°C (small mass) will give normal cure.			
Test Methods: ASTM		Aug 92	
Physical		**Mechanical**	
Colour	(Off-white, Blue)	τ (RT)	31.7 MPa
Viscosity	A=400, B=0.01 Pa.s	τ (80°C)	10.3 MPa
S.G.	1.36	τ (wet)	-
Shelf life	1 yr./RT	Peel (-)	3.5 N mm^{-1}
Pot/Work life	25 min.	σ_f	34.5 MPa
Tg	80°C	σ_c	55.2 MPa
$T_{Service}$	80°C	E	2280 MPa
CTE (°C)	-	G	1000 MPa
Thermal cond.	-	Elongation	9.0 %
Resistivity: Vol.	-	Tear strength	-
Surf.	-	H (Shore D)	81
Notes: Tensile lap shear strength and Service temperature - ASTM D 1002. Tensile properties - ASTM D 638. Compressive - ASTM D 638. Peel - ASTM D 1876. Electrical ASTM D 149/150: Dielectric constant @1kHz RT: 4.39. Dissipation factor @1kHz RT: 0.024.			

Dexter Hysol	EA 9317		Paste
Epoxy - two part			
Manufacturers Comments: RT cure. Thixotropic, two-part paste with good preload resistance at elevated temperatures up to 175°C.			
Typical Use: Good performance in 'preload' <175°C and meets the requirements of General Dynamics specification FMS 1048C.			
Cure: 5 - 7 days /RT or 1hr/80°C (small mass) will give normal cure.			
Test Methods: ASTM		July 90	
Physical		**Mechanical**	
Colour	Grey	τ (RT)	19.31 MPa
Viscosity	A=400, B=20 Pa.s	τ (175°C)	6.9 MPa
S.G.	1.36	τ (wet)	-
Shelf life	3 months/RT	Peel ( )	-
Pot/Work life	35 min.	σ_f	-
Tg	-	σ_c	82.74 MPa
$T_{Service}$	175°C	E	-
CTE (°C)	-	G	-
Thermal cond.	-	Elongation	-
Resistivity: Vol.	-	Tear strength	-
Surf.	-	H (Shore D)	90
Notes: Tensile lap shear strength and Service temperature - ASTM D 1002. Tensile properties - ASTM D 638. Compressive - ASTM D 638. Peel - ASTM D 1876. Electrical ASTM D 149/150: Dielectric constant @1kHz RT: 6.78. Dissipation factor @1kHz RT: 0.017.			

Dexter Hysol	EA 9321		Paste
Epoxy - two part, thixotropic			
Manufacturers Comments: RT cure. Tough, two-part paste with good elevated temperature strength.			
Typical Use: Gives good bonds over a wide temperature range.			
Cure: 5 - 7 days /RT or 1hr/80°C (small mass) will give normal cure.			
Test Methods: ASTM		Aug 92	
Physical		**Mechanical**	
Colour	Grey	τ (RT)	27.6 MPa
Viscosity	A=500, B=40 Pa.s	τ (205°C)	3.45 MPa
S.G.	1.36	τ (wet)	25.5 MPa
Shelf life	3 months/RT	Peel (-)	1.1 N mm^{-1}
Pot/Work life	40 min.	σ_f	49.0 MPa
Tg	110°C (88°C wet)	σ_c	116.4 MPa
$T_{Service}$	121°C	E	2900 MPa
CTE (°C)	-	G	1540 MPa
Thermal cond.	-	Elongation	6.0 %
Resistivity: Vol.	-	Tear strength	-
Surf.	-	H (Shore D)	84
Notes: Tensile lap shear strength and Service temperature - ASTM D 1002. Tensile properties - ASTM D 638. Compressive - ASTM D 638. Peel - ASTM D 1876. Electrical ASTM D 149/150: Dielectric constant @1kHz RT: 4.81. Dissipation factor @1kHz RT: 0.018.			

See Key to Data Sheets - Pages 63 to 65

Dexter Hysol — EA 9323 — Paste

Epoxy - two part, low viscosity

Manufacturers Comments: RT cure. Unfilled, two-part paste with good elevated temperature strength and some toughness.

Typical Use: Repair of metal and composite structures. Laminating and injecting.

Cure: 5 - 7 days /RT or 1hr/80°C (small mass) will give normal cure.

Test Methods: ASTM		Aug 92	
Physical		**Mechanical**	
Colour	Amber	τ (RT)	29.0 MPa
Viscosity	A=70, B=2 Pa.s	τ (175°C)	4.14 MPa
S.G.	A=1.25, B=0.96	τ (wet)	19.3 MPa
Shelf life	3 months/RT	Peel (-)	0.7 N mm^{-1}
Pot/Work life	30 min.	σ_t	24.1 MPa
Tg	97°C (69°C wet)	σ_c	60.7 MPa
$T_{Service}$	121°C	E	2590 MPa
CTE (°C)	-	G	930 MPa
Thermal cond.	-	Elongation	9.0 %
Resistivity: Vol.	-	Tear strength	-
Surf.	-	H (Shore D)	86

Notes: Tensile lap shear strength and Service temperature - ASTM D 1002. Tensile properties - ASTM D 638. Compressive - ASTM D 638. Peel - ASTM D 1876. Electrical ASTM D 149/150: Dielectric constant @1kHz RT: 3.43. Dissipation factor @1kHz RT: 0.017.

Dexter Hysol — EA 9330.3 — Paste

Epoxy - two part, thixotropic

Manufacturers Comments: RT cure. Two-part paste with good peel strength and low slump. Thixotropic, sag resistant version of EA 9330.

Typical Use: General purpose thixotropic.

Cure: 5 - 7 days /RT or <95°C. e.g. 1hr/80°C (small mass) will give normal cure.

Test Methods: ASTM		Aug 92	
Physical		**Mechanical**	
Colour	Grey	τ (RT)	33.8 MPa
Viscosity	A=400, B=30 Pa.s	τ (82°C)	6.9 MPa
S.G.	1.11	τ (wet)	25.5 MPa
Shelf life	1 yr./RT	Peel (-)	6.13 N mm^{-1}
Pot/Work life	60 min.	σ_t	42.0 MPa
Tg	55°C	σ_c	-
$T_{Service}$	80°C	E	2690 MPa
CTE (°C)	-	G	-
Thermal cond.	-	Elongation	9.0 %
Resistivity: Vol.	-	Tear strength	-
Surf.	-	H (Shore D)	81

Notes: Tensile lap shear strength and Service temperature - ASTM D 1002. Tensile properties - ASTM D 638. Compressive - ASTM D 638. Peel - ASTM D 1876. Electrical ASTM D 149/150: Dielectric constant @1kHz RT: 4.46. Dissipation factor @1kHz RT: 0.013.

Dexter Hysol — EA 9330 — Paste

Epoxy - two part

Manufacturers Comments: RT cure. Two-part paste with good peel strength and excellent environmental resistance.

Typical Use: General Purpose. Bonds to a variety of adherends and tolerates bondline thickness variations.

Cure: 5 - 7 days /RT or <93°C. e.g. 2hr/82°C (small mass) will give normal cure.

Test Methods: ASTM		March 92	
Physical		**Mechanical**	
Colour	Cream	τ (RT)	34.5 MPa
Viscosity	A=200, B=1.5 Pa.s	τ (82°C)	6.9 MPa
S.G.	1.15	τ (wet)	25.5 MPa
Shelf life	1 yr./RT	Peel (-)	6.13 N mm^{-1}
Pot/Work life	60 min.	σ_t	38.6 MPa
Tg	57°C	σ_c	53.1 MPa
$T_{Service}$	82°C	E	2650 MPa
CTE (°C)	-	G	970 MPa
Thermal cond.	-	Elongation	2.4 %
Resistivity: Vol.	1.59 x 10^{14} Ω.cm	Tear strength	-
Surf.	1.59 x 10^{14} Ω	H	(65 Barcol #935)

Notes: Tensile lap shear strength and Service temperature - ASTM D 1002. Tensile properties - ASTM D 638. Compressive - ASTM D 638. Peel - ASTM D 1876. Electrical ASTM D 149/150: Dielectric constant @1kHz RT: 4.38. Dissipation factor @1kHz RT: 0.020.

Dexter Hysol — EA 9346.2 — Paste

Epoxy - one part

Manufacturers Comments: One-component paste with 0.13mm glass beads for bondline control. Good strength under hot/wet conditions.

Typical Use: Recommended for thermoplastic and composite bonding.

Cure: 1 hr/120°C or 1hr/175°C.

Test Methods: ASTM		Aug 92	
Physical		**Mechanical**	
Colour	Burgundy	τ (RT)	41.4 MPa
Viscosity	1000 Pa.s	τ (120°C)	17.9 MPa
S.G.	1.2	τ (wet)	31.1 MPa
Shelf life	1 yr/-20°C	Peel (Bell)	(10.5 N mm^{-1})
Pot/Work life	14 days/RT	σ_t	34.5 MPa
Tg	120°C (95°C wet)	σ_c	103 MPa
$T_{Service}$	150°C	E	1790 MPa
CTE (°C)	-	G	760 MPa
Thermal cond.	-	Elongation	3.0 %
Resistivity: Vol.	3.9 x 10^{13} Ω.cm	Tear strength	-
Surf.	2.5 x 10^{14} Ω	H (Shore D)	80

Notes: Tensile lap shear strength and Service temperature - ASTM D 1002. Tensile properties - ASTM D 638. Compressive - ASTM D 638. Peel - ASTM D 1876. Electrical ASTM D 149/150: Dielectric constant @1kHz RT: 3.68. Dissipation factor @1kHz RT: 0.008.

Dexter Hysol	EA 9390		Paste
Epoxy - two part			
Manufacturers Comments: Low viscosity with good wetting. Cures at 95°C. Good hot/wet strength. High shear modulus. BMS 8-301 qualified.			
Typical Use: Designed for composite repair applications. Suitable for wet lay-up of carbon composite repairs.			
Cure: 220min/95°C, or 150min/120°C, or 130min/150°C.			
Test Methods: ASTM		Aug 92	
Physical		**Mechanical**	
Colour	Violet	τ (RT)	17.9 MPa
Viscosity	8 Pa.s	τ (175°C)	11.7 MPa
S.G.	~1.1	τ (-55°C)	15.2 MPa
Shelf life	1 yr/5°C (6 mth/RT)	Peel ( )	-
Pot/Work life	120 mins	σ_t	56.6 MPa
Tg	175°C (150°C wet)	σ_c	-
$T_{Service}$	175°C	*E*	2880 MPa
CTE (°C)	-	*G*(wet or dry)	1380 MPa
Thermal cond.	-	Elongation	2.5%
Resistivity: Vol.	-	Tear strength	-
Surf.	-	H (Shore D)	89
Notes: Tensile lap shear strength and Service temperature - ASTM D 1002. Tensile properties - ASTM D 638. Electrical ASTM D 149/150: Dielectric constant @1kHz RT: 4.19. Dissipation factor @1kHz RT: 0.017.			

Dexter Hysol	EA 9395		Paste
Epoxy - 2 part thixotropic			
Manufacturers Comments: RT cure. Sag resistant with good strength at 175°C. Non-metallic version of EA 9394.			
Typical Use: Can be used with composites and thermoplastics. Potting, filling, fairing and repairs.			
Cure: 5 days/RT, <65°C (small mass), e.g. 1hr/66°C.			
Test Methods: ASTM		Aug 92	
Physical		**Mechanical**	
Colour	Grey/Blue	τ (RT)	27.6 MPa
Viscosity	100 - 300 Pa.s	τ (175°C)	8.3 MPa
S.G.	1.27	τ (-55°C)	15.9 MPa
Shelf life	1 yr./RT	Peel (-)	0.9 - 1.4 N mm^{-1}
Pot/Work life	100 min.	σ_t	55.6 MPa
Tg	72°C	σ_c	96.5 MPa
$T_{Service}$	<177°C	*E*	4930 MPa
CTE (°C)	-	*G*	1550 MPa
Thermal cond.	-	Elongation	2.6 %
Resistivity: Vol.	-	Tear strength	-
Surf.	-	H (Shore D)	90
Notes: Tensile lap shear strength and Service temperature - ASTM D 1002, Peel - ASTM D 1876, Tensile properties - ASTM D 638. Electrical ASTM D 149/150. Dielectric constant @1kHz: 4.13. Dissipation factor @1kHz: 0.021.			

Dexter Hysol	EA 9394		Paste
Epoxy - 2 part paste			
Manufacturers Comments: RT cure and storage. Long pot life. Low toxicity. Good high temperature compressive strength. Qualified to MMM-A-132 Rev A, Type 1, Class 3. Accelerated cure possible for small quantities.			
Typical Use: Gap-filling, potting and liquid shim. Structural repair, composite bonding, metal bonding, thermoplastic bonding.			
Cure: 3 - 5 days/RT, <93°C (small mass), e.g. 1hr/66°C.			
Test Methods: ASTM		May 91	
Physical		**Mechanical**	
Colour	Grey	τ (RT)	29.0 MPa
Viscosity	160 Pa.s	τ (204°C)	4.1 MPa
S.G.	1.33	τ (wet)	28.3 MPa
Shelf life	1 yr./RT	Peel (-)	0.88 N mm^{-1}
Pot/Work life	90 min.	σ_t	46.04 MPa
Tg	78°C (68°C wet)	σ_c	158.6 MPa
$T_{Service}$	177°C	*E*	4240 MPa
CTE (°C)	55.6 K^{-1}	*G*	1460 MPa
Thermal cond.	0.332 $Wm^{-1}K^{-1}$	Elongation	1.66 %
Resistivity: Vol.	4.05x10^{13} Ω.cm	Tear strength	-
Surf.	4.60x10^{13} Ω	H (Shore D)	88
Notes: Tensile lap shear strength and Service temperature - ASTM D 1002, Peel - ASTM D 1876, Tensile properties - ASTM D 638. Electrical ASTM D 149/150. Dielectric constant @1kHz: 7.51. Dissipation factor @1kHz: 0.022.			

Dexter Hysol	EA 9396		Paste
Epoxy - two part			
Manufacturers Comments: Low viscosity with room temperature cure and storage. Excellent strength at high and low temperatures (-55 to +175°C). Qualified to MMM-A-132, Rev A, Type 1, Class 3. Unfilled, low-viscosity version of EA 9394.			
Typical Use: Composite bonding, wet lay-up repairs and thermoplastic bonding.			
Cure: 3 - 5 days/RT. <1hr/65°C may be used.			
Test Methods: ASTM		Aug 92	
Physical		**Mechanical**	
Colour	-	τ (RT)	24.1 MPa
Viscosity	3.5 Pa.s	τ (175°C)	8.6 MPa
S.G.	1.14	τ (-55°C)	22.8 MPa
Shelf life	1 yr./RT	Peel (Bell)	(4.4 N mm^{-1})
Pot/Work life	75-90 mins.	σ_t	-
Tg	-	σ_c	-
$T_{Service}$	175°C	*E*	-
CTE (°C)	-	*G*	-
Thermal cond.	-	Elongation	-
Resistivity: Vol.	-	Tear strength	-
Surf.	-	H (Shore)	-
Notes: Tensile lap shear strength and Service temperature - ASTM D 1002, Bell Peel - ASTM D 3167. Electrical ASTM D 149/150: Dielectric constant @1kHz: 4.12. Dissipation factor @1kHz: 0.017.			

Dexter Hysol	EA 9412		Paste
Two-part, low-viscosity			
Manufacturers Comments: Low viscosity, injectable, two-part epoxy providing flexible bonds.			
Typical Use: Flexibility makes it useful for dissimilar adherends. Can be used for laminating and potting.			
Cure: 5 days/RT, <82°C, e.g. 1hr/82°C.			
Test Methods: ASTM		May 93	
Physical		**Mechanical**	
Colour	Pink	τ (RT)	27.6 MPa
Viscosity	1.2 Pa.s	τ (60°C)	6.2 MPa
S.G.	1.10	τ (wet)	24.8 MPa
Shelf life	1 yr./RT	Peel (FR)	(10.3 N mm^{-1})
Pot/Work life	60 mins	σ_t	43.4 MPa
Tg	-	σ_c	-
$T_{Service}$	-	*E*	2280 MPa
CTE (°C)	-	*G*	-
Thermal cond.	-	Elongation	8.0 %
Resistivity: Vol.	-	Tear strength	-
Surf.	-	H (Shore D)	76
Notes: Tensile lap shear strength and Service temperature - ASTM D 1002, Tensile properties - ASTM D 638. Peel strength - Floating roller peel to ASTM D3167.			

Dexter Hysol	EA 9432NA		Paste
One-part, metal-filled, thixotropic			
Manufacturers Comments: Modified one-component, aluminium-filled epoxy. Excellent tensile shear strength and good impact resistance. Cures at moderate temperatures. Pumpable owing to low viscosity, but thixotropic enough to avoid slump on vertical surfaces.			
Typical Use: Bonds well to oily or poorly prepared surfaces. Suitable for many materials.			
Cure: 1 hr/120°C, <175°C, e.g. 0.5hr/150°C.			
Test Methods: ASTM		May 93	
Physical		**Mechanical**	
Colour	Grey	τ (RT)	26.2 MPa
Viscosity	150 - 300 Pa.s	τ (150°C)	16.5 MPa
S.G.	1.38	τ (-67°C)	24.1 MPa
Shelf life	1 yr./-20°C	Peel (FR)	(1.3 N mm^{-1})
Pot/Work life	6 months/RT	σ_t	50.7 MPa
Tg	-	σ_c	-
$T_{Service}$	~150°C	*E*	4690 MPa
CTE (°C)	-	*G*	-
Thermal cond.	-	Elongation	1.2 %
Resistivity: Vol.	-	Tear strength	-
Surf.	-	H (Shore D)	90
Notes: Tensile lap shear strength and Service temperature - ASTM D 1002, Tensile properties - ASTM D 638. Peel strength - Floating roller peel to ASTM D3167.			

Dexter Hysol	EA 9430		Paste
Two-part, modified			
Manufacturers Comments: RT cure with very high peel strength and excellent shear properties. Bonds are permanently flexible with very good environmental resistance. Tough and tolerant of high bondline thicknesses.			
Typical Use: Flexibility makes it useful for dissimilar adherends. Useful for large bondline thicknesses (<2.5 mm).			
Cure: 5 days/RT, <82°C, e.g. 1hr/82°C.			
Test Methods: ASTM		May 93	
Physical		**Mechanical**	
Colour	Off-white	τ (RT)	32.4 MPa
Viscosity	12 Pa.s	τ (60°C)	11.7 MPa
S.G.	1.12	τ (wet)	25.5 MPa
Shelf life	1 yr./RT	Peel (FR)	(10.5 N mm^{-1})
Pot/Work life	50 mins	σ_t	36.5 MPa
Tg	-	σ_c	-
$T_{Service}$	-	*E*	2620 MPa
CTE (°C)	-	*G*	-
Thermal cond.	-	Elongation	6.0 %
Resistivity: Vol.	-	Tear strength	-
Surf.	-	H (Shore D)	75
Notes: Tensile lap shear strength and Service temperature - ASTM D 1002, Tensile properties - ASTM D 638. Peel strength - Floating Roller peel to ASTM D3167.			

Dexter Hysol	EA 9460		Paste
Two-part, thixotropic			
Manufacturers Comments: RT cure with high peel strength and excellent shear properties. Long work life with a quick heat cure response. Non-sag, low-viscosity. Bonds are tough and flexible.			
Typical Use: Flexibility makes it useful for dissimilar adherends. Useful for metals, engineering thermoplastics and thermoset laminates such as sheet moulding compounds.			
Cure: 3 days/RT, <150°C, e.g. 1hr/120°C.			
Test Methods: ASTM		May 93	
Physical		**Mechanical**	
Colour	Grey	τ (RT)	24.1 MPa
Viscosity	1.5 - 2.5 Pa.s	τ (120°C)	2.07 MPa
S.G.	1.33	τ (wet)	20 MPa
Shelf life	1 yr./RT	Peel (-)	3.5 N mm^{-1}
Pot/Work life	40 - 50 mins	σ_t	30.3 MPa
Tg	-	σ_c	-
$T_{Service}$	~80°C	*E*	2760 MPa
CTE (°C)	-	*G*	-
Thermal cond.	-	Elongation	3.5 %
Resistivity: Vol.	-	Tear strength	-
Surf.	-	H (Shore D)	80
Notes: Tensile lap shear strength and Service temperature - ASTM D 1002, Tensile properties - ASTM D 638. Peel strength - ASTM D1876.			

See Key to Data Sheets - Pages 63 to 65

Dexter Hysol	**EA 9462**		**Paste**
Two-part, modified			
Manufacturers Comments: RT cure with a good balance of peel and shear properties. Medium viscosity, non-sag. Impact resistant.			
Typical Use: Bonds a variety of materials.			
Cure: 3 - 5 days/RT, <93°C, e.g. 1hr/82°C.			
Test Methods: ASTM		July 93	
Physical		**Mechanical**	
Colour	Beige	τ (RT)	24.1 MPa
Viscosity	60 - 80 Pa.s	τ (82°C)	4.1 MPa
S.G.	1.17	τ (wet)	-
Shelf life	1 yr./RT	Peel (-)	1.8 N mm^{-1}
Pot/Work life	55 mins	σ_t	-
Tg	-	σ_c	-
$T_{Service}$	-	E	-
CTE (°C)	-	G	-
Thermal cond.	-	Elongation	-
Resistivity: Vol.	-	Tear strength	-
Surf.	-	H (Shore D)	80
Notes: Tensile lap shear strength and Service temperature - ASTM D 1002, Tensile properties - ASTM D 638. Peel strength - ASTM D1876.			

Dexter Hysol	**EA 9628**		**Film**
Knitted and non-woven (EA9628NW) nylon supported, 0.293 kg.m^{-2}			
Manufacturers Comments: Modified epoxy film for structural bonds requiring toughness. Excellent resistance to most environments.			
Typical Use: Bonds many materials.			
Cure: 1hr/120°C or 1.5 hr/115°C. Pressure 0.17 - 0.34 MPa (25 - 50 psi).			
Test Methods: ASTM		August 92	
Physical		**Mechanical** (knit)	
Colour	-	τ (RT)	41.4 MPa
Viscosity	-	τ (120°C)	13.1 MPa
S.G.	-	τ (wet)	43.4 MPa
Shelf life	>6 mth/-20°C	Peel (-)	6.5 N mm^{-1}
Pot/Work life	20 days/RT.	σ_t	52.0 MPa
Tg	115°C (99°C wet)	σ_c	79.3 MPa
$T_{Service}$	120°C	E	2380 MPa
CTE (°C)	-	G	620 MPa
Thermal cond.	-	Elongation	7.5 %
Resistivity: Vol.	-	Tear strength	-
Surf.	-	H (Shore D)	83
Notes: Tensile lap shear strength and Service temperature - ASTM D 1002, Tensile properties - ASTM D 638. Electrical ASTM D 149/150. Dielectric constant @1kHz: 3.56. Dissipation factor @1kHz: 0.087.			

Dexter Hysol	**EA 9601**		**Film**
Nylon tricot supported, 0.22 - 0.39 kg.m^{-2} + unsupported 0.15 kg.m^{-2}			
Manufacturers Comments: Tacky film with moderate cure temperature and good environmental resistance. High tensile, peel and shear strength over a range of temperature. Exceeds MMM-A-132, type 1, class 2 and MIL-A-25463, type1, class 1 & 2. Optional primer: EA 9261.			
Typical Use: Recommended for metal-metal bonding and honeycomb structures.			
Cure: 1hr/121°C or 1.5 hr/107°C. Vacuum bag: 0.07 MPa (10 psi), Autoclave: 0.17 - 0.28 MPa (25 - 40 psi).			
Test Methods: ASTM		-	
Physical		**Mechanical**	
Colour	Beige	τ (RT)	36.5 MPa
Viscosity	-	τ (121°C)	15.2 MPa
S.G.	-	τ (wet)	41.4 MPa
Shelf life	2 mth/-18°C	Peel (-)	4.2 N mm^{-1}
Pot/Work life	>5 days/RT.	σ_t	-
Tg	-	σ_c	-
$T_{Service}$	121°C	E	-
CTE (°C)	-	G	-
Thermal cond.	-	Elongation	-
Resistivity: Vol.	-	Tear strength	-
Surf.	-	H (Shore)	-
Notes: Tensile lap shear strength and Service temperature - ASTM D 1002, Tensile properties - ASTM D 638.			

Dexter Hysol	**EA 9657**		**Film**
Fabric supported (0.390 kg.m^{-2}) and unsupported (0.268 kg.m^{-2})			
Manufacturers Comments: High service temperature film with high honeycomb peel strength. Reticulates on perforated sheet or honeycomb core. Companion primer: EA 9210B.			
Typical Use: Honeycomb constructions. Continuous service at 150°C.			
Cure: 1.5 hr/175°C, Autoclave 0.21 - 0.69 MPa (30 - 100 psi).			
Test Methods: ASTM		August 92	
Physical		**Mechanical** (U/S, primed)	
Colour	-	τ (RT)	36.6 MPa
Viscosity	-	τ (175°C)	16.5 MPa
S.G.	-	τ (-55°C)	34.5 MPa
Shelf life	>6 mth/-20°C	Peel (CD)	102 MPa
Pot/Work life	15 days/RT.	σ_t	-
Tg	-	σ_c	-
$T_{Service}$	205°C	E	-
CTE (°C)	-	G	-
Thermal cond.	-	Elongation	-
Resistivity: Vol.	-	Tear strength	-
Surf.	-	H (Shore)	-
Notes: Tensile lap shear strength and Service temperature - ASTM D 1002, Tensile properties - ASTM D 638. Peel - Climbing Drum.			

See Key to Data Sheets - Pages 63 to 65

Dexter Hysol	EA 9689		Film
Knit nylon (0.490 kg.m^{-2}), glass fabric (0.490 kg.m^{-2}) or unsupported (0.290 kg.m^{-2})			
Manufacturers Comments: Modified epoxy film with excellent high temperature strength and long-term thermal durability. Unsupported version reticulates. Thermally compatible with EA 9289 water-borne primer.			
Typical Use: Suitable for metal, composite and honeycomb bonding with continuous exposure up to 175°C.			
Cure: 1 hr/175°C, pressure 0.17 - 0.34 MPa (25 - 50 psi).			
Test Methods: ASTM		August 92	
Physical		**Mechanical** (uns./primed)	
Colour	-	τ (RT)	27.6 MPa
Viscosity	-	τ (175°C)	22.4 MPa
S.G.	-	τ (wet)	23.4 MPa
Shelf life	>6 mth/-20°C	Peel (CD)	(20.7 N.m m^{-1})
Pot/Work life	10 days/RT.	σ_t (flatwise)	(6.2 MPa)
Tg	225°C (175°C wet)	σ_c	-
$T_{Service}$	215°C	E	-
CTE (°C)	-	G	-
Thermal cond.	-	Elongation	-
Resistivity: Vol.	-	Tear strength	-
Surf.	-	H (Shore)	-
Notes: Tensile lap shear strength and Service temperature - ASTM D 1002, Tensile properties - ASTM D 638. Mechanical properties are for unsupported film adhesive and adherends primed with EA 9289 water borne primer. Peel test - honeycomb climbing drum.			

Dexter Hysol	EPK 1C		Paste
Epoxy-patch Kit 1C			
Manufacturers Comments: General purpose RT cure adhesive and sealant. Withstands severe environments. Machinable, sandable.			
Typical Use: Bonds, seals and repairs a variety of materials.			
Cure: 3days/RT (90% properties after 24hr), <82°C, e.g. 1hr/82°C.			
Test Methods: ASTM		July 93	
Physical		**Mechanical**	
Colour	White, grey or black	τ (RT)	12.1 MPa
Viscosity	~500 Pa.s	τ (150°C)	2.07 MPa
S.G.	1.55	τ (-55°C)	6.9 MPa
Shelf life	1 yr./RT	Peel ( )	-
Pot/Work life	20 - 25 mins	σ_t	-
Tg	98°C	σ_c	-
$T_{Service}$	-	E	-
CTE (RT)	63 x 10^{-6} °C^{-1}	G	-
Thermal cond.	0.54 W m.K^{-1}	Elongation	-
Resistivity: Vol.	-	Tear strength	-
Surf.	-	H (Shore D)	80
Notes: Tensile lap shear strength and Service temperature - ASTM D 1002, Tensile properties - ASTM D 638. Peel strength - ASTM D1876. Passes NASA outgassing standard as published in NASA reference publication 1124 Rev.2 1990. Moisture absorption: 0.16% after 24 hr/RT.			

Dexter Hysol	EA 9814NA		Film
Foaming film in various thicknesses, 0.4 - 2.54 mm.			
Manufacturers Comments: Structural intumescent foaming core splice adhesive for use up to 175°C. Low exotherm. 2.1:1 expansion for 2.54mm film.			
Typical Use: Thick sections of non-metallic honeycomb.			
Cure: 1 hr/120°C or 1 hr/175°C, pressure 0.17 - 0.34 MPa (25 - 50 psi).			
Test Methods: ASTM		August 92	
Physical		**Mechanical**	
Colour	-	τ (RT, tube)	(4.8 MPa)
Viscosity	-	τ (120°C, tube)	(3.1 MPa)
S.G.	-	τ (wet)	-
Shelf life	>6 mth/-20°C	Peel ( )	-
Pot/Work life	10 days/RT.	σ_t	-
Tg	-	σ_c	-
$T_{Service}$	-	E	-
CTE (°C)	-	G	-
Thermal cond.	-	Elongation	-
Resistivity: Vol.	-	Tear strength	-
Surf.	-	H (Shore)	-
Notes: Tube shear strength to Boeing BMS 5-90F using 2.54mm film.			

Dexter Hysol	X 9680.1		Film
Unsupported (0.171 and 0.244 kg.m^{-2})			
Manufacturers Comments: Composite bonding film with excellent environment resistance. Low cure temperature and low flow. Minimal resin intermingling during cocuring. Reticulates. X-ray opaque. Good pre-bond moisture resistance.			
Typical Use: Repair and cocuring of composites.			
Cure: 1 - 1.5 hr/120°C or 1 - 2hr/175°C. 0.17 - 0.34 MPa (25 - 50 psi).			
Test Methods: ASTM		August 92	
Physical		**Mechanical** (0.171 kg.m^{-2})	
Colour	-	τ (RT)	24.1 MPa
Viscosity	-	τ (150°C)	13.8 MPa
S.G.	-	τ (wet)	29.0 MPa
Shelf life	>6 mth/-20°C	Peel ( )	-
Pot/Work life	25 days/RT.	σ_t	-
Tg	155°C (170°C: cure 175)	σ_c	-
$T_{Service}$	150°C	E	-
CTE (°C)	-	G	-
Thermal cond.	-	Elongation	-
Resistivity: Vol.	-	Tear strength	-
Surf.	-	H (Shore)	-
Notes: Tensile lap shear strength and Service temperature - ASTM D 1002, Tensile properties - ASTM D 638. τ (wet): double overlap coupon.			

Dexter Hysol	X 9690		Film
Woven support (0.293 kg.m^{-2}) or non-woven mat.			
Manufacturers Comments: Modified epoxy film designed for high peel strength in bonds requiring toughness. Low flow and excellent performance with low pressure cure.			
Typical Use: Bonds many materials.			
Cure: 1 hr/120°C, pressure 0.17 - 0.34 MPa (25 - 50 psi).			
Test Methods: ASTM		August 92	
Physical		**Mechanical** (sup./primed)	
Colour	-	τ (RT)	42.0 MPa
Viscosity	-	τ (120°C)	14.5 MPa
S.G.	-	τ (wet)	24.2 MPa
Shelf life	>6 mth/-20°C	Peel (-)	7.88 N mm^{-1}
Pot/Work life	14 days/RT.	σ_f (flatwise)	(6.38 MPa)
Tg	115°C (95°C wet)	σ_c	-
$T_{Service}$	120°C	*E*	-
CTE (°C)	-	*G*	-
Thermal cond.	-	Elongation	-
Resistivity: Vol.	-	Tear strength	-
Surf.	-	H (Shore)	-

Notes: Tensile lap shear strength and Service temperature - ASTM D 1002, Tensile properties - ASTM D 638.
Mechanical properties are for supported film adhesive and adherends primed with EA 9210H primer.

Dexter Hysol	XEA 9361		Paste
Epoxy - two part			
Manufacturers Comments: RT cure. Two-component paste with high elongation. Good combination of high shear, peel and flexibility. Long pot life and good low temperature properties.			
Typical Use: Suitable for general purpose bonding of many adherends and applications requiring high elongation, e.g. sealing and cryogenic.			
Cure: 5 - 7 days/RT, < 1 hr/93°C. e.g. 1hr/80°C for normal cure.			
Test Methods: ASTM		Aug 92	
Physical		**Mechanical**	
Colour	Grey	τ (RT)	24.1 MPa
Viscosity	100 Pa.s	τ (71°C)	5.9 MPa
S.G.	1.28	τ (-196°C)	27.6 MPa
Shelf life	1 yr./RT	Peel (-)	4.4 N mm^{-1}
Pot/Work life	120 mins	σ_f	-
Tg	-	σ_c	-
$T_{Service}$	60°C	*E*	720 MPa
CTE (°C)	-	*G*	-
Thermal cond.	-	Elongation	>50% (>7, -55°C)
Resistivity: Vol.	-	Tear strength	-
Surf.	-	H (Shore)	-

Notes: Tensile lap shear strength and Service temperature - ASTM D 1002. Tensile properties - ASTM D 638. Compressive - ASTM D 638. Peel - ASTM D 1876.

Dexter Hysol	XEA 9359.3		Paste
Epoxy - two part			
Manufacturers Comments: RT cure. Two-component paste with good shear and peel strengths. Good environmental resistance, non-sag and bondline thickness control. Next generation from EA 9309.3NA.			
Typical Use: Metals, thermoplastics and composites. Available in 2:1 cartridge with a static mixer.			
Cure: 5 - 7 days/RT, < 1 hr/105°C for normal cure.			
Test Methods: ASTM		Aug 92	
Physical		**Mechanical**	
Colour	Green	τ (RT)	31.0 MPa
Viscosity	A=300, B=150 Pa.s	τ (95°C)	6.9 MPa
S.G.	1.12	τ (wet)	-
Shelf life	1 yr./RT	Peel (-)	6.1 N mm^{-1}
Pot/Work life	40 mins	σ_f	-
Tg	-	σ_c	-
$T_{Service}$	95°C	*E*	-
CTE (°C)	-	*G*	-
Thermal cond.	-	Elongation	-
Resistivity: Vol.	-	Tear strength	-
Surf.	-	H (Shore)	-

Notes: Tensile lap shear strength and Service temperature - ASTM D 1002. Tensile properties - ASTM D 638. Compressive - ASTM D 638. Peel - ASTM D 1876.

Dunlop

Dunlop	SAS 400		Paste
Epoxy - Two part, steel filled*			
Manufacturers Comments: RT cure. Metallic-filler (steel). Non-sag. Fast repairs or production-line uses. High tensile and impact properties. Good chemical resistance.			
Typical Use: Repair adhesive for metal components (pipes, tanks, castings). Ceramics, concrete, wood, some plastics.			
Cure: RT/3-4 hrs (handling); 24 hrs (practical cure); 7 days (full cure).			
Test Methods: Not stated		Oct 93	
Physical		**Mechanical**	
Colour	Grey + Beige	τ (RT)	-
Viscosity	-	τ (°C)	-
S.G.	-	τ (wet)	-
Shelf life	1 year	Peel ( )	-
Pot/Work life	30-50 mins	σ_f	-
Tg	-	σ_c	-
$T_{Service}$	≤120°C	*E*	-
CTE (°C)	-	*G*	-
Thermal cond.	-	Elongation	-
Resistivity: Vol.	-	Tear strength	-
Surf.	-	H (Shore D)	80

Notes: * SAS420 aluminium-filled version.

See Key to Data Sheets - Pages 63 to 65

Dunlop	SAS 510		Paste
Epoxy - One part, thixotropic			
Manufacturers Comments: Heat curing, solvent-free. Non-slump, toughened epoxy. Vibration resistance. High peel and impact strength. High heat resistance.			
Typical Use: Structural engineering assemblies - vehicles. Lifts, switch-gear enclosures.			
Cure: 150°C/ 30 mins; 160°C/20 mins; 190°C/ 10 mins.			
Test Methods: Not stated		July 92	
Physical		**Mechanical**	
Colour	Grey	τ (RT)	25-30 MPa
Viscosity	thixotropic	τ (130°C)	11 MPa
S.G.	1.25	τ (wet)	19 MPa
Shelf life	3 months/<25°C	Peel (T)	3.92 N mm^{-1}
Pot/Work life	-	σ_t	-
Tg	-	σ_c	-
$T_{Service}$	-	E	-
CTE (°C)	-	G	-
Thermal cond.	-	Elongation	-
Resistivity: Vol.	-	Tear strength	-
Surf.	-	H (Shore)	-
Notes: Gap-filling: <3mm. τ (wet): 500 hrs, 90%RH. T-peel: mild steel.			

Dunlop	SAS 520		Paste
Epoxy - Two part, thixotropic			
Manufacturers Comments: RT curing, solvent-free. Slump resistance. Toughened epoxy. High peel and impact strength. High heat resistance. Tolerance to thin oil films.			
Typical Use: Structural engineering assemblies - vehicles and transportation. Metal structures. Electronics. Repairs.			
Cure: RT/ 6 hrs (handling strength); 24 hrs (full cure).			
Test Methods: Not stated		Jan 95	
Physical		**Mechanical**	
Colour	Translucent-amber	τ (RT) steel	20 MPa
Viscosity	thixotropic	τ (°C)	-
S.G.	1.25	τ (wet)	-
Shelf life	1 year/-20+35°C	Peel (-)	5.2 N mm^{-1}
Pot/Work life	20-45 mins (100g)	σ_t	-
Tg	-	σ_c	-
$T_{Service}$	-10 to +70°C	E	-
CTE (°C)	-	G	-
Thermal cond.	-	Elongation	-
Resistivity: Vol.	-	Tear strength	-
Surf.	-	H (Shore)	-
Notes: Service: +110°C with 100°C cure, Gap-filling: <5mm. Peel: mild steel. Impact strength: 0.3J/cm.			

Dunlop	SAS 515		Paste
Epoxy - One part			
Manufacturers Comments: Heat curing, solvent-free. Toughened, good physical properties. Tolerance to thin oil films.			
Typical Use: Structural engineering assemblies.			
Cure: 150°C/ 30 mins; 160°C/20 mins; 190°C/ 10 mins.			
Test Methods: Not stated		March 95	
Physical		**Mechanical**	
Colour	Black	τ (RT)	-
Viscosity	-	τ (°C)	-
S.G.	-	τ (wet)	-
Shelf life	3 months/<25°C	Peel ( )	-
Pot/Work life	-	σ_t	-
Tg	-	σ_c	-
$T_{Service}$	-	E	-
CTE (°C)	-	G	-
Thermal cond.	-	Elongation	-
Resistivity: Vol.	-	Tear strength	-
Surf.	-	H (Shore)	-
Notes:			

Dunlop	SAS 530		Paste
Epoxy - Two part			
Manufacturers Comments: Fast RT curing, solvent-free. Chemical and moisture resistance.			
Typical Use: Light assemblies - metals, glass, GRP, ceramics, rubber and plastics.			
Cure: RT/ 10-15 mins (handling strength); 24 hrs (full cure).			
Test Methods: Not stated		Jan 93	
Physical		**Mechanical**	
Colour	-	τ (RT) steel	11-15 MPa
Viscosity	~11 Pa.s	τ (°C)	-
S.G.	1.25	τ (wet)	-
Shelf life	1 year	Peel ( )	-
Pot/Work life	<10 mins	σ_t	-
Tg	-	σ_c	-
$T_{Service}$	0 to +60°C	E	-
CTE (°C)	-	G	-
Thermal cond.	-	Elongation	-
Resistivity: Vol.	-	Tear strength	-
Surf.	-	H (Shore)	-
Notes: Gap-filling: 0.3 mm.			

Epirez

Dunlop	SAS 535		Paste
Epoxy - Two part			
Manufacturers Comments: Fast RT curing, solvent-free. Chemical and moisture resistance. Lower viscosity, more flexible than SAS 505.			
Typical Use: Light assemblies - metals, glass, GRP, ceramics, rubber and plastics (not PTFE, polyethylene).			
Cure: RT/ 10-15 mins (handling strength); 24 hrs (full cure).			
Test Methods: Not stated		Nov 90	
Physical		**Mechanical**	
Colour	-	τ (RT) steel	8 - 13 MPa
Viscosity	-	τ (°C)	-
S.G.	1.25	τ (wet)	-
Shelf life	1 year	Peel ( )	-
Pot/Work life	5-7 mins	σ_t	-
Tg	-	σ_c	-
$T_{Service}$	-20 to +60°C	E	-
CTE (°C)	-	G	-
Thermal cond.	-	Elongation	-
Resistivity: Vol.	-	Tear strength	-
Surf.	-	H (Shore)	-
Notes: Gap-filling: 0.3 mm. Negligible shrinkage.			

Epirez	8141		Paste
Epoxy - Two part			
Manufacturers Comments: RT (and underwater) curing. Structural compound which will cure and bond underwater to form a corrosion and impact resistant shield to steel, concrete and timber structures. Proven history in splash-zone of marine constructions. Excellent chemical resistance (except strong acids).			
Typical Use: Repairs to piled steel structures. Bonding steel sections to concrete piles. Reinforcement of timber piles. Pipes, underwater cables and hulls of ships. Used by divers underwater.			
Cure: RT (25°C)/ 24 hrs (dry); 48 hrs (wet).			
Test Methods: Not stated		Feb 93	
Physical		**Mechanical**	
Colour	Grey	τ (RT)	-
Viscosity	Mastic-like paste	τ (°C)	-
S.G.	-	τ (wet)	2.5 MPa*
Shelf life	2 years	Peel ( )	-
Pot/Work life	1 hr	σ_t	31 MPa
Tg	-	σ_c	75 MPa
$T_{Service}$	-	E	-
CTE (°C)	-	G	-
Thermal cond.	-	Elongation	-
Resistivity: Vol.	-	Tear strength	-
Surf.	-	H (Shore)	-
Notes: Mix ratio 1:1. τ (wet) Placed & cured in sea-water, 20°C/7 days steel/concrete (concrete failed). Concrete/concrete: 2.4 MPa (concrete failed); Steel/timber: 3.5 MPa (timber failed).			

Dyckerhoff

Dyckerhoff	DünnbettEpoxi #500		Paste
Epoxy - Two part			
Manufacturers Comments: RT curing. Thin-bed mortar for joining and laying ceramic coverings in areas with high loads, inc. under-floor heated. Resistant to water and chemicals. Indoor and outdoor use. Meets: DIN 18156-E; DIN 4030.			
Typical Use: Swimming pools, breweries, textile and paper factories, food-processing factories, chemical industries. Ceramic, glass, mosaic.			
Cure: RT (20°C)/ 16 hrs (foot-traffic); 4 days (full cure).			
Test Methods: Not stated		Mar 94	
Physical		**Mechanical**	
Colour	-	τ (RT)	-
Viscosity	-	τ (°C)	-
S.G.	-	τ (wet)	-
Shelf life	6 months/15°C	Peel ( )	-
Pot/Work life	90 mins/20°C	σ_t	-
Tg	-	σ_c	-
$T_{Service}$	+80°C	E	-
CTE (°C)	-	G	-
Thermal cond.	-	Elongation	-
Resistivity: Vol.	-	Tear strength	-
Surf.	-	H (Shore)	-
Notes: *Warm containers in water-bath to 20°C before mixing & use. Pot-life/30°C: ~20-30 mins.			

Epirez	8242		Paste
Epoxy - Two part			
Manufacturers Comments: RT curing. 100% solids, general purpose structural use in the construction industry. High strength, non-sag paste. Used in major construction projects in Asia-Pacific for 25 years. Chemical resistance. Meets: ASTM C881-83, Type 1, Grade 3.			
Typical Use: Fixing steel fixtures (inserts) into concrete. Pre-cast concrete elements. External reinforcements. Repair of honeycombed concrete structures, etc.			
Cure: RT (25°C)/ 24 hrs; 3 days (insert loading time).			
Test Methods: Not stated		Nov 91	
Physical		**Mechanical**	
Colour	Black + White	τ (RT) steel	15 MPa
Viscosity	Paste	τ (°C)	-
S.G.	-	τ (wet)	-
Shelf life	2 years	Peel ( )	-
Pot/Work life	45 mins	σ_t	-
Tg	-	σ_c	70 MPa
$T_{Service}$	+90°C	E	-
CTE (°C)	-	G	-
Thermal cond.	-	Elongation	-
Resistivity: Vol.	-	Tear strength	-
Surf.	-	H (Shore)	-
Notes: Mix ratio 1:1. τ Concrete: 2.7 MPa (substrate failed); Timber: 5.0 MPa (substrate failed).			

Epirez	8641	Paste
Epoxy - Two part		

Manufacturers Comments: RT curing. 100% solids. High strength impact resistant adhesive for construction industry use. Good adhesion to most materials. Chemical resistance.

Typical Use: Metals, glass, concrete, masonry, some plastics. Joint sealing in metal fabrications. Repair of structural elements.

Cure: RT (25°C)/ 24 hrs; 7 days (full cure).

Test Methods: Not stated		Feb 92	
Physical		**Mechanical**	
Colour	Grey	τ (RT) Al.	18 MPa
Viscosity	Paste	τ (°C)	-
S.G.	-	τ (wet)	-
Shelf life	2 years	Peel ( )	-
Pot/Work life	2 hrs/25°C	σ_t	-
Tg	-	σ_c	83 MPa
$T_{Service}$	-10 to +130°C	*E*	-
CTE (°C)	-	*G*	-
Thermal cond.	-	Elongation	-
Resistivity: Vol.	5×10^{13} Ω.cm	Tear strength	-
Surf.	-	H (Shore)	-

Notes: Mix ratio 1:3. Impact Strength: 50J
τ steel/steel: 15 MPa; FRP/FRP: 15MPa; steel/concrete: 3 MPa (substrate failed); rubber/concrete: 3MPa (substrate failed); timber/timber: 5.0MPa (substrate failed).

Epirez	Highway Marker 611	Paste
Epoxy - Two part		

Manufacturers Comments: RT curing. 100% solids, solvent-free. Bonding road markers to road surfaces. High durability. Good resistance to weathering, UV, temperature, impact, rain and snow. Meets: California State Highway Spec. 741.80.42 (pavement markers).

Typical Use: Road markers - bitumen, concrete, metal, plastics.

Cure: RT (25°C)/ 5 mins (gel); 30 mins (cure).

Test Methods: Not stated		July 91	
Physical		**Mechanical**	
Colour	Light Grey	τ (RT) Al	15 MPa
Viscosity	Paste	τ (°C)	-
S.G.	-	τ (wet)	-
Shelf life	2 years	Peel ( )	-
Pot/Work life	3 mins/25°C	σ_t	-
Tg	-	σ_c	-
$T_{Service}$	-	*E*	-
CTE (°C)	-	*G*	-
Thermal cond.	-	Elongation	-
Resistivity: Vol.	-	Tear strength	-
Surf.	-	H (Shore)	-

Notes: Mix ratio 1:1. Bond strength (7 days, 25°C):
Ceramic marker/concrete: 2 MPa (substrate failed);
Epoxy marker/ concrete: 2 MPa (substrate failed).

Epirez	Epistop Standard	Paste
Epoxy - Two part		

Manufacturers Comments: RT curing. High strength. For fast permanent repairs to most materials. Excellent adhesion. High impact resistance.

Typical Use: Repair of terracotta, iron pipes, timber, tanks, galvanised tubing.

Cure: RT (25°C)/ 1 hr; 2-3 hrs (machinable).

Test Methods: Not stated		July 91	
Physical		**Mechanical**	
Colour	Brown	τ (RT)	14 MPa
Viscosity	Paste	τ (°C)	-
S.G.	-	τ (wet)	-
Shelf life	2 years	Peel ( )	-
Pot/Work life	30 mins/25°C	σ_t	16 MPa
Tg	-	σ_c	36 MPa
$T_{Service}$	-	*E*	-
CTE (°C)	-	*G*	-
Thermal cond.	-	Elongation	-
Resistivity: Vol.	-	Tear strength	-
Surf.	-	H (Shore D)	70

Notes: Mix ratio 1:1.
Epistop Fast Set (Grey): 15 mins work-life, 30 mins/RT cure version.

Epirez	Highway Marker 712	Paste
Epoxy - Two part		

Manufacturers Comments: RT curing. 100% solids, solvent-free. Bonding road markers to road surfaces. High durability. Good resistance to weathering, UV, temperature, impact, rain and snow. Meets: California State Highway Spec. 741.80.44 (pavement markers).

Typical Use: Road markers - bitumen, concrete, metal, plastics.

Cure: RT (25°C)/30 mins (gel); 6 hrs (cure).

Test Methods: Not stated		July 91	
Physical		**Mechanical**	
Colour	Light Grey	τ (RT) Al	7 MPa
Viscosity	Paste	τ (°C)	-
S.G.	-	τ (wet)	-
Shelf life	2 years	Peel ( )	-
Pot/Work life	45 mins/25°C	σ_t	-
Tg	-	σ_c	-
$T_{Service}$	-	*E*	-
CTE (°C)	-	*G*	-
Thermal cond.	-	Elongation	-
Resistivity: Vol.	-	Tear strength	-
Surf.	-	H (Shore)	-

Notes: Mix ratio 1:1. Bond strength (7 days, 25°C):
Ceramic marker/concrete: 2 MPa (substrate failed);
Epoxy marker/ concrete: 2 MPa (substrate failed).

Epirez	Highway Marker 8242		Paste
Epoxy - Two part			
Manufacturers Comments: RT curing. 100% solids, solvent-free. Bonding road markers to road surfaces. High durability. Good resistance to weathering, UV, temperature, impact, rain and snow. Meets: California State Highway Spec. 741.80.44 (pavement markers).			
Typical Use: Road markers - bitumen, concrete, metal, plastics.			
Cure: RT (25°C)/ 2 hrs (gel); 24 hrs (cure).			
Test Methods: Not stated		July 91	
Physical		**Mechanical**	
Colour	Light Grey	τ (RT)	Al 13 MPa
Viscosity	Paste	τ (°C)	-
S.G.	-	τ (wet)	-
Shelf life	2 years	Peel ( )	-
Pot/Work life	45 mins/25°C	σ_t	-
Tg	-	σ_c	-
$T_{Service}$	-	*E*	-
CTE (°C)	-	*G*	-
Thermal cond.	-	Elongation	-
Resistivity: Vol.	-	Tear strength	-
Surf.	-	H (Shore)	-
Notes: Mix ratio 1:1. Bond strength (7 days, 25°C): Ceramic marker/concrete: 2 MPa (substrate failed); Epoxy marker/ concrete: 2 MPa (substrate failed).			

Forbo Helmitin	Helmitin A + B		Paste
Epoxy - Two part			
Manufacturers Comments: RT curing. Also, heated to accelerate cure. Solvent-free. Bonding construction materials.			
Typical Use: Metal, concrete, asbestos-based, ceramics, wood, plastic laminates, rigid-PVC, foam materials.			
Cure: RT (20°C)/10 hrs (set); 48 hrs (full cure).			
Test Methods: Not stated		July 94	
Physical		**Mechanical**	
Colour	Yellow	τ (RT)	-
Viscosity	~20 Pa.s	τ (°C)	-
S.G.	~1.08	τ (wet)	-
Shelf life	2 years	Peel ( )	-
Pot/Work life	~90 mins/25°C	σ_t	-
Tg	-	σ_c	-
$T_{Service}$	-	*E*	-
CTE (°C)	-	*G*	-
Thermal cond.	-	Elongation	-
Resistivity: Vol.	-	Tear strength	-
Surf.	-	H (Shore)	-
Notes: Mix ratio 1.2:1 (wt).			

Forbo Helmitin

Forbo Helmitin	Helmitin 670		Paste
Epoxy - Two part			
Manufacturers Comments: RT curing. Also, heated to accelerate cure. Solvent-free. Bonding construction materials.			
Typical Use: Metal, concrete, asbestos-based, ceramics, wood, plastic laminates, rigid-PVC, foam materials.			
Cure: RT (20°C)/6 hrs (set); 48 hrs (full cure).			
Test Methods: Not stated		July 94	
Physical		**Mechanical**	
Colour	Grey	τ (RT)	-
Viscosity	~90 Pa.s	τ (°C)	-
S.G.	~1.46	τ (wet)	-
Shelf life	2 years	Peel ( )	-
Pot/Work life	~60 mins/25°C	σ_t	-
Tg	-	σ_c	-
$T_{Service}$	-	*E*	-
CTE (°C)	-	*G*	-
Thermal cond.	-	Elongation	-
Resistivity: Vol.	-	Tear strength	-
Surf.	-	H (Shore)	-
Notes: Mix ratio 1:1 (wt).			

Grace Speciality Polymers

Grace	Eccobond 24		Liquid
Epoxy - Two part			
Manufacturers Comments: RT cure. Non-conductive. General purpose water-clear, low viscosity. For transparent materials and thin bondlines. Some flexibility (for dissimilar materials).			
Typical Use: General purpose. Most metals, ceramics, glass, many plastics (not plasticized PVC).			
Cure: RT/1-3 hrs (initial cure); 16-24 hrs (full cure).			
Test Methods: ASTM		(Aug 94)	
Physical		**Mechanical**	
Colour	Clear	τ (RT)	>12.2 MPa
Viscosity	0.6-1 Pa.s	τ (°C)	-
S.G.	1.72	τ (wet)	-
Shelf life	1 yr./<25°C	Peel ( )	-
Pot/Work life	~30 mins (50g mix)	σ_t	-
Tg	-	σ_c	-
$T_{Service}$	+120°C	*E*	-
CTE (°C)	-	*G*	-
Thermal cond.	-	Elongation	-
Resistivity: Vol.	10^{14} min Ωcm	Tear strength	-
Surf.	-	H (Shore)	-
Notes: Viscosity: ASTM D-2393; τ : ASTM D-1002; Service Temp: ASTM D-794; Resistivity ASTM D-257; Dielectric strength ASTM D-149: 16.5 kV/mm.			

Grace	Eccobond 45		Paste
Epoxy - Two part			
Manufacturers Comments: RT cure. Non-conductive. Controlled flexibility (by control of catalyst addition). Applications requiring peel and shock resistance.			
Typical Use: Metals, glass, ceramic, plastics.			
Cure: RT/8 hrs (initial cure); 24 hrs (full cure). 70°C/30 mins; 105°C/15mins.			
Test Methods: ASTM		(Aug 94)	
Physical		**Mechanical**	
Colour	Black*	τ (RT)	>12.0 MPa
Viscosity	-	τ (°C)	-
S.G.	-	τ (wet)	>10 MPa
Shelf life	1 yr./<25°C	Peel ( )	-
Pot/Work life	~2 hrs	σ_t	-
Tg	-	σ_{flex}	>34 MPa
$T_{Service}$	-55 to +90°C	E	-
CTE (°C)	-	G	-
Thermal cond.	-	Elongation	-
Resistivity: Vol.	$3x10^{13}$ Ωcm	Tear strength	-
Surf.	-	H (Shore A)	65-75
Notes: * other colours available. Hardness: ASTM D-2240; Service Temp: ASTM D-794 continuous: +120°C (intermittent) Viscosity: ASTM D-2393; τ : ASTM D-1002. τ(wet) 30 days water soak. Resistivity ASTM D-257; Dielectric strength ASTM D-149: 15.6 kV/mm.			

Grace	Eccobond 68C		Paste
Epoxy - one part			
Manufacturers Comments: Electrically conductive. Silver-filled. Solvent-free. For electrical connections (replacing soldering). Lower temperature curing (~80°C).			
Typical Use: Nichrome wires, conductive plastics. Heat-sensitive materials. Also glass, metals, plastics and ceramics.			
Cure: 80°C/ 5 hrs; 120°C/2 hrs; 160°C/45 mins.			
Test Methods: ASTM		(Aug 94)	
Physical		**Mechanical**	
Colour	Silver	τ (RT)	>7 MPa
Viscosity	-	τ (°C)	-
S.G.	-	τ (wet)	-
Shelf life	6 months/6°C	Peel ( )	-
Pot/Work life	-	σ_t	-
Tg	-	σ_c	-
$T_{Service}$	-	E	-
CTE (°C)	-	G	-
Thermal cond.	7.2 W m.K^{-1}	Elongation	-
Resistivity: Vol.	$1x10^{-3}$ max. Ωcm	Tear strength	-
Surf.	-	H (Shore)	-
Notes: Hardness: ASTM D-2240;τ : ASTM D-1002; Viscosity: ASTM D-2393; Thermal conductivity: ASTM D-2214; Resistivity: ASTM D-257.			

Grace	Eccobond 56C		Paste
Epoxy - Two part			
Manufacturers Comments: Electrically conductive. For electrical connections (replacing soldering).			
Typical Use: Nichrome wires, conductive plastics.			
Cure: 50°C/ 2 hrs; >65°C for low resistivity.			
Test Methods: ASTM		(Aug 94)	
Physical		**Mechanical**	
Colour	-	τ (RT)	>4.5 MPa
Viscosity	-	τ (°C)	-
S.G.	~3.3-3.6	τ (wet)	-
Shelf life	1 yr/25°C	Peel ( )	-
Pot/Work life	-	σ_t	-
Tg	70-80°C	σ_{flex}	>75 MPa
$T_{Service}$	-60 to +120°C	E	-
CTE (°C)	36x10^{-6} K^{-1}	G	-
Thermal cond.	7.1 W m.K^{-1}	Elongation	-
Resistivity: Vol.	-	Tear strength	-
Surf.	-	H (Shore D)	>85
Notes: Hardness: ASTM D-2240;τ : ASTM D-1002; Viscosity: ASTM D-2393; Thermal conductivity: ASTM D-2214; CTE: ASTM D-696; Resistivity: ASTM D-257; σflex ASTM D-790.			

Grace	Eccobond 72C		Paste
Epoxy - two part			
Manufacturers Comments: Electrically conductive (not silver-filled). Resistivity higher than silver-filled. For electrical connections (replacing soldering). RT curing (accelerated by heating).			
Typical Use: Nichrome wires, conductive plastics. Heat-sensitive materials. Also glass, metals, plastics and ceramics.			
Cure: RT/8 hrs; 65°C/ 2 hrs; 90°C/30 mins.			
Test Methods: ASTM		(Aug 94)	
Physical		**Mechanical**	
Colour	-	τ (RT)	>4.5 MPa
Viscosity	-	τ (°C)	-
S.G.	-	τ (wet)	-
Shelf life	1 yr./<25°C	Peel ( )	-
Pot/Work life	-	σ_t	-
Tg	-	σ_c	-
$T_{Service}$	-54 to +150°C	E	-
CTE (°C)	45 x 10^{-6} K^{-1}	G	-
Thermal cond.	2.2 W m.K^{-1}	Elongation	-
Resistivity: Vol.	$6x10^{-3}$ max. Ωcm	Tear strength	-
Surf.	-	H (Shore)	-
Notes: Hardness: ASTM D-2240;τ : ASTM D-1002; Viscosity: ASTM D-2393; Thermal conductivity: ASTM D-2214; Resistivity: ASTM D-257; Service temp: ASTM D-794.			

See Key to Data Sheets - Pages 63 to 65

Grace	Eccobond 104	Paste
	Epoxy - Two part	

Manufacturers Comments: Heat cure. Solvent-free, low volatiles. Non-conductive. High service temperature use. Excellent physical and dielectric properties at high temperatures. Chemical resistance.

Typical Use: Metals, glass, ceramic, many thermoset plastics.

Cure: 120°C/6 hrs; 150°C/3 hrs; 180°C/2 hrs; 200°C/1 hr. Postcure (260°C/12 hrs) for >205°C use.

Test Methods: ASTM		(Aug 94)	
Physical		**Mechanical**	
Colour	Black	τ (RT) Al	>10 MPa
Viscosity	-	τ (230°C)	>6.8 MPa
S.G.	1.4-1.6	τ (wet)	-
Shelf life	6 months/<25°C	Peel ( )	-
Pot/Work life	~12 hrs	σ_t	-
Tg	-	σ_c	-
$T_{Service}$	+230°C	E	-
CTE (°C)	$49x10^{-6}$ K^{-1}	G	-
Thermal cond.	-	Elongation	-
Resistivity: Vol.	10^{15} Ω.cm (RT)	Tear strength	-
Surf.	-	H (Shore D)	90 min

Notes: Hardness: ASTM D-2240; CTE ASTM D-696. Service Temp: ASTM D-794 continuous: +290°C (intermittent) Viscosity: ASTM D-2393; τ : ASTM D-1002. Resistivity ASTM D-257; Dielectric strength ASTM D-149: 15.6 kV/mm.

Grace	Eccobond 285	Paste
	Epoxy - two part	

Manufacturers Comments: RT or heat curing (depends on catalyst selected - choice of 4). 100% solids. Thixotropic, sag resistance. General purpose, high strength bonds. Low CTE, close to Al-, brass and copper. High thermal conductivity. Thermal cycling resistance.

Typical Use: Metals, glass, plastics.

Cure: Depends on catalyst used, e.g. RT/24 hrs. Heat cure: 40 -190°C

Test Methods: ASTM		(Aug 94)	
Physical		**Mechanical**	
Colour	-	τ (RT)	>14 MPa
Viscosity	-	τ (93°C)	>9 MPa
S.G.	1.4-1.6	τ (wet)	>14 MPa
Shelf life	1 yr./<25°C	Peel ( )	-
Pot/Work life	30-45 mins*	σ_t	-
Tg	-	σ_c	-
$T_{Service}$	+180°C max.*	E	-
CTE (°C)	$27x10^{-6}$ K^{-1}	G	-
Thermal cond.	1.3 W m.K^{-1}	Elongation	-
Resistivity: Vol.	>1x10^{15} Ω.cm RT	Tear strength	-
Surf.	-	H (Shore)	-

Notes: * depends on catalyst used. Hardness: ASTM D-2240; CTE ASTM D-696; τ : ASTM D-1002. τ(wet) 30 day water soak. Service Temp: ASTM D-794; Viscosity: ASTM D-2393; Resistivity ASTM D-257. Thermal Conductivity ASTM D-2214.

Grace	Eccobond 144A	Paste
	Epoxy - One part	

Manufacturers Comments: Heat cure. Fast-curing. Solvent-free. Good chemical resistance.

Typical Use: Metals, glass, plastics.

Cure: Gel Temp: (120 to 180°C) + Cure: 80°C/4 hrs; 120°C/1 hr; 150°C/30 mins; 180°C/8 mins.

Test Methods: ASTM		(Aug 94)	
Physical		**Mechanical**	
Colour	Black	τ (RT) Al	>15 MPa
Viscosity	75-110 Pa.s	τ (°C)	-
S.G.	1.4-1.6	τ (wet)	-
Shelf life	3 months/<25°C	Peel ( )	-
Pot/Work life	-	σ_t	-
Tg	-	σ_c	-
$T_{Service}$	-62 to +230°C	E	-
CTE (°C)	$48x10^{-6}$ K^{-1}	G	-
Thermal cond.	-	Elongation	-
Resistivity: Vol.	1x10^{14} min. Ω.cm	Tear strength	-
Surf.	-	H (Shore)	-

Notes: Hardness: ASTM D-2240; CTE ASTM D-696; τ : ASTM D-1002. Service Temp: ASTM D-794 continuous; Viscosity: ASTM D-2393; Resistivity ASTM D-257.

Grace	Eccobond 286	Paste
	Epoxy - two part	

Manufacturers Comments: RT curing. Solvent-free, no volatiles, non-corrosive. Controlled-flow (thixotropic). General purpose to bond wide number of materials. For production and maintenance uses, inc. pipe-work assembly (replacing solder joints). Chemical and solvent resistant.

Typical Use: Metals inc. (Al-, steels, copper, brass, iron), wood products, porcelain, brick, plastics (inc. PVC, polycarbonate). Teflon-polypropylene- and polyethylene-treated surfaces.

Cure: RT/4 hrs (initial cure); 24 hrs (full cure)

Test Methods: ASTM		(Aug 94)	
Physical		**Mechanical**	
Colour	White	τ (RT)	>12 MPa
Viscosity	-	τ (°C)	-
S.G.	-	τ (wet)	-
Shelf life	1 yr./<25°C	Peel ( )	-
Pot/Work life	-	σ_t	-
Tg	-	σ_{flex}	>80 MPa
$T_{Service}$	-	E	-
CTE (°C)	$36x10^{-6}$ K^{-1}	G	-
Thermal cond.	1.0 W m.K^{-1}	Elongation	-
Resistivity: Vol.	10^{14} min. Ω.cm	Tear strength	-
Surf.	-	H (Shore)	-

Notes: Hardness: ASTM D-2240; σflex ASTM D1184; CTE ASTM D-696; τ : ASTM D-1002; Service Temp: ASTM D-794; Viscosity: ASTM D-2393; Resistivity ASTM D-257. Thermal Conductivity ASTM D-2214.

Grace	Eccobond 927-10E		Liquid
Epoxy - one part			
Manufacturers Comments: Heat curing (90-120°C). Rigid, pourable, fast curing. Long-shelf life. Low viscosity.			
Typical Use: For bonding and sealing syringes (needle to hub).			
Cure: 90°C/ 20 mins; 100°C/15 mins; 120°C/5 mins.			
Test Methods: ASTM		(Aug 94)	
Physical		**Mechanical**	
Colour	Light Yellow	τ (RT)	>7.4 MPa
Viscosity	10-12 Pa.s	τ (°C)	-
S.G.	~1.32-1.34	τ (wet)	-
Shelf life	<3 month/20°C	Peel ( )	-
Pot/Work life	-	σ_f	-
Tg	-	σ_c	-
$T_{Service}$	-	E	-
CTE (°C)	-	G	-
Thermal cond.	-	Elongation	-
Resistivity: Vol.	-	Tear strength	-
Surf.	-	H (Shore D)	>80/RT; >50/100°C
Notes: Hardness: ASTM D-2240; τ : ASTM D-1002; Viscosity: ASTM D-2393; Cure varies with needle size and hub configuration. Similar product: Eccobond 933-39.			

Grace	Eccobond 2332-17		Paste
Epoxy - one part			
Manufacturers Comments: Heat curing (~100°C). Solvent-free. Thixotropy. High bond strengths. Low temperature flexibility (-30°C). High peel and shear strength over wide temperature range. Can be cured in paint-baking ovens.			
Typical Use: Metals (copper, Al-, nickel, oily steel); Rigid plastics, FRP			
Cure: 100°C/ 90 mins; 120°C/ 30 mins; 150°C/ 20 mins. Paint-baking ovens >130°C.			
Test Methods: ASTM		(Aug 94)	
Physical		**Mechanical**	
Colour	Black	τ (RT) Al	>20 MPa
Viscosity	90-120 Pa.s	τ (°C)	-
S.G.	~1.17-1.23	τ (wet)	-
Shelf life	<2 month/25°C	Peel ( )	-
Pot/Work life	-	σ_f	-
Tg	-	σ_c	-
$T_{Service}$	-	E	-
CTE (°C)	-	G	-
Thermal cond.	-	Elongation	-
Resistivity: Vol.	-	Tear strength	-
Surf.	-	H (Shore D)	>75/RT; >30/120°C
Notes: Hardness: ASTM D-2240; τ : ASTM D-1002; Viscosity: ASTM D-2393; Mandrel bend test at -30°C ASTM D-4338: No cracks in adhesive or separation from substrate.			

Grace	Eccobond 928-69-10		Paste
Epoxy - one part			
Manufacturers Comments: Heat curing (~80°C). Tough, pourable. Fast curing. For temperature sensitive materials or applications needing low temperature/fast curing.			
Typical Use: -			
Cure: 85°C/ 20 mins, <7 days (full cure).			
Test Methods: ASTM		(Aug 94)	
Physical		**Mechanical**	
Colour	Light Yellow	τ (RT)	-
Viscosity	33-43 Pa.s	τ (°C)	-
S.G.	~1.18-1.24	τ (wet)	-
Shelf life	<6 month/6°C	Peel ( )	-
Pot/Work life	-	σ_f	-
Tg	-	σ_c	-
$T_{Service}$	-	E	-
CTE (°C)	-	G	-
Thermal cond.	-	Elongation	-
Resistivity: Vol.	-	Tear strength	-
Surf.	-	H (Shore D)	>80/RT
Notes: Hardness: ASTM D-2240; τ : ASTM D-1002; Viscosity: ASTM D-2393.			

Grace	Eccobond 50248 F		Paste
Epoxy - one part, liquid			
Manufacturers Comments: Heat curing (~80°C). No moisture absorption (Uncured). Non-corrosive (after cure).			
Typical Use: Assembly of CD arm components.			
Cure: 80°C/ 60 mins; 90°C/ 30 mins; 100°C/ 15 mins.			
Test Methods: ASTM		(Aug 94)	
Physical		**Mechanical**	
Colour	Dark Yellow	τ (RT) Al	>11 MPa
Viscosity	10.5-14 Pa.s	τ (°C)	-
S.G.	~1.17-1.23	τ (wet)	-
Shelf life	<1 week/25°C	Peel ( )	-
Pot/Work life	-	σ_f	-
Tg	70-80°C	σ_c	-
$T_{Service}$	-	E	-
CTE (°C)	$70x10^{-6}$ K^{-1}	G	-
Thermal cond.	-	Elongation	-
Resistivity: Vol.	-	Tear strength	-
Surf.	-	H (Shore D)	>85
Notes: Hardness: ASTM D-2240; τ : ASTM D-1002; Viscosity: ASTM D-2393.			

See Key to Data Sheets - Pages 63 to 65

Grace	Eccobond A 304-6		Paste
Epoxy - one part			
Manufacturers Comments: Heat curing. Filled, pourable adhesive and sealant. Fast cure, good toughness, heat resistant. Chemical resistant.			
Typical Use: End-cap adhesive for hydraulic oil and other high-temperature industrial/aerospace filters.			
Cure: 120°C/ 10 mins(gel), 20 mins (cure); 160°C/70 secs (gel), 5 mins (cure); 180°C/15 secs (gel), 2 mins (cure).			
Test Methods: ASTM		(Aug 94)	
Physical		**Mechanical**	
Colour	Light Blue	τ (RT)	>12 MPa
Viscosity	19-24 Pa.s	τ (°C)	-
S.G.	~1.48-1.54	τ (wet)	-
Shelf life	3 months/25°C	Peel ( )	-
Pot/Work life	-	σ_t	-
Tg	-	σ_c	-
$T_{Service}$	-	E	-
CTE (°C)	-	G	-
Thermal cond.	-	Elongation	-
Resistivity: Vol.	-	Tear strength	-
Surf.	-	H (Shore D)	>80/RT; >50/125°C
Notes: Hardness: ASTM D-2240;τ : ASTM D-1002; Viscosity: ASTM D-2393; Chemical resistance ASTM D-543: No attack at 200 hrs/125°C with Skydrol500 or 7808 hydraulic oil.			

Grace	Eccobond A 329-1		Paste
Epoxy - one part			
Manufacturers Comments: Heat curing. Oxide filler. Thixotropic, sag resistant (TP-29). For high-temperature use. Good thermal stability. Chemical resistance to water, freon, acids, alkalis and solvents. Class F (155°C) dielectric properties.			
Typical Use: Metals (bonding/insulating); FRP and moulded phenolic parts.			
Cure: 100°C/ 30 mins(gel), 90 mins(cure); 120°C/ 10 mins(gel), 20 mins(cure); 160°C/1 min (gel), 3 mins (cure); 180°C/30 secs (gel), 1 mins (cure).			
Test Methods: ASTM		(Aug 94)	
Physical		**Mechanical**	
Colour	Beige	τ (RT)	15-25 MPa
Viscosity	30-55 Pa.s	τ (120°C)	13-22 MPa
S.G.	~1.2-1.4	τ (wet)	-
Shelf life	<3 months/25°C	Peel ( )	-
Pot/Work life	-	σ_t	-
Tg	-	σ_c	-
$T_{Service}$	+155°C	E	-
CTE (°C)	-	G	-
Thermal cond.	-	Elongation	-
Resistivity: Vol.	10^{14} min (RT) Ω.cm	Tear strength	-
Surf.	-	H (Shore D)	>85/RT; >65/120°C
Notes: Hardness: ASTM D-2240;τ : ASTM D-1002; Viscosity: ASTM D-2393; Resistivity: ASTM D-257. Stroke Cure 160°C (TP-12): 50-70 seconds.			

Grace	Eccobond A 316-34		Paste
Epoxy - one part			
Manufacturers Comments: Heat curing. Non-abrasive filler. Reduced wicking during cure. Shock resistance. Chemical resistance to acids, alkalis and solvents.			
Typical Use: End-cap adhesive for Skydrol and hydraulic oil filters. Heat-resistant plastics, e.g. phenolics.			
Cure: 120°C/ 5 mins(gel), 20 mins (cure); 160°C/50 secs (gel), 5 mins (cure); 180°C/20 secs (gel), 2 mins (cure).			
Test Methods: ASTM		(Aug 94)	
Physical		**Mechanical**	
Colour	Tan (cured)	τ (RT)	-
Viscosity	30-55 Pa.s	τ (°C)	-
S.G.	~1.53-1.58	τ (wet)	-
Shelf life	<2 months/25°C	Peel ( )	-
Pot/Work life	-	σ_t	-
Tg	-	σ_c	-
$T_{Service}$	-	E	-
CTE (°C)	-	G	-
Thermal cond.	-	Elongation	-
Resistivity: Vol.	-	Tear strength	-
Surf.	-	H (Shore D)	>85/RT; >65/120°C
Notes: Hardness: ASTM D-2240;τ : ASTM D-1002; Viscosity: ASTM D-2393; Stroke Cure (TP-12): 45-60 seconds.			

Grace	Eccobond A 401-12		Paste
Epoxy - one part			
Manufacturers Comments: Heat curing. Solvent-free. Aluminium oxide filler. High strength. High thermal conductivity. Excellent electrical and dielectric properties. Low viscosity version of Eccobond A 401.			
Typical Use: Metals (bonding/insulating); FRP and moulded phenolic parts.			
Cure: 120°C/ 20 mins(gel),1 hr(cure); 160°C/3 min (gel), 10 mins (cure); 180°C/1 min (gel), 5 mins (cure).			
Test Methods: ASTM		(Aug 94)	
Physical		**Mechanical**	
Colour	Black	τ (RT)	-
Viscosity	35-50 Pa.s	τ (°C)	-
S.G.	~1.7-1.8	τ (wet)	-
Shelf life	<1 month/25°C	Peel ( )	-
Pot/Work life	-	σ_t	-
Tg	-	σ_c	-
$T_{Service}$	-	E	-
CTE (°C)	-	G	-
Thermal cond.	0.54 W m.K^{-1}	Elongation	-
Resistivity: Vol.	-	Tear strength	-
Surf.	-	H (Shore D)	>80/RT; >75/120°C
Notes: Hardness: ASTM D-2240;τ : ASTM D-1002; Viscosity: ASTM D-2393; Resistivity: ASTM D-257. Thermal conductivity: ASTM D-2214.			

Grace	Eccolite 82 EP PB		Paste
Epoxy - two part			
Manufacturers Comments: RT curing. Lightweight, low flammability, syntactic adhesive for assembly of honeycomb panel components. Meets: Boeing BMS 5-28, rev.P-type18; Fokker Spec. TH 5.906/6. Flammability: Passes ASTM D-635; UL-94-V-0; Boeing BSS-7230.			
Typical Use: Sandwich bonding to shapes/containers. Fasteners.			
Cure: RT/2-3 hrs (gel), 24hrs (full cure).			
Test Methods: ASTM		(Aug 94)	
Physical		**Mechanical**	
Colour	White	τ (RT)	>7 MPa
Viscosity	Thixotropic	τ (°C)	-
S.G.	~0.65-0.75	τ (wet)	-
Shelf life	6 months/<25°C	Peel ( )	-
Pot/Work life	20 mins	σ_t	-
Tg	-	σ_c	>45 MPa
$T_{Service}$	-50 to +100°C	*E compressive*	>1800 MPa
CTE (°C)	-	*G*	-
Thermal cond.	-	Elongation	-
Resistivity: Vol.	-	Tear strength	-
Surf.	-	H (Shore)	-
Notes: Service temp: ASTM D-794; τ : ASTM D-1002; Compressive data ASTM D-1621; Moisture absorption: 1.5%.			

Hardman	Epoweld 7897		Paste
Epoxy - Two part toughened			
Manufacturers Comments: Excellent flexibility, shear properties and adhesion to many materials (including galvanised).			
Typical Use: General purpose and lamination of dissimilar materials (skis, rackets, snow boards, fishing rods, etc.)			
Cure: 15min/100°C.			
Test Methods: Not stated (USA)		(July 95)	
Physical		**Mechanical**	
Colour	Black	τ (RT)	23.5 MPa
Viscosity	2.2 Pa.s	τ (149°C)	1.5 MPa
S.G.	1.17	τ (-40°C)	21.8 MPa
Shelf life	-	Peel (-)	1.0 N mm^{-1}
Pot/Work life	<86 min.	σ_t	-
Tg	-	σ_c	-
$T_{Service}$	~60°C	*E*	-
CTE (°C)	-	*G*	-
Thermal cond.	-	Elongation	-
Resistivity: Vol.	-	Tear strength	-
Surf.	-	H (Shore D)	85
Notes:			

Hardman

Hardman	Epoweld 3243		Paste
Epoxy - Two part			
Manufacturers Comments: RT cure. Excellent pull-out strength, torque resistance and adhesion to dissimilar substrates.			
Typical Use: General purpose.			
Cure: 7 days/RT.			
Test Methods: Not stated (USA)		(July 95)	
Physical		**Mechanical**	
Colour	Ivory	τ (RT)	23.0 MPa
Viscosity	15 Pa.s	τ (149°C)	1.9 MPa
S.G.	1.17	τ (-40°C)	20.9 MPa
Shelf life	-	Peel (-)	0.63 N mm^{-1}
Pot/Work life	<150 min.	σ_t	-
Tg	-	σ_c	-
$T_{Service}$	~80°C	*E*	-
CTE (°C)	-	*G*	-
Thermal cond.	-	Elongation	-
Resistivity: Vol.	-	Tear strength	-
Surf.	-	H (Shore D)	85
Notes: Epoweld 8200 is a modified version for some thermoplastic materials (ABS, polycarbonate), Epoweld 3243A-3 is black colour version.			

Hardman	Epoweld 7929		Paste
Epoxy - Two part			
Manufacturers Comments: RT cure. Low viscosity and shrinkage. Excellent pull-out strength, impact and adhesion to dissimilar substrates.			
Typical Use: General purpose.			
Cure: 7 days/RT.			
Test Methods: Not stated (USA)		(July 95)	
Physical		**Mechanical**	
Colour	Black	τ (RT)	28.8 MPa
Viscosity	4.5 Pa.s	τ (149°C)	1.3 MPa
S.G.	1.09	τ (-40°C)	20.8 MPa
Shelf life	-	Peel (-)	0.67 N mm^{-1}
Pot/Work life	<57 min.	σ_t	-
Tg	-	σ_c	-
$T_{Service}$	~60°C	*E*	-
CTE (°C)	-	*G*	-
Thermal cond.	-	Elongation	-
Resistivity: Vol.	-	Tear strength	-
Surf.	-	H (Shore D)	83
Notes:			

Hardman	Epoweld 8173		Paste
Epoxy - Two part			
Manufacturers Comments: RT cure. Fast curing with excellent shear properties and adhesion to wood, concrete and ceramics. Cures at temperatures down to 0°C.			
Typical Use: Fast patching, fixturing and repair of many materials.			
Cure: 24hr/RT.			
Test Methods: Not stated (USA)		(July 95)	
Physical		**Mechanical**	
Colour	Amber	τ (RT)	20.7 MPa
Viscosity	40 Pa.s	τ (149°C)	1.7 MPa
S.G.	1.16	τ (-40°C)	11.7 MPa
Shelf life	-	Peel (-)	0.3 N mm^{-1}
Pot/Work life	<5 min.	σ_t	-
Tg	-	σ_c	-
$T_{Service}$	~60°C	E	-
CTE (°C)	-	G	-
Thermal cond.	-	Elongation	-
Resistivity: Vol.	-	Tear strength	-
Surf.	-	H (Shore D)	80
Notes: Slower curing versions Epoweld 8185 & 8269 available.			

Hardman	Epoweld 16469		Paste
Epoxy - Two part			
Manufacturers Comments: RT cure. Fast curing and flame retardant. Non sag with excellent flexibility and impact resistance.			
Typical Use: Metal door sandwich construction (honeycomb and foam).			
Cure: 7 days/RT.			
Test Methods: Not stated (USA)		(July 95)	
Physical		**Mechanical**	
Colour	Grey	τ (RT)	19.1 MPa
Viscosity	52 Pa.s	τ (149°C)	2.7 MPa
S.G.	1.52	τ (-40°C)	16.7 MPa
Shelf life	-	Peel (-)	0.44 N mm^{-1}
Pot/Work life	<12 min.	σ_t	-
Tg	-	σ_c	-
$T_{Service}$	~70°C	E	-
CTE (°C)	-	G	-
Thermal cond.	-	Elongation	-
Resistivity: Vol.	-	Tear strength	-
Surf.	-	H (Shore D)	90
Notes: Flame retardant: UL94V-O at 3.2mm.			

Hardman	Epoweld 13230		Paste
Epoxy - Two part			
Manufacturers Comments: RT cure. Long work life. Flexible with good impact resistance and excellent peel and shear. Good adhesion to metals, wood and many thermoplastics.			
Typical Use: Aircraft and aerospace.			
Cure: 7 days/RT.			
Test Methods: Not stated (USA)		(July 95)	
Physical		**Mechanical**	
Colour	Grey	τ (RT)	19.3 MPa
Viscosity	50 Pa.s	τ (149°C)	1.6 MPa
S.G.	1.27	τ (-40°C)	22.1 MPa
Shelf life	-	Peel (-)	4.2 N mm^{-1}
Pot/Work life	<90 min.	σ_t	-
Tg	-	σ_c	-
$T_{Service}$	~80°C	E	-
CTE (°C)	-	G	-
Thermal cond.	-	Elongation	-
Resistivity: Vol.	-	Tear strength	-
Surf.	-	H (Shore D)	65
Notes:			

Hardman	Epoweld 16480		Paste
Epoxy - Two part			
Manufacturers Comments: RT cure. Rigid, internally flexibilized adhesive long work life. High thermal conductivity with excellent impact resistance.			
Typical Use: Bonding of heatsinks.			
Cure: 7 days/RT.			
Test Methods: Not stated (USA)		(July 95)	
Physical		**Mechanical**	
Colour	Blue	τ (RT)	21.7 MPa
Viscosity	4.6 Pa.s	τ (149°C)	2.2 MPa
S.G.	2.10	τ (-40°C)	17.9 MPa
Shelf life	-	Peel (-)	0.44 N mm^{-1}
Pot/Work life	<198 min.	σ_t	-
Tg	-	σ_c	-
$T_{Service}$	~70°C	E	-
CTE (°C)	-	G	-
Thermal cond.	-	Elongation	-
Resistivity: Vol.	-	Tear strength	-
Surf.	-	H (Shore D)	85
Notes:			

Hardman	Epoweld 16530		Paste
Epoxy - Two part			
Manufacturers Comments: RT cure. Semi-rigid structural adhesive. Excellent impact properties and shear and peel strength.			
Typical Use: Aircraft and aerospace.			
Cure: 7 days/RT.			
Test Methods: Not stated (USA)		(July 95)	
Physical		**Mechanical**	
Colour	Black	τ (RT)	24.9 MPa
Viscosity	16 Pa.s	τ (149°C)	2.6 MPa
S.G.	1.17	τ (-40°C)	36.1 MPa
Shelf life	-	Peel (-)	6.53 N mm^{-1}
Pot/Work life	<45 min.	σ_f	-
Tg	-	σ_c	-
$T_{Service}$	~70°C	*E*	-
CTE (°C)	-	*G*	-
Thermal cond.	-	Elongation	-
Resistivity: Vol.	-	Tear strength	-
Surf.	-	H (Shore D)	75
Notes:			

Hardman	Epoweld 16562		Paste
Epoxy - Two part			
Manufacturers Comments: RT cure. Rigid structural adhesive. Non sag with excellent shear and peel strength.			
Typical Use: Automotive and SMC applications.			
Cure: 3 days/RT.			
Test Methods: Not stated (USA)		(July 95)	
Physical		**Mechanical**	
Colour	Grey	τ (RT)	22.7 MPa
Viscosity	126 Pa.s	τ (149°C)	3.3 MPa
S.G.	1.38	τ (-40°C)	32.8 MPa
Shelf life	-	Peel (-)	7.76 N mm^{-1}
Pot/Work life	<45 min.	σ_f	-
Tg	-	σ_c	-
$T_{Service}$	~80°C	*E*	-
CTE (°C)	-	*G*	-
Thermal cond.	-	Elongation	-
Resistivity: Vol.	-	Tear strength	-
Surf.	-	H (Shore D)	82
Notes:			

Hardman	Epoweld 16535		Paste
Epoxy - Two part			
Manufacturers Comments: RT cure. Rigid structural adhesive. Low viscosity with excellent wetting. High shear strength.			
Typical Use: Composite bonding in aircraft and aerospace.			
Cure: 7 days/RT.			
Test Methods: Not stated (USA)		(July 95)	
Physical		**Mechanical**	
Colour	Amber	τ (RT)	16.9 MPa
Viscosity	10 Pa.s	τ (149°C)	4.8 MPa
S.G.	1.11	τ (-40°C)	12.2 MPa
Shelf life	-/4°C	Peel (-)	0.18 N mm^{-1}
Pot/Work life	<42 min.	σ_f	-
Tg	-	σ_c	-
$T_{Service}$	~130°C	*E*	-
CTE (°C)	-	*G*	-
Thermal cond.	-	Elongation	-
Resistivity: Vol.	-	Tear strength	-
Surf.	-	H (Shore D)	85
Notes:			

Hardman	Epoweld 16580		Paste
Epoxy - Two part thixotropic			
Manufacturers Comments: RT cure. Fast curing with high shear strength. Non sag with good adhesion to a wide variety of materials.			
Typical Use: General purpose.			
Cure: 24hr/RT.			
Test Methods: Not stated (USA)		(July 95)	
Physical		**Mechanical**	
Colour	Translucent	τ (RT)	22.6 MPa
Viscosity	(paste)	τ (149°C)	1.2 MPa
S.G.	1.17	τ (-40°C)	5.7 MPa
Shelf life	-	Peel (-)	0.5 N mm^{-1}
Pot/Work life	<5 min.	σ_f	-
Tg	-	σ_c	-
$T_{Service}$	~70°C	*E*	-
CTE (°C)	-	*G*	-
Thermal cond.	-	Elongation	-
Resistivity: Vol.	-	Tear strength	-
Surf.	-	H (Shore D)	83
Notes:			

See Key to Data Sheets - Pages 63 to 65

Hardman	Epoweld 19157		Paste
Epoxy - Two part			
Manufacturers Comments: Moderate cure with excellent shear and peel strength. Excellent adhesion to many materials.			
Typical Use: Lamination of dissimilar materials. Honeycomb panels and sporting goods.			
Cure: 20min/80°C.			
Test Methods: Not stated (USA)		(July 95)	
Physical		**Mechanical**	
Colour	Black	τ (RT)	32.1 MPa
Viscosity	8 Pa.s	τ (149°C)	1.7 MPa
S.G.	1.18	τ (-40°C)	13.3 MPa
Shelf life	-	Peel (-)	0.7 N mm^{-1}
Pot/Work life	<74 min/RT	σ_t	-
Tg	-	σ_c	-
$T_{Service}$	~90°C	*E*	-
CTE (°C)	-	*G*	-
Thermal cond.	-	Elongation	-
Resistivity: Vol.	-	Tear strength	-
Surf.	-	H (Shore D)	80
Notes: Epoweld 13443A/19157B is neutral colour version.			

Hardman	Epoweld 20214		Paste
Epoxy - Two part			
Manufacturers Comments: RT cure. Long work life. Good impact properties and adhesion.			
Typical Use: Aircraft and aerospace.			
Cure: 7 days/RT.			
Test Methods: Not stated (USA)		(July 95)	
Physical		**Mechanical**	
Colour	Grey	τ (RT)	16.9 MPa
Viscosity	42 Pa.s	τ (149°C)	1.6 MPa
S.G.	1.30	τ (-40°C)	23.9 MPa
Shelf life	-	Peel (-)	5.1 N mm^{-1}
Pot/Work life	<116 min.	σ_t	-
Tg	-	σ_c	-
$T_{Service}$	~60°C	*E*	-
CTE (°C)	-	*G*	-
Thermal cond.	-	Elongation	-
Resistivity: Vol.	-	Tear strength	-
Surf.	-	H (Shore D)	65
Notes:			

Hardman	Epoweld 20018		Paste
Epoxy - Two part			
Manufacturers Comments: RT cure. Excellent shear strength at elevated temperatures. Asbestos free and non sag. Excellent solvent and chemical resistance. Withstands continuous exposure to temperatures in the range 149°C - 176°C with a RT cure.			
Typical Use: Aircraft and aerospace.			
Cure: 24hr/RT.			
Test Methods: Not stated (USA)		(July 95)	
Physical		**Mechanical**	
Colour	Grey	τ (RT)	30.9 MPa
Viscosity	(paste)	τ (149°C)	11.0 MPa
S.G.	1.53	τ (-40°C)	8.7 MPa
Shelf life	-	Peel (-)	0.6 N mm^{-1}
Pot/Work life	<32 min.	σ_t	-
Tg	-	σ_c	-
$T_{Service}$	<176°C	*E*	-
CTE (°C)	-	*G*	-
Thermal cond.	-	Elongation	-
Resistivity: Vol.	-	Tear strength	-
Surf.	-	H (Shore D)	95
Notes:			

Hardman	Epoweld 20215		Paste
Epoxy - Two part			
Manufacturers Comments: RT cure. Non sag with long work life. Excellent shear properties and adhesion. Good gap filling.			
Typical Use: General Purpose.			
Cure: 7 days/RT.			
Test Methods: Not stated (USA)		(July 95)	
Physical		**Mechanical**	
Colour	Amber	τ (RT)	23.6 MPa
Viscosity	408 Pa.s	τ (149°C)	2.4 MPa
S.G.	1.09	τ (-40°C)	15.7 MPa
Shelf life	-	Peel (-)	0.3 N mm^{-1}
Pot/Work life	<125 min.	σ_t	-
Tg	-	σ_c	-
$T_{Service}$	~90°C	*E*	-
CTE (°C)	-	*G*	-
Thermal cond.	-	Elongation	-
Resistivity: Vol.	-	Tear strength	-
Surf.	-	H (Shore D)	80
Notes:			

See Key to Data Sheets - Pages 63 to 65

Hardman	Monopoxy 16365		Paste
Epoxy - One part			
Manufacturers Comments: One component fast curing epoxy with low viscosity and no metallic fillers. Moderate shear and peel strength.			
Typical Use: Ferrite core bonding.			
Cure: 1.5hr/107°C.			
Test Methods: Not stated (USA)		(July 95)	
Physical		**Mechanical**	
Colour	Blue	τ (RT)	16.5 MPa
Viscosity	80 Pa.s	τ (149°C)	2.3 MPa
S.G.	1.38	τ (-40°C)	18.3 MPa
Shelf life	-/4°C	Peel (-)	0.3 N mm^{-1}
Pot/Work life	<14 min/107°C	σ_t	-
Tg	-	σ_c	-
$T_{Service}$	~125°C	*E*	-
CTE (°C)	-	*G*	-
Thermal cond.	-	Elongation	-
Resistivity: Vol.	-	Tear strength	-
Surf.	-	H (Shore D)	85
Notes:			

Hardman	Monopoxy 18512		Paste
Epoxy - One part			
Manufacturers Comments: One component, toughened, metal-filled epoxy. Moderate cure with extremely high shear strength (retained at elevated temperatures).			
Typical Use: Aircraft and aerospace.			
Cure: 40min/121°C.			
Test Methods: Not stated (USA)		(July 95)	
Physical		**Mechanical**	
Colour	Grey	τ (RT)	28.3 MPa
Viscosity	(paste)	τ (149°C)	7.6 MPa
S.G.	1.52	τ (-40°C)	23.7 MPa
Shelf life	-/4°C	Peel (-)	0.8 N mm^{-1}
Pot/Work life	<12 min/121°C	σ_t	-
Tg	-	σ_c	-
$T_{Service}$	>150°C	*E*	-
CTE (°C)	-	*G*	-
Thermal cond.	-	Elongation	-
Resistivity: Vol.	-	Tear strength	-
Surf.	-	H (Shore D)	85
Notes:			

Hardman	Monopoxy 16551		Paste
Epoxy - One part			
Manufacturers Comments: One component, toughened, metal-filled epoxy. Moderate cure with excellent shear and peel strength.			
Typical Use: Ferrite core bonding.			
Cure: 40min/121°C.			
Test Methods: Not stated (USA)		(July 95)	
Physical		**Mechanical**	
Colour	Grey	τ (RT)	29.7 MPa
Viscosity	(paste)	τ (149°C)	3.3 MPa
S.G.	1.59	τ (-40°C)	25.0 MPa
Shelf life	-/4°C	Peel (-)	0.6 N mm^{-1}
Pot/Work life	<11 min/121°C	σ_t	-
Tg	-	σ_c	-
$T_{Service}$	~135°C	*E*	-
CTE (°C)	-	*G*	-
Thermal cond.	-	Elongation	-
Resistivity: Vol.	-	Tear strength	-
Surf.	-	H (Shore D)	90
Notes:			

W.W. Henry

Henry	191		Paste
Epoxy - two part			
Manufacturers Comments: RT curing. Interior flooring adhesive.			
Typical Use: Studded rubber tiling or modules to concrete (moisture and alkali-free).			
Cure: RT(>18°C)/1-4 hrs (tack).			
Test Methods: see notes		Jan 88	
Physical		**Mechanical**	
Colour	Grey + Off-white	τ (RT)	-
Viscosity	-	τ (°C)	-
S.G.	-	τ (wet)	-
Shelf life	15-32°C storage	Peel ( )	-
Pot/Work life	30 mins	σ_t	-
Tg	-	σ_c	-
$T_{Service}$	-	*E*	-
CTE (°C)		*G*	-
Thermal cond.	-	Elongation	-
Resistivity: Vol.	-	Tear strength	-
Surf.	-	H (Shore)	-
Notes: Boiling range: 65.5-121°C; Volatile content (vol.): 4-15%.			

Hitachi-Kasei Polymer Co.

Hitachi-Kasei	Hi-Bon 3430		Paste
Epoxy - two part			
Manufacturers Comments: RT curing. Low modulus, good bond strengths.			
Typical Use: Aluminium honeycombs.			
Cure: RT/4 days.			
Test Methods: see notes		Mar 95	
Physical		**Mechanical**	
Colour	Pink + Blue	τ (RT) Al.	17.4 MPa
Viscosity	72 Pa.s (20°C)	τ (°C)	-
S.G.	1.14	τ (wet)	-
Shelf life	-	Peel (-)	7.0 N mm^{-1}
Pot/Work life	40 mins/20°C	σ_t	-
Tg	55°C	σ_c	-
$T_{Service}$	-	E	4.0 MPa
CTE (°C)		G	-
Thermal cond.	-	Elongation	10.0%
Resistivity: Vol.	-	Tear strength	-
Surf.	-	H (Shore D)	60 (20°C)

Notes: Mix ratio 10:4 (wt).τ- Al-plate (1.6x25x125mm), speed 1mm/min; Peel: Al-plate (0.5x25x150mm), speed 50mm/min.

Hitachi-Kasei	Hi-Bon 3550		Paste
Epoxy - one part			
Manufacturers Comments: Heat curing. Low modulus, good crack resistance. For assembly of small motors.			
Typical Use: Steel and ferrite.			
Cure: 150°C/40 mins.			
Test Methods: see notes		Mar 95	
Physical		**Mechanical**	
Colour	Cream	τ (RT) steel	30 MPa
Viscosity	130 Pa.s (20°C)	τ (°C)	-
S.G.	1.15	τ (wet)	-
Shelf life	6 months/20°C	Peel (-)	8.8 N mm^{-1}
Pot/Work life	-	σ_t	-
Tg	97°C	σ_c	-
$T_{Service}$	-	E	3 MPa
CTE (°C)	69x10^{-6} K^{-1}	G	-
Thermal cond.	-	Elongation	8.0%
Resistivity: Vol.	-	Tear strength	-
Surf.	-	H (Shore D)	80 (20°C)

Notes: τ Fe-plate (1.6x25x125mm), speed 1mm/min; Peel: Fe-plate (0.5x25x150mm), speed 50mm/min.

Hitachi-Kasei	Hi-Bon 3530		Paste
Epoxy - one part			
Manufacturers Comments: Heat curing. Low modulus, good bond strengths. High temperature resistance.			
Typical Use: Aluminium honeycombs.			
Cure: 120°C/1 hr.			
Test Methods: see notes		Mar 95	
Physical		**Mechanical**	
Colour	Cream	τ (RT) Al.	20 MPa
Viscosity	900 Pa.s (20°C)	τ (°C)	-
S.G.	1.10	τ (wet)	-
Shelf life	6 months/20°C	Peel (-)	6.0 N mm^{-1}
Pot/Work life	-	σ_t	-
Tg	130°C	σ_c	-
$T_{Service}$	-	E	4.0 MPa
CTE (°C)		G	-
Thermal cond.	-	Elongation	8.0%
Resistivity: Vol.	-	Tear strength	-
Surf.	-	H (Shore D)	70 (20°C)

Notes: τ- Al-plate (1.6x25x125mm), speed 1mm/min; Peel: Al-plate (0.5x25x150mm), speed 50mm/min.

Klebchemie (M.G. Becker GmbH)

Klebchemie	Kleiberit EP 554		Paste
Epoxy - Two part			
Manufacturers Comments: RT- and hot-curing. Solvent free. High bond strength. Corrosion resistant to many chemicals (acids, methanol and white spirit).			
Typical Use: Construction - bonding glass fibre reinforced plastics (pipes, shafts and sheets).			
Cure: 20°C/24 hrs; 100°C/1 hr.			
Test Methods: DIN		(Oct 95)	
Physical		**Mechanical**	
Colour	-	τ (RT, GRP)	~14 MPa
Viscosity	-	τ (GRP)*	~13 MPa
S.G.	~1.5	τ (wet)	-
Shelf life	1 yr.	Peel ( )	-
Pot/Work life	~30 mins (100g/RT)	σ_t	-
Tg	-	σ_c	-
$T_{Service}$	-	E	-
CTE (°C)	-	G	-
Thermal cond.	-	Elongation	-
Resistivity: Vol.	-	Tear strength	-
Surf.	-	H (Shore)	-

Notes: τ: similar to DIN 53283. τ (GRP)*: after 24hrs/100°C.

Klebchemie	Plastic-Mastic 528		Paste
Epoxy - Two part			
Manufacturers Comments: RT cure. Solvent free. High bond strength. Good outdoor weather resistance. Machinable. Cartridge or pots.			
Typical Use: Construction: concrete, masonry, mineral fibre cement.			
Cure: 20°C/8 hrs (handling), 48 hrs (full strength).			
Test Methods: -		(Oct 95)	
Physical		**Mechanical**	
Colour	Grey	τ (RT)	-
Viscosity	-	τ (°C)	-
S.G.	~1.93	τ (wet)	-
Shelf life	Pot*: 6 months	Peel ( )	-
Pot/Work life	~40 mins(1kg/RT)*	σ_t	-
Tg	-	σ_c	-
$T_{Service}$	-	*E*	-
CTE (°C)	-	*G*	-
Thermal cond.	-	Elongation	-
Resistivity: Vol.	-	Tear strength	-
Surf.	-	H (Shore)	-
Notes: * Cartridge: ~30mins/RT. Shelf life 2 months.			

Klebchemie	Supra-Epoxy 560.1		Paste
Epoxy - Two part			
Manufacturers Comments: RT- and hot-curing. Solvent free for high requirements.			
Typical Use: Inserts for hydraulic and industrial filters. Structural metal bonds.			
Cure: 20°C/8 hrs (handling), 72 hrs (full strength). 130°C/5 mins (handling), 2 hrs (full strength).			
Test Methods: DIN		(Oct 95)	
Physical		**Mechanical**	
Colour	-	τ (RT, Steel)	20 MPa
Viscosity	~30 Pa.s	τ (80°C)	13 MPa
S.G.	~1.1	τ (-20°C)	16 MPa
Shelf life	1 yr.	Peel ( )	-
Pot/Work life	~90 mins (100g/RT)	σ_t	-
Tg	-	σ_c	-
$T_{Service}$	-	*E*	-
CTE (°C)	-	*G*	-
Thermal cond.	-	Elongation	-
Resistivity: Vol.	-	Tear strength	-
Surf.	-	H (Shore)	-
Notes: τ: similar to DIN 53283.			

Klebchemie	Plastic-Mastic 531.1		Paste
Epoxy - Two part			
Manufacturers Comments: RT or warm curing. Solvent free. Good temperature resistance. Good chemical resistance (benzene, hydraulic and brake fluids).			
Typical Use: Industrial filters. Metal bonding.			
Cure: 20°C/6 hrs (handling), 72 hrs (full strength). 100°C/5 mins (handling, 2 hrs (full strength).			
Test Methods: -		(Oct 95)	
Physical		**Mechanical**	
Colour	Grey	τ (RT)	-
Viscosity	-	τ (°C)	-
S.G.	~1.66	τ (wet)	-
Shelf life	6 months	Peel ( )	-
Pot/Work life	~75 mins(100g/RT)	σ_t	-
Tg	-	σ_c	-
$T_{Service}$	-50 to +100°C	*E*	-
CTE (°C)	-	*G*	-
Thermal cond.	-	Elongation	-
Resistivity: Vol.	-	Tear strength	-
Surf.	-	H (Shore)	-
Notes:			

Klebchemie	Suprapox 560.5		Paste
Epoxy - Two part			
Manufacturers Comments: RT- and hot-curing. Solvent free for high requirements.			
Typical Use: Bonding ABS to wood products (e.g. chipboard).			
Cure: 20°C/8 hrs (handling), 72 hrs (full strength). 130°C/5 mins (handling), 2 hrs (full strength).			
Test Methods: -		(Oct 95)	
Physical		**Mechanical**	
Colour	-	τ (RT)	-
Viscosity	~45 Pa.s	τ (°C)	-
S.G.	~1.2	τ (wet)	-
Shelf life	1 yr.	Peel ( )	-
Pot/Work life	~90 mins(100/RT)	σ_t	-
Tg	-	σ_c	-
$T_{Service}$	-	*E*	-
CTE (°C)	-	*G*	-
Thermal cond.	-	Elongation	-
Resistivity: Vol.	-	Tear strength	-
Surf.	-	H (Shore)	-
Notes:			

Kömmerling

Kömmerling	Corapox 733		Paste
Epoxy - Two part			
Manufacturers Comments: RT cure. Solvent free. Good strength. General assembly bonding of various materials.			
Typical Use: Metals, ceramics, wood products, some plastics.			
Cure: 20°C/~16 hrs (7-8 x pot-life), 72 hrs (full cure).			
Test Methods: -		Nov. 94	
Physical		**Mechanical**	
Colour	Opaque/Yellow tint	τ (RT)	-
Viscosity	28 Pa.s	τ (°C)	-
S.G.	1.1	τ (wet)	-
Shelf life	9 months/RT	Peel ( )	-
Pot/Work life	2 hrs (100g/RT)	σ_t	-
Tg	-	σ_c	-
$T_{Service}$	-	*E*	-
CTE (°C)	-	*G*	-
Thermal cond.	-	Elongation	-
Resistivity: Vol.	-	Tear strength	-
Surf.	-	H (Shore D)	~78
Notes:			

Le Joint Français	PR 1826		Paste
Epoxy - Two part			
Manufacturers Comments: Adhesive-sealant for rapid repairs. Permapol 3 ™. Fast curing. Low temperature curing. High temperature resistance.			
Typical Use: Aerospace.			
Cure: Low temperature.			
Test Methods: -		July 95	
Physical		**Mechanical**	
Colour	-	τ (RT)	-
Viscosity	-	τ (°C)	-
S.G.	-	τ (wet)	-
Shelf life	-	Peel ( )	-
Pot/Work life	-	σ_t	-
Tg	-	σ_c	-
$T_{Service}$	203°C/10 hrs	*E*	-
CTE (°C)	-	*G*	-
Thermal cond.	-	Elongation	-
Resistivity: Vol.	-	Tear strength	-
Surf.	-	H (Shore)	-
Notes: Permapol is a trademark of PRC- Product Research & Chemical Corporation. Product replaces PR1422.			

Le Joint Français

Le Joint Français	PR 1219		Paste
Epoxy - Two part			
Manufacturers Comments: Adhesive-sealant for filling and coating. Permapol 5 ™. Fast curing. Improved electrical characteristics. Meets hygiene and safety requirements.			
Typical Use: Aerospace: Riveted joints in skins/stringer assemblies.			
Cure: -			
Test Methods: -		July 95	
Physical		**Mechanical**	
Colour	-	τ (RT)	-
Viscosity	-	τ (°C)	-
S.G.	-	τ (wet)	-
Shelf life	-	Peel ( )	-
Pot/Work life	-	σ_t	-
Tg	-	σ_c	-
$T_{Service}$	-	*E*	-
CTE (°C)	-	*G*	-
Thermal cond.	-	Elongation	-
Resistivity: Vol.	-	Tear strength	-
Surf.	-	H (Shore)	-
Notes: Permapol is a trademark of PRC- Product Research & Chemical Corporation. Product replaces PR1210Q, PR1201RLS.			

Le Joint Français	PR 1828		Paste
Epoxy - Two part			
Manufacturers Comments: Adhesive-sealant for rapid repairs. Permapol 3 ™. Fast curing. Low temperature curing. High temperature resistance. Primerless.			
Typical Use: Aerospace.			
Cure: Low temperature.			
Test Methods: -		July 95	
Physical		**Mechanical**	
Colour	-	τ (RT)	-
Viscosity	-	τ (°C)	-
S.G.	-	τ (wet)	-
Shelf life	-	Peel ( )	-
Pot/Work life	-	σ_t	-
Tg	-	σ_c	-
$T_{Service}$	203°C/10 hrs	*E*	-
CTE (°C)	-	*G*	-
Thermal cond.	-	Elongation	-
Resistivity: Vol.	-	Tear strength	-
Surf.	-	H (Shore)	-
Notes: Permapol is a trademark of PRC- Product Research & Chemical Corporation. Product replaces PR1422.			

See Key to Data Sheets - Pages 63 to 65

LePage's Limited

Le Joint Français	PR 1829		Paste
Epoxy - Two part			
Manufacturers Comments: Adhesive-sealant for sealing and repairs to glass products. Permapol 3 ™. Fast curing. Low temperature curing.			
Typical Use: Aerospace: Glass products.			
Cure: Low temperature.			
Test Methods: -		July 95	
Physical		**Mechanical**	
Colour	-	τ (RT)	-
Viscosity	-	τ (°C)	-
S.G.	-	τ (wet)	-
Shelf life	-	Peel ( )	-
Pot/Work life	-	σ_t	-
Tg	-	σ_c	-
$T_{Service}$	-	E	-
CTE (°C)	-	G	-
Thermal cond.	-	Elongation	-
Resistivity: Vol.	-	Tear strength	-
Surf.	-	H (Shore)	-
Notes: Permapol is a trademark of PRC- Product Research & Chemical Corporation. Product replaces PR1222, PR1425.			

LePage's	5 Minute Epoxy		Paste
Epoxy - Two part			
Manufacturers Comments: RT cure. Solvent-free, non-flammable. Vapour-free cure. General assembly adhesive. Non-shrink for gap fillings and surface repairs/patching. Resistance to water and most common solvents. Inert and safe for food contact (when cured).			
Typical Use: Metals, glass, ceramics, masonry, wood products, most rigid plastics (not nylon, polyethylene, polypropylene, Teflon).			
Cure: 21°C/1 hr (service), 24 hrs (full cure).			
Test Methods: ASTM		Oct. 91	
Physical		**Mechanical**	
Colour	Clear (yellow-tint)	τ (RT, Al-alloy)	15.9 (25.1) MPa
Viscosity	~16 Pa.s	τ (°C)	-
S.G.	~1.1	τ_c	20.7 MPa
Shelf life	-	Peel ( )	-
Pot/Work life	<5 mins (21°C)	σ_t	-
Tg	-	σ_c	-
$T_{Service}$	<60°C	E	-
CTE (°C)	-	G	-
Thermal cond.	-	Elongation	-
Resistivity: Vol.	~10^{12} Ω.cm	Tear strength	-
Surf.	-	H (Shore)	-
Notes: τ data to ASTM D1002: 1 hr cure (24 hr cure). τ compressive to ASTM D905: wood, 24 hr cure. Dielectric strength: >19MV/m.			

Le Joint Français	PR 1875		Paste
Epoxy - Two part			
Manufacturers Comments: Adhesive-sealant for corrosion-inhibiting assemblies. Solvent-free. Fast cure. Chromate-free. Meets: DMS 2013.			
Typical Use: Aerospace.			
Cure: Low temperature.			
Test Methods: -		July 95	
Physical		**Mechanical**	
Colour	-	τ (RT)	-
Viscosity	-	τ (°C)	-
S.G.	-	τ (wet)	-
Shelf life	-	Peel ( )	-
Pot/Work life	-	σ_t	-
Tg	-	σ_c	-
$T_{Service}$	-	E	-
CTE (°C)	-	G	-
Thermal cond.	-	Elongation	-
Resistivity: Vol.	-	Tear strength	-
Surf.	-	H (Shore)	-
Notes: Permapol is a trademark of PRC- Product Research & Chemical Corporation. Product replaces PR1431, PR1431G, PS 870C.			

LePage's	Epoxy Putty		Paste
Epoxy - Two part			
Manufacturers Comments: RT cure. 100% solids. Solvent-free, non-flammable. Vapour-free cure. Adhesive/filler for tough, waterproof bonds/seals. Non-shrink for gap filling. Resistance to water and most common solvents, oil and gas. Will cure underwater.			
Typical Use: Metal, masonry, wood products, rigid plastics. Repairs/seals: Tooling, pipes, castings, tanks. Pools and spa repairs.			
Cure: 21°C/12 hrs (full strength).			
Test Methods: ASTM		(Apr. 95)	
Physical		**Mechanical**	
Colour	Brown	τ (RT, Al-alloy)	2.6-4.6 MPa
Viscosity	Putty (mouldable)	τ (°C)	-
S.G.	~1.8	τ (wet)	-
Shelf life	-	Peel ( )	-
Pot/Work life	<1-2 hrs (24°C)	σ_t	27.5 MPa
Tg	-	σ_c	83.0 MPa
$T_{Service}$	121°C (continuous)	E	-
CTE (°C)	-	G	-
Thermal cond.	-	Elongation	-
Resistivity: Vol.	-	Tear strength	-
Surf.	-	H (Shore)	-
Notes: τ ASTM D1002, aluminium. Dielectric strength: 11.8 MV/m.			

LePage's	Epoxy (Regular)		Paste
Epoxy - Two part			
Manufacturers Comments: RT cure. 100% solids. Solvent-free, non-flammable. Vapour-free cure. General assembly adhesive. Non-shrink for gap fillings and surface repairs/patching. Resistance to water and most common solvents. Inert and safe for food contact (when cured).			
Typical Use: Metals, glass, ceramics, masonry, wood products, most rigid plastics (not nylon, polyethylene, polypropylene, Teflon).			
Cure: 21°C/24 hrs (service), 7 days (full cure).			
Test Methods: ASTM		(April 95)	
Physical		**Mechanical**	
Colour	Translucent (tinted)	τ (RT, Al-alloy)	17.2 (18.6) MPa
Viscosity	~12 Pa.s	τ (°C)	-
S.G.	~1.2	τ_c	20.7 MPa
Shelf life	-	Peel ( )	-
Pot/Work life	<2-3 hrs(21°C)	σ_f	-
Tg	-	σ_c	-
$T_{Service}$	<70°C	*E*	-
CTE (°C)	-	*G*	-
Thermal cond.	-	Elongation	-
Resistivity: Vol.	~10^{14} Ω.cm	Tear strength	-
Surf.	-	H (Shore)	-
Notes: τ data to ASTM D1002: 24 hr cure (7 day cure). τ compressive to ASTM D905: wood, 24 hr cure. Dielectric strength: >19MV/m.			

LePage's	LEP 502 Epoxy Glue		Paste
Epoxy - Two part			
Manufacturers Comments: RT cure. 100% solids. Solvent-free, non-flammable. Vapour-free cure. Adhesive/filler for wide range of materials. Non-shrink for gap filling. Resistance to water, dilute acids and alkalis, most solvents, oil and gas.			
Typical Use: Metals, masonry, glass, ceramics, most rigid plastics. Crack filling. Electrical insulation (cast handles, knobs, etc.)			
Cure: RT(>15°C)/6-7hrs (handling), 1 day (cure), 7 days(max. strength)			
Test Methods: ASTM		March 86	
Physical		**Mechanical**	
Colour	Opaque (yellow tint)	τ (RT, Al-alloy)	17.2 (18.6) MPa
Viscosity	~900 Pa.s	τ (°C)	-
S.G.	-	τ (wet)	-
Shelf life	-	Peel (CD)	3 N mm^{-1}
Pot/Work life	~2 hrs (min)	σ_f	-
Tg	-	σ_c	-
$T_{Service}$	71°C (continuous)	*E*	-
CTE (°C)	-	*G*	-
Thermal cond.	-	Elongation	-
Resistivity: Vol.	-	Tear strength	-
Surf.	-	H (Shore)	-
Notes: τ data: ASTM D1002, Al-alloy, phosphoric acid etch, 24 hr cure (7 day cure). Climbing Drum Peel ASTM D1781, Al-alloy, phosphoric acid etch, 30 day cure.			

LePage's	Epoxy Steel #400		Paste
Epoxy - Two part			
Manufacturers Comments: RT cure. Steel-filled. Solvent-free, non-flammable. Vapour-free cure. Adhesive/filler for metal repairs. Non-shrink for gap filling and surface repairs/patching. Resistance to water and most common solvents.			
Typical Use: Metal repairs: tooling, pipes, castings, sheet metal parts. Fixing fasteners into masonry.			
Cure: 21°C/24 hrs (full strength).			
Test Methods: -		Oct. 91	
Physical		**Mechanical**	
Colour	Clear (yellow-tint)	τ (RT)	15.0 MPa
Viscosity	~500 Pa.s	τ (°C)	-
S.G.	~1.1	τ (wet)	13.0 MPa
Shelf life	-	Peel ( )	-
Pot/Work life	<2-3 hrs (20°C)	σ_f	-
Tg	-	σ_c	-
$T_{Service}$	<60°C	*E*	-
CTE (°C)	-	*G*	-
Thermal cond.	-	Elongation	-
Resistivity: Vol.	-	Tear strength	-
Surf.	-	H (Shore)	-
Notes: τ(wet): after 1 month water immersion.			

LePage's	LEP 503 - 5 minute Epoxy		Paste
Epoxy - Two part			
Manufacturers Comments: RT cure. 100% solids. Solvent-free, non-flammable. Vapour-free cure. Adhesive/filler for wide range of materials. Non-shrink for gap filling. Resistance to water, dilute acids and alkalis, most solvents, oil and gas.			
Typical Use: Metals, masonry, glass, ceramics, most rigid plastics. Crack filling. Electrical insulation (cast handles, knobs, etc.)			
Cure: RT(>15°C)/15-20 min (handling), 1 hr (cure), 1 day(max. strength).			
Test Methods: ASTM		March 86	
Physical		**Mechanical**	
Colour	Opaque (yellow tint)	τ (RT, Al-alloy)	15.9 (25.1) MPa
Viscosity	~900 Pa.s	τ (-20°C)	8.7 MPa
S.G.	-	τ (wet)	10.2 MPa
Shelf life	-	Peel ( )	-
Pot/Work life	-	σ_f	-
Tg	-	σ_c	12.4 MPa
$T_{Service}$	60°C (continuous)	*E*	-
CTE (°C)	-	*G*	-
Thermal cond.	-	Elongation	-
Resistivity: Vol.	1 x 10^{12} Ωcm	Tear strength	-
Surf.	-	H (Rockwell L)	28-33
Notes: τ data: ARP 1524, Al-alloy, phosphoric acid etch, 1 hr cure (28 day cure). τ(wet): 7 day cure, 60 hrs, water immersion. τ(-20): 7 day cure, 60hrs exposure. σc: ASTM D695. Hardness: ASTM D785. Volume resistivity: ASTM D2S7. Dielectric Strength ASTM D149: 500V/mil.			

LePage's	Permabond 1700		Paste
Epoxy - Two part			
Manufacturers Comments: RT cure. High performance flooring adhesive. Internal and external use. High resistance to plasticisers. Impervious to water. High adhesion (peel and shear).			
Typical Use: Flooring: vinyl, rubber, wood parquet and strip, rubber sports coverings. Suitable for PVC.			
Cure: RT(>5<40°C)/~ 24 hrs			
Test Methods: -		(Apr 95)	
Physical		**Mechanical**	
Colour	Beige	τ (RT)	-
Viscosity	-	τ (°C)	-
S.G.	-	τ (wet)	-
Shelf life	1year/RT	Peel ( )	-
Pot/Work life	60-90 mins	σ_t	-
Tg	-	σ_c	-
$T_{Service}$	-	*E*	-
CTE (°C)	-	*G*	-
Thermal cond.	-	Elongation	-
Resistivity: Vol.	-	Tear strength	-
Surf.	-	H (Shore)	-
Notes: Open time: 30-45 mins. Apply within 30 mins.			

Lugato

Lugato	Kraftkleber		Paste
Epoxy - Two part			
Manufacturers Comments: RT curing, building adhesive. Bonds metals, masonry, wood products, some plastics. High strength and chemical resistance (grease, food, dilute acids, fruit juice, dairy products, household cleaners). Interior and external use.			
Typical Use: Industrial tiling (floor and wall), food processing factories, laboratories, swimming pools.			
Cure: RT (8-35°C). Open Time: ~45 mins.			
Test Methods: -		(Mar 95)	
Physical		**Mechanical**	
Colour	Grey	τ (RT, Steel)	12 MPa
Viscosity	Paste	τ (°C)	-
S.G.	-	τ (wet)	-
Shelf life	1 year	Peel ( )	-
Pot/Work life	~90 mins	σ_t	-
Tg	-	σ_c	45 MPa
$T_{Service}$	-50 to +70°C	*E*	-
CTE (°C)		*G*	-
Thermal cond.		Elongation	-
Resistivity: Vol.	-	Tear strength	-
Surf.	-	H (Shore)	-
Notes: -			

Loctite

Loctite	Structural Epoxy 344		Paste
Epoxy - One part, toughened			
Manufacturers Comments: Heat curing (possible during paint-stoving at 230°C). High shear strength.			
Typical Use: Steels (mild, hot-dipped galvanised). Aluminium alloys. Fibre reinforced plastics (FRP), e.g. external cladding to buildings.			
Cure: ~30 mins.			
Test Methods: -		(Feb 95)	
Physical		**Mechanical**	
Colour	-	τ (RT)	-
Viscosity	Paste	τ (°C)	-
S.G.	-	τ (wet)	-
Shelf life	-	Peel ( )	-
Pot/Work life	-	σ_t	-
Tg	-	σ_c	-
$T_{Service}$	100°C max.	*E*	-
CTE (°C)		*G*	-
Thermal cond.		Elongation	-
Resistivity: Vol.	-	Tear strength	-
Surf.	-	H (Shore)	-
Notes: Gap filling: 5 mm max.			

Panacol Elosol

Panacol Elosol	Vitralit 1500 LI		Paste
Epoxy - One part, cationic UV curing			
Manufacturers Comments: Rapid cure. 100% solids. High Tg.			
Typical Use: Electronics industry.			
Cure: UV-A, 60 mW.cm^{-2}: 90 sec. Final cure: 24 hrs.			
Test Methods: DIN (ASTM)		Jun 94	
Physical		**Mechanical**	
Colour	-	τ (RT)	-
Viscosity	8 Pa.s	τ (°C)	-
S.G.	1.45	τ (wet)	-
Shelf life	6 months/RT	Peel ( )	-
Pot/Work life	-	σ_t	-
Tg	150°C	σ_c	-
$T_{Service}$	<175°C	*E*	-
CTE (°C)	35 x 10^{-6} K^{-1}	*G*	-
Thermal cond.	0.6 W m.K^{-1}	Elongation	-
Resistivity: Vol.	-	Tear strength	-
Surf.	-	H (Shore D)	85
Notes: Dielectric constant(10 kHz, DIN 53483): 3.15. Breakdown potential(DIN 53481): 16.7 kV.mm^{-1}. Flash point (DIN 51755): >100°C. Shrinkage: 1.4%. Water absorption (ASTM 570): 0.52%. CTE: ASTM D695.			

See Key to Data Sheets - Pages 63 to 65

Panacol Elosol	Vitralit 1505		Paste
Epoxy - One part, cationic UV curing			
Manufacturers Comments: Rapid cure. Very high Tg. 100% solids.			
Typical Use: Electronics industry.			
Cure: UV-A, 45 $mW.cm^{-2}$: 90 sec. Final cure: 24 hrs.			
Test Methods: DIN (ASTM)		Jun 94	
Physical		**Mechanical**	
Colour	-	τ (RT)	-
Viscosity	0.4 - 0.5 Pa.s	τ (°C)	-
S.G.	1.12	τ (wet)	-
Shelf life	6 months/RT	Peel ( )	-
Pot/Work life	-	σ_t	-
Tg	150°C	σ_c	-
$T_{Service}$	<175°C	E	-
CTE (°C)	75 x 10^{-6} K^{-1}	G	-
Thermal cond.	-	Elongation	-
Resistivity: Vol.	-	Tear strength	-
Surf.	-	H (Shore D)	85 - 87

Notes: Breakdown potential(DIN 53481): 16.7 $kV.mm^{-1}$. Flash point (DIN 51755): >100°C. Shrinkage: 1.5%. Water absorption (ASTM 570): 0.52%. CTE: ASTM D695.

Panacol Elosol	Vitralit 1555		Paste
Epoxy - One part, cationic UV curing			
Manufacturers Comments: Rapid cure. 100% solids.			
Typical Use: Electronics industry.			
Cure: UV-A, 60 $mW.cm^{-2}$: 90 sec. Final cure: 24 hrs.			
Test Methods: DIN (ASTM)		Jun 94	
Physical		**Mechanical**	
Colour	-	τ (RT)	-
Viscosity	0.28 Pa.s	τ (°C)	-
S.G.	1.1	τ (wet)	-
Shelf life	6 months/RT	Peel ( )	-
Pot/Work life	-	σ_t	-
Tg	30 - 40°C	σ_c	-
$T_{Service}$	<150°C	E	-
CTE (°C)	45 x 10^{-6} K^{-1}	G	-
Thermal cond.	-	Elongation	-
Resistivity: Vol.	-	Tear strength	-
Surf.	-	H (Shore A)	76

Notes: Flash point (DIN 51755): >100°C. Shrinkage: 1.2%. Water absorption (ASTM 570): 0.35%. CTE: ASTM D695.

Panacol Elosol	Vitralit 1550		Paste
Epoxy - One part, cationic UV curing			
Manufacturers Comments: Rapid cure. Flexible. 100% solids.			
Typical Use: Electronics industry.			
Cure: UV-A, 60 $mW.cm^{-2}$: 90 sec. Final cure: 24 hrs.			
Test Methods: DIN (ASTM)		Jun 94	
Physical		**Mechanical**	
Colour	-	τ (RT)	-
Viscosity	6 Pa.s	τ (°C)	-
S.G.	1.46	τ (wet)	-
Shelf life	6 months/RT	Peel ( )	-
Pot/Work life	-	σ_t	-
Tg	30°C	σ_c	-
$T_{Service}$	<135°C	E	-
CTE (°C)	37 x 10^{-6} K^{-1}	G	-
Thermal cond.	0.8 W $m.K^{-1}$	Elongation	-
Resistivity: Vol.	-	Tear strength	-
Surf.	-	H (Shore D)	60

Notes: Dielectric constant(10 kHz, DIN 53483): 3.4. Breakdown potential(DIN 53481): 14.3 $kV.mm^{-1}$. Flash point (DIN 51755): >100°C. Shrinkage: 1.2%. Water absorption (ASTM 570): 0.8%. CTE: ASTM D695.

Panacol Elosol	Vitralit 1722		Paste
Epoxy - One part, cationic UV curing			
Manufacturers Comments: Rapid cure. 100% solids.			
Typical Use: Plastics, most thermoplastics and glass.			
Cure: UV-A, 60 $mW.cm^{-2}$: 120 sec. Final cure: 24 hrs.			
Test Methods: DIN (ASTM)		Jun 94	
Physical		**Mechanical**	
Colour	-	τ (RT)	-
Viscosity	5.5 Pa.s	τ (°C)	-
S.G.	1.14	τ (wet)	-
Shelf life	6 months/RT	Peel ( )	-
Pot/Work life	-	σ_t	-
Tg	40°C	σ_c	-
$T_{Service}$	<120°C	E	-
CTE (°C)	85 x 10^{-6} K^{-1}	G	-
Thermal cond.	0.4 W $m.K^{-1}$	Elongation	-
Resistivity: Vol.	-	Tear strength	-
Surf.	-	H (Shore D)	75

Notes: Dielectric constant(10 kHz, DIN 53483): 3.3. Breakdown potential(DIN 53481): 8.3 $kV.mm^{-1}$. Flash point (DIN 51755): >100°C. Shrinkage: 1.8%. Water absorption (ASTM 570): 1.4%. CTE: ASTM D695.

Permabond

Panacol Elosol	Vitralit 1731		Paste
Epoxy - One part, cationic UV curing			
Manufacturers Comments: Rapid cure. 100% solids.			
Typical Use: Plastics and glass. Optics.			
Cure: UV-A, 60 mW.cm^{-2}: 120 sec. Final cure: 24 hrs.			
Test Methods: DIN (ASTM)		Jun 94	
Physical		**Mechanical**	
Colour	Transparent	τ (RT)	-
Viscosity	0.65 Pa.s	τ (°C)	-
S.G.	1.12	τ (wet)	-
Shelf life	6 months/RT	Peel ( )	-
Pot/Work life	-	σ_t	-
Tg	40°C	σ_c	-
$T_{Service}$	<80°C	E	-
CTE (°C)	75 x 10^{-6} K^{-1}	G	-
Thermal cond.	0.4 W m.K^{-1}	Elongation	-
Resistivity: Vol.	-	Tear strength	-
Surf.	-	H (Shore D)	65
Notes: Dielectric constant(10 kHz, DIN 53483): 3.2. Breakdown potential(DIN 53481): 6.4 kV.mm^{-1}. Flash point (DIN 51755): >100°C. Shrinkage: 1.9%. Water absorption (ASTM 570): 1.3%. CTE: ASTM D695.			

Permabond	Double Bubble		Paste
Epoxy - Two part rapid repair kit			
Manufacturers Comments: Fast RT cure - handling strength: 5 mins.			
Typical Use: Rapid industrial repairs.			
Cure: Full cure: 24 hr/RT.			
Test Methods: Not stated		(Feb 95)	
Physical		**Mechanical**	
Colour	Transparent	τ (RT, steel)	20 MPa
Viscosity	-	τ (°C)	-
S.G.	-	τ (wet)	-
Shelf life	-	Peel ( )	-
Pot/Work life	-	σ_t	-
Tg	-	σ_c	-
$T_{Service}$	-	E	-
CTE (°C)	-	G	-
Thermal cond.	-	Elongation	-
Resistivity: Vol.	-	Tear strength	-
Surf.	-	H (Shore)	-
Notes: Gap filling: < 1mm.			

Panacol Elosol	Vitralit 4102		Paste
Epoxy - One part, cationic UV curing, heat			
Manufacturers Comments: Rapid cure. Low shrinkage and complete initial surface curing. 100% solids. Combined thermal curing required for larger quantities/thicknesses.			
Typical Use: -			
Cure: UV-A, 60 mW.cm^{-2}: 60 sec. Final cure: 12 hrs. 120°C/60 min.			
Test Methods: DIN (ASTM)		Jun 94	
Physical		**Mechanical**	
Colour	-	τ (RT)	-
Viscosity	32 Pa.s	τ (°C)	-
S.G.	1.23	τ (wet)	-
Shelf life	6 months/RT	Peel ( )	-
Pot/Work life	-	σ_t	-
Tg	-	σ_c	-
$T_{Service}$	-50 to +180°C	E	872 MPa
CTE (°C)	45 x 10^{-6} K^{-1}	G	-
Thermal cond.	0.31 W m.K^{-1}	Elongation	-
Resistivity: Vol.	-	Tear strength	-
Surf.	-	H (Shore D)	83
Notes: Dielectric constant(10 kHz, DIN 53483): 3.2. Breakdown potential(DIN 53481): 26.3 kV.mm^{-1}. Flash point (DIN 51755): >100°C. Shrinkage: 1.5%. Water absorption (ASTM 570): 0.8%. Refractive Index: 1.486. CTE: ASTM D695.			

Permabond	E01		Paste
Epoxy - Two part			
Manufacturers Comments: Fast RT cure - handling strength in 5 minutes.			
Typical Use: General purpose. Wood, metal, plastics, ceramic, glass.			
Cure: 24hr/RT or 30min/60°C.			
Test Methods: Not stated		Mar 94	
Physical		**Mechanical**	
Colour	Transparent	τ (RT, Al)	20 MPa
Viscosity	35 Pa.s	τ (°C)	-
S.G.	1.16	τ (wet)	-
Shelf life	1 yr/RT	Peel ( )	-
Pot/Work life	3 min/RT	σ_t	-
Tg	-	σ_c	-
$T_{Service}$	-55°C to +100°C	E	-
CTE (°C)	-	G	-
Thermal cond.	-	Elongation	-
Resistivity: Vol.	2 x 10^{15} Ω.cm	Tear strength	-
Surf.	-	H (Shore)	-
Notes: Dielectric constant(1 kHz): 4.5, Dielectric strength: 14.4 kV/mm. Gap filling: < 1mm.			

Permabond	E04		Paste
Epoxy - Two part modified			
Manufacturers Comments: RT cure - handling strength in 3 hours. Good peel and impact strength.			
Typical Use: Particularly suitable for composites and dissimilar substrates.			
Cure: 24hr/RT or 20min/60°C.			
Test Methods: Not stated		Mar 94	
Physical		**Mechanical**	
Colour	Grey	τ (RT, Al)	16 MPa
Viscosity	(soft paste)	τ (RT, steel)	16 MPa
S.G.	1.34	τ (wet, Al)	~17 MPa
Shelf life	1 yr./RT	Peel ( )	-
Pot/Work life	8 min/RT	σ_t	-
Tg	-	σ_c	-
$T_{Service}$	-55°C to +70°C	E	-
CTE (°C)	-	G	-
Thermal cond.	-	Elongation	-
Resistivity: Vol.	-	Tear strength	-
Surf.	-	H (Shore)	-
Notes: Gap filling: 1 - 2mm.			

Permabond	E32		Paste
Epoxy - Two part thixotropic			
Manufacturers Comments: Slow RT cure - handling strength in 8 hours. High strength and toughness.			
Typical Use: Large components or surfaces. Suitable for composites and dissimilar substrates.			
Cure: 72hr/RT or 1 hr/60°C.			
Test Methods: Not stated		Mar 94	
Physical		**Mechanical**	
Colour	Grey	τ (RT, Al)	28 MPa
Viscosity	(soft paste)	τ (RT, Steel)	29 MPa
S.G.	1.22	τ (wet, Al)	~25 MPa
Shelf life	1 yr./RT	Peel ( )	-
Pot/Work life	2 hrs/RT	σ_t	-
Tg	-	σ_c	-
$T_{Service}$	-55°C to +80°C	E	-
CTE (°C)	-	G	-
Thermal cond.	-	Elongation	-
Resistivity: Vol.	-	Tear strength	-
Surf.	-	H (Shore)	-
Notes: Gap filling: <1mm.			

Permabond	E18		Paste
Epoxy - Two part			
Manufacturers Comments: Fast RT cure - handling strength in 15 minutes.			
Typical Use: General purpose. Wood, metal, plastics, ceramic, glass.			
Cure: 24hr/RT or 30min/60°C.			
Test Methods: Not stated		Mar 94	
Physical		**Mechanical**	
Colour	Transparent	τ (RT, Al)	20 MPa
Viscosity	35 Pa.s	τ (°C)	-
S.G.	1.16	τ (wet)	-
Shelf life	1 yr./RT	Peel ( )	-
Pot/Work life	3 min/RT	σ_t	-
Tg	-	σ_c	-
$T_{Service}$	-55°C to +100°C	E	-
CTE (°C)	-	G	-
Thermal cond.	-	Elongation	-
Resistivity: Vol.	2×10^{15} Ω.cm	Tear strength	-
Surf.	-	H (Shore)	-
Notes: Dielectric constant(1 kHz): 4.5, Dielectric strength: 14.4 kV/mm. Gap filling: < 0.75mm.			

Permabond	Fe Rapid		Paste
Epoxy - Two part metal-filled			
Manufacturers Comments: Fast RT cure - handling strength in 10 - 15 minutes, machinable in 2 - 3 hours. High resistance to abrasion, chemicals and weathering.			
Typical Use: Repairs to metallic components.			
Cure: 6hr/RT.			
Test Methods: Not stated		Mar 94	
Physical		**Mechanical**	
Colour	Dark metallic grey	τ (RT, steel)	15 MPa
Viscosity	(paste)	τ (°C)	-
S.G.	2.6	τ (wet)	-
Shelf life	1 yr./RT	Peel ( )	-
Pot/Work life	4 mins/RT	σ_t	13 MPa
Tg	-	σ_c	-
$T_{Service}$	-55°C to +120°C	E	-
CTE (°C)	-	G	-
Thermal cond.	-	Elongation	-
Resistivity: Vol.	-	Tear strength	-
Surf.	-	H (Shore D)	~80
Notes: Gap filling: < 5mm.			

Ricon Resins Inc.

Ricon Resins Inc.	Ricotuff 1000		Paste
Epoxy - Two part modified			
Manufacturers Comments: Elevated temperature cure, flexible, modified epoxy. High impact strength. Good electrical properties, moisture resistance and thermal stability. Components mixed at 50°C.			
Typical Use: Good adhesion to most metals, GRP, etc.			
Cure: 85°C to 130°C.			
Test Methods: Not stated		(Mar 95)	
Physical		**Mechanical**	
Colour	-	τ (RT)	-
Viscosity	168 Pa.s	τ (°C)	-
S.G.	-	τ (wet)	-
Shelf life	-	Peel ( )	-
Pot/Work life	7 days/RT	σ_t	22.8 MPa
Tg	125 - 130°C	σ_c	-
$T_{Service}$	-	*E*	860 MPa
CTE (°C)	166 x 10^{-6} K^{-1}	*G*	-
Thermal cond.	-	Elongation	5 - 7%
Resistivity: Vol.	-	Tear strength	-
Surf.	5.9 x 10^{16} Ω	H (Shore D)	82
Notes: Flexural strength: 4.7 MPa, Flexural modulus: 13100 MPa. Dielectric constant(12 - 18 GHz, RT): 2.56, Dissipation factor(12 - 18 GHz, RT): 0.01, Dielectric strength (60 Hz): >28.6 kV_{ac}/mm. Water absorption (RT): 0.49%/63 days.			

Ricon Resins Inc.	Ricotuff 1110		Paste
Epoxy - Two part modified			
Manufacturers Comments: RT and elevated temperature cure, flexible, modified epoxy. High impact strength. Good electrical properties, moisture resistance and thermal stability.			
Typical Use: Good adhesion to most metals, GRP, etc.			
Cure: 48hr/RT or 2hr/55°C or 1.5hr/75°C.			
Test Methods: Not stated		(Mar 95)	
Physical		**Mechanical**	
Colour	-	τ (RT)	64.5 MPa
Viscosity	34 Pa.s	τ (°C)	-
S.G.	-	τ (wet)	-
Shelf life	-	Peel ( )	-
Pot/Work life	<1hr/RT	σ_t	18.3 MPa
Tg	119 - 127°C	σ_c	-
$T_{Service}$	-	*E*	760 MPa
CTE (°C)	130 x 10^{-6} K^{-1}	*G*	-
Thermal cond.	-	Elongation	5 - 7%
Resistivity: Vol.	-	Tear strength	-
Surf.	4.7 x 10^{16} Ω	H (Shore D)	75
Notes: Dielectric constant(12 - 18 GHz, RT): 2.55, Dissipation factor(12 - 18 GHz, RT): 0.01, Dielectric strength (60 Hz): >31.7 kV_{ac}/mm. Water absorption (RT): 0.45%/63 days.			

Ricon Resins Inc.	Ricotuff 1100		Paste
Epoxy - Two part modified			
Manufacturers Comments: Elevated temperature cure, flexible, modified epoxy. High impact strength. Good electrical properties, moisture resistance and thermal stability. Components mixed at 50°C.			
Typical Use: Good adhesion to most metals, GRP, etc.			
Cure: 85°C to 130°C.			
Test Methods: Not stated		(Mar 95)	
Physical		**Mechanical**	
Colour	-	τ (RT)	64.5 MPa
Viscosity	40 Pa.s	τ (°C)	-
S.G.	-	τ (wet)	-
Shelf life	-	Peel ( )	-
Pot/Work life	7 days/RT	σ_t	18.3 MPa
Tg	119 - 127°C	σ_c	-
$T_{Service}$	-	*E*	760 MPa
CTE (°C)	130 x 10^{-6} K^{-1}	*G*	-
Thermal cond.	-	Elongation	5 - 7%
Resistivity: Vol.	-	Tear strength	-
Surf.	4.7 x 10^{16} Ω	H (Shore D)	75
Notes: Dielectric constant(12 - 18 GHz, RT): 2.55, Dissipation factor(12 - 18 GHz, RT): 0.01, Dielectric strength (60 Hz): >31.7 kV_{ac}/mm. Water absorption (RT): 0.45%/63 days.			

SEA
(Société d'Exploitation des Adhésifs)

SEA	Epox		Paste
Epoxy - Two part			
Manufacturers Comments: 100 % solids.			
Typical Use: Construction industry.			
Cure: Initial cure: 24 hr/RT, Full cure: 5 days.			
Test Methods: Not stated		May 90	
Physical		**Mechanical**	
Colour	Light Brown	τ (RT)	-
Viscosity	100 Pa.s	τ (°C)	-
S.G.	1.65	τ (wet)	-
Shelf life	-	Peel ( )	-
Pot/Work life	120 min/RT	σ_t	-
Tg	-	σ_c	-
$T_{Service}$	-	*E*	-
CTE (°C)	-	*G*	-
Thermal cond.	-	Elongation	-
Resistivity: Vol.	-	Tear strength	-
Surf.	-	H (Shore)	-
Notes:			

Structil

SEA	SIKI 440		Paste
Epoxy - Two part			
Manufacturers Comments: -			
Typical Use: Construction industry.			
Cure: Initial cure: 24 hr/RT or 12 hr/40°C, Full cure: 7days.			
Test Methods: Not stated		Sept. 91	
Physical		**Mechanical**	
Colour	Grey	τ (RT)	-
Viscosity	48 Pa.s	τ (°C)	-
S.G.	1.75	τ (wet)	-
Shelf life	-	Peel ( )	-
Pot/Work life	120 min/RT	σ_t	-
Tg	-	σ_c	-
$T_{Service}$	-	E	-
CTE (°C)	-	G	-
Thermal cond.	-	Elongation	-
Resistivity: Vol.	-	Tear strength	-
Surf.	-	H (Shore)	-
Notes:			

Structil	ST-1013		Paste
Epoxy - Two part			
Manufacturers Comments: RT cure, bondline control by glass micro-spheres. For assembly and structural repairs. High shear and peel strength. Environmental (moisture) resistance.			
Typical Use: Composite materials and thermoplastics. Structural repairs.			
Cure: RT.			
Test Methods: Not stated		(Jul 95)	
Physical		**Mechanical**	
Colour	-	τ (RT)	-
Viscosity	-	τ (°C)	-
S.G.	-	τ (wet)	-
Shelf life	-	Peel ( )	-
Pot/Work life	35 min (100g)	σ_t	-
Tg	-	σ_c	-
$T_{Service}$	-55 to +80°C	E	-
CTE (°C)	-	G	-
Thermal cond.	-	Elongation	-
Resistivity: Vol.	-	Tear strength	-
Surf.	-	H (Shore)	-
Notes:			

Sicomin

Sicomin	Aerobond 7303		Paste
Epoxy - Two part modified			
Manufacturers Comments: RT cure, flexible, modified epoxy. Good peel and fatigue resistance.			
Typical Use: Bonding of metals, wood, composites, etc. Honeycomb bonding in aircraft, skis.			
Cure: 48hr/RT or 70min/70°C or 20min/130°C.			
Test Methods: Not stated		(Mar 95)	
Physical		**Mechanical**	
Colour	Grey	τ (RT)	25 - 35 MPa
Viscosity	48 Pa.s	τ (°C)	-
S.G.	1.11	τ (wet)	-
Shelf life	1 yr./RT	Peel ( )	-
Pot/Work life	150 min	σ_t	-
Tg	65°C	σ_c	-
$T_{Service}$	-	E	-
CTE (°C)	-	G	-
Thermal cond.	-	Elongation	-
Resistivity: Vol.	-	Tear strength	-
Surf.	-	H (Shore)	-
Notes:			

Structil	ST-1015		Paste
Epoxy - Two part, filled			
Manufacturers Comments: RT cure, aluminium filled. Non-flow. For structural repairs. Moisture resistance.			
Typical Use: Repair adhesive: potting, filling, surfacing.			
Cure: RT.			
Test Methods: Not stated		(Jul 95)	
Physical		**Mechanical**	
Colour	-	τ (RT)	-
Viscosity	-	τ (°C)	-
S.G.	-	τ (wet)	-
Shelf life	-	Peel ( )	-
Pot/Work life	40 min (100g)	σ_t	-
Tg	-	σ_c	-
$T_{Service}$	-250 to +150°C	E	-
CTE (°C)	-	G	-
Thermal cond.	-	Elongation	-
Resistivity: Vol.	-	Tear strength	-
Surf.	-	H (Shore)	-
Notes:			

Structil	ST-1016		Paste
Epoxy - Two part			
Manufacturers Comments: RT cure, unfilled. Low viscosity. Can be injected. For structural repairs to composites. Moisture resistance.			
Typical Use: Repairs to composite materials.			
Cure: RT.			
Test Methods: Not stated		(Jul 95)	
Physical		**Mechanical**	
Colour	-	τ (RT)	-
Viscosity	-	τ (°C)	-
S.G.	-	τ (wet)	-
Shelf life	-	Peel ( )	-
Pot/Work life	30 min (100g)	σ_f	-
Tg	-	σ_c	-
$T_{Service}$	-55 to +150°C	E	-
CTE (°C)	-	G	-
Thermal cond.	-	Elongation	-
Resistivity: Vol.	-	Tear strength	-
Surf.	-	H (Shore)	-
Notes:			

Structil	ST-1035		Film
Epoxy - One part unfilled film. Areal wt: 100-450 gm^{-2}			
Manufacturers Comments: Heat cure (95 to 200°C), unfilled. Can be co-cured. High shear and peel resistance. Good moisture resistance. Supported (glass or nylon) or non-supported versions available.			
Typical Use: Fabrication of honeycomb panels with composite or metal skins. Surfacing composites before painting. Plastics.			
Cure: 120°C/~1hr/~0.3 MPa pressure. (95 to 200°C when co-cured with composites).			
Test Methods: Not stated		(Nov 95)	
Physical		**Mechanical**	
Colour	Ivory	τ (RT)	40 MPa
Viscosity	-	τ (120°C)	10 MPa
S.G.	~1.2	τ (wet)	-
Shelf life	1 yr. (-18°C)	Peel (-)	8 N mm^{-1}
Pot/Work life	3 weeks (25°C)	σ_f	-
Tg	-	σ_c	-
$T_{Service}$	120°C	E	-
CTE (°C)	-	G	-
Thermal cond.	-	Elongation	-
Resistivity: Vol.	-	Tear strength	-
Surf.	-	H (Shore)	-
Notes:			

Structil	ST-1021		Paste
Epoxy - Two part			
Manufacturers Comments: RT cure, thixotropic. Low viscosity. Can be injected. For structural repairs and assembly of composites. Retains properties at elevated temperatures. Moisture resistance.			
Typical Use: Assembly of composite materials. Structural repairs.			
Cure: RT.			
Test Methods: Not stated		(Jul 95)	
Physical		**Mechanical**	
Colour	-	τ (RT)	-
Viscosity	Thixotropic	τ (°C)	-
S.G.	-	τ (wet)	-
Shelf life	-	Peel ( )	-
Pot/Work life	40 min (100g)	σ_f	-
Tg	-	σ_c	-
$T_{Service}$	-55 to +120°C	E	-
CTE (°C)	-	G	-
Thermal cond.	-	Elongation	-
Resistivity: Vol.	-	Tear strength	-
Surf.	-	H (Shore)	-
Notes:			

Structil	ST-1039		Paste
Epoxy - One part			
Manufacturers Comments: Heat cure. Non-flow. Resistance to corrosives (also at high temperatures). Moisture resistance. Qualifications: SNECMA.			
Typical Use: Rapid repair of motor components. Assembly of composites and structural repairs.			
Cure: 150°C/ 1hr; 200°C/30 mins.			
Test Methods: prEN2243-1		(Nov 95)	
Physical		**Mechanical**	
Colour	Dark Grey	τ (RT)	20 MPa
Viscosity	300 Pa.s	τ (180°C)	20 MPa
S.G.	-	τ (wet)	20 (12) MPa
Shelf life	3 months (RT)	Peel ( )	-
Pot/Work life	-	σ_f	-
Tg	-	σ_c	-
$T_{Service}$	-55 to +200°C	E	-
CTE (°C)	-	G	-
Thermal cond.	-	Elongation	-
Resistivity: Vol.	-	Tear strength	-
Surf.	-	H (Shore)	-
Notes: τ data: 2024-T3, sulpho-chromic etch. τ (wet): after moisture ageing/RT, (after moisture ageing/180°C).			

Structil	ST-1050		Paste
Epoxy - Two part			
Manufacturers Comments: RT cure. Compatible with rocket propellants. Qualifications: SEP, SNPE.			
Typical Use: Aerospace engines/rockets.			
Cure: RT/7days; 60°C/1 day.			
Test Methods: prEN2243-1		(Jul 95)	
Physical		**Mechanical**	
Colour	-	τ (RT)	20 MPa
Viscosity	-	τ (80°C)	6 MPa
S.G.	-	τ (wet)	-
Shelf life	6 months (RT)	Peel ( )	-
Pot/Work life	>6 hrs (100g)	σ_t	-
Tg	-	σ_c	-
$T_{Service}$	-55 to +80°C	*E*	-
CTE (°C)	-	*G*	-
Thermal cond.	-	Elongation	-
Resistivity: Vol.	-	Tear strength	-
Surf.	-	H (Shore)	-
Notes: τ data: 2024-T3, sulpho-chromic etch, 60°C cure.			

Structil	ST-1060 F		Paste
Epoxy - Two part			
Manufacturers Comments: Faster RT cure version of ST-1060. High peel strength. Remains flexible at low temperatures (-40°C). Qualifications: SNPE.			
Typical Use: Assembly of EDPM rubber.			
Cure: RT/7 days; 60°C/1 day.			
Test Methods: Not stated		(Nov 95)	
Physical		**Mechanical**	
Colour	-	τ (RT)	17 MPa
Viscosity	-	τ (80°C)	5 MPa
S.G.	-	τ (wet)	-
Shelf life	6 months (RT)	Peel (Bell)	6 N mm^{-1}
Pot/Work life	2 hrs (100g)	σ_t	-
Tg	-	σ_c	-
$T_{Service}$	-50 to +60°C	*E*	-
CTE (°C)	-	*G*	-
Thermal cond.	-	Elongation	85%
Resistivity: Vol.	-	Tear strength	-
Surf.	-	H (Shore)	-
Notes: τ data: 2024-T3, sulpho-chromic etch, 20°C cure.			

Structil	ST-1060		Paste
Epoxy - Two part			
Manufacturers Comments: RT cure. High peel strength. Remains flexible at low temperatures (-40°C).Qualifications: SNPE.			
Typical Use: Assembly of EDPM rubber.			
Cure: RT/7 days; 60°C/1 day.			
Test Methods: Not stated		(Nov 95)	
Physical		**Mechanical**	
Colour	-	τ (RT)	17 MPa
Viscosity	-	τ (80°C)	5 MPa
S.G.	-	τ (wet)	-
Shelf life	6 months (RT)	Peel (Bell)	6 N mm^{-1}
Pot/Work life	2 hrs (100g)	σ_t	-
Tg	-	σ_c	-
$T_{Service}$	-50 to +60°C	*E*	-
CTE (°C)	-	*G*	-
Thermal cond.	-	Elongation	85%
Resistivity: Vol.	-	Tear strength	-
Surf.	-	H (Shore)	-
Notes: τ data: 2024-T3, sulpho-chromic etch, 20°C cure.			

Structil	ST-1110		Paste
Epoxy - One part			
Manufacturers Comments: Heat cure. Low viscosity (in solution).			
Typical Use: Fabrication of aluminium honeycomb (node bonding).			
Cure: Heat.			
Test Methods: Not stated		(Jul 95)	
Physical		**Mechanical**	
Colour	-	τ (RT)	-
Viscosity	-	τ (°C)	-
S.G.	-	τ (wet)	-
Shelf life	-	Peel ( )	-
Pot/Work life	-	σ_t	-
Tg	-	σ_c	-
$T_{Service}$	-	*E*	-
CTE (°C)	-	*G*	-
Thermal cond.	-	Elongation	-
Resistivity: Vol.	-	Tear strength	-
Surf.	-	H (Shore)	-
Notes:			

Structural Adhesives Ltd.

Structural Adhesives Ltd. S - 106 E			Paste
Epoxy - One part modified			
Manufacturers Comments: Rubber modified with good peel and impact resistance. Oil tolerant.			
Typical Use: Metal bonding and plastics (those unaffected by cure temperature).			
Cure: 140 - 200°C. e.g. 30mins/160°C.			
Test Methods: Not stated		(Mar 95)	
Physical		**Mechanical**	
Colour	Grey	τ (RT)	23 - 28 MPa
Viscosity	(paste)	τ (120°C)	17 MPa
S.G.	1.3	τ (wet)	22 MPa
Shelf life	1 yr./RT	Peel (T)	11 N mm^{-1}
Pot/Work life	-	σ_t	-
Tg	-	σ_c	-
$T_{Service}$	-	*E*	-
CTE (°C)	-	*G*	-
Thermal cond.	-	Elongation	-
Resistivity: Vol.	-	Tear strength	-
Surf.	-	H (Shore)	-
Notes:			

Structural Adhesives Ltd. S - 206 E			Paste
Epoxy - Two part, rubber modified			
Manufacturers Comments: RT or elevated cure. Self-levelling if hot cured. High impact and shear. Long pot life.			
Typical Use: Metals and some rigid plastics.			
Cure: 24hr/RT or e.g. 15 - 20mins/100°C, 2.5mins/180°C.			
Test Methods: Not stated		(Mar 95)	
Physical		**Mechanical**	
Colour	Grey	τ (RT, steel)	15 - 21 MPa
Viscosity	~90 Pa.s	τ (°C)	-
S.G.	~1.0	τ (wet)	-
Shelf life	1 yr./RT	Peel ( )	-
Pot/Work life	3 - 5hr/RT	σ_t	-
Tg	-	σ_c	-
$T_{Service}$	-	*E*	-
CTE (°C)	-	*G*	-
Thermal cond.	-	Elongation	-
Resistivity: Vol.	-	Tear strength	-
Surf.	-	H (Shore)	-
Notes:			

Structural Adhesives Ltd. S - 203 E			Paste
Epoxy - Two part, polyamide modified			
Manufacturers Comments: RT cure, weakly thixotropic. 100% solids.			
Typical Use: Metal and wood-product bonding. Industrial flooring.			
Cure: RT.			
Test Methods: Not stated		(Mar 95)	
Physical		**Mechanical**	
Colour	Grey	τ (RT, steel)	19.7 MPa
Viscosity	~400 Pa.s	τ (°C)	-
S.G.	~1.5	τ (wet)	-
Shelf life	1 yr./RT	Peel ( )	-
Pot/Work life	~5.5hr/RT	σ_t	-
Tg	-	σ_c	-
$T_{Service}$	-30 to +120°C	*E*	-
CTE (°C)	-	*G*	-
Thermal cond.	-	Elongation	-
Resistivity: Vol.	-	Tear strength	-
Surf.	-	H (Shore)	-
Notes:			

Structural Adhesives Ltd. S - 241 E			Paste
Epoxy - Two part			
Manufacturers Comments: RT cure. Long pot life.			
Typical Use: Metals, wood and plastics. Marine applications (e.g. decking) and hand-tool handles.			
Cure: RT.			
Test Methods: Not stated		Feb 91	
Physical		**Mechanical**	
Colour	Grey	τ (RT, steel)	15.5 MPa
Viscosity	~100 Pa.s	τ (°C)	-
S.G.	~1.7	τ (wet)	-
Shelf life	-	Peel ( )	-
Pot/Work life	1.5 - 2 hr/RT	σ_t	-
Tg	-	σ_c	-
$T_{Service}$	-	*E*	-
CTE (°C)	-	*G*	-
Thermal cond.	-	Elongation	-
Resistivity: Vol.	-	Tear strength	-
Surf.	-	H (Shore)	-
Notes:			

See Key to Data Sheets - Pages 63 to 65

Structural Adhesives Ltd. S - 255 E			Paste
Epoxy - Two part			
Manufacturers Comments: RT cure. Long pot life. Flexible and impact resistant.			
Typical Use: Metals, wood and plastics. Marine applications (e.g. decking), machine bedding and hand-tool handles.			
Cure: RT.			
Test Methods: Not stated		Oct 91	
Physical		**Mechanical**	
Colour	Grey	τ (RT, steel)	8.4 MPa
Viscosity	~100 Pa.s	τ (°C)	-
S.G.	~1.45	τ (wet)	-
Shelf life	-	Peel ( )	-
Pot/Work life	1.5 - 2 hr/RT	σ_t	-
Tg	-	σ_c	-
$T_{Service}$	-	*E*	-
CTE (°C)	-	*G*	-
Thermal cond.	-	Elongation	-
Resistivity: Vol.	-	Tear strength	-
Surf.	-	H (Shore)	-
Notes:			

Structural Adhesives Ltd. S - 288 E			Paste
Epoxy - Two part			
Manufacturers Comments: RT cure. Long pot life with rapid cure to handling strength (4 hours). Cures to a matt finish.			
Typical Use: Brickwork, steel, building blocks.			
Cure: RT.			
Test Methods: Not stated		Apr 94	
Physical		**Mechanical**	
Colour	Off-white (see note)	τ (RT, steel)	10.3 MPa
Viscosity	(stiff paste)	τ (°C)	-
S.G.	~1.55	τ (wet)	-
Shelf life	1 yr./RT	Peel ( )	-
Pot/Work life	1 hr/RT	σ_t	-
Tg	-	σ_c	-
$T_{Service}$	-	*E*	-
CTE (°C)	-	*G*	-
Thermal cond.	-	Elongation	-
Resistivity: Vol.	-	Tear strength	-
Surf.	-	H (Shore)	-
Notes: Also available in buff, blue and red for colour blending to match application.			

Structural Adhesives Ltd. S - 264 E			Paste
Epoxy - Two part			
Manufacturers Comments: RT cure. Long pot life with rapid cure to handling strength (4 hours).			
Typical Use: Brickwork, steel, concrete.			
Cure: RT.			
Test Methods: Not stated		Dec 93	
Physical		**Mechanical**	
Colour	Grey (see note)	τ (RT, steel)	11.7 MPa
Viscosity	~200 Pa.s	τ (°C)	-
S.G.	1.34	τ (wet)	-
Shelf life	1 yr./RT	Peel ( )	-
Pot/Work life	1 hr/RT	σ_t	-
Tg	-	σ_c	-
$T_{Service}$	-	*E*	-
CTE (°C)	-	*G*	-
Thermal cond.	-	Elongation	-
Resistivity: Vol.	-	Tear strength	-
Surf.	-	H (Shore)	-
Notes: Also available in buff, blue and red for colour blending to match application.			

Structural Adhesives Ltd. S - 305 E			Paste
Epoxy - Two part thixotropic			
Manufacturers Comments: RT cure. Non sag with a medium pot life.			
Typical Use: Metals and plastics.			
Cure: ~24hr/RT.			
Test Methods: Not stated		Oct 91	
Physical		**Mechanical**	
Colour	Off-white	τ (RT, steel)	13.4 MPa
Viscosity	(paste)	τ (°C)	-
S.G.	1.07	τ (wet)	-
Shelf life	-	Peel ( )	-
Pot/Work life	~1 hr/RT	σ_t	-
Tg	-	σ_c	-
$T_{Service}$	-	*E*	-
CTE (°C)	-	*G*	-
Thermal cond.	-	Elongation	-
Resistivity: Vol.	-	Tear strength	-
Surf.	-	H (Shore)	-
Notes:			

Structural Adhesives Ltd.	**S - 307 E**		**Paste**
Epoxy - Two part			
Manufacturers Comments: RT cure. Very slow cure to allow repositioning and adjustment of structures.			
Typical Use: Structural metalwork.			
Cure: ~24hr/RT.			
Test Methods: ASTM		Jan 94	
Physical		**Mechanical**	
Colour	Khaki	τ (RT, steel)	7.2 MPa
Viscosity	(viscous paste)	τ (°C)	-
S.G.	-	τ (wet)	-
Shelf life	1 yr./RT	Peel ( )	-
Pot/Work life	~1 hr/RT	σ_t	-
Tg	-	σ_c	-
$T_{Service}$	-	E	-
CTE (°C)	-	G	-
Thermal cond.	-	Elongation	-
Resistivity: Vol.	-	Tear strength	-
Surf.	-	H (Shore)	-
Notes: Shear strength: ASTM 1002.			

Structural Adhesives Ltd.	**S - 360 E**		**Paste**
Epoxy - Two part thixotropic			
Manufacturers Comments: RT cure. Fast-curing, non sag.			
Typical Use: Designed for bonding cast-iron (guttering and drain pipes). Also suitable for most metals, wood and rigid plastics.			
Cure: RT.			
Test Methods: Not stated		Jan 93	
Physical		**Mechanical**	
Colour	Grey	τ (RT, steel)	8.6 MPa
Viscosity	~40 Pa.s	τ (RT, Al)	7.7 MPa
S.G.	~1.0	τ (wet)	-
Shelf life	1 yr./RT	Peel ( )	-
Pot/Work life	12.5 mins/RT	σ_t	-
Tg	-	σ_c	-
$T_{Service}$	-	E	-
CTE (°C)	-	G	-
Thermal cond.	-	Elongation	-
Resistivity: Vol.	-	Tear strength	-
Surf.	-	H (Shore)	-
Notes:			

Structural Adhesives Ltd.	**S - 334 E**		**Paste**
Epoxy - Two part			
Manufacturers Comments: RT cure. Long pot life. Tough and resilient.			
Typical Use: Marine applications. General purpose adhesive for the pleasure craft industry.			
Cure: RT.			
Test Methods: Not stated		Oct 91	
Physical		**Mechanical**	
Colour	Grey	τ (RT, steel)	19.3 MPa
Viscosity	~50 Pa.s	τ (RT, Al)	10.9 MPa
S.G.	1.55	τ (RT, GRP)	6.9 MPa
Shelf life	1 yr./RT	Peel ( )	-
Pot/Work life	3 hr/RT	σ_t	-
Tg	-	σ_c	-
$T_{Service}$	-	E	-
CTE (°C)	-	G	-
Thermal cond.	-	Elongation	-
Resistivity: Vol.	-	Tear strength	-
Surf.	-	H (Shore)	-
Notes:			

Structural Adhesives Ltd.	**S - 361 E**		**Paste**
Epoxy - Two part highly thixotropic			
Manufacturers Comments: RT cure. High peel and impact strength. Good gap-filling.			
Typical Use: Construction. Wall-panelling and stiffeners. Steel, wood, GRP, etc.			
Cure: RT.			
Test Methods: Not stated		Jun 94	
Physical		**Mechanical**	
Colour	Off-white	τ (RT, steel)	8.4 MPa
Viscosity	~80 Pa.s	τ (°C)	-
S.G.	~1.5	τ (wet)	-
Shelf life	1 yr./RT	Peel ( )	-
Pot/Work life	40 mins/RT	σ_t	-
Tg	-	σ_c	-
$T_{Service}$	-	E	-
CTE (°C)	-	G	-
Thermal cond.	-	Elongation	-
Resistivity: Vol.	-	Tear strength	-
Surf.	-	H (Shore)	-
Notes:			

Structural Adhesives Ltd.	**S - 384 E**		**Paste**
Epoxy - Two part highly thixotropic			
Manufacturers Comments: RT cure. Tough and flexible.			
Typical Use: Bonding of plastics, wood and metal.			
Cure: RT.			
Test Methods: Not stated		Mar 94	
Physical		**Mechanical**	
Colour	Cream	τ (RT, steel)	10.7 MPa
Viscosity	~65 Pa.s	τ (RT, Al)	10.2 MPa
S.G.	1.03	τ (RT, SMC)	6.3 MPa
Shelf life	1 yr./RT	Peel ( )	-
Pot/Work life	2 hr/RT	σ_t	-
Tg	-	σ_c	-
$T_{Service}$	-	E	-
CTE (°C)	-	G	-
Thermal cond.	-	Elongation	-
Resistivity: Vol.	-	Tear strength	-
Surf.	-	H (Shore)	-
Notes:			

Structural Adhesives Ltd.	**S - 2022 E**		**Paste**
Epoxy - Two part highly thixotropic			
Manufacturers Comments: RT cure. Good gap-filling and bonds well to damp substrates.			
Typical Use: Bonding of brick and 'clayware'. Also bonds metals, concrete, etc.			
Cure: RT.			
Test Methods: Not stated		Dec 94	
Physical		**Mechanical**	
Colour	Grey	τ (RT, steel)	10.3 MPa
Viscosity	~20 Pa.s	τ (°C)	-
S.G.	-	τ (wet)	-
Shelf life	1 yr./RT	Peel ( )	-
Pot/Work life	50 - 60 min/RT	σ_t	-
Tg	-	σ_c	-
$T_{Service}$	-	E	-
CTE (°C)	-	G	-
Thermal cond.	-	Elongation	-
Resistivity: Vol.	-	Tear strength	-
Surf.	-	H (Shore)	-
Notes:			

Structural Adhesives Ltd.	**S - 386 E**		**Paste**
Epoxy - Two part			
Manufacturers Comments: RT cure. Acid resistant.			
Typical Use: Bonding of hard plastics, battery cases. Metals and wood.			
Cure: RT.			
Test Methods: Not stated		July 93	
Physical		**Mechanical**	
Colour	Pale pink	τ (RT, steel)	4.6 MPa
Viscosity	20 - 30 Pa.s	τ (°C)	-
S.G.	1.15	τ (wet)	-
Shelf life	1 yr./RT	Peel ( )	-
Pot/Work life	25 min/RT	σ_t	-
Tg	-	σ_c	-
$T_{Service}$	-	E	-
CTE (°C)	-	G	-
Thermal cond.	-	Elongation	-
Resistivity: Vol.	-	Tear strength	-
Surf.	-	H (Shore)	-
Notes:			

Structural Adhesives Ltd.	**S - 2029 E**		**Paste**
Epoxy - Two part			
Manufacturers Comments: RT cure. High compressive strength with good high temperature and boiling water resistance.			
Typical Use: Bonding of metals and ceramics.			
Cure: RT.			
Test Methods: Not stated		Dec 94	
Physical		**Mechanical**	
Colour	Off-white	τ (RT, steel)	4.7 MPa
Viscosity	~20 Pa.s	τ (100°C, steel)	1.4 MPa
S.G.	~1.5	τ (wet)	-
Shelf life	1 yr./RT	Peel ( )	-
Pot/Work life	30 min/RT	σ_t	-
Tg	-	σ_c	75 MPa
$T_{Service}$	-	E	-
CTE (°C)	-	G	-
Thermal cond.	-	Elongation	-
Resistivity: Vol.	-	Tear strength	-
Surf.	-	H (Shore)	-
Notes:			

See Key to Data Sheets - Pages 63 to 65

Structural Adhesives Ltd.	S - 2045 E		Paste
Epoxy - Two part			
Manufacturers Comments: RT cure. Fast-curing giving a tough resilient bond.			
Typical Use: General purpose, household and DIY.			
Cure: 24 hr/RT.			
Test Methods: Not stated		Feb 95	
Physical		**Mechanical**	
Colour	Transparent	τ (RT, steel)	7.9 MPa
Viscosity	~8.5 Pa.s	τ (RT, Al)	8.4 MPa
S.G.	~1.15	τ (RT, ABS)	3 MPa
Shelf life	1 yr./RT	Peel ( )	-
Pot/Work life	~5 min/RT	σ_t	-
Tg	-	σ_c	-
$T_{Service}$	-	*E*	-
CTE (°C)	-	*G*	-
Thermal cond.	-	Elongation	-
Resistivity: Vol.	-	Tear strength	-
Surf.	-	H (Shore)	-
Notes:			

Sunstar Engineering	Penguin Cement 1064		Paste
Epoxy - One part			
Manufacturers Comments: Good adhesion to oily steel.			
Typical Use: Automotive. Bonding of body panels.			
Cure: e.g. 20mins/170°C.			
Test Methods: Not stated		(Mar 95)	
Physical		**Mechanical**	
Colour	Black	τ (RT)	24.5 MPa
Viscosity	450 - 950 Pa.s	τ (°C)	-
S.G.	-	τ (wet)	-
Shelf life	-	Peel (T)	6 N mm^{-1}
Pot/Work life	-	σ_t	-
Tg	-	σ_c	-
$T_{Service}$	-	*E*	-
CTE (°C)	-	*G*	-
Thermal cond.	-	Elongation	-
Resistivity: Vol.	-	Tear strength	-
Surf.	-	H (Shore)	-
Notes: Shear strength and peel: Oily steel to oily steel.			

Sunstar Engineering

Sunstar Engineering	Penguin Cement 1063		Paste
Epoxy - One part			
Manufacturers Comments: Good adhesion to oily steel.			
Typical Use: Automotive. Bonding of body panels.			
Cure: e.g. 20mins/170°C.			
Test Methods: Not stated		(Mar 95)	
Physical		**Mechanical**	
Colour	Brown	τ (RT)	24.5 MPa
Viscosity	800 - 1400 Pa.s	τ (°C)	-
S.G.	-	τ (wet)	-
Shelf life	-	Peel (T)	6 N mm^{-1}
Pot/Work life	-	σ_t	-
Tg	-	σ_c	-
$T_{Service}$	-	*E*	-
CTE (°C)	-	*G*	-
Thermal cond.	-	Elongation	-
Resistivity: Vol.	-	Tear strength	-
Surf.	-	H (Shore)	-
Notes: Shear strength and peel: Oily steel to oily steel.			

Sunstar Engineering	Penguin Cement 1065T		Paste
Epoxy - One part			
Manufacturers Comments: Good adhesion to oily steel.			
Typical Use: Automotive. Bonding of body panels.			
Cure: e.g. 20mins/170°C.			
Test Methods: Not stated		(Mar 95)	
Physical		**Mechanical**	
Colour	Brown	τ (RT)	24.5 MPa
Viscosity	250 - 600 Pa.s	τ (°C)	-
S.G.	-	τ (wet)	-
Shelf life	-	Peel (T)	6 N mm^{-1}
Pot/Work life	-	σ_t	-
Tg	-	σ_c	-
$T_{Service}$	-	*E*	-
CTE (°C)	-	*G*	-
Thermal cond.	-	Elongation	-
Resistivity: Vol.	-	Tear strength	-
Surf.	-	H (Shore)	-
Notes: Shear strength and peel: Oily steel to oily steel.			

See Key to Data Sheets - Pages 63 to 65

Sunstar Engineering	**Penguin Cement 1069**		**Paste**
Epoxy - One part			
Manufacturers Comments: Good adhesion to oily steel.			
Typical Use: Automotive. Bonding of body panels.			
Cure: e.g. 20mins/170°C.			
Test Methods: Not stated		(Mar 95)	
Physical		**Mechanical**	
Colour	Black	τ (RT)	13 MPa
Viscosity	1400 - 2000 Pa.s	τ (°C)	-
S.G.	-	τ (wet)	-
Shelf life	-	Peel (T)	6 N mm^{-1}
Pot/Work life	-	σ_t	-
Tg	-	σ_c	-
$T_{Service}$	-	E	-
CTE (°C)	-	G	-
Thermal cond.	-	Elongation	-
Resistivity: Vol.	-	Tear strength	-
Surf.	-	H (Shore)	-
Notes: Shear strength and peel: Oily steel to oily steel.			

Sunstar Engineering	**Penguin Cement 1081L**		**Paste**
Epoxy - One part			
Manufacturers Comments: Good adhesion to oily steel.			
Typical Use: Automotive. Bonding of body panels.			
Cure: e.g. 20mins/170°C.			
Test Methods: Not stated		(Mar 95)	
Physical		**Mechanical**	
Colour	Brown	τ (RT)	24.5 MPa
Viscosity	400 - 600 Pa.s	τ (°C)	-
S.G.	-	τ (wet)	-
Shelf life	-	Peel (T)	6 N mm^{-1}
Pot/Work life	-	σ_t	-
Tg	-	σ_c	-
$T_{Service}$	-	E	-
CTE (°C)	-	G	-
Thermal cond.	-	Elongation	-
Resistivity: Vol.	-	Tear strength	-
Surf.	-	H (Shore)	-
Notes: Shear strength and peel: Oily steel to oily steel.			

Sunstar Engineering	**Penguin Cement 1069B**		**Paste**
Epoxy - One part			
Manufacturers Comments: Good adhesion to oily steel.			
Typical Use: Automotive. Bonding of body panels.			
Cure: e.g. 20mins/170°C.			
Test Methods: Not stated		(Mar 95)	
Physical		**Mechanical**	
Colour	Black	τ (RT)	13 MPa
Viscosity	800 - 1200 Pa.s	τ (°C)	-
S.G.	-	τ (wet)	-
Shelf life	-	Peel (T)	6 N mm^{-1}
Pot/Work life	-	σ_t	-
Tg	-	σ_c	-
$T_{Service}$	-	E	-
CTE (°C)	-	G	-
Thermal cond.	-	Elongation	-
Resistivity: Vol.	-	Tear strength	-
Surf.	-	H (Shore)	-
Notes: Shear strength and peel: Oily steel to oily steel.			

Sunstar Engineering	**Penguin Cement 1086**		**Paste**
Epoxy - One part			
Manufacturers Comments: Good adhesion to oily steel.			
Typical Use: Automotive. Bonding of body panels.			
Cure: e.g. 20mins/170°C.			
Test Methods: Not stated		(Mar 95)	
Physical		**Mechanical**	
Colour	Brown	τ (RT)	24.5 MPa
Viscosity	400 - 600 Pa.s	τ (°C)	-
S.G.	-	τ (wet)	-
Shelf life	-	Peel (T)	8 N mm^{-1}
Pot/Work life	-	σ_t	-
Tg	-	σ_c	-
$T_{Service}$	-	E	-
CTE (°C)	-	G	-
Thermal cond.	-	Elongation	-
Resistivity: Vol.	-	Tear strength	-
Surf.	-	H (Shore)	-
Notes: Shear strength and peel: Oily steel to oily steel.			

TACC
(Transworld Adhesive and Chemical Corp.)

TACC	EK 2241		Paste
Epoxy - One-part, induction curing			
Manufacturers Comments: Non-sagging, induction curing, structural epoxy. Exceeds the requirements of (USA) MMM-A-132 Type 2. Excellent thermal stability and water/solvent resistance.			
Typical Use: Permanent magnet assembly. Metal-metal bonding and structural assembly.			
Cure: 45min/125°C or 30min/160°C or 15min/175°C (conventional heating).			
Test Methods: Not stated		October 93	
Physical		**Mechanical**	
Colour	Metallic grey	τ (RT)	21.8 MPa
Viscosity	700 - 900 Pa.s	τ (°C)	-
S.G.	1.35	τ (wet)	-
Shelf life	1 yr./5°C	Peel ( )	-
Pot/Work life	-	σ_t	47.6 MPa
Tg	-	σ_c	-
$T_{Service}$	-50°C to +150°C	E	-
CTE (RT)	62 x 10^{-6} °C^{-1}	G	-
Thermal cond.	-	Elongation	2%
Resistivity: Vol.	-	Tear strength	-
Surf.	-	H (Shore D)	84
Notes: Impact strength: 0.56 N.m/m.			

TACC	ER 1100		Paste
Epoxy - Two-part			
Manufacturers Comments: Fast-curing, high strength epoxy.			
Typical Use: General purpose.			
Cure: 5min/RT.			
Test Methods: Not stated		January 95	
Physical		**Mechanical**	
Colour	Transparent	τ (RT)	20.7 MPa
Viscosity	19 Pa.s	τ (°C)	-
S.G.	1.2	τ (wet)	-
Shelf life	-	Peel ( )	-
Pot/Work life	-	σ_t	-
Tg	-	σ_c	-
$T_{Service}$	-40°C to +105°C	E	-
CTE (°C)	-	G	-
Thermal cond.	-	Elongation	-
Resistivity: Vol.	-	Tear strength	-
Surf.	-	H (Shore D)	80
Notes:			

TACC	ER 1007		Paste
Epoxy - Two-part			
Manufacturers Comments: Chemical resistant, non-burning. Designed to reduce heat loss and protect lagging from fuel, oils, saltwater, etc.			
Typical Use: Superheaters, Insulated evaporators, Forced draft blowers, Turbine crossover pipes, Exhaust lines, steam lines, etc.			
Cure: 2hr/100°C or 5min/160°C.			
Test Methods: Not stated		June 93	
Physical		**Mechanical**	
Colour	Grey	τ (RT)	-
Viscosity	(paste)	τ (°C)	-
S.G.	-	τ (wet)	-
Shelf life	1 yr./RT	Peel ( )	-
Pot/Work life	4 hrs.	σ_t	-
Tg	-	σ_c	62 MPa
$T_{Service}$	-	E	-
CTE (°C)	-	G	-
Thermal cond.	-	Elongation	-
Resistivity: Vol.	-	Tear strength	-
Surf.	-	H (Shore D)	80
Notes: Flexural strength: 20.6 MPa.			

TACC	ER 1218		Paste
Epoxy - Two-part thixotropic			
Manufacturers Comments: RT cure. High tensile shear. Does not sag or drip. 100% solids.			
Typical Use: General purpose.			
Cure: 24hr/RT or 2hr/60°C.			
Test Methods: Not stated		June 94	
Physical		**Mechanical**	
Colour	Amber	τ (RT)	-
Viscosity	(paste)	τ (°C)	-
S.G.	1.2	τ (wet)	-
Shelf life	-	Peel ( )	-
Pot/Work life	35 - 40 mins/RT	σ_t	26.2 MPa
Tg	-	σ_c	14.1 MPa
$T_{Service}$	-	E	-
CTE (RT)	45 x 10^{-6} °C^{-1}	G	-
Thermal cond.	-	Elongation	-
Resistivity: Vol.	4 x 10^{14} Ω.cm	Tear strength	-
Surf.	-	H (Shore D)	75
Notes: Shrinkage: 0.2%. Dielectric strength: 40 V.mm^{-1}. Water absorption (240hr, 96% RH): 0.8%.			

See Key to Data Sheets - Pages 63 to 65

TACC	ER 1650		Paste
Epoxy - Two-part thixotropic			
Manufacturers Comments: RT cure. Good resistance to water, solvents, fuel oils and harsh chemicals.			
Typical Use: General purpose.			
Cure: RT, <100°C.			
Test Methods: Not stated		June 94	
Physical		**Mechanical**	
Colour	Grey	τ (RT)	-
Viscosity	(paste)	τ (°C)	-
S.G.	1.4	τ (wet)	-
Shelf life	-	Peel ( )	-
Pot/Work life	40 mins/RT	σ_t	26.2 MPa
Tg	-	σ_c	14.1 MPa
$T_{Service}$	-60°C to +160°C	*E*	-
CTE (°C)	-	*G*	-
Thermal cond.	-	Elongation	-
Resistivity: Vol.	-	Tear strength	-
Surf.	-	H (Shore D)	75
Notes: Dielectric strength: 19400 V.mm^{-1}.			

TACC	ER 3126		Paste
Epoxy - One-part			
Manufacturers Comments: Highly filled, structural epoxy. Excellent heat resistance and Class H (180°C) properties. Resistant to strong solvents, gasoline, freon, salt water and mild acids. High thermal conductivity. 100% solids.			
Typical Use: Continuous use at 180°C. Engine compartment applications in motor industry. Refrigeration control potting.			
Cure: 2hr/120°C or 1hr/150°C or 30min/175°C.			
Test Methods: Not stated		December 94	
Physical		**Mechanical**	
Colour	Black	τ (RT)	21.8 MPa
Viscosity	100 - 160 Pa.s	τ (°C)	-
S.G.	1.57	τ (wet)	-
Shelf life	4 mnth/RT; 1yr./0°C	Peel ( )	-
Pot/Work life	-	σ_t	44.1 MPa
Tg	-	σ_c	-
$T_{Service}$	-	*E*	-
CTE (RT)	70 x 10^{-6} °C^{-1}	*G*	-
Thermal cond.	1.7 W.m^{-1}K^{-1}	Elongation	-
Resistivity: Vol.	-	Tear strength	-
Surf.	-	H (Shore D)	86
Notes: Dielectric strength: 18000 V.mm^{-1}.			

TACC	ER 2241		Paste
Epoxy - One-part thixotropic paste			
Manufacturers Comments: Fast curing, aluminium filled, structural epoxy. Excellent thermal stability and high strength. Resistant to water, solvents and salt. No irritating vapours during rapid cure. No Exotherm. Contains no Butyl Glycidyl Ether. SPI rating: 2. 100% solids.			
Typical Use: Structural bonding: metal-metal and metal-other.			
Cure: 45min/125°C or 30min/160°C or 15min/175°C.			
Test Methods: Not stated		January 94	
Physical		**Mechanical**	
Colour	Metallic grey	τ (RT)	-
Viscosity	-	τ (°C)	-
S.G.	1.44	τ (wet)	-
Shelf life	1 yr./5°C	Peel ( )	-
Pot/Work life	-	σ_t	44.1 MPa
Tg	-	σ_c	-
$T_{Service}$	-50°C to +150°C	*E*	-
CTE (RT)	70 x 10^{-6} °C^{-1}	*G*	-
Thermal cond.	0.81 W.m^{-1}K^{-1}	Elongation	-
Resistivity: Vol.	-	Tear strength	-
Surf.	-	H (Shore D)	90
Notes:			

TACC	ER 4765		Paste
Epoxy - Two-part thixotropic liquid			
Manufacturers Comments: RT cure. For use as adhesive or semi-thixotropic potting compound. Very tough at low temperatures and excellent strength at 130°C. Good impact strength.			
Typical Use: Bonding and potting.			
Cure: 24 hr/RT or 1hr/65°C.			
Test Methods: Not stated		August 94	
Physical		**Mechanical**	
Colour	Black	τ (RT)	-
Viscosity	-	τ (°C)	-
S.G.	~1.1	τ (wet)	-
Shelf life	1 yr./RT	Peel ( )	-
Pot/Work life	-	σ_t	-
Tg	-	σ_c	-
$T_{Service}$	-	*E*	-
CTE (°C)	-	*G*	-
Thermal cond.	-	Elongation	-
Resistivity: Vol.	-	Tear strength	-
Surf.	-	H (Shore)	-
Notes: Wide range of other colours available.			

TACC	ER 7040	Paste	
Epoxy - One-part, glass bonding			
Manufacturers Comments: Designed specifically for glass-glass bonding in headlight assemblies. Low glass transition temperature.			
Typical Use: Glass-glass bonding.			
Cure: 1hr/120°C or 30min/150°C or 20min/160°C or 10min/175°C.			
Test Methods: Not stated		June 94	
Physical		**Mechanical**	
Colour	-	τ (RT)	-
Viscosity	-	τ (°C)	-
S.G.	-	τ (wet)	-
Shelf life	-	Peel ( )	-
Pot/Work life	-	σ_t	-
Tg	-50°C	σ_c	-
$T_{Service}$	-	*E*	-
CTE (°C)	-	*G*	-
Thermal cond.	-	Elongation	-
Resistivity: Vol.	-	Tear strength	-
Surf.	-	H (Shore D)	50
Notes: Volatiles (150°C): 1% max.			

TACC	ER 7093	Paste	
Epoxy - Two-part			
Manufacturers Comments: RT cure. 100% solids.			
Typical Use: -			
Cure: 7 days/RT.			
Test Methods: Not stated		June 94	
Physical		**Mechanical**	
Colour	Grey	τ (RT)	-
Viscosity	(paste)	τ (°C)	-
S.G.	-	τ (wet)	-
Shelf life	1 yr./RT	Peel ( )	-
Pot/Work life	-	σ_t	-
Tg	-	σ_c	-
$T_{Service}$	-	*E*	-
CTE (°C)	-	*G*	-
Thermal cond.	-	Elongation	-
Resistivity: Vol.	-	Tear strength	-
Surf.	-	H (Shore)	-
Notes:			

TACC	ER 7045	Paste	
Epoxy - One-part, thixotropic			
Manufacturers Comments: Designed for glass and plastics. Tough with good chemical and moisture resistance. 100% solids.			
Typical Use: Glass-glass/plastic bonding (headlight assemblies)			
Cure: 1hr/120°C or 30min/150°C or 20min/160°C or 10min/175°C.			
Test Methods: Not stated		June 94	
Physical		**Mechanical**	
Colour	Grey	τ (RT)	-
Viscosity	(non sag)	τ (°C)	-
S.G.	1.64	τ (wet)	-
Shelf life	1 yr./-18°C	Peel ( )	-
Pot/Work life	-	σ_t	-
Tg	-	σ_c	-
$T_{Service}$	-	*E*	-
CTE (°C)	-	*G*	-
Thermal cond.	-	Elongation	-
Resistivity: Vol.	-	Tear strength	-
Surf.	-	H (Shore)	-
Notes:			

TACC	ER 7124	Paste	
Epoxy - One-part			
Manufacturers Comments: Semi-rigid, moderate viscosity. Tough with good chemical and moisture resistance.			
Typical Use: Versatile general purpose adhesive for structural uses.			
Cure: 30min/125°C or 15min/150°C.			
Test Methods: Not stated		February 94	
Physical		**Mechanical**	
Colour	Green	τ (RT)	20.7 MPa
Viscosity	350 Pa.s	τ (°C)	-
S.G.	1.33	τ (wet)	-
Shelf life	1 yr./RT	Peel ( )	-
Pot/Work life	-	σ_t	-
Tg	-	σ_c	-
$T_{Service}$	-40°C to +155°C	*E*	-
CTE (°C)	-	*G*	-
Thermal cond.	-	Elongation	-
Resistivity: Vol.	-	Tear strength	-
Surf.	-	H (Shore D)	75
Notes:			

TACC	ER 7138		Paste
Epoxy - One-part			
Manufacturers Comments: High strength, very tough epoxy. Fast low temperature curing. Non-sag with good dielectric properties. 100% solids.			
Typical Use: -			
Cure: 1hr/100°C or 20min/120°C or 10min/160°C.			
Test Methods: Not stated		(95)	
Physical		**Mechanical**	
Colour	Light grey	τ (RT)	-
Viscosity	-	τ (°C)	-
S.G.	1.25	τ (wet)	-
Shelf life	1 yr./RT	Peel (-)	2.45 N mm^{-1}
Pot/Work life	-	σ_t	28.3 MPa
Tg	-	σ_c	-
$T_{Service}$	-	*E*	-
CTE (°C)	-	*G*	-
Thermal cond.	-	Elongation	-
Resistivity: Vol.	-	Tear strength	-
Surf.	-	H (Shore D)	75
Notes:			

TACC	ER 7821		Paste
Epoxy - One-part			
Manufacturers Comments: Self-levelling epoxy. Fast moderate temperature curing. Good temperature and chemical resistance.			
Typical Use: Bonding/sealing of filter materials, ceramics and metals.			
Cure: 2hr/100°C or 5min/160°C.			
Test Methods: Not stated		January 95	
Physical		**Mechanical**	
Colour	Beige	τ (RT)	-
Viscosity	20 Pa.s	τ (°C)	-
S.G.	1.49	τ (wet)	-
Shelf life	1 yr./RT	Peel ( )	-
Pot/Work life	-	σ_t	-
Tg	-	σ_c	-
$T_{Service}$	-40°C to +200°C	*E*	-
CTE (°C)	-	*G*	-
Thermal cond.	0.76 $W.m^{-1}K^{-1}$	Elongation	-
Resistivity: Vol.	4 x 10^{14} Ω.cm	Tear strength	-
Surf.	-	H (Shore D)	80
Notes: Dielectric strength: 172000 $V.mm^{-1}$, Dielectric constant (1 MHz): 4.3, Dissipation factor (1 MHz): 0.012.			

TACC	ER 7820		Paste
Epoxy - One-part			
Manufacturers Comments: Self-levelling epoxy. Fast moderate temperature curing. Good temperature and chemical resistance.			
Typical Use: Bonding/sealing of filter materials, ceramics and metals.			
Cure: 2hr/100°C or 15min/120°C or 5min/160°C or 2min/175°C.			
Test Methods: Not stated		June 94	
Physical		**Mechanical**	
Colour	Beige	τ (RT)	-
Viscosity	60 Pa.s	τ (°C)	-
S.G.	1.49	τ (wet)	-
Shelf life	1 yr./RT	Peel ( )	-
Pot/Work life	-	σ_t	-
Tg	-	σ_c	-
$T_{Service}$	-	*E*	-
CTE (°C)	-	*G*	-
Thermal cond.	-	Elongation	-
Resistivity: Vol.	-	Tear strength	-
Surf.	-	H (Shore D)	80
Notes:			

TACC	KR 2241		Paste
Epoxy - One-part, induction cure			
Manufacturers Comments: Fast induction curing, aluminium filled, structural epoxy. Excellent thermal stability and high strength in 120°C to 180°C range. Resistant to water, solvents and salt. Flexible and impact resistant. Exceeds requirements of (USA) MMM-A-132 type 2. 100% solids.			
Typical Use: Bonding of permanent magnets to metallic housing in electric motors. Bonds most magnet materials.			
Cure: <10s. cure with induction/dielectric process (2 - 3s. possible).			
Test Methods: Not stated		February 93	
Physical		**Mechanical**	
Colour	Metallic grey	τ (RT)	21.8 MPa
Viscosity	700 - 900 Pa.s	τ (°C)	-
S.G.	1.4	τ (wet)	-
Shelf life	3mnth/RT; 1yr./0°C	Peel ( )	-
Pot/Work life	-	σ_t	44.1 MPa
Tg	-	σ_c	-
$T_{Service}$	-	*E*	-
CTE (RT)	70 x 10^{-6} $°C^{-1}$	*G*	-
Thermal cond.	-	Elongation	-
Resistivity: Vol.	-	Tear strength	-
Surf.	-	H (Shore D)	78
Notes: Impact strength: 0.56 N.m/m.			

Teroson (Henkel)

Teroson	Terokal 3919 S		Paste
Epoxy - One part, modified epoxy			
Manufacturers Comments: Flexible, heat-curing paste with good temperature stability. Free from solvents and PVC with good sag resistance. Pre-gellable. 100% solids.			
Typical Use: Automotive (hem-flange). Bonding of body panels (together with spot-welding). Good adhesion to oily steel (2 - 3 g.m^{-2}).			
Cure: (Pre-gel: e.g. 6 - 10 sec at 170°C - induction heat). Final cure: 160 - 230°C, e.g. 35 - 90 mins/160°C, 12 - 25mins/230°C.			
Test Methods: DIN		May 93	
Physical		**Mechanical**	
Colour	Black	τ (RT)	~15 MPa
Viscosity	50 - 65 Pa.s	τ (°C)	-
S.G.	1.2	τ (wet)	-
Shelf life	3 months/10-20°C	Peel (T)	~30 N mm^{-1}
Pot/Work life	-	σ_t	-
Tg	-	σ_c	-
$T_{Service}$	-40 to +90°C	*E*	-
CTE (°C)	-	*G*	-
Thermal cond.	-	Elongation	-
Resistivity: Vol.	-	Tear strength	-
Surf.	-	H (Shore)	-
Notes: Shear strength: DIN 53283/DIN 1623. 'T' Peel: DIN 53282.			

Teroson	Terokal 5026		Paste
Epoxy - One part			
Manufacturers Comments: Flexible, heat-curing paste with good sag resistance. Free from solvents and PVC with good sag resistance. Pre-gellable. >99% solids.			
Typical Use: Automotive (hem-flange and structural bonds). Bonding of body panels (together with spot-welding). Good adhesion to oily steel (2 - 3 g.m^{-2}).			
Cure: (Pre-gel: e.g. 5 sec at 180°C - induction heat). Final cure: e.g. 30mins/180°C.			
Test Methods: DIN		Dec 92	
Physical		**Mechanical**	
Colour	Black	τ (RT, steel)	~20 MPa
Viscosity	65 Pa.s	τ (80°C, steel)	~15 MPa
S.G.	1.3	τ (-30°C)	~16 MPa
Shelf life	3 months/10-20°C	Peel ( )	-
Pot/Work life	-	σ_t	-
Tg	-	σ_c	-
$T_{Service}$	-40 to +90°C	*E*	-
CTE (°C)	-	*G*	-
Thermal cond.	-	Elongation	-
Resistivity: Vol.	-	Tear strength	-
Surf.	-	H (Shore D)	65 - 70
Notes: Shear strength: DIN 53283/DIN 1623.			

Teroson	Terokal 4520-34		Paste
Epoxy - One part			
Manufacturers Comments: Heat-curing paste. Free from solvents and PVC with good sag resistance. >99% solids.			
Typical Use: Automotive (hem-flange and structural bonds). Bonding of body panels (together with spot-welding). Good adhesion to oily steel (2 - 3 g.m^{-2}).			
Cure: e.g. 30 mins/180°C.			
Test Methods: DIN		Dec 92	
Physical		**Mechanical**	
Colour	Black	τ (RT, steel)	~20 MPa
Viscosity	65 Pa.s	τ (80°C, steel)	~15 MPa
S.G.	1.3	τ (-30°C)	~16 MPa
Shelf life	3 months/10-20°C	Peel ( )	-
Pot/Work life	-	σ_t	~38 MPa
Tg	-	σ_c	-
$T_{Service}$	-40 to +90°C	*E*	2700 MPa
CTE (°C)	-	*G*	530 MPa
Thermal cond.	-	Elongation	~13%
Resistivity: Vol.	-	Tear strength	-
Surf.	-	H (Shore D)	~64
Notes: Shear strength: DIN 53283/DIN 1623. Young's Modulus: DIN 53457. Shear modulus: DIN 54451.			

Teroson	Terokal 5027		Paste
Epoxy - One part			
Manufacturers Comments: Flexible, heat-curing paste with good sag resistance. Free from solvents and PVC with good sag resistance. Pre-gellable. >99% solids.			
Typical Use: Automotive (hem-flange and structural bonds). Bonding of body panels (together with spot-welding). Good adhesion to oily steel (2 - 3 g.m^{-2}).			
Cure: (Pre-gel: e.g. 5 sec at 180°C - induction heat). Final cure: e.g. 30mins/180°C.			
Test Methods: DIN		Dec 92	
Physical		**Mechanical**	
Colour	Black	τ (RT, steel)	~20 MPa
Viscosity	65 Pa.s	τ (80°C, steel)	~15 MPa
S.G.	1.4	τ (-30°C)	~20 MPa
Shelf life	3 months/10-20°C	Peel ( )	-
Pot/Work life	-	σ_t	-
Tg	-	σ_c	-
$T_{Service}$	-40 to +90°C	*E*	-
CTE (°C)	-	*G*	-
Thermal cond.	-	Elongation	-
Resistivity: Vol.	-	Tear strength	-
Surf.	-	H (Shore D)	72 - 78
Notes: Shear strength: DIN 53283/DIN 1623.			

Teroson | **Terokal 5045** | **Paste**

Epoxy - Two part, thixotropic (Experimental Product)

Manufacturers Comments: RT cure. Free from solvents. 100% solids.

Typical Use: Metal bonding (steel and aluminium).

Cure: 6hr/RT or e.g. 30mins/100°C.

Test Methods: DIN		Jan 94	
Physical		**Mechanical**	
Colour	Red brown	τ (RT, steel)	16 - 20 MPa
Viscosity	~100 Pa.s	τ (80°C, steel)	~15 MPa
S.G.	1.25	τ (-30°C)	~20 MPa
Shelf life	1 yr/10 - 20°C	Peel ( )	-
Pot/Work life	~30 mins/RT	σ_t	-
Tg	-	σ_c	-
$T_{Service}$	-40 to +80°C	E	~1300 MPa
CTE (°C)	-	G	-
Thermal cond.	-	Elongation	-
Resistivity: Vol.	-	Tear strength	-
Surf.	-	H (Shore D)	~70

Notes: Shear strength: DIN 53283. Young's modulus: DIN 53457.

Teroson | **Terokal 5811** | **Paste**

Epoxy - One part, PVC-modified epoxy

Manufacturers Comments: Flexible, tough, heat-curing paste. Free from solvents. Pre-gellable. >99% solids.

Typical Use: Automotive (hem-flange). Bonding of body panels. Good adhesion to oily steel.

Cure: (Pre-gel: e.g. 6 - 10 sec at 170°C - induction heat). Final cure: 160 - 230°C, e.g. 35 - 90 mins/160°C, 12 - 25mins/230°C.

Test Methods: DIN		May 93	
Physical		**Mechanical**	
Colour	Black	τ (RT)	~8.5 MPa
Viscosity	(paste)	τ (90°C)	~2.5 MPa
S.G.	1.5	τ (-40°C)	>13 MPa
Shelf life	3 months/15-20°C	Peel ( )	-
Pot/Work life	-	σ_t	~10 MPa
Tg	-	σ_c	-
$T_{Service}$	-40 to +80°C	E	-
CTE (°C)	-	G	-
Thermal cond.	-	Elongation	15%
Resistivity: Vol.	-	Tear strength	-
Surf.	-	H (Shore)	-

Notes: Shear strength: DIN 53283. Tensile: DIN 53455.

Teroson | **Terokal 5050** | **Paste**

Epoxy - Two part, thixotropic

Manufacturers Comments: RT cure. Free from solvents. 100% solids.

Typical Use: Metal, hard plastics, ceramics concrete and wood.

Cure: 7hr/RT or 45mins/75°C or 25mins/100°C or 5mins/150°C.

Test Methods: DIN		Dec 92	
Physical		**Mechanical**	
Colour	Light yellow	τ (RT)	25 - 30 MPa
Viscosity	~50 Pa.s	τ (°C)	-
S.G.	1.12	τ (wet)	-
Shelf life	1 yr/10 - 20°C	Peel ( )	-
Pot/Work life	~90 mins/RT	σ_t	-
Tg	-	σ_c	-
$T_{Service}$	-40 to +80°C	E	-
CTE (°C)	-	G	-
Thermal cond.	-	Elongation	-
Resistivity: Vol.	-	Tear strength	-
Surf.	-	H (Shore)	-

Notes: Shear strength: DIN 53283.

Teroson | **Terokal 6049** | **Paste**

Epoxy - One part epoxy

Manufacturers Comments: Flexible heat-curing paste with good sag resistance. Free from solvents and PVC. >99% solids.

Typical Use: Automotive (hem-flange). Bonding of body panels. Good adhesion to oily steel.

Cure: e.g. 30mins/180°C.

Test Methods: DIN		Sept 93	
Physical		**Mechanical**	
Colour	Yellow	τ (RT)	-
Viscosity	(paste)	τ (°C)	-
S.G.	1.67	τ (wet)	-
Shelf life	2 months/10 - 20°C	Peel ( )	-
Pot/Work life	-	σ_t	-
Tg	-	σ_c	-
$T_{Service}$	-40 to +90°C	E	-
CTE (°C)	-	G	-
Thermal cond.	-	Elongation	-
Resistivity: Vol.	-	Tear strength	-
Surf.	-	H (Shore)	-

Notes:

See Key to Data Sheets - Pages 63 to 65

Teroson	Terokal 8554		Paste
Epoxy - One part, PVC modified epoxy			
Manufacturers Comments: Heat-curing paste (170-220°C). Free from solvents with good sag resistance. Pre-gellable (>80°C). >99% solids.			
Typical Use: Automotive (hem-flange & structural bonds). Bonding of body panels (together with spot-welding). Good adhesion to oily steel.			
Cure: (Pre-gel: e.g. 10 sec/160°C, induction heat). Final cure: >15min/180°C or <15min/220°C.			
Test Methods: DIN		Mar 94	
Physical		**Mechanical**	
Colour	White	τ (RT, steel)	~5.4 MPa
Viscosity	(paste)	τ (80°C, steel)	~2.2 MPa
S.G.	~1.5	τ (-35°C)	>13 MPa
Shelf life	6 months/10 - 20°C	Peel (T)	4.8 N mm^{-1}
Pot/Work life	-	σ_t	~1 MPa
Tg	-	σ_c	-
$T_{Service}$	-40 to +80°C	*E*	-
CTE (°C)	-	*G*	-
Thermal cond.	-	Elongation	~45%
Resistivity: Vol.	-	Tear strength	-
Surf.	-	H (Shore)	-
Notes: Shear strength: DIN 53283. 'T' Peel: DIN 53282. Tensile: DIN 53455.			

Teroson	Terorehm 2000		Paste
Epoxy - One part, Hot-application, Heat-curing, Modified			
Manufacturers Comments: Hot applied (80-110°C), heat-curing (>170°C). Flexible, >99% solids. Semi-solid at RT, softens as temperature rises and cross-links (cures) at high temperature.			
Typical Use: Automotive. Body panel bonding. Adheres to oily steel.			
Cure: e.g. 30min/180°C.			
Test Methods: DIN		Oct 94	
Physical		**Mechanical**	
Colour	Black	τ (RT, steel)	~24 MPa
Viscosity	-	τ (90°C, steel)	~11 MPa
S.G.	~1.35	τ (-30°C)	~25 MPa
Shelf life	6 months/10 - 30°C	Peel ( )	-
Pot/Work life	-	σ_t	-
Tg	-	σ_c	-
$T_{Service}$	-40 to +90°C	*E*	-
CTE (°C)	-	*G*	-
Thermal cond.	-	Elongation	-
Resistivity: Vol.	-	Tear strength	-
Surf.	-	H (Shore)	-
Notes: Shear strength: DIN 53283.			

Tivoli

Teroson	Terokal 8554-3		Paste
Epoxy - One part, PVC modified epoxy			
Manufacturers Comments: Heat-curing paste (170-220°C). Free from solvents with good sag resistance. Pre-gellable (>80°C). >99% solids.			
Typical Use: Automotive (hem-flange & structural bonds). Bonding of body panels (together with spot-welding). Good adhesion to oily steel.			
Cure: (Pre-gel: e.g. 10 sec/160°C, induction heat). Final cure: ≥15min/180°C or ≤15min/190°C.			
Test Methods: DIN		Apr 94	
Physical		**Mechanical**	
Colour	Yellow	τ (RT, steel)	~11 MPa
Viscosity	(paste)	τ (80°C, steel)	~4.5 MPa
S.G.	~1.5	τ (-35°C)	>15 MPa
Shelf life	6 months/10 - 20°C	Peel ( )	-
Pot/Work life	-	σ_t	-
Tg	-	σ_c	-
$T_{Service}$	-40 to +80°C	*E*	-
CTE (°C)	-	*G*	-
Thermal cond.	-	Elongation	-
Resistivity: Vol.	-	Tear strength	-
Surf.	-	H (Shore)	-
Notes: Shear strength: DIN 53283.			

Tivoli	Tivo 9870/2		Paste
Epoxy - One-part, modified			
Manufacturers Comments: Flexible. Good solvent resistance. 100% solids.			
Typical Use: Automotive.			
Cure: 30min/180°C.			
Test Methods: DIN		December 93	
Physical		**Mechanical**	
Colour	Black	τ (RT, steel)	14 MPa
Viscosity	300 Pa.s	τ (°C)	-
S.G.	1.5	τ (wet)	-
Shelf life	3months/RT	Peel ( )	-
Pot/Work life	-	σ_t	-
Tg	-	σ_c	-
$T_{Service}$	-	*E*	-
CTE (°C)	-	*G*	-
Thermal cond.	-	Elongation	-
Resistivity: Vol.	-	Tear strength	-
Surf.	-	H (Shore)	-
Notes:			

UHU

Tivoli	Tivo 9870/3		Paste
Epoxy - One-part, modified			
Manufacturers Comments: Flexible. Good solvent resistance. 100% solids.			
Typical Use: Automotive.			
Cure: 30min/180°C.			
Test Methods: DIN		December 93	
Physical		**Mechanical**	
Colour	Black	τ (RT, steel)	14 MPa
Viscosity	500 Pa.s	τ (°C)	-
S.G.	1.5	τ (wet)	-
Shelf life	3months/RT	Peel ( )	-
Pot/Work life	-	σ_t	-
Tg	-	σ_c	-
$T_{Service}$	-	*E*	-
CTE (°C)	-	*G*	-
Thermal cond.	-	Elongation	-
Resistivity: Vol.	-	Tear strength	-
Surf.	-	H (Shore)	-
Notes:			

UHU	Plus		Paste
Two-part epoxy kit			
Manufacturers Comments: General purpose RT cure epoxy. Resistant to water, oils and domestic fluids. Strength: ~13 MPa between -60°C and +150°C.			
Typical Use: Bond and repair hard materials (domestic).			
Cure: RT/6 hrs.			
Test Methods: ASTM		February 94	
Physical		**Mechanical**	
Colour	Yellow	τ (RT)	-
Viscosity	~35 Pa.s	τ (°C)	-
S.G.	~1.1	τ (wet)	-
Shelf life	-	Peel ( )	-
Pot/Work life	60 mins	σ_t	-
Tg	-	σ_c	-
$T_{Service}$	-	*E*	-
CTE (°C)	-	*G*	-
Thermal cond.	-	Elongation	-
Resistivity: Vol.	-	Tear strength	-
Surf.	-	H (Shore)	-
Notes:			

Tremco

Tremco	ES 1122		Paste
Epoxy - Two-part, modified			
Manufacturers Comments: RT cure for bonding 'difficult' materials. Slightly flexible. Interior and exterior applications. Self-levelling. Good moisture and chemical resistance.			
Typical Use: Partitions and floor panels in trains and buses, structural laminated panels. Synthetic rubbers, mild and stainless steel sheeting, copper, aluminium, lead, ceramic and mosaic tiles and certain plastics including polycarbonate, PVC and ABS.			
Cure: 7 - 10hr/RT, <80°C, e.g. 20 - 30min/80°C.			
Test Methods: Not stated		Apr 91	
Physical		**Mechanical**	
Colour	Off white	τ (RT, steel)	8.3 - 10.3 MPa
Viscosity	100 - 180 Pa.s	τ (°C)	-
S.G.	-	τ (wet)	-
Shelf life	9 months/RT	Peel ( )	-
Pot/Work life	1hr/RT	σ_t	-
Tg	-	σ_c	-
$T_{Service}$	-20°C to +80°C	*E*	-
CTE (°C)	-	*G*	-
Thermal cond.	-	Elongation	-
Resistivity: Vol.	-	Tear strength	-
Surf.	-	H (Shore D)	80
Notes: Coverage: <4.5 m^2kg^{-1} (smooth), 1.5 - 1.8 m^2kg^{-1} (rough/uneven).			

UHU	Ultra Rapid		Paste
Two-part epoxy kit			
Manufacturers Comments: General purpose RT cure epoxy. Resistant to water, fuels and domestic fluids.			
Typical Use: General bond and repair (domestic).			
Cure: RT/5mins (initial cure); RT/24 hrs (full cure).			
Test Methods: ASTM		February 94	
Physical		**Mechanical**	
Colour	Transparent	τ (RT)	-
Viscosity	~30 Pa.s	τ (°C)	-
S.G.	~1.2	τ (wet)	-
Shelf life	-	Peel ( )	-
Pot/Work life	2 mins	σ_t	-
Tg	-	σ_c	-
$T_{Service}$	<100°C	*E*	-
CTE (°C)	-	*G*	-
Thermal cond.	-	Elongation	-
Resistivity: Vol.	-	Tear strength	-
Surf.	-	H (Shore)	-
Notes:			

Witco

Witco	Eurepox 710 + Euredur 10		Paste
Witco produce a range of epoxy resins and hardeners which can be used in a variety of combinations			
Manufacturers Comments: Low viscosity. Short pot life and cure.			
Typical Use:			
Cure: 8 hr/RT.			
Test Methods: DIN		(February 95)	
Physical		**Mechanical** (Steel, 7 days/RT)	
Colour	-	τ (RT)	12 MPa
Viscosity	7.5 Pa.s	τ (°C)	-
S.G.	1.1 - 1.2	τ (wet)	-
Shelf life	-	Peel (-)	2 - 4 N mm^{-1}
Pot/Work life	10 mins	σ_t	-
Tg	-	σ_c	-
$T_{Service}$	-	*E*	2000-4000 MPa
CTE (RT)	70 - 90 x 10^{-6} K^{-1}	*G*	-
Thermal cond.	0.1 - 0.2 W m.K^{-1}	Elongation	-
Resistivity: Vol.	10^{16} - 10^{17} Ω.cm	Tear strength	-
Surf.	10^{12} - 10^{13} Ω	H (Shore)	-
Notes: Tensile lap shear strength - DIN 53283, steel adherends, 7 days. cure at RT, Peel strength - DIN 53282 steel adherends, cure: 7 days at RT.			

Witco	Eurepox 710 + Euredur 26		Paste
Witco produce a range of epoxy resins and hardeners which can be used in a variety of combinations			
Manufacturers Comments: Low viscosity. Short to medium pot life and medium cure. Good higher temperature strength.			
Typical Use: Metals and hard plastics.			
Cure: 8 hr/RT or 20 min/80°C.			
Test Methods: DIN		(February 95)	
Physical		**Mechanical** (Steel, 30min/120°C)	
Colour	-	τ (RT)	21 MPa
Viscosity	0.9 Pa.s	τ (°C)	-
S.G.	1.1 - 1.2	τ (wet)	-
Shelf life	-	Peel (-)	2 - 4 N mm^{-1}
Pot/Work life	35 mins	σ_t	-
Tg	-	σ_c	-
$T_{Service}$	-	*E*	2000-4000 MPa
CTE (RT)	70 - 90 x 10^{-6} K^{-1}	*G*	-
Thermal cond.	0.1 - 0.2 W m.K^{-1}	Elongation	-
Resistivity: Vol.	10^{16} - 10^{17} Ω.cm	Tear strength	-
Surf.	10^{12} - 10^{13} Ω	H (Shore)	-
Notes: Tensile lap shear strength - DIN 53283, steel adherends, 30min. cure at 120°C, Peel strength - DIN 53282 steel adherends, cure: 7 days at RT.			

Witco	Eurepox 710 + Euredur 12		Paste
Witco produce a range of epoxy resins and hardeners which can be used in a variety of combinations			
Manufacturers Comments: Low viscosity. Short pot life and cure.			
Typical Use:			
Cure: 8 hr/RT.			
Test Methods: DIN		(February 95)	
Physical		**Mechanical** (Steel, 7 days/RT)	
Colour	-	τ (RT)	23 MPa
Viscosity	1.5 Pa.s	τ (°C)	-
S.G.	1.1 - 1.2	τ (wet)	-
Shelf life	-	Peel (-)	2 - 4 N mm^{-1}
Pot/Work life	12 mins	σ_t	-
Tg	-	σ_c	-
$T_{Service}$	-	*E*	2000-4000 MPa
CTE (RT)	70 - 90 x 10^{-6} K^{-1}	*G*	-
Thermal cond.	0.1 - 0.2 W m.K^{-1}	Elongation	-
Resistivity: Vol.	10^{16} - 10^{17} Ω.cm	Tear strength	-
Surf.	10^{12} - 10^{13} Ω	H (Shore)	-
Notes: Tensile lap shear strength - DIN 53283, steel adherends, 7 days. cure at RT, Peel strength - DIN 53282 steel adherends, cure: 7 days at RT.			

Witco	Eurepox 710 + Euredur 27		Paste
Witco produce a range of epoxy resins and hardeners which can be used in a variety of combinations			
Manufacturers Comments: Low viscosity. Short pot life and fast cure.			
Typical Use: Metals and hard plastics.			
Cure: 8 hr/RT or 1.5 hr/40°C.			
Test Methods: DIN		(February 95)	
Physical		**Mechanical** (Steel, 30min/120°C)	
Colour	-	τ (RT)	33 MPa
Viscosity	0.5 Pa.s	τ (°C)	-
S.G.	1.1 - 1.2	τ (wet)	-
Shelf life	-	Peel (-)	6 - 9 N mm^{-1}
Pot/Work life	70 mins	σ_t	-
Tg	-	σ_c	-
$T_{Service}$	-	*E*	2000-4000 MPa
CTE (RT)	70 - 90 x 10^{-6} K^{-1}	*G*	-
Thermal cond.	0.1 - 0.2 W m.K^{-1}	Elongation	-
Resistivity: Vol.	10^{16} - 10^{17} Ω.cm	Tear strength	-
Surf.	10^{12} - 10^{13} Ω	H (Shore)	-
Notes: Tensile lap shear strength - DIN 53283, steel adherends, 30min. cure at 120°C, Peel strength - DIN 53282 steel adherends, cure: 7 days at RT.			

See Key to Data Sheets - Pages 63 to 65

Witco	Eurepox 710 + Euredur 115		Paste
Witco produce a range of epoxy resins and hardeners which can be used in a variety of combinations			
Manufacturers Comments: High viscosity unfilled epoxy. Long pot life, slow cure.			
Typical Use: Metals and hard plastics.			
Cure: 2 - 3days/RT or 45min/80°C or 10min/150°C.			
Test Methods: DIN		(February 95)	
Physical		**Mechanical** (Steel, 30min/120°C)	
Colour	-	τ (RT)	30 MPa
Viscosity	130 Pa.s	τ (80°C)	~7 MPa
S.G.	1.1 - 1.2	τ (wet)	-
Shelf life	-	Peel	4 - 7 N mm^{-1}
Pot/Work life	160 mins	σ_f	-
Tg	-	σ_c	-
$T_{Service}$	~80°C	E	2000-4000 MPa
CTE (RT)	70 - 90 x 10^{-6} K^{-1}	G	-
Thermal cond.	0.1 - 0.2 W m.K^{-1}	Elongation	-
Resistivity: Vol.	10^{16} - 10^{17} Ω.cm	Tear strength	-
Surf.	10^{12} - 10^{13} Ω	H (Shore)	-
Notes: Tensile lap shear strength - DIN 53283, steel adherends, 30min. cure at 120°C, Peel strength - DIN 53282 steel adherends, cure: 7 days at RT.			

Witco	Eurepox 710 + Euredur 140		Paste
Witco produce a range of epoxy resins and hardeners which can be used in a variety of combinations			
Manufacturers Comments: Medium viscosity unfilled epoxy. Short pot life and fast cure.			
Typical Use: Metals and hard plastics.			
Cure: 8 hr/RT or 1.5 hr/40°C or 20min/60°C.			
Test Methods: DIN		(February 95)	
Physical		**Mechanical** (Steel, 30min/120°C)	
Colour	-	τ (RT)	30 MPa
Viscosity	12 Pa.s	τ (80°C)	~21 MPa
S.G.	1.1 - 1.2	τ (140°C)	~5 MPa
Shelf life	-	Peel (-)	4 - 7 N mm^{-1}
Pot/Work life	100 mins	σ_f	-
Tg	-	σ_c	-
$T_{Service}$	~130°C	E	2000-4000 MPa
CTE (RT)	70 - 90 x 10^{-6} K^{-1}	G	-
Thermal cond.	0.1 - 0.2 W m.K^{-1}	Elongation	-
Resistivity: Vol.	10^{16} - 10^{17} Ω.cm	Tear strength	-
Surf.	10^{12} - 10^{13} Ω	H (Shore)	-
Notes: Tensile lap shear strength - DIN 53283, steel adherends, 30min. cure at 120°C, Peel strength - DIN 53282 steel adherends, cure: 7 days at RT.			

Witco	Eurepox 710 + Euredur 125		Paste
Witco produce a range of epoxy resins and hardeners which can be used in a variety of combinations			
Manufacturers Comments: Medium viscosity unfilled epoxy. Medium pot life and cure.			
Typical Use: Metals and hard plastics.			
Cure: 2 days/RT or 20min/80°C or 5min/150°C.			
Test Methods: DIN		(February 95)	
Physical		**Mechanical** (Steel, 30min/120°C)	
Colour	-	τ (RT)	29 MPa
Viscosity	21 Pa.s	τ (90°C)	~7 MPa
S.G.	1.1 - 1.2	τ (100°C)	~4 MPa
Shelf life	-	Peel (-)	4 - 7 N mm^{-1}
Pot/Work life	100 mins	σ_f	-
Tg	-	σ_c	-
$T_{Service}$	~90°C	E	2000-4000 MPa
CTE (RT)	70 - 90 x 10^{-6} K^{-1}	G	-
Thermal cond.	0.1 - 0.2 W m.K^{-1}	Elongation	-
Resistivity: Vol.	10^{16} - 10^{17} Ω.cm	Tear strength	-
Surf.	10^{12} - 10^{13} Ω	H (Shore)	-
Notes: Tensile lap shear strength - DIN 53283, steel adherends, 30min. cure at 120°C, Peel strength - DIN 53282 steel adherends, cure: 7 days at RT.			

Witco	Eurepox 710 + Euredur 200		Paste
Witco produce a range of epoxy resins and hardeners which can be used in a variety of combinations			
Manufacturers Comments: Low viscosity. Short pot life and fast cure.			
Typical Use:			
Cure: 8 hr/RT.			
Test Methods: DIN		(February 95)	
Physical		**Mechanical** (Steel, 30min/120°C)	
Colour	-	τ (RT)	14 MPa
Viscosity	3.5 Pa.s	τ (°C)	-
S.G.	1.1 - 1.2	τ (wet)	-
Shelf life	-	Peel (-)	2 - 5 N mm^{-1}
Pot/Work life	35 mins	σ_f	-
Tg	-	σ_c	-
$T_{Service}$	-	E	2000-4000 MPa
CTE (RT)	70 - 90 x 10^{-6} K^{-1}	G	-
Thermal cond.	0.1 - 0.2 W m.K^{-1}	Elongation	-
Resistivity: Vol.	10^{16} - 10^{17} Ω.cm	Tear strength	-
Surf.	10^{12} - 10^{13} Ω	H (Shore)	-
Notes: Tensile lap shear strength - DIN 53283, steel adherends, 30min. cure at 120°C, Peel strength - DIN 53282 steel adherends, cure: 7 days at RT.			

See Key to Data Sheets - Pages 63 to 65

Witco	Eurepox 710 + Euredur 250		Paste
Witco produce a range of epoxy resins and hardeners which can be used in a variety of combinations			
Manufacturers Comments: Low viscosity. Medium pot life and cure.			
Typical Use:			
Cure: 16 hr/RT.			
Test Methods: DIN		(February 95)	
Physical		**Mechanical** (Steel, 30min/120°C)	
Colour	-	τ (RT)	28 MPa
Viscosity	3.0 Pa.s	τ (°C)	-
S.G.	1.1 - 1.2	τ (wet)	-
Shelf life	-	Peel (-)	4 - 6 N mm^{-1}
Pot/Work life	60 mins	σ_t	-
Tg	-	σ_c	-
$T_{Service}$	-	E	2000-4000 MPa
CTE (RT)	70 - 90 x 10^{-6} K^{-1}	G	-
Thermal cond.	0.1 - 0.2 W m.K^{-1}	Elongation	-
Resistivity: Vol.	10^{16} - 10^{17} Ω.cm	Tear strength	-
Surf.	10^{12} - 10^{13} Ω	H (Shore)	-
Notes: Tensile lap shear strength - DIN 53283, steel adherends, 30min. cure at 120°C, Peel strength - DIN 53282 steel adherends, cure: 7 days at RT.			

Witco	Eurepox 710 + Euredur 370		Paste
Witco produce a range of epoxy resins and hardeners which can be used in a variety of combinations			
Manufacturers Comments: Low viscosity. Medium pot life and cure.			
Typical Use:			
Cure: 24 hr/RT.			
Test Methods: DIN		(February 95)	
Physical		**Mechanical** (Steel, 30min/120°C)	
Colour	-	τ (RT)	27 MPa
Viscosity	2.5 Pa.s	τ (°C)	-
S.G.	1.1 - 1.2	τ (wet)	-
Shelf life	-	Peel (-)	2 - 5 N mm^{-1}
Pot/Work life	100 mins	σ_t	-
Tg	-	σ_c	-
$T_{Service}$	-	E	2000-4000 MPa
CTE (RT)	70 - 90 x 10^{-6} K^{-1}	G	-
Thermal cond.	0.1 - 0.2 W m.K^{-1}	Elongation	-
Resistivity: Vol.	10^{16} - 10^{17} Ω.cm	Tear strength	-
Surf.	10^{12} - 10^{13} Ω	H (Shore)	-
Notes: Tensile lap shear strength - DIN 53283, steel adherends, 30min. cure at 120°C, Peel strength - DIN 53282 steel adherends, cure: 7 days at RT.			

Witco	Eurepox 710 + Euredur 350		Paste
Witco produce a range of epoxy resins and hardeners which can be used in a variety of combinations			
Manufacturers Comments: Low viscosity. Long pot life and medium cure.			
Typical Use:			
Cure: 48 hr/RT.			
Test Methods: DIN		(February 95)	
Physical		**Mechanical** (Steel, 30min/120°C)	
Colour	-	τ (RT)	30 MPa
Viscosity	2.5 Pa.s	τ (°C)	-
S.G.	1.1 - 1.2	τ (wet)	-
Shelf life	-	Peel (-)	4 - 6 N mm^{-1}
Pot/Work life	360 mins	σ_t	-
Tg	-	σ_c	-
$T_{Service}$	-	E	2000-4000 MPa
CTE (RT)	70 - 90 x 10^{-6} K^{-1}	G	-
Thermal cond.	0.1 - 0.2 W m.K^{-1}	Elongation	-
Resistivity: Vol.	10^{16} - 10^{17} Ω.cm	Tear strength	-
Surf.	10^{12} - 10^{13} Ω	H (Shore)	-
Notes: Tensile lap shear strength - DIN 53283, steel adherends, 30min. cure at 120°C, Peel strength - DIN 53282 steel adherends, cure: 7 days at RT.			

Witco	Eurepox 783 + Euredur 115		Paste
Witco produce a range of epoxy resins and hardeners which can be used in a variety of combinations			
Manufacturers Comments:			
Typical Use:			
Cure: 4 days/RT.			
Test Methods: DIN		(February 95)	
Physical		**Mechanical** (Steel, 30min/120°C)	
Colour	-	τ (RT)	28 MPa
Viscosity	39 Pa.s	τ (°C)	-
S.G.	1.1 - 1.2	τ (wet)	-
Shelf life	-	Peel (-)	4 - 7 N mm^{-1}
Pot/Work life	190 mins	σ_t	-
Tg	-	σ_c	-
$T_{Service}$	-	E	2000-4000 MPa
CTE (RT)	70 - 90 x 10^{-6} K^{-1}	G	-
Thermal cond.	0.1 - 0.2 W m.K^{-1}	Elongation	-
Resistivity: Vol.	10^{16} - 10^{17} Ω.cm	Tear strength	-
Surf.	10^{12} - 10^{13} Ω	H (Shore)	-
Notes: Tensile lap shear strength - DIN 53283, steel adherends, 30min. cure at 120°C, Peel strength - DIN 53282 steel adherends, cure: 7 days at RT.			

See Key to Data Sheets - Pages 63 to 65

Witco	**Eurepox 783 + Euredur 125**		**Paste**
Witco produce a range of epoxy resins and hardeners which can be used in a variety of combinations			
Manufacturers Comments:			
Typical Use:			
Cure: 1.5 days/RT.			
Test Methods: DIN		(February 95)	
Physical		**Mechanical** (Steel, 30min/120°C)	
Colour	-	τ (RT)	31 MPa
Viscosity	6 Pa.s	τ (°C)	-
S.G.	1.1 - 1.2	τ (wet)	-
Shelf life	-	Peel (-)	4 - 7 N mm^{-1}
Pot/Work life	120 mins	σ_t	-
Tg	-	σ_c	-
$T_{Service}$	-	E	2000-4000 MPa
CTE (RT)	70 - 90 x 10^{-6} K^{-1}	G	-
Thermal cond.	0.1 - 0.2 W m.K^{-1}	Elongation	-
Resistivity: Vol.	10^{16} - 10^{17} Ω.cm	Tear strength	-
Surf.	10^{12} - 10^{13} Ω	H (Shore)	-
Notes: Tensile lap shear strength - DIN 53283, steel adherends, 30min. cure at 120°C, Peel strength - DIN 53282 steel adherends, cure: 7 days at RT.			

Witco	**Eurepox 783 + Euredur 200**		**Paste**
Witco produce a range of epoxy resins and hardeners which can be used in a variety of combinations			
Manufacturers Comments:			
Typical Use:			
Cure: 12 hr/RT.			
Test Methods: DIN		(February 95)	
Physical		**Mechanical** (Steel, 30min/120°C)	
Colour	-	τ (RT)	13 MPa
Viscosity	1.5 Pa.s	τ (°C)	-
S.G.	1.1 - 1.2	τ (wet)	-
Shelf life	-	Peel (-)	4 - 6 N mm^{-1}
Pot/Work life	40 mins	σ_t	-
Tg	-	σ_c	-
$T_{Service}$	-	E	2000-4000 MPa
CTE (RT)	70 - 90 x 10^{-6} K^{-1}	G	-
Thermal cond.	0.1 - 0.2 W m.K^{-1}	Elongation	-
Resistivity: Vol.	10^{16} - 10^{17} Ω.cm	Tear strength	-
Surf.	10^{12} - 10^{13} Ω	H (Shore)	-
Notes: Tensile lap shear strength - DIN 53283, steel adherends, 30min. cure at 120°C, Peel strength - DIN 53282 steel adherends, cure: 7 days at RT.			

Witco	**Eurepox 783 + Euredur 140**		**Paste**
Witco produce a range of epoxy resins and hardeners which can be used in a variety of combinations			
Manufacturers Comments:			
Typical Use:			
Cure: 1.5 days/RT.			
Test Methods: DIN		(February 95)	
Physical		**Mechanical** (Steel, 30min/120°C)	
Colour	-	τ (RT)	31 MPa
Viscosity	4 Pa.s	τ (°C)	-
S.G.	1.1 - 1.2	τ (wet)	-
Shelf life	-	Peel (-)	4 - 7 N mm^{-1}
Pot/Work life	120 mins	σ_t	-
Tg	-	σ_c	-
$T_{Service}$	-	E	2000-4000 MPa
CTE (RT)	70 - 90 x 10^{-6} K^{-1}	G	-
Thermal cond.	0.1 - 0.2 W m.K^{-1}	Elongation	-
Resistivity: Vol.	10^{16} - 10^{17} Ω.cm	Tear strength	-
Surf.	10^{12} - 10^{13} Ω	H (Shore)	-
Notes: Tensile lap shear strength - DIN 53283, steel adherends, 30min. cure at 120°C, Peel strength - DIN 53282 steel adherends, cure: 7 days at RT.			

Witco	**Eurepox 783 + Euredur 250**		**Paste**
Witco produce a range of epoxy resins and hardeners which can be used in a variety of combinations			
Manufacturers Comments:			
Typical Use:			
Cure: 24 hr/RT.			
Test Methods: DIN		(February 95)	
Physical		**Mechanical** (Steel, 30min/120°C)	
Colour	-	τ (RT)	27 MPa
Viscosity	0.8 Pa.s	τ (°C)	-
S.G.	1.1 - 1.2	τ (wet)	-
Shelf life	-	Peel (-)	4 - 6 N mm^{-1}
Pot/Work life	70 mins	σ_t	-
Tg	-	σ_c	-
$T_{Service}$	-	E	2000-4000 MPa
CTE (RT)	70 - 90 x 10^{-6} K^{-1}	G	-
Thermal cond.	0.1 - 0.2 W m.K^{-1}	Elongation	-
Resistivity: Vol.	10^{16} - 10^{17} Ω.cm	Tear strength	-
Surf.	10^{12} - 10^{13} Ω	H (Shore)	-
Notes: Tensile lap shear strength - DIN 53283, steel adherends, 30min. cure at 120°C, Peel strength - DIN 53282 steel adherends, cure: 7 days at RT.			

Witco	Eurepox 783 + Euredur 350		Paste
Witco produce a range of epoxy resins and hardeners which can be used in a variety of combinations			
Manufacturers Comments:			
Typical Use:			
Cure: 3 days/RT.			
Test Methods: DIN		(February 95)	
Physical		**Mechanical** (Steel, 30min/120°C)	
Colour	-	τ (RT)	30 MPa
Viscosity	0.5 Pa.s	τ (°C)	-
S.G.	1.1 - 1.2	τ (wet)	-
Shelf life	-	Peel (-)	2 - 5 N mm^{-1}
Pot/Work life	540 mins	σ_f	-
Tg	-	σ_c	-
$T_{Service}$	-	E	2000-4000 MPa
CTE (RT)	70 - 90 x 10^{-6} K^{-1}	G	-
Thermal cond.	0.1 - 0.2 W m.K^{-1}	Elongation	-
Resistivity: Vol.	10^{16} - 10^{17} Ω.cm	Tear strength	-
Surf.	10^{12} - 10^{13} Ω	H (Shore)	-
Notes: Tensile lap shear strength - DIN 53283, steel adherends, 30min. cure at 120°C, Peel strength - DIN 53282 steel adherends, cure: 7 days at RT.			

Witco	Eurepox 783 + Euredur 370		Paste
Witco produce a range of epoxy resins and hardeners which can be used in a variety of combinations			
Manufacturers Comments:			
Typical Use:			
Cure: 1.5 days/RT.			
Test Methods: DIN		(February 95)	
Physical		**Mechanical** (Steel, 30min/120°C)	
Colour	-	τ (RT)	29 MPa
Viscosity	0.6 Pa.s	τ (°C)	-
S.G.	1.1 - 1.2	τ (wet)	-
Shelf life	-	Peel (-)	2 - 5 N mm^{-1}
Pot/Work life	110 mins	σ_f	-
Tg	-	σ_c	-
$T_{Service}$	-	E	2000-4000 MPa
CTE (RT)	70 - 90 x 10^{-6} K^{-1}	G	-
Thermal cond.	0.1 - 0.2 W m.K^{-1}	Elongation	-
Resistivity: Vol.	10^{16} - 10^{17} Ω.cm	Tear strength	-
Surf.	10^{12} - 10^{13} Ω	H (Shore)	-
Notes: Tensile lap shear strength - DIN 53283, steel adherends, 30min. cure at 120°C, Peel strength - DIN 53282 steel adherends, cure: 7 days at RT.			

IMIDE-BASED ADHESIVES

NOTES

This section contains data sheets for bismaleimide and polyimide adhesives. Both classes have been developed for high-temperature applications which exceed the capabilities of epoxy-based polymers.

With the exception of the Grace 'Amicon' adhesive (a silver-loaded, conductive product for electronics assembly) all are aimed at applications primarily, but not exclusively, in the aerospace industries. Their use generally involves the bonding of high-performance, fibre-reinforced composites having a bismaleimide or polyimide resin matrix. The adhesive processing is often matched to that of the matrix resin to allow co-curing (both the bond and composite are consolidated and cured at the same time).

Secondary bonding (assembly) of metals is also possible. The metal adherends are normally those with moderate to high-temperature capabilities, such as titanium alloys, either as thin sheets or honeycomb cores. Foaming versions are designed for honeycomb structural panel assembly and edge-closures.

Currently there are few commercially available products, their development is on-going and evaluation studies are being conducted in the aerospace and other industry sectors.

COMPARISON TABLE - IMIDE-BASED ADHESIVES

Company	Adhesive	Base	Type	Colour	LSS	SG	Visc (Pa.s)	Service (°C)	Tg (°C)	Cure	Pg. No.
Ciba Polymers	Redux 326	Mod. BMI	F	Red/brown	◑	-	-	>200	-	175 + 245	237
Cytec	FM 55	PI (a)	F	Tan	○	-	-	260	299	177 + 310	237
	FM 57	PI (c)	F	Brown	●	-	-	-55/+288	298	177 + 288	237
	FM 680-1	PI (c)	F	Dk. Brown	●	-	-	371	342	176 + 371	237
	FM 680-2	PI (c)	F foam	Dk. Brown	-	0.4	-	316	-	177 + 400	238
	Metlbond 2550G	Mod. BMI	F	-	●	-	-	250	250	177 + 277	238
	Metlbond 6604-1	Mod. BMI	F foam	-	-	0.45	-	177	-	177 + 225	238
Dexter Hysol	EA 9369	BMI	1	Yellow	○	1.17	1500	290	-	177 + 246	238
	EA 9673	Mod. BMI	F	-	○	-	-	230	-	175 + 245	239
	XEA 9351 MB	BMI	1	Yellow	-	0.6	-	-	-	177 + 246	239
	XEA 9833	Mod. BMI	F foam	-	-	-	-	<230	-	175 + 245	239
Grace	Amicon C 940-4	PI	1	Metallic	-	2.9	68	200	-	280	239
Monsanto	Skybond 700	PI	1	-	-	1.17	4.5	370	-	-	240
	Skybond 703	PI	1	-	-	1.17	4.5	340	-	-	240
	Skybond 705	PI	1	-	-	1.07	1.8	-	-	-	240

See Key to Comparison Tables - Pages 63 to 65

Ciba Polymers

Ciba Polymers	Redux 326		Film
Modified bismaleimide			
Manufacturers Comments: High temperature (>200°C) requirements.			
Typical Use: Metal and composite bonding.			
Cure: 175°C/2 hrs/0.2-0.7 MPa + Post cure: 245°C/2hr (free-standing).			
Test Methods: Not stated		October 91	
Physical		**Mechanical**	
Colour	Red/Brown	τ (RT)	18 MPa
Viscosity (170°C)	0.0005 Pa.s	τ (220°C)	18 MPa
SG	-	τ (wet)	19 MPa
Shelf life	6 months/-18°C	Peel (CD)	5.92 N mm^{-1}
Pot/Work life	3 weeks/RT	σ_t	-
Tg	-	σ_c	-
$T_{Service}$	-	E	-
CTE (°C)	-	G	-
Thermal cond.	-	Elongation	-
Resistivity: Vol.	-	Tear strength	-
Surf.	-	H (Shore)	-
Notes: Areal wt. available: 0.2 kg/m², 0.5 kg/m². τ(wet) 48hr/water boil/ RT test. τ 2024-T3, clad, CSA etched to DTD 915b(ii); Peel - Climbing Drum - composite skin/metal core, RT.			

Cytec	FM 57		Film
Polyimide (condensation cure) Supported Areal wt. 0.5 kg/m²			
Manufacturers Comments: High service temperature applications. MDA-free. Radar transparent. Repairs.			
Typical Use: Aerospace - metals (with BR57 primer) and composites.			
Cure: 177°C/90 mins/0.3 MPa + Post-cure: 288°C/ 2 hrs.			
Test Methods: Not stated		June 94	
Physical		**Mechanical**	
Colour	Brown	τ (RT)	24.8 MPa
Viscosity	-	τ (288°C)	11.4 MPa
SG	-	τ (wet)	-
Shelf life	6 months/-18°C	Peel (S/wich)	32.6 Nm.m^{-1}
Pot/Work life	15 days/24°C	σ_t	-
Tg	298°C	σ_c	-
$T_{Service}$	-55 to +288°C	E	-
CTE (°C)	-	G	-
Thermal cond.	-	Elongation	-
Resistivity: Vol.	-	Tear strength	-
Surf.	-	H (Shore)	-
Notes: τ Ti-6Al-4V. Peel - sandwich Al-alloy 2024-T3, FPL etch.			

Cytec

Cytec	FM 55		Film
Polyimide (addition cure) Supported Areal wt. 0.5 kg/m²			
Manufacturers Comments: High service temperature applications. PMR-15 chemistry. Contains <1% free MDA. Co-cures with PMR-15 resin composites.			
Typical Use: Aerospace - metals (with BR55 primer) and composites.			
Cure: 177°C/2hrs/0.3 MPa applied at 204°C + Post-cure: 310°C/ 2 hrs.			
Test Methods: Not stated		1993	
Physical		**Mechanical**	
Colour	Tan	τ (RT)	15.9 MPa
Viscosity	-	τ (260°C)	16.2 MPa
SG	-	τ (wet)	15.3 MPa
Shelf life	6 months/-18°C	Peel ( )	-
Pot/Work life	7 days/RT (in bag)	σ_t	-
Tg	299°C	σ_c	-
$T_{Service}$	+260°C- continuous	E	-
CTE (°C)	-	G	-
Thermal cond.	-	Elongation	-
Resistivity: Vol.	-	Tear strength	-
Surf.	-	H (Shore)	-
Notes: τ Ti-6Al-4V, 1.27 mm, sand blasted & solvent-wipe. τ(wet) 30 days, 71°C,100%RH. Test temp. 24°C. Similar performance to FM 35-1 film.			

Cytec	FM 680-1		Film
Polyimide (condensation cure) Supported Areal wt. 0.5 kg/m²			
Manufacturers Comments: High service temperature applications. Thermal oxidation stability. Radar transparent.			
Typical Use: Aerospace - metals (with BRX680 primer): Ti-alloys, Stainless steel. Composites: Co-cure & secondary bonding.			
Cure: 176-316°C/ 2 hrs/0.7 MPa + Post-cure: 316-371°C/ 2-8 hrs.			
Test Methods: Not stated		May 90	
Physical		**Mechanical**	
Colour	Dark Brown	τ (RT)	20.7 MPa
Viscosity	-	τ (371°C)	8.3 MPa
SG	-	τ (wet)	11.2 MPa
Shelf life	1 year/-18°C	Peel ( )	-
Pot/Work life	30 days/24°C	σ_t	-
Tg	342°C	σ_c	-
$T_{Service}$	+371°C	E	-
CTE (°C)	-	G	-
Thermal cond.	-	Elongation	-
Resistivity: Vol.	-	Tear strength	-
Surf.	-	H (Shore)	-
Notes: τ Ti-6Al-4V. τ(wet) 30 days, 71°C, 100%RH.			

Cytec	FM 680-2		Film
Polyimide (condensation cure), foaming Areal wt. 1.5 kg/m²			
Manufacturers Comments: High service temperature applications. No metallic filler, radar transparent. Co-curable with 177°C composites.			
Typical Use: Aerospace - Composites. Core splicing and edge sealing honeycomb panels.			
Cure: 177°C/ 2 hrs; Post-cure: 400°C/ 2 hrs (free-standing).			
Test Methods: Not stated		Nov 93	
Physical		**Mechanical**	
Colour	Dark Brown	τ (RT) Tube	1.7 MPa
Viscosity	-	τ (316°C)	1.4 MPa
SG	~0.4 (cured)	τ (wet)	-
Shelf life		Peel ( )	-
Pot/Work life		σ_t	-
Tg		σ_c	-
$T_{Service}$	+316°C	*E*	-
CTE (°C)	-	*G*	-
Thermal cond.	-	Elongation	-
Resistivity: Vol.	-	Tear strength	-
Surf.	-	H (Shore)	-
Notes: Film thickness: 1.27 mm. Expansion ratio: 1-1.5x thickness. τ tube shear.			

Cytec	Metlbond 6604-1		Film
Modified bismaleimide, foaming (unsupported)			
Manufacturers Comments: Low density, high service temperature. Co-cures with most BMI prepregs. Radar transparent.			
Typical Use: Aerospace - core splicing, edge sealing honeycomb panels.			
Cure: 177°C/4 hrs + Post cure: 225°C/6 hrs.			
Test Methods: Not stated		(May 95)	
Physical		**Mechanical**	
Colour	-	τ (RT)	-
Viscosity	-	τ (°C)	-
SG	~0.3 - 0.6 (cured)	τ (wet)	-
Shelf life	-	Peel ( )	-
Pot/Work life	-	σ_t	-
Tg		σ_c	-
$T_{Service}$	+177°C	*E*	-
CTE (°C)	-	*G*	-
Thermal cond.	-	Elongation	-
Resistivity: Vol.	-	Tear strength	-
Surf.	-	H (Shore)	-
Notes: ex BASF product. Expansion ratio 1.7 - 2.3 x thickness.			

Cytec	Metlbond 2550G		Film
Modified bismaleimide (supported)*			
Manufacturers Comments: High temperature requirements. Tough and high strength.			
Typical Use: Aerospace - Metal and composites (co-cure and secondary bonding).			
Cure: 177°C/4 hrs/0.2 MPa + Post cure: 277°C/6 hrs (free-standing).			
Test Methods: ASTM		November 91	
Physical		**Mechanical**	
Colour	-	τ (RT)	20.7 MPa
Viscosity	-	τ (232°C)	15.9 MPa
SG	-	τ (wet)	-
Shelf life	6 mth/-18°C	Peel (S/wich)	1233 Nm.m^{-1}
Pot/Work life	21 days/24°C	σ_t	-
Tg	250°C	σ_c	-
$T_{Service}$ °C	+250(dry);+191(wet)	*E*	-
CTE (°C)	-	*G*	-
Thermal cond.	-	Elongation	-
Resistivity: Vol.	-	Tear strength	-
Surf.	-	H (Shore)	-
Notes: ex BASF product. * glass support, unsupported version available. Other areal wts. available. Thickness: 0.5, other thickness' available. τ ASTM D-1002 Al alloy 2024-T3, FPL etch. Peel - Sandwich.			

Dexter Hysol

Hysol	EA 9369		Paste
Modified Bismaleimide - One part			
Manufacturers Comments: One part, thixotropic paste. Structural properties up to 290°C. No outgassing. Subject to minimum order quantity.			
Typical Use: Sufficiently thixotropic for use in potting and edge filling. May be handled and trowelled at room temperature.			
Cure: 177°C/1 hr (at 0.17 MPa/25 psi pressure) + Post cure 246°C/2 hr non-pressurised.			
Test Methods: ASTM		(1994)	
Physical		**Mechanical**	
Colour	Yellow	τ (RT)	15.2 MPa
Viscosity	1500 Pa.s	τ (°C)	10.3 MPa
SG	1.17	τ (wet)	-
Shelf life	1 yr/-18°C	Peel ( )	-
Pot/Work life	30 days/RT	σ_t	-
Tg	-	σ_c	92.4 MPa
$T_{Service}$	290°C	*E*	-
CTE (RT)	5.4 x 10^{-5} °C^{-1}	*G*	-
Thermal cond.	-	Elongation	-
Resistivity: Vol.	-	Tear strength	-
Surf.	-	H (Shore)	-
Notes: Tensile lap shear strength & Service temperature - ASTM D 1002, Compressive properties - ASTM D 695. CTE - Perkin Elmer TMA.			

Hysol	EA 9673		Film
Modified bismaleimide, glass fabric supported (0.490 kg.m^{-2})			
Manufacturers Comments: Good strength to 290°C. Moisture resistant. No volatiles during cure.			
Typical Use: Honeycomb constructions. Continuous service at 230°C.			
Cure: 175°C/1 hr, 0.17 - 0.34 MPa (25 - 50 psi) + Post cure 245°C/2hr. Step cure of 1 - 1.5 hr/120°C at low pressure (0.03 - 0.1 MPa) + 1hr/175°C at full pressure will reduce adhesive flow.			
Test Methods: ASTM		August 92	
Physical		**Mechanical**	
Colour	-	τ (RT)	13.8 MPa
Viscosity	-	τ (230°C)	15.9 MPa
SG	-	τ (315°C)	4.1 MPa
Shelf life	>6 mth/-20°C	Peel (CD)	(12.0 N.m m^{-1})
Pot/Work life	30 days/RT.	σ_t (flatwise)	(4.1 MPa)
Tg	-	σ_c	-
$T_{Service}$	230°C	*E*	-
CTE (°C)	-	*G*	-
Thermal cond.	-	Elongation	-
Resistivity: Vol.	-	Tear strength	-
Surf.	-	H (Shore)	-
Notes: Tensile lap shear strength & Service temperature - ASTM D 1002, Tensile properties - ASTM D 638. Peel - Honeycomb Climbing Drum Peel - 6.4mm cell. Flatwise tensile strength - 6.4mm cell.			

Hysol	XEA 9833		Film
Modified bismaleimide, foaming (1.25 or 2.5 mm thick)			
Manufacturers Comments: Structural intumescent foaming core-splice adhesive for use up to 230°C. Expansion ratio: x2 - x3 at 175°C.			
Typical Use: Honeycomb core splices.			
Cure: 175°C/1 hr + Post cure 245°C/2 hr.			
Test Methods: -		August 92	
Physical		**Mechanical**	
Colour	-	τ (RT, tube)	(5.5 MPa)
Viscosity	-	τ (260°C, tube)	(3.8 MPa)
SG	-	τ (wet)	-
Shelf life	>6 mth/5°C	Peel ( )	-
Pot/Work life	-	σ_t	-
Tg	-	σ_c	-
$T_{Service}$	<230°C	*E*	-
CTE (40°C)	-	*G*	-
Thermal cond.	-	Elongation	-
Resistivity: Vol.	-	Tear strength	-
Surf.	-	H (Shore)	-
Notes: Tube shear strength to Boeing BMS 5-90F.			

Hysol	XEA 9351 MB		Paste
Bismaleimide - One part			
Manufacturers Comments: Low density paste with 24.5% glass microspheres. Excellent high temperature properties, no outgassing. Preliminary data sheet. Subject to minimum order quantity.			
Typical Use: Sufficiently thixotropic for use in potting and edge filling. Best handled, and may be trowelled, at 50°C.			
Cure: 177°C/1hr (at 0.17 MPa/25 psi pressure) + Post cure 246°C/2hr non-pressurised.			
Test Methods: ASTM		March 94	
Physical		**Mechanical**	
Colour	Yellow	τ (RT)	-
Viscosity	Putty	τ (°C)	-
SG	0.6	τ (wet)	-
Shelf life	1 yr/-18°C	Peel ( )	-
Pot/Work life	30 days/RT.	σ_t	-
Tg	-	σ_c	-
$T_{Service}$	-	*E*	-
CTE (°C)	-	*G*	-
Thermal cond.	-	Elongation	-
Resistivity: Vol.	-	Tear strength	-
Surf.	-	H (Shore)	-
Notes: Tensile lap shear strength & Service temperature - ASTM D 1002, Tensile properties - ASTM D 638.			

Grace Speciality Polymers

Grace	Amicon C 940-4		Paste
Polyimide - one part			
Manufacturers Comments: Silver-filled conductive adhesive (82-84% silver). High bond strengths. High temperature electronic uses. Low extractable corrosive ion content. Meets: MIL-883 C Method 2019.			
Typical Use: Electronics assemblies. Die attachment - plastic & ceramic.			
Cure: 160-180°C/ 10-40 mins (drying); 260-280°C/ 10-40 mins (cure).			
Test Methods: ASTM		(Aug 94)	
Physical		**Mechanical**	
Colour	Silver	τ (RT)	-
Viscosity	56-80 Pa.s	τ (°C)	-
SG	~2.8-3.0	τ (wet)	-
Shelf life	6 month/0°C	Peel ( )	-
Pot/Work life	-	σ_t	-
Tg	-	σ_c	-
$T_{Service}$	+200°C (+400°C)	*E*	-
CTE (°C)	-	*G*	-
Thermal cond.	-	Elongation	-
Resistivity: Vol.	1x10^{-4} max. Ωcm	Tear strength	-
Surf.	-	H (Shore)	-
Notes: Service temp. data: continuous (intermittent). Viscosity: ASTM D-2393. Resistivity: ASTM D-257.			

See Key to Data Sheets - Pages 63 to 65

Monsanto

Monsanto	Skybond 700		Liquid
Polyimide - one part			
Manufacturers Comments: Heat curing aromatic polyimide for applications up to 370°C. 45 - 48% Solids.			
Typical Use: -			
Cure: -			
Test Methods: Not Stated		(Apr 95)	
Physical		**Mechanical**	
Colour	-	τ (RT)	-
Viscosity	3 - 7 Pa.s	τ (°C)	-
SG	1.15 - 1.18	τ (wet)	-
Shelf life	3 month, 5°C	Peel ( )	-
Pot/Work life	-	σ_t	-
Tg	-	σ_c	-
$T_{Service}$	<370°C	E	-
CTE (°C)	-	G	-
Thermal cond.	-	Elongation	-
Resistivity: Vol.	-	Tear strength	-
Surf.	-	H (Shore)	-
Notes:			

Monsanto	Skybond 705		Liquid
Polyimide - one part			
Manufacturers Comments: Heat curing aromatic polyimide for high temperature applications. 16.5 - 19.5% Solids.			
Typical Use: -			
Cure: -			
Test Methods: Not Stated		(Apr 95)	
Physical		**Mechanical**	
Colour	-	τ (RT)	-
Viscosity	1.1 - 2.6 Pa.s	τ (°C)	-
SG	1.05 - 1.08	τ (wet)	-
Shelf life	3 month, 5°C	Peel ( )	-
Pot/Work life	-	σ_t	-
Tg	-	σ_c	-
$T_{Service}$	-	E	-
CTE (°C)	-	G	-
Thermal cond.	-	Elongation	-
Resistivity: Vol.	-	Tear strength	-
Surf.	-	H (Shore)	-
Notes:			

Monsanto	Skybond 703		Liquid
Polyimide - one part			
Manufacturers Comments: Heat curing aromatic polyimide for applications up to 340°C. 48.5 - 51.5% Solids.			
Typical Use: -			
Cure: -			
Test Methods: Not Stated		(Apr 95)	
Physical		**Mechanical**	
Colour	-	τ (RT)	-
Viscosity	3 - 7 Pa.s	τ (°C)	-
SG	1.15 - 1.18	τ (wet)	-
Shelf life	3 month, 5°C	Peel ( )	-
Pot/Work life	-	σ_t	-
Tg	-	σ_c	-
$T_{Service}$	<340°C	E	-
CTE (°C)	-	G	-
Thermal cond.	-	Elongation	-
Resistivity: Vol.	-	Tear strength	-
Surf.	-	H (Shore)	-
Notes:			

PHENOLIC ADHESIVES

NOTES

The phenolic family of adhesives are very diverse in their formulations and uses. Some are filled or modified with other polymers (vinyl, nitrile or epoxy). The composition depends largely on the intended use, e.g. for temperature and chemical resistance, or for water-proof wood bonding (phenol-resorcinol-formaldehyde). The majority are heat-curing although some wood-bonding adhesives can be cured at room temperature (RT). The comparison table includes 'Typical Use' to differentiate between the types of adhesives. Phenolic adhesives are generally poorly represented by mechanical property data.

Main uses

Depends on formulation:

- Friction materials (clutch and break linings).
- Metal bonding in hostile environments (temperature and chemicals).
- Wood bonding (plywood manufacture, heavy-duty wood structures, e.g. construction of wood frame buildings and structural members for boats).

COMPARISON TABLE - PHENOLIC ADHESIVES

Company	Adhesive	Type	Typical Use	Colour	LSS	SG	Visc (Pa.s)	Service (°C)	Pot (mins)	Cure	Pg. No.
Ciba Polymers	Redux 755	F	Metal	-	●	-	-	<70	-	150	243
	Redux 755 L/P	2	Metal	-	●	-	-	<70	-	150	243
Cytec	HT 424	F	Metals & honeycombs	Grey	●	-	-	<150	(12 days)	107 - 170	243
BF Goodrich	Plastilock 605-1	-	Friction materials. Metal.	Black	◑	0.85	-	-	-	-	243
	Plastilock 605-4	-	Friction materials. Metal.	Black	◑	0.91	5	-	-	-	244
	Plastilock 605-15	-	Friction materials. Metal.	Black	◑	0.95	15	-	-	-	244
	Plastilock 605-35	-	Friction materials. Metal.	Black	◑	0.95	30	-	-	-	244
	Plastilock 605-51	-	Friction materials. Metal.	Black	◑	0.97	43	-	-	-	244
	Aqualock 6000 (AL-6000)	-	Friction materials. Metal.	Black	●	1.04	6	>260	-	150 - 220	245
	Aqualock 6002 (EXP605W)	-	Friction materials. Metal.	Black	●	1.06	16	-	-	150 - 220	245
	Aqualock 6700 (AL-6700)	-	Friction materials. Metal.	Black	○	1.06	0.45	-	-	190 - 205	245
Huntsman Chemical Co	Lauxite PP-1597	-	Plywood manufacture	Dk. Red	-	-	0.33	-	120	135 - 155	245
	Lauxite PR-148	-	Wood. Boat, building units	Dk. Red	-	-	2	-	165	RT	246
	Lauxite PR-900	-	Wood. Boat, building units	Dk. Red	-	-	2.8	-	150	RT - 127	246

See Key to Comparison Tables - Pages 63 to 65

Ciba Polymers

Ciba Polymers	Redux 755		Nov 90
Vinyl-phenolic Film (unsupported)* Areal wt. 0.37 kg/m²			
Manufacturers Comments: Structural bonding in hostile environments. Good for large bond areas, horizontal surfaces.			
Typical Use: Metals (with Redux 101 primer).			
Cure: 150°C /30mins/0.4- 0.7 MPa. (conditions may be varied).			
Test Methods: MMM-A-132			
Physical		**Mechanical**	
Colour	-	τ (RT)	30 MPa
Viscosity	-	τ (82°C)	15 MPa
SG	-	τ (wet)	-
Shelf life	6 months/0°C	Peel (-)	8.20 N mm^{-1}
Pot/Work life	-	σ_f	-
Tg	-	Elongation	-
$T_{Service}$	+70°C max.	H (Shore)	-
Notes: * Redux 755RN supported film version. Areal wt. 0.37 kg/m². Peel - RT, metal/metal.			

Ciba Polymers	Redux 755 L/P		Nov 90
Two part - phenolic (liquid) + vinyl (powder)			
Manufacturers Comments: Structural bonding in hostile environments. Can be used on non-horizontal surfaces.			
Typical Use: Metals (with Redux 101 primer).			
Cure: 150°C /30mins/0.4- 0.7 MPa. (conditions may be varied).			
Test Methods: MMM-A-132			
Physical		**Mechanical**	
Colour	-	τ (RT)	30 MPa
Viscosity	-	τ (82°C)	12 MPa
SG	-	τ (wet)	-
Shelf life	6 months/RT	Peel (-)	6.00 N mm^{-1}
Pot/Work life	-	σ_f	-
Tg	-	Elongation	-
$T_{Service}$	+70°C max.	H (Shore)	-
Notes: Mix ratio 1:1 to 1: 2. Peel - RT, metal/metal.			

Cytec

Cytec	HT 424		Nov 82
Phenolic/mod. epoxy Film (supported)* Areal wt. 0.66 kg/m²			
Manufacturers Comments: High service temperatures. Good thermal resistance. X-ray opaque. Meets: MMM-A-132, Type II & Type III; MIL-A-25463, Type II & Type III.			
Typical Use: Aerospace - metals & honeycombs (with HT424 primer).			
Cure: (107-170°C) 170°C /80mins/0.3 MPa.			
Test Methods: MMM-A-132			
Physical		**Mechanical**	
Colour	Grey	τ (RT)	24.5 MPa
Viscosity	-	τ (150°C)	19.0 MPa
SG	-	τ (wet)	18.0 MPa
Shelf life	6 months/-18°C	Peel (CD)	1.93 N mm^{-1}
Pot/Work life	12 days/<24°C	σ_f	-
Tg	181°C	Elongation	-
$T_{Service}$	+150°C	H (Shore)	-
Notes: * unsupported & 2-part paste versions. Also without Al-filler (electrical transparency). Other areal wts. available. T service +260°C, short-term. Peel -climbing drum (no primer). τ 2024-T3; τ(wet) 30days, tap water, 24°C.			

BF Goodrich

BF Goodrich	Plastilock 605-1		Jul 92
Nitrile phenolic - solvent-based - thixotropic			
Manufacturers Comments: Black, heat curing adhesive with excellent resistance to thermal shock, chemicals and water. High thermal stability. 15% min. total solids - Roll coat, spray and brush grade.			
Typical Use: Brake, clutch and other friction materials. Metal bonding.			
Cure: Elevated temperature after 24hr dry or 30 min/85°C prebake. 0.5 - 1.0 MPa (75 - 150 psi) pressure during cure.			
Test Methods: See notes			
Physical		**Mechanical**	
Colour	Black	τ (RT, disc)	25.9 MPa
Viscosity	-	τ (260°C)	1.7 MPa
SG	0.85 (uncured)	τ (wet)	-
Shelf life	12 months/4°C	Peel ( )	-
Pot/Work life	-	σ_f	-
Tg	-	Elongation	-
$T_{Service}$	-	H (Shore)	-
Notes: Shear test: BFG ADTD #8/SAE J840C button and bar disc shear. Calculated coverage: 5 m²/litre. Thinners: Methyl Ethyl Ketone. Tack time (0.5mm/RT): 4 min.			

BF Goodrich	Plastilock 605-4		Jul 92
Nitrile phenolic - solvent-based - thixotropic			
Manufacturers Comments: Black, heat curing adhesive with excellent resistance to thermal shock, chemicals and water. High thermal stability. 29% min. total solids - Roll coat, spray and brush grade.			
Typical Use: Brake, clutch and other friction materials. Metal bonding.			
Cure: Elevated temperature after 24hr dry or 30 min/85°C prebake. 0.5 - 1.0 MPa (75 - 150 psi) pressure during cure.			
Test Methods: See notes			
Physical		**Mechanical**	
Colour	Black	τ (RT, disc)	25.9 MPa
Viscosity	4 - 6 Pa.s	τ (260°C)	1.7 MPa
SG	0.91 (uncured)	τ (wet)	-
Shelf life	12 months/4°C	Peel ( )	-
Pot/Work life	-	σ_f	-
Tg	-	Elongation	-
$T_{Service}$	-	H (Shore)	-
Notes: Shear test: BFG ADTD #8/SAE J840C button and bar disc shear. Calculated coverage: 11.5 m²/litre. Thinners: Methyl Ethyl Ketone. Tack time (0.5mm/RT): 6.5 min.			

BF Goodrich	Plastilock 605-35		Jul 92
Nitrile phenolic - solvent-based - thixotropic			
Manufacturers Comments: Black, heat curing adhesive with excellent resistance to thermal shock, chemicals and water. High thermal stability. 40% min. total solids - Extrusion grade.			
Typical Use: Brake, clutch and other friction materials. Metal bonding.			
Cure: Elevated temperature after 24hr dry or 30 min/85°C prebake. 0.5 - 1.0 MPa (75 - 150 psi) pressure during cure.			
Test Methods: See notes			
Physical		**Mechanical**	
Colour	Black	τ (RT, disc)	25.9 MPa
Viscosity	30 Pa.s	τ (260°C)	1.7 MPa
SG	0.95 (uncured)	τ (wet)	
Shelf life	12 months/4°C	Peel ( )	
Pot/Work life	-	σ_f	
Tg	-	Elongation	
$T_{Service}$		H (Shore)	
Notes: Shear test: BFG ADTD #8/SAE J840C button and bar disc shear. Calculated coverage: 13 m²/litre. Thinners: Methyl Ethyl Ketone. Tack time (0.5mm/RT): 9.5 min.			

BF Goodrich	Plastilock 605-15		Jul 92
Nitrile phenolic - solvent-based - thixotropic			
Manufacturers Comments: Black, heat curing adhesive with excellent resistance to thermal shock, chemicals and water. High thermal stability. 32% min. total solids - Special brush, roll coat grade.			
Typical Use: Brake, clutch and other friction materials. Metal bonding.			
Cure: Elevated temperature after 24hr dry or 30 min/85°C prebake. 0.5 - 1.0 MPa (75 - 150 psi) pressure during cure.			
Test Methods: See notes			
Physical		**Mechanical**	
Colour	Black	τ (RT, disc)	25.9 MPa
Viscosity	14 - 16 Pa.s	τ (260°C)	1.7 MPa
SG	0.94 (uncured)	τ (wet)	
Shelf life	12 months/4°C	Peel ( )	
Pot/Work life	-	σ_f	
Tg	-	Elongation	
$T_{Service}$		H (Shore)	
Notes: Shear test: BFG ADTD #8/SAE J840C button and bar disc shear. Calculated coverage: 11.5 m²/litre. Thinners: Methyl Ethyl Ketone. Tack time (0.5mm/RT): 5 min.			

BF Goodrich	Plastilock 605-51		Jul 92
Nitrile phenolic - solvent-based - thixotropic			
Manufacturers Comments: Black, heat curing adhesive with excellent resistance to thermal shock, chemicals and water. High thermal stability. 40% min. total solids - Extrusion grade.			
Typical Use: Brake, clutch and other friction materials. Metal bonding.			
Cure: Elevated temperature after 24hr dry or 30 min/85°C prebake. 0.5 - 1.0 MPa (75 - 150 psi) pressure during cure.			
Test Methods: See notes			
Physical		**Mechanical**	
Colour	Black	τ (RT, disc)	25.9 MPa
Viscosity	43 Pa.s	τ (260°C)	1.7 MPa
SG	0.97 (uncured)	τ (wet)	-
Shelf life	12 months/4°C	Peel ( )	-
Pot/Work life	-	σ_f	-
Tg	-	Elongation	-
$T_{Service}$	-	H (Shore)	-
Notes: Shear test: BFG ADTD #8/SAE J840C button and bar disc shear. Calculated coverage: 13.8 m²/litre. Thinners: Methyl Ethyl Ketone. Tack time (0.5mm/RT): 4 min.			

BF Goodrich	Aqualock 6000 (AL-6000)		Mar 93
Nitrile phenolic - water-based - thermosetting			
Manufacturers Comments: Heat curing, water-based adhesive with excellent resistance to thermal shock, chemicals and water. High thermal stability. Very low volatile content. 27% total solids.			
Typical Use: Brake, clutch and other friction materials. Steel bonding. Roll coating, spraying, silk screening and brushing.			
Cure: 5 - 30mins/150°C - 220°C, apply pressure during cure.			
Test Methods: See notes			
Physical		**Mechanical**	
Colour	Black	τ (RT, disc)	30.3 MPa
Viscosity	5 - 7 Pa.s	τ (260°C)	4.8 MPa
SG	1.04 (uncured)	τ (wet)	-
Shelf life	TBD/4 - 16°C	Peel ( )	-
Pot/Work life	-	σ_t	-
Tg	-	Elongation	-
$T_{Service}$	>260°C	H (Shore)	-
Notes: Shear test: SAE J840C button and bar disc shear. Calculated coverage: 5 m²/litre.			

BF Goodrich	Aqualock 6700 (AL-6700)		Mar 93
Nitrile phenolic - water-based - thermosetting			
Manufacturers Comments: Heat curing, water-based adhesive with good thermal stability. Resistant to solvents, oils and water. High bond strength. 29% total solids.			
Typical Use: Bonding metals and non-metallic compositions to metallic substrates. Airless spray, curtain coating and silk screening.			
Cure: 24 - 30mins/190°C - 205°C, 14 - 28 MPa (200 - 400 psi).			
Test Methods: See notes			
Physical		**Mechanical**	
Colour	Black	τ (RT, disc)	19 MPa
Viscosity	0.3 - 0.6 Pa.s	τ (205°C)	3.4 MPa
SG	1.055 (uncured)	τ (wet)	-
Shelf life	12 months/4 - 21°C	Peel ( )	-
Pot/Work life	-	σ_t	-
Tg	-	Elongation	-
$T_{Service}$	-	H (Shore)	-
Notes: Shear test: SAE J840C button and bar disc shear. Calculated coverage: ~11 m²/litre. Tack time (0.5mm/RT): 48 min.			

BF Goodrich	Aqualock 6002 (EXP605W)		Mar 93
Nitrile phenolic - water-based - thermosetting			
Manufacturers Comments: Heat curing, water-based adhesive with excellent resistance to thermal shock, chemicals and water. High thermal stability. Very low volatile content. Good melt flow during cure. Gap filling. 26% total solids.			
Typical Use: Brake, clutch and other friction materials. Steel bonding. Roll coating, spraying, silk screening and brushing.			
Cure: 5 - 30mins/150°C - 220°C, apply pressure during cure.			
Test Methods: See notes			
Physical		**Mechanical**	
Colour	Black	τ (RT, disc)	30.3 MPa
Viscosity	14 - 18 Pa.s	τ (260°C)	4.8 MPa
SG	1.06 (uncured)	τ (wet)	-
Shelf life	TBD/4 - 21°C	Peel ( )	-
Pot/Work life	-	σ_t	-
Tg	-	Elongation	-
$T_{Service}$	-	H (Shore)	-
Notes: Shear test: SAE J840C button and bar disc shear. Calculated coverage: 10.5 m²/litre.			

Huntsman Chemical Company

Huntsman	Lauxite PP-1597		Aug 85
Phenolic resorcinol-formaldehyde (filled)			
Manufacturers Comments: RT curing. 40-42.5% solids. Contains Macadamia NSF/Wheatflour. For use in hot, dry conditions to avoid dry-out. Will meet 72 hr boil. For hot-pressing. Retains tack in hot weather.			
Typical Use: Plywood manufacture.			
Cure: 135-155°C/12mins (assembly), 1-2 hrs (closed),1.2 MPa (press time varies with thickness).			
Test Methods: Not stated			
Physical		**Mechanical**	
Colour	Amber - Dark Red	τ (RT)	-
Viscosity	0.28-0.38 Pa.s	τ (°C)	-
SG	-	τ (wet)	-
Shelf life	1 yr/25°C	Peel ( )	-
Pot/Work life	2-4 hrs/24°C	σ_t	-
Tg	-	Elongation	-
$T_{Service}$	-	H (Shore)	-
Notes: Sunshine Gel time (100°C): 28-40 mins. Total alkalinity: 6.0-6.5%. Veneer moisture content: 0-8% (5-7% preferred).			

See Key to Data Sheets - Pages 63 to 65

Huntsman	Lauxite PR-148		Apr 91
Phenolic resorcinol-formaldehyde			
Manufacturers Comments: RT curing. 50-54% solids. Water & boil-proof. Weather resistant, salt water resistance. Fungus proof. For hard & soft wood assemblies. Also laminating, bag moulding veneering.			
Typical Use: Boat-building (laminated keels, ribs, arches, beams, spars). Prefabricated structural building units.			
Cure: RT/6 hrs (initial cure), 6 days (full cure). Pressure: 1-1.3MPa.			
Test Methods: Not stated			
Physical		**Mechanical**	
Colour	Amber - Dark Red	τ (RT)	-
Viscosity	1.5-2.5 Pa.s	τ (°C)	-
SG	-	τ (wet)	-
Shelf life	1 yr/25°C	Peel ( )	-
Pot/Work life	165 min/25°C*	σ_t	-
Tg	-	Elongation	-
$T_{Service}$	-	H (Shore)	-
Notes: * working temperature <30°C recommended. Gel time (43.5°C): 25-55 mins. Assembly time ~1hr/20°C. Wood moisture content: 8-14%. Low pressure for softwoods.			

Huntsman	Lauxite PR-900		Aug 85
Phenolic resorcinol-formaldehyde			
Manufacturers Comments: RT curing. 54-57% solids. Water & boil-proof. Weather resistant, salt water resistance. Fungus proof. For hard & soft wood assemblies. Also laminating, bag moulding veneering.			
Typical Use: Boat-building (laminated keels, ribs, arches, beams, spars). Prefabricated structural building units.			
Cure: RT/6 hrs (initial cure), 6 days (full cure). Pressure: 1.1-1.4MPa Example hot assembly press times: 60°C/15 mins; 127°C/12 seconds.			
Test Methods: Not stated			
Physical		**Mechanical**	
Colour	Amber - Dark Red	τ (RT)	-
Viscosity	2.6-3.1 Pa.s	τ (°C)	-
SG	-	τ (wet)	-
Shelf life	1 yr/25°C	Peel ( )	-
Pot/Work life	150 min/24°C*	σ_t	-
Tg	-	Elongation	-
$T_{Service}$	-	H (Shore)	-
Notes: * working temperature <30°C recommended. Gel time (100°C): ~1-2 mins. Wood moisture content: 8-14%. Low pressure for soft-woods.			

POLYURETHANE ADHESIVES

NOTES

Polyurethane adhesives fall into two main groups with different curing mechanisms:

- One-part adhesives are generally moisture curing.
- Two-part adhesives are resin and hardener mixes.

Moisture curing varieties may require water misting before assembly (for large bond areas). Most one- and two-part varieties are room temperature (RT) curing, although heating may be used to accelerate the cure. Few are solely heat-cured.

Some of the one-part polyurethanes are supplied as solids and require melting before application. These then moisture cure on cooling (often over several days). These are variously referred to by manufacturers as hot-melt, reaction hot-melt, warm-melt or melt-applied.

Some adhesives have recommended primers for certain materials (e.g. galvanised or anodised metals).

There are a wide range of viscosities - from thick pastes, to very thin liquids. Some are thixotropic. The very high viscosity versions are used as adhesive/sealants.

Main Uses

- Building and construction: Assembly of structural panels (wood, metal, plastics and insulation materials), laminated panels, assembly of glazing systems, high-duty flooring, adhesive/sealants for woodwork and masonry.
- Automotive: vehicle body work and trim assembly (metals, plastics, composites and painted surfaces); assembly of filters.
- Boat-building: assembly and sealing.
- Wood and wood-products: laminating with metals and plastics, joints.
- Footwear industry: assembly and sole attachment.

COMPARISON TABLE - POLYURETHANE ADHESIVES

Company	Adhesive	Type	Colour	LSS	TS	TM	Hs	El. (%)	SG	Visc (Pa.s)	Service (°C)	Usable (mins)	Cure	Pg. No.
3E	Andex 950	1m	-	○	-	-	85D	-	-	M	-	-	RT	252
	Polarjoint - Super	2	Grey	-	◕	-	78A	-	1.05	-	<93	-	RT	252
	Polarjoint - T	2	Black	-	○	-	46A	-	0.95	-	<93	-	RT	252
3M	3532 B/A	2	Brown	◕	-	-	-	-	-	30	-	7	RT	252
	3535 B/A	2	Off-white	◕	-	-	-	-	-	30	-	3	RT	253
	3549 B/A	2	Brown	◕	-	-	-	-	-	30	-	60	RT	253
	DP-605 NS	2	Off-white	◕	-	-	-	-	-	M	-	4	RT	253
	TE-030	1m*	Off-white	-	●	○	60D	725	-	17	-	1	121 + RT	253
	TE-031	1m*	Off-white	-	●	○	50D	-	-	13	-	2	121 + RT	254
	TE-100	1m*	Off-white	-	●	○	-	-	-	7	-	2	121 + RT	254
	TE-230	1m*	Off-white	-	◕	○	-	-	-	9	-	4	121 + RT	254
Apollo	Astrolok A 77	1m	-	-	-	-	-	-	-	0.2	-40/+140	20	RT	254
	Astrolok A 84	1m	-	-	-	-	-	-	-	0.35	-40/+120	1	RT	255
	Astrolok A 7166	1m	Brown	-	-	-	-	-	-	0.13	-	1	RT	255
	Astrolok A 7504	1m	Brown	-	-	-	-	-	-	4	-	55	RT	255
	Astrolok A 7515	1m	Lt. Brown	-	-	-	-	-	-	13	-	10	RT	255
	Astrolok A 7521	1m	Lt. Brown	-	-	-	-	-	-	4	-	40	RT	256
	Astrolok AX 9106	2	Brown	-	-	-	-	-	-	-	-	60	RT	256
Ashland Chemicals	Isogrip 3030D	1m	Grey	-	-	-	-	-	1.1	6	-	25	RT	256
	Isogrip WD3-A312	2	White	-	-	-	-	-	1.2	3.25	-	180	RT/RF + Heat	256
	Isogrip WD3-A320	2	White	-	-	-	-	-	1.1	10	-	180	RT/RF + Heat	257
	Isogrip WD3-A322	2	White	-	-	-	-	-	1.1	6.25	-	180	RT/RF + Heat	257
	Isogrip WD3-CM402	2	White	-	-	-	-	-	1.1	3.25	-	120	RT/RF + Heat	257
	Pliogrip 6600/6610	2	Green	-	◕	◕	60D	55	1.2	30	-	5	RT - 110	257
	Pliogrip 6600/6612	2	Green	-	◕	◕	60D	55	1.2	30	-	2	RT - 110	258
	Pliogrip 6600/6620	2	Green	-	◕	◕	60D	55	1.2	30	-	15	RT - 121	258
	Pliogrip 7000/7020	2	Green	○	◕	◕	65D	45	1.2	18	-	15	149	258
	Pliogrip 7000/7021	2	Green	○	◕	◕	65D	45	1.2	18	-	30	149	258
	Pliogrip 7000/7024	2	Green	○	◕	◕	65D	45	1.2	18	-	25	149	259
	Pliogrip 7300/7305	2	Green	○	◕	◕	-	46	1.25	24	-	4	RT - 120	259
	Pliogrip 7300/7308	2	Green	○	◕	◕	-	46	1.25	24	-	9	RT - 148	259
	Pliogrip 7400/7406C	2	Green	○	●	◕	69D	70	1.3	20	-	20	100 - 149	259
	Pliogrip 7400/7410	2	Green	○	●	◕	69D	70	1.3	20	-	10	100 - 149	260
	Pliogrip 7400/7420	2	Green	○	◕	○	69D	120	1.3	15	-	12	135 - 149	260
Bostik	2639	1m	Various	-		○	41A	>250	1.2	-	-5/+35		RT	260
	Matrix 500	1m	-	-	○	-	35A	>700	-	H	-40/+90	60	RT	260
	Matrix 600FC	1m	Various	-	○	-	-	>600	-	H	-40/+90	45	RT	261
	Matrix 700 Marine	1m	-	-	○	-	-	>600	-	H	-40/+90	45	RT	261
	Parcol	1	Lt. Beige	-	-	-	-	-	1.3	-	-5/+35	120	RT	261
	PU 505	2	Bk. Beige	-	-	-	-	-	1.2	100	<150	60	RT	261
Wm. Canning	Hylomar Polyurethane Sealant	1	Various	-	○	○	40A	450	-	H	-25/+80	-	RT	262
Casco Nobel	1801 PUR 2K	2	Lt. Brown	-	-	-	90D	-	1.8	14	-	90	RT	262
	1829 PUR 1K	1m	Brown	-	-	-	-	-	1.5	10	-	10	RT	262
	1833 PUR 1K	1m	Brown	-	-	-	-	-	1.6	15	-	45	RT	262
	1846 PUR 2K	2	White	-	-	-	80D	-	1.5	8.5	-	20	RT	263
	1894 PUR 1K	1m	Brown	-	-	-	-	-	1.1	11	-	45	RT	263
	1897 PUR 2K	2	White	-	-	-	80D	-	1.5	550	-	90	RT	263
	1899 PUR 2K	2	Lt. Brown	-	-	-	85D	-	1.5	8	-	135	RT	263
	1959 PUR 2K	2	Red	-	◕	●	83D	4.3	1.5	4	-	20	RT	264
CECA	Colle 455	2	Yellow	-	-	-	-	-	1.58	25	<120	60	RT	264
	Mastic Colle	1m	Various	-	-	-	-	-	1.2	H	-30/+80	120	5 -40	264

Continued...

Company	Adhesive	Type	Colour	LSS	TS	TM	Hs	El. (%)	SG	Visc (Pa.s)	Service (°C)	Usable (mins)	Cure	Pg. No.
CFPI	Galfix A 835	2	Lt. Green	◑	-	○	55D	25	1.34	H	-	15	RT - 80	264
	Galfix A 838	2	Dk. Grey	○	-	-	60D	25	1.75	H	-	30	RT	265
	Lidec A 223 D	2	Lt. Green	○	-	-	60A	150	1.42	H	-40/+100	10	RT	265
	Lidec A 250 D	2	Green	○	-	○	80A	60	1.24	H	-40/+100	5	RT - 80	265
Chryso	Tectane 1000	1m	Various	○	-	-	-	-	-	H	-40/+120	-	RT	265
	Tectane 1020	1m	-	○	-	-	-	-	-	H	-40/+120	20	RT	266
Ciba Polymers	Araldite 2018	2	Off-white	◑	-	-	-	35	1.13	8	-55/+80	40	RT - 60	266
	Uralane 5774 A/B	2	Beige	◑	-	-	60D		1.15	M	-55/+80	20	RT - 95	266
Conap	Conastic AD-20	2	Amber	-	◑	-	90A	500	0.99	5.5	-55/+80	15	RT - 80	266
Degussa	Agomet PU 540 AB	2	-	●	-	-	65D	-	1.3	20	-30/+80	40	RT	267
Dunlop	HM 4200	1m*	Straw	-	-	-	-	-	-	12.5	<100	3	110-120+RT	267
	HM 4201	1m*	Straw	-	-	-	-	-	-	45	-	30	150-160+RT	267
	HM 4202	1m*	Straw	-	-	-	-	-	-	10	<100	2	110-130+RT	267
Dyckerhoff	PU-Kleber #503	2	-	-	-	-	-	-	-	-	-	45	RT	268
EMFI	Isole 3100 Mono	1m	Brown	-	-	-	-	-	1.1	11	-	-	RT	268
	Isole S 35/521	2	Brown	-	-	-	-	-	1.45	350	-70/+150	50	RT	268
	Mastic PU 15	1m	Various	-	-	○	15A	>250	1.2	H	-30/+80	15	RT	268
	Mastic PU 25 Bâtiment	1m	Various	-	-	○	25A	>250	1.2	H	-30/+80	15	RT	269
	Mastic PU 40 Bâtiment	1m	Various	-	-	○	40A	>250	1.2	H	-30/+80	15	RT	269
	Mastic PU 50 Industrie	1m	Various	-	-	○	50A	80	1.17	H	-40/+90	15	RT	269
	Sapocol DK 4001	1m	Black	-	●	-	-	-	1.2	H	-	15	RT	269
Forbo Helmitin	Helmipur 700T	2	Beige	◑	-	-	-	-	1.5	100	-	40	RT - 120	270
	Helmipur 701T	2	Beige	◑	-	-	-	-	1.5	30	-	40	RT - 100	270
	Helmipur 702T	2	White	◑	-	-	-	-	1.67	H	-	90	RT	270
	Helmipur 15065	1 or 2	Transp.	-	-	-	-	-	0.86	4	-	20	RT - 70	270
	Helmipur 15104	2	Cl. Yellow	-	-	-	-	-	0.93	5	-	(1 day)	RT	271
	Helmipur 15141/2	2	Cl. Yellow	-	-	-	-	-	0.88	3.3	-	(1 day)	RT	271
	Helmipur 46010	2	Grey	◑	-	-	-	-	1.53	12	-	40	RT	271
	Helmipur 46018	2	Brown	◑	-	-	-	-	1.6	4	-	120	RT	271
	Helmipur 46019/2	2	Grey	◑	-	-	-	-	1.45	10	-	4	RT	272
	Helmipur 46508	1m	Brown	-	-	-	-	-	1.1	30	-	120	RT	272
	Helmipur 46516	1m	Yellow	-	-	-	-	-	1.1	H	-	120	RT	272
	Helmipur 46519	1m	Brown	-	-	-	-	-	1.15	4.5	-	40	RT - 40	272
	Helmipur DS/18	2	White	-	-	-	-	-	1.05	2	-	30	RT - 80	273
	Helmipur GPV/B	1 or 2	Transp.	-	-	-	-	-	0.86	3.3	-	20	RT - 80	273
	Helmipur SH100	1m	Various	-	○	-	53A	630	1.26	H	-40/+90	-	RT	273
	Helmipur UN neu	2	Transp.	-	-	-	-	-	0.98	6	-	(12 hrs)	RT	273
	Helmitherm RK 99	1m*	White	○	-	-	-	-	~1	-	-	<1	130+RT	274
	Helmitin-kleber 34331	2	White	-	-	-	-	-	1.04	1	90 - 120	(8 hrs)	RT - 90	274
Ed. Geistlich Söhne AG	Miracol 9505	1 or 2	Clear	-	-	-	-	-	1.13	2	-20/+150	30	RT	274
	Mirapur 9510 Rapid	1m	Brown	-	-	-	-	-	1.14	7	<120	15	RT	274
Hardman	Kalex 15036	2	Off-white	●	-	-	85D	-	1.24	11	<70	5	RT	275
	Kalex 15038	2	Beige	●	-	-	70D	-	1.26	12	<70	5	RT	275
	Kalex 16308	2	Black	○	-	-	40D	-	1.22	40	<20	7	RT	275
	Kalex 18515	2	Off-white	○	-	-	72A	-	1.89	11	-	85	RT	275
Hitachi-Kasei Polymer Co.	Hi-Bon 4610	1	Black	○	○	◑	32D	38	1.58	710	-20/+150	30	180	276
	Hi-Bon 4620	1	Black	○	○	◑	32D	57	1.58	680	-20/+150	30	180	276
	Hi-Bon 4811	1m*	Clear	-	◑	○	-	420	-	-	-	(10 sec)	100-200+RT	276
	Hi-Bon 4812	1m*	Clear	-	◑	○	-	510	-	-	-	(10 sec)	100-200+RT	276
	Hi-Bon 4820	1m*	Yellow	-	●	○	-	580	-	-	-	(7 sec)	100-200+RT	277
	Hi-Bon 4830	1m*	Opaque	-	●	◑	-	480	-	-	-	(2 sec)	100-200+RT	277
	Hi-Bon 4832	1m*	Yellow	-	●	◑	-	700	-	-	-	(<1 sec)	100-200+RT	277

Continued...

See Key to Comparison Tables - Pages 63 to 65

Company	Adhesive	Type	Colour	LSS	TS	TM	Hs	El. (%)	SG	Visc (Pa.s)	Service (°C)	Usable (mins)	Cure	Pg. No.
Kemira Polymers	Hyperlast 2874/019	2	Brown	○	◑	-	70A	10	1.5	9	-	10	70	277
	Hyperlast 2875-2874/022	2	Brown	-	◑	-	60D	150	1.1	H	-	8	RT	278
	Hyperlast 2875-2874/701	2	Brown	-	◑	-	73D	35	1.3	3	-	4.5	RT	278
	Hyperlast 2875-2874/7003	2	Cream	-	-	-	88D		1.6	12	-	6	RT - 50	278
	Hyperlast 5000-2874/08	2	Brown	-	-	-	77D	10	1.5	7	-	1.5	RT	278
	Hyperlast 5000-2874/701	2	Brown	-	-	-	85D		1.3	3	-	2	RT	279
	Hyperlast 5000-3283	2	Brown	-	○	-	60D	35	1.5	8	-	<1	RT	279
Klebchemie (M.G Becker)	Klebschaum 576.1	2	-	-	-	-	-	-	1.4	-	-	2	RT - 40	279
	Kleiberit PU-501	1m	Brown	-	-	-	-	-	1.2	8	-	25	RT - 70	279
	Kleiberit PU-520	2	-	-	●	-	-	25	1.2	H	-	1.5	RT - 60	280
	Kleiberit PU-543.1	2	Various	-	-	-	77A		1.2	H	-	2	RT - 60	280
	Supramelt PU 703.6	1*	-	-	-	-	-	-	1.15	-	-40/+100	<1	120-140+RT	280
	Supramelt PU 704.1	1*	-	-	-	-	-	-	1.1	-	-40/+130	-	120-140+RT	280
	Supramelt PU 706.1	1*	-	-	-	-	-	-	1.1	-	-40/+130	1	120-140+RT	281
Kömmerling	Corapur 646	2	Lt. Brown	◑	-	-	75D	-	1.4	3	-40/+130	20	RT	281
Loctite	Forteflex 3951	1m	Black	○	○	-	60A	725	1.1	H	<75	60	RT	281
Mydrin	Mybond 66102/66500	2	Brown	◑	●	-	-	-	1.4	8.5	<190	30	RT	281
	Mybond 66103/66500	2	Brown	○	○	-	-	-	1.5	8.5	-	30	RT	282
	Mybond 66115/66500	2	Brown	◑	◑	-	-	-	1.4	H	-	30	RT	282
	Mybond 66116/66500	2	Brown	◑	◑	-	-	-	1.4	100	-	30	RT	282
National Casein	LU-3123A/B	2	Lt. Grey	-	-	-	-	-	1.2	4	-	240	RT + Heat	282
SLC Rinaldi	Gutoid PU70	2	-	-	-	-	-	-	-	-	-30/+80	120	RT	283
SABA	Sabaplast 5130	1	Yell.-brown	-	-	-	-	-	1.13	8.5	-40/+150	120	RT	283
	Sabaplast 5503	1	Various	-	○	-	43A	400	1.2	H	-40/+90	-	RT	283
	Sabaplast 5538	1m	Lt. Brown	-	-	-	-	-	1.18	0.08	-20/+120	30	RT	283
	Sabaplast 5817	1 or 2	Transp.	-	-	-	-	-	0.89	2	-20/+90	10	RT	284
Scholten Lijmen	Dynol U 462	1m	Brown	-	-	-	-	-	1.12	7	-	60	RT	284
Scott Bader	Crestomer 1152PA	2	-	◑	●	◑	65D	100	1.03	-	-	<50	RT	284
	Crestomer 1186PA	2	-	◑	◑	●	70D	6	1.32	H	-	<50	RT	284
	Crestomer 1196PA	2	-	◑	◑	●	70D	4	0.58	-	-	<50	RT	285
SEA	Mastic PU	1	Various	-	-	-	-	-	-	H	-	-	RT	285
Sika	Sikaflex 252	1m	-	○	○	-	55A	>300	1.14	H	-40/+90	<40	RT	285
	Sikaflex 255FC	1m	-	○	○	-	60A	>400	1.18	-	-40/+90	<20	RT	285
	Sikaflex 260	1m	-	○	○	-	55A	>400	1.22	-	-40/+90	<40	RT	286
	Sikaflex 352HC	1m	-	○	○	-	55A	>300	1.18	-	-40/+90	<120	RT - 130	286
	Sikaflex 360H	1m	-	○	○	-	55A	>400	1.22	-	-40/+90	<120	RT - 130	286
	SikaTack Ultrafast	1m*	-	○	○	-	60A	>400	1.13	-	-40/+90	<10	Melt+RT-80	286
Sunkyong Industries	Skybon U0231T	1	-	-	●	○	-	900	-	33	-	-	RT	287
	Skybon U0251M	1	-	-	●	○	-	900	-	40	-	-	RT	287
	Skybon U0841H	1	-	-	○	○	-	1100	-	0.04	-	-	RT	287
	Skybon U1351H	1	-	-	●	○	-	300	-	1.75	-	-	RT	287
Sunstar Engineering	Penguin Cement 930	-	Lt. Brown	-	-	-	-	-	-	125	-	-	RT	288
	Penguin Cement 965	-	Black	○	-	-	-	-	-	49	-	-	RT	288
	Penguin Seal 551	-	Black	○	-	-	-	-	-	46	-	-	RT	288
	Penguin Seal 915	-	Grey	○	-	-	-	-	-	500	-	-	RT	288
Swift Adhesives	EC2 3710R/EC2 3700D	2	Beige	○	-	-	60D	-	-	5	<120	8	RT	289
	EC2 3810R/EC2 3700D	2	Grey	○	-	-	50D	-	-	3	<130	15	RT	289
	HMPUR 146	1*	White	○	◑	◑	63D	-	-	-	-	-	120-150+RT	289
	HMPUR 200	1*	White	○	○	◑	50D	>100	-	-	-	-	120-150+RT	289
	HMPUR 205/1	1*	White	○	○	◑	55D	>100	-	-	-	-	120-150+RT	290
	Polyfree 161	1m	Cream	-	-	-	-	-	1.02	2	-	30	RT - 70	290

Continued...

See Key to Comparison Tables - Pages 63 to 65

Company	Adhesive	Type	Colour	LSS	TS	TM	Hs	El. (%)	SG	Visc (Pa.s)	Service (°C)	Usable (mins)	Cure	Pg. No.
Teroson	Terokal 6015	2	Grey	◑	-	-	65D	-	1.75	170	-40/+80	20	RT -100	290
(Henkel)	Terostat 8590	1m	Black	○	○	-	50A	425	1.2	H	-40/+90	20	RT + heat	290
	Terostat 8690	1m	Black	○	○	-	70A	250	1.2	7000	-40/+90	15	RT + heat	291
	Terostat 8890	1m	Black	○	○	○	65A	425	1.2	H	-40/+90	10	RT + heat	291
Tivoli	Tivo 9551/43	1	Black	○	-	-	-	-	1.5	400	-	-	140 -200	291
	Tivomelt 9617/11	1m*	-	-	-	-	-	-	-	-	-	5	90-150+RT	291
	Tivomelt 9625	1m*	-	-	-	-	-	-	-	-	-	2	90-160+RT	292
	Tivomelt 9626	1m*	-	-	-	-	-	-	-	-	-	1	90-140+RT	292
	Tivomelt 9627/1	1m*	-	-	-	-	-	-	-	-	-	<1	120-140+RT	292
	Tivomelt 9627/10	1m*	-	-	-	-	-	-	-	-	-	1	130-150+RT	292
	Tivomelt 9628/28	1m*	-	-	-	-	-	-	-	-	-	<1	130-160+RT	293
	Tivomelt 9630	1m*	-	-	-	-	-	-	-	-	-	1	100-140+RT	293
	Tivomelt 9630/10	1m*	-	-	-	-	-	-	-	-	-	<1	130-150+RT	293
	Tivomelt 9635/3	1m*	-	-	-	-	-	-	-	-	-	<1	120-150+RT	293
	Tivomelt 9635/4	1m*	-	-	-	-	-	-	-	-	-	<1	120-150+RT	294
	Tivomelt 9645	1m*	-	-	-	-	-	-	-	-	-	<1	130-150+RT	294
	Tivopur 1609	2	Beige	●	●	-	100A	3	1.4	100	-	60	RT - 40	294
	Tivopur 1636	2	Beige	●	◑	-	98A	10	1.4	80	-	70	RT - 40	294
	Tivopur 1657	2	Beige	○	●	-	95A	5	1.4	5	-	2	RT - 40	295
	Tivopur 1672	2	Beige	●	○	-	95A	30	1.4	66	-	170	RT - 40	295
	Tivopur 1672/20	2	Beige	◑	○	-	95A	55	1.3	1.5	-	10	RT - 40	295
	Tivopur 1672/70	2	Beige	◑	◑	-	98A	35	1.5	5	-	32	RT - 40	295
	Tivopur 1673	2	Beige	◑	◑	-	95A	5	1.4	8	-	4	RT - 40	296
	Tivopur 1717	2	Black	○	○	-	40A	285	1.4	8	-	100	RT + heat	296
	Tivopur 1768	2	Grey	○	○	-	73A	55	1.5	30	-	45	RT + heat	296
	Tivopur 1782/40	2	White	○	●	-	90A	5	1.5	15	-	<1	RT + heat	296
	Tivopur 1802/10	2	Black	○	○	-	75A	150	1.1	30	-	14	RT + heat	297
	Tivopur 1815	2	Brown	○	●	-	95A	5	1.4	4	-	1	RT + heat	297
	Tivopur 1823/30	2	Beige	○	○	-	70A	200	1.0	35	-	2	RT + heat	297
	Tivopur 1840	2	Beige	●	◑	-	98A	5	1.4	7	-	14	RT + heat	297
	Tivopur 1841/10	2	Ochre	○	○	-	95A	20	1.5	8	-	<1	RT + heat	298
Wulff	Puro-Tack	2	-	-	-	-	-	-	1.2	70	<60	30	RT	298

See Key to Comparison Tables - Pages 63 to 65

3E

3E	Andex 950		(Feb 95)
One part polyurethane, moisture curing			
Manufacturers Comments: Bonds all materials used in the construction industry. Fills voids. Rapid cure to full strength. Non-flammable. Two versions with different viscosity: Type A & Type B.			
Typical Use: Bridges, piers, foundations - metals, wood & masonry.			
Cure: RT Pressure injection: 8-12 hrs; Crack-sealing: 6 hrs (min.).			
Test Methods: Not stated			
Physical		**Mechanical**	
Colour	-	τ (RT)	4.1 MPa
Viscosity	A: Reg. B: Low	τ (°C)	-
SG	-	τ (wet)	-
Shelf life	-	Peel (-)	0.2 MPa
Pot/Work life	-	σ_t	-
Tg	-	σ_c	-
$T_{Service}$	-	E	-
CTE (°C)	-	G	-
Thermal cond.	-	Elongation	-
Resistivity: Vol.	-	Tear strength	-
Surf.	-	H (Shore D)	85
Notes:			

3E	Polarjoint - T		(Mar 95)
Two part polyurethane			
Manufacturers Comments: 100% solids, coal-tar modified. Pourable. Forms a hard, durable joint. High resistance to chemicals.			
Typical Use: Construction industry. Joints with moderate movement.			
Cure: RT.			
Test Methods: ASTM			
Physical		**Mechanical**	
Colour	Black	τ (RT)	-
Viscosity	-	τ (°C)	-
SG	0.95	τ (wet)	-
Shelf life	1 year	Peel ( )	-
Pot/Work life	-	σ_t	2.8 MPa
Tg	-	σ_c	-
$T_{Service}$	+93°C max.	E	-
CTE (°C)	-	G	-
Thermal cond.	-	Elongation	-
Resistivity: Vol.	-	Tear strength	-
Surf.	-	H (Shore A)	46
Notes: Mix ratio 1:1. Hardness ASTM D-2240. SG - ASTM D-792. σ ASTM D-417. Water absorption ASTM D-570-77: zero. Flash point >55°C.			

3E	Polarjoint - Super		(Mar 95)
Two part polyurethane			
Manufacturers Comments: 100% solids. Forms a flexible, but highly durable bond. High resistance to chemicals.			
Typical Use: Construction industry. Joints with extensive movement.			
Cure: RT.			
Test Methods: ASTM			
Physical		**Mechanical**	
Colour	Grey	τ (RT)	-
Viscosity	-	τ (°C)	-
SG	1.05	τ (wet)	-
Shelf life	1 year	Peel ( )	-
Pot/Work life	-	σ_t	15.2 MPa
Tg	-	σ_c	-
$T_{Service}$	+93°C max.	E	-
CTE (°C)	-	G	-
Thermal cond.	-	Elongation	-
Resistivity: Vol.	-	Tear strength	-
Surf.	-	H (Shore A)	78
Notes: Mix ratio 2:1. Hardness ASTM D-2240. SG - ASTM D-792. σ ASTM D-417. Water absorption ASTM D-570-77: zero. Flash point >93°C.			

3M

3M	3532 B/A		(Mar 95)
Two part polyurethane			
Manufacturers Comments: Rapid RT curing.			
Typical Use: Wood, plastics & painted metals.			
Cure: RT Handling strength: 25-30 mins; Full cure: 2 days.			
Test Methods: Not stated			
Physical		**Mechanical**	
Colour	Brown	τ (RT)	13.8 MPa
Viscosity	30 Pa.s	τ (121°C)	1.0 MPa
SG	-	τ (wet)	15.9 MPa
Shelf life	-	Peel (T)	4.40 N mm^{-1}
Pot/Work life	6 - 8 mins	σ_t	-
Tg	-	σ_c	-
$T_{Service}$	-	E	-
CTE (°C)	-	G	-
Thermal cond.	-	Elongation	-
Resistivity: Vol.	-	Tear strength	-
Surf.	-	H (Shore)	-
Notes: τ(wet) 30 days, tap water, RT using Scotch-weld 1945B/A primer.			

See Key to Data Sheets - Pages 63 to 65

3M	**3535 B/A**		(Mar 95)
Two part polyurethane			
Manufacturers Comments: More rapidly curing version of 3532B/A.			
Typical Use: Metal, plastics & painted metals.			
Cure: RT Handling strength: 15 - 20 mins; Full cure: 1 day.			
Test Methods: Not stated			
Physical		**Mechanical**	
Colour	Off-white	τ (RT)	13.8 MPa
Viscosity	30 Pa.s	τ (121°C)	1.0 MPa
SG	-	τ (wet)	14.5 MPa
Shelf life	-	Peel (T)	4.38 N mm^{-1}
Pot/Work life	2 - 4 mins	σ_t	-
Tg	-	σ_c	-
$T_{Service}$	-	*E*	-
CTE (°C)	-	*G*	-
Thermal cond.	-	Elongation	-
Resistivity: Vol.	-	Tear strength	-
Surf.	-	H (Shore)	-
Notes: τ(wet) 30 days, tap water, RT using Scotch-weld 1945B/A primer.			

3M	**DP-605 NS**		(Mar 95)
Two part polyurethane			
Manufacturers Comments: Fast cure, non-sag, good gap-filler. Duo-pak self metering & mixing cartridge.			
Typical Use: General industrial assembly: Metal, plastics, ceramics & glass.			
Cure: RT Handling strength:15 - 20 mins; Full cure: 1 day.			
Test Methods: Not stated			
Physical		**Mechanical**	
Colour	Off-white	τ (RT)	13.8 MPa
Viscosity	(A) 10 (B) 100 Pa.s	τ (82°C)	2.8 MPa
SG	-	τ (wet)	-
Shelf life	RT	Peel (T)	4.73 N mm^{-1}
Pot/Work life	3- 5 mins	σ_t	-
Tg	-	σ_c	-
$T_{Service}$	-	*E*	-
CTE (50-110°C)	219×10^{-6} °C^{-1}	*G*	-
Thermal cond.	-	Elongation	-
Resistivity: Vol.	1×10^{14} Ω.cm	Tear strength	-
Surf.	-	H (Shore)	-
Notes:			

3M	**3549 B/A**		(Mar 95)
Two part polyurethane			
Manufacturers Comments: Longer work-life, slower cure version of 3532B/A.			
Typical Use: Metal, plastics & painted metals.			
Cure: RT Handling strength: 8 - 12 hours; Full cure: 7 days.			
Test Methods: Not stated			
Physical		**Mechanical**	
Colour	Brown	τ (RT)	13.8 MPa
Viscosity	30 Pa.s	τ (121°C)	1.0 MPa
SG	-	τ (wet)	11.7 MPa
Shelf life	-	Peel (T)	4.38 N mm^{-1}
Pot/Work life	1 hr	σ_t	-
Tg	-	σ_c	-
$T_{Service}$	-	*E*	-
CTE (°C)	-	*G*	-
Thermal cond.	-	Elongation	-
Resistivity: Vol.	-	Tear strength	-
Surf.	-	H (Shore)	-
Notes: τ(wet) 30 days, tap water, RT using Scotch-weld 1945B/A primer.			

3M	**TE-030**		(Mar 95)
One part polyurethane, moisture curing			
Manufacturers Comments: Jet-Weld series. 100% solids, thermoset. Extrudable. Fast set time. Furniture, vehicle interiors & trims, electronic cabinets, sports goods.			
Typical Use: Wood, some plastics.			
Cure: Application Temperature: 121°C, 30 seconds.			
Test Methods: Not stated			
Physical		**Mechanical**	
Colour	Off-white	τ (RT)	-
Viscosity	17 Pa.s	τ (°C)	-
SG	-	τ (wet)	-
Shelf life	-	Peel ( )	-
Pot/Work life	1 min	σ_t	26.2 MPa
Tg	-	σ_c	-
$T_{Service}$	-	*E*	76 MPa
CTE (°C)	-	*G*	-
Thermal cond.	-	Elongation	725%
Resistivity: Vol.	-	Tear strength	-
Surf.	-	H (Shore D)	60
Notes:			

3M	TE-031		(Mar 95)
One part polyurethane, moisture curing			
Manufacturers Comments: Jet-Weld series. 100% solids, thermoset. Extrudable. Fast set time. General assembly of plastics.			
Typical Use: Plastics, inc. polystyrene & polyacrylic.			
Cure: Application Temperature: 121°C, 30 seconds.			
Test Methods: Not stated			
Physical		**Mechanical**	
Colour	Off-white	τ (RT)	-
Viscosity	13 Pa.s	τ (°C)	-
SG	-	τ (wet)	-
Shelf life	-	Peel ( )	-
Pot/Work life	2 mins	σ_t	26.9 MPa
Tg	-	σ_c	-
$T_{Service}$	-	E	38.6 MPa
CTE (°C)	-	G	-
Thermal cond.	-	Elongation	-
Resistivity: Vol.	-	Tear strength	-
Surf.	-	H (Shore D)	50
Notes:			

3M	TE-230		(Mar 95)
One part polyurethane, moisture curing			
Manufacturers Comments: Jet-Weld series. 100% solids, thermoset. Spray/extrudable. Long set time. General assembly.			
Typical Use: Plastics, inc. polystyrene & polyacrylic. Aluminium, glass to plastic & wood.			
Cure: Application Temperature: 121°C, 2.5 mins.			
Test Methods: Not stated			
Physical		**Mechanical**	
Colour	Off-white	τ (RT)	-
Viscosity	9 Pa.s	τ (°C)	-
SG	-	τ (wet)	-
Shelf life	-	Peel ( )	-
Pot/Work life	4 mins	σ_t	22.8 MPa
Tg	-	σ_c	-
$T_{Service}$	-	E	37.2 MPa
CTE (°C)	-	G	-
Thermal cond.	-	Elongation	-
Resistivity: Vol.	-	Tear strength	-
Surf.	-	H (Shore)	-
Notes:			

3M	TE-100		(Mar 95)
One part polyurethane, moisture curing			
Manufacturers Comments: Jet-Weld series. 100% solids, thermoset. Extrudable. Medium set time, low viscosity. Thin bondlines.			
Typical Use: Wood, some plastics.			
Cure: Application Temperature: 121°C, 1 min.			
Test Methods: Not stated			
Physical		**Mechanical**	
Colour	Off-white	τ (RT)	-
Viscosity	7 Pa.s	τ (°C)	-
SG	-	τ (wet)	-
Shelf life	-	Peel ( )	-
Pot/Work life	2 mins	σ_t	29.0 MPa
Tg	-	σ_c	-
$T_{Service}$	-	E	84.1 MPa
CTE (°C)	-	G	-
Thermal cond.	-	Elongation	-
Resistivity: Vol.	-	Tear strength	-
Surf.	-	H (Shore)	-
Notes:			

Apollo

Apollo	Astrolok A 77		(Feb 95)
One part polyurethane, moisture curing			
Manufacturers Comments: Non-flammable, high solids (~50%), low viscosity (sprayable). Assembly of vehicle panels & building units.			
Typical Use: Panel laminating - plastics, metals, plywood, GRP, decorative laminates.			
Cure: RT (>15°C)/ 0.03 MPa.			
Test Methods: Not stated			
Physical		**Mechanical**	
Colour	Tinted*	τ (RT)	-
Viscosity	0.1 - 0.3 Pa.s	τ (°C)	-
SG	-	τ (wet)	-
Shelf life	90 days	Peel ( )	-
Pot/Work life	<20 mins	σ_t	-
Tg	-	σ_c	-
$T_{Service}$	-40 to +140°C	E	-
CTE (°C)	-	G	-
Thermal cond.	-	Elongation	-
Resistivity: Vol.	-	Tear strength	-
Surf.	-	H (Shore)	-
Notes: * Red, blue, etc. Apply adhesive to one adherend only. Assemble within 20 mins.			

Apollo	Astrolok A 84		(Feb 95)
One part polyurethane, moisture curing			
Manufacturers Comments: Non-flammable, high solids (~40%), low viscosity (sprayable). Good chemical resistance. Assembly of vehicle panels & building units.			
Typical Use: Panel laminating - plastics, metals, wood composites, GRP, insulation & asbestos sheets.			
Cure: RT (>14°C)/ Pressure from rollers or press.			
Test Methods: Not stated			
Physical		**Mechanical**	
Colour	Tinted*	τ (RT)	-
Viscosity	0.3 - 0.4 Pa.s	τ (°C)	-
SG	-	τ (wet)	-
Shelf life	12 weeks	Peel ( )	-
Pot/Work life	1 - 4 mins	σ_t	-
Tg	-	σ_c	-
$T_{Service}$	-40 to +120°C	*E*	-
CTE (°C)	-	*G*	-
Thermal cond.	-	Elongation	-
Resistivity: Vol.	-	Tear strength	-
Surf.	-	H (Shore)	-
Notes: * Also clear & amber. Modified formulations (AX) available. Apply adhesive to one adherend only.			

Apollo	Astrolok A 7504		Mar 93
One part polyurethane, moisture curing			
Manufacturers Comments: 100% solids. Requires water misting. Assembly of vehicle panels e.g. mobile homes, caravans, and insulated building units.			
Typical Use: Insulating materials to flexible or rigid skins. Some Al-alloys, galvanised steel & plastics may need a primer.			
Cure: RT (20°C) 55 mins (handling).			
Test Methods: Not stated			
Physical		**Mechanical**	
Colour	Brown	τ (RT)	-
Viscosity	4±0.5 Pa.s	τ (°C)	-
SG	-	τ (wet)	-
Shelf life	3 months	Peel ( )	-
Pot/Work life	55 mins	σ_t	-
Tg	-	σ_c	-
$T_{Service}$	-	*E*	-
CTE (°C)	-	*G*	-
Thermal cond.	-	Elongation	-
Resistivity: Vol.	-	Tear strength	-
Surf.	-	H (Shore)	-
Notes: * based on: 0.08 to 0.14 kg/m² adhesive applied + ~0.01 kg/m² water mist. Modified formulations available. Flash point >150°C			

Apollo	Astrolok A 7166		Nov 93
One part polyurethane, moisture curing			
Manufacturers Comments: Non-flammable, low viscosity (sprayable). Assembly of vehicle panels & building units.			
Typical Use: Galvanised & non-galvanised metals (without primer), wood, decorative laminates.			
Cure: RT (>12°C)/ Pressure from rollers or press.			
Test Methods: Not stated			
Physical		**Mechanical**	
Colour	Clear Brown	τ (RT)	-
Viscosity	0.13 Pa.s	τ (°C)	-
SG	-	τ (wet)	-
Shelf life	3 months	Peel ( )	-
Pot/Work life	1 min	σ_t	-
Tg	-	σ_c	-
$T_{Service}$	-	*E*	-
CTE (°C)	-	*G*	-
Thermal cond.	-	Elongation	-
Resistivity: Vol.	-	Tear strength	-
Surf.	-	H (Shore)	-
Notes: Apply adhesive to one adherend only.			

Apollo	Astrolok A 7515		Aug 92
One part polyurethane, moisture curing			
Manufacturers Comments: 100% solids, reinforced. Requires water misting. Roller-coatable. Assembly of vehicle panels & building units, e.g. cold-stores, caravans, etc.			
Typical Use: Insulation materials. Metals & decorative laminates. May need a primer.			
Cure: RT (20°C) 25 mins (handling).			
Test Methods: Not stated			
Physical		**Mechanical**	
Colour	Light Brown	τ (RT)	-
Viscosity	13±2 Pa.s	τ (°C)	-
SG	-	τ (wet)	-
Shelf life	3 months <25°C	Peel ( )	-
Pot/Work life	10 mins	σ_t	-
Tg	-	σ_c	-
$T_{Service}$	-	*E*	-
CTE (°C)	-	*G*	-
Thermal cond.	-	Elongation	-
Resistivity: Vol.	-	Tear strength	-
Surf.	-	H (Shore)	-
Notes: * based on: 0.10 to 0.2 kg/m² adhesive applied + ~0.01 kg/m² water mist. Modified formulations available. Flash point >150°C.			

Ashland Chemicals

Apollo	Astrolok A 7521		Nov 93
One part polyurethane, moisture curing			
Manufacturers Comments: High solids (~95%). Requires water misting. Roller-coatable. Assembly of vehicle panels & building units.			
Typical Use: Metals. Some Al-grades, galvanised steels & plastics may need a primer.			
Cure: RT (20°C) 90 mins (handling); 1 day (full cure). ~0.09MPa pressure.			
Test Methods: Not stated			
Physical		**Mechanical**	
Colour	Light Brown	τ (RT)	-
Viscosity	3 - 5 Pa.s	τ (°C)	-
SG	-	τ (wet)	-
Shelf life	3 months <25°C	Peel ( )	-
Pot/Work life	40 mins/20°C	σ_t	-
Tg	-	σ_c	-
$T_{Service}$	-	E	-
CTE (°C)	-	G	-
Thermal cond.	-	Elongation	-
Resistivity: Vol.	-	Tear strength	-
Surf.	-	H (Shore)	-
Notes: * based on: 0.10 to 0.13 kg/m² adhesive applied + ~0.01 kg/m² water mist. Flash point >55°C.			

Ashland Chemicals	Isogrip 3030D		May 94
One part polyurethane, moisture curing			
Manufacturers Comments: 100% solids. Laminating building structural panels. Water-based, no formaldehyde to replace current solvent-based adhesives. Requires water-misting.			
Typical Use: Porous & non-porous materials. Insulation materials to wood, FRP, vinyl, metals.			
Cure: Metal or plastic to wood:* 24°C/20-95%RH/30mins assembly/75 mins/0.2MPa in press.			
Test Methods: Not Stated			
Physical		**Mechanical**	
Colour	Grey	τ (RT)	-
Viscosity	3.5-8 Pa.s	τ (°C)	-
SG	~1.1	τ (wet)	-
Shelf life	6 months (7-24°C)	Peel ( )	-
Pot/Work life	25 mins/27°C	σ_t	-
Tg	-	σ_c	-
$T_{Service}$	-	E	-
CTE (°C)	-	G	-
Thermal cond.	-	Elongation	-
Resistivity: Vol.	-	Tear strength	-
Surf.	-	H (Shore)	-
Notes: Adhesive applied to one surface only. Water-mist before assembly. * Varies for substrates. Lower %RH increases press time. Flash point >93°C.			

Apollo	Astrolok AX 9106		Jan 94
Two part polyurethane, thixotropic paste			
Manufacturers Comments: 100% solids. Forms tough, resilient bond. Resists water, solvents & heat.			
Typical Use: Metals, plastics, concrete, asbestos boards.			
Cure: RT (20°C)/ 4 - 6 hrs (handling); 1 day (full cure)			
Test Methods: Not stated			
Physical		**Mechanical**	
Colour	Brown	τ (RT)	-
Viscosity	-	τ (°C)	-
SG	-	τ (wet)	-
Shelf life	6 months <25°C	Peel ( )	-
Pot/Work life	1 hr/20°C	σ_t	-
Tg	-	σ_c	-
$T_{Service}$	-	E	-
CTE (°C)	-	G	-
Thermal cond.	-	Elongation	-
Resistivity: Vol.	-	Tear strength	-
Surf.	-	H (Shore)	-
Notes: Apply adhesive to one adherend only. Mix ratio 6.6:1 (wt).			

Ashland Chemicals	Isogrip WD3-A312		Dec 89
Two part polyurethane			
Manufacturers Comments: 45% solids. Water-based. Laminating building structural panels (rail doors). Wood bonding (face, edge & finger joints). Specifications: NLGA SpS1-81; ANSI/NWWDA I.S .6-86; ASTM D-3110-50; WWPA 101.86-1.70.			
Typical Use: Wood & plastics. Joints in wood products.			
Cure: RF, cold & hot pressing			
Test Methods: Not Stated			
Physical		**Mechanical**	
Colour	White + Opaque	τ (RT)	-
Viscosity	2 - 4.5 Pa.s	τ (°C)	-
SG	~1.2	τ (wet)	-
Shelf life	3 months (7-24°C)	Peel ( )	-
Pot/Work life	3 hrs*/24°C	σ_t	-
Tg	-	σ_c	-
$T_{Service}$	-	E	-
CTE (°C)	-	G	-
Thermal cond.	-	Elongation	-
Resistivity: Vol.	-	Tear strength	-
Surf.	-	H (Shore)	-
Notes: * Varies with hardener content. Value for 10% Isoset CX. Mix ratio 5-20:100(wt). Flash point >93°C; pH ~6.5 Mixed adhesive evolves CO_2, do not keep in closed containers.			

Ashland Chemicals	Isogrip WD3-A320		June 87
Two part polyurethane			
Manufacturers Comments: ~45% solids. Water-based. Laminating building structural panels & wood bonding (X-ray doors). Creep-resistant bonds (with 20:100 mix). Specifications: ANSI/AITC A 190.1-1983; ANSI/NWWDA I.S 1.6.AITC 200.83; ASTM D-2559-84.			
Typical Use: Wood, metal (inc. coated), & plastics			
Cure: RF, cold & hot pressing.			
Test Methods: Not Stated			
Physical		**Mechanical**	
Colour	White + Opaque	τ (RT)	-
Viscosity	8 - 12 Pa.s	τ (°C)	-
SG	~1.1	τ (wet)	-
Shelf life	6 months (7-24°C)	Peel ( )	-
Pot/Work life	3 hrs*/21°C	σ_t	-
Tg	-	σ_c	-
$T_{Service}$	-	*E*	-
CTE (°C)	-	*G*	-
Thermal cond.	-	Elongation	-
Resistivity: Vol.	-	Tear strength	-
Surf.	-	H (Shore)	-

Notes: * Varies with hardener content. Value for 10% Isoset CX-11. Mix ratio 5-20:100(wt). Flash point >93°C; pH ~6.5 Mixed adhesive evolves CO_2, do not keep in closed containers.

Ashland Chemicals	Isogrip WD3-CM402		June 87
Two part polyurethane			
Manufacturers Comments: ~45% solids. Water-based. Laminating building structural panels (porous & non-porous materials). Ocean containers, trailer & storm doors. Specifications: Fruehauf FES-18; Sealand M-008-metal.			
Typical Use: Wood, metal (inc. coated), mill-finished Al foils (having water emulsifiable oils).			
Cure: RF, cold & hot pressing.			
Test Methods: Not Stated			
Physical		**Mechanical**	
Colour	White + Opaque	τ (RT)	-
Viscosity	2 - 4.5 Pa.s	τ (°C)	-
SG	~1.1	τ (wet)	-
Shelf life	3 months (7-24°C)	Peel ( )	-
Pot/Work life	2 hrs*/21°C	σ_t	-
Tg	-	σ_c	-
$T_{Service}$	-	*E*	-
CTE (°C)	-	*G*	-
Thermal cond.	-	Elongation	-
Resistivity: Vol.	-	Tear strength	-
Surf.	-	H (Shore)	-

Notes: * Varies with hardener content. Value for 10% Isoset CX-11. Mix ratio 10-20:100(wt). Flash point >93°C; pH ~7 Mixed adhesive evolves CO_2, do not keep in closed containers.

Ashland Chemicals	Isogrip WD3-A322		Nov 90
Two part polyurethane			
Manufacturers Comments: 45% solids. Laminating building structural panels. Water-based. Specifications: ICBO; ASTM; CABO-BOCA-ICBO SBCC; ANSI/NWWDA I.S .6-87; WHI 013 UL.			
Typical Use: I-beams, fire-doors. Wood & plastics.			
Cure: Cold RF + hot pressing.			
Test Methods: Not Stated			
Physical		**Mechanical**	
Colour	White + Opaque	τ (RT)	-
Viscosity	5 - 7.5 Pa.s	τ (°C)	-
SG	~1.1	τ (wet)	-
Shelf life	3 months (7-24°C)	Peel ( )	-
Pot/Work life	3 hrs*/24°C	σ_t	-
Tg	-	σ_c	-
$T_{Service}$	-	*E*	-
CTE (°C)	-	*G*	-
Thermal cond.	-	Elongation	-
Resistivity: Vol.	-	Tear strength	-
Surf.	-	H (Shore)	-

Notes: * Varies with hardener content. Value for 10% Isoset CX-47. Mix ratio 5-20:100(wt). Flash point >93°C; pH ~6.5 Mixed adhesive evolves CO_2, do not keep in closed containers.

Ashland Chemicals	Pliogrip 6600/6610		(Apr 95)
Two part polyurethane, 100% reactive			
Manufacturers Comments: High strength, temperature & chemical resistant. Automotive (Chevrolet) & heavy trucks. Marine & other industries. Requires primer.			
Typical Use: Automotive body parts - composite plastics & metals.			
Cure: RT(23°C)/ 15 - 20 mins; 100 - 110°C/ 4 - 10 mins.			
Test Methods: Not stated			
Physical		**Mechanical**	
Colour	Tan + Dark Green	τ (RT)	-
Viscosity	~30 Pa.s	τ (°C)	-
SG	1.2	τ (wet)	-
Shelf life	6 months	Peel ( )	-
Pot/Work life	4 - 6 mins /23°C	σ_t	16.9 MPa
Tg	-25°C	σ_c	-
$T_{Service}$	-	*E*	388 MPa
CTE (°C)	-	*G*	-
Thermal cond.	-	Elongation	55%
Resistivity: Vol.	-	Tear strength	-
Surf.	-	H (Shore D)	60

Notes: Water absorption <1.5% Pliogrip 6600/6611 - slower cure version.

Ashland Chemicals	Pliogrip 6600/6612		(Apr 95)
Two part polyurethane, 100% reactive			
Manufacturers Comments: Low-sag, short work-life. Temperature & chemical resistant. Automotive , marine & other industries. Requires primer.			
Typical Use: Automotive body parts - composite plastics & metals.			
Cure: RT(23°C)/ 5 - 6 mins; 100 - 110°C/ 3 - 5 mins.			
Test Methods: Not stated			
Physical		**Mechanical**	
Colour	Dark Tan + Green	τ (RT)	-
Viscosity	~30 Pa.s	τ (°C)	-
SG	1.2	τ (wet)	-
Shelf life	6 months	Peel ( )	-
Pot/Work life	2 -3 mins /23°C	σ_t	16.9 MPa
Tg	-25°C	σ_c	-
$T_{Service}$	-	E	388 MPa
CTE (°C)	-	G	-
Thermal cond.	-	Elongation	55%
Resistivity: Vol.	-	Tear strength	-
Surf.	-	H (Shore D)	60
Notes: -			

Ashland Chemicals	Pliogrip 7000/7020		(Apr 95)
Two part polyurethane, 100% reactive			
Manufacturers Comments: Solvent-free. High strength, chemical resistant. Primerless & minimal surface preparation. Meets General Motors GM3629M.			
Typical Use: Automotive body parts - metals & SMC plastics.			
Cure: Initial: 149°C/ 1 min (handling); Post cure: 149°C/ 30 mins.			
Test Methods: See notes			
Physical		**Mechanical**	
Colour	Tan + Green	τ (RT)	>3.4 MPa
Viscosity	~18 Pa.s (mixed)	τ (82°C)	>2.4 MPa
SG	1.2	τ (wet)	>3.3 MPa
Shelf life	6 months	Peel ( )	-
Pot/Work life	15 mins /23°C	σ_t	17.2 MPa
Tg	-35°C	σ_c	-
$T_{Service}$	-	E	400 MPa
CTE (°C)	-	G	193 MPa
Thermal cond.	-	Elongation	45%
Resistivity: Vol.	-	Tear strength	-
Surf.	-	H (Shore D)	65
Notes: τ min value for automotive grade SMCs, dry wipe, 0.76mm bondline. 2.54 x 10.16 cm lap shear, 6.45 cm² bond area. τ(wet) 14 days, water, 23°C. Water absorption (cured adhesive) <1.5%.			

Ashland Chemicals	Pliogrip 6600/6620		(Apr 95)
Two part polyurethane, 100% reactive			
Manufacturers Comments: Longer work-life. Temperature & chemical resistant. Automotive, marine & other industries. Requires primer.			
Typical Use: Automotive body parts - composite plastics & metals.			
Cure: RT(23°C)/ 1 hr; 104 - 121°C/ 4 - 6 mins.			
Test Methods: Not stated			
Physical		**Mechanical**	
Colour	Tan + Dark Green	τ (RT)	-
Viscosity	~30 Pa.s	τ (°C)	-
SG	1.2	τ (wet)	-
Shelf life	6 months	Peel ( )	-
Pot/Work life	15 -20 mins /23°C	σ_t	16.9 MPa
Tg	-25°C	σ_c	-
$T_{Service}$	-	E	388 MPa
CTE (°C)	-	G	-
Thermal cond.	-	Elongation	55%
Resistivity: Vol.	-	Tear strength	-
Surf.	-	H (Shore D)	60
Notes: Pliogrip 6600/6621 similar product.			

Ashland Chemicals	Pliogrip 7000/7021		(Apr 95)
Two part polyurethane, 100% reactive			
Manufacturers Comments: Solvent-free. High strength, chemical resistant. Primerless & minimal surface preparation. Meets General Motors GM3629M.			
Typical Use: Automotive body parts - metals & SMC plastics.			
Cure: Initial: 149°C/ 2 mins (handling); Post cure: 149°C/ 30 mins.			
Test Methods: See notes			
Physical		**Mechanical**	
Colour	Tan + Green	τ (RT)	>3.4 MPa
Viscosity	~18 Pa.s (mixed)	τ (82°C)	>2.4 MPa
SG	1.2	τ (wet)	>3.3 MPa
Shelf life	6 months	Peel ( )	-
Pot/Work life	30 mins /23°C	σ_t	17.2 MPa
Tg	-35°C	σ_c	-
$T_{Service}$	-	E	400 MPa
CTE (°C)	-	G	193 MPa
Thermal cond.	-	Elongation	45%
Resistivity: Vol.	-	Tear strength	-
Surf.	-	H (Shore D)	65
Notes: τ min value for automotive grade SMCs, dry wipe, 0.76mm bondline. 2.54 x 10.16 cm lap shear, 6.45 cm² bond area. τ(wet) 14 days, water, 23°C. Water absorption (cured adhesive) <1.5%.			

Ashland Chemicals	**Pliogrip 7000/7024**		(Apr 95)
Two part polyurethane, 100% reactive			
Manufacturers Comments: Solvent-free. High strength, chemical resistant. Primerless & minimal surface preparation. Meets General Motors GM3629M.			
Typical Use: Automotive body parts - metals & SMC plastics.			
Cure: Initial: 149°C/ 2 mins (handling); Post cure: 149°C/ 30 mins.			
Test Methods: See notes			
Physical		**Mechanical**	
Colour	Tan + Green	τ (RT)	>3.4 MPa
Viscosity	~18 Pa.s (mixed)	τ (82°C)	>2.4 MPa
SG	1.2	τ (wet)	>3.3 MPa
Shelf life	6 months	Peel ( )	-
Pot/Work life	25 mins /23°C	σ_t	17.2 MPa
Tg	-35°C	σ_c	-
$T_{Service}$	-	E	400 MPa
CTE (°C)	-	G	193 MPa
Thermal cond.	-	Elongation	45%
Resistivity: Vol.	-	Tear strength	-
Surf.	-	H (Shore D)	65
Notes: τ min value for automotive grade SMCs, dry wipe, 0.76mm bondline. 2.54 x 10.16 cm lap shear, 6.45 cm² bond area. τ(wet) 14 days, water, 23°C. Water absorption (cured adhesive) <1.5%.			

Ashland Chemicals	**Pliogrip 7300/7308**		(Apr 95)
Two part polyurethane, 100% reactive			
Manufacturers Comments: High performance, temperature resistant. Primerless. Automotive applications.			
Typical Use: Automotive body parts - metals & SMC plastics.			
Cure: Handling: RT/ 15-20 mins; 37°C/5-6 mins. Cure: 70-148°C/1-4 mins. Post cure: 93°C/ 30 mins.			
Test Methods: See notes			
Physical		**Mechanical**	
Colour	Tan + Green	τ (RT)	>3.3 MPa
Viscosity	~24 Pa.s	τ (82°C)	>2.7 MPa
SG	1.25	τ (wet)	>3.1 MPa
Shelf life	6 months	Peel ( )	-
Pot/Work life	9-13 mins /23°C	σ_t	21.0 MPa
Tg	-53°C	σ_c	-
$T_{Service}$	-	E	378 MPa
CTE (°C)	-	G	-
Thermal cond.	-	Elongation	46%
Resistivity: Vol.	-	Tear strength	-
Surf.	-	H (Shore)	-
Notes: τ min value for automotive grade SMCs, dry wipe, 0.76mm bondline. 2.54 x 10.16 cm lap shear, 6.45 cm² bond area. τ(wet) 14 days, water, 23°C. Water absorption 0%. Shrinkage <1.			

Ashland Chemicals	**Pliogrip 7300/7305**		(Apr 95)
Two part polyurethane, 100% reactive			
Manufacturers Comments: High performance, heat & chemical resistant. Primerless. Automotive, marine applications.			
Typical Use: Automotive body parts - metals & SMC plastics.			
Cure: Handling: RT/ 13-18 mins; 40°C/8-10 mins. Cure: 120°C/3-4 mins. Post cure: 120°C/ 30 mins; 85°C/1 hr.			
Test Methods: See notes			
Physical		**Mechanical**	
Colour	Tan + Green	τ (RT)	>3.3 MPa
Viscosity	~24 Pa.s	τ (82°C)	>2.7 MPa
SG	1.25	τ (wet)	>3.1 MPa
Shelf life	6 months	Peel ( )	-
Pot/Work life	4-5 mins /23°C	σ_t	21.0 MPa
Tg	-53°C	σ_c	-
$T_{Service}$	-	E	378 MPa
CTE (°C)	-	G	-
Thermal cond.	-	Elongation	46%
Resistivity: Vol.	-	Tear strength	-
Surf.	-	H (Shore)	-
Notes: τ min value for automotive grade SMCs, dry wipe, 0.76mm bondline. 2.54 x 10.16 cm lap shear, 6.45 cm² bond area. τ(wet) 14 days, water, 23°C. Water absorption 0%. Shrinkage <1.			

Ashland Chemicals	**Pliogrip 7400/7406C**		(Apr 95)
Two part polyurethane, 100% reactive			
Manufacturers Comments: High performance, chemical & temperature resistant. Primerless. Automotive. RF or heat curing. Meets: Ford Motors WSB-11P27B & General Motors GM 3639M.			
Typical Use: Automotive thermosetting & thermoplastics, metals.			
Cure: Handling: 100-110°C/4-6 mins. Cure: 149°C/90 secs. Post cure: 149°C/ 30 mins.			
Test Methods: See notes			
Physical		**Mechanical**	
Colour	Tan +Dark Green	τ (RT)	>3.8 MPa
Viscosity	~20 Pa.s	τ (82°C)	>3.4 MPa
SG	1.3	τ (wet)	>3.0 MPa
Shelf life	6 months	Peel ( )	-
Pot/Work life	20-23 mins /23°C	σ_t	25.6 MPa
Tg	-35°C	σ_c	-
$T_{Service}$	-	E	452 MPa
CTE (°C)	-	G	-
Thermal cond.	-	Elongation	70%
Resistivity: Vol.	-	Tear strength	-
Surf.	-	H (Shore D)	69
Notes: τ min value for automotive grade SMCs, dry wipe, 0.76mm bondline. 2.54 x 10.16 cm lap shear, 6.45 cm² bond area. τ(wet) 14 days, water, 23°C. Water absorption <1.0%. Shrinkage: Not seen.			

Bostik

Ashland Chemicals	Pliogrip 7400/7410		(Apr 95)
Two part polyurethane, 100% reactive			
Manufacturers Comments: High performance, chemical & temperature resistant. Primerless. Automotive. Meets: Ford Motors WSB-11P27B & General Motors GM 3639M.			
Typical Use: Automotive thermosetting & thermoplastics, metals.			
Cure: Handling: 100-110°C/2-3 mins. Cure: 149°C/1 min. Post cure: 149°C/ 30 mins.			
Test Methods: See notes			
Physical		**Mechanical**	
Colour	Tan +Dark Green	τ (RT)	>3.8 MPa
Viscosity	~20 Pa.s	τ (82°C)	>3.4 MPa
SG	1.3	τ (wet)	>3.0 MPa
Shelf life	6 months	Peel ( )	-
Pot/Work life	10-12 mins /23°C	σ_f	25.6 MPa
Tg	-35°C	σ_c	-
$T_{Service}$	-	E	452 MPa
CTE (°C)	-	G	-
Thermal cond.	-	Elongation	70%
Resistivity: Vol.	-	Tear strength	-
Surf.	-	H (Shore D)	69
Notes: τ min value for automotive grade SMCs, dry wipe, 0.76mm bondline. 2.54 x 10.16 cm lap shear, 6.45 cm² bond area. τ(wet) 14 days, water, 23°C. Water absorption <1.0%. Shrinkage: <1.			

Bostik	2639		July 92
One part polyurethane, moisture curing			
Manufacturers Comments: 93% solids. Good moisture resistance, poor solvent resistance. Adhesive sealant for motor vehicle body assembly, air conditioning conduits.			
Typical Use:, Metals (inc. anodised & galvanised steel, stainless), wood, plastics, glass & masonry.			
Cure: RT (23°C)/ ~2 hrs/65%RH (handling).			
Test Methods: NF			
Physical		**Mechanical**	
Colour	White*	τ (RT)	-
Viscosity	-	τ (°C)	-
SG	~1.2	τ (wet)	-
Shelf life	9 months/10 - 25°C	Peel ( )	-
Pot/Work life	-	σ_f	-
Tg	-	σ_c	-
$T_{Service}$	-5 to +35°C	E	0.6 MPa
CTE (°C)	-	G	-
Thermal cond.	-	Elongation	>250%
Resistivity: Vol.	-	Tear strength	-
Surf.	-	H (Shore A)	40 - 42
Notes: * Grey & black versions available. Cured adhesive can be painted.			

Ashland Chemicals	Pliogrip 7400/7420		(Apr 95)
Two part polyurethane, 100% reactive			
Manufacturers Comments: High performance, chemical & temperature resistant. Primerless. Automotive. Meets: Ford Motors WSB-11P27B & General Motors GM 3639M.			
Typical Use: Automotive thermosetting & thermoplastics, metals.			
Cure: Handling: 135°C/2 mins. Cure: 149°C/1 min. Post cure: 149°C/ 30 mins.			
Test Methods: See notes			
Physical		**Mechanical**	
Colour	Tan +Dark Green	τ (RT)	>4.5 MPa
Viscosity	~15 Pa.s	τ (82°C)	>2.9 MPa
SG	1.3	τ (wet)	>3.7 MPa
Shelf life	6 months	Peel ( )	-
Pot/Work life	12-15 mins /23°C	σ_f	17.0 MPa
Tg	-5°C	σ_c	-
$T_{Service}$	-	E	48 MPa
CTE (°C)	-	G	-
Thermal cond.	-	Elongation	120%
Resistivity: Vol.	-	Tear strength	-
Surf.	-	H (Shore D)	69
Notes: τ min value for automotive grade SMCs, dry wipe, 0.76mm bondline. 2.54 x 10.16 cm lap shear, 6.45 cm² bond area. τ(wet) 14 days, water, 23°C. Water absorption <1.0%. Shrinkage: <1.			

Bostik	Matrix 500		Nov 94
One part polyurethane, moisture curing			
Manufacturers Comments: General purpose adhesive sealant for engineering uses where joint movement expected. Non-slump, non-corrosive. Paintable. Good ageing resistance. Cartridge pack.			
Typical Use: Metals, concrete, masonry, GRP & wood.			
Cure: RT (10-35°C). Cure rate: 3-4 mm/24 hrs (23°C/50%RH).			
Test Methods: Not Stated			
Physical		**Mechanical**	
Colour	-	τ (RT)	-
Viscosity	-	τ (°C)	-
SG	-	τ (wet)	-
Shelf life	9 months/10-25°C	Peel ( )	-
Pot/Work life	1 hr (tack-free time)	σ_f	> 1.5 MPa
Tg	-	σ_c	-
$T_{Service}$	-40 to +90°C*	E	-
CTE (°C)	-	G	-
Thermal cond.	-	Elongation	>700%
Resistivity: Vol.	-	Tear strength	-
Surf.	-	H (Shore A)	35
Notes: Intermittent use +120°C.			

Bostik	Matrix 600FC		Nov 94
One part polyurethane, moisture curing			
Manufacturers Comments: General purpose adhesive sealant for engineering uses. Fast curing, non-slump, non-corrosive. Sandable & paintable. Water & chemical resistance. Cartridge pack.			
Typical Use: Metals, concrete, masonry, GRP & wood.			
Cure: RT (10-35°C). Cure rate: 4 mm/ hr (23°C/50%RH).			
Test Methods: Not Stated			
Physical		**Mechanical**	
Colour	(Various)	τ (RT)	-
Viscosity	-	τ (°C)	-
SG	-	τ (wet)	-
Shelf life	9 months/10-25°C	Peel ( )	-
Pot/Work life	45 mins (tack-free)	σ_t	> 1.6 MPa
Tg	-	σ_c	-
$T_{Service}$	-40 to +90°C*	E	-
CTE (°C)	-	G	-
Thermal cond.	-	Elongation	>600%
Resistivity: Vol.	-	Tear strength	-
Surf.	-	H (Shore)	-
Notes: Intermittent use +120°C.			

Bostik	Parcol		July 95
One part polyurethane, thixotropic paste			
Manufacturers Comments: Non-flammable. Wood floor bonding (inc. heated floors). Good temperature, humidity & impact resistance. High bond strengths.			
Typical Use: Wood flooring (untreated & finished) to concrete.			
Cure: RT/ 1 hr (handling); 1-2 days (full cure).			
Test Methods: Not Stated			
Physical		**Mechanical**	
Colour	Light Beige	τ (RT)	-
Viscosity	-	τ (°C)	-
SG	~1.3	τ (wet)	-
Shelf life	1 year/ RT	Peel ( )	-
Pot/Work life	~ 2 hrs	σ_t	-
Tg	-	σ_c	-
$T_{Service}$	-5 to +35°C	E	-
CTE (°C)	-	G	-
Thermal cond.	-	Elongation	-
Resistivity: Vol.	-	Tear strength	-
Surf.	-	H (Shore)	-
Notes:			

Bostik	Matrix 700 Marine		Nov 94
One part polyurethane, moisture curing			
Manufacturers Comments: Marine adhesive sealant for engineering uses. Non-slump, non-corrosive. Sandable & paintable. Good resistance to fresh, salt & chlorinated water. Cartridge pack.			
Typical Use: Wood, GRP, metal, polycarbonate & acrylic panels.			
Cure: RT (10-35°C). Cure rate: 4 mm/ hr (23°C/50%RH).			
Test Methods: Not Stated			
Physical		**Mechanical**	
Colour	-	τ (RT)	-
Viscosity	-	τ (°C)	-
SG	-	τ (wet)	-
Shelf life	9 months/10-25°C	Peel ( )	-
Pot/Work life	45 mins (tack-free)	σ_t	> 1.6 MPa
Tg	-	σ_c	-
$T_{Service}$	-40 to +90°C*	E	-
CTE (°C)	-	G	-
Thermal cond.	-	Elongation	>600%
Resistivity: Vol.	-	Tear strength	-
Surf.	-	H (Shore)	-
Notes: Intermittent use +120°C.			

Bostik	PU 505		Jan 91
Two part polyurethane			
Manufacturers Comments: For wide range of materials. Building uses & laminated panels. High performance (mechanical & chemical).			
Typical Use: Wood, floor tiles, metals, glass, plastics (PVC, vinyl, polystyrene), masonry.			
Cure: RT (20°C)/ 4 hrs (handling); 2 days (full cure).			
Test Methods: Not Stated			
Physical		**Mechanical**	
Colour	Dark Beige	τ (RT)	-
Viscosity	100 Pa.s	τ (°C)	-
SG	1.2	τ (wet)	-
Shelf life	-	Peel ( )	-
Pot/Work life	1 hr (1 kg/20°C)	σ_t	-
Tg	-	σ_c	-
$T_{Service}$	<150°C*	E	-
CTE (°C)	-	G	-
Thermal cond.	-	Elongation	-
Resistivity: Vol.	-	Tear strength	-
Surf.	-	H (Shore)	-
Notes: Mix ratio 25:100. * Creep (beech-beech) after 7 day cure.			

See Key to Data Sheets - Pages 63 to 65

Wm. Canning (Hylomar)

W. Canning	Hylomar Polyurethane Sealant		(Feb 95)
One part polyurethane			
Manufacturers Comments: Permanently flexible. Excellent chemical and weathering resistance. Bonds well to a wide range of materials.			
Typical Use: Automotive, coach-building, construction and industrial. Bonding and sealing of metal panels, etc.			
Cure: RT / 4mm in 24 hrs.			
Test Methods: DIN			
Physical		**Mechanical**	
Colour	Black, white or grey	τ (RT)	-
Viscosity	-	τ (°C)	-
SG	-	τ (wet)	-
Shelf life	-	Peel ( )	-
Pot/Work life	-	σ_t	1.74 MPa
Tg	-	σ_c	-
$T_{Service}$	-25 to +80°C	E	0.75 MPa
CTE (°C)	-	G	-
Thermal cond.	-	Elongation	450 %
Resistivity: Vol.	-	Tear strength	-
Surf.	-	H (Shore A)	40
Notes: Skin time: 10 - 15 min. Tensile: DIN 53504.			

Casco Nobel	1829 PUR 1K		Aug 94
One part polyurethane, thixotropic, moisture curing			
Manufacturers Comments: 99% solids. Manufacture of building structural panels. Water-proof wood bonding. High heat resistance, foaming (gap-filling). Requires water-misting.			
Typical Use: Metals, plastics, GRP, insulating materials, glass & wood.			
Cure: RT (23°C/50%RH)/ 2.5 hrs. Press: 20°C/ >0.2 kg/cm²/ 30 min.			
Test Methods: Not Stated			
Physical		**Mechanical**	
Colour	Brown	τ (RT)	-
Viscosity	10 Pa.s	τ (°C)	-
SG	~1.5	τ (wet)	-
Shelf life	6 months/10-18°C	Peel ( )	-
Pot/Work life	10 mins/20°C*	σ_t	-
Tg	-	σ_c	-
$T_{Service}$	-	E	-
CTE (°C)	-	G	-
Thermal cond.	-	Elongation	-
Resistivity: Vol.	-	Tear strength	-
Surf.	-	H (Shore)	-
Notes: * after water misting. Flash point: >200°C. Inflammability test (bulkheads) approved by: See-Berufsgenossenschaft Ref 119.050.			

Casco Nobel

Casco Nobel	1801 PUR 2K		March 94
Two part polyurethane (1821 curing agent)			
Manufacturers Comments: 100% solids. Manufacture of building structural panels. High strength at high temperatures.			
Typical Use: Metals, asbestos sheets, plastics, GRP, insulating materials, glass & wood.			
Cure: RT (20°C)/ 3 hrs (handling); ~1 week (full cure). Press: 20°C/ >0.02 MPa/ 90 mins.			
Test Methods: Not Stated			
Physical		**Mechanical**	
Colour	Light Brown	τ (RT)	-
Viscosity	14 Pa.s	τ (°C)	-
SG	~1.8	τ (wet)	-
Shelf life	1 year/10-18°C	Peel ( )	-
Pot/Work life	~ 90 mins/20°C	σ_t	-
Tg	-	σ_c	-
$T_{Service}$	-	E	-
CTE (°C)	-	G	-
Thermal cond.	-	Elongation	-
Resistivity: Vol.	-	Tear strength	-
Surf.	-	H (Shore D)	90
Notes: Mix ratio 100:20(wt). Flash point: >200°C.			

Casco Nobel	1833 PUR 1K		June 93
One part polyurethane, moisture curing			
Manufacturers Comments: 98% solids. Manufacture of building structural panels. Water-proof wood bonding. High heat resistance, foaming (gap-filling). Requires water-misting.			
Typical Use: Metals (inc. galvanised steel), plastics, GRP, insulating materials, glass & wood.			
Cure: RT (23°C/50%RH)/ 5 hrs. Press: 20°C/ >0.02 MPa/ 1 hr.			
Test Methods: Not Stated			
Physical		**Mechanical**	
Colour	Brown	τ (RT)	-
Viscosity	15 Pa.s	τ (°C)	-
SG	~1.6	τ (wet)	-
Shelf life	6 months/10-18°C	Peel ( )	-
Pot/Work life	~45 mins/20°C*	σ_t	-
Tg	-	σ_c	-
$T_{Service}$	-	E	-
CTE (°C)	-	G	-
Thermal cond.	-	Elongation	-
Resistivity: Vol.	-	Tear strength	-
Surf.	-	H (Shore)	-
Notes: * after water misting. Flash point: >200°C. Inflammability test (bulkheads) approved by: See-Berufsgenossenschaft Ref 119.050.			

See Key to Data Sheets - Pages 63 to 65

Casco Nobel	1846 PUR 2K		March 94
Two part polyurethane (1821 curing agent)			
Manufacturers Comments: Manufacture of building structural panels. Open time not affected by humidity.			
Typical Use: Metals, asbestos sheets, plastics, GRP to PVC, insulating materials, glass & wood.			
Cure: RT (20°C)/ 90 min (handling); ~2 weeks (full cure). Press: 20°C/ >0.02 MPa/ 2 hrs.			
Test Methods: Not Stated			
Physical		**Mechanical**	
Colour	White	τ (RT)	-
Viscosity	8.5 Pa.s	τ (°C)	-
SG	~1.5	τ (wet)	-
Shelf life	1 year/10-18°C	Peel ( )	-
Pot/Work life	~ 20 mins/20°C	σ_f	-
Tg	-	σ_c	-
$T_{Service}$	-	*E*	-
CTE (°C)	-	*G*	-
Thermal cond.	-	Elongation	-
Resistivity: Vol.	-	Tear strength	-
Surf.	-	H (Shore D)	80
Notes: Mix ratio 100:22 (wt). Flash point: >200ºC.			

Casco Nobel	1897 PUR 2K		June 94
Two part polyurethane (1802 curing agent)			
Manufacturers Comments: Assembly of profiles & structural panels. Thixotropic, gap-filling. Heat resistance (sanded metals).			
Typical Use: GRP, wood, metals, ceramics & pre-treated plastics.			
Cure: RT (20°C)/ 4 hrs (handling); 1 month (full cure). Press: 20°C/ >0.02 MPa/ 2 hrs.			
Test Methods: Not Stated			
Physical		**Mechanical**	
Colour	White	τ (RT)	-
Viscosity	550 Pa.s	τ (°C)	-
SG	~1.5	τ (wet)	-
Shelf life	1 year/10-18°C	Peel ( )	-
Pot/Work life	~ 90 min/20°C	σ_f	-
Tg	-	σ_c	-
$T_{Service}$	-	*E*	-
CTE (°C)	-	*G*	-
Thermal cond.	-	Elongation	-
Resistivity: Vol.	-	Tear strength	-
Surf.	-	H (Shore D)	80
Notes: Mix ratio 100:20 (wt). Flash point: >200ºC. Inflammability test (bulkheads) approved by: See-Berufsgenossenschaft Ref. 119.050.			

Casco Nobel	1894 PUR 1K		Aug 93
One part polyurethane, moisture curing			
Manufacturers Comments: 100% solids. Manufacture of building structural panels. Water-proof wood bonding. High heat resistance, slight foaming (gap-filling). Requires water-misting.			
Typical Use: Metals, plastics, GRP, insulating materials, glass & wood.			
Cure: RT (23°C/50%RH)/ 10 hrs. Press: 20°C/ >0.02 MPa/ 90 min.			
Test Methods: Not Stated			
Physical		**Mechanical**	
Colour	Brown	τ (RT)	-
Viscosity	11 Pa.s	τ (°C)	-
SG	~1.1	τ (wet)	-
Shelf life	1 year/10-18°C	Peel ( )	-
Pot/Work life	~45 mins/20°C*	σ_f	-
Tg	-	σ_c	-
$T_{Service}$	-	*E*	-
CTE (°C)	-	*G*	-
Thermal cond.	-	Elongation	-
Resistivity: Vol.	-	Tear strength	-
Surf.	-	H (Shore)	-
Notes: * after water misting. Flash point: >200ºC.			

Casco Nobel	1899 PUR 2K		Mar 94
Two part polyurethane (1821 curing agent)			
Manufacturers Comments: 100% solids. Manufacturing building structural panels. Long open-time, low press time. Not affected by humidity.			
Typical Use: Metals, asbestos sheets, insulation materials, wood, plastics, GRP.			
Cure: RT (20°C)/ 4 hrs (handling); 1 month (full cure). Press: 20°C/ >0.02 MPa/ 1 hr.			
Test Methods: Not Stated			
Physical		**Mechanical**	
Colour	Light Brown	τ (RT)	-
Viscosity	8 Pa.s	τ (°C)	-
SG	~1.5	τ (wet)	-
Shelf life	1 year/10-18°C	Peel ( )	-
Pot/Work life	~ 135 min/20°C	σ_f	-
Tg	-	σ_c	-
$T_{Service}$	-	*E*	-
CTE (°C)	-	*G*	-
Thermal cond.	-	Elongation	-
Resistivity: Vol.	-	Tear strength	-
Surf.	-	H (Shore D)	85
Notes: Mix ratio 100:22 (wt). Flash point: >200ºC. Inflammability test (bulkheads) approved by: See-Berufsgenossenschaft Ref. 119.050.			

Casco Nobel	1959 PUR 2K		Nov 94
Two part polyurethane (1875 curing agent)			
Manufacturers Comments: 100% solids. Assembly adhesive. High impact strength.			
Typical Use: Wood, metal & special constructions (bone).			
Cure: RT (20°C/50%RH)/ 1 week; Pressure:0.02 MPa.			
Test Methods: ISO, ASTM			
Physical		**Mechanical**	
Colour	Red	τ (RT)	-
Viscosity	4 Pa.s	τ (°C)	-
SG	~1.5	τ (wet)	-
Shelf life	1 year/10-18°C	Peel ( )	-
Pot/Work life	20-30mins (100g)	σ_t	17.5 MPa
Tg	-	σ_c	56.3 MPa
$T_{Service}$	-	E	2239 MPa
CTE (°C)	-	G	-
Thermal cond.	-	Elongation	4.3%
Resistivity: Vol.	-	Tear strength	-
Surf.	-	H (Shore D)	80-85
Notes: Mix ratio 100:25 (wt). Mechanical data after 3 weeks: σt, elong ISO527; σc ISO604 (cylinder); *E* ASTM D-790 (3-point rod bend)			

CECA	Mastic Colle		Sept 92
One part polyurethane, moisture curing			
Manufacturers Comments: Adhesive sealant for building uses. Expansion joints materials & facades. Good environmental resistance (inc. corrosive atmospheres). Specifications: SNJF, CEBTP, DTU 53.2.			
Typical Use: Concrete, glass, metals & wood.			
Cure: RT(5 to 40°C). Cure rate: 5mm/24 hrs.			
Test Methods: Not stated			
Physical		**Mechanical**	
Colour	Grey or White	τ (RT)	-
Viscosity	-	τ (°C)	-
SG	~1.2	τ (wet)	-
Shelf life	1 year (5-25°C)	Peel ( )	-
Pot/Work life	2 hrs	σ_t	-
Tg	-	σ_c	-
$T_{Service}$	-30 to +80°C	E	-
CTE (°C)	-	G	-
Thermal cond.	-	Elongation	-
Resistivity: Vol.	-	Tear strength	-
Surf.	-	H (Shore)	-
Notes:			

CECA

CECA	Colle 455		July 94
Two part polyurethane			
Manufacturers Comments: 98% solids. Flooring adhesive (inc. heated floors). Interior/exterior use. Heat & humidity resistance.			
Typical Use: Flooring - wood, PVC, carpet, synthetic grass surfaces.			
Cure: RT(12 to 30°C)/ 24 hrs; 72 hrs (full cure).			
Test Methods: NF			
Physical		**Mechanical**	
Colour	Yellow	τ (RT)	-
Viscosity	25 Pa.s	τ (°C)	-
SG	1.58	τ (wet)	-
Shelf life	6 months (10-30°C)	Peel ( )	-
Pot/Work life	1-1.5 hrs/7.8 kg(RT)	σ_t	-
Tg	-	σ_c	-
$T_{Service}$	<120°C	E	-
CTE (°C)	-	G	-
Thermal cond.	-	Elongation	-
Resistivity: Vol.	-	Tear strength	-
Surf.	-	H (Shore)	-
Notes:			

CFPI

CFPI	Galfix A 835		(Sept 95)
Two part polyurethane			
Manufacturers Comments: Solvent-free, >99% solids. Thixotropic. Assembly adhesive developed for bonding GRP without surface preparation. Also coated metal parts. Resistance to temperature, moisture, mineral oils & hydrocarbons. Meets: RVI 45-14-200; Jeanneau Div. Microcar (with surface prep.).			
Typical Use: Automotive assembly: GRP, metals.			
Cure: RT / <10 secs. (gel); ~12hrs (23°C/50%RH); 80°C/~20 mins.			
Test Methods: See notes			
Physical		**Mechanical**	
Colour	Pale Green	τ (RT, steel)	10-12 MPa
Viscosity	Thixotropic	τ (°C)	-
SG	1.34	τ (wet)	-
Shelf life	3 months/10-30°C	Peel ( )	-
Pot/Work life	15-20 mins	σ_t	-
Tg	-	σ_c	-
$T_{Service}$	-	E	6-7 MPa (rupture)
CTE (°C)	-	G	-
Thermal cond.	-	Elongation	25 %
Resistivity: Vol.	-	Tear strength	-
Surf.	-	H (Shore D)	55 (95 Shore A)
Notes: Flash point >200°C. τ data (ME RNUR 1180), 0.2mm sheet, phos. coated. 2mm thick: 5-6 MPa; SMC: 10 MPa (0.2 mm thick), 4-5 MPa (2mm thick).			

See Key to Data Sheets - Pages 63 to 65

CFPI	Galfix A 838		(Sept 95)
Two part polyurethane			
Manufacturers Comments: Solvent-free, >99% solids. Thixotropic. Assembly adhesive-sealant for automotive industry. Resistance to temperature, moisture, mineral oils & hydrocarbons, dilute acids & alkalis. Meets: Renault 45 14 508.			
Typical Use: Automotive materials: ABS, PC, PMMA, SMC, metal, wood.			
Cure: RT/10-20 mins (application); Full cure: 48 hrs (23°C/50-60%RH).			
Test Methods: See notes			
Physical		**Mechanical**	
Colour	Dk. Grey	τ (RT, steel)	≥8 (≥4) MPa
Viscosity	Thixotropic	τ (°C)	-
SG	1.75	τ (wet)	-
Shelf life	6 months/10-35°C	Peel ( )	-
Pot/Work life	30-35 mins	σ_t	-
Tg	-	σ_c	-
$T_{Service}$	-	E	-
CTE (°C)	-	G	-
Thermal cond.	-	Elongation	20-25 %
Resistivity: Vol.	-	Tear strength	-
Surf.	-	H (Shore D)	60 (80 Shore A)
Notes: τ data (ME RNUR 1180), steel sheet, phos.: 0.2 mm thick, (2mm thick); SMC: ≥4 MPa (2 mm thick); Wood: ≥4 MPa (2mm thick).			

CFPI	Lidec A 250 D		(Sept 95)
Two part polyurethane			
Manufacturers Comments: Solvent-free, flexible, elastic adhesive for automotive assembly. Thixotropic. For bonding GRP without surface preparation. Also coated metal parts. Resistance to temperature, moisture, mineral oils & hydrocarbons. Meets requirements of: Renault 03 70 160 (RNUR).			
Typical Use: Polyester-based composites (SMC, BZMC, RTM, etc), Metal (with primer). Also thermoplastics (ABS, PC, PP).			
Cure: RT/<30 secs (gel); Full cure: 12 hrs (RT), 15 mins (80°C).			
Test Methods: See notes			
Physical		**Mechanical**	
Colour	Green	τ (RT, steel)	6-7 MPa
Viscosity	Thixotropic	τ (°C)	-
SG	1.24	τ (wet) SMC	3-5 MPa
Shelf life	-	Peel ( )	-
Pot/Work life	5-10 min (work)	σ_t	-
Tg	-	σ_c	-
$T_{Service}$	-40 to +100°C	E	8 MPa (rupture)
CTE (°C)	-	G	-
Thermal cond.	-	Elongation	60 %
Resistivity: Vol.	-	Tear strength	-
Surf.	-	H (Shore A)	80
Notes: τ data (ME RNUR 1180): steel; SMC: 4-5 MPa; ABS: 5-6 MPa. τ (wet): After moisture ageing (ME RNUR 1165); Elong. & E: ME RNUR 1545.			

CFPI	Lidec A 223 D		(Sept 95)
Two part polyurethane			
Manufacturers Comments: Flexible, elastic adhesive for automotive assembly. Meets: Matra specification (coach-building); Renault 03 70 160 (RNUR) without surface preparation.			
Typical Use: Polyester-based composites, metal (galvanised or painted). Also glass, thermoplastics (ABS, PC), wood.			
Cure: RT/15 mins (open); Full cure: 7 days (RT), 1 day (80°C).			
Test Methods: See notes			
Physical		**Mechanical**	
Colour	Pale Green	τ (RT, steel)	3 MPa
Viscosity	Thixotropic	τ (°C)	-
SG	1.42	τ (wet)	-
Shelf life	3 months/15-35°C	Peel ( )	-
Pot/Work life	10 min (pot)	σ_t	-
Tg	-	σ_c	-
$T_{Service}$	-40 to +100°C	E	-
CTE (°C)	-	G	-
Thermal cond.	-	Elongation	150 %
Resistivity: Vol.	-	Tear strength	-
Surf.	-	H (Shore A)	60
Notes: τ data : steel sheet 2mm thick; Composite: 2.5 MPa.			

Chryso

Chryso	Tectane 1000		(Aug 95)
One part polyurethane, moisture curing			
Manufacturers Comments: Adhesive sealant for boat building & automotive body-work. High adhesion. Fungus & bacteria resistance. Low UV resistance. Good ageing.			
Typical Use: Polyester, polycarbonate & aluminium.			
Cure: RT.			
Test Methods: Not stated			
Physical		**Mechanical**	
Colour	Grey*	τ (RT)	~1.5 MPa
Viscosity	High	τ (°C)	-
SG	-	τ (wet)	-
Shelf life	-	Peel ( )	-
Pot/Work life	-	σ_t	-
Tg	-	σ_c	-
$T_{Service}$	-40 to +120°C	E	-
CTE (°C)	-	G	-
Thermal cond.	-	Elongation	-
Resistivity: Vol.	-	Tear strength	-
Surf.	-	H (Shore)	-
Notes: *Also black & white. Similar products: Tectane 1030 (high UV resistance); Tectane 1035 (slow cure, decking compound); Tectane 1010 (bubble-free).			

Chryso	Tectane 1020		(Aug 95)
One part polyurethane, moisture curing			
Manufacturers Comments: Adhesive sealant for boat building & automotive body-work.			
Typical Use: Glass to metal structures.			
Cure: RT.			
Test Methods: Not stated			
Physical		**Mechanical**	
Colour	-	τ (RT)	2.5 MPa
Viscosity	High	τ (°C)	-
SG	-	τ (wet)	-
Shelf life	-	Peel ( )	-
Pot/Work life	20 mins	σ_t	-
Tg	-	σ_c	-
$T_{Service}$	-40 to +120°C	E	-
CTE (°C)	-	G	-
Thermal cond.	-	Elongation	-
Resistivity: Vol.	-	Tear strength	-
Surf.	-	H (Shore)	-
Notes: Similar product: Tectane 1050 (fast cure), τ: 5 MPa, black.			

Ciba Polymers	Uralane 5774A/B		Aug 92
Two part polyurethane			
Manufacturers Comments: Solvent-free, self extinguishing. Tough, impact resistant bonds. High peel strength. Meets: BMS5-105, Type 5. Cartridge pack.			
Typical Use: Metals, engineering plastics.			
Cure: RT/ 7 days; RT(gel) + 65°C/1 - 2 hrs or 95°C/1 - 2 hrs.			
Test Methods: ASTM			
Physical		**Mechanical**	
Colour	Beige	τ (RT)	15 MPa
Viscosity	Semi-paste*	τ (80°C)	9 MPa
SG	1.15	τ (wet)	13 MPa
Shelf life	6 months/25°C	Peel (T)	8.00 N mm^{-1}
Pot/Work life	15-25mins(100g/RT)	σ_t	-
Tg	-	σ_c	-
$T_{Service}$	-55 to +80°C	E	-
CTE (°C)	-	G	-
Thermal cond.	-	Elongation	-
Resistivity: Vol.	-	Tear strength	-
Surf.	-	H (Shore D)	60 (RT)
Notes: Mix ratio 100:55 (wt). Optimum bondline thickness 0.05-0.1mm * Viscosity ASTM D-2393; T-service ASTM D-648; Hardness ASTM D-2240; τ Al-alloy, acid etched; τ(wet) 30 days/60°C/95%RH; Flash point (ASTM D-92).			

Ciba Polymers

Ciba Polymers	Araldite 2018		Nov 93
Two part polyurethane, thixotropic, liquid			
Manufacturers Comments: Bonds well to most thermoplastics. Flexible bond, pale colour.			
Typical Use: Thermoplastics (SMC, polycarbonate, ABS, etc.).			
Cure: RT(23°C) / 16 hrs; 60°C/ 30 mins.			
Test Methods: ISO			
Physical		**Mechanical**	
Colour	Pale Opaque	τ (RT)	12 MPa
Viscosity	~ 8 Pa.s	τ (80°C)	3 MPa
SG	~1.13	τ (wet)	4 MPa
Shelf life	1 year (18-25°C)	Peel (FR)	4.00 N mm^{-1}
Pot/Work life	40 mins(100g/RT)	σ_t	-
Tg	-	σ_c	-
$T_{Service}$	-55 to +80°C	E	-
CTE (°C)	-	G	-
Thermal cond.	-	Elongation	35%
Resistivity: Vol.	-	Tear strength	-
Surf.	-	H (Shore)	-
Notes: Mix ratio 100:95 (wt). τ ISO4587; τ(wet) 60 days/23°C/water; Peel - Roller ISO4578.			

Conap

Conap	Conastic AD-20		(April 95)
Two part polyurethane			
Manufacturers Comments: 100% solids. Dielectric adhesive. Flexible, high strength for bonding thermoplastics.			
Typical Use: Polycarbonate, Kapton®, Hystrel®, PVC, ABS, nylon.			
Cure: RT(25°C) / 5-7 days; 80°C/ 4-6 hrs.			
Test Methods: ASTM			
Physical		**Mechanical**	
Colour	Amber Opaque	τ (RT)	-
Viscosity	5.5 Pa.s	τ (°C)	-
SG	0.99	τ (wet)	-
Shelf life	-	Peel ( )	-
Pot/Work life	15 mins(225g/RT)	σ_t	13.8 MPa
Tg	-	σ_c	-
$T_{Service}$	-55 to +80°C	E	-
CTE (°C)	-	G	-
Thermal cond.	-	Elongation	500%
Resistivity: Vol.	2.1 × 10^{15} Ω.cm	Tear strength	48 N mm^{-1}
Surf.	-	H (Shore A)	90
Notes: Mix ratio 100:17.5 (wt).SG ASTM D-792-66; σ, Elong ASTM D-214-80; Hardness ASTM D-2240-81. Resistivity ASTM D-257-78. Electrical Properties, RT: Dielectric Strength ASTM D-149-81: 24×10^3 V/mm; Arc Resistance: >120 secs; Dielectric Constants ASTM D-150-81: 3.14 (100Hz), 2.87 (1 MHz); Dissipation Factors ASTM D-150-81: 0.029 (100Hz), 0.017 (1 MHz).			

Degussa

Degussa	Agomet PU 540 AB		July 94
Two part polyurethane			
Manufacturers Comments: High shear strength, humidity resistant, neutral odour. For manufacture of aluminium-based products in the building/construction industry.			
Typical Use: Metals - Al-alloys, steels (inc. stainless), copper, brass. Plastics - rigid PVC, ABS, polyacrylic, polycarbonate. Wood, ceramic.			
Cure: RT(10 - 35°C).			
Test Methods: DIN			
Physical		**Mechanical**	
Colour	-	τ (RT)	22.6 MPa
Viscosity	~20 Pa.s	τ (40°C)	15.3 MPa
SG	~1.3	τ (wet)	20.4 MPa
Shelf life	9 months	Peel ( )	-
Pot/Work life*	40 + 210 mins/20°C	σ_f	-
Tg	-	σ_c	-
$T_{Service}$	-30 to +80°C	E	-
CTE (°C)	-	G	-
Thermal cond.	-	Elongation	-
Resistivity: Vol.	-	Tear strength	-
Surf.	-	H (Shore D)	65

Notes: * Pot life + worklife, these, plus viscosity, decrease with increasing temperature. Mix ratio 0.8:1 (wt). Hardness DIN 53505. τ(wet) DIN 50017, 28 days. Flash point: >100°C.

Dunlop	HM 4201		Jan 94
One part, warm-melt, moisture-curing polyurethane			
Manufacturers Comments: Solid at RT, warmed before applying. Reacts with atmospheric moisture to cure. Fast curing. Tough, heat resistant adhesive.			
Typical Use: Automotive assembly. Wood - edge-banding.			
Cure: Application Temp. 150-160°C. RT/~4 days (cure).			
Test Methods: Not Stated			
Physical		**Mechanical**	
Colour	White→Straw	τ (RT)	-
Viscosity	45 Pa.s (160°C)	τ (°C)	-
SG	-	τ (wet)	-
Shelf life	-	Peel ( )	-
Pot/Work life	30-60 secs (open)	σ_f	-
Tg	-	σ_c	-
$T_{Service}$	-	E	-
CTE (°C)	-	G	-
Thermal cond.	-	Elongation	-
Resistivity: Vol.	-	Tear strength	-
Surf.	-	H (Shore)	-

Notes: Softening point (ring & ball): ~80°C. Application temperature must not exceed 170°C.

Dunlop

Dunlop	HM 4200		June 92
One part, warm-melt, moisture-curing polyurethane			
Manufacturers Comments: Solid at RT, warmed before applying. Reacts with atmospheric moisture to cure. Excellent adhesion to a variety of materials. Tough, heat resistant adhesive.			
Typical Use: Leather-cloth, painted metals, wood.			
Cure: Application Temp. 110-120°C. RT/3-7 days (cure).			
Test Methods: Not Stated			
Physical		**Mechanical**	
Colour	White→Straw	τ (RT)	-
Viscosity	10-15 Pa.s (120°C)	τ (°C)	-
SG	-	τ (wet)	-
Shelf life	-	Peel ( )	-
Pot/Work life	3 mins (open)	σ_f	-
Tg	-	σ_c	-
$T_{Service}$	+100°C (cured)	E	-
CTE (°C)	-	G	-
Thermal cond.	-	Elongation	-
Resistivity: Vol.	-	Tear strength	-
Surf.	-	H (Shore)	-

Notes: Softening point (ring & ball): ~60-65°C . Application temperature must not exceed 120°C.

Dunlop	HM 4202		June 93
One part, warm-melt, moisture-curing polyurethane			
Manufacturers Comments: Solid at RT, warmed before applying. Reacts with atmospheric moisture to cure. Adhesion to a variety of substrates. Tough, heat resistant adhesive.			
Typical Use: Leather-cloth, painted metals, plastics, GRP, wood.			
Cure: Application Temp. 110-130°C. RT/3-7 days (cure).			
Test Methods: Not Stated			
Physical		**Mechanical**	
Colour	White→Straw	τ (RT)	-
Viscosity	10 Pa.s (120°C)	τ (°C)	-
SG	-	τ (wet)	-
Shelf life	-	Peel ( )	-
Pot/Work life	~2 mins (open)	σ_f	-
Tg	-	σ_c	-
$T_{Service}$	+100°C	E	-
CTE (°C)	-	G	-
Thermal cond.	-	Elongation	-
Resistivity: Vol.	-	Tear strength	-
Surf.	-	H (Shore)	-

Notes: Softening point (ring & ball): ~62°C. Set time 10 secs (metal) Application temperature must not exceed 160°C.

Dyckerhoff

Dyckerhoff	PU-Kleber #503		(Feb 95)
Two-part, reaction, polyurethane			
Manufacturers Comments: Fast RT cure. Flexible, water-resistant. For joining & laying ceramic coverings (floor/wall). Suitable for use on moisture-sensitive surfaces (wood products, gypsum board). Use with coverings to: DIN EN 159; DIN E 176; DIN 18 158 & equivalents.			
Typical Use: Housing. Ceramic coverings to wood, concrete & plaster.			
Cure: RT (20°C)/24 hrs (foot-traffic); 48 hrs (full cure).			
Test Methods: Not Stated			
Physical		**Mechanical**	
Colour	-	τ (RT)	-
Viscosity	-	τ (°C)	-
SG	-	τ (wet)	-
Shelf life	-	Peel ( )	-
Pot/Work life	45 mins (open)	σ_t	-
Tg	-	σ_c	-
$T_{Service}$	-	*E*	-
CTE (°C)	-	*G*	-
Thermal cond.	-	Elongation	-
Resistivity: Vol.	-	Tear strength	-
Surf.	-	H (Shore)	-
Notes:			

EMFI	Isole S 35 / 521		(Nov 95)
Two part polyurethane			
Manufacturers Comments: Solvent-free. For assembly of building panels, etc.			
Typical Use: Insulation materials (plastic foams and mineral wool) to plaster-board, wood, decorative laminates, sheet metal (steel, aluminium, St. steel), inc. galvanised and painted.			
Cure: RT/5 hrs (assembly); 2-3 days (full strength).			
Test Methods: Not stated			
Physical		**Mechanical**	
Colour	Brown	τ (RT)	-
Viscosity	~350 Pa.s	τ (°C)	-
SG	1.45	τ (wet)	-
Shelf life	6 months	Peel ( )	-
Pot/Work life	~50 mins	σ_t	-
Tg	-	σ_c	-
$T_{Service}$	-70 to +150°C	*E*	-
CTE (°C)	-	*G*	-
Thermal cond.	-	Elongation	-
Resistivity: Vol.	-	Tear strength	-
Surf.	-	H (Shore)	-
Notes:			

EMFI

EMFI	Isole 3100 Mono		(Nov 95)
One part polyurethane (moisture curing)			
Manufacturers Comments: High cohesion, flexibility. Water resistance, low temperatures and ageing. For assembly of exterior woodwork and boat-building. Assembly of building materials and panels (insulation to skins).			
Typical Use: Wood, plaster-board, metals, masonry, mineral insulation materials.			
Cure: RT/ 8-12 hrs (initial cure); 24-48 hrs (full cure).			
Test Methods: Not stated			
Physical		**Mechanical**	
Colour	Brown/translucent	τ (RT)	-
Viscosity	11 Pa.s	τ (°C)	-
SG	1.10	τ (wet)	-
Shelf life	6 months (RT)	Peel ( )	-
Pot/Work life	-	σ_t	-
Tg	-	σ_c	-
$T_{Service}$	-	*E*	-
CTE (°C)	-	*G*	-
Thermal cond.	-	Elongation	-
Resistivity: Vol.	-	Tear strength	-
Surf.	-	H (Shore)	-
Notes:			

EMFI	Mastic PU 15		(Nov 95)
One part polyurethane (moisture curing)			
Manufacturers Comments: Non-slump, adhesive-sealant for building uses, e.g. expansion joints in prefabricated units. Adheres to most building materials. Developed for PVC-based materials. UV and moisture resistance. Meets: SNJF Spec. (NFP 85 507 and 518) for PVC; DIN 18 540 (Inst. Würzburg)			
Typical Use: Joints (windows, doors, etc). PVC frames, etc.			
Cure: RT (5-40°C): 6 hrs (skin). Cure rate: 1mm/24 hrs (23°C/65%RH)			
Test Methods: ISO			
Physical		**Mechanical**	
Colour	White*	τ (RT)	-
Viscosity	Thixotropic paste	τ (°C)	-
SG	1.20	τ (wet)	-
Shelf life	9 months (5-25°C)	Peel ( )	-
Pot/Work life	15 mins (20°C)	σ_t	-
Tg	-	σ_c	-
$T_{Service}$	-30 to +80°C	*E*	0.2 MPa
CTE (°C)	-	*G*	-
Thermal cond.	-	Elongation	>250%
Resistivity: Vol.	-	Tear strength	-
Surf.	-	H (Shore A)	15
Notes: * Other colours available: Grey, black, beige. Slump: ISO 7390; Hardness: ISO 868; E: ISO 37; Elong. ISO 8339.			

See Key to Data Sheets - Pages 63 to 65

EMFI	**Mastic PU 25 Bâtiment**		(Nov 95)
One part polyurethane (moisture curing)			
Manufacturers Comments: Non-slump, flexible adhesive-sealant for building uses, e.g. expansion joints in prefabricated units. Rapid cure with atmospheric moisture. Adheres to most building materials. UV and moisture resistance. Meets: SNJF Spec.			
Typical Use: Joints (wood, aluminium, PVC). Concrete & ceramic tiles.			
Cure: RT (5-40°C): 2-3 hrs (skin). Cure rate: 3mm/24 hrs (23°C/65%RH).			
Test Methods: ISO			
Physical		**Mechanical**	
Colour	White*	τ (RT)	-
Viscosity	Thixotropic paste	τ (°C)	-
SG	1.20	τ (wet)	-
Shelf life	9 months (5-25°C)	Peel ( )	-
Pot/Work life	15 mins (20°C)	σ_t	-
Tg	-	σ_c	-
$T_{Service}$	-30 to +80°C	E	0.3 MPa
CTE (°C)	-	G	-
Thermal cond.	-	Elongation	>250%
Resistivity: Vol.	-	Tear strength	-
Surf.	-	H (Shore A)	25
Notes: * Other colours available: Grey, black, bronze, brown, beige. Slump: ISO 7390; Hardness: ISO 868; E: ISO 37; Elong. ISO 8339.			

EMFI	**Mastic PU 50 Industrie**		(Nov 95)
One part polyurethane (moisture curing)			
Manufacturers Comments: Non-slump, tough, flexible adhesive-sealant for assembly and sealing materials. Fast skinning, fast cure. Adheres to most building materials and general engineering uses. Excellent tear resistance, good weatherproofing. UV and moisture resistance.			
Typical Use: Joints (concrete, wood, aluminium, painted metals, polyester, glass, PVC tiles (concrete & ceramic).			
Cure: RT (5-40°C): 1.5-2 hrs (skin). Cure rate: 4mm/24 hrs (23°C/65%RH).			
Test Methods: ISO			
Physical		**Mechanical**	
Colour	White*	τ (RT)	-
Viscosity	Thixotropic paste	τ (°C)	-
SG	1.17	τ (wet)	-
Shelf life	9 months (5-25°C)	Peel ( )	-
Pot/Work life	15 mins (20°C)	σ_t	-
Tg	-	σ_c	-
$T_{Service}$	-40 to +90°C	E	0.8 MPa
CTE (°C)	-	G	-
Thermal cond.	-	Elongation	80%
Resistivity: Vol.	-	Tear strength	-
Surf.	-	H (Shore A)	50
Notes: * Other colours available: Grey, black. Slump: ISO 7390; Hardness: ISO 868; E: ISO 37; Elong: ISO 8339.			

EMFI	**Mastic PU 40 Bâtiment**		(Nov 95)
One part polyurethane (moisture curing)			
Manufacturers Comments: Non-slump, tough, flexible adhesive-sealant for assembly and sealing materials in buildings, car, caravan, ship-building, cooling systems and air conditioning units. Rapid cure with atmospheric moisture. Adheres to most building materials. UV and moisture resistance. Meets: SNJF Spec.			
Typical Use: Joints (concrete, wood, aluminium, painted metals, polyester, glass, PVC tiles (concrete & ceramic).			
Cure: RT (5-40°C): 2-3 hrs (skin). Cure rate: 4mm/24 hrs (23°C/65%RH).			
Test Methods: ISO			
Physical		**Mechanical**	
Colour	White*	τ (RT)	-
Viscosity	Thixotropic paste	τ (°C)	-
SG	1.20	τ (wet)	-
Shelf life	9 months (5-25°C)	Peel ( )	-
Pot/Work life	15 mins (20°C)	σ_t	-
Tg	-	σ_c	-
$T_{Service}$	-30 to +80°C	E	0.6 MPa
CTE (°C)	-	G	-
Thermal cond.	-	Elongation	>250%
Resistivity: Vol.	-	Tear strength	-
Surf.	-	H (Shore A)	40
Notes: * Other colours available: Grey, black, bronze, brown, beige. Slump: ISO 7390; Hardness: ISO 868; E: ISO 37; Elong. ISO 8339.			

EMFI	**Sapocol DK 4001**		(Nov 95)
One part polyurethane (moisture curing)			
Manufacturers Comments: High viscosity, non-slump, flexible adhesive. High strength. Good resistance to ageing (thermal, moisture, UV). Recommends primer Sapocol GR 2001.			
Typical Use: Assembly adhesive, inc. glass.			
Cure: RT (assemble & clamp for 15 mins/23°C/50%RH).			
Test Methods: Not stated			
Physical		**Mechanical**	
Colour	Black	τ (RT)	-
Viscosity	-	τ (°C)	-
SG	1.2	τ (wet)	-
Shelf life	6 months (0-35°C)	Peel ( )	-
Pot/Work life	<15 mins (20°C)	σ_t	560 MPa
Tg	-	σ_c	-
$T_{Service}$	-	E	-
CTE (°C)	-	G	-
Thermal cond.	-	Elongation	-
Resistivity: Vol.	-	Tear strength	-
Surf.	-	H (Shore)	-
Notes: σ data: after 7 days, 23°C, 50%RH. After 4 ageing cycles (40/100%RH, -40°C, 80°C/100%RH): 510 MPa. After 100 days UV exposure: 380 MPa.			

Forbo-Helmitin

Forbo- Helmitin	Helmipur 700T		Feb 93
Two part polyurethane			
Manufacturers Comments: Solvent-free. Assembly of most materials, laminated panels. Metal roof joints. Structural joints with low flexibility. High resistance to petrol, oils, greases, dilute acids & alkalis. Cure can be accelerated by heat. May require primer.			
Typical Use: Foams - polyurethane, PVC, phenolic-resin. Polyester, thermosetting plastics. Laminated sheets. Wood products. Ceramics, concrete, masonry. Metals (aluminium, steel, etc.).			
Cure: RT/30 mins (wet-life); 24 hrs (set time); 20-22°C/5-8 hrs; 120°C/10-15 mins.			
Test Methods: Not stated			
Physical		**Mechanical**	
Colour	Beige	τ (RT)	12.4 MPa
Viscosity	~100 Pa.s	τ (°C)	-
SG	~1.50	τ (wet)	-
Shelf life	1 yr/RT	Peel ( )	-
Pot/Work life	~40 mins	σ_t	-
Tg	-	σ_c	-
$T_{Service}$	-	*E*	-
CTE (°C)	-	*G*	-
Thermal cond.	-	Elongation	-
Resistivity: Vol.	-	Tear strength	-
Surf.	-	H (Shore)	-
Notes: Hardener (20% addition): 500LF.			

Forbo- Helmitin	Helmipur 702T		Feb 93
Two part polyurethane			
Manufacturers Comments: Solvent-free, non-sag. Assembly of most materials, laminated panels & elements. High resistance to weathering, petrol, oils, greases, dilute acids & alkalis. May require primer.			
Typical Use: Foams - polyurethane, PVC, phenolic-resin. Polyester, thermosetting plastics. Laminated sheets (inc. foam-to-glass & wood). Wood products. Ceramics, concrete, masonry. Metals.			
Cure: RT/70 mins (wet-life); 24 hrs (set time).			
Test Methods: Not stated			
Physical		**Mechanical**	
Colour	White	τ (RT)	9.3 MPa
Viscosity	Paste	τ (°C)	-
SG	~1.67	τ (wet)	-
Shelf life	1 yr/RT	Peel ( )	-
Pot/Work life	~90 mins	σ_t	-
Tg	-	σ_c	-
$T_{Service}$	-	*E*	-
CTE (°C)	-	*G*	-
Thermal cond.	-	Elongation	-
Resistivity: Vol.	-	Tear strength	-
Surf.	-	H (Shore)	-
Notes: Hardener (15% addition): 500LF.			

Forbo- Helmitin	Helmipur 701T		Feb 93
Two part polyurethane			
Manufacturers Comments: Solvent-free. Assembly of most materials, laminated panels. Metal roof joints. Structural joints with low flexibility. High resistance to weathering, petrol, oils, greases, dilute acids & alkalis. Cure can be accelerated by heat. May require primer.			
Typical Use: Foams - polyurethane, PVC, phenolic-resin. Polyester, thermosetting plastics. Laminated sheets. Wood products. Ceramics, concrete, masonry. Metals (aluminium, steel, etc.).			
Cure: RT/30 mins (wet-life); 24 hrs (set time); 20 - 22°C/5 - 8 hrs; 100°C/15 - 20 mins.			
Test Methods: Not stated			
Physical		**Mechanical**	
Colour	Beige	τ (RT)	11.1 MPa
Viscosity	~30 Pa.s	τ (°C)	-
SG	~1.50	τ (wet)	-
Shelf life	1 yr/RT	Peel ( )	-
Pot/Work life	~40 mins	σ_t	-
Tg	-	σ_c	-
$T_{Service}$	-	*E*	-
CTE (°C)	-	*G*	-
Thermal cond.	-	Elongation	-
Resistivity: Vol.	-	Tear strength	-
Surf.	-	H (Shore)	-
Notes: Hardener (20% addition): 500LF.			

Forbo- Helmitin	Helmipur 15065		Jul 92
Two part polyurethane (also used as 1-part)			
Manufacturers Comments: High performance for footwear assembly. Low distortion during sole pressing light-materials. Good heat, moisture & oil resistance. Low temperature flexibility. May need primer.			
Typical Use: Sole attaching - leather, PVC, polyurethane, primed rubber soles, thermoplastic-rubber soles, unprimed composition rubber & most synthetic upper materials.			
Cure: RT/24 hrs (set time) + 70°C heat-activation (sole press). 2 days (full cure).			
Test Methods: Not stated			
Physical		**Mechanical**	
Colour	Transparent	τ (RT)	-
Viscosity	~4 Pa.s	τ (°C)	-
SG	~0.86	τ (wet)	-
Shelf life	1 yr/RT	Peel ( )	-
Pot/Work life	20-30 mins (drying)*	σ_t	-
Tg	-	σ_c	-
$T_{Service}$	-	*E*	-
CTE (°C)	-	*G*	-
Thermal cond.	-	Elongation	-
Resistivity: Vol.	-	Tear strength	-
Surf.	-	H (Shore)	-
Notes: Choice of hardener (all 5% addition): Helmitin 500 FB, 500 FE or 49631. *Hardener affects pot-life & open time: 500 FB - 24 hrs (pot), 6 hrs (open); 500 FE - 24 hrs (pot), 3 hrs (open); 49631 48 hrs (pot), 15 hrs (open). Inflammable.			

See Key to Data Sheets - Pages 63 to 65

Forbo- Helmitin	Helmipur 15104		July 92
Two part polyurethane			
Manufacturers Comments: Contact adhesive for footwear assembly.			
Typical Use: Lasting - uppers to moulded PUR & PVC insoles. Covering moulded PUR shoe components. Leather sock lining onto PUR injected soles.			
Cure: RT/20 - 30 mins (contact life)*.			
Test Methods: Not stated			
Physical		**Mechanical**	
Colour	Yellow-transparent	τ (RT)	-
Viscosity	~5 Pa.s	τ (°C)	-
SG	~0.93	τ (wet)	-
Shelf life	1 yr/RT	Peel ( )	-
Pot/Work life	~24 hrs	σ_t	-
Tg	-	σ_c	-
$T_{Service}$	-	*E*	-
CTE (°C)	-	*G*	-
Thermal cond.	-	Elongation	-
Resistivity: Vol.	-	Tear strength	-
Surf.	-	H (Shore)	-
Notes: Hardener: 5% 500 FB. Drying time: 10-15 mins. *Slight heating required if contact time is exceeded. Inflammable.			

Forbo- Helmitin	Helmipur 46010		Feb 93
Two part polyurethane			
Manufacturers Comments: Solvent-free. Manufacture of laminated panels. High resistance to weathering, petrol, oils, greases, dilute acids & alkalis. May require primer.			
Typical Use: Foam-to-wood. Asbestos cements. Wood products. Polyester. Metals (inc. aluminium & steels, etc).			
Cure: RT/30 mins (wet-life); 24 hrs (set time).			
Test Methods: Not stated			
Physical		**Mechanical**	
Colour	Grey	τ (RT)	12.6 MPa
Viscosity	~12 Pa.s	τ (°C)	-
SG	~1.53	τ (wet)	-
Shelf life	6 months/RT	Peel ( )	-
Pot/Work life	~40 mins	σ_t	-
Tg	-	σ_c	-
$T_{Service}$	-	*E*	-
CTE (°C)	-	*G*	-
Thermal cond.	-	Elongation	-
Resistivity: Vol.	-	Tear strength	-
Surf.	-	H (Shore)	-
Notes: Hardener (20% addition): 500LF			

Forbo- Helmitin	Helmipur 15141/2		May 92
Two part polyurethane			
Manufacturers Comments: Brushable. For assembly of car side panels & dashboards.			
Typical Use: Leather with ABS & fibre-moulding materials.			
Cure: RT/5 mins (drying time).			
Test Methods: Not stated			
Physical		**Mechanical**	
Colour	Yellow-transparent	τ (RT)	-
Viscosity	~3.3 Pa.s	τ (°C)	-
SG	~0.88	τ (wet)	-
Shelf life	1 yr/RT	Peel ( )	-
Pot/Work life	24-48 hrs*	σ_t	-
Tg	-	σ_c	-
$T_{Service}$	-	*E*	-
CTE (°C)	-	*G*	-
Thermal cond.	-	Elongation	-
Resistivity: Vol.	-	Tear strength	-
Surf.	-	H (Shore)	-
Notes: Helmipur 151142/2 - sprayable version (lower viscosity) Choice of two hardeners (10% addition): 500 FL or 49631. *pot life depends on hardener. Inflammable.			

Forbo- Helmitin	Helmipur 46018		Mar 91
Two part polyurethane			
Manufacturers Comments: Solvent-free. Assembly & manufacture of laminated panels. High resistance to weathering, petrol, oils, greases, dilute acids & alkalis. May require primer.			
Typical Use: Foams - polyurethane, PVC, phenolic-resin. Polyester, thermosetting plastics. Laminated sheets (inc. foam-to-glass & wood). Wood products. Ceramics, concrete, masonry. Metals.			
Cure: RT/90 mins (wet-life); 36 hrs (set time).			
Test Methods: Not stated			
Physical		**Mechanical**	
Colour	Brown	τ (RT)	11.1 MPa
Viscosity	~4 Pa.s	τ (°C)	-
SG	~1.60	τ (wet)	-
Shelf life	6 months/RT	Peel ( )	-
Pot/Work life	~2 hrs	σ_t	-
Tg	-	σ_c	-
$T_{Service}$	-	*E*	-
CTE (°C)	-	*G*	-
Thermal cond.	-	Elongation	-
Resistivity: Vol.	-	Tear strength	-
Surf.	-	H (Shore)	-
Notes: Hardener (20% addition): 500LF.			

Forbo- Helmitin	**Helmipur 46019/2**		Feb 91
Two part polyurethane			
Manufacturers Comments: Solvent-free. Assembly & manufacture of laminated panels. High resistance to weathering, petrol, oils, greases, dilute acids & alkalis. May require primer.			
Typical Use: Foams - polyurethane, PVC, phenolic-resin. Polyester, thermosetting plastics. Laminated sheets (inc. foam-to-glass & wood). Wood products. Ceramics, concrete, masonry. Metals.			
Cure: RT/20 mins (wet-life); 24 hrs (set time); 20°C/1 hr (cure).			
Test Methods: Not stated			
Physical		**Mechanical**	
Colour	Grey	τ (RT)	10.7 MPa
Viscosity	~10 Pa.s	τ (°C)	-
SG	~1.45	τ (wet)	-
Shelf life	6 months/RT	Peel ( )	-
Pot/Work life	~4 mins	σ_t	-
Tg	-	σ_c	-
$T_{Service}$	-	E	-
CTE (°C)	-	G	-
Thermal cond.	-	Elongation	-
Resistivity: Vol.	-	Tear strength	-
Surf.	-	H (Shore)	-
Notes: Hardener (25% addition): 500LF.			

Forbo- Helmitin	**Helmipur 46516**		Oct 90
One part polyurethane, moisture curing			
Manufacturers Comments: Assembly of building materials. Applied by cartridge.			
Typical Use: Wood products, asbestos cements, hard foams, polyester, laminates, metals.			
Cure: RT/48 hrs (varies with temperature & moisture).			
Test Methods: Not stated			
Physical		**Mechanical**	
Colour	Yellow	τ (RT)	-
Viscosity	Paste	τ (°C)	-
SG	~1.1	τ (wet)	-
Shelf life	6 months/15-25°C	Peel ( )	-
Pot/Work life	~ 2 hrs	σ_t	-
Tg	-	σ_c	-
$T_{Service}$	-	E	-
CTE (°C)	-	G	-
Thermal cond.	-	Elongation	-
Resistivity: Vol.	-	Tear strength	-
Surf.	-	H (Shore)	-
Notes:			

Forbo- Helmitin	**Helmipur 46508**		Feb 93
One part polyurethane, moisture curing			
Manufacturers Comments: Assembly of building materials. Applied by brush, roller or trowel.			
Typical Use: Wood products, hard foams, polyester, laminates, metals.			
Cure: RT/12 hrs (varies with temperature & moisture).			
Test Methods: Not stated			
Physical		**Mechanical**	
Colour	Brown	τ (RT)	-
Viscosity	~30 Pa.s	τ (°C)	-
SG	~1.1	τ (wet)	-
Shelf life	6 months/RT	Peel ( )	-
Pot/Work life	~ 2 hrs	σ_t	-
Tg	-	σ_c	-
$T_{Service}$	-	E	-
CTE (°C)	-	G	-
Thermal cond.	-	Elongation	-
Resistivity: Vol.	-	Tear strength	-
Surf.	-	H (Shore)	-
Notes:			

Forbo- Helmitin	**Helmipur 46519**		Jan 91
One part polyurethane, moisture curing			
Manufacturers Comments: Assembly of building materials. Applied by brush, roller or trowel.			
Typical Use: Wood products, asbestos cements, hard foams, polyester, laminates, metals.			
Cure: RT/2-3 hrs (varies with temperature & moisture). Accelerated cure: 40°C/10 mins in press.			
Test Methods: Not stated			
Physical		**Mechanical**	
Colour	Brown	τ (RT)	-
Viscosity	~4.5 Pa.s	τ (°C)	-
SG	~1.15	τ (wet)	-
Shelf life	6 months/15-25°C	Peel ( )	-
Pot/Work life	~40 mins	σ_t	-
Tg	-	σ_c	-
$T_{Service}$	-	E	-
CTE (°C)	-	G	-
Thermal cond.	-	Elongation	-
Resistivity: Vol.	-	Tear strength	-
Surf.	-	H (Shore)	-
Notes:			

Forbo- Helmitin	Helmipur DS/18		Aug 92
Two part polyurethane			
Manufacturers Comments: Dispersion-based, high performance for footwear assembly. Good heat, moisture & oil resistance. Low temperature flexibility. May need primer.			
Typical Use: Sole attaching - leather, PVC, polyurethane, primed rubber soles, thermoplastic-rubber soles, unprimed composition rubber & most synthetic upper materials.			
Cure: RT/48 hrs (set time); 24 hrs (assembly time) + 70-80°C heat-activation (sole press). 2 days (full cure).			
Test Methods: Not stated			
Physical		**Mechanical**	
Colour	White	τ (RT)	-
Viscosity	~2 Pa.s	τ (°C)	-
SG	~1.05	τ (wet)	-
Shelf life	6 months/RT	Peel ( )	-
Pot/Work life	30-60mins(drying)	σ_t	-
Tg	-	σ_c	-
$T_{Service}$	-	E	-
CTE (°C)	-	G	-
Thermal cond.	-	Elongation	-
Resistivity: Vol.	-	Tear strength	-
Surf.	-	H (Shore)	-
Notes: Hardener (2.5-5% addition): Helmitin D improves heat resistance, adhesion to some materials, grease, hydrolysis and plasticiser resistance. Frost sensitive.			

Forbo- Helmitin	Helmipur SH100		Nov 86
One part polyurethane, moisture curing			
Manufacturers Comments: High grade, fast curing, non-sag adhesive sealant. Metal-curtain walls. Ventilation, air conditioning units. Refrigerating equipment - vehicle, container, ship construction. Tiling. Shop-fitting. Can be lacquered. Resistant to sea water, water, weak acids & alkalis. Limited stability to fuels, oils, grease.			
Typical Use: Metals (inc. galvanised & coated), FRP, wood products, glass, ceramic, building (masonry), rubber, some plastics*.			
Cure: RT. Cure rate: 3mm/24 hrs (20°C/65%RH)			
Test Methods: DIN			
Physical		**Mechanical**	
Colour	Various	τ (RT)	-
Viscosity	Paste	τ (°C)	-
SG	~1.26	τ (wet)	-
Shelf life	6 months/RT	Peel ( )	-
Pot/Work life	-	σ_t	2.1 MPa
Tg	-	σ_c	-
$T_{Service}$	-40 to +90°C	E	-
CTE (°C)	-	G	-
Thermal cond.	-	Elongation	630%
Resistivity: Vol.	-	Tear strength	-
Surf.	-	H (Shore A)	50-55
Notes: * No adhesion to polyethylene, polypropylene, silicone, EPDM. Not for sealing glass grooves. Colours: beige, grey, white, black, brown. σ, elong DIN 53504.			

Forbo- Helmitin	Helmipur GPV/B		Feb 95
Two part polyurethane (also used as 1-part)			
Manufacturers Comments: High performance for footwear assembly. Good heat, moisture & oil resistance. Low temperature flexibility. May need primer.			
Typical Use: Sole attaching - leather, PVC, polyurethane, primed rubber soles, thermoplastic-rubber soles, unprimed composition rubber & most synthetic upper materials.			
Cure: RT/48 hrs (set time) + 70-80°C heat-activation (sole press). 2 days (full cure).			
Test Methods: Not stated			
Physical		**Mechanical**	
Colour	Transparent	τ (RT)	-
Viscosity	~3.3 Pa.s	τ (°C)	-
SG	~0.86	τ (wet)	-
Shelf life	1 yr/RT	Peel ( )	-
Pot/Work life	20-30 mins (drying)*	σ_t	-
Tg	-	σ_c	-
$T_{Service}$	-	E	-
CTE (°C)	-	G	-
Thermal cond.	-	Elongation	-
Resistivity: Vol.	-	Tear strength	-
Surf.	-	H (Shore)	-
Notes: Choice of hardener (all 5% addition): Helmitin 500 FB, 500 FE/1 or 49631. *Hardener affects pot-life & open time: 500 FB - 24 hrs (pot), 6 hrs (open); 500 FE/1 - 24 hrs (pot), 3 hrs (open); 49631 - 48 hrs (pot), 15 hrs (open). Inflammable.			

Forbo- Helmitin	Helmipur UN neu		July 94
Two part polyurethane			
Manufacturers Comments: Assembly adhesive. High resistance to grease, oil, petrol and solvents.			
Typical Use: Bonding PUR materials (elastomer, foam, etc.).			
Cure: RT/10-20 mins (drying time); ~48 hrs (set time).			
Test Methods: Not stated			
Physical		**Mechanical**	
Colour	Transparent	τ (RT)	-
Viscosity	~6 Pa.s	τ (°C)	-
SG	~0.98	τ (wet)	-
Shelf life	1 yr/RT	Peel ( )	-
Pot/Work life	12 or 24 hrs*	σ_t	-
Tg	-	σ_c	-
$T_{Service}$	-	E	-
CTE (°C)	-	G	-
Thermal cond.	-	Elongation	-
Resistivity: Vol.	-	Tear strength	-
Surf.	-	H (Shore)	-
Notes: Choice of two hardeners (5-8% addition): 500 I or 500 FB (better adhesion). * pot life depends on hardener: 500 I /24 hrs; 500FB/12 hrs. Inflammable.			

Ed. Geistlich Söhne AG

Forbo- Helmitin — **Helmitherm RK 99** — July 94

One part, melt applied, moisture cure

Manufacturers Comments: Applied in molten form, then moisture cures (atmospheric or from substrate). Assembly adhesive.

Typical Use: Wood, sheet materials, textiles, leather, plastics (automotive industry).

Cure: Applied: 130°C. RT/5 - 7days (cure).

Test Methods: Not stated

Physical		Mechanical	
Colour	White	τ (RT)	2 - 8 MPa
Viscosity	30 Pa.s (120°C)	τ (°C)	-
SG	~1.0	τ (wet)	-
Shelf life	~1 yr/RT	Peel ( )	-
Pot/Work life	15 - 60 secs (open)	σ_t	-
Tg	-	σ_c	-
$T_{Service}$	-	E	-
CTE (°C)	-	G	-
Thermal cond.	-	Elongation	-
Resistivity: Vol.	-	Tear strength	-
Surf.	-	H (Shore)	-

Notes: Softening temperature: ~70°C (Ring and Ball).
Similar product: Helmitherm RK 99/1.

Geistlich — **Miracol 9505** — May 91

One part polyurethane

Manufacturers Comments: 100% solids, solvent-free. Assembly adhesive for wood products requiring water- & moisture resistance. Can be mixed with primer to accelerate cure.
Meets: DIN 68705 section 2/AW & EN-204-D4 (water resistance)

Typical Use: Wood products - water-resistance. Windows, doors, building elements, saunas, swimming-pools. Boat-building.

Cure: RT(15-25°C)/~24 hrs (handling); 24 hrs (full cure).
Press: 5-10°C/10 hrs/0.2-0.7MPa; 20°C/4-6hrs/0.2-0.7MPa

Test Methods: DIN

Physical		Mechanical	
Colour	Colourless	τ (RT)	-
Viscosity	2 Pa.s	τ (°C)	-
SG	~1.13	τ (wet)	-
Shelf life	6 months/RT	Peel ( )	-
Pot/Work life	30-60 mins (open)	σ_t	*
Tg	-	σ_c	-
$T_{Service}$	-20 to +150°C	E	-
CTE (°C)	-	G	-
Thermal cond.	-	Elongation	-
Resistivity: Vol.	-	Tear strength	-
Surf.	-	H (Shore)	-

Notes: All assemblies require pressing. Flash point (DIN51758, 53213): +200°C.
Wood moisture content: 18%. max.; * DIN 204-D1.

Forbo- Helmitin — **Helmitin-kleber 34331** — May 92

Two part polyurethane

Manufacturers Comments: Dispersion-based, for laminating thermoplastic foils on pre-profiled work pieces by deep-drawing & heat-sealing.

Typical Use: Laminating plastic foils.

Cure: RT/30-60mins(drying); 48 hrs (set time); + 70-90°C/10-20 secs deep-drawing activation.

Test Methods: Not stated

Physical		Mechanical	
Colour	White	τ (RT)	-
Viscosity	~1 Pa.s	τ (°C)	-
SG	~1.04	τ (wet)	-
Shelf life	6 months/RT	Peel ( )	-
Pot/Work life	~ 8 hrs	σ_t	-
Tg	-	σ_c	-
$T_{Service}$	+90 to +120°C	E	-
CTE (°C)	-	G	-
Thermal cond.	-	Elongation	-
Resistivity: Vol.	-	Tear strength	-
Surf.	-	H (Shore)	-

Notes: Hardener (2.5-5% addition): Helmitin D. Frost sensitive.
Similar product: Helmitin-kleber 34333 (higher viscosity), also sprayable.

Geistlich — **Mirapur 9510 Rapid** — Jan 92

One part polyurethane, moisture curing

Manufacturers Comments: 100% solids. Assembly adhesive for wood products requiring water- & moisture resistance. Meets: EN 204-D1 (strength) & EN 204-D4 (moisture resistance).

Typical Use: Wood products - water-resistance.

Cure: RT(10-30°C)/3 hrs (handling); 7 days (full cure).
Press: >30mins/0.2-0.7MPa.

Test Methods: EN

Physical		Mechanical	
Colour	Brown	τ (RT)	-
Viscosity	7 Pa.s	τ (°C)	-
SG	~1.14	τ (wet)	-
Shelf life	6 months/RT	Peel ( )	-
Pot/Work life	<15 mins (open)	σ_t	*
Tg	-	σ_c	-
$T_{Service}$	+120°C	E	-
CTE (°C)	-	G	-
Thermal cond.	-	Elongation	-
Resistivity: Vol.	-	Tear strength	-
Surf.	-	H (Shore)	-

Notes: Wood moisture content: 6-18%. * DIN 204-D1.

Hardman

Hardman	Kalex 15036		(July 95)
Two part polyurethane			
Manufacturers Comments: Fast curing, rigid, structural adhesive. Excellent shear strength and good peel. Excellent adhesion.			
Typical Use:			
Cure: 24hr/RT.			
Test Methods: Not stated (USA)			
Physical		**Mechanical**	
Colour	Off-white	τ (RT)	28.1 MPa
Viscosity	11 Pa.s	τ (149°C)	1.9 MPa
SG	1.24	τ (-40°C)	28.3 MPa
Shelf life	-	Peel (-)	1.8 N mm^{-1}
Pot/Work life	<5 mins/RT	σ_t	-
Tg	-	σ_c	-
$T_{Service}$	~70°C	*E*	-
CTE (°C)	-	*G*	-
Thermal cond.	-	Elongation	-
Resistivity: Vol.	-	Tear strength	-
Surf.	-	H (Shore D)	85
Notes:			

Hardman	Kalex 16308		(July 95)
Two part polyurethane			
Manufacturers Comments: Fast curing, flexible adhesive. Excellent peel strength. Good adhesion to leather, rubbers and thermoplastics.			
Typical Use: Good for low temperature applications.			
Cure: 7 days/RT.			
Test Methods: Not stated (USA)			
Physical		**Mechanical**	
Colour	Black	τ (RT)	6.6 MPa
Viscosity	40 Pa.s	τ (149°C)	1.2 MPa
SG	1.22	τ (-40°C)	29.7 MPa
Shelf life	-	Peel (-)	4.2 N mm^{-1}
Pot/Work life	<7 mins/RT	σ_t	-
Tg	-	σ_c	-
$T_{Service}$	~20°C	*E*	-
CTE (°C)	-	*G*	-
Thermal cond.	-	Elongation	-
Resistivity: Vol.	-	Tear strength	-
Surf.	-	H (Shore D)	40
Notes:			

Hardman	Kalex 15038		(July 95)
Two part polyurethane			
Manufacturers Comments: Fast curing, semi-rigid, structural adhesive. Excellent shear strength and exceptional peel. Excellent adhesion.			
Typical Use: Thermoplastics and SMC.			
Cure: 24hr/RT.			
Test Methods: Not stated (USA)			
Physical		**Mechanical**	
Colour	Beige	τ (RT)	27.2 MPa
Viscosity	12 Pa.s	τ (149°C)	2.5 MPa
SG	1.26	τ (-40°C)	27.3 MPa
Shelf life	-	Peel	12.3 N mm^{-1}
Pot/Work life	<5 mins/RT	σ_t	-
Tg	-	σ_c	-
$T_{Service}$	~70°C	*E*	-
CTE (°C)	-	*G*	-
Thermal cond.	-	Elongation	-
Resistivity: Vol.	-	Tear strength	-
Surf.	-	H (Shore D)	70
Notes:			

Hardman	Kalex 18515		(July 95)
Two part polyurethane			
Manufacturers Comments: Very flexible adhesive/sealant. Thermally conductive. Excellent impact strength and good low-temperature peel and shear strength.			
Typical Use: Thermal printers and heatsinks.			
Cure: 7 days/RT.			
Test Methods: Not stated (USA)			
Physical		**Mechanical**	
Colour	Off-white	τ (RT)	4.7 MPa
Viscosity	11 Pa.s	τ (-40°C)	24.8 MPa
SG	1.89	τ (wet)	-
Shelf life	-	Peel (-)	0.93 N mm^{-1}
Pot/Work life	<85 mins/RT	σ_t	-
Tg	-	σ_c	-
$T_{Service}$	-	*E*	-
CTE (°C)	-	*G*	-
Thermal cond.	-	Elongation	-
Resistivity: Vol.	-	Tear strength	-
Surf.	-	H (Shore A)	72
Notes:			

See Key to Data Sheets - Pages 63 to 65

Hitachi-Kasei Polymer Co.

Hitachi-Kasei	Hi-Bon 4610		Mar 95
One part polyurethane-epoxy			
Manufacturers Comments: Heat curing. Bonding reinforcements to car bodies. For use with oily steel.			
Typical Use: Automotive - steel bonding.			
Cure: 180°C/20 mins.			
Test Methods: See notes			
Physical		**Mechanical**	
Colour	Black	τ (RT)	5.5 MPa
Viscosity	710 (170) Pa.s	τ (°C)	-
SG	1.58	τ (wet)	-
Shelf life	+6 months/RT	Peel (-)	5.9 N mm^{-1}
Pot/Work life	30-60 mins (open)	σ_t	4.9 MPa
Tg	100°C	σ_c	-
$T_{Service}$	-20 to +150°C	*E*	100 MPa
CTE (°C)	-	*G*	-
Thermal cond.	-	Elongation	38%
Resistivity: Vol.	-	Tear strength	-
Surf.	-	H (Shore D)	32 (RT)
Notes: Viscosity Flowtester, shear rate 20/s data: 20°C (60°C). σ: 50mm/min; τ oily steel (0.8x25x125mm), adhesive 1mm thick, Tensile speed 50mm/min; Peel- oily steel (0.8x25x125mm), adhesive 1mm thick, 200mm/min.			

Hitachi-Kasei	Hi-Bon 4811		Mar 95
One part polyurethane. Melt applied, moisture curing			
Manufacturers Comments: Applied in molten form, then moisture curing (atmospheric or from substrate). Tacky before curing. Good flexibility. Fast setting, high green strength & temperature resistance.			
Typical Use: Laminating: Films, textiles, carpeting.			
Processing: Application Temp:100 - 200°C. + Moisture curing RT/48 hrs.			
Test Methods: Not stated			
Physical		**Mechanical**	
Colour	Clear	τ (RT)	-
Viscosity	3 Pa.s (120°C)	τ (°C)	-
SG	-	τ (wet)	-
Shelf life	-	Peel ( )	-
Pot/Work life	>10 sec (open)	σ_t	17 (40) MPa
Tg	5°C	σ_c	-
$T_{Service}$	-	*E*	10 (500) MPa
CTE (°C)	-	*G*	-
Thermal cond.	-	Elongation	420 (280) %
Resistivity: Vol.	-	Tear strength	-
Surf.	-	H (Shore)	-
Notes: Tensile property data format: 20°C (-10°C). Tensile shear strengths (MPa): Wood (Japanese birch)- 8.3 MPa, ABS- 6.6 MPa, PMMA-2.1 MPa PVC- 7.3 MPa, steel/wood- 10.2 MPa. Softening Temp: 160°C.			

Hitachi-Kasei	Hi-Bon 4620		Mar 95
One part polyurethane-epoxy			
Manufacturers Comments: Heat curing. Bonding reinforcements to car bodies. For use with Electrodeposition (EDP) coated steel.			
Typical Use: Automotive - steel bonding.			
Cure: 180°C/20 mins.			
Test Methods: See notes			
Physical		**Mechanical**	
Colour	Black	τ (RT)	6.0 MPa
Viscosity	680 (150) Pa.s	τ (°C)	-
SG	1.58	τ (wet)	-
Shelf life	+6 months/RT	Peel (-)	5.8 N mm^{-1}
Pot/Work life	30-60 mins (open)	σ_t	5.2 MPa
Tg	100°C	σ_c	-
$T_{Service}$	-20 to +150°C	*E*	100 MPa
CTE (°C)	-	*G*	-
Thermal cond.	-	Elongation	57%
Resistivity: Vol.	-	Tear strength	-
Surf.	-	H (Shore D)	32 (RT)
Notes: Viscosity Flowtester, shear rate 20/s data: 20°C (60°C). σ: 50mm/min; τ EDP steel (0.8x25x125mm), adhesive 1mm thick, Tensile speed 50mm/min; Peel- EDP steel (0.8x25x125mm), adhesive 1mm thick, 200mm/min.			

Hitachi-Kasei	Hi-Bon 4812		Mar 95
One part polyurethane. Melt applied, moisture curing			
Manufacturers Comments: Applied in molten form, then moisture curing (atmospheric or from substrate). Tacky before curing. Good flexibility. Fast setting, high green strength & temperature resistance.			
Typical Use: Assembly of loudspeakers.			
Processing: Application Temp:100 - 200°C. + Moisture curing RT/48 hrs.			
Test Methods: Not stated			
Physical		**Mechanical**	
Colour	Clear	τ (RT)	-
Viscosity	15 Pa.s (120°C)	τ (°C)	-
SG	-	τ (wet)	-
Shelf life	-	Peel ( )	-
Pot/Work life	>10 sec (open)	σ_t	19 (39) MPa
Tg	5°C	σ_c	-
$T_{Service}$	-	*E*	3 (300) MPa
CTE (°C)	-	*G*	-
Thermal cond.	-	Elongation	510 (300) %
Resistivity: Vol.	-	Tear strength	-
Surf.	-	H (Shore)	-
Notes: Tensile property data format: 20°C (-10°C). Tensile shear strengths (MPa): Wood (Japanese birch)- 5.5 MPa, ABS- 6.5 MPa, PMMA-2.2 MPa PVC- 7.1 MPa, steel/wood- 7.5 MPa. Softening Temp: 160°C.			

Hitachi-Kasei	Hi-Bon 4820		Mar 95
One part polyurethane. Melt applied, moisture curing			
Manufacturers Comments: Applied in molten form, then moisture curing (atmospheric or from substrate). Open + set time. Fast setting, high green strength & temperature resistance.			
Typical Use: Door panels: Textiles to PVC.			
Processing: Application Temp:100 - 200°C. + Moisture curing RT/48 hrs.			
Test Methods: Not stated			
Physical		**Mechanical**	
Colour	Opaque/yellow	τ (RT)	-
Viscosity	25 Pa.s (120°C)	τ (°C)	-
SG	-	τ (wet)	-
Shelf life	-	Peel ( )	-
Pot/Work life	7 (open);15(set) sec	σ_t	25 (45) MPa
Tg	0°C	σ_c	-
$T_{Service}$	-	*E*	20 (300) MPa
CTE (°C)	-	*G*	-
Thermal cond.	-	Elongation	580 (350) %
Resistivity: Vol.	-	Tear strength	-
Surf.	-	H (Shore)	-
Notes: Tensile property data format: 20°C (-10°C). Tensile shear strengths (MPa): Wood (Japanese birch)- 10.5 MPa, ABS- 6.2 MPa, PMMA-3.0 MPa PVC- 7.2 MPa, steel/wood- 7.8 MPa. Softening Temp: 160°C.			

Hitachi-Kasei	Hi-Bon 4832		Mar 95
One part polyurethane. Melt applied, moisture curing			
Manufacturers Comments: Applied in molten form, then moisture curing (atmospheric or from substrate). Rapid set time. Fast setting, high green strength & temperature resistance.			
Typical Use: Assembly of honeycomb core panels. Textile bonding to polyurethane foams.			
Processing: Application Temp:100 - 200°C. + Moisture curing RT/48 hrs.			
Test Methods: Not stated			
Physical		**Mechanical**	
Colour	Opaque/yellow	τ (RT)	-
Viscosity	35 Pa.s (120°C)	τ (°C)	-
SG	-	τ (wet)	-
Shelf life	-	Peel ( )	-
Pot/Work life	0.5 (open);1(set)sec	σ_t	46 (>29) MPa
Tg	-30°C	σ_c	-
$T_{Service}$	-	*E*	300 (400) MPa
CTE (°C)	-	*G*	-
Thermal cond.	-	Elongation	700 (>400) %
Resistivity: Vol.	-	Tear strength	-
Surf.	-	H (Shore)	-
Notes: Tensile property data format: 20°C (-10°C). Tensile shear strengths (MPa): Wood (Japanese birch)- 10.2 MPa, ABS- 4.0 MPa, PMMA-3.5 MPa PVC- 7.2 MPa, steel/wood- 8.7 MPa. Softening Temp: 170°C.			

Hitachi-Kasei	Hi-Bon 4830		Mar 95
One part polyurethane. Melt applied, moisture curing			
Manufacturers Comments: Applied in molten form, then moisture curing (atmospheric or from substrate). Open + set time. Fast setting, high green strength & temperature resistance.			
Typical Use: Assembly: Lacquered surfaces, aluminium, mirrors.			
Processing: Application Temp:100 - 200°C. + Moisture curing RT/48 hrs.			
Test Methods: Not stated			
Physical		**Mechanical**	
Colour	Opaque	τ (RT)	-
Viscosity	18 Pa.s (120°C)	τ (°C)	-
SG	-	τ (wet)	-
Shelf life	-	Peel ( )	-
Pot/Work life	2 (open); 3(set) sec	σ_t	34 (>26) MPa
Tg	-30°C	σ_c	-
$T_{Service}$	-	*E*	200 (300) MPa
CTE (°C)	-	*G*	-
Thermal cond.	-	Elongation	480 (>400) %
Resistivity: Vol.	-	Tear strength	-
Surf.	-	H (Shore)	-
Notes: Tensile property data format: 20°C (-10°C). Tensile shear strengths (MPa): Wood (Japanese birch)- 12.1 MPa, ABS- 5.5 MPa, PMMA-7.4 MPa PVC- 7.6 MPa, steel/wood- 10.3 MPa. Softening Temp: 170°C.			

Kemira Polymers

Kemira	Hyperlast 2874/019		Nov 88
Two part polyurethane			
Manufacturers Comments: Heat curing adhesive for bonding Sheet Moulding Compound (SMC). Initial mixing provides non-slump compound.			
Typical Use: SMC.			
Cure: ~70°C/30 - 60 mins (for optimum cure).			
Test Methods: See notes			
Physical		**Mechanical**	
Colour	Brown	τ (RT)	4.8 MPa
Viscosity	~9 Pa.s	τ (70°C)	2.7 MPa
SG	~1.5	τ (wet)	4.8 MPa
Shelf life	>6 month/RT	Peel ( )	-
Pot/Work life	10 mins (100g/RT)	σ_t	14 MPa
Tg	-	σ_c	-
$T_{Service}$	-	*E*	-
CTE (°C)	-	*G*	-
Thermal cond.	-	Elongation	10%
Resistivity: Vol.	-	Tear strength	-
Surf.	-	H (Shore A)	70
Notes: σ, elong: BS 903 Pt A2; Hardness: BS 2782 Pt 3; τ data: Automotive grade SMC, bond area 25x25mm, adhesive 1 mm thick. Post-cure 70°C/1 hr + 72 hrs @ RT. All failed in SMC. τ(wet) 14 days, water immersion, RT. τ after 14 days @ 70°C, RT test: 4.8 MPa.			

Kemira	Hyperlast 2875-2874/022		Aug 92
Two part polyurethane			
Manufacturers Comments: Highly thixotropic elastomer. Non-slump when applied in thick layers.			
Typical Use: Filter assembly.			
Cure: RT.			
Test Methods: See notes			
Physical		**Mechanical**	
Colour	Brown	τ (RT)	-
Viscosity	Thixotropic	τ (°C)	-
SG	~1.1	τ (wet)	-
Shelf life	-	Peel ( )	-
Pot/Work life	~8 mins (100g/RT)	σ_t	18 MPa
Tg	-	σ_c	-
$T_{Service}$	-	E	-
CTE (°C)	-	G	-
Thermal cond.	-	Elongation	150%
Resistivity: Vol.	-	Tear strength	45 N mm^{-1}
Surf.	-	H (Shore D)	60
Notes: σ,elong: BS903 Pt A2; Hardness: BS 2782 Pt 3; Angle Tear: BS 903 Pt A3			

Kemira	Hyperlast 2875-2874/7003		Aug 93
Two part polyurethane			
Manufacturers Comments: Rigid, heat resistant. Hafting adhesive for cutlery assembly. Meets: UK Cutlery standard BS5577:1978.			
Typical Use: Cutlery assembly.			
Cure: RT/30 mins (handling); 40-50°C (for accelerated cure).			
Test Methods: See notes			
Physical		**Mechanical**	
Colour	Cream	τ (RT)	-
Viscosity	~12 Pa.s	τ (°C)	-
SG	~1.6	τ (wet)	-
Shelf life	~6 months/RT	Peel ( )	-
Pot/Work life	6-7 mins (100g/RT)	σ_t	-
Tg	-	σ_{flex}	90 MPa
$T_{Service}$	-	E_{flex}	4500 MPa
CTE (°C)	-	G	-
Thermal cond.	-	Elongation	-
Resistivity: Vol.	-	Tear strength	-
Surf.	-	H (Shore D)	88
Notes: σ flex, *E* flex: ASTM D-790; Hardness: BS 2782 Pt 3; Linear shrinkage (200g/RT): 0.7%.			

Kemira	Hyperlast 2875-2874/701		Aug 93
Two part polyurethane			
Manufacturers Comments: Semi-rigid. For filter industry & moulding use.			
Typical Use: Filter assembly.			
Cure: RT/24 hrs.			
Test Methods: See notes			
Physical		**Mechanical**	
Colour	Brown	τ (RT)	-
Viscosity	~3 Pa.s	τ (°C)	-
SG	~1.3	τ (wet)	-
Shelf life	~6 months/RT	Peel ( )	-
Pot/Work life	4.5 mins (100g/RT)	σ_t	21 MPa
Tg	-	σ_{flex}	8 MPa
$T_{Service}$	-	E_{flex}	435 MPa
CTE (°C)	-	G	-
Thermal cond.	-	Elongation	35 %
Resistivity: Vol.	-	Tear strength	-
Surf.	-	H (Shore D)	73
Notes: σt, elong: BS 903 Pt A2; σ flex, *E* flex: ASTM D-790; Hardness: BS 2782 Pt 3.			

Kemira	Hyperlast 5000-2874/08		Aug 92
Two part polyurethane			
Manufacturers Comments: Semi-rigid. Good oil/fuel resistance. General purpose adhesive. For filters & bonding paper filter cartridges to tin plate or Zintec end-caps.			
Typical Use: Filter assembly.			
Cure: RT.			
Test Methods: See notes			
Physical		**Mechanical**	
Colour	Brown	τ (RT)	-
Viscosity	~7 Pa.s	τ (°C)	-
SG	~1.5	τ (wet)	-
Shelf life	>6 month/RT	Peel ( )	-
Pot/Work life	~1.5 mins (100g/RT)	σ_t	-
Tg	-	σ_{flex}	30 MPa
$T_{Service}$	-	E_{flex}	1200 MPa
CTE (°C)	-	G	-
Thermal cond.	-	Elongation	10%
Resistivity: Vol.	-	Tear strength	-
Surf.	-	H (Shore D)	77
Notes: σ flex, *E* flex: ASTM D-790; Hardness: BS 2782 Pt 3.			

Klebchemie (M.G. Becker GmbH)

Kemira	Hyperlast 5000-2874/701		Aug 93
Two part polyurethane			
Manufacturers Comments: Rigid. Good resistance to oils & fuels. For filter industry & consolidation of geological core samples.			
Typical Use: Filter assembly (end-caps).			
Cure: RT/24 hrs.			
Test Methods: See notes			
Physical		**Mechanical**	
Colour	Blue+Brown	τ (RT)	-
Viscosity	~3 Pa.s	τ (°C)	-
SG	~1.3	τ (wet)	-
Shelf life	~6 months/RT	Peel ( )	-
Pot/Work life	2 mins (100g/RT)	σ_t	-
Tg	-	σ_{flex}	45 MPa
$T_{Service}$	-	E_{flex}	2500 MPa
CTE (°C)	-	G	-
Thermal cond.	-	Elongation	-
Resistivity: Vol.	-	Tear strength	-
Surf.	-	H (Shore D)	85
Notes: σ flex, *E* flex: ASTM D-790; Hardness: BS 2782 Pt 3; Heat distortion temperature BS 2782 Pt 1: 60°C; Impact strength (notched Izod) BS2782 Pt 3: 3.2 kJ/m².			

Klebchemie	Klebeschaum 576.1		(Oct. 95)
Two-part, polyurethane			
Manufacturers Comments: Solvent-free. Foaming. Short cycle time.			
Typical Use: Automotive industry: Filter constructions (air, diesel, etc).			
Cure: RT. Warm (35-40°C).			
Test Methods: Not stated			
Physical		**Mechanical**	
Colour	-	τ (RT)	-
Viscosity	-	τ (°C)	-
SG	~1.4	τ (wet)	-
Shelf life	3 month/RT	Peel ( )	-
Pot/Work life	~2 mins(~50g/RT)	σ_t	-
Tg	-	σ_c	-
$T_{Service}$	-	E	-
CTE (°C)	-	G	-
Thermal cond.	-	Elongation	-
Resistivity: Vol.	-	Tear strength	-
Surf.	-	H (Shore)	-
Notes:			

Kemira	Hyperlast 5000-3283		Aug 93
Two part polyurethane			
Manufacturers Comments: Semi-rigid. Fast cure. Good adhesive properties.			
Typical Use: Filter assembly.			
Cure: RT.			
Test Methods: See notes			
Physical		**Mechanical**	
Colour	Brown	τ (RT)	-
Viscosity	~8 Pa.s	τ (°C)	-
SG	~1.5	τ (wet)	-
Shelf life	~6 months/RT	Peel ( )	-
Pot/Work life	30 sec (100g/RT)	σ_t	9 MPa
Tg	-	σ_c	-
$T_{Service}$	-	E	-
CTE (°C)	-	G	-
Thermal cond.	-	Elongation	35 %
Resistivity: Vol.	-	Tear strength	48 N mm^{-1}
Surf.	-	H (Shore D)	60
Notes: σt, elong: BBS 903 Pt A2; Hardness: BS 2782 Pt 3; Tear Strength (nicked crescent): ASTM D-624-73.			

Klebchemie	Kleiberit PU-501		(Oct. 95)
One-part, polyurethane, reaction (humidity) cure			
Manufacturers Comments: Moisture curing, requires water spray. High water & temperature resistance. Meets: DIN 68602 - B4 class.			
Typical Use: Construction: wood products (boards, windows, door parts). Laminating. Mineral-based boards, concrete & hard foams.			
Cure: RT. Cold and hot (70°C max.) pressing. RT/2-3 hrs (handling), 24 hrs (final strength).			
Test Methods: Not stated			
Physical		**Mechanical**	
Colour	Brown	τ (RT)	-
Viscosity	~8 Pa.s (medium)	τ (°C)	-
SG	~1.2	τ (wet)	-
Shelf life	6 month/RT	Peel ( )	-
Pot/Work life	~25 mins/RT (open)	σ_t	-
Tg	-	σ_c	-
$T_{Service}$	-	E	-
CTE (°C)	-	G	-
Thermal cond.	-	Elongation	-
Resistivity: Vol.	-	Tear strength	-
Surf.	-	H (Shore)	-
Notes: Single-sided application. Wood moisture content: 8 - 12%.			

See Key to Data Sheets - Pages 63 to 65

Klebchemie	Kleiberit PU-520	(Oct. 95)	
Two-part, polyurethane			
Manufacturers Comments: Solvent-free. Short cycle time. Thixotropic paste-like when mixed.			
Typical Use: Automotive industry: Glass fibre reinforced plastics, e.g. back-rest shells for car seats.			
Cure: RT/~15 mins (clamped); 60°C/~4-5 mins (clamped)			
Test Methods: Not stated			
Physical		**Mechanical**	
Colour	-	τ (RT)	-
Viscosity	Thixotropic paste	τ (°C)	-
SG	~1.2	τ (wet)	-
Shelf life	3 month/RT	Peel ( )	-
Pot/Work life*	~1.5 mins(~60g/RT)	σ_t (RT)	~35 MPa
Tg	-	σ_c	-
$T_{Service}$	-	*E*	-
CTE (°C)	-	*G*	-
Thermal cond.	-	Elongation	~25 %
Resistivity: Vol.	-	Tear strength	-
Surf.	-	H (Shore)	-
Notes: * Machine operation: ~70 seconds/50g mix. Tensile strength: DIN 53455-6-3. Elong: DIN 53455-6-3.			

Klebchemie	Supramelt PU 703.6	(Oct. 95)	
One-part, polyurethane (cross-linking, hot melt)			
Manufacturers Comments: Low melt temperature. Good adhesion to many plastics. No-flow on slanted surfaces. High joint strength after cure. Good temperature resistance.			
Typical Use: Automotive industry: Interior assembly (doors, textiles)			
Cure: Application Temperature: 120-140°C. Setting: >1min (clamped). Cure (crosslinking): ~1-3 days			
Test Methods: Not stated			
Physical		**Mechanical**	
Colour	-	τ (RT)	-
Viscosity	15 Pa.s (140°C)	τ (°C)	-
SG	~1.15	τ (wet)	-
Shelf life	3 month/RT	Peel ( )	-
Pot/Work life	<30 secs (open)	σ_t	-
Tg	-	σ_c	-
$T_{Service}$	-40 to ~100°C	*E*	-
CTE (°C)	-	*G*	-
Thermal cond.	-	Elongation	-
Resistivity: Vol.	-	Tear strength	-
Surf.	-	H (Shore)	-
Notes:			

Klebchemie	Kleiberit PU-543.1	(Oct. 95)	
Two-part, polyurethane			
Manufacturers Comments: Elastic structural adhesive. Solvent-free. Short cycle time. Thixotropic paste-like when mixed.			
Typical Use: Automotive industry: Glass fibre reinforced plastics, e.g. back-rest shells for car seats.			
Cure: RT/~15 mins (clamped); 60°C/~4-5 mins (clamped).			
Test Methods: Not stated			
Physical		**Mechanical**	
Colour	Black*	τ (RT)	-
Viscosity	Thixotropic paste	τ (°C)	-
SG	~1.2	τ (wet)	-
Shelf life	3 month/RT	Peel ( )	-
Pot/Work life	~2 mins(~50g/RT)	σ_t	-
Tg	-	σ_c	-
$T_{Service}$	-	*E*	-
CTE (°C)	-	*G*	-
Thermal cond.	-	Elongation	-
Resistivity: Vol.	-	Tear strength	-
Surf.	-	H (Shore A)	~77
Notes: *Other colours possible.			

Klebchemie	Supramelt PU 704.1	(Oct. 95)	
One-part, polyurethane (cross-linking, hot melt)			
Manufacturers Comments: Low melt temperature. Very high tack. Temperature resistance. Resistance to steaming & boiling. Exterior use.			
Typical Use: Highly stressed wrapping of PVC profiles. Wood wrapping with plastics (laminates, veneers, etc).			
Cure: Application Temperature: 120 - 140°C. Cure (crosslinking): ~1 - 2 days.			
Test Methods: Not stated			
Physical		**Mechanical**	
Colour	-	τ (RT)	-
Viscosity	15 Pa.s (140°C)	τ (°C)	-
SG	~1.1	τ (wet)	-
Shelf life	>6 month/RT	Peel ( )	-
Pot/Work life	-	σ_t	-
Tg	-	σ_c	-
$T_{Service}$	-40 to >130°C	*E*	-
CTE (°C)	-	*G*	-
Thermal cond.	-	Elongation	-
Resistivity: Vol.	-	Tear strength	-
Surf.	-	H (Shore)	-
Notes:			

See Key to Data Sheets - Pages 63 to 65

Loctite

Klebchemie	Supramelt PU 706.1		(Oct. 95)
One-part, polyurethane (cross-linking, hot melt)			
Manufacturers Comments: Low melt temperature. Very high tack. Temperature resistance. Resistance to steaming & boiling. Exterior use.			
Typical Use: Highly stressed wrapping of PVC profiles, wood & pre-heated aluminium profiles with plastics, laminates, veneers, foils, etc.			
Cure: Application Temperature: 120-140°C. Press (>10 secs). Cure (crosslinking): ~1-2 days.			
Test Methods: Not stated			
Physical		**Mechanical**	
Colour	-	τ (RT)	-
Viscosity	20 Pa.s (140°C)	τ (°C)	-
SG	~1.1	τ (wet)	-
Shelf life	>6 month/RT	Peel ( )	-
Pot/Work life	1 min max. (open)	σ_t	-
Tg	-	σ_c	-
$T_{Service}$	-40 to >130°C	*E*	-
CTE (°C)	-	*G*	-
Thermal cond.	-	Elongation	-
Resistivity: Vol.	-	Tear strength	-
Surf.	-	H (Shore)	-
Notes:			

Loctite	Forteflex 3951		Feb 93
One-part, polyurethane (moisture curing)			
Manufacturers Comments: RT cure. General purpose, elastomeric adhesive. Bonds wide range of materials, good gap filling. Flexible, durable. Exposed adhesive can be painted. Three primers available.			
Typical Use: GRP mouldings & reinforcing struts to GRP.			
Cure: Tack-free: 30-60 mins. Full cure: ~7 days (22°C/50%RH).			
Test Methods: ASTM/DIN			
Physical		**Mechanical**	
Colour	Black	τ (RT, steel)	1-2.5 (6.0) MPa
Viscosity	Thixotropic paste	τ (°C)	-
SG	1.1	τ (wet)	-
Shelf life	1 year/6-20°C	Peel (T)	3 (23) N mm^{-1}
Pot/Work life	60 mins (open)	σ_t	9-16 MPa
Tg	-	σ_c	-
$T_{Service}$	~75°C	*E*	-
CTE (°C)	-	*G*	-
Thermal cond.	-	Elongation	650-800%
Resistivity: Vol.	-	Tear strength	-
Surf.	-	H (Shore A)	60
Notes: Gap filling: 0.5-5 mm. Flash point (ASTM D93/DIN 51758) : >60°C. τ data ASTM D1002/DIN 53283 (modified): grit-blasted mild steel, unprimed (7253 primer). Polycarbonate: 1-3 MPa (unprimed), 6.6 MPa (7251 primer); σ_t & Elong.: ASTM D638; T-peel ASTM D1876 data: unprimed (primer 7253).			

Kömmerling

Kömmerling	Corapur 646		Jan 95
Two-part, polyurethane			
Manufacturers Comments: Solvent-free. Sprayable with slight thixotropy. Assembly of sandwich panels for building uses.			
Typical Use: Metal, wood products, laminates & insulation materials.			
Cure: RT. Press/clamping: 7-8 x pot life; Full cure: 24-48 hrs.			
Test Methods: Not stated			
Physical		**Mechanical**	
Colour	Beige-brown	τ (RT)	16 MPa
Viscosity	3 Pa.s	τ (°C)	-
SG	~1.4	τ (wet)	-
Shelf life	9 month/RT	Peel ( )	-
Pot/Work life	~20 mins*	σ_t	-
Tg	-	σ_c	-
$T_{Service}$	-	*E*	-
CTE (°C)	-	*G*	-
Thermal cond.	-	Elongation	-
Resistivity: Vol.	-	Tear strength	-
Surf.	-	H (Shore D)	75 ± 4
Notes: Other pot-life versions available as standard: ~50 mins, ~120 mins. τ : aluminium coupons with epoxy primer.			

Mydrin

Mydrin	Mybond 66102 / 66500		Dec 94
Two-part, polyurethane			
Manufacturers Comments: Solvent-free. Isocyanate cure.			
Typical Use: Vehicle and construction industry. Bonding of sandwich elements for cryogenic use.			
Cure: Initial: 5 - 8 hrs, Full: 7 days.			
Test Methods: Not stated			
Physical		**Mechanical**	
Colour	Brown	τ (RT)	10 - 12 MPa
Viscosity	8 - 9 Pa.s	τ (°C)	-
SG	~1.4	τ (wet)	-
Shelf life	6 month/RT	Peel ( )	-
Pot/Work life	30 - 60 mins.	σ_t	25 MPa
Tg	-	σ_c	-
$T_{Service}$	-190°C	*E*	-
CTE (°C)	-	*G*	-
Thermal cond.	-	Elongation	-
Resistivity: Vol.	-	Tear strength	-
Surf.	-	H (Shore)	-
Notes: -			

Mydrin	Mybond 66103 / 66500		Dec 94
Two-part, polyurethane			
Manufacturers Comments: Solvent-free. Isocyanate cure.			
Typical Use: Vehicle and construction industry. Bonding of sandwich elements with low strength core materials. Oil filter construction.			
Cure: Initial: 5 - 8 hrs, Full: 7 days.			
Test Methods: Not stated			
Physical		**Mechanical**	
Colour	Brown	τ (RT)	8 - 9 MPa
Viscosity	8 - 9 Pa.s	τ (°C)	-
SG	~1.5	τ (wet)	-
Shelf life	6 month/RT	Peel ( )	-
Pot/Work life	30 - 60 mins.	σ_t	10 MPa
Tg	-	σ_c	-
$T_{Service}$	-	E	-
CTE (°C)	-	G	-
Thermal cond.	-	Elongation	-
Resistivity: Vol.	-	Tear strength	-
Surf.	-	H (Shore)	-
Notes: -			

Mydrin	Mybond 66116 / 66500		Dec 94
Two-part, polyurethane			
Manufacturers Comments: Solvent-free. Isocyanate cure.			
Typical Use: Vehicle and construction industry. Bonding of sandwich elements for cryogenic use.			
Cure: Initial: 5 - 8 hrs, Full: 7 days.			
Test Methods: Not stated			
Physical		**Mechanical**	
Colour	Brown	τ (RT)	9 - 11 MPa
Viscosity	100 Pa.s	τ (°C)	-
SG	~1.4	τ (wet)	-
Shelf life	6 month/RT	Peel ( )	-
Pot/Work life	30 - 60 mins.	σ_t	22 - 25 MPa
Tg	-	σ_c	-
$T_{Service}$	-	E	-
CTE (°C)	-	G	-
Thermal cond.	-	Elongation	-
Resistivity: Vol.	-	Tear strength	-
Surf.	-	H (Shore)	-
Notes: -			

Mydrin	Mybond 66115 / 66500		Dec 94
Two-part, polyurethane			
Manufacturers Comments: Solvent-free. Non-slumping. Isocyanate cure.			
Typical Use: Vehicle and construction industry. Bonding of sandwich elements for cryogenic use. Vertical application.			
Cure: Initial: 5 - 8 hrs, Full: 7 days.			
Test Methods: Not stated			
Physical		**Mechanical**	
Colour	Brown	τ (RT)	10 - 12 MPa
Viscosity	(paste)	τ (°C)	-
SG	~1.4	τ (wet)	-
Shelf life	6 month/RT	Peel ( )	-
Pot/Work life	30 - 60 mins.	σ_t	20 - 23 MPa
Tg	-	σ_c	-
$T_{Service}$	-	E	-
CTE (°C)	-	G	-
Thermal cond.	-	Elongation	-
Resistivity: Vol.	-	Tear strength	-
Surf.	-	H (Shore)	-
Notes: -			

National Casein of California

National Casein	LU-3123A / B		Dec. 83
Two-part, polyurethane aqueous/polymer emulsion			
Manufacturers Comments: 44 - 48% solids.			
Typical Use: Wood products, metals and plastics. Laminating.			
Cure: RT, cold and hot pressing.			
Test Methods: Not stated			
Physical		**Mechanical**	
Colour	Pale grey	τ (RT)	-
Viscosity	~4 Pa.s	τ (°C)	-
SG	~1.2	τ (wet)	-
Shelf life	6 month/RT	Peel ( )	-
Pot/Work life	4 hrs.	σ_t	-
Tg	-	σ_c	-
$T_{Service}$	-	E	-
CTE (°C)	-	G	-
Thermal cond.	-	Elongation	-
Resistivity: Vol.	-	Tear strength	-
Surf.	-	H (Shore)	-
Notes: Flash point: >90°C.			

Rinaldi (SLC Rinaldi)

Rinaldi	Gutoid PU70		(Feb 95)
Two-part, polyurethane			
Manufacturers Comments: Single coat. Range of colours. Excellent strength in permanently wet conditions.			
Typical Use: Interior and exterior flooring.			
Cure: 24 hr/RT.			
Test Methods: Not Stated			
Physical		**Mechanical**	
Colour	-	τ (RT)	-
Viscosity	-	τ (°C)	-
SG	-	τ (wet)	-
Shelf life	-	Peel ( )	-
Pot/Work life	2 hr/RT	σ_t	-
Tg	-	σ_c	-
$T_{Service}$	-30°C to +80°C	E	-
CTE (°C)	-	G	-
Thermal cond.	-	Elongation	-
Resistivity: Vol.	-	Tear strength	-
Surf.	-	H (Shore)	-
Notes: 30 min. open time.			

SABA	Sabaplast 5503		(Sept 95)
One-part polyurethane			
Manufacturers Comments: Multi-purpose adhesive/sealant. Fast skinning, low shrinkage. Thixotropic. Excellent adhesion to many materials: metals, polyester, glass, wood, plastics, concrete, etc. 93% solids.			
Typical Use: Construction industry, boat-building, etc.			
Cure: Cure rate: 3mm per day.			
Test Methods: Not stated			
Physical		**Mechanical**	
Colour	Black, grey, white	τ (RT)	-
Viscosity	-	τ (°C)	-
SG	1.2	τ (wet)	-
Shelf life	6 month/RT	Peel ( )	-
Pot/Work life	-	σ_t (yield)	0.96 MPa
Tg	-	σ_c	-
$T_{Service}$	-40 to +90°C	$E_{100\%}$	0.55 MPa
CTE (°C)	-	G	-
Thermal cond.	-	Elongation	400 %
Resistivity: Vol.	-	Tear strength	-
Surf.	-	H (Shore A)	40 - 45
Notes: Flash point >40°C. Skin time: 20 mins/RT.			

SABA

SABA	Sabaplast 5130		(Sept 95)
One-part moisture curing polyurethane			
Manufacturers Comments: One-sided application. Sprayable. 98% solids. Solvent content: <0.5%.			
Typical Use: Construction industry. Wood-wood, wood-steel, etc.			
Cure: 4 hrs/RT.			
Test Methods: Not stated			
Physical		**Mechanical**	
Colour	Yellow-brown	τ (RT)	-
Viscosity	8.5 Pa.s	τ (°C)	-
SG	1.13	τ (wet)	-
Shelf life	6 month/RT	Peel ( )	-
Pot/Work life	120 mins (open)	σ_t	-
Tg	-	σ_c	-
$T_{Service}$	-40 to +150°C	E	-
CTE (°C)	-	G	-
Thermal cond.	-	Elongation	-
Resistivity: Vol.	-	Tear strength	-
Surf.	-	H (Shore)	-
Notes: Flash point 60°C.			

SABA	Sabaplast 5538		(Sept 95)
One-part moisture curing polyurethane			
Manufacturers Comments: One-sided application. Sprayable. 50% solids.			
Typical Use: Construction industry. Sandwich panel manufacture.			
Cure: 50 mins/RT.			
Test Methods: Not stated			
Physical		**Mechanical**	
Colour	Light brown	τ (RT)	-
Viscosity	0.075 Pa.s	τ (°C)	-
SG	1.18	τ (wet)	-
Shelf life	6 month/RT	Peel ( )	-
Pot/Work life	30 mins (open)	σ_t	-
Tg	-	σ_c	-
$T_{Service}$	-20 to +120°C	E	-
CTE (°C)	-	G	-
Thermal cond.	-	Elongation	-
Resistivity: Vol.	-	Tear strength	-
Surf.	-	H (Shore)	-
Notes: Flash point 40°C.			

See Key to Data Sheets - Pages 63 to 65

Scott Bader

SABA	Sabaplast 5817		(Sept 95)
One- or Two-part solvent-based polyurethane			
Manufacturers Comments: Excellent adhesion to PVC. Long open time. 17% solids. Can be used with 8% SABA Vulk SU 95 to increase crosslinking and service temperature limit (to 90°C).			
Typical Use: Shoe industry.			
Cure: RT. Dry 10 - 45 mins (both surfaces) then press together. Full cure: 24 hr. (8 hr. with Vulk SU 95).			
Test Methods: Not stated			
Physical		**Mechanical**	
Colour	Transparent	τ (RT)	-
Viscosity	2 Pa.s	τ (°C)	-
SG	0.89	τ (wet)	-
Shelf life	6 month/RT	Peel ( )	-
Pot/Work life	10 - 30 mins (open)	σ_t	-
Tg	-	σ_c	-
$T_{Service}$ (Vulk)	-20 to +50°C (+90)	*E*	-
CTE (°C)	-	*G*	-
Thermal cond.	-	Elongation	-
Resistivity: Vol.	-	Tear strength	-
Surf.	-	H (Shore)	-
Notes: Flash point <0°C.			

Scott Bader	Crestomer 1152PA		Feb 94
Two part modified polyurethane (acrylate), highly thixotropic.			
Manufacturers Comments: RT cure. Flexible with good shock and impact resistance. 53% Non-volatiles.			
Typical Use: Marine. FRP bonding and many other substrates.			
Cure: RT, Gel time: 50 mins/RT.			
Test Methods: Not stated.			
Physical		**Mechanical**	
Colour	-	τ (RT) GRP	15 MPa
Viscosity	-	τ (°C)	-
SG	1.03	τ (wet)	-
Shelf life	-	Peel ( )	-
Pot/Work life	(See cure)	σ_t	26 MPa
Tg	-	σ_c	-
$T_{Service}$	-	*E*	500 MPa
CTE (°C)	-	*G*	-
Thermal cond.	-	Elongation	100%
Resistivity: Vol.	-	Tear strength	-
Surf.	-	H (Shore D)	65
Notes: Shrinkage: 5%, Gardiner impact: 200 N/mm.			

Scholten Lijmen

Scholten Lijmen	Dynol U 462		Apr 88
One part moisture curing polyurethane			
Manufacturers Comments: RT cure. Coverage: 0.09 - 0.25 kg.m^{-2}			
Typical Use: Construction industry materials. Sandwich panel bonding for a wide range of core and facesheet materials.			
Cure: RT, pressure: 0.01 - 0.1 MPa (4 hrs).			
Test Methods: Not stated.			
Physical		**Mechanical**	
Colour	Brown	τ (RT)	-
Viscosity	7 Pa.s	τ (°C)	-
SG	1.12	τ (wet)	-
Shelf life	6 month/RT	Peel ( )	-
Pot/Work life	~60 mins/RT	σ_t	-
Tg	-	σ_c	-
$T_{Service}$	-	*E*	-
CTE (°C)	-	*G*	-
Thermal cond.	-	Elongation	-
Resistivity: Vol.	-	Tear strength	-
Surf.	-	H (Shore)	-
Notes: Flash point: 100°C.			

Scott Bader	Crestomer 1186PA		Feb 94
Two part modified polyurethane (acrylate).			
Manufacturers Comments: RT cure. Non-slumping filled adhesive with excellent gap-filling properties. Also exists in fast-cure (PD7962PA) and slow cure (1186A) versions. 74% Non-volatiles.			
Typical Use: Marine applications - FRP, metals, plastics, etc.			
Cure: RT, Gel time: 50 mins (1186PA), 120 mins (1186A), 20 mins (PD7962PA).			
Test Methods: Not stated.			
Physical		**Mechanical**	
Colour	-	τ (RT) GRP	9 - 10 MPa
Viscosity	-	τ (°C)	-
SG	1.32	τ (wet)	-
Shelf life	-	Peel ( )	-
Pot/Work life	(See cure)	σ_t	14 MPa
Tg	-	σ_c	-
$T_{Service}$	-	*E*	800 MPa
CTE (°C)	-	*G*	-
Thermal cond.	-	Elongation	6%
Resistivity: Vol.	-	Tear strength	-
Surf.	-	H (Shore D)	70
Notes: Shrinkage: 5%.			

Sika

Scott Bader	Crestomer 1196PA		Feb 94
Two part modified polyurethane (acrylate), thixotropic.			
Manufacturers Comments: RT cure. Spreadable with good gap-filling properties. Very low density. 53% Non-volatiles.			
Typical Use: Marine. Bonding primed PVC foam and balsa core to cured FRP panels in sandwich constructions.			
Cure: RT, Gel time: 50 mins/RT.			
Test Methods: Not stated.			
Physical		**Mechanical**	
Colour	-	τ (RT) GRP	9 - 10 MPa
Viscosity	-	τ (°C)	-
SG	0.58	τ (wet)	-
Shelf life	-	Peel ( )	-
Pot/Work life	(See cure)	σ_f	19 MPa
Tg	-	σ_c	-
$T_{Service}$	-	E	1300 MPa
CTE (°C)	-	G	-
Thermal cond.	-	Elongation	4%
Resistivity: Vol.	-	Tear strength	-
Surf.	-	H (Shore D)	70
Notes: Shrinkage: 4.5%			

Sika	Sikaflex 252		Apr 88
One part polyurethane, moisture curing, thixotropic.			
Manufacturers Comments: Knife-grade adhesive. Flexible with good aqueous chemical resistance - short-exposure resistance to fuels and oils and poor resistance to organic solvents, acids, strong alkalis, etc.			
Typical Use: Automotive - panel bonding. Repairs, e.g. boat hulls.			
Cure: Rate of cure: ~4 mm/day.			
Test Methods: DIN			
Physical		**Mechanical**	
Colour	-	τ (RT)	3 MPa
Viscosity	-	τ (°C)	-
SG	1.14	τ (wet)	-
Shelf life	9 month/RT	Peel ( )	-
Pot/Work life	(See note below)	σ_f	4 MPa
Tg	-	σ_c	-
$T_{Service}$	-40 to +90°C	E	-
CTE (°C)	-	G	-
Thermal cond.	-	Elongation	>300%
Resistivity: Vol.	-	Tear strength	9 N mm^{-1}
Surf.	-	H (Shore A)	55
Notes: Tack-free time: 40 min. Shrinkage: ~6% (DIN 52451). Tensile: DIN 53504, Tear DIN 53515			

SEA

(Société d'Exploitation des Adhésifs)

SEA	Mastic PU		May 94
One-part polyurethane			
Manufacturers Comments: Good flexibility. Good adhesion to PVC, wood, zinc, steel, stainless steel, glass, concrete, fibre-reinforced cement, aluminium, brick and stone.			
Typical Use: Construction industry.			
Cure: RT. Cure rate: 1mm per day under normal conditions.			
Test Methods: Not stated			
Physical		**Mechanical**	
Colour	Grey, brown	τ (RT)	-
Viscosity	(Paste)	τ (°C)	-
SG	-	τ (wet)	-
Shelf life	-	Peel ( )	-
Pot/Work life	-	σ_f	-
Tg	-	σ_c	-
$T_{Service}$	-	E	-
CTE (°C)	-	G	-
Thermal cond.	-	Elongation	-
Resistivity: Vol.	-	Tear strength	-
Surf.	-	H (Shore)	-
Notes: -			

Sika	Sikaflex 255 FC		Apr 88
One part polyurethane, moisture curing			
Manufacturers Comments: Fast-curing and flexible with good aqueous chemical resistance - short-exposure resistance to fuels and oils and poor resistance to organic solvents, acids, strong alkalis, etc.			
Typical Use: Automotive - glass bonding.			
Cure: Rate of cure: ~4 mm/day.			
Test Methods: DIN			
Physical		**Mechanical**	
Colour	-	τ (RT)	4.5 MPa
Viscosity	-	τ (°C)	-
SG	1.18	τ (wet)	-
Shelf life	9 month/RT	Peel ( )	-
Pot/Work life	(See note below)	σ_f	5 MPa
Tg	-	σ_c	-
$T_{Service}$	-40 to +90°C	E	-
CTE (°C)	-	G	-
Thermal cond.	-	Elongation	>400%
Resistivity: Vol.	-	Tear strength	12 N mm^{-1}
Surf.	-	H (Shore A)	60
Notes: Tack-free time: 20 min. Shrinkage: ~4% (DIN 52451). Tensile: DIN 53504, Tear DIN 53515.			

See Key to Data Sheets - Pages 63 to 65

Sika — Sikaflex 260 — Apr 88

One part polyurethane, moisture curing

Manufacturers Comments: Fast-curing and flexible with good aqueous chemical resistance - short-exposure resistance to fuels and oils and poor resistance to organic solvents, acids, strong alkalis, etc.

Typical Use: Automotive and construction. Bonding and sealing of steel and plastic components.

Cure: Rate of cure: ~4 mm/day.

Test Methods: DIN

Physical		Mechanical	
Colour	-	τ (RT)	4 MPa
Viscosity	-	τ (°C)	-
SG	1.22	τ (wet)	-
Shelf life	9 month/RT	Peel ( )	-
Pot/Work life	(See note below)	σ_t	6 MPa
Tg	-	σ_c	-
$T_{Service}$	-40 to +90°C	*E*	-
CTE (°C)	-	*G*	-
Thermal cond.	-	Elongation	>400%
Resistivity: Vol.	-	Tear strength	8 $N\ mm^{-1}$
Surf.	-	H (Shore A)	55

Notes: Tack-free time: 40 min. Shrinkage: ~6% (DIN 52451). Tensile: DIN 53504, Tear DIN 53515.

Sika — Sikaflex 360 HC — Apr 88

One part polyurethane, heat & moisture curing

Manufacturers Comments: High-strength. Variable rate cure and flexible with good aqueous chemical resistance - short-exposure resistance to fuels and oils and poor resistance to organic solvents, acids, strong alkalis, etc.

Typical Use: Automotive and construction. Bonding and sealing of steel and plastic components, pre-fabricated panels, etc.

Cure: 50 min/80°C, <5 min/140°C (<200°C). Rate of cure, RT: ~5mm/2days, 90°C (full cure): 25 min, 130°C (full cure): 5 min.

Test Methods: DIN

Physical		Mechanical	
Colour	-	τ (RT)	4 MPa
Viscosity	-	τ (°C)	-
SG	1.22	τ (wet)	-
Shelf life	9 month/RT	Peel ( )	-
Pot/Work life	(See note below)	σ_t	~6 MPa
Tg	-	σ_c	-
$T_{Service}$	-40 to +90°C	*E*	-
CTE (°C)	-	*G*	-
Thermal cond.	-	Elongation	>400%
Resistivity: Vol.	-	Tear strength	12 $N\ mm^{-1}$
Surf.	-	H (Shore A)	55

Notes: Tack-free time: 2 hrs. Shrinkage: <5% (DIN 52451). Tensile: DIN 53504, Tear DIN 53515.

Sika — Sikaflex 352 HC — Oct 89

One part polyurethane, heat & moisture curing

Manufacturers Comments: Variable rate cure and flexible with good aqueous chemical resistance - short-exposure resistance to fuels and oils and poor resistance to organic solvents, acids, strong alkalis, etc. Good resistance to impact, shock-loading and vibration.

Typical Use: Automotive - body panel bonding.

Cure: Rate of cure, RT: ~5 mm/2 days, 90°C (full cure): 25 min, 130°C (full cure): 5 min.

Test Methods: DIN

Physical		Mechanical	
Colour	-	τ (RT)	3.5 MPa
Viscosity	-	τ (°C)	-
SG	1.18	τ (wet)	-
Shelf life	9 month/RT	Peel ( )	-
Pot/Work life	(See note below)	σ_t	~4.5 MPa
Tg	-	σ_c	-
$T_{Service}$	-40 to +90°C	*E*	-
CTE (°C)	-	*G*	-
Thermal cond.	-	Elongation	>300%
Resistivity: Vol.	-	Tear strength	12 $N\ mm^{-1}$
Surf.	-	H (Shore A)	55

Notes: Tack-free time: 2 hrs. Shrinkage: ~8% (DIN 52451). Tensile: DIN 53504, Tear DIN 5076.

Sika — SikaTack Ultrafast — Aug 89

One part polyurethane, moisture curing

Manufacturers Comments: Very fast-curing and flexible.

Typical Use: Automotive - glass bonding. (Vehicle driveable after 30 mins using hot application, 60 mins at RT).

Cure: Rate of cure: ~4 mm/day.

Test Methods: DIN

Physical		Mechanical	
Colour	-	τ (RT)	5 MPa
Viscosity	-	τ (°C)	-
SG	1.13	τ (wet)	-
Shelf life	9 month/RT	Peel ( )	-
Pot/Work life	(See note below)	σ_t	7 MPa
Tg	-	σ_c	-
$T_{Service}$	-40 to +90°C	*E*	-
CTE (°C)	-	*G*	-
Thermal cond.	-	Elongation	>400%
Resistivity: Vol.	-	Tear strength	7 $N\ mm^{-1}$
Surf.	-	H (Shore A)	60

Notes: Tack-free time: 10 - 15 min (80°C), 15 - 20 min (20°C). Shrinkage: ~4% (DIN 52451). Tensile: DIN 53504, Tear DIN 53507B.

Sunkyong Industries

Sunkyong Industries	Skybon U0231T		(Oct 95)
One-part, solvent-borne polyurethane			
Manufacturers Comments: 24% solids. MEK/Toluene solvent.			
Typical Use: Automotive and shoe industries.			
Cure: RT.			
Test Methods: Not stated			
Physical		**Mechanical**	
Colour	-	τ (RT)	-
Viscosity	25 - 40 Pa.s	τ (°C)	-
SG	-	τ (wet)	-
Shelf life	-	Peel ( )	-
Pot/Work life	-	σ_t (yield)	39.2 MPa
Tg	-46°C	σ_c	-
$T_{Service}$	-	*E*	6.4 MPa
CTE (°C)	-	*G*	-
Thermal cond.	-	Elongation	900 %
Resistivity: Vol.	-	Tear strength	-
Surf.	-	H (Shore)	-
Notes: -			

Sunkyong Industries	Skybon U0841H		(Oct 95)
One-part, aqueous polyurethane emulsion			
Manufacturers Comments: 30% solids.			
Typical Use: Woodworking and general. (Wood and wood products, glass, PVC, etc.).			
Cure: RT.			
Test Methods: Not stated			
Physical		**Mechanical**	
Colour	-	τ (RT)	-
Viscosity	0.03 - 0.05 Pa.s	τ (°C)	-
SG	-	τ (wet)	-
Shelf life	-	Peel ( )	-
Pot/Work life	-	σ_t (yield)	10.8 MPa
Tg	-	σ_c	-
$T_{Service}$	-	*E*	1 MPa
CTE (°C)	-	*G*	-
Thermal cond.	-	Elongation	1100 %
Resistivity: Vol.	-	Tear strength	-
Surf.	-	H (Shore)	-
Notes: -			

Sunkyong Industries	Skybon U0251M		(Oct 95)
One-part, solvent-borne polyurethane			
Manufacturers Comments: 40% solids. MEK solvent.			
Typical Use: Automotive and shoe industries.			
Cure: RT.			
Test Methods: Not stated			
Physical		**Mechanical**	
Colour	-	τ (RT)	-
Viscosity	30 - 50 Pa.s	τ (°C)	-
SG	-	τ (wet)	-
Shelf life	-	Peel ( )	-
Pot/Work life	-	σ_t (yield)	39.2 MPa
Tg	-46°C	σ_c	-
$T_{Service}$	-	*E*	6.4 MPa
CTE (°C)	-	*G*	-
Thermal cond.	-	Elongation	900 %
Resistivity: Vol.	-	Tear strength	-
Surf.	-	H (Shore)	-
Notes: -			

Sunkyong Industries	Skybon U1351H		(Oct 95)
One-part, aqueous polyurethane dispersion			
Manufacturers Comments: 40% solids.			
Typical Use: Woodworking and general. (Wood and wood products, glass, PVC, etc.).			
Cure: RT.			
Test Methods: Not stated			
Physical		**Mechanical**	
Colour	-	τ (RT)	-
Viscosity	1 - 2.5 Pa.s	τ (°C)	-
SG	-	τ (wet)	-
Shelf life	-	Peel ( )	-
Pot/Work life	-	σ_t (yield)	53.9 MPa
Tg	-	σ_c	-
$T_{Service}$	-	*E*	24.5 MPa
CTE (°C)	-	*G*	-
Thermal cond.	-	Elongation	300 %
Resistivity: Vol.	-	Tear strength	-
Surf.	-	H (Shore)	-
Notes: -			

Sunstar Engineering

Sunstar Engineering	Penguin Cement 930		(Mar 95)
-			
Manufacturers Comments: RT cure.			
Typical Use: Construction, wood bonding.			
Cure: 16 hr/RT.			
Test Methods: Not stated.			
Physical		**Mechanical**	
Colour	Light brown	τ (RT) comp.	19.6 MPa
Viscosity	100 - 150 Pa.s	τ (°C)	-
SG	-	τ (wet)	-
Shelf life	-	Peel ( )	-
Pot/Work life	-	σ_t	-
Tg	-	σ_c	-
$T_{Service}$	-	*E*	-
CTE (°C)	-	*G*	-
Thermal cond.	-	Elongation	-
Resistivity: Vol.	-	Tear strength	-
Surf.	-	H (Shore)	-
Notes: τ compressive			

Sunstar Engineering	Penguin Seal 551		(Mar 95)
-			
Manufacturers Comments: RT cure.			
Typical Use: Automotive, glass bonding.			
Cure: 3 day/RT.			
Test Methods: Not stated.			
Physical		**Mechanical**	
Colour	Black	τ (RT)	4.7 MPa
Viscosity	46 Pa.s	τ (°C)	-
SG	-	τ (wet)	-
Shelf life	-	Peel ( )	-
Pot/Work life	-	σ_t	-
Tg	-	σ_c	-
$T_{Service}$	-	*E*	-
CTE (°C)	-	*G*	-
Thermal cond.	-	Elongation	-
Resistivity: Vol.	-	Tear strength	-
Surf.	-	H (Shore)	-
Notes:			

Sunstar Engineering	Penguin Cement 965		(Mar 95)
-			
Manufacturers Comments: RT cure.			
Typical Use: Automotive, glass bonding.			
Cure: 3 day/RT.			
Test Methods: Not stated.			
Physical		**Mechanical**	
Colour	Black	τ (RT)	4.6 MPa
Viscosity	49 Pa.s (60°C)	τ (°C)	-
SG	-	τ (wet)	-
Shelf life	-	Peel ( )	-
Pot/Work life	-	σ_t	-
Tg	-	σ_c	-
$T_{Service}$	-	*E*	-
CTE (°C)	-	*G*	-
Thermal cond.	-	Elongation	-
Resistivity: Vol.	-	Tear strength	-
Surf.	-	H (Shore)	-
Notes:			

Sunstar Engineering	Penguin Seal 915		(Mar 95)
-			
Manufacturers Comments: RT cure.			
Typical Use: Plastic bonding.			
Cure: 7day/RT.			
Test Methods: Not stated.			
Physical		**Mechanical**	
Colour	Grey	τ (RT)	7 MPa
Viscosity	500 Pa.s	τ (°C)	-
SG	-	τ (wet)	-
Shelf life	-	Peel ( )	-
Pot/Work life	-	σ_t	-
Tg	-	σ_c	-
$T_{Service}$	-	*E*	-
CTE (°C)	-	*G*	-
Thermal cond.	-	Elongation	-
Resistivity: Vol.	-	Tear strength	-
Surf.	-	H (Shore)	-
Notes:			

Swift Adhesives

Swift Adhesives	EC2 3710R / EC2 3700D		(Sept 95)
Two-part, polyurethane			
Manufacturers Comments: Fast setting. Good chemical resistance. High hot tack. Good adhesion to PVC, aluminium and metals.			
Typical Use: Panels, doors and containers.			
Cure: RT. Handling time: 2 hrs.			
Test Methods: Not stated			
Physical		**Mechanical**	
Colour	Beige	τ (Al/Al)	7.3 MPa
Viscosity	5 Pa.s	τ (Al/wood)	3.5 MPa
SG	-	τ (wet)	-
Shelf life	-	Peel ( )	-
Pot/Work life	8 mins.	σ_t	-
Tg	-	σ_c	-
$T_{Service}$	<120°C	E	-
CTE (°C)	-	G	-
Thermal cond.	-	Elongation	-
Resistivity: Vol.	-	Tear strength	-
Surf.	-	H (Shore D)	60
Notes: -			

Swift Adhesives	HMPUR 146		(Sept 95)
One-part, reactive hot-melt polyurethane			
Manufacturers Comments: Fast setting time. High hot tack. Good adhesion to PVC and wood substrates.			
Typical Use: Woodworking industry.			
Cure: Applied: 120 - 150°C, cure 2 mins/RT.			
Test Methods: Not stated			
Physical		**Mechanical**	
Colour	White	τ (wood/wood)	1.5 MPa
Viscosity	65 Pa.s (140°C)	τ (°C)	-
SG	-	τ (wet)	-
Shelf life	-	Peel ( )	-
Pot/Work life	-	σ_t (yield)	18 MPa
Tg	-	σ_c	-
$T_{Service}$	-	E	330 MPa
CTE (°C)	-	G	-
Thermal cond.	-	Elongation	-
Resistivity: Vol.	-	Tear strength	-
Surf.	-	H (Shore D)	63
Notes: Softening point: 65°C.			

Swift Adhesives	EC2 3810R / EC2 3700D		(Sept 95)
Two-part, polyurethane			
Manufacturers Comments: Low viscosity. Flexible. Good chemical resistance. High hot tack. Good adhesion to PVC, aluminium and metals.			
Typical Use: Panels, doors and containers.			
Cure: RT. Handling time: 2 hrs.			
Test Methods: Not stated			
Physical		**Mechanical**	
Colour	Grey	τ (Al/Al)	5 MPa
Viscosity	3 Pa.s	τ (Al/wood)	3 MPa
SG	-	τ (wet)	-
Shelf life	-	Peel ( )	-
Pot/Work life	15 mins.	σ_t	-
Tg	-	σ_c	-
$T_{Service}$	<130°C*	E	-
CTE (°C)	-	G	-
Thermal cond.	-	Elongation	-
Resistivity: Vol.	-	Tear strength	-
Surf.	-	H (Shore D)	50
Notes: * Temperature excursions to 200°C for 6 mins.			

Swift Adhesives	HMPUR 200		(Sept 95)
One-part, reactive hot-melt polyurethane			
Manufacturers Comments: Long open time. High initial strength. Good adhesion to PVC, ABS, SMC and primed metals.			
Typical Use: Automotive industry.			
Cure: Applied: 120 - 150°C, cure 6 mins/RT.			
Test Methods: Not stated			
Physical		**Mechanical**	
Colour	White	τ (SMC/ABS)	1.1 MPa
Viscosity	22 Pa.s (140°C)	τ (°C)	-
SG	-	τ (wet)	-
Shelf life	-	Peel ( )	-
Pot/Work life	-	σ_t (yield)	9.8 MPa
Tg	-	σ_c	-
$T_{Service}$	-	E	190 MPa
CTE (°C)	-	G	-
Thermal cond.	-	Elongation	>100 %
Resistivity: Vol.	-	Tear strength	-
Surf.	-	H (Shore D)	50
Notes: Softening point: 65°C.			

Teroson (Henkel)

Swift Adhesives	HMPUR 205 / 1		(Sept 95)
One-part, reactive hot-melt polyurethane			
Manufacturers Comments: Long open time, fast cure. High initial strength. Good adhesion to PVC, ABS, SMC and primed metals.			
Typical Use: Automotive industry.			
Cure: Applied: 120 - 150°C, cure 6 mins/RT.			
Test Methods: Not stated			
Physical		**Mechanical**	
Colour	White	τ (SMC/ABS)	1.3 MPa
Viscosity	30 Pa.s (140°C)	τ (°C)	-
SG	-	τ (wet)	-
Shelf life	-	Peel ( )	-
Pot/Work life	-	σ_t (yield)	10.5 MPa
Tg	-	σ_c	-
$T_{Service}$	-	E	210 MPa
CTE (°C)	-	G	-
Thermal cond.	-	Elongation	>100 %
Resistivity: Vol.	-	Tear strength	-
Surf.	-	H (Shore D)	55
Notes: Softening point: 65°C.			

Teroson	Terokal 6015		Aug 93
Two part polyurethane			
Manufacturers Comments: RT cure - elevated temperature cure possible. Paintable. >98% solids.			
Typical Use: Automotive. Metals/fibre-reinforced hard plastics (SMC). Plastic repair.			
Cure: 6hr/RT(90% final strength), 80-100°C (preferred), e.g. 8min/80°C			
Test Methods: DIN			
Physical		**Mechanical**	
Colour	Grey	τ (RT)	~10 MPa
Viscosity	~170 Pa.s	τ (40°C)	15.3 MPa
SG	~1.75	τ (wet)	20.4 MPa
Shelf life	12mnth/RT	Peel ( )	-
Pot/Work life	20 mins/RT	σ_t	-
Tg	-	σ_c	-
$T_{Service}$	-40 to +80°C	E	-
CTE (°C)	-	G	-
Thermal cond.	-	Elongation	-
Resistivity: Vol.	-	Tear strength	-
Surf.	-	H (Shore D)	65
Notes:			

Swift Adhesives	Polyfree 161		(Sept 95)
One-part, polyurethane, moisture curing			
Manufacturers Comments: Solvent-free. Spray application. Open time: 30 mins. 100% solids.			
Typical Use: Bonding aluminium and coated steel to wood and rigid insulation materials.			
Cure: Hot or cold pressing. (Pressure during initial cure). Initial cure: 90 mins/RT or 10 min/60 - 70°C. Full cure: 12 hrs.			
Test Methods: Not stated			
Physical		**Mechanical**	
Colour	Cream	τ (RT)	-
Viscosity	1.5 - 2.5 Pa.s	τ (°C)	-
SG	1.02	τ (wet)	-
Shelf life	6 month/RT	Peel ( )	-
Pot/Work life	-	σ_t	-
Tg	-	σ_c	-
$T_{Service}$	-	E	-
CTE (°C)	-	G	-
Thermal cond.	-	Elongation	-
Resistivity: Vol.	-	Tear strength	-
Surf.	-	H (Shore)	-
Notes: -			

Teroson	Terostat 8590		Aug 94
One part polyurethane, moisture curing,			
Manufacturers Comments: High viscosity adhesive/sealant. Cure accelerated by increasing temperature and humidity. Paintable. 100% solids. Very good sag resistance.			
Typical Use: Automotive, ships, lorries, trains. Glass bonding.			
Cure: Cure rate: 4 - 5 mm/day (RT, 50%RH).			
Test Methods: DIN			
Physical		**Mechanical**	
Colour	Black	τ (RT)	5 - 6 MPa
Viscosity	(See notes below)	τ (40°C)	-
SG	~1.2	τ (wet)	-
Shelf life	6mnth/RT	Peel (-)	6.5 N mm^{-1}
Pot/Work life	20 mins/RT	σ_t	7.5 MPa
Tg	-	σ_c	-
$T_{Service}$	-40 to +90°C	E	-
CTE (°C)	-	G	-
Thermal cond.	-	Elongation	350 - 500%
Resistivity: Vol.	-	Tear strength	-
Surf.	-	H (Shore A)	50
Notes: Available in four viscosity grades (values at RT): 8590N - low viscosity(~2800 Pa.s), 8590DK - medium viscosity(~2800 Pa.s), 8590HV - high viscosity(~4000 Pa.s), 8590UHV - very high viscosity(~7000 Pa.s). Shrinkage: <1%. Tensile: DIN 53504, shear: DIN 53283.			

See Key to Data Sheets - Pages 63 to 65

Tivoli

Teroson	Terostat 8690		July 94
One part polyurethane, moisture curing			
Manufacturers Comments: High viscosity, flexible adhesive/sealant. Cure accelerated by increasing temperature and humidity. Paintable. 100% solids. Very good sag resistance.			
Typical Use: Automotive, ships, lorries, trains. Glass bonding.			
Cure: Cure rate: 4 mm/day (RT, 50%RH).			
Test Methods: DIN			
Physical		**Mechanical**	
Colour	Black	τ (RT)	~1.5 MPa
Viscosity	7000 Pa.s (RT)	τ (°C)	-
SG	~1.2	τ (wet)	-
Shelf life	6mnth/RT	Peel ( )	-
Pot/Work life	15 mins/RT	σ_t	~8.5 MPa
Tg	-	σ_c	-
$T_{Service}$	-40 to +90°C	E	-
CTE (°C)	-	G	1.5 - 3 MPa
Thermal cond.	-	Elongation	~250%
Resistivity: Vol.	-	Tear strength	-
Surf.	-	H (Shore A)	70
Notes: Shrinkage: <1%. Tensile: DIN 53504, shear: DIN 53283.			

Tivoli	Tivo 9551/43		May 95
One part modified polyurethane			
Manufacturers Comments: Solvent-free thixotropic adhesive sealant. Bonding is insensitive to oils and over-bake. 99% solids.			
Typical Use: Automotive.			
Cure: 140°C - 200°C, e.g. 30min/180°C.			
Test Methods: DIN			
Physical		**Mechanical**	
Colour	Black	τ (RT)	3.5 MPa
Viscosity	400 Pa.s	τ (°C)	-
SG	~1.5	τ (wet)	-
Shelf life	6 month/RT	Peel ( )	-
Pot/Work life	-	σ_t	-
Tg	-	σ_c	-
$T_{Service}$	-	E	-
CTE (°C)	-	G	-
Thermal cond.	-	Elongation	-
Resistivity: Vol.	-	Tear strength	-
Surf.	-	H (Shore)	-
Notes: A range of properties are available with Tivo 9551/6, /7, /40, /44, /47, /48, /55 and /91.			

Teroson	Terostat 8890		Apr 94
One part polyurethane, moisture curing			
Manufacturers Comments: High viscosity adhesive/sealant. Cure accelerated by increasing temperature and humidity. Paintable. 100% solids. Good sag resistance.			
Typical Use: Automotive, ships, lorries, trains. Glass bonding.			
Cure: Cure rate: 3 - 4 mm/day (RT, 50%RH).			
Test Methods: DIN			
Physical		**Mechanical**	
Colour	Black	τ (RT)	4 - 5 MPa
Viscosity	(See note below)	τ (°C)	-
SG	~1.2	τ (wet)	-
Shelf life	6mnth/RT	Peel ( )	-
Pot/Work life	10 mins/RT	σ_t	8.5 MPa
Tg	-	σ_c	-
$T_{Service}$	-40 to +90°C	E	(See note below)
CTE (°C)	-	G	-
Thermal cond.	-	Elongation	350 - 500%
Resistivity: Vol.	-	Tear strength	-
Surf.	-	H (Shore A)	60 - 70
Notes: Available in two modulus grades: 8890 - normal modulus (E = ~1.5 MPa, visc = ~7000 Pa.s) and 8890HM - high modulus (E = 2.5 - 3 MPa, visc = ~14500 Pa.s). Shrinkage: <1%. Tensile: DIN 53504, shear: DIN 53283.			

Tivoli	Tivomelt 9617/11		May 92
One part, moisture-curing, hot-melt polyurethane			
Manufacturers Comments: Very good adhesion to many plastics and enamelled/primed steel. Bead, slit nozzle or spray application.			
Typical Use: -			
Cure: Work temperature: 90°C - 150°C, cure time: <16hr/RT/50%RH.			
Test Methods: Not stated.			
Physical		**Mechanical**	
Colour	-	τ (RT)	-
Viscosity	7.5 Pa.s/130°C	τ (°C)	-
SG	-	τ (wet)	-
Shelf life	-	Peel ( )	-
Pot/Work life	(very long)	σ_t	-
Tg	-	σ_c	-
$T_{Service}$	-	E	-
CTE (°C)	-	G	-
Thermal cond.	-	Elongation	-
Resistivity: Vol.	-	Tear strength	-
Surf.	-	H (Shore)	-
Notes: Tack time: 5 - 10 mins.			

Tivoli	Tivomelt 9625	May 92

One part, moisture-curing, hot-melt polyurethane

Manufacturers Comments: Very good adhesion to many plastics and enamelled/primed steel. Bead, slit nozzle or spray application.

Typical Use: -

Cure: Work temperature: 90°C - 160°C, cure time: <10hr/RT/50%RH.

Test Methods: Not stated.

Physical		Mechanical	
Colour	-	τ (RT)	-
Viscosity	14 Pa.s/130°C	τ (°C)	-
SG	-	τ (wet)	-
Shelf life	-	Peel ( )	-
Pot/Work life	(long)	σ_t	-
Tg	-	σ_c	-
$T_{Service}$	-	E	-
CTE (°C)	-	G	-
Thermal cond.	-	Elongation	-
Resistivity: Vol.	-	Tear strength	-
Surf.	-	H (Shore)	-

Notes: Tack time: 2 - 5 mins.

Tivoli	Tivomelt 9627/1	May 92

One part, moisture-curing, hot-melt polyurethane

Manufacturers Comments: Extremely fast initial set. Bead or slit nozzle application.

Typical Use: Wood and fibre products.

Cure: Work temperature: 120°C - 140°C, cure time: <16hr/RT/50%RH.

Test Methods: Not stated.

Physical		Mechanical	
Colour	-	τ (RT)	-
Viscosity	7.5 Pa.s/130°C	τ (°C)	-
SG	-	τ (wet)	-
Shelf life	-	Peel ( )	-
Pot/Work life	(very short)	σ_t	-
Tg	-	σ_c	-
$T_{Service}$	-	E	-
CTE (°C)	-	G	-
Thermal cond.	-	Elongation	-
Resistivity: Vol.	-	Tear strength	-
Surf.	-	H (Shore)	-

Notes: Tack time: <1 min.

Tivoli	Tivomelt 9626	May 92

One part, moisture-curing, hot-melt polyurethane

Manufacturers Comments: Fast initial set. Very good adhesion to many plastics and enamelled/primed steel. Bead or slit nozzle application.

Typical Use: -

Cure: Work temperature: 90°C - 140°C, cure time: <16hr/RT/50%RH.

Test Methods: Not stated.

Physical		Mechanical	
Colour	-	τ (RT)	-
Viscosity	6.5 Pa.s/130°C	τ (°C)	-
SG	-	τ (wet)	-
Shelf life	-	Peel ( )	-
Pot/Work life	(short)	σ_t	-
Tg	-	σ_c	-
$T_{Service}$	-	E	-
CTE (°C)	-	G	-
Thermal cond.	-	Elongation	-
Resistivity: Vol.	-	Tear strength	-
Surf.	-	H (Shore)	-

Notes: Tack time: 1 - 2 mins.

Tivoli	Tivomelt 9627/10	May 92

One part, moisture-curing, hot-melt polyurethane

Manufacturers Comments: Long reactivation time. Bead or slit nozzle application.

Typical Use: Wood and fibre products.

Cure: Work temperature: 130°C - 150°C, cure time: ~48hr/RT/50%RH.

Test Methods: Not stated.

Physical		Mechanical	
Colour	-	τ (RT)	-
Viscosity	14 Pa.s/130°C	τ (°C)	-
SG	-	τ (wet)	-
Shelf life	-	Peel ( )	-
Pot/Work life	(short)	σ_t	-
Tg	-	σ_c	-
$T_{Service}$	-	E	-
CTE (°C)	-	G	-
Thermal cond.	-	Elongation	-
Resistivity: Vol.	-	Tear strength	-
Surf.	-	H (Shore)	-

Notes: Tack time: 1 - 2 mins.

Tivoli	Tivomelt 9628/28		May 92
One part, moisture-curing, hot-melt polyurethane			
Manufacturers Comments: Fast amorphous setting adhesive with good adhesion to many plastics, wood and fibre products and enamelled/primed steel. Bead, slit nozzle or spray application.			
Typical Use: Wood and fibre products.			
Cure: Work temperature: 130°C - 160°C, cure time: <10hr/RT/50%RH.			
Test Methods: Not stated.			
Physical		**Mechanical**	
Colour	-	τ (RT)	-
Viscosity	25 Pa.s/130°C	τ (°C)	-
SG	-	τ (wet)	-
Shelf life	-	Peel ( )	-
Pot/Work life	(short)	σ_t	-
Tg	-	σ_c	-
$T_{Service}$	-	E	-
CTE (°C)	-	G	-
Thermal cond.	-	Elongation	-
Resistivity: Vol.	-	Tear strength	-
Surf.	-	H (Shore)	-
Notes: Tack time: <1 min.			

Tivoli	Tivomelt 9630/10		May 92
One part, moisture-curing, hot-melt polyurethane			
Manufacturers Comments: Highly accelerated, fast curing adhesive. Bead application.			
Typical Use: Plastics, rigid foamed plastics, enamelled/primed steel.			
Cure: Work temperature: 130°C - 150°C, cure time: ~3hr/RT/50%RH.			
Test Methods: Not stated.			
Physical		**Mechanical**	
Colour	-	τ (RT)	-
Viscosity	(thixotropic)	τ (°C)	-
SG	-	τ (wet)	-
Shelf life	-	Peel ( )	-
Pot/Work life	(very short)	σ_t	-
Tg	-	σ_c	-
$T_{Service}$	-	E	-
CTE (°C)	-	G	-
Thermal cond.	-	Elongation	-
Resistivity: Vol.	-	Tear strength	-
Surf.	-	H (Shore)	-
Notes: Tack time: <1 min.			

Tivoli	Tivomelt 9630		May 92
One part, moisture-curing, hot-melt polyurethane			
Manufacturers Comments: Highly accelerated, fast curing adhesive. Bead application.			
Typical Use: Plastics, rigid foamed plastics, enamelled/primed steel.			
Cure: Work temperature: 100°C - 140°C, cure time: ~3hr/RT/50%RH.			
Test Methods: Not stated.			
Physical		**Mechanical**	
Colour	-	τ (RT)	-
Viscosity	12.5 Pa.s (130°C)	τ (°C)	-
SG	-	τ (wet)	-
Shelf life	-	Peel ( )	-
Pot/Work life	(short)	σ_t	-
Tg	-	σ_c	-
$T_{Service}$	-	E	-
CTE (°C)	-	G	-
Thermal cond.	-	Elongation	-
Resistivity: Vol.	-	Tear strength	-
Surf.	-	H (Shore)	-
Notes: Tack time: 1 - 2 mins.			

Tivoli	Tivomelt 9635/3		May 92
One part, moisture-curing, hot-melt polyurethane			
Manufacturers Comments: Bead or slit nozzle application.			
Typical Use: Plastics, wood, fibre products, textiles.			
Cure: Work temperature: <120°C - 150°C, cure time: <16hr/RT/50%RH			
Test Methods: Not stated.			
Physical		**Mechanical**	
Colour	-	τ (RT)	-
Viscosity	25 Pa.s (130°C)	τ (°C)	-
SG	-	τ (wet)	-
Shelf life	-	Peel ( )	-
Pot/Work life	(very short)	σ_t	-
Tg	-	σ_c	-
$T_{Service}$	-	E	-
CTE (°C)	-	G	-
Thermal cond.	-	Elongation	-
Resistivity: Vol.	-	Tear strength	-
Surf.	-	H (Shore)	-
Notes: Tack time: <1 min.			

See Key to Data Sheets - Pages 63 to 65

Tivoli	Tivomelt 9635/4	May 92

One part, moisture-curing, hot-melt polyurethane

Manufacturers Comments: Fast curing adhesive. Bead or slit nozzle application.

Typical Use: Plastics, wood, fibre products, textiles.

Cure: Work temperature: <120°C - 150°C, cure time: <3hr/RT/50%RH.

Test Methods: Not stated.

Physical		Mechanical	
Colour	-	τ (RT)	-
Viscosity	30 Pa.s (130°C)	τ (°C)	-
SG	-	τ (wet)	-
Shelf life	-	Peel ( )	-
Pot/Work life	(very short)	σ_t	-
Tg	-	σ_c	-
$T_{Service}$	-	*E*	-
CTE (°C)	-	*G*	-
Thermal cond.	-	Elongation	-
Resistivity: Vol.	-	Tear strength	-
Surf.	-	H (Shore)	-

Notes: Tack time: <1 min.

Tivoli	Tivopur 1609	(July 95)

Two part polyurethane

Manufacturers Comments: Rigid.

Typical Use: General purpose. Polycarbonate, polyester, phenolics, foamed polymers, steel and wood.

Cure: 6hr/RT or 1.5hr/40°C.

Test Methods: DIN

Physical		Mechanical	
Colour	Beige	τ (RT)	18 MPa
Viscosity	~100 Pa.s	τ (°C)	-
SG	~1.4	τ (wet)	-
Shelf life	-	Peel ( )	-
Pot/Work life	60 mins/RT	σ_t	30 MPa
Tg	-	σ_c	-
$T_{Service}$	-	*E*	-
CTE (°C)	-	*G*	-
Thermal cond.	-	Elongation	3%
Resistivity: Vol.	-	Tear strength	-
Surf.	-	H (Shore A)	100

Notes: Tensile: DIN 53455, Shear: DIN 53283. Cured: 24hr/RT + 3hr/105°C. Data is for /09 hardener - /08 hardener gives slightly higher viscosity and shorter pot life.

Tivoli	Tivomelt 9645	May 92

One part, moisture-curing, hot-melt polyurethane

Manufacturers Comments: Bead or slit nozzle application.

Typical Use: Plastics, wood, fibre products.

Cure: Work temperature: 130°C - 150°C, cure time: <10hr/RT/50%RH.

Test Methods: Not stated.

Physical		Mechanical	
Colour	-	τ (RT)	-
Viscosity	22 Pa.s (130°C)	τ (°C)	-
SG	-	τ (wet)	-
Shelf life	-	Peel ( )	-
Pot/Work life	(very short)	σ_t	-
Tg	-	σ_c	-
$T_{Service}$	-	*E*	-
CTE (°C)	-	*G*	-
Thermal cond.	-	Elongation	-
Resistivity: Vol.	-	Tear strength	-
Surf.	-	H (Shore)	-

Notes: Tack time: <1 min.

Tivoli	Tivopur 1636	(July 95)

Two part polyurethane

Manufacturers Comments: Tough and rigid.

Typical Use: General purpose. ABS, PVC, polycarbonate, PBTP, polyester, polyurethane, polyetherester, phenolics, foamed polymers, steel, aluminium and wood.

Cure: 7hr/RT or 100min/40°C.

Test Methods: DIN

Physical		Mechanical	
Colour	Beige	τ (RT)	17 MPa
Viscosity	~80 Pa.s	τ (°C)	-
SG	~1.4	τ (wet)	-
Shelf life	-	Peel ( )	-
Pot/Work life	70 mins/RT	σ_t	20 MPa
Tg	-	σ_c	-
$T_{Service}$	-	*E*	-
CTE (°C)	-	*G*	-
Thermal cond.	-	Elongation	10%
Resistivity: Vol.	-	Tear strength	-
Surf.	-	H (Shore A)	98

Notes: Tensile: DIN 53455, Shear: DIN 53283. Cured: 24hr/RT + 3hr/105°C. Data is for /09 hardener - /08 hardener gives slightly higher viscosity and shorter pot life.

Tivoli	**Tivopur 1657**		(July 95)
Two part, fast curing polyurethane			
Manufacturers Comments: Rigid.			
Typical Use: Bonding of wood and wood products.			
Cure: 10 min/RT or 8min/40°C.			
Test Methods: DIN			
Physical		**Mechanical**	
Colour	Beige	τ (RT)	>6 MPa
Viscosity	~5 Pa.s	τ (°C)	-
SG	~1.4	τ (wet)	-
Shelf life	-	Peel ( )	-
Pot/Work life	2 mins/RT	σ_t	35 MPa
Tg	-	σ_c	-
$T_{Service}$	-	E	-
CTE (°C)	-	G	-
Thermal cond.	-	Elongation	5%
Resistivity: Vol.	-	Tear strength	-
Surf.	-	H (Shore A)	95
Notes: Tensile: DIN 53455, Shear: DIN 53283 (failure in wood adherend). Cured: 24hr/RT + 3hr/105°C.			

Tivoli	**Tivopur 1672/20**		(July 95)
Two part polyurethane			
Manufacturers Comments: Tough and rigid.			
Typical Use: General purpose. ABS, PVC, polycarbonate, PBTP, polyester, polyurethane, polyetherester, phenolics, foamed polymers, steel, aluminium and wood.			
Cure: 2hr/RT or 45min/40°C.			
Test Methods: DIN			
Physical		**Mechanical**	
Colour	Beige	τ (RT)	12 MPa
Viscosity	~1.5 Pa.s	τ (°C)	-
SG	~1.3	τ (wet)	-
Shelf life	-	Peel ( )	-
Pot/Work life	10 mins/RT	σ_t	10 MPa
Tg	-	σ_c	-
$T_{Service}$	-	E	-
CTE (°C)	-	G	-
Thermal cond.	-	Elongation	55%
Resistivity: Vol.	-	Tear strength	-
Surf.	-	H (Shore A)	95
Notes: Tensile: DIN 53455, Shear: DIN 53283. Cured: 24hr/RT + 3hr/105°C.			

Tivoli	**Tivopur 1672**		(July 95)
Two part polyurethane			
Manufacturers Comments: Tough and rigid.			
Typical Use: General purpose. ABS, PVC, polycarbonate, PBTP, polyester, polyurethane, polyetherester, phenolics, foamed polymers, steel, aluminium and wood.			
Cure: 18hr/RT or 4hr/40°C.			
Test Methods: DIN			
Physical		**Mechanical**	
Colour	Beige	τ (RT)	17 MPa
Viscosity	~66 Pa.s	τ (°C)	-
SG	~1.4	τ (wet)	-
Shelf life	-	Peel ( )	-
Pot/Work life	170 mins/RT	σ_t	10 MPa
Tg	-	σ_c	-
$T_{Service}$	-	E	-
CTE (°C)	-	G	-
Thermal cond.	-	Elongation	30%
Resistivity: Vol.	-	Tear strength	-
Surf.	-	H (Shore A)	95
Notes: Tensile: DIN 53455, Shear: DIN 53283. Cured: 24hr/RT + 3hr/105°C. Data is for /09 hardener - /08 hardener gives slightly higher viscosity and shorter pot life.			

Tivoli	**Tivopur 1672/70**		(July 95)
Two part polyurethane			
Manufacturers Comments: Tough and rigid.			
Typical Use: General purpose. ABS, PVC, polycarbonate, PBTP, polyester, polyurethane, polyetherester, phenolics, foamed polymers, steel, aluminium and wood.			
Cure: 3hr/RT or 40min/40°C.			
Test Methods: DIN			
Physical		**Mechanical**	
Colour	Beige	τ (RT)	13 MPa
Viscosity	~5 Pa.s	τ (°C)	-
SG	~1.5	τ (wet)	-
Shelf life	-	Peel ( )	-
Pot/Work life	32 mins/RT	σ_t	15 MPa
Tg	-	σ_c	-
$T_{Service}$	-	E	-
CTE (°C)	-	G	-
Thermal cond.	-	Elongation	35%
Resistivity: Vol.	-	Tear strength	-
Surf.	-	H (Shore A)	98
Notes: Tensile: DIN 53455, Shear: DIN 53283. Cured: 24hr/RT + 3hr/105°C.			

Tivoli	Tivopur 1673		(July 95)
Two part polyurethane			
Manufacturers Comments: Tough and rigid.			
Typical Use: General purpose. ABS, polycarbonate, PBTP, polyester, polyurethane, polyetherester, phenolics, foamed polymers, steel, aluminium and wood.			
Cure: 1hr/RT or 20min/40°C.			
Test Methods: DIN			
Physical		**Mechanical**	
Colour	Beige	τ (RT)	14 MPa
Viscosity	~8 Pa.s	τ (°C)	-
SG	~1.4	τ (wet)	-
Shelf life	-	Peel ( )	-
Pot/Work life	4 mins/RT	σ_t	20 MPa
Tg	-	σ_c	-
$T_{Service}$	-	*E*	-
CTE (°C)	-	*G*	-
Thermal cond.	-	Elongation	5%
Resistivity: Vol.	-	Tear strength	-
Surf.	-	H (Shore A)	95
Notes: Tensile: DIN 53455, Shear: DIN 53283. Cured: 24hr/RT + 3hr/105°C.			

Tivoli	Tivopur 1768		(July 95)
Two part polyurethane			
Manufacturers Comments: Tough with moderate rigidity.			
Typical Use: General purpose. ABS, polycarbonate, polyamide, rubber, steel, aluminium, glass and ceramics.			
Cure: 3hr/RT or 1hr/40°C.			
Test Methods: DIN			
Physical		**Mechanical**	
Colour	Grey	τ (RT)	7 MPa
Viscosity	~30 Pa.s	τ (°C)	-
SG	~1.5	τ (wet)	-
Shelf life	-	Peel ( )	-
Pot/Work life	45 mins/RT	σ_t	3 MPa
Tg	-	σ_c	-
$T_{Service}$	-	*E*	-
CTE (°C)	-	*G*	-
Thermal cond.	-	Elongation	55%
Resistivity: Vol.	-	Tear strength	-
Surf.	-	H (Shore A)	73
Notes: Tensile: DIN 53455, Shear: DIN 53283. Cured: 24hr/RT + 3hr/105°C.			

Tivoli	Tivopur 1717		(July 95)
Two part polyurethane			
Manufacturers Comments: Soft, high-flexibility.			
Typical Use: General purpose. ABS, polycarbonate, PBTP, polyamide, rubber, foamed polymers, steel and aluminium.			
Cure: -			
Test Methods: DIN			
Physical		**Mechanical**	
Colour	Black	τ (RT)	3 MPa
Viscosity	~8 Pa.s	τ (°C)	-
SG	~1.4	τ (wet)	-
Shelf life	-	Peel ( )	-
Pot/Work life	100 mins/RT	σ_t	1.5 MPa
Tg	-	σ_c	-
$T_{Service}$	-	*E*	-
CTE (°C)	-	*G*	-
Thermal cond.	-	Elongation	285%
Resistivity: Vol.	-	Tear strength	-
Surf.	-	H (Shore A)	40
Notes: Tensile: DIN 53455, Shear: DIN 53283. Cured: 24hr/RT + 3hr/105°C. Data is for /09 hardener - /08 hardener gives slightly higher viscosity and shorter pot life.			

Tivoli	Tivopur 1782/40		(July 95)
Two part polyurethane			
Manufacturers Comments: Rigid.			
Typical Use: Bonding of wood and wood products.			
Cure: 2min/RT.			
Test Methods: DIN			
Physical		**Mechanical**	
Colour	White	τ (RT)	>6 MPa
Viscosity	~15 Pa.s	τ (°C)	-
SG	~1.5	τ (wet)	-
Shelf life	-	Peel ( )	-
Pot/Work life	0.8 min/RT	σ_t	30 MPa
Tg	-	σ_c	-
$T_{Service}$	-	*E*	-
CTE (°C)	-	*G*	-
Thermal cond.	-	Elongation	5%
Resistivity: Vol.	-	Tear strength	-
Surf.	-	H (Shore A)	90
Notes: Tensile: DIN 53455, Shear: DIN 53283 (failure in wood adherend). Cured: 24hr/RT + 3hr/105°C.			

Tivoli	**Tivopur 1802/10**		(July 95)
Two part polyurethane			
Manufacturers Comments: High flexibility.			
Typical Use: General purpose. ABS, polycarbonate, foamed polymers, steel, aluminium, glass and ceramics.			
Cure: 6hr/RT or 1hr/40°C.			
Test Methods: DIN			
Physical		**Mechanical**	
Colour	Black	τ (RT)	8 MPa
Viscosity	~30 Pa.s	τ (°C)	-
SG	~1.1	τ (wet)	-
Shelf life	-	Peel ( )	-
Pot/Work life	14 mins/RT	σ_t	5 MPa
Tg	-	σ_c	-
$T_{Service}$	-	E	-
CTE (°C)	-	G	-
Thermal cond.	-	Elongation	150%
Resistivity: Vol.	-	Tear strength	-
Surf.	-	H (Shore A)	75
Notes: Tensile: DIN 53455, Shear: DIN 53283. Cured: 24hr/RT + 3hr/105°C.			

Tivoli	**Tivopur 1823/30**		(July 95)
Two part polyurethane			
Manufacturers Comments: Flexible.			
Typical Use: General purpose. ABS, polycarbonate, phenolics, foamed polymers, steel, aluminium, glass and ceramics.			
Cure: 2hr/40°C.			
Test Methods: DIN			
Physical		**Mechanical**	
Colour	Beige	τ (RT)	8 MPa
Viscosity	~35 Pa.s	τ (°C)	-
SG	~1.0	τ (wet)	-
Shelf life	-	Peel ( )	-
Pot/Work life	2 mins/RT	σ_t	7 MPa
Tg	-	σ_c	-
$T_{Service}$	-	E	-
CTE (°C)	-	G	-
Thermal cond.	-	Elongation	200%
Resistivity: Vol.	-	Tear strength	-
Surf.	-	H (Shore A)	70
Notes: Tensile: DIN 53455, Shear: DIN 53283. Cured: 24hr/RT + 3hr/105°C.			

Tivoli	**Tivopur 1815**		(July 95)
Two part polyurethane			
Manufacturers Comments: Rigid.			
Typical Use: Bonding of wood and wood products.			
Cure: 15min/RT or 10min/40°C			
Test Methods: DIN			
Physical		**Mechanical**	
Colour	Brown	τ (RT)	>6 MPa
Viscosity	~4 Pa.s	τ (°C)	-
SG	~1.4	τ (wet)	-
Shelf life	-	Peel ( )	-
Pot/Work life	1 min/RT	σ_t	35 MPa
Tg	-	σ_c	-
$T_{Service}$	-	E	-
CTE (°C)	-	G	-
Thermal cond.	-	Elongation	5%
Resistivity: Vol.	-	Tear strength	-
Surf.	-	H (Shore A)	95
Notes: Tensile: DIN 53455, Shear: DIN 53283 (failure in wood adherend). Cured: 24hr/RT + 3hr/105°C.			

Tivoli	**Tivopur 1840**		(July 95)
Two part polyurethane			
Manufacturers Comments: Tough and rigid.			
Typical Use: General purpose. ABS, polycarbonate, PBTP, polyester, polyurethane, polyetherester, phenolics, foamed polymers, steel, aluminium and wood.			
Cure: 3hr/RT or 40min/40°C			
Test Methods: DIN			
Physical		**Mechanical**	
Colour	Beige	τ (RT)	17 MPa
Viscosity	~7 Pa.s	τ (°C)	-
SG	~1.4	τ (wet)	-
Shelf life	-	Peel ( )	-
Pot/Work life	14 mins/RT	σ_t	20 MPa
Tg	-	σ_c	-
$T_{Service}$	-	E	-
CTE (°C)	-	G	-
Thermal cond.	-	Elongation	5%
Resistivity: Vol.	-	Tear strength	-
Surf.	-	H (Shore A)	98
Notes: Tensile: DIN 53455, Shear: DIN 53283. Cured: 24hr/RT + 3hr/105°C.			

Tivoli	Tivopur 1841/10		(July 95)
Two part polyurethane			
Manufacturers Comments: Tough and rigid.			
Typical Use: Bonding of wood and wood products.			
Cure: 2min/RT.			
Test Methods: DIN			
Physical		**Mechanical**	
Colour	Ochre	τ (RT)	>6 MPa
Viscosity	~8 Pa.s	τ (°C)	-
SG	~1.5	τ (wet)	-
Shelf life	-	Peel ( )	-
Pot/Work life	0.5 min/RT	σ_f	10 MPa
Tg	-	σ_c	-
$T_{Service}$	-	E	-
CTE (°C)	-	G	-
Thermal cond.	-	Elongation	20%
Resistivity: Vol.	-	Tear strength	-
Surf.	-	H (Shore A)	95
Notes: Tensile: DIN 53455, Shear: DIN 53283 (failure in wood adherend). Cured: 24hr/RT + 3hr/105°C.			

Wulff

Wulff (Naturalan)	Puro - Tack		(Feb 95)
Two component, liquid			
Manufacturers Comments: Solvent free, two part polyurethane. Castor wheel and radiant floor heating resistant.			
Typical Use: Floor coverings, mineral wool, metal, wood & ceramic tiles.			
Cure: 72 hr.			
Test Methods:			
Physical		**Mechanical**	
Colour	-	τ (RT)	-
Viscosity	~70 Pa.s	τ (°C)	-
SG	~1.2	τ (wet)	-
Shelf life	-	Peel ( )	-
Pot/Work life	30 min/20°C,75%rh	σ_f	-
Tg	-	σ_c	-
$T_{Service}$	~60°C	E	-
CTE (°C)	-	G	-
Thermal cond.	-	Elongation	-
Resistivity: Vol.	-	Tear strength	-
Surf.	-	H (Shore)	-
Notes: Conforms to DIN 53 269 (tensile shear and peel strength) and DIN 68 131 (castor wheel use). Coverage: 350 $g.m^{-2}$ - non-porous, <800 $g.m^{-2}$ - porous or textured.			

SILICONE ADHESIVES

NOTES

Silicone adhesives fall into two main groups:

- One-part. Within these, some liberate acetic acid during cure - these are not appropriate for corrosion sensitive applications. Others are described as 'neutral' cure, indicating the absence of acid cure products.
- Two-part mixed systems with a variety of cure mechanisms.

The majority of products are RT curing (RTV - room temperature vulcanisation), although heating may accelerate the cure.

With the exception of a few products designed to be poured (for potting or casting), viscosities are normally high to allow use as adhesive/sealants.

The particular advantages of silicone adhesives are:

- High, permanent flexibility to accommodate movement of parts and thermal expansion/contraction.
- Upper and lower service temperatures are very wide (maximum range: -115 to +265°C); specific products have service ranges within this band.
- Good environmental resistance.

Main Uses

Most silicone-based products are used as adhesive/sealants and are often available in a variety of colours.

- Building/Construction: Glazing assembly and fitting (wood, metals, plastics, glass), gap filling and expansion joints in facades and masonry, cold-storage units, water-tight sealing (industrial and household).
- Automotive: Bonding/sealing for hostile environments.
- Shipbuilding and aircraft: Bonding/sealing.
- Low viscosity (non-conductive) versions are generally used for potting of electronic components or electrical applications.
- Special grades exist for:
  - high vacuum applications (industrial and space use) in which volatile substances liberated during cure and exposure are restricted (controlled volatiles).
  - radiation resistance (nuclear installations).
  - electrical conductivity (metal filled).

Comparison Table

Company	Adhesive	Type	Colour	TS	Hs	El. (%)	SG	Visc (Pa.s)	Service (°C)	Pot (mins)	Cure	Skin Time (mins)	Cure Rate (mm/day)	Pg. No.
Bostik	Bondflex 100HMA	1a	-	○	22A	>500	1	-	-60/+200	-	RT	7	1.5	302
	Bondflex 200LMA	1a	-	◑	17A	850	1	-	-60/+200	-	RT	7	1.5	302
	Bondflex 300LMN	1n	-	○	25A	800	1.1	-	-60/+180	-	RT	5	1.5	302
	Bondflex 400LMN	1n	Various	○	25A	800	1.1	-	-60/+180	-	RT	5	1.5	302
W. Canning (Hylomar)	Hylosil 101	1n	Ivory	◑	25A	350	1.16	-	<240	-	-	15	-	303
	Hylosil 301	1a	White	◑	28A	400	1.05	-	<240	-	-	15	-	303
	Hylosil 310	1a	Red	◑	43A	380	1.3	-	<280	-	-	15	-	303
CECA	Mastic Bâtiment	1a	Various	-	-	-	1.05	-	-40/+150	-	RT	5	1.5	303
	Mastic Construction	1n	Various	-	-	-	1.04	-	-40/+150	-	RT	<45	1.5	304
	Mastic Isolation	1a	Transp.	-	-	-	1.05	-	-40/+150	-	RT	10	1.5	304
	Mastic Sanitaire	1a	Transp.	-	-	-	1.05	-	-40/+160	-	RT	10	1.5	304
Delo	Delomet 2302	1a	Transl.	-	28A	500	1.03	350	-40/+180	-	RT	20	2	304
Dow Corning	6-1104	1n	Transl.	◑	28A	575	1.11	H	-65/+200	90	RT	25	-	305
	6-1125	1n	White	◑	30A	500	1.12	H	-65/+260	90	RT	25	-	305
	895	1n	-	◑	30A	560	-	H	-55/+150	45	RT	-	-	305
	938	2n	Black	◑	45A	200	-	H	-55/+150	20	RT	-	-	305
	993	2n	Black	○	35A	130	1.36	34	-55/+150	90	RT	-	-	306
Dunlop	SAS 600	1a	White	◑	23A	400	-	H	-50/+150	10	RT	-	-	306
	SAS 640	1n	White	○	16A	500	-	H	-55/+120	10	RT	30	-	306
Dyckerhoff	SiliconFuge Karat	1n	Various	-	-	-	-	-	-55/+120	-	RT	15	1	306
EMFI	Mastic BN1	1n	Transl.	-	22A	>250	1.04	-	-50/+150	10	RT	60	2	307
	Mastic BTN	1n	Various	-	22A	>250	1.39	-	-60/+180	15	RT	<120	1	307
	Mastic MS1	1a	Various	-	21A	>250	1.04	-	-50/+200	5	RT	5	2	307
	Mastic S	1a	Various	-	24A	>250	1.25	-	-50/+200	10	RT	-	1.5	307
Forbo Helmitin	Hemisil BO	1n	Transp.	-	22A		1.0	H	-45/+150	-	RT	-	-	308
	Hemisil E	1a	Various	○	25A	600	1.05	H	-45/+150	-	RT	15	5	308
GE Silicones	FRV1102	1a	White	◑	38A	245	1.46	-	-60/+204	-	RT	-	-	308
	IS808	1a	Transl.	◑	23A	400	1.04	-	-60/+204	-	RT	-	-	308
	RTV 31	2	Red	●	54A	170	1.42	25	-54/+260	-	RT	150	-	309
	RTV 142	1n	White	◑	34A	400	1.09	H	-60/+204	-	RT	240	1.5	309
	RTV 167	1n	Grey	●	37A	650	1.12	H	-60/+204	-	RT	-	-	309
	RTV 560	2	Red	●	55A	120	1.42	30	-115/+260	-	RT	150	-	309
	RTV 566	2	Red-brown	●	63A	120	1.4	M	-115/+260	90	RT	150	-	310
	RTV 655	2	Transp.	●	45A	120	1.04	5.7	-110/+204	-	RT	150	-	310
	RTV 5818	1n	Transl.	◑	21A	500	1.04	-	-60/+204	-	RT	-	-	310
	RTV 6424	1	White	●	30A	500	1.17	-	-60/+204	-	RT + Heat	-	-	310
	RTV 6708	1n	Transl.	◑	18A	450	1.04	-	-60/+204	-	RT	-	-	311
	RTV 6808	1n	Transl.	◑	18A	400	1.04	-	-60/+204	-	RT	-	-	311
Le Joint Français	PR 1991	-	-	-	-	-	-	-	-	-	-	-	-	311
Loctite	Superflex	1n	Various	◑	25A	450	1.04	M	-60/+250		RT	15	3	311
McGhan NuSil	CCF1-2566	2	Red	◑	45A	100	1.42	35	-115/+260	60	RT	-	-	312
	CF2-1130	1n	Transl.	●	45A	300	1.09		-65/+232	180	RT	-	-	312
	CV-1142	1n	Transl.	◑	35A	200	1.12	H	-115/+260	30	RT	-	-	312
	CV-1143	1n	Transl.	◑	30A	300	1.12	H	-115/+260	30	RT	-	-	312
	CV-1500	1n	Black	◑	70A	25	1.25	H	-115/+260	30	RT	-	-	313
	CV-2520	1	White	○	55A	75	0.75		-115/+260	60	RT	-	-	313
	CV-2564	1	Transl.	●	50A	150	1.11	90	-115/+260	60	RT	-	-	313
	CV-2566	1	Red	●	60A	120	1.5	50	-115/+260	60	RT	-	-	313
	CV-2568	1	Red	○	45A	40	0.64	-	-115/+300	60	RT	-	-	314

Continued...

Company	Adhesive	Type	Colour	TS	Hs	El. (%)	SG	Visc (Pa.s)	Service (°C)	Pot (mins)	Cure	Skin Time (mins)	Cure Rate (mm/day)	Pg. No.
McGhan Nusil	CV1-1142	1n	Transl.	◐	35A.	100	1.1	M	-115/+260	30	RT	-	-	314
(continued)	CV2-1142-1	1n	White	◐	35A	200	1.13	M	-115/+260	30	RT	-	-	314
	FS 3730	1a	White	◐	35A	250	1.4	-	-60/+204	30	RT	-	-	314
	R 1140	1	Transl.	◐	30A	550	1.07	-	-65/+232	30	-	-	-	315
	R 1505	1n	Black	◐	70A	25	1.24	H	-115/+260	30	RT	-	-	315
	R 2076	2	Grey	●	40A	400	1.12	-	-65/+260	-	RT	-	-	315
	R 2560	2	Red	●	50A	140	1.42	30	-115/+260	90	RT	-	-	315
	R 2630	2	Black	◐	60A	50	1.09	20	-60/+204	120	RT - 150	-	-	316
Panacol Elosol	Penloc Silicon Blau	1	-	-	35A	460	-	-	-60/+265	-	-	-	-	316
SEA	Mastic Silicone	1m	Various	○	-	500	-	-	-40/+150	30	-	-	-	316
ThreeBond	1207B	1a	Black	○	(30)	425	1.01	H	-60/+250	-	RT - 120	7	3	316
	1207C	1a	Red-brown	●	(60)	170	1.5	H	-60/+250	-	RT - 120	5	3	317
	1208D	1a	Grey	○	(40)	350	1.47	H	-40/+120	-	RT - 120	9	3	317
	1211	1	White	○	(20)	230	1.01	70	-60/+250	-	RT	90	-	317
	1212	1	White	◐	(35)	260	1.05	H	-60/+250	-	RT	15	-	317
	1215	1	Grey	○	(45)	320	1.5	M	-60/+250	-	RT	60	2.5	318
	1216	1	Grey	◐	(60)	200	1.4	H	-60/+250	-	RT	20	2.5	318
Wacker Chemie	E104	1a	Transp.	○	16A	600	1.02	H	-40/+180	-	RT	18	1.5	318
	Elastosil IG 21	2	Black	◐	32A	350	1.39	H	-50/+150	30	RT	90	-	318
	Elastosil N182	1n	Transp.	○	18A	500	1.04	H	-40/+150	-	RT	15	1.5	319
	Elastosil N183	1n	Grey	○	30A	430	1.39	H	-40/+150	-	RT	15	1	319
	Elastosil N184	1n	Grey	○	21A	700	1.38	H	-40/+150	-	RT	20	1	319
	Elastosil N192	1n	Transp.	○	20A	400	1.03	H	-50/+100	-	RT	10	1.5	319
	Elastosil N193	1n	White	○	22A	550	1.38	H	-50/+150	-	RT	20	1	320
	Elastosil N195	1n	Transp.	○	25A	260	1.04	H	-40/+150	-	RT	20	-	320
	N180	1n	Grey	○	27A	500	1.34	H	-60/+150	-	RT	30	1.5	320
	N185	1n	Transp.	○	19A	550	1.04	H	-40/+150	-	RT	15	1.5	320
	N186	1n	White	○	25A	550	1.37	H	-40/+150	-	RT	15	1	321
	RTV-S 691	2	Red	●	55A	130	1.42	22	-40/+150	100	RT - 150	-	-	321
	RTV-S 695	2	Transp.	-	10A	-	-	5.9	-	(8 hrs)	RT - 100	-	-	321

Bostik

Bostik	Bondflex 100HMA		Oct 93
Silicone - One part, acetoxy cure.			
Manufacturers Comments: General purpose adhesive-sealant (cartridge pack). Accommodates movement. Meets USA FDA Title 21, Reg. No. 177.2600 & UK BS 5889(89) Type B.			
Typical Use: Waterproof seals. Building materials. Internal & external.			
Cure: RT(1°C - 30°C, 20°C opt.)		**Cure rate:** 1.5mm/day RT/50%RH	
Test Methods: Not stated.			
Physical		**Mechanical**	
Colour	-	τ (RT)	-
Viscosity	-	τ (°C)	-
SG	~1.0	τ (wet)	-
Shelf life	-	Peel ( )	-
Pot/Work life	-	σ_t	1.5 MPa
Tg	-	σ_c	-
$T_{Service}$	-60 to +200°C	E	-
Skin time	7 mins.	G	-
Cure shrinkage	-	Elongation	>500%
Resistivity: Vol.	-	Tear strength	-
Surf.	-	H (Shore A)	22
Notes:			

Bostik	Bondflex 300LMN		Oct 93
Silicone - One part, Neutral cure.			
Manufacturers Comments: General purpose adhesive-sealant (cartridge pack). Accommodates movement. Low odour. Meets UK BS 5889(89) Type A.			
Typical Use: Waterproof seals. Building materials. Internal & external. uPVC and Acrylic plastics.			
Cure: RT(1°C - 30°C, 20°C opt.)		**Cure rate:** 1.5mm/day RT/50%RH	
Test Methods: Not stated.			
Physical		**Mechanical**	
Colour	-	τ (RT)	-
Viscosity	-	τ (°C)	-
SG	~1.1	τ (wet)	-
Shelf life	-	Peel ( )	-
Pot/Work life	-	σ_t	1.3 MPa
Tg	-	σ_c	-
$T_{Service}$	-60 to +180°C	E	-
Skin time	5 mins.	G	-
Cure shrinkage	-	Elongation	800%
Resistivity: Vol.	-	Tear strength	-
Surf.	-	H (Shore A)	25
Notes:			

Bostik	Bondflex 200LMA		Oct 93
Silicone - One part, acetoxy cure.			
Manufacturers Comments: General purpose adhesive-sealant (cartridge pack). Accommodates movement. Meets UK BS 5889(89)Type A			
Typical Use: Waterproof seals. Building materials. Internal & external.			
Cure: RT(1°C - 30°C, 20°C opt.)		**Cure rate:** 1.5mm/day RT/50%RH	
Test Methods: Not stated.			
Physical		**Mechanical**	
Colour	-	τ (RT)	-
Viscosity	-	τ (°C)	-
SG	~1.0	τ (wet)	-
Shelf life	-	Peel ( )	-
Pot/Work life	-	σ_t	1.6 MPa
Tg	-	σ_c	-
$T_{Service}$	-60 to +200°C	E	-
Skin time	7 mins.	G	-
Cure shrinkage	-	Elongation	850%
Resistivity: Vol.	-	Tear strength	-
Surf.	-	H (Shore A)	17
Notes:			

Bostik	Bondflex 400LMN		Oct 93
Silicone - One part, Neutral cure.			
Manufacturers Comments: Marine adhesive-sealant (cartridge pack). Fresh and salt water, above and below waterline. Contact with potable water & food (when cured). Meets UK BS 5889(89) Type A.			
Typical Use: Wood, GRP, plastics (including polycarbonate), metals. Hatches and portholes.			
Cure: RT(1°C - 30°C, 20°C opt.)		**Cure rate:** 1.5mm/day RT/65%RH	
Test Methods: Not stated.			
Physical		**Mechanical**	
Colour	Various	τ (RT)	-
Viscosity	-	τ (°C)	-
SG	~1.1	τ (wet)	-
Shelf life	-	Peel ( )	-
Pot/Work life	-	σ_t	1.3 MPa
Tg	-	σ_c	-
$T_{Service}$	-60 to +180°C	E	-
Skin time	5 mins.	G	-
Cure shrinkage	-	Elongation	800%
Resistivity: Vol.	-	Tear strength	-
Surf.	-	H (Shore A)	25
Notes:			

William Canning Ltd. (Hylomar)

Wm. Canning (Hylomar)	**Hylosil 101**		(Feb 95)
Silicone - One part, Amine cure.			
Manufacturers Comments: Engineering grade adhesive-sealant. Non-corrosive, low odour. Non-slump. Good gap filling.			
Typical Use: Automotive and engineering. Bonds to a wide variety of materials.			
Cure: -		**Cure rate:** -	
Test Methods: DIN			
Physical		**Mechanical**	
Colour	Ivory	τ (RT)	-
Viscosity	-	τ (°C)	-
SG	1.16	τ (wet)	-
Shelf life	-	Peel ( )	-
Pot/Work life	-	σ_t	2.5 MPa
Tg	-	σ_c	-
$T_{Service}$	<240°C	*E*	-
Skin time	10 - 20 mins.	*G*	-
Cure shrinkage	-	Elongation	350 %
Resistivity: Vol.	-	Tear strength	-
Surf.	-	H (Shore A)	25
Notes: Hardness & Tensile: DIN 53504. Also available as: Hylosil 102 - Black and Hylosil 103 - Clear.			

Wm. Canning (Hylomar)	**Hylosil 310**		(Feb 95)
Silicone - One part, Acetoxy cure.			
Manufacturers Comments: High temperature adhesive-sealant. Non-slump. Good gap filling.			
Typical Use: Automotive and engineering. Bonds to a wide variety of materials. Gas turbine applications.			
Cure: -		**Cure rate:** -	
Test Methods: DIN			
Physical		**Mechanical**	
Colour	Red	τ (RT)	-
Viscosity	-	τ (°C)	-
SG	1.30	τ (wet)	-
Shelf life	-	Peel ( )	-
Pot/Work life	-	σ_t	2.0 MPa
Tg	-	σ_c	-
$T_{Service}$	<280°C	*E*	-
Skin time	10 - 20 mins.	*G*	-
Cure shrinkage	-	Elongation	380 %
Resistivity: Vol.	-	Tear strength	-
Surf.	-	H (Shore A)	43
Notes: Hardness & Tensile: DIN 53504.			

Wm. Canning (Hylomar)	**Hylosil 301**		(Feb 95)
Silicone - One part, Acetoxy cure.			
Manufacturers Comments: General purpose adhesive-sealant. Non-slump. Good gap filling.			
Typical Use: Automotive and engineering. Bonds to a wide variety of materials.			
Cure: -		**Cure rate:** -	
Test Methods: DIN			
Physical		**Mechanical**	
Colour	White	τ (RT)	-
Viscosity	-	τ (°C)	-
SG	1.05	τ (wet)	-
Shelf life	-	Peel ( )	-
Pot/Work life	-	σ_t	2.1 MPa
Tg	-	σ_c	-
$T_{Service}$	<240°C	*E*	-
Skin time	10 - 20 mins.	*G*	-
Cure shrinkage	-	Elongation	400 %
Resistivity: Vol.	-	Tear strength	-
Surf.	-	H (Shore A)	28
Notes: Hardness & Tensile: DIN 53504. Also available as: Hylosil 302 - Black and Hylosil 303 - Clear.			

CECA

CECA	**Mastic Bâtiment**		Mar 91
Silicone - One part, Acetic cure.			
Manufacturers Comments: General purpose building adhesive-sealant (cartridge pack). Accommodates movement. Good adhesion on glass. Norm. SNJF.			
Typical Use: Glass, aluminium, ceramic, wood.			
Cure: RT (+5°C - +40°C)		**Cure rate:** 1.5mm/day	
Test Methods: Not stated.			
Physical		**Mechanical**	
Colour	Transparent	τ (RT)	-
Viscosity	-	τ (°C)	-
SG	~1.05	τ (wet)	-
Shelf life	1 yr (RT)	Peel ( )	-
Pot/Work life	-	σ_t	-
Tg	-	σ_c	-
$T_{Service}$	-40 to +150°C	*E*	-
Skin time	5mins.	*G*	-
Cure shrinkage	-	Elongation	-
Resistivity: Vol.	-	Tear strength	-
Surf.	-	H (Shore)	-
Notes: Max. gap sizes: 30mm wide x 10mm deep. Other colours: white, grey, brown, black.			

See Key to Data Sheets - Pages 63 to 65

CECA	Mastic Construction		Sept 92
Silicone - One part, Neutral cure.			
Manufacturers Comments: General purpose building adhesive-sealant (cartridge pack). No odour. Accommodates 50% movement. Environmental resistance. Norm. SNJF.			
Typical Use: Masonry. Expansion joints in facades. Cold stores.			
Cure: RT (+5°C - +40°C)		**Cure rate:** 1.5mm/day	
Test Methods: Not stated.			
Physical		**Mechanical**	
Colour	Transparent	τ (RT)	-
Viscosity	-	τ (°C)	-
SG	~1.04	τ (wet)	-
Shelf life	1 yr (RT)	Peel ( )	-
Pot/Work life	-	σ_t	-
Tg	-	σ_c	-
$T_{Service}$	-40 to +150°C	E	-
Skin time	10mins. (transp.)	G	-
Cure shrinkage	-	Elongation	-
Resistivity: Vol.	-	Tear strength	-
Surf.	-	H (Shore)	-
Notes: Max. gap sizes: 20mm wide x 10mm deep. Other colours: white, grey, brown. Skin time: 45mins for coloured varieties.			

CECA	Mastic Sanitaire		Sept 92
Silicone - One part, Acetic cure.			
Manufacturers Comments: General purpose bathroom and kitchen adhesive-sealant. Resistance to humidity, household cleaners & fungus.			
Typical Use: Glass, enamel surfaces, ceramic tiles, aluminium, stainless steel.			
Cure: RT (+5°C - +40°C)		**Cure rate:** 1.5mm/day	
Test Methods: Not stated.			
Physical		**Mechanical**	
Colour	Transparent	τ (RT)	-
Viscosity	-	τ (°C)	-
SG	~1.05	τ (wet)	-
Shelf life	1 yr (RT)	Peel ( )	-
Pot/Work life	-	σ_t	-
Tg	-	σ_c	-
$T_{Service}$	-40 to +160°C	E	-
Skin time	10mins. (transp.)	G	-
Cure shrinkage	-	Elongation	-
Resistivity: Vol.	-	Tear strength	-
Surf.	-	H (Shore)	-
Notes: Max. gap sizes: 30mm wide x 10mm deep. Also available as white.			

CECA	Mastic Isolation		Mar 91
Silicone - One part, Acetic cure.			
Manufacturers Comments: General purpose adhesive-sealant (cartridge pack). Window and door installation. Durable and accommodates movement.			
Typical Use: Glass, aluminium, wood, ceramic tiles and smooth surfaces. Aquariums.			
Cure: RT		**Cure rate:** 1.5mm/day	
Test Methods: Not stated.			
Physical		**Mechanical**	
Colour	Transparent	τ (RT)	-
Viscosity	-	τ (°C)	-
SG	~1.05	τ (wet)	-
Shelf life	1 yr (RT)	Peel ( )	-
Pot/Work life	-	σ_t	-
Tg	-	σ_c	-
$T_{Service}$	-40 to +150°C	E	-
Skin time	10mins.	G	-
Cure shrinkage	-	Elongation	-
Resistivity: Vol.	-	Tear strength	-
Surf.	-	H (Shore)	-
Notes: Max. gap sizes: 30mm wide x 10mm deep.			

Delo

Delo	Delomet 2302		(July 95)
Silicone - One part, Acetic cure.			
Manufacturers Comments: Paste-like, self-levelling adhesive-sealant. RTV-1. Fast curing. High temperature resistance, good adhesion to range of materials, high strength in tension.			
Typical Use: Machine construction. Glass. Electronics industry.			
Cure: RT (5-35°C)		**Cure rate:** 2mm/24 hrs	
Test Methods: DIN			
Physical		**Mechanical**	
Colour	Milky-transparent	τ (RT)	-
Viscosity	~350 Pa.s	τ (°C)	-
SG	~1.03	τ (wet)	-
Shelf life	1 yr (RT)	Peel ( )	-
Pot/Work life	-	σ_t	-
Tg	-	σ_c	-
$T_{Service}$	-40 to +180°C	E	-
Skin time	20 mins	G	-
Cure shrinkage	~4%	Elongation	500%
Resistivity: Vol.	-	Tear strength	3.5 MPa
Surf.	-	H (Shore A)	28
Notes: Elongation & Tear resistance DIN 53504; Hardness DIN 53505.			

Dow Corning

Dow	6-1104		July 88
Silicone - One part, neutral cure.			
Manufacturers Comments: RTV cure. Non-corrosive. Controlled volatiles, low weight loss under high vacuum. Non-slump. Good physical & electrical stability over a wide range of frequencies, temperatures & humidities. Listed by ESA for use in space. Primer recommended for high adhesion.			
Typical Use: High vacuum applications (space & equipment). Electrical, electronic & optical assemblies. Metals, most plastics.			
Cure: RT (25°C)/72 hrs; 7 days		**Cure rate:** (Not for thick sections)	
Test Methods: Not stated			
Physical		**Mechanical**	
Colour	Translucent	τ (RT)	1.6 MPa
Viscosity	Paste-like	τ (°C)	-
SG	1.11	τ (wet)	-
Shelf life	6 months <30°C	Peel (180°)	10
Pot/Work life	1.5 hrs tack-free	σ_t	3.6 MPa
Tg	-	σ_c	-
$T_{Service}$	-65 to +200°C	E	-
Skin time	~25 mins	G	-
Cure shrinkage	-	Elongation	575%
Resistivity: Vol.	3.36×10^{14} Ωcm	Tear strength	22 N mm^{-1}
Surf.	-	H (Shore A)	28
Notes: Volatiles: %TML: 0.15, %CVCM ; 0.065. CTE (volumetric) 8.5x10^{-4} /K Electric strength: 19 kV/mm; Permittivity (RT): 2.60 (100Hz & 100kHz); Dissipation Factor (RT): 1.0x10^{-3} (100Hz) 1.0x10^{-3} (100kHz). Thermal cond: 0.12 W/mK			

Dow	895		(Feb 95)
Silicone - One part, neutral cure.			
Manufacturers Comments: RT cure. Non-corrosive. Non-slump. Adhesive-sealant for structural glazing experiencing UV, heat & humid atmospheres. Good adhesion. Insignificant property change after ASTM G-53 exposures.			
Typical Use: Structural glazing: glass, metal & other building facade materials. Stiffeners to building panels.			
Cure:*RT(25°C50%RH)/21 days		**Cure rate:** Varies with thickness	
Test Methods: ASTM			
Physical		**Mechanical**	
Colour	-	τ (RT)	-
Viscosity	Paste-like	τ (°C)	-
SG	-	τ (wet)	-
Shelf life	6 months/RT	Peel (-)	4.5 N mm^{-1}
Pot/Work life	45 mins tack-free	σ_t	2.6 MPa
Tg	-	σ_c	-
$T_{Service}$	-55 to +150°C	E	-
Skin time	-	G	-
Cure shrinkage	-	Elongation	560%
Resistivity: Vol.	-	Tear strength	15.3 N mm^{-1}
Surf.	-	H (Shore A)	30
Notes: * Conditions: 8-35°C, 40-95%RH. Preferred temp range 15-30°C. σ, Elong. ASTM D-412; Tear ASTM D-624; Hardness ASTM D-676 Peel ASTM C 794-80 (al painted or anodised & float glass).			

Dow	6-1125		July 88
Silicone - One part, neutral cure.			
Manufacturers Comments: RTV cure. Non-corrosive. Controlled volatiles, low weight loss under high vacuum. Non-slump. Good physical & electrical stability over a wide range of frequencies, temperatures & humidities. Listed by ESA for use in space. Primer recommended for high adhesion.			
Typical Use: High vacuum applications (space & equipment). Electrical, electronic & optical assemblies. Metals, most plastics.			
Cure: RT (25°C)/72 hrs; 7 days		**Cure rate:** (Not for thick sections)	
Test Methods: Not stated			
Physical		**Mechanical**	
Colour	White	τ (RT)	2.1 MPa
Viscosity	Paste-like	τ (°C)	-
SG	1.12	τ (wet)	-
Shelf life	6 months <30°C	Peel (180°)	10
Pot/Work life	1.5 hrs tack-free	σ_t	3.8 MPa
Tg	-	σ_c	-
$T_{Service}$	-65 to +260°C	E	-
Skin time	~25 mins	G	-
Cure shrinkage	-	Elongation	500%
Resistivity: Vol.	1.0×10^{14} Ωcm	Tear strength	22 N mm^{-1}
Surf.	-	H (Shore A)	30
Notes: Volatiles: %TML: 0.15, %CVCM ; 0.065. CTE (volumetric) 8.5x10^{-4} /K Electric strength: 19 kV/mm; Permittivity (RT): 2.60 (100Hz & 100kHz); Thermal cond: 0.125 W/mK. Dissipation Factor (RT): 1.0x10^{-3} (100Hz) 1.0x10^{-3} (100kHz).			

Dow	983		(Feb 95)
Silicone - Two part, neutral cure.			
Manufacturers Comments: RT cure. Non-corrosive. Non-slump. Adhesive-sealant for structural glazing experiencing UV, heat & humid atmospheres. Good adhesion to wide range of construction materials.			
Typical Use: Structural glazing: coated, enamelled glasses; anodised or coated aluminium profiles; stainless steel profiles. Plastic 'glazing'.			
Cure: RT (25°C)/7 days		**Cure rate:** -	
Test Methods: ISO, ASTM			
Physical		**Mechanical**	
Colour	Black	τ (RT)	1.4 MPa
Viscosity	Paste-like	τ (°C)	-
SG	-	τ (wet)	-
Shelf life	1 year/RT	Peel ( )	-
Pot/Work life	20-25 mins	σ_t	1.9 MPa
Tg	-	σ_c	-
$T_{Service}$	-55 to +150°C	E	-
Skin time	-	G	-
Cure shrinkage	-	Elongation	200%
Resistivity: Vol.	-	Tear strength	9.8 N mm^{-1}
Surf.	-	H (Shore A)	45
Notes: τ aluminium. Same for glass. σ, Elong. ASTM D-412; Tear ASTM D-624; Hardness ASTM D-676.			

See Key to Data Sheets - Pages 63 to 65

Dow	993		(Feb 95)
Silicone - Two part, neutral cure.			
Manufacturers Comments: RT cure. Adhesive-sealant for structural glazing experiencing UV, heat & humid conditions. Good adhesion to wide range of construction materials. Non-slump, odourless, non-corrosive. Meets: CEN 129 WG 16 (European structural glazing).			
Typical Use: Structural glazing: coated, enamelled glasses; anodised or coated aluminium profiles; stainless steel profiles.			
Cure: RT (25°C/50%RH)/7 days		**Cure rate:** -	
Test Methods: ISO, ASTM			
Physical		**Mechanical**	
Colour	Black	τ (RT)	-
Viscosity	34 Pa.s	τ (°C)	-
SG	1.36	τ (wet)	-
Shelf life	8-12 months	Peel ( )	-
Pot/Work life	1.5 hrs tack-free*	σ_t	0.95 MPa
Tg	-	σ_c	-
$T_{Service}$	-55 to +150°C	*E*	-
Skin time	-	*G*	-
Cure shrinkage	-	Elongation	130%
Resistivity: Vol.	-	Tear strength	6 N mm^{-1}
Surf.	-	H (Shore A)	35
Notes: * varies with mix ratio used & temperature. σ ISO 8339; Tear ASTM D-624; Elong ISO 8339; Hardness ASTM D-676.			

Dunlop	SAS 640		Jan 92
Silicone - One part, neutral cure.			
Manufacturers Comments: RT cure. Non-corrosive, non-slump, thixotropic adhesive-sealant for glazing systems, boats, electrical uses. Good resistance to weathering. Good adhesion to masonry (no primer).			
Typical Use: Glazing: glass, metal & other building facade materials.			
Cure: RT moisture curing		**Cure rate:** -	
Test Methods: Not stated			
Physical		**Mechanical**	
Colour	White	τ (RT)	-
Viscosity	Thixotropic	τ (°C)	-
SG	-	τ (wet)	-
Shelf life	9 months/+5°C	Peel ( )	-
Pot/Work life	~10 mins	σ_t	1.0 MPa
Tg	-	σ_c	-
$T_{Service}$	-55 to +120°C	*E*	-
Skin time	30 mins	*G*	-
Cure shrinkage	-	Elongation	500%
Resistivity: Vol.	-	Tear strength	15.3 N mm^{-1}
Surf.	-	H (Shore A)	16
Notes: Joint width: 3mm (min); 25mm (max.) Movement accommodation: 50%.			

Dunlop

Dunlop	SAS 600		Jan 92
Silicone - One part, acid cure.			
Manufacturers Comments: RT cure. Thixotropic, non-sag. Good weathering resistance. Chemical resistance (dilute acids/alkalis).			
Typical Use: Glazing (primer for glass & ceramics); Joints seals (gas meters, steam lines, boilers).			
Cure:RT/1-4 days (full cure)		**Cure rate:** -	
Test Methods: Not stated			
Physical		**Mechanical**	
Colour	White*	τ (RT)	-
Viscosity	Thixotropic	τ (°C)	-
SG	-	τ (wet)	-
Shelf life	1 year/+5°C	Peel ( )	-
Pot/Work life	~10 mins	σ_t	2.5 MPa
Tg	-	σ_c	-
$T_{Service}$	-50 to +150°C	*E*	-
Skin time	-	*G*	-
Cure shrinkage	-	Elongation	400%
Resistivity: Vol.	-	Tear strength	-
Surf.	-	H (Shore A)	23
Notes: Other versions: SAS 610 (clear); SAS 620 (black); SAS630(aluminium). Joint width: 3mm (min); 25mm (max.) Movement accommodation: 12.5%.			

Dyckerhoff

Dyckerhoff	SiliconFuge Karat		Oct 93
Silicone - One part, neutral cure.			
Manufacturers Comments: RT cure. Non-corrosive, adhesive-sealant for expansion joints (frames to buildings, etc). Strong flank adhesion. Indoor-outdoor use. Good environmental resistance.			
Typical Use: Most building materials (glass, aluminium profiles, GRP). Some may require primer, e.g. brass, steels, concrete, plaster.			
Cure: RT moisture curing		**Cure rate:** 1mm/24 hrs	
Test Methods: Not stated			
Physical		**Mechanical**	
Colour	*	τ (RT)	-
Viscosity	-	τ (°C)	-
SG	-	τ (wet)	-
Shelf life	-	Peel ( )	-
Pot/Work life	-	σ_t	-
Tg	-	σ_c	-
$T_{Service}$	-55 to +120°C	*E*	-
Skin time	~15 mins	*G*	-
Cure shrinkage	-	Elongation	-
Resistivity: Vol.	-	Tear strength	-
Surf.	-	H (Shore)	-
Notes: *Product codes for range of colours: 720 (white); 721 (grey); 728 (silver-grey); 722 (transparent); 723 (sand-stone); 724 (black); 725 (mahogany); 727 (concrete-grey). Joint width: 3mm (min); 25mm (max.) Movement accommodation: 50%.			

EMFI

EMFI	Mastic BN1		(Nov 95)
Silicone - One part, neutral cure			
Manufacturers Comments: RT cure, acetic-free, adhesive-sealant for assembling building materials. No primer required. Resists temperature changes, climatic ageing (inc. UV). Meets: SNJF spec.			
Typical Use: Glazing assemblies: glass, varnished wood, metals, PVC. Expansion joints: concrete, masonry, wood.			
Cure: RT moisture (50%RH)		**Cure rate:** 2mm/24 hrs	
Test Methods: ISO			
Physical		**Mechanical**	
Colour	Translucent	τ (RT)	-
Viscosity	Thixotropic	τ (°C)	-
SG	1.04	τ (wet)	-
Shelf life	1 year (5-25°C)	Peel ( )	-
Pot/Work life	10 mins	σ_t	-
Tg	-	σ_c	-
$T_{Service}$	-50 to +150°C	E	0.3 MPa
Skin time	1 hr	G	-
Cure shrinkage	None	Elongation	>250%
Resistivity: Vol.	-	Tear strength	-
Surf.	-	H (Shore A)	22
Notes: Slump resistance (ISO 7390): good. E: ISO 37; Elong: ISO 8339. Shrinkage (DIN 52451).			

EMFI	Mastic MS1		(Nov 95)
Silicone - One part, acetic			
Manufacturers Comments: RT cure. Adhesive-sealant for assembling aquariums, cold-rooms, ship-building, cooling systems. Resists temperature changes, climatic ageing (inc. UV). Meets: SNJF spec.			
Typical Use: Glazing assemblies, building materials, metals.			
Cure: RT moisture (50%RH)		**Cure rate:** 2mm/24 hrs	
Test Methods: ISO			
Physical		**Mechanical**	
Colour	White*	τ (RT)	-
Viscosity	Thixotropic	τ (°C)	-
SG	1.04	τ (wet)	-
Shelf life	1 year (5-25°C)	Peel ( )	-
Pot/Work life	5 mins	σ_t	-
Tg	-	σ_c	-
$T_{Service}$	-50 to +200°C	E	0.5 MPa
Skin time	5 mins	G	-
Cure shrinkage	None	Elongation	>250%
Resistivity: Vol.	-	Tear strength	-
Surf.	-	H (Shore A)	~21
Notes: * Other colours available: Clear, grey, black, bronze. Slump resistance (ISO 7390): good. E: ISO 37; Elong: ISO 8339. Shrinkage (DIN 52451).			

EMFI	Mastic BTN		(Nov 95)
Silicone - One part, neutral cure			
Manufacturers Comments: RT cure, acetic-free, adhesive-sealant for assembling building materials. No primer required for non-porous materials. Resists temperature changes, climatic ageing (inc. UV). Meets: SNJF specification.			
Typical Use: Glazing assemblies: glass, varnished wood, metals, PVC. Expansion joints: concrete, masonry, wood.			
Cure: RT moisture (50%RH)		**Cure rate:** 1mm/24 hrs	
Test Methods: ISO			
Physical		**Mechanical**	
Colour	White*	τ (RT)	-
Viscosity	Thixotropic	τ (°C)	-
SG	1.39	τ (wet)	-
Shelf life	1 year (5-25°C)	Peel ( )	-
Pot/Work life	15 mins	σ_t	-
Tg	-	σ_c	-
$T_{Service}$	-60 to +180°C	E	0.39 MPa
Skin time	1 - 2 hrs	G	-
Cure shrinkage	None	Elongation	>250%
Resistivity: Vol.	-	Tear strength	-
Surf.	-	H (Shore A)	22
Notes: * Other colours available: Grey, black, bronze, brown, beige. Slump resistance (ISO 7390): good. E: ISO 37; Elong: ISO 8339. Shrinkage (DIN 52451).			

EMFI	Mastic S		(Nov 95)
Silicone - One part, acetic			
Manufacturers Comments: RT cure. Adhesive-sealant for laboratories, bathroom and plumbing uses. Resists temperature changes, moisture, fungus/bacteria, household cleaners. Not for assembly of aquariums.			
Typical Use: Porcelain, glass, enamelled materials, wood, plastics, metals.			
Cure: RT moisture (50%RH)		**Cure rate:** 1.5mm/24 hrs	
Test Methods: ISO			
Physical		**Mechanical**	
Colour	White*	τ (RT)	-
Viscosity	Thixotropic	τ (°C)	-
SG	1.25	τ (wet)	-
Shelf life	1 year (5-25°C)	Peel ( )	-
Pot/Work life	~10 mins	σ_t	-
Tg	-	σ_c	-
$T_{Service}$	-50 to +200°C	E	0.3 MPa
Skin time	-	G	-
Cure shrinkage	None	Elongation	>250%
Resistivity: Vol.	-	Tear strength	-
Surf.	-	H (Shore A)	~24
Notes: * Translucent version available. Slump resistance (ISO 7390): good. E: ISO 37; Elong: ISO 8339. Shrinkage (DIN 52451).			

Forbo Helmitin

Forbo Helmitin	Hemisil BO		Nov 89
Silicone - One part, neutral cure			
Manufacturers Comments: Elastic bonding of mirrors. Good ageing & weathering resistance.			
Typical Use: General purpose adhesive-sealant.			
Cure: RT/24 hrs		**Cure rate:** -	
Test Methods: Not stated.			
Physical		**Mechanical**	
Colour	Transparent	τ (RT)	-
Viscosity	Soft paste	τ (°C)	-
SG	1.0	τ (wet)	-
Shelf life	6 months/RT	Peel ( )	-
Pot/Work life	-	σ_t	-
Tg	-	σ_c	-
$T_{Service}$	-45 to +150°C	*E*	-
Skin time	-	*G*	-
Cure shrinkage	-	Elongation	-
Resistivity: Vol.	-	Tear strength	-
Surf.	-	H (Shore A)	22
Notes: Bead size: 10-12 mm wide, 1-1.5 mm thick. Clamp for 24 hrs.			

Forbo Helmitin	Hemisil E		Nov 89
Silicone - One part, acetic cure			
Manufacturers Comments: Adhesive-sealant for glazing units. Vehicle and containers. Good weathering, UV & thermal stability. Chemical resistance (some chemicals). May require primer.			
Typical Use: Wood, aluminium, plastics. Glass. Metals.			
Cure: RT		**Cure rate:** ~5mm/24 hrs	
Test Methods: DIN			
Physical		**Mechanical**	
Colour	Various*	τ (RT)	-
Viscosity	Soft paste	τ (°C)	-
SG	1.05	τ (wet)	-
Shelf life	1 yr/RT	Peel ( )	-
Pot/Work life	-	σ_t	1.0 MPa
Tg	-	σ_c	-
$T_{Service}$	-45 to +150°C	*E*	-
Skin time	15-20 mins	*G*	-
Cure shrinkage	-	Elongation	~600%
Resistivity: Vol.	-	Tear strength	-
Surf.	-	H (Shore A)	25
Notes: *Other colours: Transparent, white, grey, black, brown. Gap: >4mm<30mm.			

GE Silicones

GE Silicones	FRV1102		Feb 94
Silicone - One part, acetic curing			
Manufacturers Comments: Fluorosilicone adhesive-sealant. Fuel resistant. Chemical & solvent resistance.			
Typical Use: Metals & plastics.			
Cure: RT		**Cure rate:** -	
Test Methods: Not Stated			
Physical		**Mechanical**	
Colour	White*	τ (RT)	-
Viscosity	-	τ (°C)	-
SG	1.33-1.58	τ (wet)	-
Shelf life	-	Peel (-)	5.8 N mm^{-1}
Pot/Work life	-	σ_t	2.4-3.4 MPa
Tg	-	σ_c	-
$T_{Service}$	-60 to +204°C	*E*	-
Skin time	-	*G*	-
Cure shrinkage	-	Elongation	230-260%
Resistivity: Vol.	-	Tear strength	-
Surf.	-	H (Shore A)	33-42
Notes: *Other colours: FRV1103 (black): FRV1106 (red); FRV1107 (grey) SG, hardness, σ, elongation data: Range for different colours Dielectric Strength: 13.7 kV/mm; Dielectric Constant (1kHz): 6.3.			

GE Silicones	IS808		Feb 94
Silicone - One part, acetic curing			
Manufacturers Comments: General purpose adhesive-sealant. Cure-in-place electrical insulation. Meets: UL File No. E-36952; FDA; NSF.			
Typical Use: Metals & plastics.			
Cure: RT		**Cure rate:** -	
Test Methods: Not Stated			
Physical		**Mechanical**	
Colour	Translucent*	τ (RT)	0.9 MPa
Viscosity	-	τ (°C)	-
SG	1.04	τ (wet)	-
Shelf life	-	Peel ( )	-
Pot/Work life	-	σ_t	2.2 MPa
Tg	-	σ_c	-
$T_{Service}$	-60 to +204°C	*E*	-
Skin time	-	*G*	-
Cure shrinkage	-	Elongation	400%
Resistivity: Vol.	-	Tear strength	-
Surf.	-	H (Shore A)	23
Notes: *Other colours: IS802 (white), IS803 (black). Dielectric Strength: 19.5 kV/mm; Dielectric Constant (60Hz): 2.8.			

GE Silicones — RTV 31 — Feb 94

Silicone - Two part, condensation cure

Manufacturers Comments: Mechanical applications (also potting) requiring high temperature resistance & low viscosity. Pourable.

Typical Use: Wood, aluminium, plastics. Glass. Metals.

Cure: RT/24hrs(handling); 7days | **Cure rate:** -

Test Methods: Not Stated

Physical		Mechanical	
Colour	Red	τ (RT)	-
Viscosity	25 Pa.s	τ (°C)	-
SG	1.42	τ (wet)	-
Shelf life	6 months/-18°C	Peel ( )	-
Pot/Work life	-	σ_t	6.0 MPa
Tg	-	σ_c	-
$T_{Service}$	-54 to +260°C	E	-
Skin time	2.5 hrs	G	-
Cure shrinkage	0.6% linear	Elongation	170%
Resistivity: Vol.	1.6×10^{14} Ωcm	Tear strength	5.1 N mm^{-1}
Surf.	-	H (Shore A)	54

Notes: CTE: 21×10^{-4} /°C.Dielectric Strength: 17 kV/mm; Dielectric Constant (1kHz): 4.4; Dissipation Factor (1kHz): 0.03.

GE Silicones — RTV 167 — Feb 94

Silicone - One part, neutral curing

Manufacturers Comments: Alkoxy cure mechanism. Non-corrosive, high strength adhesive-sealant for mechanical & electrical applications. Use on horizontal & vertical surfaces. Electrical insulator. Meets: UL File No. E-36952; MIL-A-46146.

Typical Use: Metals & plastics.

Cure: RT | **Cure rate:** -

Test Methods: Not Stated

Physical		Mechanical	
Colour	Grey	τ (RT)	-
Viscosity	-	τ (°C)	-
SG	1.12	τ (wet)	-
Shelf life	-	Peel (-)	14.1 N mm^{-1}
Pot/Work life	-	σ_t	5.5 MPa
Tg	-	σ_c	-
$T_{Service}$	-60 to +204°C	E	-
Skin time	-	G	-
Cure shrinkage	-	Elongation	650%
Resistivity: Vol.	-	Tear strength	-
Surf.	-	H (Shore A)	37

Notes: Dielectric Strength: 19.5 kV/mm; Dielectric Constant (60Hz): 2.8.

GE Silicones — RTV 142 — Jan 91

Silicone - One part, neutral cure

Manufacturers Comments: Low volatile, adhesive-sealant. Primerless adhesion to many substrates. Good weathering, UV, ozone & thermal stability. Electrical insulation. High vacuum applications, inc. space.

Typical Use: Glass. Metals.

Cure: RT/48 hrs (handling) | **Cure rate:** ~3mm/48 hrs

Test Methods: Not Stated

Physical		Mechanical	
Colour	White	τ (RT)	2.1 MPa
Viscosity	Paste	τ (°C)	-
SG	1.09	τ (wet)	-
Shelf life	4 months/-18°C	Peel (-)	10.5 N mm^{-1}
Pot/Work life	-	σ_t	3.8 MPa
Tg	-	σ_c	-
$T_{Service}$	-60 to +204°C	E	-
Skin time	4 hrs	G	-
Cure shrinkage	<1% linear	Elongation	400%
Resistivity: Vol.	4.5×10^{15} Ωcm	Tear strength	8.8 N mm^{-1}
Surf.	-	H (Shore A)	34

Notes: Gap: <6mm thick. CTE: 2.7×10^{-4}/°C. Thermal Conductivity: 0.21 W/m°C. Dielectric Strength: 20kV/mm; Dielectric Constant (60Hz): 2.8; Dissipation Factor (60Hz): 0.001
Volatiles (Ball Aerospace VCM Test 33074): %TWL 0.22; %VCM 0.05.

GE Silicones — RTV 560 — Feb 94

Silicone - Two part, condensation cure

Manufacturers Comments: Mechanical applications (aerospace). Thermal insulation/ablative material requiring wide temperature range. Pourable.

Typical Use: Aerospace - insulation.

Cure: RT/24hrs(handling); 7days | **Cure rate:** -

Test Methods: Not Stated

Physical		Mechanical	
Colour	Red	τ (RT)	-
Viscosity	30 Pa.s	τ (°C)	-
SG	1.42	τ (wet)	-
Shelf life	6 months/-18°C	Peel ( )	-
Pot/Work life	-	σ_t	4.8 MPa
Tg	-	σ_c	-
$T_{Service}$	-115 to +260°C	E	-
Skin time	2.5 hrs	G	-
Cure shrinkage	1% linear	Elongation	120%
Resistivity: Vol.	2×10^{14} Ωcm	Tear strength	5.5 N mm^{-1}
Surf.	-	H (Shore A)	55

Notes: CTE: 21×10^{-5}/°C. Dielectric Strength: 21.2 kV/mm; Dielectric Constant (1kHz): 3.9; Dissipation Factor (1kHz): 0.02.

See Key to Data Sheets - Pages 63 to 65

GE Silicones	RTV 566		Sept 91
Silicone - Two part,			
Manufacturers Comments: Low volatile, adhesive-sealant. Adhesion to many substrates with primers. High vacuum applications, inc. space.			
Typical Use: Glass. Metals.			
Cure: RT/24hrs(handling); 7days		**Cure rate:** -	
Test Methods: Not Stated			
Physical		**Mechanical**	
Colour	Red+yellow/brown	τ (RT) Primed Al.	3.2 MPa
Viscosity	Liquid	τ (°C)	-
SG	~1.4	τ (wet)	-
Shelf life	6 months/-18°C	Peel ( )	-
Pot/Work life	1.5 hrs	σ_t	5.5 MPa
Tg	-	σ_c	-
$T_{Service}$	-115 to +260°C	E	-
Skin time	2.5 hrs	G	-
Cure shrinkage	-	Elongation	120%
Resistivity: Vol.	-	Tear strength	8.8 N mm^{-1}
Surf.	-	H (Shore A)	63
Notes: Service temp: +316°C intermittent. Post cure for +150°C use. Volatiles (After 24hrs, 125°C, 1x10^{-6} Torr): %TWL 0.11; %VCM 0.01.			

GE Silicones	RTV 5818		Feb 94
Silicone - One part, neutral curing			
Manufacturers Comments: Modified alkoxy cure mechanism. Fast curing adhesive-sealant. Primerless adhesion to metals & some plastics. High production rates. Meets: UL File No. E-36952.			
Typical Use: Metals & plastics.			
Cure: RT		**Cure rate:** -	
Test Methods: Not Stated			
Physical		**Mechanical**	
Colour	Translucent*	τ (RT)	-
Viscosity	-	τ (°C)	-
SG	1.04	τ (wet)	-
Shelf life	-	Peel (-)	6.1 N mm^{-1}
Pot/Work life	-	σ_t	1.9 MPa
Tg	-	σ_c	-
$T_{Service}$	-60 to +204°C	E	-
Skin time	-	G	-
Cure shrinkage	-	Elongation	500%
Resistivity: Vol.	-	Tear strength	-
Surf.	-	H (Shore A)	21
Notes: *Other colours: RTV5812 (white), RTV5813 (black). Dielectric Strength: 16.6 kV/mm; Dielectric Constant (1kHz): 2.8			

GE Silicones	RTV 655		Feb 94
Silicone - Two part, addition cure			
Manufacturers Comments: Mechanical applications (aerospace). Low temperature service. Glass-plastic interlayer bonding. Optical instrument applications.			
Typical Use: Aerospace - bonding, also potting.			
Cure: RT/24hrs(handling); 7days		**Cure rate:** -	
Test Methods: Not Stated			
Physical		**Mechanical**	
Colour	Clear	τ (RT)	-
Viscosity	5.7 Pa.s	τ (°C)	-
SG	1.04	τ (wet)	-
Shelf life	6 months/-18°C	Peel ( )	-
Pot/Work life	-	σ_t	6.3 MPa
Tg	-	σ_c	-
$T_{Service}$	-110 to +204°C	E	-
Skin time	2.5 hrs	G	-
Cure shrinkage	0.2% linear	Elongation	120%
Resistivity: Vol.	1.2×10^{15} Ωcm	Tear strength	-
Surf.	-	H (Shore A)	45
Notes: CTE: 33x10^{-5}/°C. Dielectric Strength: 20.9.2 kV/mm; Dielectric Constant (1kHz): 2.7; Dissipation Factor (1kHz): 0.004.			

GE Silicones	RTV 6424		Feb 94
Silicone - One part			
Manufacturers Comments: Heat-curing adhesive-sealant for automotive parts. Non-corrosive. Thermal barrier. Fabric seam.			
Typical Use: Automotive - metals & plastics			
Cure: Heat-curing		**Cure rate:** -	
Test Methods: Not Stated			
Physical		**Mechanical**	
Colour	White	τ (RT)	2.6 MPa
Viscosity	-	τ (°C)	-
SG	1.17	τ (wet)	-
Shelf life	-	Peel ( )	-
Pot/Work life	-	σ_t	4.5 MPa
Tg	-	σ_c	-
$T_{Service}$	-60 to +204°C	E	-
Skin time	-	G	-
Cure shrinkage	-	Elongation	500%
Resistivity: Vol.	-	Tear strength	-
Surf.	-	H (Shore A)	30
Notes: Dielectric Strength: 19.5 kV/mm.			

GE Silicones	RTV 6708		Feb 94
Silicone - One part, neutral curing			
Manufacturers Comments: Modified alkoxy cure mechanism. General purpose adhesive-sealant. Primerless adhesion to many plastics & metals. Long working life. Meets: UL File No. E-36952.			
Typical Use: Metals & plastics.			
Cure: RT		**Cure rate:** -	
Test Methods: Not Stated			
Physical		**Mechanical**	
Colour	Translucent*	τ (RT)	-
Viscosity	-	τ (°C)	-
SG	1.04	τ (wet)	-
Shelf life	-	Peel (-)	5.2 N mm^{-1}
Pot/Work life	-	σ_t	1.6 MPa
Tg	-	σ_c	-
$T_{Service}$	-60 to +204°C	*E*	-
Skin time	-	*G*	-
Cure shrinkage	-	Elongation	450%
Resistivity: Vol.	-	Tear strength	-
Surf.	-	H (Shore A)	18

Notes: *Other colours: RTV6702 (white), RTV6703 (black). Dielectric Strength: 16.0 kV/mm; Dielectric Constant (1kHz): 2.9

Le Joint Français

Le Joint Français	PR 1991		(Jul.95)
Silicone			
Manufacturers Comments: Adhesive-sealant for fire-protection assemblies. Resistant to hydraulic fluids. Primerless. Easy to repair.			
Typical Use: Aerospace.			
Cure:-		**Cure rate:** -	
Test Methods: -			
Physical		**Mechanical**	
Colour	-	τ (RT)	-
Viscosity	-	τ (°C)	-
SG	-	τ (wet)	-
Shelf life	-	Peel ( )	-
Pot/Work life	-	σ_t	-
Tg	-	σ_c	-
$T_{Service}$	-	*E*	-
Skin time	-	*G*	-
Cure shrinkage	-	Elongation	-
Resistivity: Vol.	-	Tear strength	-
Surf.	-	H (Shore)	-

Notes: Replaces PR812.

GE Silicones	RTV 6808		Feb 94
Silicone - One part, neutral curing			
Manufacturers Comments: Modified alkoxy cure mechanism. Industrial plastic adhesive. Primerless adhesion to many plastics. Assembly bonding & sealing. Meets: UL File No. E-36952.			
Typical Use: Plastics (inc. polycarbonate).			
Cure: RT		**Cure rate:** -	
Test Methods: Not Stated			
Physical		**Mechanical**	
Colour	Translucent*	τ (RT)	-
Viscosity	-	τ (°C)	-
SG	1.04	τ (wet)	-
Shelf life	-	Peel (-)	6.1 N mm^{-1}
Pot/Work life	-	σ_t	1.7 MPa
Tg	-	σ_c	-
$T_{Service}$	-60 to +204°C	*E*	-
Skin time	-	*G*	-
Cure shrinkage	-	Elongation	400%
Resistivity: Vol.	-	Tear strength	-
Surf.	-	H (Shore A)	18

Notes: *Other colours: RTV6802 (white), RTV6803 (black), RTV6809 (aluminium). Dielectric Strength: 19.5 kV/mm; Dielectric Constant (1kHz): 2.8.

Loctite

Loctite	Superflex		(Feb 95)
Silicone - 1 part (moisture curing)			
Manufacturers Comments: RT cure, 100% solids adhesive-sealant, good mechanical properties, chemical & temperature resistance. For applications involving noise, vibration, sealing. Non-sag, non-flow.			
Typical Use: Bonding & sealing.			
Cure: RT/24hrs(handling); 7days		**Cure rate:** 3mm/24 hrs	
Test Methods: Not Stated			
Physical		**Mechanical**	
Colour	Translucent*	τ (RT)	-
Viscosity	-	τ (°C)	-
SG	1.04	τ (wet)	-
Shelf life	9 months/<30°C	Peel ( )	-
Pot/Work life	-	σ_t	1.9 MPa
Tg	-	σ_c	-
$T_{Service}$	-60 to +250°C	*E*	-
Skin time	15 mins	*G*	-
Cure shrinkage	-	Elongation	450%
Resistivity: Vol.	1×10^{15} Ω.cm	Tear strength	-
Surf.	-	H (Shore A)	25

Notes: * Other colours available: White, Black. Other grades: Self-levelling, low odour (no acetic acid evolved during curing) Dielectric Strength (ASTM D-149): 550 V/mil.

McGhan NuSil / NuSil Technology

McGhan NuSil	CCF1 - 2566		(Apr 95)
Silicone RTV - Two part			
Manufacturers Comments: Low outgassing (controlled volatility) phenyl silicone.			
Typical Use: Space applications.			
Cure: Initial: 24hr, Full: 7 days.		**Cure rate:** -	
Test Methods: Not Stated			
Physical		**Mechanical**	
Colour	Red	τ (RT)	-
Viscosity	35 Pa.s	τ (°C)	-
SG	1.42	τ (wet)	-
Shelf life	6 month/RT	Peel ( )	-
Pot/Work life	60 mins.	σ_t	2.76 MPa
Tg	-	σ_c	-
$T_{Service}$	-115 to +260°C	*E*	-
Skin time	-	*G*	-
Cure shrinkage	-	Elongation	100%
Resistivity: Vol.	-	Tear strength	5.3 N mm^{-1}
Surf.	-	H (Shore A)	45
Notes: CVCM: 0.1%, TML: 1.0%.			

NuSil Technology	CV - 1142		(Apr 95)
Silicone RTV - One part			
Manufacturers Comments: Non-corrosive, controlled volatility phenyl silicone. Non-slump. Oxime cure.			
Typical Use: High vacuum. Industrial and space applications requiring low emission of volatiles.			
Cure: Full: 7 days.		**Cure rate:** -	
Test Methods: Not Stated			
Physical		**Mechanical**	
Colour	Translucent	τ (RT)	-
Viscosity	-	τ (°C)	-
SG	1.12	τ (wet)	-
Shelf life	6 month/RT	Peel ( )	-
Pot/Work life	30 min.	σ_t	3.8 MPa
Tg	-	σ_c	-
$T_{Service}$	-115 to +260°C	*E*	-
Skin time	-	*G*	-
Cure shrinkage	-	Elongation	200%
Resistivity: Vol.	1 x 10^{15} Ω.cm	Tear strength	10.5 N mm^{-1}
Surf.	-	H (Shore A)	35
Notes: Refractive Index: 1.43. Dielectric strength: 19.7 kV/mm. CVCM: 0.01%. TML: 0.50%. Also exists as CV - 1142-2 Black formulation.			

McGhan NuSil	CF2 - 1130		(Apr 95)
Silicone RTV - One part			
Manufacturers Comments: Non-corrosive methyl silicone dispersion (xylene). Oxime cure. Long pot life. 65% solids.			
Typical Use: Glass, aluminium, silicone, etc.			
Cure: Initial: 24hr, Full: 7 days.		**Cure rate:** -	
Test Methods: Not Stated			
Physical		**Mechanical**	
Colour	Translucent	τ (RT)	-
Viscosity	-	τ (°C)	-
SG	1.09	τ (wet)	-
Shelf life	6 month/RT	Peel ( )	-
Pot/Work life	3 hr (tack free)	σ_t	4.48 MPa
Tg	-	σ_c	-
$T_{Service}$	-65 to +232°C	*E*	-
Skin time	-	*G*	-
Cure shrinkage	-	Elongation	300%
Resistivity: Vol.	1.5 x 10^{15} Ω.cm	Tear strength	13 N mm^{-1}
Surf.	-	H (Shore A)	45
Notes: Dielectric Strength: 19.7 kV/mm.			

NuSil Technology	CV - 1143		(Apr 95)
Silicone RTV - One part			
Manufacturers Comments: Non-corrosive, controlled volatility dimethyl silicone. Non-slumping. Oxime cure.			
Typical Use: High vacuum. Industrial and space applications requiring low emission of volatiles.			
Cure: Full: 7 days.		**Cure rate:** -	
Test Methods: Not Stated			
Physical		**Mechanical**	
Colour	Translucent	τ (RT)	-
Viscosity	-	τ (°C)	-
SG	1.12	τ (wet)	-
Shelf life	6 month/RT	Peel ( )	-
Pot/Work life	30 min.	σ_t	3.8 MPa
Tg	-	σ_c	-
$T_{Service}$	-115 to +260°C	*E*	-
Skin time	-	*G*	-
Cure shrinkage	-	Elongation	300%
Resistivity: Vol.	1 x 10^{15} Ω.cm	Tear strength	10.5 N mm^{-1}
Surf.	-	H (Shore A)	30
Notes: Refractive Index: 1.41. Dielectric strength: 19.7 kV/mm. CVCM: 0.01%. TML: 0.50%.			

See Key to Data Sheets - Pages 63 to 65

NuSil Technology	**CV - 1500**		(Apr 95)
Silicone RTV - One part			
Manufacturers Comments: Electrically conductive, non-corrosive, controlled volatility phenyl silicone. Non-slumping. Oxime cure.			
Typical Use: High vacuum. Industrial and space applications requiring an electrically conductive adhesive-sealant with low emission of volatiles.			
Cure: Full: 7 days.		**Cure rate:** -	
Test Methods: Not Stated			
Physical		**Mechanical**	
Colour	Black	τ (RT)	-
Viscosity	-	τ (°C)	-
SG	1.24	τ (wet)	-
Shelf life	6 month/RT	Peel ( )	-
Pot/Work life	30 min.	σ_t	3.45 MPa
Tg	-	σ_c	-
$T_{Service}$	-115 to +260°C	E	-
Skin time	-	G	-
Cure shrinkage	-	Elongation	25%
Resistivity: Vol.	1.5 Ω.cm	Tear strength	-
Surf.	-	H (Shore A)	70
Notes: Dielectric strength: 19.7 kV/mm. CVCM: 0.01%. TML: 0.50%.			

NuSil Technology	**CV - 2564**		(Apr 95)
Silicone RTV - One part			
Manufacturers Comments: Controlled-volatility phenyl silicone. Tin condensation cure.			
Typical Use: High vacuum. Industrial and space applications requiring an adhesive-sealant with low emission of volatiles.			
Cure: Initial: 24 hr, Full: 7 days.		**Cure rate:** -	
Test Methods: Not Stated			
Physical		**Mechanical**	
Colour	Translucent	τ (RT)	-
Viscosity	90 Pa.s	τ (°C)	-
SG	1.11	τ (wet)	-
Shelf life	6 month/RT	Peel ( )	-
Pot/Work life	60 min.	σ_t	4.14 MPa
Tg	-	σ_c	-
$T_{Service}$	-115 to +260°C	E	-
Skin time	-	G	-
Cure shrinkage	-	Elongation	150%
Resistivity: Vol.	1×10^{15} Ω.cm	Tear strength	10.5 N mm^{-1}
Surf.	-	H (Shore A)	50
Notes: Dielectric strength: 19.7 kV/mm. CVCM: 0.01%. TML: 0.50%.			

NuSil Technology	**CV - 2520**		(Apr 95)
Silicone RTV - One part thixotropic			
Manufacturers Comments: Controlled-volatility, ablative phenyl silicone. Tin condensation cure.			
Typical Use: High vacuum. Industrial and space applications requiring an adhesive-sealant with low emission of volatiles.			
Cure: Initial: 24 hr, Full: 7 days.		**Cure rate:** -	
Test Methods: Not Stated			
Physical		**Mechanical**	
Colour	White	τ (RT)	-
Viscosity	-	τ (°C)	-
SG	0.75	τ (wet)	-
Shelf life	6 month/RT	Peel ( )	-
Pot/Work life	60 min.	σ_t	1.0 MPa
Tg	-	σ_c	-
$T_{Service}$	-115 to +260°C	E	-
Skin time	-	G	-
Cure shrinkage	-	Elongation	75%
Resistivity: Vol.	1×10^{15} Ω.cm	Tear strength	5.3 N mm^{-1}
Surf.	-	H (Shore A)	55
Notes: Dielectric strength: 19.7 kV/mm. CVCM: 0.01%. TML: 0.50%.			

NuSil Technology	**CV - 2566**		(Apr 95)
Silicone RTV - One part			
Manufacturers Comments: Controlled-volatility diphenyl dimethyl silicone. Tin condensation cure.			
Typical Use: High vacuum. Industrial and space applications requiring an adhesive-sealant with low emission of volatiles.			
Cure: Initial: 24 hr, Full: 7 days.		**Cure rate:** -	
Test Methods: Not Stated			
Physical		**Mechanical**	
Colour	Red	τ (RT)	-
Viscosity	50 Pa.s	τ (°C)	-
SG	1.50	τ (wet)	-
Shelf life	6 month/RT	Peel ( )	-
Pot/Work life	60 min.	σ_t	4.14 MPa
Tg	-	σ_c	-
$T_{Service}$	-115 to +260°C	E	-
Skin time	-	G	-
Cure shrinkage	-	Elongation	120%
Resistivity: Vol.	1×10^{15} Ω.cm	Tear strength	5.3 N mm^{-1}
Surf.	-	H (Shore A)	60
Notes: Dielectric strength: 19.7 kV/mm. CVCM: 0.01%. TML: 0.50%. Also exists as CV2 - 2566 thixotropic formulation.			

NuSil Technology	**CV - 2568**		(Apr 95)
Silicone RTV - One part thixotropic			
Manufacturers Comments: Low specific gravity, thixotropic. Controlled-volatility diphenyl dimethyl silicone. Tin condensation cure.			
Typical Use: High vacuum. Industrial and space applications requiring an adhesive-sealant with low emission of volatiles. (Bonding of solar cells).			
Cure: Initial: 24 hr, Full: 7 days.		**Cure rate:** -	
Test Methods: Not Stated			
Physical		**Mechanical**	
Colour	Red	τ (RT)	-
Viscosity	-	τ (°C)	-
SG	0.64	τ (wet)	-
Shelf life	6 month/RT	Peel ( )	-
Pot/Work life	60 min.	σ_t	0.9 MPa
Tg	-	σ_c	-
$T_{Service}$	-115 to +300°C	E	-
Skin time	-	G	-
Cure shrinkage	-	Elongation	40%
Resistivity: Vol.	1 x 10^{15} Ω.cm	Tear strength	4.4 N mm^{-1}
Surf.	-	H (Shore A)	45
Notes: Thermal conductivity: 0.13 W/mK. Dielectric strength: 19.7 kV/mm. CVCM: 0.01%. TML: 0.50%.			

NuSil Technology	**CV2 - 1142 - 1**		(Apr 95)
Silicone RTV - One part			
Manufacturers Comments: Non-corrosive, controlled volatility phenyl silicone. Self-levelling. Oxime cure.			
Typical Use: High vacuum. Industrial and space applications requiring low emission of volatiles.			
Cure: Full: 7 days.		**Cure rate:** -	
Test Methods: Not Stated			
Physical		**Mechanical**	
Colour	White	τ (RT)	-
Viscosity	-	τ (°C)	-
SG	1.13	τ (wet)	-
Shelf life	6 month/RT	Peel ( )	-
Pot/Work life	30 min.	σ_t	3.8 MPa
Tg	-	σ_c	-
$T_{Service}$	-115 to +260°C	E	-
Skin time	-	G	-
Cure shrinkage	-	Elongation	200%
Resistivity: Vol.	1 x 10^{15} Ω.cm	Tear strength	-
Surf.	-	H (Shore A)	35
Notes: Dielectric strength: 19.7 kV/mm. CVCM: 0.01%. TML: 0.50%.			

NuSil Technology	**CV1 - 1142**		(Apr 95)
Silicone RTV - One part			
Manufacturers Comments: Non-corrosive, controlled volatility phenyl silicone. Self-levelling. Oxime cure.			
Typical Use: High vacuum. Industrial and space applications requiring low emission of volatiles.			
Cure: Full: 7 days.		**Cure rate:** -	
Test Methods: Not Stated			
Physical		**Mechanical**	
Colour	Translucent	τ (RT)	-
Viscosity	-	τ (°C)	-
SG	1.10	τ (wet)	-
Shelf life	6 month/RT	Peel ( )	-
Pot/Work life	30 min.	σ_t	1.72 MPa
Tg	-	σ_c	-
$T_{Service}$	-115 to +260°C	E	-
Skin time	-	G	-
Cure shrinkage	-	Elongation	100%
Resistivity: Vol.	1 x 10^{15} Ω.cm	Tear strength	-
Surf.	-	H (Shore A)	35
Notes: Refractive Index: 1.43. Dielectric strength: 19.7 kV/mm. CVCM: 0.01%. TML: 0.50%. Also exists as CV1 - 1142-1 (White) and CV1 - 1142-2 (Black) formulations.			

McGhan NuSil	**FS 3730**		(Apr 95)
Silicone RTV - One part, Fluorosilicone, thixotropic			
Manufacturers Comments: Very resistant to fuels and solvents. Acetoxy cure (releases acetic acid). 100% solids.			
Typical Use: Fuel/solvent contact applications.			
Cure: Initial: 24 hr, Full: 7 days.		**Cure rate:** -	
Test Methods: Not Stated			
Physical		**Mechanical**	
Colour	White	τ (RT)	2.76 MPa
Viscosity	-	τ (°C)	-
SG	1.40	τ (wet)	-
Shelf life	6 month/RT	Peel ( )	-
Pot/Work life	30 min.	σ_t	3.45 MPa
Tg	-	σ_c	-
$T_{Service}$	-60 to +204°C	E	-
Skin time	-	G	-
Cure shrinkage	-	Elongation	250%
Resistivity: Vol.	-	Tear strength	7 N mm^{-1}
Surf.	-	H (Shore A)	35
Notes:			

See Key to Data Sheets - Pages 63 to 65

McGhan NuSil	R 1140		(Apr 95)
Silicone - One part, thixotropic			
Manufacturers Comments: Clear silicone, approved for food contact. 100% solids.			
Typical Use: Glass, ceramics, metals, natural and synthetic fibres, wood, painted surfaces, many plastics and other silicones, etc.			
Cure: Initial: 24hr, Full: 7 days.		**Cure rate:** -	
Test Methods: Not Stated			
Physical		**Mechanical**	
Colour	Translucent	τ (RT)	1.24 MPa
Viscosity	-	τ (°C)	-
SG	1.07	τ (wet)	-
Shelf life	6 month/RT	Peel (-)	7 N mm^{-1}
Pot/Work life	30 min (tack free)	σ_t	3.1 MPa
Tg	-	σ_c	-
$T_{Service}$	-65 to +232°C	E	-
Skin time	-	G	-
Cure shrinkage	-	Elongation	550%
Resistivity: Vol.	1 x 10^{15} Ω.cm	Tear strength	7 N mm^{-1}
Surf.	-	H (Shore A)	30
Notes: Dielectric Strength: 23.6 kV/mm.			

McGhan NuSil	R 2076		(Apr 95)
Silicone RTV - Two part, thixotropic			
Manufacturers Comments: High tensile and tear strength. Condensation cure. Long-term property stability.			
Typical Use: High temperature thermal barrier/insulating adhesive-sealant.			
Cure: Initial: 24hr, Full: 7 days.		**Cure rate:** -	
Test Methods: Not Stated			
Physical		**Mechanical**	
Colour	Grey	τ (RT)	-
Viscosity	-	τ (°C)	-
SG	1.12	τ (wet)	-
Shelf life	6 month/RT	Peel ( )	-
Pot/Work life	-	σ_t	5.52 MPa
Tg	-	σ_c	-
$T_{Service}$	-65 to +260°C	E	-
Skin time	-	G	-
Cure shrinkage	-	Elongation	400%
Resistivity: Vol.	1 x 10^{13} Ω.cm	Tear strength	21 N mm^{-1}
Surf.	-	H (Shore A)	40
Notes: Dielectric Strength: 17.7 kV/mm. Dissipation factor (10 kHz): 0.1. Dielectric factor (10 kHz): 3.0.			

McGhan NuSil	R 1505		(Apr 95)
Silicone RTV - One part, thixotropic			
Manufacturers Comments: Non-corrosive, electrically conductive. Non-slumping. 100% solids. Oxime cure. Good resistance to radiation, humidity and thermal/mechanical stresses.			
Typical Use: Bonding/sealing in harsh conditions. Suitable for use in nuclear power applications.			
Cure: Initial: 24hr, Full: 7 days.		**Cure rate:** -	
Test Methods: Not Stated			
Physical		**Mechanical**	
Colour	Black	τ (RT)	1.4 MPa
Viscosity	-	τ (°C)	-
SG	1.24	τ (wet)	-
Shelf life	6 month/RT	Peel ( )	-
Pot/Work life	30 min (tack free)	σ_t	3.5 MPa
Tg	-	σ_c	-
$T_{Service}$	-115 to +260°C	E	-
Skin time	-	G	-
Cure shrinkage	-	Elongation	25%
Resistivity: Vol.	1.5 x 10^{15} Ω.cm	Tear strength	-
Surf.	-	H (Shore A)	70
Notes:			

McGhan NuSil	R 2560		(Apr 95)
Silicone - Two part			
Manufacturers Comments: High temperature resistance and gamma ray resistance.			
Typical Use: Potting, bonding and sealing. Also usable as thermal insulation, firewall sealant or ablative coating.			
Cure: Initial: 24hr, Full: 7 days.		**Cure rate:** -	
Test Methods: Not Stated			
Physical		**Mechanical**	
Colour	Red	τ (RT)	2.76 MPa
Viscosity	30 Pa.s	τ (°C)	-
SG	1.42	τ (wet)	-
Shelf life	6 month/RT	Peel ( )	-
Pot/Work life	90 mins.	σ_t	4.14 MPa
Tg	-	σ_c	-
$T_{Service}$	-115 to +260°C	E	-
Skin time	-	G	-
Cure shrinkage	0.5 %	Elongation	140%
Resistivity: Vol.	1 x 10^{15} Ω.cm	Tear strength	5.3 N mm^{-1}
Surf.	-	H (Shore A)	50
Notes: Dielectric Strength: 19.7 kV/mm.			

See Key to Data Sheets - Pages 63 to 65

McGhan NuSil	R 2630		(Apr 95)
Silicone RTV - Two part, Electrically conductive			
Manufacturers Comments: Graphite-fibre filled methyl silicone.			
Typical Use: -			
Cure: 24 hr/RT or 4 hr/65°C or 1 hr/100°C or 15 mins/150°C		**Cure rate:** -	
Test Methods: Not Stated			
Physical		**Mechanical**	
Colour	Black	τ (RT)	-
Viscosity	20 Pa.s	τ (°C)	-
SG	1.09	τ (wet)	-
Shelf life	6 month/RT	Peel ( )	-
Pot/Work life	2 hrs.	σ_t	3.45 MPa
Tg	-	σ_c	-
$T_{Service}$	-60 to +204°C	E	-
Skin time	-	G	-
Cure shrinkage	-	Elongation	50%
Resistivity: Vol.	100 Ω.cm	Tear strength	-
Surf.	-	H (Shore A)	60
Notes:			

SEA
(Société d'Exploitation des Adhésifs)

SEA	Mastic Silicone		Jul 95
Silicone - One part, moisture curing			
Manufacturers Comments: Very good resistance to weathering, UV, acids, dilute alkalis and solvents.			
Typical Use: Construction industry.			
Cure: -		**Cure rate:** -	
Test Methods: Not Stated			
Physical		**Mechanical**	
Colour	Clear, white, grey	τ (RT)	-
Viscosity	-	τ (°C)	-
SG	-	τ (wet)	-
Shelf life	-	Peel ( )	-
Pot/Work life	30 min/RT	σ_t	0.3 MPa
Tg	-	σ_c	-
$T_{Service}$	-40 to +150°C	E	-
Skin time	-	G	-
Cure shrinkage	-	Elongation	500%
Resistivity: Vol.	-	Tear strength	-
Surf.	-	H (Shore)	-
Notes:			

Panacol Elosol

Panacol Elosol	Penloc Silicon Blau		Jun 94
Silicone - One part			
Manufacturers Comments: Very good chemical and radiation resistance. 99% solids. Meets the requirements of (USA) MIL-Spec 45180 B and C.			
Typical Use: Very good adhesion to a wide variety of materials.			
Cure: -		**Cure rate:** -	
Test Methods: Not Stated			
Physical		**Mechanical**	
Colour	-	τ (RT)	-
Viscosity	-	τ (°C)	-
SG	-	τ (wet)	-
Shelf life	-	Peel ( )	-
Pot/Work life	-	σ_t	-
Tg	-	σ_c	-
$T_{Service}$	-60 to +265°C	E	-
Skin time	-	G	-
Cure shrinkage	< 1% linear	Elongation	460%
Resistivity: Vol.	-	Tear strength	-
Surf.	-	H (Shore A)	35
Notes:			

ThreeBond

ThreeBond	1207B		(Aug 95)
Silicone - RTV One part, deacetonisation			
Manufacturers Comments: Oil resistant			
Typical Use: -			
Cure: RT, <120°C		**Cure rate:** 3 mm/day	
Test Methods: Not stated.			
Physical		**Mechanical**	
Colour	Black	τ (RT)	0.9 MPa
Viscosity	(paste)	τ (°C)	-
SG	1.01	τ (wet)	-
Shelf life	-	Peel ( )	-
Pot/Work life	-	σ_t	1.5 MPa
Tg	-	σ_c	-
$T_{Service}$	-60 to +250°C	E	-
Skin time	7 mins.	G	-
Cure shrinkage	-	Elongation	425%
Resistivity: Vol.	-	Tear strength	-
Surf.	-	H (Hs)	30
Notes: Hardness scale not stated			

ThreeBond	1207C	(Aug 95)
Silicone - RTV One part, deacetonisation		
Manufacturers Comments: Oil and chemical resistant.		
Typical Use: -		
Cure: RT, <120°C	**Cure rate:** 3 mm/day	
Test Methods: Not stated.		

Physical		Mechanical	
Colour	Red-Brown	τ (RT)	1.2 MPa
Viscosity	(paste)	τ (°C)	-
SG	1.50	τ (wet)	-
Shelf life	-	Peel ( )	-
Pot/Work life	-	σ_t	5.0 MPa
Tg	-	σ_c	-
$T_{Service}$	-60 to +250°C	E	-
Skin time	5 mins.	G	-
Cure shrinkage	-	Elongation	170%
Resistivity: Vol.	-	Tear strength	-
Surf.	-	H (Hs)	60

Notes: ThreeBond 1207D is a silver coloured version. Hardness scale not stated.

ThreeBond	1211	(Aug 95)
Silicone - RTV One part		
Manufacturers Comments: Low viscosity.		
Typical Use: General purpose adhesive-sealant.		
Cure: RT	**Cure rate:** -	
Test Methods: Not stated.		

Physical		Mechanical	
Colour	White	τ (RT)	1.1 MPa
Viscosity	70 Pa.s	τ (°C)	-
SG	1.01	τ (wet)	-
Shelf life	-	Peel ( )	-
Pot/Work life	-	σ_t	1.0 MPa
Tg	-	σ_c	-
$T_{Service}$	-60 to +250°C	E	-
Skin time	90 mins.	G	-
Cure shrinkage	-	Elongation	230%
Resistivity: Vol.	-	Tear strength	-
Surf.	-	H (Hs)	20

Notes: Hardness scale not stated.

ThreeBond	1208D	(Aug 95)
Silicone - RTV One part, deacetonisation		
Manufacturers Comments: Oil resistant.		
Typical Use: -		
Cure: RT, <120°C	**Cure rate:** 3 mm/day	
Test Methods: Not stated.		

Physical		Mechanical	
Colour	Grey	τ (RT)	1.3 MPa
Viscosity	(paste)	τ (°C)	-
SG	1.47	τ (wet)	-
Shelf life	-	Peel ( )	-
Pot/Work life	-	σ_t	1.5 MPa
Tg	-	σ_c	-
$T_{Service}$	-40 to +120°C	E	-
Skin time	9 mins.	G	-
Cure shrinkage	-	Elongation	350%
Resistivity: Vol.	-	Tear strength	-
Surf.	-	H (Hs)	40

Notes: ThreeBond 1207D is a silver coloured version. Hardness scale not stated.

ThreeBond	1212	(Aug 95)
Silicone - RTV One part		
Manufacturers Comments: Deoxymisation cure.		
Typical Use: General purpose adhesive-sealant.		
Cure: RT	**Cure rate:** -	
Test Methods: Not stated.		

Physical		Mechanical	
Colour	White	τ (RT)	1.0 MPa
Viscosity	(paste)	τ (°C)	-
SG	1.05	τ (wet)	-
Shelf life	-	Peel ( )	-
Pot/Work life	-	σ_t	2.0 MPa
Tg	-	σ_c	-
$T_{Service}$	-60 to +250°C	E	-
Skin time	15 mins.	G	-
Cure shrinkage	-	Elongation	260%
Resistivity: Vol.	-	Tear strength	-
Surf.	-	H (Hs)	35

Notes: Hardness scale not stated.

Wacker Chemie

ThreeBond	1215		(Aug 95)
Silicone - RTV One part, deoxymisation			
Manufacturers Comments: Oil resistant.			
Typical Use: -			
Cure: RT		**Cure rate:** 2.5 mm/day	
Test Methods: Not stated.			
Physical		**Mechanical**	
Colour	Grey	τ (RT)	0.8 MPa
Viscosity	(semi-fluid)	τ (°C)	-
SG	1.50	τ (wet)	-
Shelf life	-	Peel ( )	-
Pot/Work life	-	σ_t	1.5 MPa
Tg	-	σ_c	-
$T_{Service}$	-60 to +250°C	E	-
Skin time	60 mins.	G	-
Cure shrinkage	-	Elongation	320%
Resistivity: Vol.	-	Tear strength	-
Surf.	-	H (Hs)	45
Notes: Hardness scale not stated.			

Wacker Chemie	E104		Oct 89
Silicone - RTV-1 One part, Acetic cure.			
Manufacturers Comments: Non-slump sealant with low shrinkage and good flexibility. Paintable. Meets class E of DIN 18545 & USA TT-S-001543A. Long shelf life.			
Typical Use: General purpose adhesive-sealant. Particularly suitable for glass and ceramic.			
Cure: RT		**Cure rate:** 1 - 2 mm/day	
Test Methods: DIN & ASTM			
Physical		**Mechanical**	
Colour	Transparent	τ (RT)	-
Viscosity	(non slump)	τ (°C)	-
SG	1.02	τ (wet)	-
Shelf life	1 yr (RT)	Peel ( )	-
Pot/Work life	-	σ_t	~1.5 MPa
Tg	-	σ_c	-
$T_{Service}$	-40 to +180°C	E	~0.25 MPa
Skin time	10 - 25 mins.	G	-
Cure shrinkage	3.5%	Elongation	~600%
Resistivity: Vol.	-	Tear strength	~5 N mm^{-1}
Surf.	-	H (Shore A)	~16
Notes: Density: DIN 53479, Shrinkage: DIN 52451, Tensile properties: DIN 53504-S3A, Hardness: DIN 53505, Tear resistance: ASTM D624 rod B.			

ThreeBond	1216		(Aug 95)
Silicone - RTV One part, deoxymisation			
Manufacturers Comments: Oil resistant.			
Typical Use: -			
Cure: RT		**Cure rate:** 2.5 mm/day	
Test Methods: Not stated.			
Physical		**Mechanical**	
Colour	Grey	τ (RT)	2.0 MPa
Viscosity	(paste)	τ (°C)	-
SG	1.40	τ (wet)	-
Shelf life	-	Peel ( )	-
Pot/Work life	-	σ_t	3.0 MPa
Tg	-	σ_c	-
$T_{Service}$	-60 to +250°C	E	-
Skin time	20 mins.	G	-
Cure shrinkage	-	Elongation	200%
Resistivity: Vol.	-	Tear strength	-
Surf.	-	H (Hs)	60
Notes: Hardness scale not stated.			

Wacker Chemie	Elastosil IG 21		Apr 93
Silicone - Two part, Condensation cure.			
Manufacturers Comments: Non-sag sealant with low shrinkage. Meets requirements of NFP 78 451 & DIN 1286 part 1. Excellent adhesion to glass, aluminium and galvanised metals. Excellent UV, moisture and weathering resistance. 95% recovery after 125% extension.			
Typical Use: Insulating glass sealant for dual seal glass units.			
Cure: RT		**Cure rate:**	
Test Methods: ISO			
Physical		**Mechanical**	
Colour	Black	τ (RT)	-
Viscosity	(paste-like)	τ (°C)	-
SG	1.39	τ (wet)	-
Shelf life	1 yr (RT)	Peel (-)	10 N mm^{-1}
Pot/Work life	30 mins.	σ_t	1.8 MPa
Tg	-	σ_c	-
$T_{Service}$	-50 to +150°C	E	0.7 MPa
Skin time	90 mins.	G	-
Cure shrinkage	3.5%	Elongation	350%
Resistivity: Vol.	-	Tear strength	6 N mm^{-1}
Surf.	-	H (Shore A)	32 (3 days)
Notes: Density: ISO 1183A, Shrinkage: ISO 10563, Tensile properties: ISO 37 rod S2, Hardness: ISO 868, Tear resistance: ISO 34 Method C.			

Wacker Chemie	Elastosil N182		Nov 93
Silicone - One part, Neutral cure.			
Manufacturers Comments: Non-sag low-modulus sealant. Excellent adhesion to a variety of substrates. Short tack-free time and long shelf life. Non-corrosive to metals. Suitable for alkaline surfaces (concrete, plaster, etc.).			
Typical Use: Joint sealant for natural stone.			
Cure: RT		**Cure rate:** ~1 - 2 mm/day	
Test Methods: ISO			
Physical		**Mechanical**	
Colour	Transparent	τ (RT)	-
Viscosity	(non-sag)	τ (°C)	-
SG	1.04	τ (wet)	-
Shelf life	1 yr (RT)	Peel ( )	-
Pot/Work life	-	σ_t	1.2 MPa
Tg	-	σ_c	-
$T_{Service}$	-40 to +150°C	E	0.32 MPa
Skin time	10 - 20 mins.	G	-
Cure shrinkage	7.5%	Elongation	500%
Resistivity: Vol.	-	Tear strength	4.5 N mm^{-1}
Surf.	-	H (Shore A)	18
Notes: Density: ISO 1183A, Shrinkage: ISO 10563, Tensile properties: ISO 37 rod S2, Hardness: ISO 868, Tear resistance: ISO 34 Method C.			

Wacker Chemie	Elastosil N184		Nov 93
Silicone - One part, Neutral cure.			
Manufacturers Comments: Non-sag low-modulus sealant. Excellent adhesion to a variety of substrates. Short tack-free time and long shelf life. Non-corrosive to metals. Suitable for alkaline surfaces (concrete, plaster, etc.).			
Typical Use: Joint sealant for natural stone.			
Cure: RT		**Cure rate:** ~1 mm/day	
Test Methods: ISO			
Physical		**Mechanical**	
Colour	Grey	τ (RT)	-
Viscosity	(non-sag)	τ (°C)	-
SG	1.38	τ (wet)	-
Shelf life	1 yr (RT)	Peel ( )	-
Pot/Work life	-	σ_t	1.1 MPa
Tg	-	σ_c	-
$T_{Service}$	-40 to +150°C	E	0.33 MPa
Skin time	10 - 30 mins.	G	-
Cure shrinkage	8%	Elongation	700%
Resistivity: Vol.	-	Tear strength	4.5 N mm^{-1}
Surf.	-	H (Shore A)	21
Notes: Density: ISO 1183A, Shrinkage: ISO 10563, Tensile properties: ISO 37 rod S2, Hardness: ISO 868, Tear resistance: ISO 34 Method C. Available in a range of colours.			

Wacker Chemie	Elastosil N183		Nov 93
Silicone - One part, Neutral cure.			
Manufacturers Comments: Non-sag low-modulus sealant. Excellent adhesion to a variety of substrates. Short tack-free time and long shelf life. Non-corrosive to metals. Suitable for alkaline surfaces (concrete, plaster, etc.).			
Typical Use: Joint sealant for natural stone.			
Cure: RT		**Cure rate:** ~1 mm/day	
Test Methods: ISO			
Physical		**Mechanical**	
Colour	Grey	τ (RT)	-
Viscosity	(non-sag)	τ (°C)	-
SG	1.39	τ (wet)	-
Shelf life	1 yr (RT)	Peel ( )	-
Pot/Work life	-	σ_t	1.0 MPa
Tg	-	σ_c	-
$T_{Service}$	-40 to +150°C	E	0.55 MPa
Skin time	10 - 20 mins.	G	-
Cure shrinkage	7.5%	Elongation	430%
Resistivity: Vol.	-	Tear strength	5.4 N mm^{-1}
Surf.	-	H (Shore A)	30
Notes: Density: ISO 1183A, Shrinkage: ISO 10563, Tensile properties: ISO 37 rod S2, Hardness: ISO 868, Tear resistance: ISO 34 Method C. Available in a range of colours.			

Wacker Chemie	Elastosil N192		Nov 93
Silicone - One part, Neutral cure.			
Manufacturers Comments: Non-slump sealant with low shrinkage and good flexibility. Short tack-free time and long shelf life. Non-corrosive to metals. Suitable for alkaline surfaces (concrete, plaster, etc.). 95% recovery after 100% extension.			
Typical Use: General purpose flexible sealant.			
Cure: RT		**Cure rate:** 1 - 2 mm/day	
Test Methods: ISO			
Physical		**Mechanical**	
Colour	Transparent	τ (RT)	-
Viscosity	(non sag)	τ (°C)	-
SG	1.03	τ (wet)	-
Shelf life	1 yr (RT)	Peel ( )	-
Pot/Work life	-	σ_t	1.2 MPa
Tg	-	σ_c	-
$T_{Service}$	-50 to +100°C	E	0.38 MPa
Skin time	5 - 15 mins.	G	-
Cure shrinkage	5.5%	Elongation	400%
Resistivity: Vol.	-	Tear strength	3.0 N mm^{-1}
Surf.	-	H (Shore A)	20
Notes: Density: ISO 1183 Method A, Shrinkage: ISO 10563, Tensile properties: ISO 37 & ISO 34, Hardness: ISO 868, Tear resistance: ISO 34 Method C. Available in a range of colours.			

Wacker Chemie	**Elastosil N193**		Nov 93
Silicone - One part, Neutral cure.			
Manufacturers Comments: Non-slump, fungicide-containing, sealant with low shrinkage and good flexibility. Short tack-free time and long shelf life. Non-corrosive to metals. Suitable for alkaline surfaces (concrete, plaster, etc.). 90% recovery after 100% extension.			
Typical Use: General purpose flexible sealant.			
Cure: RT		**Cure rate:** 1 mm/day	
Test Methods: ISO			
Physical		**Mechanical**	
Colour	White	τ (RT)	-
Viscosity	(non sag)	τ (°C)	-
SG	1.38	τ (wet)	-
Shelf life	1 yr (RT)	Peel ( )	-
Pot/Work life	-	σ_t	1.1 MPa
Tg	-	σ_c	-
$T_{Service}$	-50 to +150°C	E	0.38 MPa
Skin time	10 - 30 mins.	G	-
Cure shrinkage	6%	Elongation	550%
Resistivity: Vol.	-	Tear strength	4.0 N mm^{-1}
Surf.	-	H (Shore A)	22
Notes: Density: ISO 1183 Method A, Shrinkage: ISO 10563, Tensile properties: ISO 37 & ISO 34, Hardness: ISO 868, Tear resistance: ISO 34 Method C. Available in a range of colours.			

Wacker Chemie	**N180**		Feb 92
Silicone - RTV-1 One part, Neutral cure.			
Manufacturers Comments: Non-slump, flame retardant, sealant.			
Typical Use: General purpose adhesive-sealant for construction and general industry.			
Cure: RT		**Cure rate:** 1 - 2 mm/day	
Test Methods: DIN & ASTM			
Physical		**Mechanical**	
Colour	Grey	τ (RT)	-
Viscosity	(non slump)	τ (°C)	-
SG	1.34	τ (wet)	-
Shelf life	1 yr (RT)	Peel ( )	-
Pot/Work life	-	σ_t	1.4 MPa
Tg	-	σ_c	-
$T_{Service}$	-60 to +150°C	E	0.45 MPa
Skin time	20 - 40mins.	G	-
Cure shrinkage	8%	Elongation	500%
Resistivity: Vol.	1.4 x 10^{11} Ω.cm	Tear strength	6 N mm^{-1}
Surf.	-	H (Shore A)	27
Notes: Density: DIN 53479A Method A, Shrinkage: DIN 52451, Tensile properties: DIN 53504-S3A, Hardness: DIN 53505, Tear resistance: ASTM D624B.			

Wacker Chemie	**Elastosil N195**		Feb 95
Silicone - One part, Neutral cure.			
Manufacturers Comments: Non-sag low shrinkage adhesive-sealant. Excellent adhesive strength. Short tack-free time and long shelf life. Non-corrosive to metals.			
Typical Use: General purpose adhesive-sealant. DIY, automotive, aircraft, shipbuilding, electrical, mechanical engineering, silo and container construction.			
Cure: RT		**Cure rate:** -	
Test Methods: ISO			
Physical		**Mechanical**	
Colour	Transparent	τ (RT)	-
Viscosity	(non-sag)	τ (°C)	-
SG	1.04	τ (wet)	-
Shelf life	1 yr (RT)	Peel ( )	-
Pot/Work life	-	σ_t	1.4 MPa
Tg	-	σ_c	-
$T_{Service}$	-40 to +150°C	E	0.5 MPa
Skin time	10 - 30 mins.	G	-
Cure shrinkage	5.5%	Elongation	260%
Resistivity: Vol.	-	Tear strength	3.5 N mm^{-1}
Surf.	-	H (Shore A)	25
Notes: Density: ISO 1183A, Shrinkage: ISO 10563, Tensile properties: ISO 37 rod S2, Hardness: ISO 868, Tear resistance: ISO 34 Method C.			

Wacker Chemie	**N185**		Feb 92
Silicone - RTV-1 One part, Neutral cure.			
Manufacturers Comments: Non-slump sealant with low shrinkage and good flexibility. Paintable (solvent and water-based paints). Meets class E of DIN 18545. Short tack-free time and long shelf life. Non-corrosive to metals. Suitable for alkaline surfaces (concrete, plaster, etc.)			
Typical Use: Construction and general industry. Particularly for contact with paint (e.g. window construction).			
Cure: RT		**Cure rate:** ~1 - 2mm/day	
Test Methods: DIN & ISO			
Physical		**Mechanical**	
Colour	Transparent	τ (RT)	-
Viscosity	(non slump)	τ (°C)	-
SG	1.04	τ (wet)	-
Shelf life	1 yr (RT)	Peel ( )	-
Pot/Work life	-	σ_t	1.2 MPa
Tg	-	σ_c	-
$T_{Service}$	-40 to +150°C	E	0.33 MPa
Skin time	10 - 20mins.	G	-
Cure shrinkage	6%	Elongation	550%
Resistivity: Vol.	-	Tear strength	3.5 N mm^{-1}
Surf.	-	H (Shore A)	19
Notes: Density: ISO 1183 Method A, Shrinkage: DIN 52451, Tensile properties: DIN 53504-S3A, Hardness: ISO 868, Tear resistance: ISO 34.			

See Key to Data Sheets - Pages 63 to 65

Wacker Chemie	**N186**		Feb 92
Silicone - RTV-1 One part, Neutral cure.			
Manufacturers Comments: Non-slump sealant with low shrinkage and good flexibility. Paintable (solvent and water-based paints). Meets class E of DIN 18545. Short tack-free time and long shelf life. Non-corrosive to metals. Suitable for alkaline surfaces (concrete, plaster, etc.)			
Typical Use: Construction and general industry. Particularly for contact with paint (e.g. window construction).			
Cure: RT		**Cure rate:** ~1 mm/day	
Test Methods: DIN & ISO			
Physical		**Mechanical**	
Colour	White	τ (RT)	-
Viscosity	(non slump)	τ (°C)	-
SG	1.37	τ (wet)	-
Shelf life	1 yr (RT)	Peel ( )	-
Pot/Work life	-	σ_t	1.1 MPa
Tg	-	σ_c	-
$T_{Service}$	-40 to +150°C	E	0.4 MPa
Skin time	10 - 20mins.	G	-
Cure shrinkage	8%	Elongation	550%
Resistivity: Vol.	-	Tear strength	4.0 N mm^{-1}
Surf.	-	H (Shore A)	25
Notes: Density: ISO 1183 Method A, Shrinkage: DIN 52451, Tensile properties: DIN 53504-S3A, Hardness: ISO 868, Tear resistance: ISO 34.			

Wacker Chemie	**RTV-S 695**		Mar 90
Silicone - RTV-2 Two part, Addition cure.			
Manufacturers Comments: Free-flowing adhesive-sealant with very low volatile content. High transparency and resistant to very low temperatures. Admitted for ESTEC aerospace applications (ESA PSS-01-701 Iss1).			
Typical Use: Low outgassing for space applications, e.g. bonding cover glasses of solar cells.			
Cure: 30min/100°C		**Cure rate:** -	
Test Methods: DIN			
Physical		**Mechanical**	
Colour	Transparent	τ (RT)	-
Viscosity	5 - 6.8 Pa.s(mix)	τ (°C)	-
SG	-	τ (wet)	-
Shelf life	1 yr (RT)	Peel ( )	-
Pot/Work life	7 - 9 hours	σ_t	-
Tg	-110°C	σ_c	-
$T_{Service}$	-	E	-
Skin time	-	G	-
Cure shrinkage	-	Elongation	-
Resistivity: Vol.	-	Tear strength	-
Surf.	-	H (Shore A)	10 ± 2
Notes: Hardness: DIN 53505. Total weight loss: <1% & Condensable matter: <0.1% (both to ESRO PSS-09/QRM-02T, Iss2, Aug 76).			

Wacker Chemie	**RTV-S 691**		Mar 90
Silicone - RTV-2 Two part, Addition cure.			
Manufacturers Comments: Non-slump adhesive-sealant with very low volatile content. Resistant to low temperatures. Admitted for ESTEC aerospace applications (ESA PSS-01-701). Primer G790 required for adhesive applications.			
Typical Use: Low outgassing for space applications, e.g. bonding solar cells.			
Cure: 30min/150°C, 6 hr/100°C		**Cure rate:** -	
Test Methods: DIN & ASTM			
Physical		**Mechanical**	
Colour	Red	τ (RT)	-
Viscosity	18 - 26 Pa.s(mix)	τ (°C)	-
SG	1.41 - 1.43	τ (wet)	-
Shelf life	1 yr (RT)	Peel ( )	-
Pot/Work life	90 - 110 mins.	σ_t	4.0 - 6.0 MPa
Tg	-104°C	σ_c	-
$T_{Service}$	-40 to +150°C	E	1.33 - 1.53 MPa
Skin time	-	G	0.70 - 0.73 MPa
Cure shrinkage	-	Elongation	100 - 160%
Resistivity: Vol.	>10^{14} Ω.cm	Tear strength	4.0 - 6.0 N mm^{-1}
Surf.	>10^{12} Ω	H (Shore A)	50 - 60
Notes: Density: DIN 53479, Tensile properties: DIN 53504, Shear modulus: DIN 53455, DIN 53520, Modulus of elasticity: DIN 53455, DIN 53457, DIN 53504, Hardness: DIN 53505, Tear resistance: ASTM D624B. Electrical: DIN 53482. Total weight loss (ESRO PSS-09/QRM-02T, Iss2): <1%. Condensable matter (ESRO PSS-09/QRM-02T, Iss2): <0.1%.			

THERMOPLASTIC HOT-MELT ADHESIVES

NOTES

Thermoplastic hot-melt adhesives are generally based on EVA, polyesters or polyamides (including nylon). Although there are also a small number of polyolefin- and polyurethane-based products. All consist of a base polymer blended with a variety of additives (other thermoplastics, waxes, oils, extenders, etc.) to modify their melt, flow and setting characteristics and their adhesive properties.

All are supplied in solid form and require melting, either in automatic processing machinery (feeder systems, spray-guns, hot-presses) or hand-held gun applicators. Most are available in a range of forms to match application equipment, e.g. granules, sticks of a certain diameter, etc.

The viscosity of a molten thermoplastic decreases with increasing temperature. So, for reproducible production, the viscosity of the adhesive, hence the temperature, must be closely controlled. Most manufacturers produce a wide range temperature/viscosity variants with suggested temperature ranges for each. Many also suggest blending of their product grades to produce intermediate properties for specific applications.

Assemblies are made whilst the adhesive is still in a hot, flowable condition; normally within a few seconds. Most of the available data is concerned with the processing of the adhesives; the property data quoted here varies with the intended use.

Main Uses

- ❑ Packaging: assembly of cartons and other paper, cardboard or plastic-based products, canning (seals).
- ❑ Textile industries: bonding (fabrics, leather, vinyl cloth, plastics, foams and padding) to other materials, heat-sealing, clothing.
- ❑ Footwear: assembly (folding and lasting).
- ❑ Wood-working: furniture finishing, bonding laminates or veneers to wood products, or metals.
- ❑ Automotive: filter assembly (air, oil and fuel).
- ❑ Electrical industry: assembly.
- ❑ Special grades for aircraft trim (flame-retardant).

COMPARISON TABLE - THERMOPLASTIC HOT-MELT ADHESIVES

Company	Adhesive	Base	Form	Colour (Gardner)	Soft. Temp (°C)	Application Temp (°C)	Assembly Time (sec)	Service (°C)	SG	TS	El. (%)	Pg. No.
3M	4060	-	Film	Transp.	-	149	-	-	-	-	-	325
American Tech. Int.	ATI 555	Urethane	Film	-	-	121	-	-	-	-	-	325
	ATI 786	Urethane	Film	-	-	105	-	-	-	-	-	325
Borden	HMR 50	EVA	-	Straw	88 - 118	-	M	-	-	-	-	325
	HMR 55	EVA	-	Straw	85 - 130	-	M	-	-	-	-	326
	HMR 68	EVA	-	Straw	85 - 98	-	M	-	-	-	-	326
Bostik	Thermogrip 6330	EVA	Sticks	Transl.	80	-	45	-25/+65	1.0	○	1800	326
Dexter Hysol	Hysol 7804	Polyamide	Granules/Rods	-	137	-	-	<104	1.01	○	900	326
	Hysol 7901	Polyamide	Various	-	171	-	-	<149	0.99	○	80	327
Electro-Seal	E-510	EVA	Film	Clear	93 - 104	107 - 135	-	-	-	-	-	327
	E-520	EVA	Film	Clear	99 - 110	121 - 149	-	-	-	-	-	327
	E-540	EVA	Film	Clear	82 - 93	93 - 121	-	-	-	-	-	327
	E-560	EVA	Film	Transl.	135 - 141	163 - 191	-	-	-	-	-	328
	N-610	Nylon	Film	Clear	127 - 134	177 - 204	-	-	-	-	-	328
	N-620	Nylon	Film	Clear	149 - 160	232 - 260	-	-	-	-	-	328
	N-625	Nylon	Film	Clear	187 - 199	246 - 274	-	-	-	-	-	328
	P-210	Polyester	Film	Transl.	127 - 138	163 - 191	-	-	-	-	-	329
	P-212	Polyester	Film	Transl.	149 - 160	191 - 218	-	-	-	-	-	329
	P-220	Polyester	Film	Transl.	110 - 121	163 - 191	-	-	-	-	-	329
	P-230	Polyester	Film	Transl.	143 - 154	191 - 218	-	-	-	-	-	329
	P-250	Polyester	Film	Cl. Tint	177 - 204	191 - 218	-	-	-	-	-	330
	U-810	Urethane	Film	Cl. Tint	135 - 143	177 - 218	-	-	-	-	-	330
	U-820	Urethane	Film	Transl.	71 - 82	121 - 177	-	-	-	-	-	330
	U-840	Urethane	Film	Clear	179 - 185	204 - 246	-	-	-	-	-	330
	U-854	Urethane	Film	Clear	177 - 182	204 - 246	-	-	-	-	-	331
EMS	Griltex 1	Polyamide	Powder/Granules	-	105 - 115	130 - 150	-	-	-	-	-	331
	Griltex 2	Polyamide	Powder/Granules	-	120 - 130	140 - 160	-	-	-	-	-	331
	Griltex 3	Polyamide	Powder/Granules	-	110 - 120	120 - 150	-	-	-	-	-	331
	Griltex 4	Polyamide	Powder/Granules	-	100 - 110	115 - 135	-	-	-	-	-	332
	Griltex 5	Polyamide	Powder/Granules	-	78 - 85	95 - 120	-	-	-	-	-	332
	Griltex 6	Polyester	Powder/Granules	-	125 - 130	140 - 170	-	-	-	-	-	332
	Griltex 8	Polyester	Powder/Granules	-	111 - 116	120 - 140	-	-	-	-	-	332
	Griltex 9	Polyester	Powder/Granules	-	118 - 123	130 - 160	-	-	-	-	-	333
	Griltex 11	Polyamide	Powder/Granules	-	100 - 110	110 - 140	-	-	-	-	-	333
	Griltex D1260	Polyamide	Powder/Granules	-	110 - 125	115 - 150	-	-	-	-	-	333
Evacor	Evacor 1	Polyamide	Cylinder	-	165	200 - 220	-	-	-	-	-	333
	Evacor 2	Polyamide	Cylinder	-	136	170 - 190	-	-	-	-	-	334
	Evacor 230	Polyamide	Pellets/Cylinder	<8	105 - 115	150 - 170	-	-	-	-	-	334
	Evacor 816	Polyamide	Cylinder/Pieces	<8	133 - 138	-	-	-	-	-	-	334
	Evacor 823	Polyamide	Pellets/Various	<8	165 - 170	-	-	-	-	-	-	334
	Evacor 824	Polyamide	Pieces	<8	87 - 104	-	-	-	-	-	-	335
	Evacor 1232	Polyamide	Pellets	<6	105 - 115	145 - 165	-	-	-	-	-	335
	Evacor 1315	Polyamide	Pellets	<6	150 - 155	-	-	-	-	-	-	335
	Evacor 1352	Polyamide	Pellets	<6	170 - 175	190 - 205	-	-	-	-	-	335
	Evacor 1385	Polyamide	Pellets	<6	180 - 186	205 - 220	-	-	-	-	-	336
	Evacor 2272	Polyamide	Pellets	<6	132 - 138	160 - 190	-	-	-	-	-	336
	Evacor 2410	Polyamide	Pellets	<6	88 - 95	-	-	-	-	-	-	336
	Evacor 2462	Polyamide	Pellets/Cylinder	<6	128 - 132	-	-	-	-	-	-	336
	Evacor 2472	Polyamide	Pellets	<6	130 - 136	-	-	-	-	-	-	337
	Evacor 2482	Polyamide	Pellets	<6	132 - 138	-	-	-	-	-	-	337
	Evacor 2492	Polyamide	Pellets	<6	136 - 140	-	-	-	-	-	-	337
	Evacor 80833	Polyamide	Pellets	<6	98 - 104	-	-	-	-	-	-	337
	Evacor O Rod 63	Polyamide	Rod	-	168	180 - 210	-	-	-	-	-	338
	Evacor O Rod 65	Polyamide	Rod	-	164	190 - 210	-	-	-	-	-	338
	Evacor O Rod 70	Polyamide	Rod	-	184	195 - 210	-	-	-	-	-	338
	Evacor O/V Rod 10	Polyamide	Rod	-	184	195 - 210	-	-	-	-	-	338
	Evacor O/V Rod 30A	Polyamide	Rod	-	135	160 - 180	-	-	-	-	-	339
	Evacor O/V Rod 2278	Polyamide	Rod	-	163	200 - 230	-	-	-	-	-	339
Forbo Helmitin	Helmidur 862/6	Polyamide	Granules	Cl. Yellow	110	160 - 180	-	-	0.98	-	-	339
	Helmidur 40023	Polyamide	-	Cl. Yellow	144	170 - 190	F	-	0.98	-	-	339
	Helmidur 40026	Polyamide	-	Cl. Yellow	134	170 - 190	F	-	0.98	-	-	340
	Helmidur 40029	Polyamide	Rod	Cl. Yellow	172	170 - 190	<5	-	0.98	-	-	340

Continued...

See Key to Comparison Tables - Pages 63 to 65

Company	Adhesive	Base	Form	Colour (Gardner)	Soft. Temp (°C)	Application Temp (°C)	Assembly Time (sec)	Service (°C)	SG	TS	El. (%)	Pg. No.
Klebchemie	Hot-Melt 725.1	Polyolefin	Paste(Foaming)	Lt. Beige	140 - 150	150 - 200	<60	<90	1.1	-	-	340
Mydrin (TRL)	Thermelt 099	EVA	Pellets/Forms	Ivory	102 - 112	170 - 190	F	<90	1.0	◐	250	340
	Thermelt 102 CH	Polyamide	Pellets	<2	139 - 149	-	-	-	1.0	-	-	341
	Thermelt 159	Polyamide	Pellets	<6	170 - 180	195 - 210	F	-	1.0	◐	550	341
	Thermelt 168 SP	Polyamide	Pellets	<4	177	190 - 220	F	-	0.98	●	300	341
	Thermelt 187	Polyamide	Pellets	<6	185 - 195	210 - 240	F	-	0.98	◐	750	341
	Thermelt 190	Polyamide	Pellets	4	203	210 - 240	F	<180	0.98	●	550	342
	Thermelt 811	Polyamide	Pellets	<9	163 - 173	180 - 210	-	~100	1.0	○	48	342
	Thermelt 814	Polyamide	Pellets	<8	145 - 155	160 - 200	-	~100	0.98	○	-	342
	Thermelt 817	Polyamide	Pellets/Forms	<9	163 - 173	180 - 210	F	-	1.0	○	48	342
	Thermelt 866	Polyamide	Pellets/Forms	4	157	180 - 210	F	-	0.98	○	900	343
	Thermelt 869	Polyamide	Pellets/Forms	<6	154 - 166	192 - 210	F	<130	1.0	○	900	343
	Thermelt 870	Polyamide	Pellets/Forms	4	144	180 - 210	F	-	0.98	◐	800	343
	Thermelt 874	Polyamide	Pellets	6	135 - 145	190 - 200	F	-	1.0	◐	70	343
	Thermelt 895	Polyamide	Pellets	4	163 - 173	180 - 200	F	-	0.98	◐	-	344
	Thermelt 897	Polyamide	Pellets/Forms	<6	158 - 168	150 - 180	-	-	0.98	○	20	344
	Thermelt 1564	Polyamide	Pellets	<6	136 - 146	180 - 210	M	-	1.0	○	275	344
	Thermelt 1570	Polyamide	Pellets/Forms	4	144	180 - 210	F	<110	0.98	◐	800	344
	Thermelt 2147	Polyolefin	Pellets	White	161	190 - 200	F	<90	0.9	○	1300	345
	Thermelt 9172	Polyester	Pellets	White	171	210 - 230	F	-	1.3	-	-	345
	Thermelt OROD 28	Polyester	Rod	White	170 - 190	230 - 250	F	-	1.3	●	500	345
	Thermelt OROD 168	Polyamide	Rod	4	170	190 - 200	F	-	0.98	●	550	345
Power Adhesives	Tec-Bond 213	-	Stick/Cylinder	Brown	78	160 - 200	-	<70	-	-	-	346
	Tec-Bond 214	-	Stick/Cylinder	Lt. Brown	96	160 - 200	-	<70	-	-	-	346
	Tec-Bond 240	-	Stick/Cylinder	Transp.	80	160 - 200	-	<67	-	-	-	346
	Tec-Bond 260	-	Stick/Cylinder	Lt. Brown	84	160 - 200	-	<70	-	-	-	346
	Tec-Bond 420	-	Cylinder	Brown	98	200	-	<75	-	-	-	347
	Tec-Bond 430	-	Cylinder	Lt. Brown	157	200	-	<75	-	-	-	347
	Tec-Bond 6002	EVA	Stick/Cylinder	Brown	101	160 - 200	-	<65	-	-	-	347
	Tec-Bond LM 44	-	Stick/Cylinder	Transl.	98	120 - 140	-	<60	-	-	-	347
Quilosa	Komelt C-18	EVA	Granules	Amber	109 - 119	160 - 180	-	-	-	-	-	348
	Sintex DM-15	EVA	Granules	Beige	101 - 105	190 - 210	-	<55	-	-	-	348
	Sintex DM-43	EVA	Granules	Beige	80 - 82	180 - 210	-	<55	-	-	-	348
	Sintex DM-64	EVA	Granules	Beige	93 - 97	190 - 210	-	<65	-	-	-	348
Sunkyong Industries	EH 105	Polyester	Pellet/Rod/Film	-	173 - 183	-	F	-40/+150	-	●	50	349
	EH 500	Polyester	Pellet/Rod/Film	-	172 - 182	-	F	-40/+150	-	●	400	349
	NH 430	Polyamide	Pellet/Sheet	Various	130 - 140	-	-	-	-	◐	-	349
	NH 540	Polyamide	Pellet/Sheet	Various	150 - 160	-	-	-	-	◐	-	349
	NH 550	Polyamide	Pellet/Sheet	Various	180 - 190	-	-	-	-	●	-	350
Witco	Euremelt 900	Polyamide	Pellets/Various	Cl. Amber	180 - 190	200 - 210	<1	-	0.98	-	<5	350
	Euremelt 930	Polyamide	Pellets/Various	Cl. Amber	105 - 115	150 - 180	<1	-	0.98	◐	<5	350
	Euremelt 940	Polyamide	Pellets/Various	Cl. Amber	105 - 115	140 - 170	<1	-	0.98	◐	<5	350
	Euremelt 1155	Polyamide	Pellets/Various	Cl. Amber	155 - 165	180 - 200	<1	-	0.95	◐	15	351
	Euremelt 1166	Polyamide	Pellets/Various	Cl. Amber	172 - 178	175 - 200	-	-	0.98	◐	15	351
	Euremelt 1174	Polyamide	Pellets/Various	Cl. Amber	170 - 180	200 - 230	-	-	-	◐	10	351
	Euremelt 1175	Polyamide	Pellets/Various	Cl. Amber	175 - 185	190 - 220	<3	-	-	●	20	351
	Euremelt 2095	Polyamide	Pellets/Various	Cl. Amber	90 - 100	120 - 160	-	-	0.95	○	70	352
	Euremelt 2096	Polyamide	Pellets/Various	Cl. Amber	92 - 102	140 - 180	-	-	-	◐	70	352
	Euremelt 2110	Polyamide	Pellets/Various	Cl. Amber	100 - 110	180 - 200	-	-	-	○	90	352
	Euremelt 2125	Polyamide	Pellets/Various	Cl. Amber	125 - 135	180 - 200	-	-	-	○	60	352
	Euremelt 2130	Polyamide	Pellets/Various	Cl. Amber	125 - 135	160 - 200	<8	-	0.98	○	225	353
	Euremelt 2135	Polyamide	Pellets/Various	Cl. Amber	130 - 136	160 - 200	-	-	-	○	320	353
	Euremelt 2137	Polyamide	Pellets/Various	Cl. Amber	140 - 150	190 - 220	-	-	0.95	○	350	353
	Euremelt 2138	Polyamide	Pellets/Various	Cl. Amber	138 - 148	200 - 230	-	-	0.95	◐	600	353
	Euremelt 2140	Polyamide	Pellets/Various	Cl. Amber	135 - 145	200 - 230	8	-	0.95	○	600	354
	Euremelt 2145	Polyamide	Pellets/Various	Cl. Amber	138 - 148	200 - 230	-	-	0.95	◐	150	354
	Euremelt 2166	Polyamide	Pellets/Various	Cl. Amber	163 - 169	180 - 200	-	-	-	◐	200	354
	Euremelt 2170	Polyamide	Pellets/Various	Cl. Amber	-	180 - 210	<1	-	-	-	-	354

3M

3M	4060		(Mar 95)
Thermoplastic Type: Not stated. Film ~0.73mm thick.			
Manufacturers Comments: Short bond cycle with moderate temperature & low pressure.			
Typical Use: Metals, textiles, wood, some plastics.			
Typ. Processing: 149°C/5 mins/0.07MPa (10 psi).			
Test Methods: Not stated			
Physical		**Mechanical**	
Colour	Transparent	τ (RT)	13.8 MPa
Softening Temp	-	τ (82°C)	2.8 MPa
Temp. Range	-	τ (wet)	14.5 MPa
Visc. (°C)	-	Peel (T)	0.64 N mm^{-1}
Visc. (°C)	-	σ_t	-
SG	-	σ_c	-
Assembly time	-	E	-
Tg	-	G	-
$T_{Service}$	-	Elongation	-
Notes: τ(wet) 30 days, tap water, 24°C.			

Am. Tech. Int.	ATI 786		Sept 94
Thermoplastic Type: Urethane Film. Areal wt. 0.08 kg/m².			
Manufacturers Comments: Solvent-based, heat activated, flame retardant film. Low heat release, low smoke release. No emission of toxic gases. Meets: Airbus ATS1000.001/ABD 0031-Issue 5; FAR 25.853 (a) (b) (a-1).			
Typical Use: Aircraft interiors. Automotive interiors. Furniture & luggage (PVC shells).			
Typ. Processing: 70°C + 105°C/4 - 15 mins/0.04 MPa (vacuum).			
Test Methods: Not stated			
Physical		**Mechanical**	
Colour	-	τ (RT)	-
Softening Temp	-	τ (°C)	-
Temp. Range	-	τ (wet)	-
Visc. (°C)	-	Peel (70°C)*	> 30N
Visc. (°C)	-	σ_t	-
SG	-	σ_c	-
Assembly time	-	E	-
Tg	-	G	-
$T_{Service}$	-	Elongation	-
Notes: * Peel (RT) - substrate failure.			

American Technologies International

Am. Tech. Int.	ATI 555		Sept 94
Thermoplastic Type: Urethane Film. Areal wts. 0.12 - 0.14 kg/m².			
Manufacturers Comments: Solvent-based, texture retaining, flame retardant film. Low heat release, low smoke release. No emission of toxic gases. Meets: Airbus ATS 1000.001/ABD 0031-Issue 5; Boeing BMS 8-254F (Type 4); FAR 25.853 (a) (b) (a-1).			
Typical Use: Aircraft interiors. Automotive interiors. Furniture & luggage (PVC shells).			
Typ. Processing: 100°C + 121°C/4 mins/0.04 MPa (vacuum).			
Test Methods: Not stated			
Physical		**Mechanical**	
Colour	-	τ (RT)	>substrate
Softening Temp	-	τ (°C)	-
Temp. Range	-	τ (wet)	-
Visc. (°C)	-	Peel ( )	-
Visc. (°C)	-	σ_t	-
SG	-	σ_c	-
Assembly time	-	E	-
Tg	-	G	-
$T_{Service}$	-	Elongation	-
Notes: * Shear test (RT) - substrate failure.			

Borden

Borden	HMR 50		(July 95)
Thermoplastic Type: EVA.			
Manufacturers Comments: Woodworking, edge-banding.			
Typical Use: Wood, some plastics, impregnated paper.			
Typ. Processing: -			
Test Methods: Not stated			
Physical		**Mechanical**	
Colour	Straw*	τ (RT)	-
Softening Temp	88 - 118°C	τ (°C)	-
Temp. Range	-	τ (wet)	-
Visc. (200°C)	0.9 Pa.s	Peel ( )	-
Visc. (°C)	-	σ_t	-
SG	-	σ_c	-
Assembly time	Medium	E	-
Tg	-	G	-
$T_{Service}$	-	Elongation	-
Notes: * other colours with similar softening range & viscosity: HMR-42 (white); HMR 56 (black); HMR 71(brown); HMR 74(straw).			

Borden **HMR 55** (July 95)

Thermoplastic Type: EVA.

Manufacturers Comments: Woodworking, edge-banding.

Typical Use: Wood, some plastics, impregnated paper.

Typ. Processing: -

Test Methods: Not stated

Physical		**Mechanical**	
Colour	Straw*	τ (RT)	-
Softening Temp	85 - 130°C	τ (°C)	-
Temp. Range	-	τ (wet)	-
Visc. (200°C)	9 Pa.s	Peel ( )	-
Visc. (°C)	-	σ_f	-
SG	-	σ_c	-
Assembly time	Medium	E	-
Tg	-	G	-
$T_{Service}$	-	Elongation	-

Notes: * other colours with similar softening range & viscosity: HMR-63 (straw); HMR 64 (white); HMR 77(black).

Bostik

Bostik **Thermogrip 6330** Oct 94

Thermoplastic Type: EVA. Sticks (11.5Ø x 300mm).

Manufacturers Comments: 100% solids. Good strength. General assembly, flexible packaging. Use with Bostik Hot Glue guns (preset application temperature). Meets: FDA Section 175.105 (food packaging); UK MoD Def. Std 80-167 Issue 1.

Typical Use: Paper products, textile, leather, foam rubbers, some plastics, metals, flowers.

Typ. Processing: Glue gun applied to one surface. Full strength: 2 hrs.

Test Methods: Not stated

Physical		**Mechanical**	
Colour	Translucent	τ (RT)	2.4 MPa*
Softening Temp	80°C	τ (°C)	-
Temp. Range	-	τ (wet)	-
Visc. (190°C)	16 Pa.s	Peel (-)	4.50 N mm^{-1}
Visc. (°C)	-	σ_f	3 MPa
SG	1.0	σ_c	-
Assembly time	~ 45 seconds	E	-
Tg	-	G	-
$T_{Service}$	-25 to + 65°C	Elongation	1800%

Notes: τ aluminium. Other materials: Rigid PVC: 2.6 MPa; ABS: 1.9 MPa; Galvanised steel: 1.8 MPa. Peel: canvas/canvas. Good water, acid & alkali resistance. Poor solvent/oil resistance.

Borden **HMR 68** (July 95)

Thermoplastic Type: EVA.

Manufacturers Comments: Woodworking, edge-banding.

Typical Use: Wood, some plastics, impregnated paper.

Typ. Processing: -

Test Methods: Not stated

Physical		**Mechanical**	
Colour	Straw	τ (RT)	-
Softening Temp	85 - 98°C	τ (°C)	-
Temp. Range	-	τ (wet)	-
Visc. (200°C)	6 Pa.s	Peel ( )	-
Visc. (°C)	-	σ_f	-
SG	-	σ_c	-
Assembly time	Medium	E	-
Tg	-	G	-
$T_{Service}$	-	Elongation	-

Notes: Similar softening range & viscosity: HMR-68 (straw).

Dexter Hysol

Dexter Hysol **Hysol 7804** (Sept. 95)

Thermoplastic Type: Polyamide. Granules/Rod.

Manufacturers Comments: Tough, flexible with good low temperature impact properties. Good adhesion on difficult substrates. Meets requirements of (USA) FDA CFR21 175 - 105 (Adhesives).

Typical Use: -

Typ. Processing: Melt-on-demand equipment.

Test Methods: ASTM

Physical		**Mechanical**	
Colour	-	τ (RT)	-
Softening Temp	137°C	τ (°C)	-
Temp. Range	-	τ (wet)	-
Visc. (204°C)	2.6 Pa.s	Peel ( )	-
Visc. (232°C)	1.0 Pa.s	σ_f	2.3 MPa
SG	1.01	σ_c	-
Assembly time	-	E	-
Tg	-	G	-
$T_{Service}$	104°C	Elongation	900 %

Notes: Dielectric constant (1 kHz): 6.7. Dielectric strength: 23.6 kV/mm. Volume resistivity: 1.6 x 10^{12}. Surface resistivity: 2.2 x 10^{12}.

Dexter Hysol	Hysol 7901		(Sept. 95)
Thermoplastic Type: Polyamide.		Various forms.	
Manufacturers Comments: Low viscosity with a sharp melting point and good toughness. Meets requirements of (USA) FDA CFR21 175 - 105 (Adhesives).			
Typical Use: Electronics.			
Typ. Processing: Melt-on-demand equipment.			
Test Methods: ASTM			
Physical		**Mechanical**	
Colour	-	τ (RT)	-
Softening Temp	171°C	τ (°C)	-
Temp. Range	-	τ (wet)	-
Visc. (177°C)	2 Pa.s	Peel ( )	-
Visc. (232°C)	~0.5 Pa.s	σ_t	1.8 MPa
SG	0.99	σ_c	-
Assembly time	-	E	-
Tg	-	G	-
$T_{Service}$	149°C	Elongation	80 %
Notes: Dielectric constant (1 kHz): 3.00. Dielectric strength: 29.1 kV/mm. Volume resistivity: 2 x 10^{11}. Surface resistivity: 2.2 x 10^{11}.			

Electro-Seal	E-520		Aug 93
Thermoplastic Type: EVA.		Film.	
Manufacturers Comments: Heat-seal adhesive films. Metallised films & porous materials without degradation to substrates. Fair washability, poor dry-clean ability.			
Typical Use: Textiles. Metals.			
Typ. Processing: 121 - 149°C/2 - 5 secs/0.3 - 0.4 MPa.			
Test Methods: Not stated			
Physical		**Mechanical**	
Colour	Clear	τ (RT)	-
Softening Temp	99 - 110°C	τ (°C)	-
Temp. Range	121 - 149°C	τ (wet)	-
Visc. (°C)	-	Peel ( )	-
Visc. (°C)	-	σ_t	-
SG	-	σ_c	-
Assembly time	-	E	-
Tg	-	G	-
$T_{Service}$	-	Elongation	-
Notes: Melt Flow Index ASTM 1238-Condition E: 120-150. Hardness Shore A: 86 Standard film thickness: 0.13mm.			

Electro-Seal

Electro-Seal	E-510		Aug 93
Thermoplastic Type: EVA.		Film.	
Manufacturers Comments: Heat-seal adhesive films. Children's clothing. Flame retardant. Poor washability & dry-cleanability.			
Typical Use: Textiles. Metals. Non-porous substrates.			
Typ. Processing: 107 - 135°C/2 - 5 secs/0.3 - 0.4 MPa.			
Test Methods: Not stated			
Physical		**Mechanical**	
Colour	Cloudy → Clear	τ (RT)	-
Softening Temp	93 - 104°C	τ (°C)	-
Temp. Range	107 - 135°C	τ (wet)	-
Visc. (°C)	-	Peel ( )	-
Visc. (°C)	-	σ_t	-
SG	-	σ_c	-
Assembly time	-	E	-
Tg	-	G	-
$T_{Service}$	-	Elongation	-
Notes: Melt Flow Index ASTM 1238-Condition E: 500-600. Hardness Shore A: 85 Standard film thickness: 0.05mm Similar product: E-510W (whitened version).			

Electro-Seal	E-540		Aug 93
Thermoplastic Type: EVA.		Film.	
Manufacturers Comments: Heat-seal adhesive films. Low melt, fast flow. Poor washability, poor dry-cleanability.			
Typical Use: Textiles. Metals.			
Typ. Processing: 93 - 121°C/2 - 5 secs/0.3 - 0.4 MPa.			
Test Methods: Not stated			
Physical		**Mechanical**	
Colour	Clear	τ (RT)	-
Softening Temp	82 - 93°C	τ (°C)	-
Temp. Range	93 - 121°C	τ (wet)	-
Visc. (°C)	-	Peel ( )	-
Visc. (°C)	-	σ_t	-
SG	-	σ_c	-
Assembly time	-	E	-
Tg	-	G	-
$T_{Service}$	-	Elongation	-
Notes: Melt Flow Index ASTM 1238-Condition E: 550-650. Hardness Shore A: 80 Standard film thickness: 0.05mm.			

Electro-Seal	E-560		Aug 93
Thermoplastic Type: EVA.	Film.		
Manufacturers Comments: Heat-seal adhesive films. Adhesion to metallised Mylar & polar, non-porous surfaces. Barrier properties & grease resistance. Fair washability, poor dry-clean ability.			
Typical Use: Textiles. Metals.			
Typ. Processing: 163 - 191°C/3 - 8 secs/0.4 - 0.6 MPa.			
Test Methods: Not stated			
Physical		**Mechanical**	
Colour	Cloudy	τ (RT)	-
Softening Temp	135 - 141°C	τ (°C)	-
Temp. Range	163 - 191°C	τ (wet)	-
Visc. (°C)	-	Peel ( )	-
Visc. (°C)	-	σ_t	-
SG	-	σ_c	-
Assembly time	-	E	-
Tg	-	G	-
$T_{Service}$	-	Elongation	-
Notes: Melt Flow Index ASTM 1238-Condition E: 10-15. Hardness Shore A: 88 Standard film thickness: 0.08mm.			

Electro-Seal	N-620		Aug 93
Thermoplastic Type: Nylon.	Film.		
Manufacturers Comments: Heat-seal adhesive films. Good bond strengths. Washable. Dry-cleanable.			
Typical Use: Textiles, leather, foam, rubbers, various plastics. Metals. Cork. Wood.			
Typ. Processing: 232 - 260°C/5 - 10 secs/0.4 - 0.7 MPa.			
Test Methods: Not stated			
Physical		**Mechanical**	
Colour	Clear	τ (RT)	-
Softening Temp	149 - 160°C	τ (°C)	-
Temp. Range	232 - 260°C	τ (wet)	-
Visc. (°C)	-	Peel ( )	-
Visc. (°C)	-	σ_t	-
SG	-	σ_c	-
Assembly time	-	E	-
Tg	-	G	-
$T_{Service}$	-	Elongation	-
Notes: Melt Flow Index ASTM 1238-Condition E: 3-6. Hardness Shore A: 96 Standard film thickness: 0.08 mm, 0.13mm.			

Electro-Seal	N-610		Aug 93
Thermoplastic Type: Nylon.	Film.		
Manufacturers Comments: Heat-seal adhesive films. Dielectric & ultrasonic 'activated'. Washable. Dry-cleanable.			
Typical Use: Textiles (inc. rayon, wool, polyester blends), leather, foam, rubbers, various plastics.			
Typ. Processing: 177 - 204°C/3 - 8 secs/0.3 - 0.6 MPa.			
Test Methods: Not stated			
Physical		**Mechanical**	
Colour	Clear	τ (RT)	-
Softening Temp	127 - 134°C	τ (°C)	-
Temp. Range	177 - 204°C	τ (wet)	-
Visc. (°C)	-	Peel ( )	-
Visc. (°C)	-	σ_t	-
SG	-	σ_c	-
Assembly time	-	E	-
Tg	-	G	-
$T_{Service}$	-	Elongation	-
Notes: Melt Flow Index ASTM 1238-Condition E: 50-60. Hardness Shore A: 97 Standard film thickness: 0.03mm, 0.08 mm, 0.13mm.			

Electro-Seal	N-625		Aug 93
Thermoplastic Type: Nylon.	Film.		
Manufacturers Comments: Heat-seal adhesive films. Heat resistance. Solvent resistance. Washable. Dry-cleanable.			
Typical Use: Textiles, leather, foam, rubbers, various plastics.			
Typ. Processing: 246 - 274°C/5 - 15 secs/0.4 - 0.7 MPa.			
Test Methods: Not stated			
Physical		**Mechanical**	
Colour	Clear	τ (RT)	-
Softening Temp	187 - 199°C	τ (°C)	-
Temp. Range	246 - 274°C	τ (wet)	-
Visc. (°C)	-	Peel ( )	-
Visc. (°C)	-	σ_t	-
SG	-	σ_c	-
Assembly time	-	E	-
Tg	-	G	-
$T_{Service}$	-	Elongation	-
Notes: Melt Flow Index ASTM 1238-Condition E: <1. Hardness Shore A: 98 Standard film thickness: 0.10.			

Electro-Seal	**P-210**		Aug 93
Thermoplastic Type: Polyester.	Film.		
Manufacturers Comments: Heat-seal adhesive films. Excellent bonding to textiles. Washable. Dry-cleanable.			
Typical Use: Cotton, polycotton, polyesters, wood, some metals & plastics.			
Typ. Processing: 163 - 191°C/3 - 8 secs/0.4 - 0.6 MPa (60 - 80 psi).			
Test Methods: Not stated			
Physical		**Mechanical**	
Colour	Milky translucent	τ (RT)	-
Softening Temp	127 - 138°C	τ (°C)	-
Temp. Range	163 - 191°C	τ (wet)	-
Visc. (°C)	-	Peel ( )	-
Visc. (°C)	-	σ_t	-
SG	-	σ_c	-
Assembly time	-	E	-
Tg	-	G	-
$T_{Service}$	-	Elongation	-
Notes: Melt Flow Index ASTM 1238-Condition E: 50-60. Hardness Shore A: 95 Standard film thickness: 0.04mm, 0.08mm, 0.13mm.			

Electro-Seal	**P-220**		Aug 93
Thermoplastic Type: Polyester.	Film.		
Manufacturers Comments: Heat-seal adhesive films. Low softening point, high melt flow allows fast processing. Fair washability. Poor/fair dry-cleanability.			
Typical Use: Textiles (woven & non-woven), leather, vinyl. Metals, wood, most plastics.			
Typ. Processing: 163 - 191°C/5 - 10 secs/0.3 - 0.4 MPa (40 - 60 psi).			
Test Methods: Not stated			
Physical		**Mechanical**	
Colour	Milky amber tint	τ (RT)	-
Softening Temp	110 - 121°C	τ (°C)	-
Temp. Range	163 - 191°C	τ (wet)	-
Visc. (°C)	-	Peel ( )	-
Visc. (°C)	-	σ_t	-
SG	-	σ_c	-
Assembly time	-	E	-
Tg	-	G	-
$T_{Service}$	-	Elongation	-
Notes: Melt Flow Index ASTM 1238-Condition E: 120-130. Hardness Shore A: 95 Standard film thickness: 0.08mm, 0.13mm, 0.18mm.			

Electro-Seal	**P-212**		Aug 93
Thermoplastic Type: Polyester.	Film.		
Manufacturers Comments: Heat-seal adhesive films. Excellent bonding to textiles. More flexible than P-210 & higher heat-resistance. Washable. Dry-cleanable.			
Typical Use: Cotton, polycotton, polyesters, wood, some metals & plastics.			
Typ. Processing: 191 - 218°C/5 - 10 secs/0.4 - 0.6 MPa (60 - 80 psi).			
Test Methods: Not stated			
Physical		**Mechanical**	
Colour	Milky translucent	τ (RT)	-
Softening Temp	149 - 160°C	τ (°C)	-
Temp. Range	191 - 218°C	τ (wet)	-
Visc. (°C)	-	Peel ( )	-
Visc. (°C)	-	σ_t	-
SG	-	σ_c	-
Assembly time	-	E	-
Tg	-	G	-
$T_{Service}$	-	Elongation	-
Notes: Melt Flow Index ASTM 1238-Condition E: 5-15. Hardness Shore A: 87 Standard film thickness: 0.08mm.			

Electro-Seal	**P-230**		Aug 93
Thermoplastic Type: Polyester.	Film.		
Manufacturers Comments: Heat-seal adhesive films. Washable. Dry-cleanable.			
Typical Use: Label end-seal & spot-coating. Textiles, metals, wood, most plastics.			
Typ. Processing: 191 - 218°C/3 - 8 secs/0.-0.6 MPa (60 - 80 psi).			
Test Methods: Not stated			
Physical		**Mechanical**	
Colour	Cloudy	τ (RT)	-
Softening Temp	143 - 154°C	τ (°C)	-
Temp. Range	191 - 218°C	τ (wet)	-
Visc. (°C)	-	Peel ( )	-
Visc. (°C)	-	σ_t	-
SG	-	σ_c	-
Assembly time	-	E	-
Tg	-	G	-
$T_{Service}$	-	Elongation	-
Notes: Melt Flow Index ASTM 1238-Condition E: 15-25. Hardness Shore A: 97 Standard film thickness: 0.10mm.			

Electro-Seal **P-250** Aug 93

Thermoplastic Type: Polyester. Film.

Manufacturers Comments: Heat-seal adhesive films. High-duty applications. Washable. Dry-cleanable.

Typical Use: Textiles, metals (inc. steel, copper & aluminium), wood, most plastics.

Typ. Processing: 191 - 218°C/3 - 8 secs/0.-0.6 MPa (60 - 80 psi).

Test Methods: Not stated

Physical		Mechanical	
Colour	Clear green tint	τ (RT)	-
Softening Temp	177 - 204°C	τ (°C)	-
Temp. Range	191 - 218°C	τ (wet)	-
Visc. (°C)	-	Peel ( )	-
Visc. (°C)	-	σ_t	-
SG	-	σ_c	-
Assembly time	-	E	-
Tg	-	G	-
$T_{Service}$	-	Elongation	-

Notes: Melt Flow Index ASTM 1238-Condition E: 120-130. Hardness Shore A: 85 Standard film thickness: 0.08 mm, 0.13mm.

Electro-Seal **U-820** Aug 93

Thermoplastic Type: Modified urethane. Film.

Manufacturers Comments: Heat-seal adhesive films. Low temperature 'activation'. High peel strength. Wash & dry resistance. Dry-cleanable.

Typical Use: Textiles, plastics. Urethane, some vinyls. Cotton & polycotton.

Typ. Processing: 121 - 177°C/2 - 5 secs/0.4 - 0.6 MPa.

Test Methods: Not stated

Physical		Mechanical	
Colour	Translucent	τ (RT)	-
Softening Temp	71 - 82°C	τ (°C)	-
Temp. Range	121 - 177°C	τ (wet)	-
Visc. (°C)	-	Peel ( )	-
Visc. (°C)	-	σ_t	-
SG	-	σ_c	-
Assembly time	-	E	-
Tg	-	G	-
$T_{Service}$	-	Elongation	-

Notes: Melt Flow Index ASTM 1238-Condition E: 15-25. Hardness Shore A: 95 Standard film thickness: 0.05mm, 0.08mm, 0.15mm.

Electro-Seal **U-810** Aug 93

Thermoplastic Type: Modified urethane. Film.

Manufacturers Comments: Heat-seal adhesive films. Withstands industrial laundering and dry cleaning.

Typical Use: Textiles, plastics. Urethane, some vinyls. Cotton & polycotton.

Typ. Processing: 177 - 218°C/5 - 10 secs/0.4 - 0.6 MPa.

Test Methods: Not stated

Physical		Mechanical	
Colour	Clear amber tint	τ (RT)	-
Softening Temp	135 - 143°C	τ (°C)	-
Temp. Range	177 - 218°C	τ (wet)	-
Visc. (°C)	-	Peel ( )	-
Visc. (°C)	-	σ_t	-
SG	-	σ_c	-
Assembly time	-	E	-
Tg	-	G	-
$T_{Service}$	-	Elongation	-

Notes: Melt Flow Index ASTM 1238-Condition E: 45-55. Hardness Shore A: 80 Standard film thickness: 0.08mm, 0.15mm.

Electro-Seal **U-840** Aug 93

Thermoplastic Type: Modified urethane. Film.

Manufacturers Comments: Heat-seal adhesive films. Abrasion resistance, high tensile strength. Withstands industrial laundering. Wash & dry resistance. Dry-cleanable.

Typical Use: Textiles, plastics. Urethane, some vinyls. Cotton & polycotton.

Typ. Processing: 204 - 246°C/5 - 15 secs/0.4 - 0.7 MPa.

Test Methods: Not stated

Physical		Mechanical	
Colour	Clear	τ (RT)	-
Softening Temp	179 - 185°C	τ (°C)	-
Temp. Range	204 - 246°C	τ (wet)	-
Visc. (°C)	-	Peel ( )	-
Visc. (°C)	-	σ_t	-
SG	-	σ_c	-
Assembly time	-	E	-
Tg	-	G	-
$T_{Service}$	-	Elongation	-

Notes: Melt Flow Index ASTM 1238-Condition E: 5-15. Hardness Shore A: 95 Standard film thickness: 0.05mm, 0.09mm.

Electro-Seal	**U-854**		Aug 93
Thermoplastic Type: Modified urethane. Film.			
Manufacturers Comments: Heat-seal adhesive films. Hydrolysis resistance. Low temperature flexibility.			
Typical Use: Textiles, plastics.			
Typ. Processing: 204 - 246°C/5 - 15 secs/0.4 - 0.7 MPa.			
Test Methods: Not stated			
Physical		**Mechanical**	
Colour	Clear	τ (RT)	-
Softening Temp	177 - 182°C	τ (°C)	-
Temp. Range	204 - 246°C	τ (wet)	-
Visc. (°C)	-	Peel ( )	-
Visc. (°C)	-	σ_t	-
SG	-	σ_c	-
Assembly time	-	E	-
Tg	-	G	-
$T_{Service}$	-	Elongation	-
Notes: Melt Flow Index ASTM 1238-Condition E: 10-15. Hardness Shore A: 85 Standard film thickness: 0.14mm.			

EMS	**Griltex 2**		Jul 94
Thermoplastic Type: Copolyamide. Powder/Granules.			
Manufacturers Comments: Textiles & industrial laminating. Washing resistance (<95°C). Dry cleanable. Steam resistance.			
Typical Use: Outdoor clothing textiles (uniforms).			
Typ. Processing: 140 - 160°C/12 - 20 secs/0.3 - 0.5 MPa.			
Test Methods: DIN/ISO			
Physical		**Mechanical**	
Colour	-	τ (RT)	-
Softening Temp	120 - 130°C	τ (°C)	-
Temp. Range	140 - 160°C	τ (wet)	-
Visc. (160°C)	400 Pa.s	Peel ( )	-
Visc. (°C)	-	σ_t	-
SG	-	σ_c	-
Assembly time	-	E	-
Tg	-	G	-
$T_{Service}$	-	Elongation	-
Notes: Viscosity DIN/ISO1133. Melt Volume Rate (MVR) DIN/ISO 1133: 26 cm³/10 mins.			

EMS

EMS	**Griltex 1**		Jul 94
Thermoplastic Type: Copolyamide. Powder/Granules.			
Manufacturers Comments: Textiles & industrial laminating. Steam-fusible. Washing resistance (<60°C). Dry cleanable. Limited steam resistance.			
Typical Use: Outdoor clothing textiles (front-fusing).			
Typ. Processing: 130 - 150°C/12 - 20 secs/0.3 - 0.5 MPa.			
Test Methods: DIN/ISO			
Physical		**Mechanical**	
Colour	-	τ (RT)	-
Softening Temp	105 - 115°C	τ (°C)	-
Temp. Range	130 - 150°C	τ (wet)	-
Visc. (160°C)	450 Pa.s	Peel ( )	-
Visc. (°C)	-	σ_t	-
SG	-	σ_c	-
Assembly time	-	E	-
Tg	-	G	-
$T_{Service}$	-	Elongation	-
Notes: Viscosity: DIN/ISO 1133. Melt Volume Rate (MVR) DIN/ISO 1133: 23 cm³/10 mins.			

EMS	**Griltex 3**		Jul 94
Thermoplastic Type: Copolyamide. Powder/Granules.			
Manufacturers Comments: Textiles & industrial laminating. Wide fusing range. Washing resistance (<60°C). Dry cleanable. Limited steam resistance.			
Typical Use: Outdoor clothing textiles (front-fusing).			
Typ. Processing: 120 - 150°C/12 - 20 secs/0.3 - 0.5 MPa.			
Test Methods: DIN/ISO			
Physical		**Mechanical**	
Colour	-	τ (RT)	-
Softening Temp	110 - 120°C	τ (°C)	-
Temp. Range	120 - 150°C	τ (wet)	-
Visc. (160°C)	500 Pa.s	Peel ( )	-
Visc. (°C)	-	σ_t	-
SG	-	σ_c	-
Assembly time	-	E	-
Tg	-	G	-
$T_{Service}$	-	Elongation	-
Notes: Viscosity DIN/ISO1133. Melt Volume Rate (MVR) DIN/ISO 1133: 21 cm³/10 mins.			

See Key to Data Sheets - Pages 63 to 65

EMS	Griltex 4	Jul 94

Thermoplastic Type: Copolyamide. Powder/Granules.

Manufacturers Comments: Textiles & industrial laminating. Wide fusing range (steam-fusible). Washing resistance (<40°C). Dry cleanable.

Typical Use: Leather & furs (heat- & pressure-sensitive fabrics).

Typ. Processing: 115 - 135°C/10 - 15 secs/0.1 - 0.4 MPa.

Test Methods: DIN/ISO

Physical		Mechanical	
Colour	-	τ (RT)	-
Softening Temp	100 - 110°C	τ (°C)	-
Temp. Range	115 - 135°C	τ (wet)	-
Visc. (160°C)	100 Pa.s	Peel ( )	-
Visc. (°C)	-	σ_f	-
SG	-	σ_c	-
Assembly time	-	*E*	-
Tg	-	*G*	-
$T_{Service}$	-	Elongation	-

Notes: Viscosity DIN/ISO1133.
Melt Volume Rate (MVR) DIN/ISO 1133: 105 cm³/10 mins.

EMS	Griltex 6	Jul 94

Thermoplastic Type: Copolyester. Powder/Granules.

Manufacturers Comments: Textiles & industrial laminating. Wide fusing range. Washing resistance (<95°C). Limited dry cleaning resistance. Steam resistance.

Typical Use: Lightweight clothing, non-woven interlining.

Typ. Processing: 140 - 170°C/12 - 20 secs/0.5 - 1.5 MPa.

Test Methods: DIN/ISO

Physical		Mechanical	
Colour	-	τ (RT)	-
Softening Temp	125 - 130°C	τ (°C)	-
Temp. Range	140 - 170°C	τ (wet)	-
Visc. (160°C)	750 Pa.s	Peel ( )	-
Visc. (°C)	-	σ_f	-
SG	-	σ_c	-
Assembly time	-	*E*	-
Tg	-	*G*	-
$T_{Service}$	-	Elongation	-

Notes: Viscosity DIN/ISO1133.
Melt Volume Rate (MVR) DIN/ISO 1133: 14 cm³/10 mins.

EMS	Griltex 5	Jul 94

Thermoplastic Type: Copolyamide. Powder/Granules.

Manufacturers Comments: Textiles & industrial laminating. Low melting temperature range. Washing resistance (<40°C). Dry cleanable. Limited steam resistance.

Typical Use: Leather, furs, padding, shoes.

Typ. Processing: 95 - 120°C/10 - 15 secs/0.1 - 0.4 MPa.

Test Methods: DIN/ISO

Physical		Mechanical	
Colour	-	τ (RT)	-
Softening Temp	78 - 85°C	τ (°C)	-
Temp. Range	95 - 120°C	τ (wet)	-
Visc. (160°C)	150 Pa.s	Peel ( )	-
Visc. (°C)	-	σ_f	-
SG	-	σ_c	-
Assembly time	-	*E*	-
Tg	-	*G*	-
$T_{Service}$	-	Elongation	-

Notes: Viscosity DIN/ISO1133.
Melt Volume Rate (MVR) DIN/ISO 1133: 70 cm³/10 mins.

EMS	Griltex 8	Jul 94

Thermoplastic Type: Copolyester. Powder/Granules.

Manufacturers Comments: Textiles & industrial laminating. Wide fusing range. Washing resistance (<60°C). Limited dry cleaning resistance. Steam resistance.

Typical Use: Lightweight clothing.

Typ. Processing: 120 - 140°C/12 - 20 secs/0.5 - 1.5 MPa.

Test Methods: DIN/ISO

Physical		Mechanical	
Colour	-	τ (RT)	-
Softening Temp	111 - 116°C	τ (°C)	-
Temp. Range	120 - 140°C	τ (wet)	-
Visc. (160°C)	750 Pa.s	Peel ( )	-
Visc. (°C)	-	σ_f	-
SG	-	σ_c	-
Assembly time	-	*E*	-
Tg	-	*G*	-
$T_{Service}$	-	Elongation	-

Notes: Viscosity DIN/ISO1133.
Melt Volume Rate (MVR) DIN/ISO 1133: 14 cm³/10 mins.

EMS	Griltex 9		Jul 94
Thermoplastic Type: Copolyester. Powder/Granules.			
Manufacturers Comments: Textiles & industrial laminating. High bond strength. Washing resistance (<75°C). Limited dry cleaning resistance. Steam resistance.			
Typical Use: Lightweight clothing.			
Typ. Processing: 130 - 160°C/12 - 20 secs/0.5 - 1.5 MPa.			
Test Methods: DIN/ISO			
Physical		**Mechanical**	
Colour	-	τ (RT)	-
Softening Temp	118 - 123°C	τ (°C)	-
Temp. Range	130 - 160°C	τ (wet)	-
Visc. (160°C)	300 Pa.s	Peel ( )	-
Visc. (°C)	-	σ_t	-
SG	-	σ_c	-
Assembly time	-	E	-
Tg	-	G	-
$T_{Service}$	-	Elongation	-
Notes: Viscosity DIN/ISO1133. Melt Volume Rate (MVR) DIN/ISO 1133: 35 cm³/10 mins.			

EMS	Griltex D1260		Jul 94
Thermoplastic Type: Copolyamide. Powder/Granules.			
Manufacturers Comments: Textiles & industrial laminating. Wide fusing range. Washing resistance (<60°C). Dry cleanable. Steam resistance.			
Typical Use: Outdoor clothing - strike-back protected.			
Typ. Processing: 115 - 150°C/10 - 20 secs/0.3 - 0.5 MPa.			
Test Methods: DIN/ISO			
Physical		**Mechanical**	
Colour	-	τ (RT)	-
Softening Temp	110 - 125°C	τ (°C)	-
Temp. Range	115 - 150°C	τ (wet)	-
Visc. (160°C)	250 Pa.s	Peel ( )	-
Visc. (°C)	-	σ_t	-
SG	-	σ_c	-
Assembly time	-	E	-
Tg	-	G	-
$T_{Service}$	-	Elongation	-
Notes: Viscosity DIN/ISO1133. Melt Volume Rate (MVR) DIN/ISO 1133: 42 cm³/10 mins.			

EMS	Griltex 11		Jul 94
Thermoplastic Type: Copolyamide. Powder/Granules.			
Manufacturers Comments: Textiles & industrial laminating. Low melting temperature range. Steam fusible. Washing resistance (<60°C). Dry cleanable. Limited steam resistance.			
Typical Use: Light-weight outdoor clothing.			
Typ. Processing: 110 - 140°C/10 - 20 secs/0.1 - 0.5 MPa.			
Test Methods: DIN/ISO			
Physical		**Mechanical**	
Colour	-	τ (RT)	-
Softening Temp	100 - 110°C	τ (°C)	-
Temp. Range	110 - 140°C	τ (wet)	-
Visc. (160°C)	300 Pa.s	Peel ( )	-
Visc. (°C)	-	σ_t	-
SG	-	σ_c	-
Assembly time	-	E	-
Tg	-	G	-
$T_{Service}$	-	Elongation	-
Notes: Viscosity DIN/ISO1133. Melt Volume Rate (MVR) DIN/ISO 1133: 35 cm³/10 mins.			

Evacor

Evacor	Evacor 1		(Mar 95)
Thermoplastic Type: Polyamide. Cylinders.			
Manufacturers Comments: 100% solids, solvent-free. Foot wear assembly. High melting resin for injection moulded shoes.			
Typical Use: Toe-puffs.			
Typ. Processing: -			
Test Methods: Not stated			
Physical		**Mechanical**	
Colour	-	τ (RT)	-
Softening Temp	165°C	τ (°C)	-
Temp. Range	200 - 220°C	τ (wet)	-
Visc. (°C)	-	Peel ( )	-
Visc. (°C)	-	σ_t	-
SG	-	σ_c	-
Assembly time	-	E	-
Tg	-	G	-
$T_{Service}$	-	Elongation	-
Notes: Softening temperature - Ring & Ball.			

Evacor | **Evacor 2** | (Mar 95)

Thermoplastic Type: Polyamide. Cylinder.

Manufacturers Comments: 100% solids, solvent-free. Foot wear assembly. Standard version.

Typical Use: Toe-puffs.

Typ. Processing: -

Test Methods: Not stated

Physical		Mechanical	
Colour	-	τ (RT)	-
Softening Temp	136°C	τ (°C)	-
Temp. Range	170 - 190°C	τ (wet)	-
Visc. (°C)	-	Peel ( )	-
Visc. (°C)	-	σ_t	-
SG	-	σ_c	-
Assembly time	-	*E*	-
Tg	-	*G*	-
$T_{Service}$	-	Elongation	-

Notes: Softening temperature - Ring & Ball.

Evacor | **Evacor 816** | (Mar 95)

Thermoplastic Type: Polyamide. Cylinders/Pieces.

Manufacturers Comments: 100% solids, solvent-free. Medium performance. For textiles, footwear, automotive components, large specialist air-filter assembly. Electrical industry.

Typical Use: Good adhesion to metals, plastics & wood.

Typ. Processing: -

Test Methods: Not stated

Physical		Mechanical	
Colour	Gardner 8 max.	τ (RT)	-
Softening Temp	133 - 138°C	τ (°C)	-
Temp. Range	-	τ (wet)	-
Visc. (193°C)	0.5 - 0.7 Pa.s	Peel ( )	-
Visc. (°C)	-	σ_t	-
SG	-	σ_c	-
Assembly time	-	*E*	-
Tg	-	*G*	-
$T_{Service}$	-	Elongation	-

Notes: Softening temperature - Ring & Ball.

Evacor | **Evacor 230** | (Mar 95)

Thermoplastic Type: Polyamide. Pellets/Cylinders.

Manufacturers Comments: 100% solids, solvent-free. Medium performance. For textiles, footwear, automotive components, large specialist air-filter assembly. Electrical industry. Good adhesion to metals & wood.

Typical Use: Footwear (folding, side-lasting & insole binders).

Typ. Processing: -

Test Methods: Not stated

Physical		Mechanical	
Colour	Gardner 8 max.	τ (RT)	-
Softening Temp	105 - 115°C	τ (°C)	-
Temp. Range	150 - 170°C	τ (wet)	-
Visc. (149°C)	5 - 6 Pa.s	Peel ( )	-
Visc. (°C)	-	σ_t	-
SG	-	σ_c	-
Assembly time	-	*E*	-
Tg	-	*G*	-
$T_{Service}$	-	Elongation	-

Notes: Softening temperature - Ring & Ball.

Evacor | **Evacor 823** | (Mar 95)

Thermoplastic Type: Polyamide. Pellets/Cylinders/Pieces.

Manufacturers Comments: 100% solids, solvent-free. Medium performance. For textiles, footwear, automotive components, large specialist air-filter assembly. Electrical industry.

Typical Use: Good adhesion to metals, plastics & wood.

Typ. Processing: -

Test Methods: Not stated

Physical		Mechanical	
Colour	Gardner 8 max.	τ (RT)	-
Softening Temp	165 - 170°C	τ (°C)	-
Temp. Range	-	τ (wet)	-
Visc. (193°C)	0.6 - 0.8 Pa.s	Peel ( )	-
Visc. (°C)	-	σ_t	-
SG	-	σ_c	-
Assembly time	-	*E*	-
Tg	-	*G*	-
$T_{Service}$	-	Elongation	-

Notes: Softening temperature - Ring & Ball.

Evacor	Evacor 824		(Mar 95)
Thermoplastic Type: Polyamide.		Pieces.	
Manufacturers Comments: 100% solids, solvent-free. Medium performance. For textiles, footwear, automotive components, large specialist air-filter assembly. Electrical industry.			
Typical Use: Good adhesion to metals, plastics & wood.			
Typ. Processing: -			
Test Methods: Not stated			
Physical		**Mechanical**	
Colour	Gardner 8 max.	τ (RT)	-
Softening Temp	87 - 104°C	τ (°C)	-
Temp. Range	-	τ (wet)	-
Visc. (149°C)	0.3 - 0.5 Pa.s	Peel ( )	-
Visc. (°C)	-	σ_t	-
SG	-	σ_c	-
Assembly time	-	*E*	-
Tg	-	*G*	-
$T_{Service}$	-	Elongation	-
Notes: Softening temperature - Ring & Ball.			

Evacor	Evacor 1315		(Mar 95)
Thermoplastic Type: Polyamide.		Pellets.	
Manufacturers Comments: 100% solids, solvent-free. High performance. For textiles, footwear, automotive components, large specialist air-filter assembly. Electrical industry. Adhesion to metals & wood.			
Typical Use: Footwear (toe-puffs).			
Typ. Processing: -			
Test Methods: Not stated			
Physical		**Mechanical**	
Colour	Gardner 6 max.	τ (RT)	-
Softening Temp	150 - 155°C	τ (°C)	-
Temp. Range	-	τ (wet)	-
Visc. (180°C)	3 - 5 Pa.s	Peel ( )	-
Visc. (°C)	-	σ_t	-
SG	-	σ_c	-
Assembly time	-	*E*	-
Tg	-	*G*	-
$T_{Service}$	-	Elongation	-
Notes: Softening temperature - Ring & Ball.			

Evacor	Evacor 1232		(Mar 95)
Thermoplastic Type: Polyamide.		Pellets.	
Manufacturers Comments: 100% solids, solvent-free. High performance. For textiles, footwear, automotive components, large specialist air-filter assembly. Electrical industry. Adhesion to metals, & wood.			
Typical Use: Footwear (folding for difficult uppers, inc. leather).			
Typ. Processing: -			
Test Methods: Not stated			
Physical		**Mechanical**	
Colour	Gardner 6 max.	τ (RT)	-
Softening Temp	105 - 115°C	τ (°C)	-
Temp. Range	145 - 165°C	τ (wet)	-
Visc. (149°C)	5 - 6 Pa.s	Peel ( )	-
Visc. (°C)	-	σ_t	-
SG	-	σ_c	-
Assembly time	-	*E*	-
Tg	-	*G*	-
$T_{Service}$	-	Elongation	-
Notes: Softening temperature - Ring & Ball.			

Evacor	Evacor 1352		(Mar 95)
Thermoplastic Type: Polyamide.		Pellets.	
Manufacturers Comments: 100% solids, solvent-free. High performance. For textiles, footwear, automotive components, large specialist air-filter assembly. Electrical industry. Adhesion to metals & wood.			
Typical Use: Footwear (toe-lasting, ideal for leather).			
Typ. Processing: -			
Test Methods: Not stated			
Physical		**Mechanical**	
Colour	Gardner 6 max.	τ (RT)	-
Softening Temp	170 - 175°C	τ (°C)	-
Temp. Range	190 - 205°C	τ (wet)	-
Visc. (190°C)	7 - 9 Pa.s	Peel ( )	-
Visc. (°C)	-	σ_t	-
SG	-	σ_c	-
Assembly time	-	*E*	-
Tg	-	*G*	-
$T_{Service}$	-	Elongation	-
Notes: Softening temperature - Ring & Ball.			

Evacor	Evacor 1385		(Mar 95)
Thermoplastic Type: Polyamide.		Pellets.	
Manufacturers Comments: 100% solids, solvent-free. High performance. For textiles, footwear, automotive components, large specialist air-filter assembly. Electrical industry. Adhesion to metals & wood.			
Typical Use: Footwear (lasting poromeric uppers).			
Typ. Processing: -			
Test Methods: Not stated			
Physical		**Mechanical**	
Colour	Gardner 6 max.	τ (RT)	-
Softening Temp	180 - 186°C	τ (°C)	-
Temp. Range	205 - 220°C	τ (wet)	-
Visc. (210°C)	2.5 - 5 Pa.s	Peel ( )	-
Visc. (°C)	-	σ_t	-
SG	-	σ_c	-
Assembly time	-	*E*	-
Tg	-	*G*	-
$T_{Service}$	-	Elongation	-
Notes: Softening temperature - Ring & Ball.			

Evacor	Evacor 2410		(Mar 95)
Thermoplastic Type: Polyamide.		Pellets.	
Manufacturers Comments: 100% solids, solvent-free. Flexible, high performance. For textiles, footwear, automotive components, large specialist air-filter assembly. Electrical industry.			
Typical Use: Good adhesion to metals, plastics & wood.			
Typ. Processing: -			
Test Methods: Not stated			
Physical		**Mechanical**	
Colour	Gardner 6 max.	τ (RT)	-
Softening Temp	88 - 95°C	τ (°C)	-
Temp. Range	-	τ (wet)	-
Visc. (204°C)	2 - 3 Pa.s	Peel ( )	-
Visc. (°C)	-	σ_t	-
SG	-	σ_c	-
Assembly time	-	*E*	-
Tg	-	*G*	-
$T_{Service}$	-	Elongation	-
Notes: Softening temperature - Ring & Ball.			

Evacor	Evacor 2272		(Mar 95)
Thermoplastic Type: Polyamide.		Pellets.	
Manufacturers Comments: 100% solids, solvent-free. High performance. For textiles, footwear, automotive components, large specialist air-filter assembly. Electrical industry. Adhesion to metals & wood.			
Typical Use: Footwear (toe-puffs).			
Typ. Processing: -			
Test Methods: Not stated			
Physical		**Mechanical**	
Colour	Gardner 6 max.	τ (RT)	-
Softening Temp	132 - 138°C	τ (°C)	-
Temp. Range	160 - 190°C	τ (wet)	-
Visc. (190°C)	5 - 7.5 Pa.s	Peel ( )	-
Visc. (°C)	-	σ_t	-
SG	-	σ_c	-
Assembly time	-	*E*	-
Tg	-	*G*	-
$T_{Service}$	-	Elongation	-
Notes: Softening temperature - Ring & Ball.			

Evacor	Evacor 2462		(Mar 95)
Thermoplastic Type: Polyamide.		Pellets/Cylinders.	
Manufacturers Comments: 100% solids, solvent-free. Flexible, high performance. For textiles, footwear, automotive components, large specialist air-filter assembly. Electrical industry.			
Typical Use: Good adhesion to metals, plastics & wood.			
Typ. Processing: -			
Test Methods: Not stated			
Physical		**Mechanical**	
Colour	Gardner 6 max.	τ (RT)	-
Softening Temp	128 - 132°C	τ (°C)	-
Temp. Range	-	τ (wet)	-
Visc. (204°C)	3 - 4.5 Pa.s	Peel ( )	-
Visc. (°C)	-	σ_t	-
SG	-	σ_c	-
Assembly time	-	*E*	-
Tg	-	*G*	-
$T_{Service}$	-	Elongation	-
Notes: Softening temperature - Ring & Ball.			

See Key to Data Sheets - Pages 63 to 65

Evacor	Evacor 2472		(Mar 95)
Thermoplastic Type: Polyamide.		Pellets.	
Manufacturers Comments: 100% solids, solvent-free. Flexible, high performance. For textiles, footwear, automotive components, large specialist air-filter assembly. Electrical industry.			
Typical Use: Good adhesion to metals, plastics & wood.			
Typ. Processing: -			
Test Methods: Not stated			
Physical		**Mechanical**	
Colour	Gardner 6 max.	τ (RT)	-
Softening Temp	130 - 136°C	τ (°C)	-
Temp. Range	-	τ (wet)	-
Visc. (170°C)	6 - 8 Pa.s	Peel ( )	-
Visc. (°C)	-	σ_t	-
SG	-	σ_c	-
Assembly time	-	*E*	-
Tg	-	*G*	-
$T_{Service}$	-	Elongation	-
Notes: Softening temperature - Ring & Ball.			

Evacor	Evacor 2492		(Mar 95)
Thermoplastic Type: Polyamide.		Pellets.	
Manufacturers Comments: 100% solids, solvent-free. Flexible, high performance. For textiles, footwear, automotive components, large specialist air-filter assembly. Electrical industry.			
Typical Use: Good adhesion to metals, plastics & wood.			
Typ. Processing: -			
Test Methods: Not stated			
Physical		**Mechanical**	
Colour	Gardner 6 max.	τ (RT)	-
Softening Temp	136 - 140°C	τ (°C)	-
Temp. Range	-	τ (wet)	-
Visc. (200°C)	11.5 - 13.5 Pa.s	Peel ( )	-
Visc. (°C)	-	σ_t	-
SG	-	σ_c	-
Assembly time	-	*E*	-
Tg	-	*G*	-
$T_{Service}$	-	Elongation	-
Notes: Softening temperature - Ring & Ball.			

Evacor	Evacor 2482		(Mar 95)
Thermoplastic Type: Polyamide.		Pellets.	
Manufacturers Comments: 100% solids, solvent-free. Flexible, high performance. For textiles, footwear, automotive components, large specialist air-filter assembly. Electrical industry.			
Typical Use: Good adhesion to metals, plastics & wood.			
Typ. Processing: -			
Test Methods: Not stated			
Physical		**Mechanical**	
Colour	Gardner 6 max.	τ (RT)	-
Softening Temp	132 - 138°C	τ (°C)	-
Temp. Range	-	τ (wet)	-
Visc. (215°C)	4 - 8 Pa.s	Peel ( )	-
Visc. (°C)	-	σ_t	-
SG	-	σ_c	-
Assembly time	-	*E*	-
Tg	-	*G*	-
$T_{Service}$	-	Elongation	-
Notes: Softening temperature - Ring & Ball.			

Evacor	Evacor 80833		(Mar 95)
Thermoplastic Type: Polyamide.		Pellets.	
Manufacturers Comments: 100% solids, solvent-free. Flexible, high performance. For textiles, footwear, automotive components, large specialist air-filter assembly. Electrical industry. Good adhesion to metals, plastics & wood.			
Typical Use: Good adhesion to metals, plastics & wood.			
Typ. Processing: -			
Test Methods: Not stated			
Physical		**Mechanical**	
Colour	Gardner 6 max.	τ (RT)	-
Softening Temp	98 - 104°C	τ (°C)	-
Temp. Range	-	τ (wet)	-
Visc. (190°C)	3 - 5 Pa.s	Peel ( )	-
Visc. (°C)	-	σ_t	-
SG	-	σ_c	-
Assembly time	-	*E*	-
Tg	-	*G*	-
$T_{Service}$	-	Elongation	-
Notes: Softening temperature - Ring & Ball.			

Evacor	Evacor O Rod 63		(Mar 95)
Thermoplastic Type: Polyamide.	Rod.		
Manufacturers Comments: 100% solids, solvent-free. Foot wear assembly. General purpose toe-lasting (some side-lasting machines)			
Typical Use: Footwear - lasting.			
Typ. Processing: -			
Test Methods: Not stated			
Physical		**Mechanical**	
Colour	-	τ (RT)	-
Softening Temp	168°C	τ (°C)	-
Temp. Range	180 - 210°C	τ (wet)	-
Visc. (°C)	-	Peel ( )	-
Visc. (°C)	-	σ_t	-
SG	-	σ_c	-
Assembly time	-	E	-
Tg	-	G	-
$T_{Service}$	-	Elongation	-
Notes: Softening temperature - Ring & Ball.			

Evacor	Evacor O Rod 70		(Mar 95)
Thermoplastic Type: Polyamide.	Rod.		
Manufacturers Comments: 100% solids, solvent-free. Foot wear assembly. Excellent heat resistance. Recommended for poromeric uppers & for high heat setting.			
Typical Use: Footwear - lasting.			
Typ. Processing: -			
Test Methods: Not stated			
Physical		**Mechanical**	
Colour	-	τ (RT)	-
Softening Temp	184°C	τ (°C)	-
Temp. Range	195 - 210°C	τ (wet)	-
Visc. (°C)	-	Peel ( )	-
Visc. (°C)	-	σ_t	-
SG	-	σ_c	-
Assembly time	-	E	-
Tg	-	G	-
$T_{Service}$	-	Elongation	-
Notes: Softening temperature - Ring & Ball.			

Evacor	Evacor O Rod 65		(Mar 95)
Thermoplastic Type: Polyamide.	Rod.		
Manufacturers Comments: 100% solids, solvent-free. Foot wear assembly. For automatic side-lasting machines.			
Typical Use: Footwear - lasting.			
Typ. Processing: -			
Test Methods: Not stated			
Physical		**Mechanical**	
Colour	-	τ (RT)	-
Softening Temp	164°C	τ (°C)	-
Temp. Range	190 - 210°C	τ (wet)	-
Visc. (°C)	-	Peel ( )	-
Visc. (°C)	-	σ_t	-
SG	-	σ_c	-
Assembly time	-	E	-
Tg	-	G	-
$T_{Service}$	-	Elongation	-
Notes: Softening temperature - Ring & Ball.			

Evacor	Evacor O/V Rod 10		(Mar 95)
Thermoplastic Type: Polyamide.	Rod.		
Manufacturers Comments: 100% solids, solvent-free. Foot wear assembly. Standard version in O & V format (for TCF machines)			
Typical Use: Footwear - folding.			
Typ. Processing: -			
Test Methods: Not stated			
Physical		**Mechanical**	
Colour	-	τ (RT)	-
Softening Temp	184°C	τ (°C)	-
Temp. Range	195 - 210°C	τ (wet)	-
Visc. (°C)	-	Peel ( )	-
Visc. (°C)	-	σ_t	-
SG	-	σ_c	-
Assembly time	-	E	-
Tg	-	G	-
$T_{Service}$	-	Elongation	-
Notes: Softening temperature - Ring & Ball.			

Forbo Helmitin

Evacor	Evacor O/V Rod 30A		(Mar 95)
Thermoplastic Type: Polyamide.	Rod.		
Manufacturers Comments: 100% solids, solvent-free. Foot wear assembly. Print-on adhesive combining strength & flexibility (use with Tru-line machines).			
Typical Use: Footwear - toe-puffs.			
Typ. Processing: -			
Test Methods: Not stated			
Physical		**Mechanical**	
Colour	-	τ (RT)	-
Softening Temp	135°C	τ (°C)	-
Temp. Range	160 - 180°C	τ (wet)	-
Visc. (°C)	-	Peel ( )	-
Visc. (°C)	-	σ_t	-
SG	-	σ_c	-
Assembly time	-	*E*	-
Tg	-	*G*	-
$T_{Service}$	-	Elongation	-
Notes: Softening temperature - Ring & Ball.			

Forbo Helmitin	Helmidur 862 / 6		Dec 92
Thermoplastic Type: Polyamide.	Granules.		
Manufacturers Comments: Automatic assembly of footwear. Low melting flow, flexible film.			
Typical Use: Footwear - folding and side-lasting.			
Typ. Processing: -			
Test Methods: -			
Physical		**Mechanical**	
Colour	Yellow-transparent	τ (RT)	-
Softening Temp	~110°C	τ (°C)	-
Temp. Range	160 - 180°C	τ (wet)	-
Visc. (180°C)	1.8 Pa.s	Peel ( )	-
Visc. (°C)	-	σ_t	-
SG	~0.98	σ_c	-
Assembly time	-	*E*	-
Tg	-	*G*	-
$T_{Service}$	-	Elongation	-
Notes: Softening temperature: Ring and Ball. Shelf life: ~ 1 year.			

Evacor	Evacor O/V Rod 2278		(Mar 95)
Thermoplastic Type: Polyamide.	Rod.		
Manufacturers Comments: 100% solids, solvent-free. Foot wear assembly. Firmer & tougher adhesive for heavier types of footwear.			
Typical Use: Footwear - toe-puffs.			
Typ. Processing: -			
Test Methods: Not stated			
Physical		**Mechanical**	
Colour	-	τ (RT)	-
Softening Temp	163°C	τ (°C)	-
Temp. Range	200 - 230°C	τ (wet)	-
Visc. (°C)	-	Peel ( )	-
Visc. (°C)	-	σ_t	-
SG	-	σ_c	-
Assembly time	-	*E*	-
Tg	-	*G*	-
$T_{Service}$	-	Elongation	-
Notes: Softening temperature - Ring & Ball.			

Forbo Helmitin	Helmidur 40023		Jan 92
Thermoplastic Type: Polyamide.			
Manufacturers Comments: High heat resistance. Heat sealing and bonding for HF welding of non-thermoplastics.			
Typical Use: PVC. HF welding: textiles, leather, paste-board.			
Typ. Processing: -			
Test Methods: -			
Physical		**Mechanical**	
Colour	Yellow-transparent	τ (RT)	-
Softening Temp	~144°C	τ (°C)	-
Temp. Range	170 - 190°C	τ (wet)	-
Visc. (190°C)	29 Pa.s	Peel ( )	-
Visc. (°C)	-	σ_t	-
SG	~0.98	σ_c	-
Assembly time	Fast	*E*	-
Tg	-	*G*	-
$T_{Service}$	-	Elongation	-
Notes: Softening temperature: Ring and Ball. Shelf life: ~ 1 year.			

See Key to Data Sheets - Pages 63 to 65

Klebchemie (M.G. Becker GmbH)

Forbo Helmitin	Helmitherm 40026		Jan 92
Thermoplastic Type: Polyamide.			
Manufacturers Comments: High heat resistance. Heat sealing and bonding for HF welding of non-thermoplastics.			
Typical Use: PVC. HF welding: textiles, leather, paste-board.			
Typ. Processing: -			
Test Methods: -			
Physical		**Mechanical**	
Colour	Yellow-transparent	τ (RT)	-
Softening Temp	~134°C	τ (°C)	-
Temp. Range	170 - 190°C	τ (wet)	-
Visc. (190°C)	12 Pa.s	Peel ( )	-
Visc. (°C)	-	σ_t	-
SG	~0.98	σ_c	-
Assembly time	Fast	*E*	-
Tg	-	*G*	-
$T_{Service}$	-	Elongation	-
Notes: Softening temperature: Ring and Ball. Shelf life: ~ 1 year.			

Klebchemie	Hot-Melt 725.1		(Oct. 95)
Thermoplastic Type: Polyolefin.			
Manufacturers Comments: Tough, elastic bond. High strength. Sprayable. Foaming (FoamMelt™).			
Typical Use: Automotive interiors: Foamed plastics and textiles to wood products, plastics (sheets and foams), and lacquered metals.			
Typ. Processing: Working Temp: 150 - 200°C. Press Time: 3-10 secs.			
Test Methods: Not stated			
Physical		**Mechanical**	
Colour	Light beige	τ (RT)	-
Softening Temp	145±5°C	τ (°C)	-
Temp. Range	150 - 200°C	τ (wet)	-
Visc. (180°C)	4.5 Pa.s	Peel ( )	-
Visc. (°C)	-	σ_t	-
SG	~1.1	σ_c	-
Assembly time	<1 min (open)	*E*	-
Tg	-	*G*	-
$T_{Service}$	90°C	Elongation	-
Notes: Softening Temp: Ring and Ball. FoamMelt is a registered trademark of Nordson GmbH.			

Mydrin (TRL)

Forbo Helmitin	Helmidur 40029		Apr 91
Thermoplastic Type: Polyamide.		Rod 4mm∅.	
Manufacturers Comments: Footwear assembly.			
Typical Use: Footwear: lasting - leather, textiles, PVC and poromeric materials. Composition insoles.			
Typ. Processing: Automatic lasting machines with rod feeders. 225 - 230°C injection head temp.			
Test Methods: -			
Physical		**Mechanical**	
Colour	Yellow-transparent	τ (RT)	-
Softening Temp	~172°C	τ (°C)	-
Temp. Range	170 - 190°C	τ (wet)	-
Visc. (230°C)	9 Pa.s	Peel ( )	-
Visc. (°C)	-	σ_t	-
SG	~0.98	σ_c	-
Assembly time	3 - 5 sec (wipe in)	*E*	-
Tg	-	*G*	-
$T_{Service}$	-	Elongation	-
Notes: Softening temperature: Ring and Ball. Shelf life: ~ 1 year.			

Mydrin (TRL)	Thermelt 099		Apr 94
Thermoplastic Type: EVA.		Pellets/Forms.	
Manufacturers Comments: Very short setting time. Good wetting and temperature resistance (<90°C). Long skin time.			
Typical Use: Packaging.			
Typ. Processing: Working Temp: 170 - 190°C.			
Test Methods: ASTM			
Physical		**Mechanical**	
Colour	Ivory	τ (RT)	-
Softening Temp	102 - 112°C	τ (°C)	-
Temp. Range	170 - 190°C	τ (wet)	-
Visc. (170°C)	3.6 Pa.s	Peel ( )	-
Visc. (190°C)	2 Pa.s	σ_t	4.5 - 5.5 MPa
SG	~1.0	σ_c	-
Assembly time	Very short set time.	*E*	-
Tg	-	*G*	-
$T_{Service}$	<90	Elongation	200 - 300 %
Notes: Softening Temp: ASTM D 3461, Density: ASTM D 792, Viscosity: ASTM D 3236, Tensile ASTM D 638. Skin time at 204°C (TRL): 9 hr.			

Mydrin (TRL)	**Thermelt 102 CH**		Aug 94
Thermoplastic Type: Copolyamide.	Pellets.		
Manufacturers Comments: Good resistance to chlorinated solvents, laundry cleaning at 60°C and sunlight.			
Typical Use: Textiles.			
Typ. Processing: -			
Test Methods: ASTM			
Physical		**Mechanical**	
Colour	Gardner 2 max.	τ (RT)	-
Softening Temp	139 - 149°C	τ (°C)	-
Temp. Range	-	τ (wet)	-
Visc. (°C)	-	Peel ( )	-
Visc. (°C)	-	σ_t	-
SG	~1.0	σ_c	-
Assembly time	-	*E*	-
Tg	-	*G*	-
$T_{Service}$	-	Elongation	-
Notes: Softening Temp: ASTM D 3461, Density: ASTM D 792.			

Mydrin (TRL)	**Thermelt 168 SP**		93
Thermoplastic Type: Copolyamide.	Pellets.		
Manufacturers Comments: Very short open time and setting time. Good wetting at application temperature. Very good temperature resistance.			
Typical Use: Woodworking industry.			
Typ. Processing: Working Temp: 190 - 220°C.			
Test Methods: ASTM			
Physical		**Mechanical**	
Colour	Gardner 4	τ (RT)	-
Softening Temp	177°C	τ (°C)	-
Temp. Range	190 - 220°C	τ (wet)	-
Visc. (190°C)	17.5 Pa.s	Peel ( )	-
Visc. (220°C)	6.7 Pa.s	σ_t	16 MPa
SG	0.98	σ_c	-
Assembly time	Very fast set time.	*E*	-
Tg	-	*G*	-
$T_{Service}$	-	Elongation	300 %
Notes: Softening Temp: ASTM D 3461, Colour: ASTM D 1544, Density: ASTM D 792, Viscosity: ASTM D3236, Tensile ASTM D638. Skin time at 204°C (TRL): 8 hr.			

Mydrin (TRL)	**Thermelt 159**		Aug 94
Thermoplastic Type: Copolyamide.	Pellets.		
Manufacturers Comments: Flexible with short open time and short setting time. Good wetting at the application temperature. Good solvent resistance.			
Typical Use: Automotive fuel and oil filter construction.			
Typ. Processing: Working Temp: 195 - 210°C.			
Test Methods: ASTM			
Physical		**Mechanical**	
Colour	Gardner 6 max.	τ (RT)	-
Softening Temp	170 - 180°C	τ (°C)	-
Temp. Range	195 - 210°C	τ (wet)	-
Visc. (195°C)	4.7 Pa.s	Peel ( )	-
Visc. (210°C)	2.6 Pa.s	σ_t	6 - 7 MPa
SG	~1	σ_c	-
Assembly time	Fast set time.	*E*	-
Tg	-	*G*	-
$T_{Service}$	-	Elongation	400 - 700 %
Notes: Softening Temp: ASTM D 3461, Colour: ASTM D 1544, Density: ASTM D 792, Viscosity: ASTM D3236, Tensile ASTM D638. Skin time at 204°C (TRL): 8 hr.			

Mydrin (TRL)	**Thermelt 187**		90
Thermoplastic Type: Copolyamide.	Pellets.		
Manufacturers Comments: Very short open time and setting time. Very good temperature resistance.			
Typical Use: Automotive fuel and oil filter construction.			
Typ. Processing: Working Temp: 210 - 240°C.			
Test Methods: ASTM			
Physical		**Mechanical**	
Colour	Gardner 6 max.	τ (RT)	-
Softening Temp	190±5°C	τ (°C)	-
Temp. Range	210 - 240°C	τ (wet)	-
Visc. (210°C)	4.9 Pa.s	Peel ( )	-
Visc. (240°C)	2.2 Pa.s	σ_t	8 - 9 MPa
SG	0.98	σ_c	-
Assembly time	Very fast set time.	*E*	-
Tg	-	*G*	-
$T_{Service}$	-	Elongation	700 - 800 %
Notes: Softening Temp: ASTM D 3461, Colour: ASTM D 1544, Density: ASTM D 792, Viscosity: ASTM D3236, Tensile ASTM D638. Skin time at 204°C (TRL): 6 hr.			

Mydrin (TRL) **Thermelt 190** 93

Thermoplastic Type: Copolyamide.

Manufacturers Comments: Flexible. Very high softening temperature. Very short open time and setting time. Very good temperature resistance (<180°C).

Typical Use: Automotive fuel and oil filter construction.

Typ. Processing: Working Temp: 210 - 240°C.

Test Methods: ASTM

Physical		Mechanical	
Colour	Gardner 4	τ (RT)	-
Softening Temp	203°C	τ (°C)	-
Temp. Range	210 - 240°C	τ (wet)	-
Visc. (220°C)	19 Pa.s	Peel ( )	-
Visc. (240°C)	12 Pa.s	σ_t	18.5 MPa
SG	0.98	σ_c	-
Assembly time	Very fast set time.	E	-
Tg	-	G	-
$T_{Service}$	<180	Elongation	550 %

Notes: Softening Temp: ASTM D 3461, Colour: ASTM D 1544, Density: ASTM D 792, Viscosity: ASTM D3236, Tensile ASTM D638. Skin time at 232°C (TRL): 1 hr.

Mydrin (TRL) **Thermelt 814** 89

Thermoplastic Type: Copolyamide. Pellets.

Manufacturers Comments: Low viscosity at the application temperature. Good adhesion on plastics and porous substrates.

Typical Use: Plastics. Electrical industry.

Typ. Processing: Working Temp: 160 - 200°C.

Test Methods: ASTM

Physical		Mechanical	
Colour	Gardner 8 max.	τ (RT)	-
Softening Temp	150±5°C	τ (°C)	-
Temp. Range	160 - 200°C	τ (wet)	-
Visc. (177°C)	0.2 - 0.4 Pa.s	Peel ( )	-
Visc. (°C)	-	σ_t	2.5 - 3.5 MPa
SG	0.98	σ_c	-
Assembly time	-	E	-
Tg	-	G	-
$T_{Service}$	~100	Elongation	-

Notes: Softening Temp: ASTM D 3461, Colour: ASTM D 1544, Density: ASTM D 792, Viscosity: ASTM D3236, Tensile ASTM D638. Skin time at 204°C (TRL): 9 hr.

Mydrin (TRL) **Thermelt 811** Mar 95

Thermoplastic Type: Copolyamide. Pellets.

Manufacturers Comments: Low viscosity at the application temperature. No creep at 100°C. Meets requirements of UL 94 VO.

Typical Use: Electrical industry.

Typ. Processing: Working Temp: 180 - 210°C.

Test Methods: ASTM

Physical		Mechanical	
Colour	Gardner 9 max.	τ (RT)	-
Softening Temp	163 - 173°C	τ (°C)	-
Temp. Range	180 - 210°C	τ (wet)	-
Visc. (180°C)	1.5 Pa.s	Peel ( )	-
Visc. (210°C)	0.3 Pa.s	σ_t	3.5 - 4.5 MPa
SG	~1	σ_c	-
Assembly time	-	E	-
Tg	-	G	-
$T_{Service}$	~100	Elongation	40 - 55 %

Notes: Softening Temp: ASTM D 3461, Colour: ASTM D 1544, Density: ASTM D 792, Viscosity: ASTM D 3236, Tensile ASTM D 638. Skin time at 205°C (TRL): 9 hr.

Mydrin (TRL) **Thermelt 817** Mar 95

Thermoplastic Type: Copolyamide. Pellets/Forms.

Manufacturers Comments: Short open time and setting time. Good wetting at the application temperature.

Typical Use: Electrical and automotive industry.

Typ. Processing: Working Temp: 180 - 210°C.

Test Methods: ASTM

Physical		Mechanical	
Colour	Gardner 9 max.	τ (RT)	-
Softening Temp	163 - 173°C	τ (°C)	-
Temp. Range	180 - 210°C	τ (wet)	-
Visc. (180°C)	1.7 Pa.s	Peel ( )	-
Visc. (210°C)	0.3 Pa.s	σ_t	3.5 - 4.5 MPa
SG	~1	σ_c	-
Assembly time	Short set time.	E	-
Tg	-	G	-
$T_{Service}$	-	Elongation	40 - 55 %

Notes: Softening Temp: ASTM D 3461, Colour: ASTM D 1544, Density: ASTM D 792, Viscosity: ASTM D 3236, Tensile ASTM D 638. Skin time at 205°C (TRL): 8 hr.

Mydrin (TRL)	**Thermelt 866**		93
Thermoplastic Type: Copolyamide.		Pellets/Forms.	
Manufacturers Comments: Medium open time, fast setting time. Good wetting at the application temperature.			
Typical Use: Polyester packaging materials.			
Typ. Processing: Working Temp: 180 - 210°C.			
Test Methods: ASTM			
Physical		**Mechanical**	
Colour	Gardner 4	τ (RT)	-
Softening Temp	157°C	τ (°C)	-
Temp. Range	180 - 210°C	τ (wet)	-
Visc. (180°C)	5.1 Pa.s	Peel ( )	-
Visc. (210°C)	1.8 Pa.s	σ_t	3 MPa
SG	0.98	σ_c	-
Assembly time	Fast set time.	*E*	-
Tg	-	*G*	-
$T_{Service}$	-	Elongation	900 %
Notes: Softening Temp: ASTM D 3461, Colour: ASTM D 1544, Density: ASTM D 792, Viscosity: ASTM D 3236, Tensile ASTM D 638. Skin time at 204°C (TRL): 8 hr.			

Mydrin (TRL)	**Thermelt 870**		93
Thermoplastic Type: Copolyamide.		Pellets/Forms.	
Manufacturers Comments: Medium open time, fast setting time. Good wetting at the application temperature.			
Typical Use: Plastics. Electrical industry.			
Typ. Processing: Working Temp: 180 - 210°C.			
Test Methods: ASTM			
Physical		**Mechanical**	
Colour	Gardner 4	τ (RT)	-
Softening Temp	144°C	τ (°C)	-
Temp. Range	180 - 210°C	τ (wet)	-
Visc. (180°C)	14.5 Pa.s	Peel ( )	-
Visc. (210°C)	5.2 Pa.s	σ_t	5.5 MPa
SG	0.98	σ_c	-
Assembly time	Fast set time.	*E*	-
Tg	-	*G*	-
$T_{Service}$	-	Elongation	800 %
Notes: Softening Temp: ASTM D 3461, Colour: ASTM D 1544, Density: ASTM D 792, Viscosity: ASTM D3236, Tensile ASTM D638. Skin time at 204°C (TRL): 8 hr.			

Mydrin (TRL)	**Thermelt 869**		Sept 94
Thermoplastic Type: Copolyamide.		Pellets/Forms.	
Manufacturers Comments: Very short setting time. Very good temperature resistance (<130°C). Good wetting and low viscosity at the application temperature.			
Typical Use: Automotive industry.			
Typ. Processing: Working Temp: 190 - 210°C.			
Test Methods: ASTM			
Physical		**Mechanical**	
Colour	Gardner 6 max.	τ (RT)	-
Softening Temp	154 - 166°C	τ (°C)	-
Temp. Range	190 - 210°C	τ (wet)	-
Visc. (190°C)	4.5 Pa.s	Peel ( )	-
Visc. (210°C)	2.9 Pa.s	σ_t	2.5 - 4.5 MPa
SG	~1	σ_c	-
Assembly time	Very short set time.	*E*	-
Tg	-	*G*	-
$T_{Service}$	<130	Elongation	600 - 1200 %
Notes: Softening Temp: ASTM D 3461, Colour: ASTM D 1544, Density: ASTM D 792, Viscosity: ASTM D3236, Tensile ASTM D638. Skin time at 204°C (TRL): 7 hr. Hardness (Shore A): 85. Low temperature test (ASTM D 3111): 35°C.			

Mydrin (TRL)	**Thermelt 874**		Jan 95
Thermoplastic Type: Copolyamide.		Pellets.	
Manufacturers Comments: Medium open time, quick setting time. Good wetting and low viscosity at the application temperature.			
Typical Use: Packaging for hot products (<80°C).			
Typ. Processing: Working Temp: 150 - 200°C.			
Test Methods: ASTM			
Physical		**Mechanical**	
Colour	Gardner 6 max.	τ (RT)	-
Softening Temp	135 - 145°C	τ (°C)	-
Temp. Range	190 - 200°C	τ (wet)	-
Visc. (150°C)	7.3 Pa.s	Peel ( )	-
Visc. (200°C)	~1 Pa.s	σ_t	5 - 6 MPa
SG	~1.0	σ_c	-
Assembly time	Quick set time.	*E*	-
Tg	-	*G*	-
$T_{Service}$	-	Elongation	65 - 75 %
Notes: Softening Temp: ASTM D 3461, Colour: ASTM D 1544, Density: ASTM D 792, Viscosity: ASTM D 3236, Tensile ASTM D 638. Skin time at 205°C (TRL): 8 hr.			

Mydrin (TRL) **Thermelt 895** 93

Thermoplastic Type: Copolyamide. Pellets.

Manufacturers Comments: Short open time and fast setting time. Good wetting and low viscosity at the application temperature.

Typical Use: Electrical and automotive industry.

Typ. Processing: Working Temp: 180 - 200°C.

Test Methods: ASTM

Physical		**Mechanical**	
Colour	Gardner 4	τ (RT)	-
Softening Temp	163 - 173°C	τ (°C)	-
Temp. Range	180 - 200°C	τ (wet)	-
Visc. (190°C)	0.095 Pa.s	Peel ( )	-
Visc. (200°C)	<0.09 Pa.s	σ_t	5 MPa
SG	0.98	σ_c	-
Assembly time	Fast set time.	E	-
Tg	-	G	-
$T_{Service}$	-	Elongation	-

Notes: Softening Temp: ASTM D 3461, Colour: ASTM D 1544, Density: ASTM D 792, Viscosity: ASTM D3236, Tensile ASTM D638. Skin time at 204°C (TRL): 24 hr.

Mydrin (TRL) **Thermelt 1564** Jan 95

Thermoplastic Type: Copolyamide. Pellets.

Manufacturers Comments: Medium open time and setting time. Good wetting at application temperature.

Typical Use: Woodworking industry.

Typ. Processing: Working Temp: 180 - 210°C.

Test Methods: ASTM

Physical		**Mechanical**	
Colour	Gardner 6 max.	τ (RT)	-
Softening Temp	136 - 146°C	τ (°C)	-
Temp. Range	180 - 210°C	τ (wet)	-
Visc. (180°C)	5.2 Pa.s	Peel ( )	-
Visc. (210°C)	1.9 Pa.s	σ_t	3.0 - 4.5 MPa
SG	~1	σ_c	-
Assembly time	Medium set time.	E	-
Tg	-	G	-
$T_{Service}$	-	Elongation	200 - 350 %

Notes: Softening Temp: ASTM D 3461, Colour: ASTM D 1544, Density: ASTM D 792, Viscosity: ASTM D3236, Tensile ASTM D638. Skin time at 205°C (TRL): 9 hr.

Mydrin (TRL) **Thermelt 897** 91

Thermoplastic Type: Copolyamide. Pellets/Forms.

Manufacturers Comments: High melting point and very low viscosity at the application temperature. Good adhesion on plastics and porous substrates.

Typical Use: Plastics. Electrical industry.

Typ. Processing: Working Temp: 150 - 180°C.

Test Methods: ASTM

Physical		**Mechanical**	
Colour	Gardner 6 max.	τ (RT)	-
Softening Temp	163±5°C	τ (°C)	-
Temp. Range	150 - 180°C	τ (wet)	-
Visc. (163°C)	0.5 - 1.5 Pa.s	Peel ( )	-
Visc. (°C)	-	σ_t	2.5 - 3.5 MPa
SG	0.98	σ_c	-
Assembly time	-	E	-
Tg	-	G	-
$T_{Service}$	-	Elongation	10 - 30%

Notes: Softening Temp: ASTM D 3461, Colour: ASTM D 1544, Density: ASTM D 792, Viscosity: ASTM D3236, Tensile ASTM D638. Skin time at 204°C (TRL): 9 hr.

Mydrin (TRL) **Thermelt 1570** 93

Thermoplastic Type: Copolyamide. Pellets/Forms.

Manufacturers Comments: Medium open time and fast setting time. Good wetting at application temperature. Very good temperature resistance (<110°C).

Typical Use: Woodworking industry.

Typ. Processing: Working Temp: 180 - 210°C.

Test Methods: ASTM

Physical		**Mechanical**	
Colour	Gardner 4	τ (RT)	-
Softening Temp	144°C	τ (°C)	-
Temp. Range	180 - 210°C	τ (wet)	-
Visc. (180°C)	14.5 Pa.s	Peel ( )	-
Visc. (210°C)	5.2 Pa.s	σ_t	5.5 MPa
SG	0.98	σ_c	-
Assembly time	Fast set time	E	-
Tg	-	G	-
$T_{Service}$	<110	Elongation	800 %

Notes: Softening Temp: ASTM D 3461, Colour: ASTM D 1544, Density: ASTM D 792, Viscosity: ASTM D3236, Tensile ASTM D638. Skin time at 204°C (TRL): 8 hr.

Mydrin (TRL)	Thermelt 2147		93
Thermoplastic Type: Polyolefin.		Pellets.	
Manufacturers Comments: Flexible with medium open time and short setting time. Very good temperature resistance (<90°C). Good wetting at the application temperature.			
Typical Use: Automotive industry.			
Typ. Processing: Working Temp: 190 - 200°C.			
Test Methods: ASTM			
Physical		**Mechanical**	
Colour	White	τ (RT)	-
Softening Temp	161°C	τ (°C)	-
Temp. Range	190 - 200°C	τ (wet)	-
Visc. (190°C)	13.8 Pa.s	Peel ( )	-
Visc. (200°C)	10.7 Pa.s	σ_t	1.2 MPa
SG	0.9	σ_c	-
Assembly time	Short set time.	*E*	-
Tg	-	*G*	-
$T_{Service}$	<90	Elongation	1300 %
Notes: Softening Temp: ASTM D 3461, Density: ASTM D 792, Viscosity: ASTM D3236, Tensile ASTM D638. Skin time at 204°C (TRL): 2 hr.			

Mydrin (TRL)	Thermelt OROD 28		91
Thermoplastic Type: Copolyester.		Rod.	
Manufacturers Comments: Very short setting time. Good wetting and flexibility. Good adhesion on porous surfaces.			
Typical Use: Shoe-making.			
Typ. Processing: Working Temp: 230 - 250°C.			
Test Methods: ASTM			
Physical		**Mechanical**	
Colour	White	τ (RT)	-
Softening Temp	180±10°C	τ (°C)	-
Temp. Range	230 - 250°C	τ (wet)	-
Visc. (232°C)	~7 Pa.s	Peel ( )	-
Visc. (°C)	-	σ_t	15 - 18 MPa
SG	1.3	σ_c	-
Assembly time	Very fast set time.	*E*	-
Tg	-	*G*	-
$T_{Service}$	-	Elongation	400 - 600 %
Notes: Softening Temp: ASTM D 3461, Density: ASTM D 792, Viscosity: ASTM D 3236, Tensile ASTM D 638. Skin time at 232°C (TRL): 9 hr.			

Mydrin (TRL)	Thermelt 9172		93
Thermoplastic Type: Copolyester.		Pellets.	
Manufacturers Comments: Very fast setting time. Good wetting at application temperature.			
Typical Use: Automotive fuel and oil filter construction.			
Typ. Processing: Working Temp: 210 - 230°C.			
Test Methods: ASTM			
Physical		**Mechanical**	
Colour	White	τ (RT)	-
Softening Temp	171°C	τ (°C)	-
Temp. Range	210 - 230°C	τ (wet)	-
Visc. (210°C)	13.8 Pa.s	Peel ( )	-
Visc. (230°C)	2.0 Pa.s	σ_t	-
SG	1.3	σ_c	-
Assembly time	Very fast set time.	*E*	-
Tg	-	*G*	-
$T_{Service}$	-	Elongation	-
Notes: Softening Temp: AFNOR T 76106, Density: ASTM D 792, Viscosity: ASTM D3236.			

Mydrin (TRL)	Thermelt OROD 168		Jan 94
Thermoplastic Type: Polyamide.		Rod.	
Manufacturers Comments: Very short setting time. Good wetting and flexibility.			
Typical Use: Shoe-making.			
Typ. Processing: Working Temp: 190 - 200°C.			
Test Methods: ASTM			
Physical		**Mechanical**	
Colour	Gardner 4	τ (RT)	-
Softening Temp	170°C	τ (°C)	-
Temp. Range	190 - 200°C	τ (wet)	-
Visc. (190°C)	12.5 Pa.s	Peel ( )	-
Visc. (200°C)	8.3 Pa.s	σ_t	13.5 MPa
SG	0.98	σ_c	-
Assembly time	Very short set time.	*E*	-
Tg	-	*G*	-
$T_{Service}$	-	Elongation	550 %
Notes: Softening Temp: ASTM D 3461, Colour: ASTM D 1544, Density: ASTM D 792, Viscosity: ASTM D 3236, Tensile ASTM D 638. Skin time at 204°C (TRL): 8 hr.			

Power Adhesives Ltd.

Power Adhesives Ltd.	**Tec-Bond 213**		(Sept 95)
Thermoplastic Type: -	Sticks/Cylinders.		
Manufacturers Comments: High performance, multi purpose adhesive with a long open time. Very high tack. Meets requirements of (USA) FDA CFR21 175 - 105 (Adhesives).			
Typical Use: General purpose.			
Typ. Processing: 160 - 200°C.			
Test Methods: ASTM / BS			
Physical		**Mechanical**	
Colour	Brown	τ (RT)	-
Softening Temp	78°C	τ (°C)	-
Temp. Range	160°C - 200°C	τ (wet)	-
Visc. (160°C)	14 Pa.s	Peel ( )	-
Visc. (200°C)	4.45 Pa.s	σ_t	-
SG	-	σ_c	-
Assembly time	45 sec. (open time)	E	-
Tg	-	G	-
$T_{Service}$	<70°C	Elongation	-
Notes: Softening point: Ring and ball (ASTM E28). Heat resistance and open time: BS 5350. Viscosity: Brookfield (ASTM D 3236). Available in several physical forms: 213/12, 213/43, 213/20/60.			

Power Adhesives Ltd.	**Tec-Bond 240**		(Sept 95)
Thermoplastic Type: -	Sticks/Cylinders.		
Manufacturers Comments: Flexible, transparent, general purpose adhesive with a medium open time. Very high tack. Meets requirements of (USA) FDA CFR21 175 - 105 (Adhesives).			
Typical Use: General purpose.			
Typ. Processing: 160 - 200°C.			
Test Methods: ASTM / BS			
Physical		**Mechanical**	
Colour	Transparent	τ (RT)	-
Softening Temp	80°C	τ (°C)	-
Temp. Range	160°C - 200°C	τ (wet)	-
Visc. (160°C)	8.3 Pa.s	Peel ( )	-
Visc. (200°C)	2.9 Pa.s	σ_t	-
SG	-	σ_c	-
Assembly time	45 sec. (open time)	E	-
Tg	-	G	-
$T_{Service}$	<67°C	Elongation	-
Notes: Softening point: Ring and ball (ASTM E28). Heat resistance and open time: BS 5350. Viscosity: Brookfield (ASTM D 3236). Available in several physical forms: 240/12, 240/43, 240/20/60.			

Power Adhesives Ltd.	**Tec-Bond 214**		(Sept 95)
Thermoplastic Type: EVA.	Sticks/Cylinders.		
Manufacturers Comments: Fast setting EVA. High tack. Meets requirements of (USA) FDA CFR21 175 - 105 (Adhesives).			
Typical Use: Packaging industry.			
Typ. Processing: 160 - 200°C.			
Test Methods: ASTM / BS			
Physical		**Mechanical**	
Colour	Light Brown	τ (RT)	-
Softening Temp	96°C	τ (°C)	-
Temp. Range	160°C - 200°C	τ (wet)	-
Visc. (160°C)	8.35 Pa.s	Peel ( )	-
Visc. (200°C)	2.8 Pa.s	σ_t	-
SG	-	σ_c	-
Assembly time	15 sec. (open time)	E	-
Tg	-	G	-
$T_{Service}$	<70°C	Elongation	-
Notes: Softening point: Ring and ball (ASTM E28). Heat resistance and open time: BS 5350. Viscosity: Brookfield (ASTM D 3236). Available in several physical forms: 214/12, 214/43, 214/20/60.			

Power Adhesives Ltd.	**Tec-Bond 260**		(Sept 95)
Thermoplastic Type: -	Sticks/Cylinders.		
Manufacturers Comments: High performance adhesive. Tough and resilient. Very high tack. Meets requirements of (USA) FDA CFR21 175 - 105 and CFR21 176 - 180 (Adhesives).			
Typical Use: Excellent on plastics, glass, ceramics and metals.			
Typ. Processing: 160 - 200°C.			
Test Methods: ASTM / BS			
Physical		**Mechanical**	
Colour	Light Brown	τ (RT)	-
Softening Temp	84°C	τ (°C)	-
Temp. Range	160°C - 200°C	τ (wet)	-
Visc. (160°C)	11.7 Pa.s	Peel ( )	-
Visc. (200°C)	3.9 Pa.s	σ_t	-
SG	-	σ_c	-
Assembly time	40 sec. (open time)	E	-
Tg	-	G	-
$T_{Service}$	<70°C	Elongation	-
Notes: Softening point: Ring and ball (ASTM E28). Heat resistance and open time: BS 5350. Viscosity: Brookfield (ASTM D 3236). Available in several physical forms: 260/12, 260/43, 260/20/60.			

Power Adhesives Ltd.	**Tec-Bond 420**		(Sept 95)
Thermoplastic Type: -	Cylinders.		
Manufacturers Comments: High performance sprayable adhesive. Extended open time. Very high tack.			
Typical Use: Packaging.			
Typ. Processing: 200°C.			
Test Methods: ASTM / BS			
Physical		**Mechanical**	
Colour	Brown	τ (RT)	-
Softening Temp	98°C	τ (°C)	-
Temp. Range	-	τ (wet)	-
Visc. (160°C)	15 Pa.s	Peel ( )	-
Visc. (200°C)	5.5 Pa.s	σ_f	-
SG	-	σ_c	-
Assembly time	3 mins. (open time)	E	-
Tg	-	G	-
$T_{Service}$	<75°C	Elongation	-
Notes: Softening point: Ring and ball (ASTM E28). Heat resistance and open time: BS 5350. Viscosity: Brookfield (ASTM D 3236).			

Power Adhesives Ltd.	**Tec-Bond 6002**		(Sept 95)
Thermoplastic Type: EVA.	Sticks/Cylinders.		
Manufacturers Comments: Very fast setting EVA. Very high tack. Meets requirements of (USA) FDA CFR21 175 - 105 (Adhesives).			
Typical Use: Packaging industry.			
Typ. Processing: 160 - 205°C.			
Test Methods: ASTM / BS			
Physical		**Mechanical**	
Colour	Brown	τ (RT)	-
Softening Temp	101°C	τ (°C)	-
Temp. Range	160°C - 200°C	τ (wet)	-
Visc. (160°C)	5.25 Pa.s	Peel ( )	-
Visc. (200°C)	1.5 Pa.s	σ_f	-
SG	-	σ_c	-
Assembly time	10 sec. (open time)	E	-
Tg	-	G	-
$T_{Service}$	<65°C	Elongation	-
Notes: Softening point: Ring and ball (ASTM E28). Heat resistance and open time: BS 5350. Viscosity: Brookfield (ASTM D 3236). Available in several physical forms: 6002/12, 6002/43, 6002/20/60.			

Power Adhesives Ltd.	**Tec-Bond 430**		(Sept 95)
Thermoplastic Type: -	Cylinders.		
Manufacturers Comments: High performance sprayable adhesive. Extended open time. Very high tack.			
Typical Use: Packaging.			
Typ. Processing: 200°C.			
Test Methods: ASTM / BS			
Physical		**Mechanical**	
Colour	Light Brown	τ (RT)	-
Softening Temp	157°C	τ (°C)	-
Temp. Range	-	τ (wet)	-
Visc. (180°C)	10.7 Pa.s	Peel ( )	-
Visc. (200°C)	7.3 Pa.s	σ_f	-
SG	-	σ_c	-
Assembly time	8 mins. (open time)	E	-
Tg	-	G	-
$T_{Service}$	<75°C	Elongation	-
Notes: Softening point: Ring and ball (ASTM E28). Heat resistance and open time: BS 5350. Viscosity: Brookfield (ASTM D 3236).			

Power Adhesives Ltd.	**Tec-Bond LM44**		(Sept 95)
Thermoplastic Type: -	Sticks/Cylinders.		
Manufacturers Comments: Low-melt adhesive for use in low temperature glue guns. Very fast setting. Very high tack. Meets requirements of (USA) FDA CFR21 175 - 105 (Adhesives).			
Typical Use: Packaging - suitable for delicate materials (expanded polystyrene, films, etc.).			
Typ. Processing: 120 - 140°C.			
Test Methods: ASTM / BS			
Physical		**Mechanical**	
Colour	Translucent white	τ (RT)	-
Softening Temp	98°C	τ (°C)	-
Temp. Range	120°C - 140°C	τ (wet)	-
Visc. (120°C)	7.1 Pa.s	Peel ( )	-
Visc. (140°C)	3.4 Pa.s	σ_f	-
SG	-	σ_c	-
Assembly time	10 sec. (open time)	E	-
Tg	-	G	-
$T_{Service}$	<60°C	Elongation	-
Notes: Softening point: Ring and ball (ASTM E28). Heat resistance and open time: BS 5350. Viscosity: Brookfield (ASTM D 3236). Available in several physical forms.			

See Key to Data Sheets - Pages 63 to 65

Quilosa

(Industrias Quimicas Löwenberg SA)

Quilosa	Komelt C-18		Feb 92
Thermoplastic Type: EVA and synthetic resins. Granules.			
Manufacturers Comments: Good thermal stability. Short open time. Medium setting time.			
Typical Use: Assembly and closing of compact cardboard cartons.			
Typ. Processing: -			
Test Methods: -			
Physical		**Mechanical**	
Colour	Amber	τ (RT)	-
Softening Temp	114±5°C	τ (°C)	-
Temp. Range	160 - 180°C	τ (wet)	-
Visc. (160°C)	0.8 Pa.s	Peel ( )	-
Visc. (°C)	-	σ_f	-
SG	-	σ_c	-
Assembly time	-	E	-
Tg	-	G	-
$T_{Service}$	-	Elongation	-
Notes: Softening temperature: ASTM D3461-85. Shelf life: <1 year. Komelt C-11 is a faster setting version.			

Quilosa	Sintex DM-43		Sept 89
Thermoplastic Type: EVA and synthetic resins. Granules.			
Manufacturers Comments: 100% solids.			
Typical Use: Veneering and moulding of wood and wood products. Wood veneer and polyester.			
Typ. Processing: -			
Test Methods: -			
Physical		**Mechanical**	
Colour	Beige	τ (RT)	-
Softening Temp	81±1°C	τ (°C)	-
Temp. Range	180 - 210°C	τ (wet)	-
Visc. (200°C)	16 Pa.s	Peel ( )	-
Visc. (°C)	-	σ_f	-
SG	-	σ_c	-
Assembly time	-	E	-
Tg	-	G	-
$T_{Service}$	<55°C	Elongation	-
Notes: Softening temperature: ASTM D3461-85. Shelf life: <1 year.			

Quilosa	Sintex DM-15		May 90
Thermoplastic Type: EVA and synthetic resins. Granules.			
Manufacturers Comments: Fast setting.			
Typical Use: Wood and wood products.			
Typ. Processing: -			
Test Methods: -			
Physical		**Mechanical**	
Colour	Beige	τ (RT)	-
Softening Temp	103±2°C	τ (°C)	-
Temp. Range	190 - 210°C	τ (wet)	-
Visc. (190°C)	66.5 Pa.s	Peel ( )	-
Visc. (210°C)	39.5 Pa.s	σ_f	-
SG	-	σ_c	-
Assembly time	-	E	-
Tg	-	G	-
$T_{Service}$	<55°C	Elongation	-
Notes: Softening temperature: ASTM D3461-85. Shelf life: <1 year.			

Quilosa	Sintex DM-64		(Jul 95)
Thermoplastic Type: EVA and synthetic resins. Granules.			
Manufacturers Comments: Very flexible with good surface adhesion.			
Typical Use: Furniture. Wood and wood products, polyester, melamine, PVC and rigid materials.			
Typ. Processing: -			
Test Methods: -			
Physical		**Mechanical**	
Colour	Beige	τ (RT)	-
Softening Temp	95±2°C	τ (°C)	-
Temp. Range	190 - 210°C	τ (wet)	-
Visc. (190°C)	90 Pa.s	Peel ( )	-
Visc. (210°C)	30 Pa.s	σ_f	-
SG	-	σ_c	-
Assembly time	-	E	-
Tg	-	G	-
$T_{Service}$	<65°C	Elongation	-
Notes: Softening temperature: ASTM D3461-85. Shelf life: <1 year.			

Sunkyong Industries

Sunkyong Industries	EH 105		(Oct 95)
Thermoplastic Type: Copolyester. Pellets/Rods/Films.			
Manufacturers Comments: Fast setting time. Short open time. Good adhesion and chemical resistance.			
Typical Use: Shoe industry.			
Typ. Processing: -			
Test Methods: ASTM			
Physical		**Mechanical**	
Colour	-	τ (RT)	-
Softening Temp	178 ± 5°C	τ (°C)	-
Temp. Range	-	τ (wet)	-
Visc. (°C)	-	Peel ( )	-
Visc. (°C)	-	σ_t	22 MPa
SG	-	σ_c	-
Assembly time	Fast set time.	*E*	-
Tg	-	*G*	-
$T_{Service}$	-40 to +150°C	Elongation	50 %
Notes: Softening point: Ring and ball.			

Sunkyong Industries	NH 430		(Oct 95)
Thermoplastic Type: Copolyamide. Pellets/Sheet.			
Manufacturers Comments: Flexible. Good low temperature properties. Long open time.			
Typical Use: -			
Typ. Processing: -			
Test Methods: ASTM			
Physical		**Mechanical**	
Colour	Various	τ (RT)	-
Softening Temp	135 ± 5°C	τ (°C)	-
Temp. Range	-	τ (wet)	-
Visc. (200°C)	6 Pa.s	Peel ( )	-
Visc. (°C)	-	σ_t	9.8 MPa
SG	-	σ_c	-
Assembly time	-	*E*	-
Tg	-	*G*	-
$T_{Service}$	-	Elongation	-
Notes: -			

Sunkyong Industries	EH 500		(Oct 95)
Thermoplastic Type: Copolyester. Pellets/Rods/Films.			
Manufacturers Comments: Fast setting time. Short open time. Good adhesion and chemical resistance.			
Typical Use: Automotive filters.			
Typ. Processing: -			
Test Methods: ASTM			
Physical		**Mechanical**	
Colour	-	τ (RT)	-
Softening Temp	177 ± 5°C	τ (°C)	-
Temp. Range	-	τ (wet)	-
Visc. (°C)	-	Peel ( )	-
Visc. (°C)	-	σ_t	25.5 MPa
SG	-	σ_c	-
Assembly time	Fast set time.	*E*	-
Tg	-	*G*	-
$T_{Service}$	-40 to +150°C	Elongation	400 %
Notes: Softening point: Ring and ball.			

Sunkyong Industries	NH 540		(Oct 95)
Thermoplastic Type: Copolyamide. Pellets/Sheet.			
Manufacturers Comments: Flexible and tough. Medium open time.			
Typical Use: -			
Typ. Processing: -			
Test Methods: ASTM			
Physical		**Mechanical**	
Colour	Various	τ (RT)	-
Softening Temp	155 ± 5°C	τ (°C)	-
Temp. Range	-	τ (wet)	-
Visc. (200°C)	5 Pa.s	Peel ( )	-
Visc. (°C)	-	σ_t	10.3 MPa
SG	-	σ_c	-
Assembly time	-	*E*	-
Tg	-	*G*	-
$T_{Service}$	-	Elongation	-
Notes: -			

Sunkyong Industries	**NH 550**		(Oct 95)
Thermoplastic Type: Copolyamide.	Pellets/Sheet.		
Manufacturers Comments: Flexible and very tough. High strength. Short open time.			
Typical Use: -			
Typ. Processing: -			
Test Methods: ASTM			
Physical		**Mechanical**	
Colour	Various	τ (RT)	-
Softening Temp	185 ± 5°C	τ (°C)	-
Temp. Range	-	τ (wet)	-
Visc. (200°C)	8 Pa.s	Peel ( )	-
Visc. (°C)	-	σ_t	12.7 MPa
SG	-	σ_c	-
Assembly time	-	*E*	-
Tg	-	*G*	-
$T_{Service}$	-	Elongation	-
Notes: -			

Witco	**Euremelt 930**		(Feb 95)
Thermoplastic Type: Polyamide.	Pellets/Various.		
Manufacturers Comments: Hard and brittle. Euremelt series resins may be blended to produce a range of properties.			
Typical Use: Furniture, electrical, packaging, canning and automotive.			
Typ. Processing: -			
Test Methods: DIN			
Physical		**Mechanical**	
Colour	Clear Amber	τ (RT)	6.5 MPa
Softening Temp	110±5°C	τ (90°C)	2.2 MPa
Temp. Range	150 - 180°C	τ (-20°C)	4.0 MPa
Visc. (150°C)	6.0 Pa.s	Peel (T)	2 N mm^{-1}
Visc. (180°C)	1.7 Pa.s	σ_t	5.0 MPa
SG	0.98	σ_c	-
Assembly time	<1 sec.	*E*	-
Tg	-	*G*	-
$T_{Service}$	-	Elongation	<5%
Notes: Tensile shear strength (steel adherends): DIN 53283. Peel strength (steel adherends): DIN 53282. Tensile strength & Elongation: DIN 53455, Setting time: In-house. Melt stability: 16 hrs to start of skinning at 180°C. Water absorption (DIN 53495): 0.3% after 14 days. Application temperature range is only an indication by the manufacturer.			

Witco

Witco	**Euremelt 900**		(Feb 95)
Thermoplastic Type: Polyamide.	Pellets/Various.		
Manufacturers Comments: Hard and brittle. Euremelt series resins may be blended to produce a range of properties.			
Typical Use: Furniture, electrical, packaging, canning and automotive.			
Typ. Processing: -			
Test Methods: DIN			
Physical		**Mechanical**	
Colour	Clear Amber	τ (RT)	4.3 MPa
Softening Temp	185±5°C	τ (90°C)	2.0 MPa
Temp. Range	200 - 210°C	τ (-20°C)	3.3 MPa
Visc. (200°C)	0.45 Pa.s	Peel (T)	2 N mm^{-1}
Visc. (210°C)	0.3 Pa.s	σ_t	-
SG	0.98	σ_c	-
Assembly time	<1 sec.	*E*	-
Tg	-	*G*	-
$T_{Service}$	-	Elongation	<5%
Notes: Tensile shear strength (steel adherends): DIN 53283. Peel strength (steel adherends): DIN 53282. Tensile strength & Elongation: DIN 53455, Setting time: In-house. Melt stability: 10 hrs to start of skinning at 200°C. Water absorption (DIN 53495): 0.6% after 14 days. Application temperature range is only an indication by the manufacturer.			

Witco	**Euremelt 940**		(Feb. 95)
Thermoplastic Type: Polyamide.	Pellets/Various.		
Manufacturers Comments: Hard and brittle. Euremelt series resins may be blended to produce a range of properties.			
Typical Use: Furniture, electrical, packaging, canning and automotive.			
Typ. Processing: -			
Test Methods: DIN			
Physical		**Mechanical**	
Colour	Clear Amber	τ (RT)	5.8 MPa
Softening Temp	110±5°C	τ (90°C)	2.0 MPa
Temp. Range	140 - 170°C	τ (-20°C)	3.9 MPa
Visc. (140°C)	4.6 Pa.s	Peel (T)	2 N mm^{-1}
Visc. (170°C)	1.3 Pa.s	σ_t	5.0 MPa
SG	0.98	σ_c	-
Assembly time	<1 sec.	*E*	-
Tg	-	*G*	-
$T_{Service}$	-	Elongation	<5%
Notes: Tensile shear strength (steel adherends): DIN 53283. Peel strength (steel adherends): DIN 53282. Tensile strength & Elongation: DIN 53455, Setting time: In-house. Melt stability: 200 hrs to start of skinning at 160°C. Water absorption (DIN 53495): 0.4% after 14 days. Application temperature range is only an indication by the manufacturer.			

Witco	Euremelt 1155		(Feb 95)
Thermoplastic Type: Polyamide.	Pellets/Various.		
Manufacturers Comments: Hard and tough. Unaffected by oxidation. Euremelt series resins may be blended to produce a range of properties.			
Typical Use: Furniture, electrical, packaging, canning and automotive.			
Typ. Processing: -			
Test Methods: DIN			
Physical		**Mechanical**	
Colour	Clear Amber	τ (RT)	8.6 MPa
Softening Temp	160±5°C	τ (°C)	-
Temp. Range	180 - 200°C	τ (wet)	-
Visc. (180°C)	4.8 Pa.s	Peel (T)	3 N mm^{-1}
Visc. (200°C)	2.3 Pa.s	σ_t	10.0 MPa
SG	0.95	σ_c	-
Assembly time	<1 sec.	E	-
Tg	-5°C	G	-
$T_{Service}$	-	Elongation	15%
Notes: Tensile shear strength (steel adherends): DIN 53283. Peel strength (steel adherends): DIN 53282. Tensile strength, Elongation & Tg: DIN 53455, Setting time: In-house. Melt stability: 75 hrs to start of skinning at 190°C. Water absorption (DIN 53495): 0.4% after 14 days. Application temperature range is only an indication by the manufacturer.			

Witco	Euremelt 1174		(Feb 95)
Thermoplastic Type: Polyamide.	Pellets/Various.		
Manufacturers Comments: Hard and tough. Euremelt series resins may be blended to produce a range of properties.			
Typical Use: Furniture, electrical, packaging, canning and automotive.			
Typ. Processing: -			
Test Methods: DIN			
Physical		**Mechanical**	
Colour	Clear Amber	τ (RT)	10.0 MPa
Softening Temp	170 - 180°C	τ (°C)	-
Temp. Range	200 - 230°C	τ (wet)	-
Visc. (200°C)	9.3 Pa.s	Peel (T)	2 N mm^{-1}
Visc. (210°C)	7.0 Pa.s	σ_t	11.3 MPa
SG	-	σ_c	-
Assembly time	-	E	-
Tg	-	G	-
$T_{Service}$	-	Elongation	10%
Notes: Tensile shear strength (steel adherends): DIN 53283. Peel strength (steel adherends): DIN 53282. Tensile strength & Elongation: DIN 53455. Melt stability: 3.5 hrs to start of skinning at 200°C. Water absorption (DIN 53495): 0.47% after 14 days. Application temperature range is only an indication by the manufacturer.			

Witco	Euremelt 1166		(Feb 95)
Thermoplastic Type: Polyamide.	Pellets/Various.		
Manufacturers Comments: Hard and tough. Unaffected by oxidation. Extremely low melt viscosity and good thermal stability. Euremelt series resins may be blended to produce a range of properties.			
Typical Use: Furniture, electrical, packaging, canning and automotive.			
Typ. Processing:			
Test Methods: DIN			
Physical		**Mechanical**	
Colour	Clear Amber	τ (RT)	3.7 MPa
Softening Temp	175±3°C	τ (°C)	-
Temp. Range	175 - 200°C	τ (wet)	-
Visc. (180°C)	0.28 Pa.s	Peel (T)	3 N mm^{-1}
Visc. (200°C)	0.16 Pa.s	σ_t	8.0 MPa
SG	0.98	σ_c	-
Assembly time	-	E	-
Tg	0°C	G	-
$T_{Service}$	-	Elongation	15%
Notes: Tensile shear strength (steel adherends): DIN 53283. Peel strength (steel adherends): DIN 53282. Tensile strength, Elongation & Tg: DIN 53455. Melt stability: 80 hrs to start of skinning at 190°C. Water absorption (DIN 53495): 0.5% after 14 days. Application temperature range is only an indication by the manufacturer.			

Witco	Euremelt 1175		(Feb 95)
Thermoplastic Type: Polyamide.	Pellets/Various.		
Manufacturers Comments: Hard and tough. Euremelt series resins may be blended to produce a range of properties.			
Typical Use: Furniture, electrical, packaging, canning and automotive.			
Typ. Processing:			
Test Methods: DIN			
Physical		**Mechanical**	
Colour	Clear Amber	τ (RT)	10.6 MPa
Softening Temp	180±5°C	τ (90°C)	2.2 MPa
Temp. Range	190 - 220°C	τ (-20°C)	6.3 MPa
Visc. (190°C)	8.5 Pa.s	Peel (T)	4 N mm^{-1}
Visc. (220°C)	3.0 Pa.s	σ_t	12.5 MPa
SG	-	σ_c	-
Assembly time	2 - 3 sec.	E	-
Tg	0°C	G	-
$T_{Service}$	-	Elongation	20%
Notes: Tensile shear strength (steel adherends): DIN 53283. Peel strength (steel adherends): DIN 53282. Tensile strength, Elongation & Tg: DIN 53455, Setting time: In-house. Melt stability: 2.5 hrs to start of skinning at 200°C. Water absorption (DIN 53495): 0.6% after 14 days. Application temperature range is only an indication by the manufacturer.			

Witco	Euremelt 2095		(Feb 95)
Thermoplastic Type: Polyamide.	Pellets/Various.		
Manufacturers Comments: Flexible with low melt viscosity and good thermal stability. Euremelt series resins may be blended to produce a range of properties. The flexible, series 2000 resins have much better adhesion to rubbers and plastics than Witco Euremelt 900 and 1000 series.			
Typical Use: Furniture, electrical, packaging, canning and automotive.			
Typ. Processing: -			
Test Methods: DIN			
Physical		**Mechanical**	
Colour	Clear Amber	τ (RT)	4.0 MPa
Softening Temp	95±5°C	τ (50°C)	1.0 MPa
Temp. Range	120 - 160°C	τ (-20°C)	7.0 MPa
Visc. (120°C)	7.5 Pa.s	Peel (T)	4 N mm^{-1}
Visc. (160°C)	1.0 Pa.s	σ_t	3.5 MPa
SG	0.95	σ_c	-
Assembly time	-	E	-
Tg	-5°C	G	-
$T_{Service}$	-	Elongation (RT)	70%

Notes: Tensile shear strength (steel adherends): DIN 53283. Peel strength (steel adherends): DIN 53282. Tensile strength, Elongation & Tg: DIN 53455. Melt stability: 170 hrs to start of skinning at 160°C. Water absorption (DIN 53495): 0.9% after 14 days. Application temperature range is only an indication by the manufacturer.

Witco	Euremelt 2110		(Feb 95)
Thermoplastic Type: Polyamide.	Pellets/Various.		
Manufacturers Comments: Flexible. Euremelt series resins may be blended to produce a range of properties. The flexible, series 2000 resins have much better adhesion to rubbers and plastics than Witco Euremelt 900 and 1000 series.			
Typical Use: Furniture, electrical, packaging, canning and automotive.			
Typ. Processing: -			
Test Methods: DIN			
Physical		**Mechanical**	
Colour	Clear Amber	τ (RT)	6.6 MPa
Softening Temp	105±5°C	τ (°C)	-
Temp. Range	180 - 200°C	τ (wet)	-
Visc. (180°C)	19.5 Pa.s	Peel (T)	9.6 N mm^{-1}
Visc. (200°C)	11.1 Pa.s	σ_t	2.3 MPa
SG	-	σ_c	-
Assembly time	-	E	-
Tg	-	G	-
$T_{Service}$	-	Elongation (RT)	90%

Notes: Tensile shear strength (steel adherends): DIN 53283. Peel strength (steel adherends): DIN 53282. Tensile strength & Elongation: DIN 53455. Melt stability: 7 hrs to start of skinning at 200°C. Water absorption (DIN 53495): 0.5% after 14 days. Application temperature range is only an indication by the manufacturer.

Witco	Euremelt 2096		(Feb 95)
Thermoplastic Type: Polyamide.	Pellets/Various.		
Manufacturers Comments: Flexible. Euremelt series resins may be blended to produce a range of properties. The flexible, series 2000 resins have much better adhesion to rubbers and plastics than Witco Euremelt 900 and 1000 series.			
Typical Use: Furniture, electrical, packaging, canning and automotive.			
Typ. Processing: -			
Test Methods: DIN			
Physical		**Mechanical**	
Colour	Clear Amber	τ (RT)	7.2 MPa
Softening Temp	97±5°C	τ (°C)	-
Temp. Range	140 - 180°C	τ (wet)	-
Visc. (150°C)	2.9 Pa.s	Peel (T)	3.5 N mm^{-1}
Visc. (180°C)	1.0 Pa.s	σ_t	5.4 MPa
SG	-	σ_c	-
Assembly time	-	E	-
Tg	-	G	-
$T_{Service}$	-	Elongation (RT)	70%

Notes: Tensile shear strength (steel adherends): DIN 53283. Peel strength (steel adherends): DIN 53282. Tensile strength & Elongation: DIN 53455. Melt stability: 25 hrs to start of skinning at 160°C. Water absorption (DIN 53495): 0.74% after 14 days. Application temperature range is only an indication by the manufacturer.

Witco	Euremelt 2125		(Feb 95)
Thermoplastic Type: Polyamide.	Pellets/Various.		
Manufacturers Comments: Flexible. Euremelt series resins may be blended to produce a range of properties. The flexible, series 2000 resins have much better adhesion to rubbers and plastics than Witco Euremelt 900 and 1000 series.			
Typical Use: Furniture, electrical, packaging, canning and automotive.			
Typ. Processing: -			
Test Methods: DIN			
Physical		**Mechanical**	
Colour	Clear Amber	τ (RT)	8.5 MPa
Softening Temp	130±5°C	τ (°C)	-
Temp. Range	180 - 200°C	τ (wet)	-
Visc. (180°C)	16.9 Pa.s	Peel (T)	8.4 N mm^{-1}
Visc. (200°C)	8.5 Pa.s	σ_t	4.0 MPa
SG	-	σ_c	-
Assembly time	-	E	-
Tg	-	G	-
$T_{Service}$	-	Elongation (RT)	60%

Notes: Tensile shear strength (steel adherends): DIN 53283. Peel strength (steel adherends): DIN 53282. Tensile strength & Elongation: DIN 53455. Melt stability: 7 hrs to start of skinning at 200°C. Water absorption (DIN 53495): 0.5% after 14 days. Application temperature range is only an indication by the manufacturer.

Witco	Euremelt 2130		(Feb 95)
Thermoplastic Type: Polyamide.	Pellets/Various.		
Manufacturers Comments: Flexible. Euremelt series resins may be blended to produce a range of properties. The flexible, series 2000 resins have much better adhesion to rubbers and plastics than Witco Euremelt 900 and 1000 series.			
Typical Use: Furniture, electrical, packaging, canning and automotive.			
Typ. Processing: -			
Test Methods: DIN			
Physical		**Mechanical**	
Colour	Clear Amber	τ (RT)	5.5 MPa
Softening Temp	130±5°C	τ (50°C)	1.5 MPa
Temp. Range	160 - 200°C	τ (-10°C)	13.0 MPa
Visc. (160°C)	8.5 Pa.s	Peel (T)	10.0 N mm^{-1}
Visc. (200°C)	3.0 Pa.s	σ_t	2.5 MPa
SG	0.98	σ_c	-
Assembly time	7 - 8 sec.	E	-
Tg	0°C	G	-
$T_{Service}$	-	Elongation (RT)	225%
Notes: Tensile shear strength (steel adherends): DIN 53283. Peel strength (steel adherends): DIN 53282. Tensile strength, Elongation & Tg: DIN 53455, Setting time: In-house. Melt stability: 5 hrs to start of skinning at 180°C. Water absorption (DIN 53495): 1.8% after 14 days. Application temperature range is only an indication by the manufacturer.			

Witco	Euremelt 2137		(Feb 95)
Thermoplastic Type: Polyamide.	Pellets/Various.		
Manufacturers Comments: Flexible. Outstanding adhesion to PVC, ABS and rubber. Euremelt series resins may be blended to produce a range of properties. The flexible, series 2000 resins have much better adhesion to rubbers and plastics than Witco Euremelt 900 and 1000 series.			
Typical Use: Furniture, electrical, packaging, canning and automotive.			
Typ. Processing: -			
Test Methods: DIN			
Physical		**Mechanical**	
Colour	Clear Amber	τ (RT)	9.7 MPa
Softening Temp	145±5°C	τ (°C)	-
Temp. Range	190 - 220°C	τ (wet)	-
Visc. (190°C)	10.8 Pa.s	Peel (T)	12.0 N mm^{-1}
Visc. (220°C)	3.6 Pa.s	σ_t	4.0 MPa
SG	0.95	σ_c	-
Assembly time	-	E	-
Tg	10°C	G	-
$T_{Service}$	-	Elongation (RT)	350%
Notes: Tensile shear strength (steel adherends): DIN 53283. Peel strength (steel adherends): DIN 53282. Tensile strength, Elongation & Tg: DIN 53455. Melt stability: 12 hrs to start of skinning at 200°C. Water absorption (DIN 53495): 2.1% after 14 days. Application temperature range is only an indication by the manufacturer.			

Witco	Euremelt 2135		(Feb 95)
Thermoplastic Type: Polyamide.	Pellets/Various.		
Manufacturers Comments: Flexible. Euremelt series resins may be blended to produce a range of properties. The flexible, series 2000 resins have much better adhesion to rubbers and plastics than Witco Euremelt 900 and 1000 series.			
Typical Use: Furniture, electrical, packaging, canning and automotive.			
Typ. Processing: -			
Test Methods: DIN			
Physical		**Mechanical**	
Colour	Clear Amber	τ (RT)	7.0 MPa
Softening Temp	133±3°C	τ (°C)	-
Temp. Range	160 - 200°C	τ (wet)	-
Visc. (160°C)	9.3 Pa.s	Peel (T)	10.1 N mm^{-1}
Visc. (200°C)	2.4 Pa.s	σ_t	4.5 MPa
SG	-	σ_c	-
Assembly time	-	E	-
Tg	0°C	G	-
$T_{Service}$	-	Elongation (RT)	320%
Notes: Tensile shear strength (steel adherends): DIN 53283. Peel strength (steel adherends): DIN 53282. Tensile strength, Elongation & Tg: DIN 53455. Melt stability: 10 hrs to start of skinning at 200°C. Water absorption (DIN 53495): 1.34% after 14 days. Application temperature range is only an indication by the manufacturer.			

Witco	Euremelt 2138		(Feb 95)
Thermoplastic Type: Polyamide.	Pellets/Various.		
Manufacturers Comments: Flexible. Outstanding adhesion to PVC, ABS and rubber. Euremelt series resins may be blended to produce a range of properties. The flexible, series 2000 resins have much better adhesion to rubbers and plastics than Witco Euremelt 900 and 1000 series.			
Typical Use: Furniture, electrical, packaging, canning and automotive.			
Typ. Processing: -			
Test Methods: DIN			
Physical		**Mechanical**	
Colour	Clear Amber	τ (RT)	9.5 MPa
Softening Temp	143±5°C	τ (70°C)	1.5 MPa
Temp. Range	200 - 230°C	τ (-20°C)	15.0 MPa
Visc. (200°C)	15.0 Pa.s	Peel (T)	11.0 N mm^{-1}
Visc. (230°C)	6.2 Pa.s	σ_t	5.0 MPa
SG	0.95	σ_c	-
Assembly time	-	E	-
Tg	5°C	G	-
$T_{Service}$	-	Elongation (RT)	600%
Notes: Tensile shear strength (steel adherends): DIN 53283. Peel strength (steel adherends): DIN 53282. Tensile strength, Elongation & Tg: DIN 53455. Melt stability: 6.5 hrs to start of skinning at 200°C. Water absorption (DIN 53495): 2.0% after 14 days. Application temperature range is only an indication by the manufacturer.			

Witco	Euremelt 2140		(Feb 95)
Thermoplastic Type: Polyamide. Pellets/Various.			
Manufacturers Comments: Flexible. Outstanding adhesion to PVC, ABS and rubber. Euremelt series resins may be blended to produce a range of properties. The flexible, series 2000 resins have much better adhesion to rubbers and plastics than Witco Euremelt 900 and 1000 series.			
Typical Use: Furniture, electrical, packaging, canning and automotive.			
Typ. Processing: -			
Test Methods: DIN			
Physical		**Mechanical**	
Colour	Clear Amber	τ (RT)	9.7 MPa
Softening Temp	140±5°C	τ (90°C)	0.7 MPa
Temp. Range	200 - 230°C	τ (-20°C)	16.7 MPa
Visc. (200°C)	14.4 Pa.s	Peel (T)	11.0 N mm^{-1}
Visc. (230°C)	5.4 Pa.s	σ_t	3.0 MPa
SG	0.95	σ_c	-
Assembly time	8 sec.	E	-
Tg	10°C	G	-
$T_{Service}$	-	Elongation (RT)	600%
Notes: Tensile shear strength (steel adherends): DIN 53283. Peel strength (steel adherends): DIN 53282. Tensile strength, Elongation & Tg: DIN 53455, Setting time: In-house. Melt stability: 1.5 hrs to start of skinning at 220°C. Water absorption (DIN 53495): 1.2% after 14 days. Application temperature range is only an indication by the manufacturer.			

Witco	Euremelt 2166		(Feb 95)
Thermoplastic Type: Polyamide. Pellets/Various.			
Manufacturers Comments: Flexible. Euremelt series resins may be blended to produce a range of properties. The flexible, series 2000 resins have much better adhesion to rubbers and plastics than Witco Euremelt 900 and 1000 series.			
Typical Use: Furniture, electrical, packaging, canning and automotive.			
Typ. Processing: -			
Test Methods: DIN			
Physical		**Mechanical**	
Colour	Clear Amber	τ (RT)	8.3 MPa
Softening Temp	166±3°C	τ (°C)	-
Temp. Range	180 - 200°C	τ (wet)	-
Visc. (180°C)	2.1 Pa.s	Peel (T)	2.5 N mm^{-1}
Visc. (200°C)	1.3 Pa.s	σ_t	5.4 MPa
SG	-	σ_c	-
Assembly time	-	E	-
Tg	-	G	-
$T_{Service}$	-	Elongation (RT)	200%
Notes: Tensile shear strength (steel adherends): DIN 53283. Peel strength (steel adherends): DIN 53282. Tensile strength & Elongation: DIN 53455. Melt stability: 5 hrs to start of skinning at 180°C. Water absorption (DIN 53495): 1.6% after 14 days. Application temperature range is only an indication by the manufacturer.			

Witco	Euremelt 2145		(Feb 95)
Thermoplastic Type: Polyamide. Pellets/Various.			
Manufacturers Comments: Flexible. Euremelt series resins may be blended to produce a range of properties. The flexible, series 2000 resins have much better adhesion to rubbers and plastics than Witco Euremelt 900 and 1000 series.			
Typical Use: Furniture, electrical, packaging, canning and automotive.			
Typ. Processing: -			
Test Methods: DIN			
Physical		**Mechanical**	
Colour	Clear Amber	τ (RT)	10.0 MPa
Softening Temp	143±5°C	τ (70°C)	3.0 MPa
Temp. Range	200 - 230°C	τ (-20°C)	13.0 MPa
Visc. (200°C)	6.5 Pa.s	Peel (T)	5.0 N mm^{-1}
Visc. (220°C)	3.0 Pa.s	σ_t	9.0 MPa
SG	0.95	σ_c	-
Assembly time	-	E	-
Tg	0°C	G	-
$T_{Service}$	-	Elongation (RT)	150%
Notes: Tensile shear strength (steel adherends): DIN 53283. Peel strength (steel adherends): DIN 53282. Tensile strength, Elongation & Tg: DIN 53455. Melt stability: 4 hrs to start of skinning at 200°C. Water absorption (DIN 53495): 0.8% after 14 days. Application temperature range is only an indication by the manufacturer.			

Witco	Euremelt 2170		(Feb 95)
Thermoplastic Type: Polyamide. Pellets/Various.			
Manufacturers Comments: Flexible. Euremelt series resins may be blended to produce a range of properties. The flexible, series 2000 resins have much better adhesion to rubbers and plastics than Witco Euremelt 900 and 1000 series.			
Typical Use: Furniture, electrical, packaging, canning and automotive.			
Typ. Processing: -			
Test Methods: DIN			
Physical		**Mechanical**	
Colour	Clear Amber	τ (RT)	8.0 MPa
Softening Temp	-	τ (70°C)	2.5 MPa
Temp. Range	180 - 210°C	τ (-20°C)	6.5
Visc. (180°C)	13 Pa.s	Peel (T)	7.0 N mm^{-1}
Visc. (210°C)	3.3 Pa.s	σ_t	-
SG	-	σ_c	-
Assembly time	<1 sec.	E	-
Tg	-	G	-
$T_{Service}$	-	Elongation (RT)	-
Notes: Tensile shear strength (steel adherends): DIN 53283. Peel strength (steel adherends): DIN 53282. Setting time: In-house. Melt stability: 8 hrs to start of skinning at 200°C. Application temperature range is only an indication by the manufacturer.			

MISCELLANEOUS ADHESIVES

Notes

This section contains a collection of data sheets for adhesives used for the structural bonding of various materials, but whose various chemical bases lie outside the broad classes used elsewhere in this book. Included here are:

- **Cyanate Ester** - developed for bonding the new high-performance, fibre-reinforced composites with cyanate ester resin matrices. These materials have been proposed for aerospace applications requiring improved thermal performance, electrical properties (radomes) and moisture resistance.
- **Vinyl-containing** - (either as a principal or secondary phase). These are mainly water-based dispersions. Included are examples of polyvinylacetate (PVA and PVAC -copolymer) adhesives. Commonly known as 'white glue', these are widely used in the woodworking industry. Those given here are considered the higher performance, 'industrial' types rather than products for household use.
- **Plastisol** - commonly used in automotive assembly for large bond areas, low-load applications (bonding of bonnet and panel reinforcements to reduce flutter). Capable of bonding oily surfaces; some absorb the oil film during cure. Properties are semi-structural (low peel, low cleavage and will creep under prolonged static loading, but have good resistance to cyclic loads).
- **Ceramic-based** - adhesives for very high temperature applications, i.e. 400 to ~1600°C. Temperatures which exceed the capabilities of any polymer-based materials. Used for bonding refractory materials (mainly ceramics) to themselves or other materials.
- **Unknown chemical base** - details of some formulations are not stated by the manufacturers. In some cases a general description is used by the manufacturer, e.g. 'synthetic resin'. Some of these are UV curing adhesives.

Their disparate characteristics and intended uses make a direct comparison of properties meaningless. In this section the comparison table gives a broad description of each adhesive and its typical applications.

COMPARISON TABLE - MISCELLANEOUS ADHESIVES

Company	Adhesive	Description: Typical Uses	Cure (°C)	Service (°C)	Pg. No.
Cytec	Metlbond 2555	1 part - Cyanate ester. Film: Aerospace metals & composites, inc. radome applications	~177	232	358
	Metlbond 6555	1 part - Cyanate ester. Foaming film: Aerospace composites, inc. radome applications	~177+225	177	358
EMFI	Bois 860	1 part - vinyl polymer dispersion: Wood products	RT	-	358
	Bois 863 B3	1 part - vinyl polymer dispersion: Wood products	RT	-	358
	Bois Rapide 885	1 part - vinyl polymer dispersion: Wood products. Furniture. Edge-banding.	RT		359
	Colle PVC/PVC	1 part - synthetic resin: PVC glazing frames, water & waste pipes	RT	-	359
Forbo	Helmitin 34026	1 part - EVA dispersion: Plastics, foams, metal foils to porous materials (inc. textiles)	RT	-	359
Helmitin	PF/S	1 part - EVA dispersion: PVC sheets to wood products & other porous materials	RT	-	359
	Rapid 33207	1 part - PVAC water-based disp.: Woodworking. Construction, joints, laminates to wood	RT	-	360
	Rapid 33511	1 part - PVAC water-based disp.: Woodworking. Construction, joints, laminates to wood	RT	-	360
	Rapid 39533	1 part - PVAC dispersion: Wood products (inc. paper & cardboard), textiles.	RT	-	360
	Rapid A	1 part - PVAC water-based disp.: Wood working. Wood products	RT	-	360
	Rapid D	1 part - PVAC water-based disp.: Wood working. Doweling (automatic)	RT	-	361
	Rapid KA	1 part - PVAC water-based disp.: Wood working, inc. profiles (soft-forming) & edge-banding	RT	-	361
	Rapid S	1 part - PVAC water-based disp.: Wood working, inc. joints, dowels, laminates	RT	-	361
Fortafix	AL/CS	2 part - ceramic: High temperature. Ceramics, glass, metals	RT/ ~90	1600	361
	Chromix	2 part - ceramic: High temperature. Ceramics, glass, metals	RT/ ~90	1600	362
	Chromix Ready-to-Use	1 part - water-based ceramic: High temperature. Ceramics, glass, metals	RT/~90	1400	362
	Fiborclad Grade	1 part - water-based ceramic: Fire-resistant insulation (ceramic- & glass-fibre) materials.	RT/~85	850	362
	Heavy Grade	1 part - water-based ceramic: High temperature. Ceramics, glass, metals	RT/~85	1200	362
	Light Grade	1 part - water-based ceramic: High temperature. Ceramics, glass, metals	RT/~85	950	363
	LQ/S6	2 part - ceramic: High temperature. Ceramics, glass, metals	RT/~90	1000	363
	Medium Grade	1 part - water-based ceramic: High temperature. Ceramics, glass, metals	RT/~85	1000	363
	QS/B4	2 part - ceramic: High temperature. Ceramics, glass, metals	RT/~90	1000	363
	SBC & SBC/FRB	1 part - water-based ceramic: Fire-resistant insulation (ceramic- & glass-fibre) materials.	RT/~85	850	364
	TC Caulking Grade	1 part - water-based ceramic: High temperature. Ceramics, glass, metals	RT/~85	1250	364
	Zircon RM/3	1 part - water-based Zirconia: High temperature. Ceramics, glass, metals	RT/~85	1600	364
Geistlich	Miracol 8F1	1 part - PVA dispersion: Wood working. Wood products	RT	-20/+120	364
Henry	411 Grip Rite	1 part - PVA: Wood working. Wood products	RT	-	365
Loctite	Glass Bonding Adhesive 350	1 part - base unknown: High optical clarity glass bonding. Glass & metals	UV	120	365
Lugato	Montier mit Mir	1 part - vinyl-acryl-ester disp.: Building materials (wood-panels, isolation panels, metals)	RT	-	365
Panacol	Vitralit 3033	1 part - base unknown: Plastics	UV	-40/+120	365
Elosol	Vitralit 4010	1 part - base unknown: Glass laminating. Plastics (inc. polycarbonate, epoxy).	UV	-55/+125	366
	Vitralit 4011	1 part - base unknown: Glass laminating. Plastics (inc. polycarbonate, epoxy).	UV	-55/+125	366
	Vitralit 5634	1 part - base unknown: Glass, wood, plastics (inc. PMMA, ABS, PC)	UV	-40/+100	366
	Vitralit 5634 LV	1 part - base unknown: Glass, wood, plastics (inc. PMMA, ABS, PC)	UV	-40/+100	366
	Vitralit 6102	1 part - base unknown: Glass, metals, plastics	UV	-40/+150	367
	Vitralit 6127	1 part - base unknown: Glass, metals, plastics. Optical components	UV	-40/+125	367
	Vitralit 6128	1 part - base unknown: Glass, metals, plastics	UV, Heat, Cat.	-40/+125	367
	Vitralit 6128 VT	1 part - base unknown: Glass, metals, plastics	UV, Heat, Cat.	-40/+125	367
	Vitralit 6129	1 part - base unknown: Glass, aluminium, plastics	UV, Heat, Cat.	-50/+180	368
	Vitralit 6140	1 part - base unknown: Glass, aluminium, plastics	UV	-40/+130	368
	Vitralit 6143	1 part - base unknown: Glass, aluminium, plastics	UV	-40/+130	368
	Vitralit 6164	1 part - base unknown: Glass bonding. Repair	UV	-20/+120	368
	Vitralit 7256	1 part - base unknown: Glass bonding	UV	-40/+125	369
	Vitralit 7512	1 part - base unknown: Glass bonding	UV	-40/+125	369
	Vitralit 7515	1 part - base unknown: Glass bonding	UV	-40/+120	369
	Vitralit 8101	1 part - base unknown: Glass, metal & plastic medical equipment	UV	-50/+150	369
	Vitralit 8103	1 part - base unknown: Glass, metal & plastic medical equipment	UV	-50/+150	370
	Vitralit 8106	1 part - base unknown: Glass, metal & plastic medical equipment	UV	-50/+150	370

Continued...

Company	Adhesive	Description: Typical Uses	Cure (°C)	Service (°C)	Pg. No.
Quilosa	Aklesil L-32	1 part - water-based vinyl copolymers: Wood panel fixing	RT	-	370
	Aklesil L-51	1 part - water-based vinyl copolymers: Laminating wood products & PVC	RT	+70	370
	Sintex POL K-52	1 part - modified polyoxyfines: General industry adhesive-sealant. Vehicles, building, marine.	RT	-40/+90	371
	Unifix M-02	1 part - modified vinyl polyacetate water disp.: Wood products. Furniture	RT	-	371
	Unifix M-09	1 part - water-dispersion modified vinyl polyacetate: Laminating melamine products	RT	-	371
	Unifix M-20	1 part - water-dispersion modified vinyl polyacetate: Wood products. Furniture	RT	-	371
	Unifix M-46	1 part - water-dispersion modified vinyl polyacetate: Wood products. Furniture	RT	-	372
	Unifix M-49	1 part - water-dispersion mod. vinyl polyacetate: Wood products. Furniture, construction.	RT	-	372
	Unifix M-54	1 part - water-dispersion mod. vinyl polyacetate: Wood products. Furniture, construction.	RT/<110	-	372
	Unifix M-75	1 part - water-dispersion mod. vinyl polyacetate: Wood products. Laminating. Furniture.	RT/<100	-	372
Rinaldi	Silovil Forte	Water dispersion, synthetic resin: Wood flooring	RT	-	373
	Silovil L34	2 part - base unknown: Wood flooring to ceramic or concrete	RT	-	373
	Silovil U 68L	1 part - water-based synthetic resin: Wood flooring (inc. heated sub-floors)	RT	-	373
	Silovil U 68M	1 part - water-based synthetic resin: Wood & polymer-based flooring	RT	-	373
Saba	Sabaplast 7200 UV	1 part - silane-mod. polymer: Bond/sealer for automotive industry: Metal & plastic panels	RT	-40/+100	374
SEA	ELO	1 part - vinyl: Construction industry. Cement products, plaster, wood, laminates.	RT	-	374
	Parkecol	1 part - alcohol solution - base unknown: Construction industry. Wood flooring	RT	-	374
	Siel Parquet	1 part - PVA emulsion: Construction industry. Wood flooring	RT	-	374
Teroson	Terokal 8549	1 part - PVC Plastisol: Automotive: Body panels (hem flange seams)	170-220	-40/+80	375
(Henkel)	Terokal 8552	1 part - PVC Plastisol: Automotive: Body panels (hem flange seams)	170-220	-40/+80	375
ThreeBond	3056B	base unknown: Magnetic rubber bonding	UV+Moisture	-	375
	3056C	base unknown: Metal & plastic	UV+Moisture	-	375

See Key to Comparison Tables - Pages 63 to 65

Cytec

Cytec	Metlbond 2555		(May 95)
One-part, modified cyanate ester. Film - areal wt. 0.3 kg/m²			
Manufacturers Comments: Low dielectric constant and loss tangent for radome applications. For bonding aerospace materials. Also surfacing. Can be co-cured with most 177°C composites. Supported and unsupported versions available.			
Typical Use: Aerospace: Metals and composites.			
Cure: ~177°C (co-cure).			
Test Methods: ASTM			
Physical		**Mechanical**	
Colour	-	τ (RT)	18.6 MPa
Viscosity	-	τ (232°C)	20.7 MPa
S.G.	-	τ (wet)	-
Shelf life	-	Peel ( )	-
Pot/Work life	-	σ_t	-
Tg	232°C	σ_c	-
$T_{Service}$	232°C	*E*	-
CTE (°C)	-	Elongation	-
Flash Point	-	H (Shore)	-
Notes: τ data (ASTM D-1002): metal-to-metal lap.			

Cytec	Metlbond 6555		(May 95)
One-part, cyanate ester. Syntactic Film			
Manufacturers Comments: Foaming, low dielectric constant and loss tangent for radome applications.			
Typical Use: Aerospace: Composites.			
Cure: 177°C/4 hrs + 225°C/6 hrs (post-cure).			
Test Methods: Not stated			
Physical		**Mechanical**	
Colour	-	τ (RT)	-
Viscosity	-	τ (°C)	-
S.G.	-	τ (wet)	-
Shelf life	-	Peel ( )	-
Pot/Work life	-	σ_t	-
Tg	-	σ_c	-
$T_{Service}$	177°C	*E*	-
CTE (°C)	-	Elongation	-
Flash Point	-	H (Shore)	-
Notes: Foam density: 0.48-0.56 g/cm³.			

EMFI

EMFI	Bois 860		(Nov 95)
One part, vinyl polymer dispersion			
Manufacturers Comments: 60% solids. Multi-purpose wood adhesive. Easy to use. Non-staining. Heat and moisture (not water) resistance.			
Typical Use: Wood and wood products.			
Cure: RT/~1 hr; 60°C/5 mins.			
Test Methods: Not stated			
Physical		**Mechanical**	
Colour	White	τ (RT)	-
Viscosity	13 Pa.s	τ (°C)	-
S.G.	1.15	τ (wet)	-
Shelf life	9 months/RT	Peel ( )	-
Pot/Work life	10-15 mins (open)	σ_t	-
Tg	-	σ_c	-
$T_{Service}$	-	*E*	-
CTE (°C)	-	Elongation	-
Flash Point	-	H (Shore)	-
Notes: Frost sensitive.			

EMFI	Bois 863 B3		(Nov 95)
One part, vinyl polymer dispersion			
Manufacturers Comments: 49% solids. Multi-purpose wood adhesive for exterior and humid environments. Non-staining. Heat and moisture (inc. water) resistance.			
Typical Use: Wood and wood products.			
Cure: RT/15-20 mins; 80°C/2-3 mins.			
Test Methods: Not stated			
Physical		**Mechanical**	
Colour	White	τ (RT)	-
Viscosity	~9 Pa.s	τ (°C)	-
S.G.	1.15	τ (wet)	-
Shelf life	9 months/RT	Peel ( )	-
Pot/Work life	10 mins (open)	σ_t	-
Tg	-	σ_c	-
$T_{Service}$	-	*E*	-
CTE (°C)	-	Elongation	-
Flash Point	-	H (Shore)	-
Notes: Frost sensitive.			

Forbo Helmitin

EMFI	Bois Rapide 885		(Nov 95)
One part, vinyl polymer dispersion			
Manufacturers Comments: Fast cure, multi-purpose wood adhesive for interior use.			
Typical Use: Wood and wood products. Furniture, inc. edge banding.			
Cure: RT/5-15 mins (clamping).			
Test Methods: Not stated			
Physical		**Mechanical**	
Colour	White	τ (RT)	-
Viscosity	9.5 Pa.s	τ (°C)	-
S.G.	1.13	τ (wet)	-
Shelf life	6-8 months/RT	Peel ( )	-
Pot/Work life	10-15 mins (open)	σ_t	-
Tg	-	σ_c	-
$T_{Service}$	-	E	-
CTE (°C)	-	Elongation	-
Flash Point	-	H (Shore)	-
Notes: Frost sensitive.			

Forbo Helmitin	Helmitin 34026		Oct 92
One-part, EVA dispersion			
Manufacturers Comments: Bonding plastic sheets, foams, metallic foils to porous materials (wood, leather, textiles, etc). Can be used for wet bonding or heat-sealing.			
Typical Use: PVC, polystyrene, ABS. Aluminium foils.			
Cure: RT/ 2-3 hrs (cure); >80°C (heat-sealing).			
Test Methods: Not stated			
Physical		**Mechanical**	
Colour	White	τ (RT)	-
Viscosity	~16 Pa.s	τ (°C)	-
S.G.	~1.05	τ (wet)	-
Shelf life	1 yr/RT	Peel ( )	-
Pot/Work life	2-3 mins (wet)	σ_t	-
Tg	-	σ_c	-
$T_{Service}$	-	E	-
CTE (°C)	-	Elongation	-
Flash Point	-	H (Shore)	-
Notes:			

EMFI	Colle PVC/PVC		(Nov 95)
One part, synthetic resin base			
Manufacturers Comments: For assembly of rigid-PVC parts.			
Typical Use: PVC glazing frames, water and waste pipes.			
Cure: RT/24 hrs (full cure).			
Test Methods: Not stated			
Physical		**Mechanical**	
Colour	Translucent	τ (RT)	-
Viscosity	5 Pa.s	τ (°C)	-
S.G.	~0.93	τ (wet)	-
Shelf life	6 months/RT	Peel ( )	-
Pot/Work life	-	σ_t	-
Tg	-	σ_c	-
$T_{Service}$	-	E	-
CTE (°C)	-	Elongation	-
Flash Point	-	H (Shore)	-
Notes: Flammable.			

Forbo Helmitin	PF/S		Sept 92
One-part, EVA dispersion			
Manufacturers Comments: Bonding PVC sheets to wood product substrates and other porous materials.			
Typical Use: Wood products and textiles.			
Cure: RT/ 3-4 hrs (handling); ~24 hrs (cure).			
Test Methods: Not stated			
Physical		**Mechanical**	
Colour	White	τ (RT)	-
Viscosity	~21 Pa.s	τ (°C)	-
S.G.	~1.15	τ (wet)	-
Shelf life	1 yr/RT	Peel ( )	-
Pot/Work life	10-12 mins (wet)	σ_t	-
Tg	-	σ_c	-
$T_{Service}$	-	E	-
CTE (°C)	-	Elongation	-
Flash Point	-	H (Shore)	-
Notes:			

Forbo Helmitin — **Rapid 33207** — Sept 92

One-part, PVAC water-based dispersion

Manufacturers Comments: Woodworking industry. High moisture resistance. For window and door manufacture, dividing walls, stairs. Also general assembly - joints, edging, doweling in all wood products. Bonding decorative plastics to wood product substrates.
Meets: DIN 68602, Group B3. Note: 5% addition meets B4 Group.

Typical Use: Woodworking.

Cure: RT. Press Times: 20°C/15 mins; 50°C/5 mins; 80°C/2 mins. Pressure: 0.2 - 0.5 MPa.

Test Methods: Not stated

Physical		Mechanical	
Colour	White	τ (RT)	-
Viscosity	~15 Pa.s	τ (°C)	-
S.G.	~1.08	τ (wet)	-
Shelf life	1 yr/RT	Peel ( )	-
Pot/Work life	4-6 mins (wet)*	σ_t	-
Tg	-	σ_c	-
$T_{Service}$	-	E	-
CTE (°C)	-	Elongation	-
Flash Point	-	H (Shore)	-

Notes: * Pot-life: 4-5 hrs (with addition).
Cure conditions: 8-12% moisture in wood; <60-75%RH air.
Higher moisture contents require longer pressing times.

Forbo Helmitin — **Rapid 39533** — Aug 93

One-part, PVAC dispersion

Manufacturers Comments: General purpose for bonding paper, cardboard, textiles to wood products. Transparent, flexible film.

Typical Use: Packaging. Wood products and textiles.

Cure: RT/3 hrs (cure).

Test Methods: Not stated

Physical		Mechanical	
Colour	White	τ (RT)	-
Viscosity	~2.5 Pa.s	τ (°C)	-
S.G.	~1.09	τ (wet)	-
Shelf life	6 months/RT	Peel ( )	-
Pot/Work life	5 mins (wet)	σ_t	-
Tg	-	σ_c	-
$T_{Service}$	-	E	-
CTE (°C)	-	Elongation	-
Flash Point	-	H (Shore)	-

Notes:

Forbo Helmitin — **Rapid 33511** — Sept 92

One-part, PVAC water-based dispersion

Manufacturers Comments: Woodworking industry. High shear strength, non-staining, shock-resistant. For assembly and joints, edging, doweling in all wood products. Bonding decorative plastics to wood product substrates.

Typical Use: Woodworking.

Cure: RT/4 hrs (cure). Press Times: 20°C/15 mins; 50°C/8 mins; 90°C/4 mins. Pressure: 0.2-0.5 MPa.

Test Methods: Not stated

Physical		Mechanical	
Colour	White	τ (RT)	-
Viscosity	~9 Pa.s	τ (°C)	-
S.G.	~1.18	τ (wet)	-
Shelf life	1 yr/RT	Peel ( )	-
Pot/Work life	12-15 mins (wet)	σ_t	-
Tg	-	σ_c	-
$T_{Service}$	-	E	-
CTE (°C)	-	Elongation	-
Flash Point	-	H (Shore)	-

Notes: Cure conditions: 8-12% moisture in wood; <60-75%RH air.
Higher moisture contents require longer pressing times.

Forbo Helmitin — **Rapid A** — Sept 92

One-part, PVAC water-based dispersion

Manufacturers Comments: Woodworking industry. General purpose for bonding all types of wood products. High shear strength, shock-resistant. Dries transparent, without discolouration.

Typical Use: Woodworking.

Cure: RT/24 hrs (cure). Press Times: 20°C/20 mins; 50°C/8 mins; 90°C/5 mins. Pressure: 0.2-0.5 MPa.

Test Methods: Not stated

Physical		Mechanical	
Colour	White	τ (RT)	-
Viscosity	~8 Pa.s	τ (°C)	-
S.G.	~1.05	τ (wet)	-
Shelf life	1 yr/RT	Peel ( )	-
Pot/Work life	15-20 mins (wet)	σ_t	-
Tg	-	σ_c	-
$T_{Service}$	-	E	-
CTE (°C)	-	Elongation	-
Flash Point	-	H (Shore)	-

Notes: Cure conditions: 8-12% moisture in wood; <65-70%RH air.
Higher moisture contents require longer pressing times.

Forbo Helmitin	**Rapid D**		Sept 92
One-part, PVAC water-based dispersion			
Manufacturers Comments: Non-filled, ready-to-use, wood bonding. For automatic doweling with low pressure. Tough, moisture resistant joint.			
Typical Use: Doweling wood.			
Cure: RT(18-22°C)/15-20 mins/0.2-0.5MPa.			
Test Methods: Not stated			
Physical		**Mechanical**	
Colour	White	τ (RT)	-
Viscosity	0.21 Pa.s	τ (°C)	-
S.G.	~1.06	τ (wet)	-
Shelf life	1 yr/RT	Peel ( )	-
Pot/Work life	8-10 mins (wet)	σ_t	-
Tg	-	σ_c	-
$T_{Service}$	-	E	-
CTE (°C)	-	Elongation	-
Flash Point	-	H (Shore)	-
Notes: Cure conditions: 8-12% moisture in wood; <60-75%RH air.			

Forbo Helmitin	**Rapid S**		Sept 92
One-part, PVAC water-based dispersion			
Manufacturers Comments: Woodworking industry. High shear strength, non-staining, shock-resistant. For assembly and joints, edging, doweling in all wood products. Bonding decorative plastics to wood product substrates.			
Typical Use: Woodworking.			
Cure: RT/4 hrs (cure). Press Times: 20°C/5-8 mins; 50°C/2-3 mins; 90°C/1-1.5 mins. Pressure: 0.2-0.5 MPa.			
Test Methods: Not stated			
Physical		**Mechanical**	
Colour	White	τ (RT)	-
Viscosity	~17.5 Pa.s	τ (°C)	-
S.G.	~1.13	τ (wet)	-
Shelf life	1 yr/RT	Peel ( )	-
Pot/Work life	5-6 mins (wet)	σ_t	-
Tg	-	σ_c	-
$T_{Service}$	-	E	-
CTE (°C)	-	Elongation	-
Flash Point	-	H (Shore)	-
Notes: Cure conditions: 8-12% moisture in wood; <60-75%RH air. Higher moisture contents require longer pressing times.			

Forbo Helmitin	**Rapid KA**		Sept 92
One-part, PVAC water-based dispersion			
Manufacturers Comments: Woodworking industry. Also for cold-glue activation for bonding profiles (soft forming) and edge-banding.			
Typical Use: Woodworking.			
Cure: RT/4 hrs (cure). Press Times: 20°C/4 mins; 50°C/2 mins; 90°C/1 mins.			
Test Methods: Not stated			
Physical		**Mechanical**	
Colour	White	τ (RT)	-
Viscosity	~12 Pa.s	τ (°C	-
S.G.	~1.10	τ (wet)	-
Shelf life	1 yr/RT	Peel ( )	-
Pot/Work life	5-10 mins (wet)	σ_t	-
Tg	-	σ_c	-
$T_{Service}$	-	E	-
CTE (°C)	-	Elongation	-
Flash Point	-	H (Shore)	-
Notes: Cure conditions: 8-12% moisture in wood; <60-75%RH air. Higher moisture contents require longer pressing times.			

Fortafix

Fortafix	**AL/CS**		Sept 94
Two-part, ceramic			
Manufacturers Comments: Inorganic adhesive cement for high-temperature applications. High alumina content. High temperature resistance. Dielectric properties.			
Typical Use: Ceramics, metals, glass and most organic materials. Repair of castings and moulds.			
Cure: RT/ 2-2.5 hrs; 80-95°C heat-accelerated cure.			
Test Methods: Not stated			
Physical		**Mechanical**	
Colour	White	τ (RT)	-
Viscosity	-	τ (°C)	-
S.G.	-	τ (wet)	-
Shelf life	-	Peel ( )	-
Pot/Work life	-	σ_t	-
Tg	-	σ_c	-
$T_{Service}$	+1600°C max.	E	-
CTE (°C)	-	Elongation	-
Flash Point	-	H (Shore)	-
Notes:			

Fortafix	Chromix	Sept 94
Two-part, ceramic		

Manufacturers Comments: Inorganic adhesive cement for high-temperature applications. Similar to 1-part Chromix grade - but faster curing. Good bond strength to metals. High tensile and compressive strength. Provides gas tight seals and heat resistant filler.

Typical Use: Ceramics, metals, glass and most organic materials. Repair of castings and moulds.

Cure: RT/ 2-2.5 hrs; 80-95°C heat-accelerated cure.

Test Methods: Not stated

Physical		Mechanical	
Colour	Brown/grey	τ (RT)	-
Viscosity	-	τ (°C)	-
S.G.	-	τ (wet)	-
Shelf life	-	Peel ( )	-
Pot/Work life	-	σ_f	-
Tg	-	σ_c	-
$T_{Service}$	+1400°C max.	E	-
CTE (°C)	-	Elongation	-
Flash Point	-	H (Shore)	-

Notes:

Fortafix	Fiborclad Grade	Sept 94
One-part, water-based ceramic		

Manufacturers Comments: Water based inorganic adhesive cement for high-temperature applications. Free-flowing. Fire-resistant applications. Meets: UK BS476:part 4 (1970).

Typical Use: Ceramic- and glass-fibre products. Fixing insulation materials to metals.

Cure: RT/ 24-36 hrs; 80-90°C heat-accelerated cure.

Test Methods: Not stated

Physical		Mechanical	
Colour	-	τ (RT)	-
Viscosity	-	τ (°C)	-
S.G.	-	τ (wet)	-
Shelf life	-	Peel ( )	-
Pot/Work life	-	σ_f	-
Tg	-	σ_c	-
$T_{Service}$	+850°C	E	-
CTE (°C)	-	Elongation	-
Flash Point	-	H (Shore)	-

Notes:

Fortafix	Chromix Ready-to-Use	Sept 94
One-part, water-based ceramic		

Manufacturers Comments: Water based inorganic adhesive cement for high-temperature applications. Low permeability.

Typical Use: Ceramics, metals, glass and most organic materials.

Cure: RT/ 24-36 hrs; 80-90°C heat-accelerated cure.

Test Methods: Not stated

Physical		Mechanical	
Colour	Grey or Brown	τ (RT)	-
Viscosity	-	τ (°C)	-
S.G.	-	τ (wet)	-
Shelf life	-	Peel ( )	-
Pot/Work life	-	σ_f	-
Tg	-	σ_c	-
$T_{Service}$	+1400°C	E	-
CTE (°C)	-	Elongation	-
Flash Point	-	H (Shore)	-

Notes: Special grade Chromix - higher viscosity version

Fortafix	Heavy Grade	Sept 94
One-part, water-based ceramic		

Manufacturers Comments: Water based inorganic adhesive cement for high-temperature applications. For bonding porous materials. Sealing and gap-filling, low porosity.

Typical Use: Ceramics, metals, glass and most organic materials.

Cure: RT/ 24-36 hrs; 80-90°C heat-accelerated cure.

Test Methods: Not stated

Physical		Mechanical	
Colour	White	τ (RT)	-
Viscosity	High, paste	τ (°C)	-
S.G.	-	τ (wet)	-
Shelf life	-	Peel ( )	-
Pot/Work life	-	σ_f	-
Tg	-	σ_c	-
$T_{Service}$	+1200°C	E	-
CTE (°C)	-	Elongation	-
Flash Point	-	H (Shore)	-

Notes: Continuous service temperature, without softening or degrading.

Fortafix	**Light Grade**		Sept 94
One-part, water-based ceramic			
Manufacturers Comments: Water based inorganic adhesive cement for high-temperature applications. For bonding porous materials. Pourable liquid.			
Typical Use: Ceramics, metals, glass and most organic materials.			
Cure: RT/ 24-36 hrs; 80-90°C heat-accelerated cure.			
Test Methods: Not stated			
Physical		**Mechanical**	
Colour	-	τ (RT)	-
Viscosity	-	τ (°C)	-
S.G.	-	τ (wet)	-
Shelf life	-	Peel ( )	-
Pot/Work life	-	σ_t	-
Tg	-	σ_c	-
$T_{Service}$	+950°C	*E*	-
CTE (°C)	-	Elongation	-
Flash Point	-	H (Shore)	-
Notes:			

Fortafix	**Medium Grade**		Sept 94
One-part, water-based ceramic			
Manufacturers Comments: Water based inorganic adhesive cement for high-temperature applications. For bonding porous materials. Dense paste.			
Typical Use: Ceramics, metals, glass and most organic materials.			
Cure: RT/ 24-36 hrs; 80-90°C heat-accelerated cure.			
Test Methods: Not stated			
Physical		**Mechanical**	
Colour	White	τ (RT)	-
Viscosity	-	τ (°C)	-
S.G.	-	τ (wet)	-
Shelf life	-	Peel ( )	-
Pot/Work life	-	σ_t	-
Tg	-	σ_c	-
$T_{Service}$	+1000°C	*E*	-
CTE (°C)	-	Elongation	-
Flash Point	-	H (Shore)	-
Notes: Continuous service temperature, without softening or degrading.			

Fortafix	**LQ/S6**		Sept 94
Two-part, ceramic			
Manufacturers Comments: Inorganic adhesive cement for high-temperature applications. Resistant to water, oils, acids (not hydrofluoric). Good dielectric strength, low thermal expansion. High tensile and compressive strength.			
Typical Use: Ceramics, metals, glass and most organic materials.			
Cure: RT/ 2-2.5 hrs; 80-95°C heat-accelerated cure.			
Test Methods: Not stated			
Physical		**Mechanical**	
Colour	Cream	τ (RT)	-
Viscosity	-	τ (°C)	-
S.G.	-	τ (wet)	-
Shelf life	-	Peel ( )	-
Pot/Work life	-	σ_t	-
Tg	-	σ_c	-
$T_{Service}$	+1000°C max.	*E*	-
CTE (°C)	-	Elongation	-
Flash Point	-	H (Shore)	-
Notes:			

Fortafix	**QS/B4**		Sept 94
Two-part, ceramic			
Manufacturers Comments: Inorganic adhesive cement for high-temperature applications. Similar material to 1-part Light, etc. grades. Faster curing.			
Typical Use: Ceramics, metals, glass and most organic materials.			
Cure: RT/ 2-2.5 hrs; 80-95°C heat-accelerated cure.			
Test Methods: Not stated			
Physical		**Mechanical**	
Colour	White	τ (RT)	-
Viscosity	-	τ (°C)	-
S.G.	-	τ (wet)	-
Shelf life	-	Peel ( )	-
Pot/Work life	-	σ_t	-
Tg	-	σ_c	-
$T_{Service}$	+1000°C max.	*E*	-
CTE (°C)	-	Elongation	-
Flash Point	-	H (Shore)	-
Notes:			

Fortafix	**SBC & SBC/FRB**		Sept 94
One-part, water-based ceramic			
Manufacturers Comments: Water based inorganic adhesive cement for high-temperature applications. Laminating insulation to metals. Higher tack. Fire resistance.			
Typical Use: Ceramic- and glass-fibre products. Fixing insulation materials to metals.			
Cure: RT/ 24-36 hrs; 80-90°C heat-accelerated cure.			
Test Methods: Not stated			
Physical		**Mechanical**	
Colour	-	τ (RT)	-
Viscosity	-	τ (°C)	-
S.G.	-	τ (wet)	-
Shelf life	-	Peel ( )	-
Pot/Work life	-	σ_t	-
Tg	-	σ_c	-
$T_{Service}$	+850°C	*E*	-
CTE (°C)	-	Elongation	-
Flash Point	-	H (Shore)	-
Notes:			

Fortafix	**Zircon RM/3**		Sept 94
One-part, water-based ceramic			
Manufacturers Comments: Water based inorganic adhesive cement for high-temperature applications. Zirconia-based.			
Typical Use: Ceramics, metals, glass and most organic materials.			
Cure: RT/ 24-36 hrs; 80-90°C heat-accelerated cure.			
Test Methods: Not stated			
Physical		**Mechanical**	
Colour	-	τ (RT)	-
Viscosity	High, paste	τ (°C)	-
S.G.	-	τ (wet)	-
Shelf life	-	Peel ( )	-
Pot/Work life	-	σ_t	-
Tg	-	σ_c	-
$T_{Service}$	+1600°C	*E*	-
CTE (°C)	-	Elongation	-
Flash Point	-	H (Shore)	-
Notes:			

Fortafix	**TC Caulking Grade**		Sept 94
One-part, water-based ceramic			
Manufacturers Comments: Water based inorganic adhesive cement for high-temperature applications. Thixotropic. Resistant to organic solvents, acids, etc.			
Typical Use: Ceramics, metals, glass and most organic materials.			
Cure: RT/ 24-36 hrs; 80-90°C heat-accelerated cure.			
Test Methods: Not stated			
Physical		**Mechanical**	
Colour	-	τ (RT)	-
Viscosity	Thixotropic	τ (°C)	-
S.G.	-	τ (wet)	-
Shelf life	-	Peel ( )	-
Pot/Work life	-	σ_t	-
Tg	-	σ_c	-
$T_{Service}$	+1250°C	*E*	-
CTE (°C)	-	Elongation	-
Flash Point	-	H (Shore)	-
Notes:			

Ed. Geistlich Söhne AG

Geistlich	**Miracol 8F1**		Jul 92
One-part, PVA dispersion			
Manufacturers Comments: ~51% solids, solvent-free. General purpose wood-working assembly. Moisture-resistance. HF use. Can be mixed with hardener. Meets: FFEM/SZFF & VSSM/FFF recommendations; EN 204-D3 (moisture); EN204-D4 (with hardener addition).			
Typical Use: Wood products - windows, facades.			
Cure: RT/ 1-10 hrs (handling); 7 days (full cure); Press: 20°C/10-60 mins (solid wood); 30-40°C/6-8 mins (laminates); 80°C/~1.5-2 mins (edge-finishing).			
Test Methods: DIN			
Physical		**Mechanical**	
Colour	White	τ (RT)	-
Viscosity	~4.4 Pa.s	τ (°C)	-
S.G.	~1.07	τ (wet)	-
Shelf life	9 months/RT	Peel ( )	-
Pot/Work life	<8 mins (open)	σ_t	*
Tg	-	σ_c	-
$T_{Service}$	-20 to +120°C max.	*E*	-
CTE (°C)	-	Elongation	-
Flash Point	-	H (Shore)	-
Notes: All assemblies require pressing. Wood humidity: 12% max. * DIN 204-D1.			

W. W. Henry

Henry	411 Grip Rite	Mar 92
One-part, PVA		
Manufacturers Comments: Fast setting carpenters wood adhesive for interior use. 45 % solids. Strong dependable joints. Heat resistance and sandable. For a variety of porous and semi-porous materials.		
Typical Use: Wood products (hard and soft wood, chipboard, laminates, etc). Leather. Paper products.		
Cure: RT (>10°C)/15 mins (closed assembly); several hours (full cure). Pressure: ~1 MPa (pine).		
Test Methods: Not stated		

Physical		Mechanical	
Colour	Yellow→Translucent	τ (RT)	*
Viscosity	5.4 Pa.s	τ (°C)	-
S.G.	1.1	τ (wet)	-
Shelf life	1 year/RT	Peel ( )	-
Pot/Work life	5 mins (open)	σ_t	-
Tg	-	σ_c	-
$T_{Service}$	-	E	-
CTE (°C)	-	Elongation	-
Flash Point	-	H (Shore)	-

Notes: pH: 4.7. Bond strength (hard maple):
21°C/overnight @ 23.4 MPa: 45% wood failure.
65.5°C/overnight @ 10.4 MPa: 15% wood failure.
Some oily or exotic woods may require special surface preparation.

Lugato

Lugato	Montier mit Mir	(Mar 95)
One-part, vinyl-acryl-ester copolymer dispersion		
Manufacturers Comments: Solvent-free, high viscosity, non-sag adhesive for assembly of various building materials. Cartridge pack.		
Typical Use: Wood panelling, insulation panels, metals.		
Cure: RT (10 -30°C) /24 hrs (full cure).		
Test Methods: DIN		

Physical		Mechanical	
Colour	Beige	τ (RT, wood)	2.8 MPa
Viscosity	200 Pa.s	τ (°C)	-
S.G.	~1.4	τ (wet)	-
Shelf life	1 year (frost-free)	Peel ( )	-
Pot/Work life	15 mins (assembly)	σ_t	-
Tg	-	σ_c	-
$T_{Service}$	-	E	-
CTE (°C)	-	Elongation	-
Flash Point	18°C	H (Shore)	-

Notes: Gap filling: 0.5 mm max. Flash point (DIN 55680).
Ignition Temperature: >400°C.

Loctite

Loctite	Glass Bonding Adhesive 350	(Feb 95)
One-part, UV cure.		
Manufacturers Comments: Glass bonds requiring good optical clarity and high resistance to humidity (inc. wash-up and boil cycles), solvents, ageing, thermal shock. Silane primer not required.		
Typical Use: Glass, glass-to-metal.		
Cure: UV, 10mW/cm²: ~40 secs. (0.05mm bondline); ~100 secs (0.5mm bondline). UV, 50mW/ cm²: ~30 secs (0.5mm bondline).		
Test Methods: Not stated		

Physical		Mechanical	
Colour	Transparent	τ (RT)	-
Viscosity	-	τ (°C)	-
S.G.	-	τ (wet)	-
Shelf life	-	Peel ( )	-
Pot/Work life	-	σ_t	-
Tg	-	σ_c	-
$T_{Service}$	120°C max.	E	-
CTE (°C)	-	Elongation	-
Flash Point	-	H (Shore)	-

Notes: Gap filling: 0.5 mm max. Refractive index: 1.51.
Typical light transmittance: 98%

Panacol Elosol

Panacol Elosol	Vitralit 3033	Jun 94
One-part, UV cure.		
Manufacturers Comments: Rapid cure. Good adhesion to Polyester, Polyurethane and other plastics. Solvent-free, 100% solids.		
Typical Use: Plastics.		
Cure: UV-A, 60 mW.cm²: 45 sec. Final strength: 24 hrs.		
Test Methods: DIN (ASTM)		

Physical		Mechanical	
Colour	Transparent	τ (RT)	3.9 MPa
Viscosity	0.7 Pa.s	τ (°C)	-
S.G.	1.22	τ (wet)	-
Shelf life	6 months/RT	Peel ( )	-
Pot/Work life	-	σ_t	-
Tg	-	σ_c	-
$T_{Service}$	-40°C to +120°C	E	347 MPa
CTE (°C)	$72 \times 10^{-6}\ K^{-1}$	Elongation	-
Flash Point	105 °C	H (Shore D)	82

Notes: Shrinkage: 10.4%. Refractive index: 1.5. E (DIN 53457). Breakdown potential (DIN 53481): 26.5 $kV.mm^{-1}$. Moisture absorption (ASTM 570): 1.4%. Thermal conductivity (DIN 52612): 0.31 $W.m^{-1}K^{-1}$. Dielectric constant (DIN 53483): 4.9. Shear: ASTM D1002. CTE: (ASTM D695).

Panacol Elosol	Vitralit 4010		Jun 94
One-part, UV cure			
Manufacturers Comments: Rapid cure, flexible. Good adhesion to Polycarbonate and epoxy. Solvent-free, 100% solids.			
Typical Use: Glass laminating, etc.			
Cure: UV-A, 60 mW.cm^2: 15 sec. Final strength: 12 hrs.			
Test Methods: DIN (ASTM)			
Physical		**Mechanical**	
Colour	Transparent	τ (RT)	12.3 MPa
Viscosity	1 Pa.s	τ (°C)	-
S.G.	1.15	τ (wet)	-
Shelf life	6 months/RT	Peel ( )	-
Pot/Work life	-	σ_t	-
Tg	-	σ_c	-
$T_{Service}$	-55°C to +125°C	E	182 MPa
CTE (°C)	34 x 10^{-6} K^{-1}	Elongation	-
Flash Point	98 °C	H (Shore D)	60
Notes: Shrinkage: 4.2%. Refractive index: 1.495. E (DIN 53457). Breakdown potential (DIN 53481): 13.5 kV.mm^{-1}. Moisture absorption (ASTM 570): 3.5%. Thermal conductivity (DIN 52612): 0.18 W.m^{-1}K^{-1}. Dielectric constant (DIN 53483): 6.1. Shear: ASTM D1002. CTE: (ASTM D695).			

Panacol Elosol	Vitralit 5634		Jun 94
One-part, UV cure			
Manufacturers Comments: Rapid cure. Solvent-free, 100% solids.			
Typical Use: Glass, wood, plastics (PMMA, ABS, PC), etc.			
Cure: UV-A, 60 mW.cm^2: 10 sec. Final strength: 12 hrs.			
Test Methods: DIN (ASTM)			
Physical		**Mechanical**	
Colour	Transparent	τ (RT)	13.5 MPa
Viscosity	8 Pa.s	τ (°C)	-
S.G.	1.15	τ (wet)	-
Shelf life	6 months/RT	Peel ( )	-
Pot/Work life	-	σ_t	-
Tg	-	σ_c	-
$T_{Service}$	-40°C to +100°C	E	535 MPa
CTE (°C)	58 x 10^{-6} K^{-1}	Elongation	-
Flash Point	95 °C	H (Shore D)	75
Notes: Shrinkage: 5.2%. Refractive index: 1.52. E (DIN 53457). Breakdown potential (DIN 53481): 12.3 kV.mm^{-1}. Moisture absorption (ASTM 570): 3.2%. Thermal conductivity (DIN 52612): 0.21 W.m^{-1}K^{-1}. Dielectric constant (DIN 53483): 4.8. Shear: ASTM D1002. CTE: ASTM D695.			

Panacol Elosol	Vitralit 4011		Jun 94
One-part, UV cure			
Manufacturers Comments: Rapid cure, flexible. Good adhesion to Polycarbonate and epoxy. Solvent-free, 100% solids.			
Typical Use: Glass laminating, etc.			
Cure: UV-A, 60 mW.cm^2: 15 sec. Final strength: 12 hrs.			
Test Methods: DIN (ASTM)			
Physical		**Mechanical**	
Colour	Transparent	τ (RT)	12.3 MPa
Viscosity	10 Pa.s	τ (°C)	-
S.G.	1.15	τ (wet)	-
Shelf life	6 months/RT	Peel ( )	-
Pot/Work life	-	σ_t	-
Tg	-	σ_c	-
$T_{Service}$	-55°C to +125°C	E	182 MPa
CTE (°C)	34 x 10^{-6} K^{-1}	Elongation	-
Flash Point	98 °C	H (Shore D)	46
Notes: Shrinkage: 3.8%. Refractive index: 1.495. E (DIN 53457). Breakdown potential (DIN 53481): 13.5 kV.mm^{-1}. Moisture absorption (ASTM 570): 3.5%. Thermal conductivity (DIN 52612): 0.19 W.m^{-1}K^{-1}. Dielectric constant (DIN 53483): 6.1. Shear: ASTM D1002. CTE: (ASTM D695).			

Panacol Elosol	Vitralit 5634 LV		Jun 94
One-part, UV cure			
Manufacturers Comments: Rapid cure. Lower viscosity variant. Solvent-free, 100% solids.			
Typical Use: Glass, wood, plastics (PMMA, ABS, PC), etc.			
Cure: UV-A, 60 mW.cm^2: 10 sec. Final strength: 12 hrs.			
Test Methods: DIN (ASTM)			
Physical		**Mechanical**	
Colour	Transparent	τ (RT)	13.5 MPa
Viscosity	2 Pa.s	τ (°C)	-
S.G.	1.15	τ (wet)	-
Shelf life	6 months/RT	Peel ( )	-
Pot/Work life	-	σ_t	-
Tg	-	σ_c	-
$T_{Service}$	-40°C to +100°C	E	475 MPa
CTE (°C)	54 x 10^{-6} K^{-1}	Elongation	-
Flash Point	95 °C	H (Shore D)	75
Notes: Shrinkage: 5.2%. Refractive index: 1.52. E (DIN 53457). Breakdown potential (DIN 53481): 12.3 kV.mm^{-1}. Moisture absorption (ASTM 570): 3.2%. Thermal conductivity (DIN 52612): 0.23 W.m^{-1}K^{-1}. Dielectric constant (DIN 53483): 4.8. Shear: ASTM D1002. CTE: ASTM D695.			

Panacol Elosol	**Vitralit 6102**		Jun 94
One-part, UV cure			
Manufacturers Comments: Rapid cure. Solvent-free, 100% solids.			
Typical Use: Glass, metals, plastics, etc.			
Cure: UV-A, 60 mW.cm²: 15 sec. 120°C/5 mins. Final strength: 12 hr.			
Test Methods: DIN (ASTM)			
Physical		**Mechanical**	
Colour	Translucent	τ (Al)	18.5 MPa
Viscosity	4.5 Pa.s	τ (Oily steel)	6.2 MPa
S.G.	1.15	τ (steel)	15.1 MPa
Shelf life	6 months/RT	Peel ( )	-
Pot/Work life	-	σ_t	-
Tg	-	σ_c	-
$T_{Service}$	-40°C to +150°C	*E*	726 MPa
CTE (°C)	76.3 x 10^{-6} K^{-1}	Elongation	-
Flash Point	95 °C	H (Shore D)	70

Notes: Shrinkage: 3.6%. Refractive index: 1.52. E (DIN 53457). Breakdown potential (DIN 53481): 15.6 kV.mm⁻¹. Moisture absorption (ASTM 570): 0.8%. Thermal conductivity (DIN 52612): 0.43 W.m⁻¹K⁻¹. Dielectric constant (DIN 53483): 5.6. Shear: ASTM D1002. CTE: ASTM D695.

Panacol Elosol	**Vitralit 6128**		Jun 94
One-part, UV, thermal or catalytic cure			
Manufacturers Comments: Rapid cure. Good adhesion to glass, metals and many thermoplastics. Solvent-free, 100% solids.			
Typical Use: Glass, metals, plastics, etc.			
Cure: UV-A, 60 mW.cm²: 15 sec. 120/90 mins. Catalyst 535: 25 mins. Final strength: 12 hr.			
Test Methods: DIN (ASTM)			
Physical		**Mechanical**	
Colour	Translucent	τ (RT)	26.3 MPa
Viscosity	0.8 Pa.s	τ (°C)	-
S.G.	1.12	τ (wet)	-
Shelf life	6 months/RT	Peel ( )	-
Pot/Work life	-	σ_t	-
Tg	-	σ_c	-
$T_{Service}$	-40°C to +150°C	*E*	420 MPa
CTE (°C)	62 x 10^{-6} K^{-1}	Elongation	-
Flash Point	95 °C	H (Shore D)	70

Notes: Shrinkage: 2.3%. Refractive index: 1.55. E (DIN 53457). Breakdown potential (DIN 53481): 18.7 kV.mm⁻¹. Moisture absorption (ASTM 570): 2.8%. Thermal conductivity (DIN 52612): 0.22 W.m⁻¹K⁻¹. Dielectric constant (DIN 53483): 6.2. Shear: ASTM D1002. CTE: ASTM D695.

Panacol Elosol	**Vitralit 6127**		Jun 94
One-part, UV cure			
Manufacturers Comments: Rapid cure. Good adhesion to glass, metals and many thermoplastics. Solvent-free, 100% solids.			
Typical Use: Glass, metals, plastics, optical components, etc.			
Cure: UV-A, 60 mW.cm²: 15 sec. Final strength: 12 hr.			
Test Methods: DIN (ASTM)			
Physical		**Mechanical**	
Colour	Transparent	τ (RT)	23.2 MPa
Viscosity	0.05 Pa.s	τ (°C)	-
S.G.	1.1	τ (wet)	-
Shelf life	6 months/RT	Peel ( )	-
Pot/Work life	-	σ_t	-
Tg	-	σ_c	-
$T_{Service}$	-40°C to +125°C	*E*	245 MPa
CTE (°C)	54.4 x 10^{-6} K^{-1}	Elongation	-
Flash Point	95 °C	H (Shore D)	65

Notes: Shrinkage: 3.3%. Refractive index: 1.55. E (DIN 53457). Breakdown potential (DIN 53481): 20.4 kV.mm⁻¹. Moisture absorption (ASTM 570): 0.7%. Thermal conductivity (DIN 52612): 0.18 W.m⁻¹K⁻¹. Dielectric constant (DIN 53483): 5.1. Shear: ASTM D1002. CTE: ASTM D695.

Panacol Elosol	**Vitralit 6128 VT**		Jun 94
One-part thixotropic, UV, thermal or catalytic cure			
Manufacturers Comments: Rapid cure. Good adhesion to glass, metals and many thermoplastics. Solvent-free, 100% solids.			
Typical Use: Glass, metals, plastics, etc.			
Cure: UV-A, 60 mW.cm²: 15 sec. 120/90 mins. Catalyst 535: 25 mins. Final strength: 12 hr.			
Test Methods: DIN (ASTM)			
Physical		**Mechanical**	
Colour	Translucent	τ (RT)	26.3 MPa
Viscosity	20 Pa.s	τ (°C)	-
S.G.	1.1	τ (wet)	-
Shelf life	6 months/RT	Peel ( )	-
Pot/Work life	-	σ_t	-
Tg	-	σ_c	-
$T_{Service}$	-40°C to +150°C	*E*	455 MPa
CTE (°C)	58 x 10^{-6} K^{-1}	Elongation	-
Flash Point	95 °C	H (Shore D)	72

Notes: Shrinkage: 2.3%. Refractive index: 1.55. E (DIN 53457). Breakdown potential (DIN 53481): 18.7 kV.mm⁻¹. Moisture absorption (ASTM 570): 2.8%. Thermal conductivity (DIN 52612): 0.25 W.m⁻¹K⁻¹. Dielectric constant (DIN 53483): 6.2. Shear: ASTM D1002. CTE: ASTM D695.

Panacol Elosol		**Vitralit 6129**	Jun 94
One-part, UV, thermal or catalytic cure			
Manufacturers Comments: Rapid cure. Good adhesion to glass, aluminium and many thermoplastics. Good thermal and chemical resistance. Solvent-free, 100% solids.			
Typical Use: Glass, aluminium, plastics, etc.			
Cure: UV-A, 60 mW.cm^2: 15 sec. 120/30 mins. Catalyst 535: 60 mins. Final strength: 12 hr.			
Test Methods: DIN (ASTM)			
Physical		**Mechanical**	
Colour	-	τ (Al)	16.6 MPa
Viscosity	30 Pa.s	τ (Oily steel)	10.3 MPa
S.G.	1.75	τ (Steel)	13.7 MPa
Shelf life	6 months/RT	Peel ( )	-
Pot/Work life	-	σ_t	-
Tg	-	σ_c	-
$T_{Service}$	-50°C to +180°C	E	1055 MPa
CTE (°C)	36.4 x 10^{-6} K^{-1}	Elongation	-
Flash Point	92 °C	H (Shore D)	78
Notes: Shrinkage: 2.2%. E (DIN 53457). Breakdown potential (DIN 53481): 2.1 kV.mm^{-1}. Moisture absorption (ASTM 570): 0.8%. Thermal conductivity (DIN 52612): 0.95 W.m^{-1}K^{-1}. Dielectric constant (DIN 53483): 3.2. Shear: ASTM D1002. CTE: ASTM D695.			

Panacol Elosol		**Vitralit 6143**	Jun 94
One-part thixotropic, UV cure			
Manufacturers Comments: Rapid cure, highly flexible. Good adhesion on metals and plastics. Solvent-free, 100% solids.			
Typical Use: Glass, aluminium, plastics, etc.			
Cure: UV-A, 60 mW.cm^2: 30 sec. Final strength: 12 hr.			
Test Methods: DIN (ASTM)			
Physical		**Mechanical**	
Colour	-	τ (Al)	11.3 MPa
Viscosity	32 Pa.s	τ (Oily steel)	2.1 MPa
S.G.	1.1	τ (Steel)	8.6 MPa
Shelf life	6 months/RT	Peel ( )	-
Pot/Work life	-	σ_t	-
Tg	-	σ_c	-
$T_{Service}$	-40°C to +130°C	E	324 MPa
CTE (°C)	76.3 x 10^{-6} K^{-1}	Elongation	-
Flash Point	95 °C	H (Shore D)	57
Notes: Shrinkage: 4.7%. E (DIN 53457). Breakdown potential (DIN 53481): 9.3 kV.mm^{-1}. Moisture absorption (ASTM 570): 0.8%. Thermal conductivity (DIN 52612): 0.43 W.m^{-1}K^{-1}. Dielectric constant (DIN 53483): 4.2. Shear: ASTM D1002. CTE: ASTM D695.			

Panacol Elosol		**Vitralit 6140**	Jun 94
One-part, UV cure			
Manufacturers Comments: Rapid cure, highly flexible. Good adhesion on metals and plastics. Solvent-free, 100% solids.			
Typical Use: Glass, aluminium, plastics, etc.			
Cure: UV-A, 60 mW.cm^2: 30 sec. Final strength: 12 hr.			
Test Methods: DIN (ASTM)			
Physical		**Mechanical**	
Colour	-	τ (Al)	11.3 MPa
Viscosity	1.2 Pa.s	τ (Oily steel)	2.1 MPa
S.G.	1.1	τ (Steel)	8.6 MPa
Shelf life	6 months/RT	Peel ( )	-
Pot/Work life	-	σ_t	-
Tg	-	σ_c	-
$T_{Service}$	-40°C to +130°C	E	324 MPa
CTE (°C)	76.3 x 10^{-6} K^{-1}	Elongation	-
Flash Point	95 °C	H (Shore D)	58
Notes: Shrinkage: 4.7%. E (DIN 53457). Breakdown potential (DIN 53481): 9.3 kV.mm^{-1}. Moisture absorption (ASTM 570): 0.8%. Thermal conductivity (DIN 52612): 0.43 W.m^{-1}K^{-1}. Dielectric constant (DIN 53483): 4.2. Shear: ASTM D1002. CTE: ASTM D695.			

Panacol Elosol		**Vitralit 6164**	Jun 94
One-part, UV cure			
Manufacturers Comments: Rapid cure, very low viscosity. Solvent-free, 100% solids.			
Typical Use: Glass bonding and repair.			
Cure: UV-A, 60 mW.cm^2: 15 sec. Final strength: 12 hr.			
Test Methods: DIN (ASTM)			
Physical		**Mechanical**	
Colour	-	τ (RT)	7.2 MPa
Viscosity	0.025 Pa.s	τ (°C)	-
S.G.	0.98	τ (wet)	-
Shelf life	6 months/RT	Peel ( )	-
Pot/Work life	-	σ_t	-
Tg	-	σ_c	-
$T_{Service}$	-20°C to +120°C	E	174 MPa
CTE (°C)	73 x 10^{-6} K^{-1}	Elongation	-
Flash Point	93 °C	H (Shore D)	55
Notes: Shrinkage: 6.8%. Refractive index: 1.508. E (DIN 53457). Breakdown potential (DIN 53481): 10.2 kV.mm^{-1}. Moisture absorption (ASTM 570): 0.9%. Thermal conductivity (DIN 52612): 0.14 W.m^{-1}K^{-1}. Dielectric constant (DIN 53483): 4.1. Shear: ASTM D1002. CTE: ASTM D695.			

Panacol Elosol	Vitralit 7256		Jun 94
One-part, UV cure			
Manufacturers Comments: Rapid cure. Solvent-free, 100% solids.			
Typical Use: Glass bonding.			
Cure: UV-A, 60 mW.cm^2: 15 sec. Final strength: 12 hr.			
Test Methods: DIN (ASTM)			
Physical		**Mechanical**	
Colour	-	τ (RT)	16.2 MPa
Viscosity	0.9 Pa.s	τ (°C)	-
S.G.	1.09	τ (wet)	-
Shelf life	6 months/RT	Peel ( )	-
Pot/Work life	-	σ_t	-
Tg	-	σ_c	-
$T_{Service}$	-40°C to +125°C	E	215 MPa
CTE (°C)	64 x 10^{-6} K^{-1}	Elongation	-
Flash Point	95 °C	H (Shore D)	76
Notes: Shrinkage: 5.2%. Refractive index: 1.52. E (DIN 53457). Breakdown potential (DIN 53481): 12.1 kV.mm^{-1}. Moisture absorption (ASTM 570): 1.3%. Thermal conductivity (DIN 52612): 0.15 W.m^{-1}K^{-1}. Dielectric constant (DIN 53483): 4.7. Shear: ASTM D1002. CTE: ASTM D695.			

Panacol Elosol	Vitralit 7515		Jun 94
One-part, UV cure			
Manufacturers Comments: Rapid cure. Solvent-free, 100% solids.			
Typical Use: Glass bonding.			
Cure: UV-A, 60 mW.cm^2: 25 sec. Final strength: 12 hr.			
Test Methods: DIN (ASTM)			
Physical		**Mechanical**	
Colour	-	τ (RT)	21.2 MPa
Viscosity	6 Pa.s	τ (°C)	-
S.G.	1.15	τ (wet)	-
Shelf life	6 months/RT	Peel ( )	-
Pot/Work life	-	σ_t	-
Tg .	-	σ_c	-
$T_{Service}$	-40°C to +120°C	E	274 MPa
CTE (°C)	66 x 10^{-6} K^{-1}	Elongation	-
Flash Point	95 °C	H (Shore D)	84
Notes: Shrinkage: 3.6%. Refractive index: 1.48. E (DIN 53457). Breakdown potential (DIN 53481): 14.2 kV.mm^{-1}. Moisture absorption (ASTM 570): 1.9%. Thermal conductivity (DIN 52612): 0.17 W.m^{-1}K^{-1}. Dielectric constant (DIN 53483): 4.6. Shear: ASTM D1002. CTE: ASTM D695.			

Panacol Elosol	Vitralit 7512		Jun 94
One-part, UV cure			
Manufacturers Comments: Rapid cure. Solvent-free, 100% solids.			
Typical Use: Glass bonding.			
Cure: UV-A, 60 mW.cm^2: 20 sec. Final strength: 12 hr.			
Test Methods: DIN (ASTM)			
Physical		**Mechanical**	
Colour	-	τ (RT)	14.5 MPa
Viscosity	0.3 Pa.s	τ (°C)	-
S.G.	1.10	τ (wet)	-
Shelf life	6 months/RT	Peel ( )	-
Pot/Work life	-	σ_t	-
Tg	-	σ_c	-
$T_{Service}$	-40°C to +125°C	E	237 MPa
CTE (°C)	45 x 10^{-6} K^{-1}	Elongation	-
Flash Point	95 °C	H (Shore D)	65
Notes: Shrinkage: 4.2%. Refractive index: 1.507. E (DIN 53457). Breakdown potential (DIN 53481): 13.4 kV.mm^{-1}. Moisture absorption (ASTM 570): 0.08%. Thermal conductivity (DIN 52612): 0.19 W.m^{-1}K^{-1}. Dielectric constant (DIN 53483): 4.1. Shear: ASTM D1002. CTE: ASTM D695.			

Panacol Elosol	Vitralit 8101		Jun 94
One-part, UV cure			
Manufacturers Comments: Rapid cure. Solvent-free, 100% solids.			
Typical Use: Bonding of glass, metal and plastic medical equipment.			
Cure: UV-A, 60 mW.cm^2: 5 sec. Final strength: 12 hr.			
Test Methods: DIN (ASTM)			
Physical		**Mechanical**	
Colour	-	τ (RT)	12.4 MPa
Viscosity	5 Pa.s	τ (°C)	-
S.G.	1.1	τ (wet)	-
Shelf life	6 months/RT	Peel ( )	-
Pot/Work life	-	σ_t	-
Tg	-	σ_c	-
$T_{Service}$	-50°C to +150°C	E	110 MPa
CTE (°C)	85 x 10^{-6} K^{-1}	Elongation	-
Flash Point	>93.3 °C	H (Shore D)	85
Notes: Shrinkage: 8.5%. Refractive index: 1.48. E (DIN 53457). Moisture absorption (ASTM 570): 4.7%. Thermal conductivity (DIN 52612): 0.23 W.m^{-1}K^{-1}. Dielectric constant (DIN 53483): 3.7. Shear: ASTM D1002. CTE: ASTM D695.			

Panacol Elosol	Vitralit 8103		Jun 94
One-part, UV cure			
Manufacturers Comments: Rapid cure. Solvent-free, 100% solids.			
Typical Use: Bonding of glass, metal and plastic medical equipment.			
Cure: UV-A, 60 mW.cm^2: 10 sec. Final strength: 12 hr.			
Test Methods: DIN (ASTM)			
Physical		**Mechanical**	
Colour	-	τ (RT)	12.9 MPa
Viscosity	8 Pa.s	τ (°C)	-
S.G.	1.1	τ (wet)	-
Shelf life	6 months/RT	Peel ( )	-
Pot/Work life	-	σ_t	-
Tg	-	σ_c	-
$T_{Service}$	-50°C to +150°C	E	127 MPa
CTE (°C)	87 x 10^{-6} K^{-1}	Elongation	-
Flash Point	>93.3 °C	H (Shore D)	50
Notes: Shrinkage: 6%. Refractive index: 1.48. E (DIN 53457). Moisture absorption (ASTM 570): 12.5%. Thermal conductivity (DIN 52612): 0.24 W.m^{-1}K^{-1}. Dielectric constant (DIN 53483): 3.6. Shear: ASTM D1002. CTE: ASTM D695.			

Quilosa
(Industrias Quimicas Löwenberg SA)

Quilosa	Aklesil L-32		Dec 89
One-part, water-based vinyl (homopolymers & modified copolymers)			
Manufacturers Comments: Fast RT cure. White liquid. Transparent when cured. Moderate flexibility.			
Typical Use: Wood panel fixing.			
Cure: e.g. 6 - 10min/RT.			
Test Methods: Not stated			
Physical		**Mechanical**	
Colour	Transparent	τ (RT)	-
Viscosity	-	τ (°C)	-
S.G.	-	τ (wet)	-
Shelf life	6 months/RT	Peel ( )	-
Pot/Work life	-	σ_t	-
Tg	-	σ_c	-
$T_{Service}$	-	E	-
CTE (°C)	-	Elongation	-
Flash Point	-	H (Shore)	-
Notes: Coverage: 0.11 - 0.16 kg.m^{-2}.			

Panacol Elosol	Vitralit 8106		Jun 94
One-part, UV cure			
Manufacturers Comments: Rapid cure. Solvent-free, 100% solids.			
Typical Use: Bonding of glass, metal and plastic medical equipment.			
Cure: UV-A, 60 mW.cm^2: 10 sec. Final strength: 12 hr.			
Test Methods: DIN (ASTM)			
Physical		**Mechanical**	
Colour	-	τ (RT)	14.3 MPa
Viscosity	0.16 Pa.s	τ (°C)	-
S.G.	1.1	τ (wet)	-
Shelf life	6 months/RT	Peel ( )	-
Pot/Work life	-	σ_t	-
Tg	-	σ_c	-
$T_{Service}$	-50°C to +150°C	E	112 MPa
CTE (°C)	93 x 10^{-6} K^{-1}	Elongation	-
Flash Point	>93.3 °C	H (Shore D)	85
Notes: Shrinkage: 7%. Refractive index: 1.48. E (DIN 53457). Moisture absorption (ASTM 570): 0.6%. Thermal conductivity (DIN 52612): 0.29 W.m^{-1}K^{-1}. Dielectric constant (DIN 53483): 3.8. Shear: ASTM D1002. CTE: ASTM D695.			

Quilosa	Aklesil L-51		May 93
One-part, water-based vinyl (homopolymers & modified copolymers)			
Manufacturers Comments: Fast RT cure. White fluid. Dark colour when cured. Moderate flexibility.			
Typical Use: Laminating of wood products and PVC.			
Cure: >24hr/RT.			
Test Methods: Not stated			
Physical		**Mechanical**	
Colour	(dark colour)	τ (RT)	-
Viscosity	11 Pa.s	τ (°C)	-
S.G.	-	τ (wet)	-
Shelf life	6 months/RT	Peel ( )	-
Pot/Work life	6 - 10min/RT	σ_t	-
Tg	-	σ_c	-
$T_{Service}$	60°C to 70°C	E	-
CTE (°C)	-	Elongation	-
Flash Point	-	H (Shore)	-
Notes: Coverage: 0.12 - 0.15 kg.m^{-2}			

Quilosa	Sintex POL K-52		Nov 94
One-part, adhesive/sealant based on modified polyoxyfines			
Manufacturers Comments: Flexible and good adhesion without primers. Non-slumping with good mechanical, chemical and weathering resistance. Very good UV resistance. Long shelf life. Sandable and paintable. Vibration absorbing.			
Typical Use: Automotive, construction, marine and general industry. Bonding, sealing and repair.			
Cure: Cure rate: 3 mm/day. e.g. 4 weeks/RT(55%RH).			
Test Methods: DIN, BS and ASTM			
Physical		**Mechanical**	
Colour	Black	τ (RT)	-
Viscosity	-	τ (°C)	-
S.G.	-	τ (wet)	-
Shelf life	9 months.	Peel ( )	-
Pot/Work life	7 - 9 min (open)	σ_t	3.0 - 3.3 MPa
Tg	-	σ_c	-
$T_{Service}$	-40°C to +90°C	E	-
CTE (°C)	-	Elongation	300 - 350%
Flash Point	430°C	H (Shore A)	60
Notes: Tack free time (ASTM C679-71): 7 - 12 mins. Skin time (BS 5889): 23 - 25. Hardness: DIN 53505, Tensile: DIN 53504.			

Quilosa	Unifix M-09		May 94
One-part, aqueous dispersion of modified vinyl polyacetate			
Manufacturers Comments: RT cure. White fluid.			
Typical Use: Laminating of melamine and wood products.			
Cure: 6hr/RT.			
Test Methods: DIN			
Physical		**Mechanical**	
Colour	-	τ (RT)	-
Viscosity	15 Pa.s	τ (°C)	-
S.G.	-	τ (wet)	-
Shelf life	1 yr/RT	Peel ( )	-
Pot/Work life	15 - 16mins (open)	σ_t	5 - 6 MPa
Tg	-	σ_c	-
$T_{Service}$	-	E	-
CTE (°C)	-	Elongation	-
Flash Point	-	H (Shore)	-
Notes:			

Quilosa	Unifix M-02		(Jul 95)
One-part, aqueous dispersion of modified vinyl polyacetate			
Manufacturers Comments: Fast RT cure. White fluid. Translucent when cured. Hard and resistant.			
Typical Use: Bonding of wood and wood products. Furniture.			
Cure: RT. Pressure: 6 - 9 mins/0.3 - 0.5 MPa.			
Test Methods: DIN			
Physical		**Mechanical**	
Colour	Translucent	τ (RT)	-
Viscosity	13 Pa.s	τ (°C)	-
S.G.	-	τ (wet)	-
Shelf life	1 yr/RT	Peel ( )	-
Pot/Work life	-	σ_t	16 - 17 MPa
Tg	-	σ_c	-
$T_{Service}$	-	E	-
CTE (°C)	-	Elongation	-
Flash Point	-	H (Shore)	-
Notes: Tensile: DIN 68603.			

Quilosa	Unifix M-20		May 94
One-part, aqueous dispersion of modified vinyl polyacetate			
Manufacturers Comments: Fast RT cure. White fluid.			
Typical Use: Bonding of wood and wood products. Furniture.			
Cure: RT. Pressure: 6 - 9 mins/0.3 - 0.5 MPa.			
Test Methods: DIN			
Physical		**Mechanical**	
Colour	-	τ (RT)	-
Viscosity	15 Pa.s	τ (°C)	-
S.G.	-	τ (wet)	-
Shelf life	1 yr/RT	Peel ( )	-
Pot/Work life	-	σ_t	-
Tg	-	σ_c	-
$T_{Service}$	-	E	-
CTE (°C)	-	Elongation	-
Flash Point	-	H (Shore)	-
Notes:			

Quilosa | **Unifix M-46** | May 94

One-part, aqueous dispersion of modified vinyl polyacetate

Manufacturers Comments: Very fast RT cure. White fluid.

Typical Use: Bonding of wood and wood products. Furniture.

Cure: Initial cure: <5min, Full cure: 1hr/RT.
Pressure: 6 - 9 mins/0.3 - 0.5 MPa.

Test Methods: DIN

Physical		Mechanical	
Colour	-	τ (RT)	-
Viscosity	10 Pa.s	τ (°C)	-
S.G.	-	τ (wet)	-
Shelf life	1 yr/RT	Peel ( )	-
Pot/Work life	2 - 3mins (open)	σ_t	14 MPa
Tg	-	σ_c	-
$T_{Service}$	-	*E*	-
CTE (°C)	-	Elongation	-
Flash Point	-	H (Shore)	-

Notes:

Quilosa | **Unifix M-54** | Jun 90

One-part, aqueous dispersion of modified vinyl polyacetate

Manufacturers Comments: Fast RT and elevated temperature cure. White fluid. Opaque when cured. Hard and tough.

Typical Use: Bonding of wood and wood products. Furniture, DIY, Construction.

Cure: Initial cure: 2hr/RT, Final cure 24hr/RT (cures up to 110°C).
Pressure (optional): 90 - 180 mins/RT/0.3 - 0.5 MPa.

Test Methods: DIN

Physical		Mechanical	
Colour	Opaque	τ (RT)	-
Viscosity	11 Pa.s	τ (°C)	-
S.G.	-	τ (wet)	-
Shelf life	1 yr/RT	Peel ( )	-
Pot/Work life	15 - 20 min (open)	σ_t	10 - 12 MPa
Tg	-	σ_c	-
$T_{Service}$	-	*E*	-
CTE (°C)	-	Elongation	-
Flash Point	-	H (Shore)	-

Notes: Tensile: DIN 68603.

Quilosa | **Unifix M-49** | Jun 90

One-part, aqueous dispersion of modified vinyl polyacetate

Manufacturers Comments: Fast RT cure. White fluid. Opaque when cured. Hard and tough.

Typical Use: Bonding of wood and wood products. Furniture, DIY, Construction.

Cure: Initial cure: 2hr/RT, Final cure 24hr/RT.
Pressure (optional): 90 - 180 mins/RT/0.3 - 0.5 MPa.

Test Methods: DIN

Physical		Mechanical	
Colour	Opaque	τ (RT)	-
Viscosity	11 Pa.s	τ (°C)	-
S.G.	-	τ (wet)	-
Shelf life	1 yr/RT	Peel ( )	-
Pot/Work life	15 - 20 min (open)	σ_t	10 - 12 MPa
Tg	-	σ_c	-
$T_{Service}$	-	*E*	-
CTE (°C)	-	Elongation	-
Flash Point	-	H (Shore)	-

Notes: Tensile: DIN 68603.

Quilosa | **Unifix M-75** | Dec 89

One-part, aqueous dispersion of modified vinyl polyacetate

Manufacturers Comments: Fast RT and elevated temperature cure. White fluid. Translucent when cured. Hard and tough. Good resistance to humidity.

Typical Use: Bonding of wood and wood products and laminating. Furniture. Recommended for exportable goods (moisture resistance).

Cure: Initial cure: <1hr/RT, Final cure 24hr/RT (cures up to 100°C).
Pressure (optional): 60 - 150 mins/RT/0.2 - 0.5 MPa.

Test Methods: DIN

Physical		Mechanical	
Colour	Translucent	τ (RT)	-
Viscosity	14 - 22 Pa.s	τ (°C)	-
S.G.	-	τ (wet)	-
Shelf life	1 yr/RT	Peel ( )	-
Pot/Work life	7 - 9 min (open)	σ_t	14 MPa
Tg	-	σ_c	-
$T_{Service}$	-	*E*	-
CTE (°C)	-	Elongation	-
Flash Point	-	H (Shore)	-

Notes: Tensile: DIN 68603.

Rinaldi

(SLC Rinaldi)

Rinaldi	Silovil Forte		(July 95)
Water dispersed, solvent free, synthetic glue with low water content			
Manufacturers Comments: Very fast RT cure.			
Typical Use: Wood flooring.			
Cure: RT.			
Test Methods: DIN			
Physical		**Mechanical**	
Colour	-	τ (RT)	-
Viscosity	-	τ (°C)	-
S.G.	-	τ (wet)	-
Shelf life	1 yr/RT (<5°C)	Peel ( )	-
Pot/Work life	20 min/RT	σ_f	-
Tg	-	σ_c	-
$T_{Service}$	-	E	-
CTE (°C)	-	Elongation	-
Flash Point	-	H (Shore)	-
Notes:			

Rinaldi	Silovil U 68L		(July 95)
One-part, water-based synthetic resin			
Manufacturers Comments: RT cure. Suitable for heated sub-floors.			
Typical Use: Wood flooring.			
Cure: RT.			
Test Methods: DIN			
Physical		**Mechanical**	
Colour	Brown or Beige	τ (RT)	-
Viscosity	-	τ (°C)	-
S.G.	-	τ (wet)	-
Shelf life	1 yr/RT	Peel ( )	-
Pot/Work life	30 min (open time)	σ_f	-
Tg	-	σ_c	-
$T_{Service}$	-	E	-
CTE (°C)	-	Elongation	-
Flash Point	-	H (Shore)	-
Notes:			

Rinaldi	Silovil L34		(July 95)
Water and Solvent Free, Two-part adhesive.			
Manufacturers Comments: RT cure. Flexible.			
Typical Use: Wood flooring to ceramic or concrete substrates.			
Cure: 24 - 28 hr/RT.			
Test Methods: DIN			
Physical		**Mechanical**	
Colour	-	τ (RT)	-
Viscosity	-	τ (°C)	-
S.G.	-	τ (wet)	-
Shelf life	RT	Peel ( )	-
Pot/Work life	1 - 2 hr.	σ_f	-
Tg	-	σ_c	-
$T_{Service}$	-	E	-
CTE (°C)	-	Elongation	-
Flash Point	-	H (Shore)	-
Notes:			

Rinaldi	Silovil U 68M		(July 95)
One-part, water-based synthetic resin			
Manufacturers Comments: RT cure.			
Typical Use: Wood and polymer flooring.			
Cure: RT.			
Test Methods: DIN			
Physical		**Mechanical**	
Colour	Beige	τ (RT)	-
Viscosity	-	τ (°C)	-
S.G.	-	τ (wet)	-
Shelf life	1 yr/RT	Peel ( )	-
Pot/Work life	30 min (open time)	σ_f	-
Tg	-	σ_c	-
$T_{Service}$	-	E	-
CTE (°C)	-	Elongation	-
Flash Point	-	H (Shore)	-
Notes:			

SABA

SABA	Sabaplast 7200 UV		(Sept 95)
One-part silane-modified polymer			
Manufacturers Comments: Contains no solvent, isocyanate, silicones or PVC. 100% solids. Good adhesion to many materials. No shrinkage. Paintable. Excellent UV resistance.			
Typical Use: Automotive industry - metal and plastic panel bonding/sealing.			
Cure: 3 mm per day.			
Test Methods: Not Stated			
Physical		**Mechanical**	
Colour	White	τ (RT)	-
Viscosity	(paste)	τ (°C)	-
S.G.	1.37	τ (wet)	-
Shelf life	6 months/RT	Peel ( )	-
Pot/Work life	-	σ_t	2.1 MPa
Tg	-	σ_c	-
$T_{Service}$	-40 to +100°C	E	1.1 MPa*
CTE (°C)	-	Elongation	400 %
Flash Point	-	H (Shore)	40 - 45
Notes: Modulus at 100% elongation. Skin time: 10 min. Tack-free time: ~4 hr.			

SEA	Parkecol		(Aug 95)
One-part alcohol solution of adhesive			
Manufacturers Comments: Fast drying, high tack.			
Typical Use: Construction industry - wood flooring.			
Cure: Initial tack: 1 - 5 mins.			
Test Methods: Not Stated			
Physical		**Mechanical**	
Colour	Light Beige	τ (RT)	-
Viscosity	(thick liquid)	τ (°C)	-
S.G.	-	τ (wet)	-
Shelf life	1 yr/RT	Peel ( )	-
Pot/Work life	20 min (open)	σ_t	-
Tg	-	σ_c	-
$T_{Service}$	-	E	-
CTE (°C)	-	Elongation	-
Flash Point	-	H (Shore)	-
Notes: -			

SEA
(Société d'Exploitation des Adhésifs)

SEA	ELO		Jun 91
One-part vinyl adhesive			
Manufacturers Comments: 56% solids.			
Typical Use: Construction industry - sheet fibre-reinforced cement products. For bonding to cement, plaster, wood, laminates, etc.			
Cure: 24 hr/RT or 12 mins/70°C.			
Test Methods: Not Stated			
Physical		**Mechanical**	
Colour	White	τ (RT)	-
Viscosity	(thick liquid)	τ (°C)	-
S.G.	1.1	τ (wet)	-
Shelf life	1 yr/RT	Peel ( )	-
Pot/Work life	10 - 15 min (open)	σ_t	-
Tg	-	σ_c	-
$T_{Service}$	-	E	-
CTE (°C)	-	Elongation	-
Flash Point	-	H (Shore)	-
Notes: -			

SEA	Siel Parquet		Jul 92
One-part polyvinyl acetate emulsion			
Manufacturers Comments: 59% solids.			
Typical Use: Construction industry - wood flooring.			
Cure: 24 hr/RT.			
Test Methods: Not Stated			
Physical		**Mechanical**	
Colour	White	τ (RT)	-
Viscosity	18 - 24 Pa.s	τ (°C)	-
S.G.	1.3	τ (wet)	-
Shelf life	1 yr/RT	Peel ( )	-
Pot/Work life	20 - 30 min (open)	σ_t	-
Tg	-	σ_c	-
$T_{Service}$	-	E	-
CTE (°C)	-	Elongation	-
Flash Point	-	H (Shore)	-
Notes: -			

Teroson
(Henkel)

Teroson	Terokal 8549		July 94
One-part PVC Plastisol			
Manufacturers Comments: Heat-curing (170 - 220°C), solvent free paste. Pre-gellable. >99% solids. Paintable.			
Typical Use: Automotive, bonding of body panels (hem flange seams).			
Cure: Pre-gel: e.g. 6sec/160°C (induction heat). Final cure: >15min/180°C or <15min/210°C.			
Test Methods: DIN			
Physical		**Mechanical**	
Colour	Light Grey	τ (RT)	~0.48 MPa
Viscosity	(paste)	τ (90°C)	~0.26 MPa
S.G.	1.4	τ (-40°C)	>0.55 MPa
Shelf life	6mnth/10 - 25°C	Peel (T)	2.7 N mm^{-1}
Pot/Work life	-	σ_t	~3.4 MPa
Tg	-	σ_c	-
$T_{Service}$	-40°C to +80°C	*E*	-
CTE (°C)	-	Elongation	105%
Flash Point	-	H (Shore)	-
Notes: Shear strength: DIN 53283. 'T' Peel: DIN 53282. Tensile: DIN 53504.			

Teroson	Terokal 8552		July 94
One-part PVC Plastisol			
Manufacturers Comments: Heat-curing (170 - 220°C), solvent free paste. Must be pre-gelled. >99% solids. Paintable.			
Typical Use: Automotive, bonding of body panels (hem flange seams).			
Cure: Pre-gel: e.g. 6sec/160°C (induction heat). Final cure: >15min/170°C or <15min/220°C.			
Test Methods: DIN			
Physical		**Mechanical**	
Colour	Pale yellow	τ (RT)	~0.46 MPa
Viscosity	(paste)	τ (90°C)	~0.12 MPa
S.G.	1.37	τ (-40°C)	>0.5 MPa
Shelf life	3mnth/10 - 20°C	Peel (T)	2.7 N mm^{-1}
Pot/Work life	-	σ_t	-
Tg	-	σ_c	-
$T_{Service}$	-40°C to +80°C	*E*	-
CTE (°C)	-	Elongation	-
Flash Point	-	H (Shore)	-
Notes: Shear strength: DIN 53283. 'T' Peel: DIN 53282.			

ThreeBond

ThreeBond	3056B		(Aug 95)
UV cure with moisture hardening			
Manufacturers Comments: Silicone-containing. Initial cure by UV irradiation - resin then continues to cure by contact with airborne moisture (for temperature sensitive applications). Flexible with good moisture resistance.			
Typical Use: Magnetic rubber bonding.			
Cure: 4kW mercury lamp (2 J.cm^{-2}): 20 sec, then moisture cure.			
Test Methods: Not stated.			
Physical		**Mechanical**	
Colour	Clear yellow	τ (RT)	-
Viscosity	20 Pa.s	τ (°C)	-
S.G.	1.07	τ (wet)	-
Shelf life	-	Peel ( )	-
Pot/Work life	-	σ_t	19.6 MPa
Tg	-	σ_c	98 MPa
$T_{Service}$	-	*E*	-
CTE (°C)	-	Elongation	70%
Flash Point	-	H (Shore)	95-A, 70-D
Notes: Surface resistivity: 7.7 x 10^{14} Ω.cm. Recommends 3095B primer.			

ThreeBond	3056C		(Aug 95)
UV cure with moisture hardening			
Manufacturers Comments: Silicone-containing. Initial cure by UV irradiation - resin then continues to cure by contact with airborne moisture (for temperature sensitive applications). Good moisture resistance.			
Typical Use: Metal and plastic bonding.			
Cure: 4kW mercury lamp (2 J.cm^{-2}): 20 sec, then moisture cure.			
Test Methods: Not stated.			
Physical		**Mechanical**	
Colour	Clear blue	τ (RT)	-
Viscosity	3 Pa.s	τ (°C)	-
S.G.	1.08	τ (wet)	-
Shelf life	-	Peel ( )	-
Pot/Work life	-	σ_t	36.3 MPa
Tg	-	σ_c	-
$T_{Service}$	-	*E*	255 MPa
CTE (°C)	-	Elongation	20%
Flash Point	-	H (Shore)	95-A, 80-D
Notes: Surface resistivity: 6.2 x 10^{14} Ω.cm.			

If you have any suggestions for changes or additions to the information presented here, please contact the authors, giving full details. The final page of the book is a reply form for this purpose.

Jo Wilson & Bob Hussey
RJ Technical Consultants
Bois Moreau
17770 Juicq
Charente Maritime
France
Tel: +33 46 95 35 10
Fax: +33 46 95 35 10

Part 4: APPENDICES

Appendix A: SPECIFICATIONS AND STANDARDS

Listings by reference codes and by area of applicability for international, national and industry specifications and standards. Including those specifically related to adhesives and bonding, together with applications-orientated standards found in manufacturers' data.

Appendix B: GLOSSARY

Glossary of materials and adhesive bonding terms.

Appendix C: VOCABULARY

Multilingual dictionary of materials and adhesive bonding terms.

Appendix D: CONVERSION FACTORS

Conversion factors for quantities commonly found in adhesives data sheets and manufacturers' literature.

Appendix A - SPECIFICATIONS AND STANDARDS

INTRODUCTION

Many standards and specifications have been devised for use in adhesive selection and the design of adhesively bonded joints. The majority of these have been issued by standards organisations, but some industries and major companies develop test methods and standards for their own purposes. Often these are based on well-known and established documents, but are modified to meet their specific needs. For example, tests applied to adherends other than those stated in a standard, test coupons of different dimensions or variations in the loading methods to suit different types and forms of adherend materials.

In recent years, great efforts have been made within the various interested organisations to bring the proliferation of standards into line, to allow comparisons between the results of tests to different national standards. Although progress has been made in, for example, the basic test methods and procedures, a unified international system of standards is still a long way off.

This appendix provides a collated list of commonly quoted test methods, standards and specifications, for use in R & D studies, prototyping, design, manufacture and quality assurance. These are listed first in numerical order for each standards organisation, then by industry sector and use.

Note: The listings are neither exhaustive, nor fully cross-referenced to all national standards.

IMPORTANT

All standards and specifications are reviewed and revised periodically. Always ensure that you are using the latest version

ASTM Standards

American Society for Testing Materials - USA

ASTM have developed a wide range of standards covering all aspects of adhesives and bonding technology. Tests conducted to ASTM methods are widely quoted by adhesive manufacturers and in research, design and development studies (property data to a stated specification). Those standards listed here have been selected for their relevance to structural bonding applications. For a comprehensive list of ASTM standards, refer to:

Annual Book of Standards. Part 15.06 - Adhesives

ASTM C 0297 Flatwise Tensile Strength of Metal-to-Honeycomb Core Bonds.

ASTM C 557-73 (1985) Specification for Adhesives for Fastening Gypsum Wallboard to Wood Framing.

ASTM D 0229 Test Method for Shear Strength and Shear Modulus of Structural Adhesives. *Note: Napkin Ring style test piece.*

ASTM D 570 *relating to* Moisture Absorption.

ASTM D 696 *relating to* Coefficient of Thermal Expansion.

ASTM D 816-82 (1988) Test Methods for Testing Rubber Cements.

ASTM D 896-90 Test Method for Resistance of Adhesive Bonds to Chemical Reagents.

ASTM D 997-78 (1983) Test Method for Tensile Properties of Adhesive Bonds.

ASTM D 898-90 Test Method for Applied Weight per Unit Area of Dried Adhesive Solids.

ASTM D 899-89 Test Method for Applied Weight per Unit Area of Liquid Adhesive.

ASTM D 903-49 (1983) Test Method for Peel or Stripping Strength of Adhesive Bonds.

ASTM D 904-57 (1981) Practice for Exposure of Adhesive Specimens to Artificial (Carbon-Arc Type) and Natural Light.

ASTM D 905-8911 Test Method for Strength Properties of Adhesive Bonds in Shear by Compression Loading.

ASTM D 906-82 (1987) Test Method for Strength Properties of Adhesives in Plywood Type Construction in Shear by Tension Loading.

ASTM D 907-91b Terminology of Adhesives.

ASTM D 950-82 (1987) Test Method for Impact Strength of Adhesive Bonds.

ASTM D 1002-72 (1983) Test Method for Strength Properties of Adhesives in Shear by Tension Loading (Metal-to-Metal).

ASTM D 1062-79 (1983) Test Method for Cleavage Strength of Metal-to-Metal Adhesive Bonds.

ASTM D 1084-88 Test Methods for Viscosity of Adhesives.

ASTM D 1101-89 Test Methods for Integrity of Glue Joints in Structural Laminated Wood Products for Exterior Use.

ASTM D 1144-89 Practice for Determining Strength Development of Adhesive Bonds.

ASTM D 1146-88 Test Method for Blocking Point of Potentially Adhesive Layers.

ASTM D 1151-90 Test Method for Effect of Moisture and Temperature on Adhesive Bonds.

ASTM D 1183-70 (1987) Test Methods for Resistance of Adhesives to Cyclic Laboratory Ageing Conditions.

ASTM D 1184-69 (1986) Test Method for Flexural Strength of Adhesive Bonded Laminated Assemblies.

ASTM D 1304-69 (1983) Methods of Testing Adhesives Relative to Their Use as Electrical Insulation.

ASTM D 1337-91 Test Method for Storage Life of Adhesives by Consistency and Bond Strength.

ASTM D 1338-91 Test Method for Working Life of Liquid or Paste Adhesives by Consistency and Bond Strength.

ASTM D 1344-7811 Method of Testing Cross-Lap Specimens for Tensile Properties of Adhesives (Discontinued 1987).

ASTM D 1382-64 (1987) Test Method for Susceptibility of Dry Adhesive Films to Attack by Roaches.

ASTM D 1383-64 (1987) Test Method for Susceptibility of Dry Adhesive Films to Attack by Laboratory Rats.

ASTM D 1488-86 Test Method for Amylaceous Matter in Adhesives.

ASTM D 1489-87 Test Method for Non-volatile Content of Aqueous Adhesives.

ASTM D 1490-82 (1987) Test Method for Non-volatile Content of Urea-Formaldehyde Resin Solutions.

ASTM D 1579-86 Test Method for Filler Content of Phenol, Resorcinol, and Melamine Adhesives.

ASTM D 1580-60 (1984) Specification for Liquid Adhesives for Automatic Machine Labelling of Glass Bottles (Intent to Withdraw).

ASTM D 1581-60 (1984) Test Method for Bonding Permanency of Water- or Solvent-Soluble Liquid Adhesives for Labelling Glass Bottles (Intent to Withdraw).

ASTM D 1582-86 Test Method for Non-volatile Content of Phenol, Resorcinol, and Melamine Adhesives.

ASTM D 1583-86 Test Method for Hydrogen Ion Concentration of Dry Adhesive Films.

ASTM D 1584-60 (1984) Test Method for Water Absorptiveness of Paper Labels (Intent to Withdraw).

ASTM D 1713-65 (1986) Test Method for Bonding Permanency of Water- or Solvent-Soluble Liquid Adhesives for Automatic Machine Sealing Top Flaps of Fiberboard Specimens (Intent to Withdraw).

ASTM D 1714-65 (1983) Test Method for Water Absorptiveness of Fiberboard Specimens for Adhesives (Intent to Withdraw).

ASTM D 1779-65 (1983) Specification for Adhesive for Acoustical Materials.

ASTM D 1780-72 (1983) Practice for Conducting Creep Tests of Metal-to-Metal Adhesives.

ASTM D 1781-76 (1986) Method for Climbing Drum Peel Test for Adhesives.

ASTM D 1828-70 (1987) Practice for Atmospheric Exposure of Adhesive-Bonded Joints and Structures.

ASTM D 1874-62 (1986) Specification for Water- or Solvent-Soluble Liquid Adhesives for Automatic Machine Scaling of Top Flaps of Fiberboard Shipping Cases.

ASTM D 1875-90 Test Method for Density of Adhesives in Fluid Form.

ASTM D 1876-72 (1983) Test Method for Peel Resistance of Adhesives (T-Peel Test).

ASTM D 1879-70 (1987) Practice for Exposure of Adhesive Specimens to High-Energy Radiation.

ASTM D 1916-88 Test Method for Penetration of Adhesives.

ASTM D 1994-91 Test Method for Determination of Acid Numbers of Hot Melt Adhesives.

ASTM D 1995-91 Test Method for Multi-Modal Strength of Autoadhesives (Contact Adhesives).

ASTM D 2093-84 Practice for Preparation of Surfaces of Plastics Prior to Adhesive Bonding.

ASTM D 2094-69 (1980) Practice for Preparation of Bar and Rod Specimens for Adhesion Tests.

ASTM D 2095-72 (1983) Test Method for Tensile Strength of Adhesives by Means of Bar and Rod Specimens.

ASTM D 2183-69 (1982) Test Method for Flow Properties of Adhesives (Discontinued 1991).

ASTM D 2293-69 (1980) Test Method for Creep Properties of Adhesives in Shear by Compression Loading (Metal-to-Metal).

ASTM D 2294-69 (1980) Test Method for Creep Properties of Adhesives in Shear by Tension Loading (Metal-to-Metal).

ASTM D 2295-72 (1983) Test Method for Strength Properties of Adhesives in Shear by Tension Loading at Elevated Temperatures (Metal-to-Metal).

ASTM D 2339-82 (1987) Test Method for Strength Properties of Adhesives in Two-Ply Wood Construction in Shear by Tension Loading.

ASTM D 2556-91 Test Method for Apparent Viscosity of Adhesives Having Shear-Rate-Dependent Flow Properties.

ASTM D 2557-72 (1983) Test Method for Strength Properties of Adhesives in Shear by Tension Loading in the Temperature Range from -267.8 to -55°C (-450 to -67°F).

ASTM D 2558-69 (1984) Test Method for Evaluating Peel Strength of Shoe Sole-Attaching Adhesives (Intent to Withdraw).

ASTM D 2559-84 (1990) Specification for Adhesives for Structural Laminated Wood Products for Use Under Exterior (Wet Use) Exposure Conditions.

ASTM D 2651-90 Practice for Preparation of Metal Surfaces for Adhesive Bonding.

ASTM D 2674-72 (1984) Methods of Analysis of Sulfochromate Etch Solution Used in Surface Preparation of Aluminium.

ASTM D 2739-90 Test Method for Volume Resistivity of Conductive Adhesives.

ASTM D 2851-86 Specification for Liquid Optical Adhesive.

ASTM D 2918-71 (1987) Practice for Determining Durability of Adhesive Joints Stressed in Peel.

ASTM D 2919-90 Test Method for Determining Durability of Adhesive Joints Stressed in Shear by Tension Loading.

ASTM D 2979-88 Test Method for Pressure-Sensitive Tack of Adhesives Using an Inverted Probe Machine.

ASTM D 3024-84 Specification for Protein-Base Adhesives for Structural Laminated Wood Products for Use Under Interior (Dry Use) Exposure Conditions (Discontinued 1991).

ASTM D 3110-90 Specification for Adhesives Used in Nonstructural Glued Lumber Products.

ASTM D 3111-88 Test Method for Flexibility Determination of Hot-Melt Adhesives by Mandrel Bend Test Method.

ASTM D 3121-89 Test Method for Tack of Pressure-Sensitive Adhesives by Rolling Ball.

ASTM D 3163-73 (1984) Test Method for Determining the Strength of Adhesively Bonded Rigid Plastic Lap-Shear Joints in Shear by Tension Loading.

ASTM D 3164-73 (1984) Test Method for Determining the Strength of Adhesively Bonded Plastic Lap-Shear Sandwich Joints in Shear by Tension Loading.

ASTM D 3165-73 (1979) Test Method for Strength Properties of Adhesives in Shear by Tension Loading of Laminated Assemblies.

ASTM D 3166-73 (1979) Test Method for Fatigue Properties of Adhesives in Shear by Tension Loading (Metal/Metal).

ASTM D 3167-76 (1986) Test Method for Floating Roller Peel Resistance of Adhesives.

ASTM D 3310-90 Test Method for Determining Corrosivity of Adhesive Materials.

ASTM D 3418 *relating to* Glass Transition Temperature.

ASTM D 3433-75 (1985) Practice for Fracture Strength in Cleavage of Adhesives in Bonded Joints.

ASTM D 3434-90 Practice for Multiple-Cycle Accelerated Ageing Test (Automatic Boil Test) for Exterior Wet Use Wood Adhesives.

ASTM D 3482-90 Test Method for Determining Electrolytic Corrosion of Copper by Adhesives.

ASTM D 3498-91 Specification for Adhesives for Field-Glueing Plywood to Lumber Framing for Floor Systems.

ASTM D 3528-76 (1981) Test Method for Strength Properties of Double Lap Shear Adhesive Joints by Tension Loading.

ASTM D 3535-90 Test Method for Resistance to Deformation Under Static Loading for Structural Wood Laminating Adhesives Used Under Exterior (Wet Use) Exposure Conditions.

ASTM D 3632-77 (1990) Practice for Accelerated Ageing of Adhesive Joints by the Oxygen-Pressure Method.

ASTM D 3658-90 Test Method for Determining the Torque Strength of Ultraviolet (UV) Light-Cured Glass/Metal Adhesive Joints.

ASTM D 3762-79 (1988) Test Method for Adhesive-Bonded Surface Durability of Aluminium (Wedge Test).

ASTM D 3807-79 (1984) Test Method for Strength Properties of Adhesives in Cleavage Peel by Tension Loading (Engineering Plastics-to-Engineering Plastics).

ASTM D 3808-79 (1984) Practice for Qualitative Determination of Adhesion of Adhesives to Substrates by Spot Adhesion Test Method.

ASTM D 3929-80 (1984) Practice for Evaluating the Stress Cracking of Plastics by Adhesives Using the Bent-Beam Method.

ASTM D 3930-90a Specification for Adhesives for Wood-Based Materials for Construction of Manufactured Homes.

ASTM D 3931-90 Test Method for Determining Strength of Gap-Filling Adhesive Bonds in Shear by Compression Loading.

ASTM D 3932-80 Practice for the Control of the Application of Structural Fasteners When Attached by Hot Melt Adhesives.

ASTM D 3933-80 Practice for Preparation of Aluminium Surfaces for Structural Adhesives Bonding (Phosphoric Acid Anodising).

ASTM D 3983-81 (1991) Test Method for Measuring Strength and Shear Modulus of Non-rigid Adhesives by the Thick Adherend Tensile Lap Specimen.

ASTM D 4027-81 (1991) Test Method for Measuring Shear Properties of Structural Adhesives by the Modified-Rail Test.

ASTM D 4299-83 Test Methods for Effect of Bacterial Contamination on Permanence of Adhesive Preparations and Adhesive Films (Discontinued 1990. Replaced by D 4300 and D 4783).

ASTM D 4300-88 Test Methods for Ability of Adhesive Films to Support or Resist the Growth of Fungi.

ASTM D 4317-88 Specification for Polyvinyl Acetate-Based Emulsion Adhesives.

ASTM D 4338-90 Test Method for Flexibility Determination of Supported Adhesive Films by Mandrel Bend Test Method.

ASTM D 4339-89 Test Method for Determination of the Odour of Adhesives.

ASTM D 4426-84 Test Method for Determination of Percent Non-volatile Content of Liquid Phenolic Resins Used for Wood Laminating.

ASTM D 4497-89 Test Method for Determining the Open Time of Hot Melt Adhesives (Manual Method).

ASTM D 4498-85 (1989) Test Method for Heat-Fail Temperature in Shear of Hot Melt Adhesives.

ASTM D 4499-89 Test Method for Heat Stability of Hot-Melt Adhesives.

ASTM D 4500-89 Test Method for Determining Grit, Lumps, or Undissolved Matter in Water-Borne Adhesives.

ASTM D 4501-85 Test Method for Shear Strength of Adhesive Bonds Between Rigid Substrates by the Block-Shear Method.

ASTM D 4502-85 (1990) Test Method for Heat and Moisture Resistance of Wood-Adhesive Joints.

ASTM D 4562-90 Test Method for Shear Strength of Adhesives Using Pin-and-Collar Specimen.

ASTM D 4680-87 Test Method for Creep and Time to Failure of Adhesives in Static Shear by Compression Loading (Wood-to-Wood).

ASTM D 4688-90 Test Methods for Evaluating Structural Adhesives for Fingerjointing Lumber.

ASTM D 4689-90 Specification for Adhesive, Casein-Type.

ASTM D 4690-90 Specification for Urea-Formaldehyde Resin Adhesives.

ASTM D 4783-90 Test Methods for Resistance of Adhesive Preparations in Container to Attack by Bacteria, Yeast and Fungi.

ASTM D 4800-87 Guide for Classifying and Specifying Adhesives.

ASTM D 4896-89 Guide for Use of Adhesive-Bonded Single Lap-Joint Specimen Test Results.

ASTM D 5040-90 Test Methods for Ash Content of Adhesives.

ASTM D 5041-90 Test Method for Fracture Strength in Cleavage of Adhesives in Bonded Joints.

ASTM D 5113-90 Test Method for Determining Adhesive Attack on Rigid Cellular Polystyrene Foam.

ASTM E 229-70 (1981) Test Method for Shear Strength and Shear Modulus of Structural Adhesives.

British Standards

British Standards Institute - UK

- All specifications listed are copyright (1990) by AFNOR, BSI & DIN.
- Equivalent ISO and EN standards are also given, where applicable.

3J 12:1991 Specification for pressure-sensitive adhesive identification tape.

4J 10:1991 Specification for pressure-sensitive adhesive water resistant PVC tape.

4J 11:1991 Specification for pressure-sensitive adhesive paper masking tape.

BS 647:1993; ISO 9665:1993 Adhesives. Methods for sampling and testing animal glues.

BS 1000[665]:1979 Universal Decimal Classification. Second English full edition. Oils. Fats. Waxes. Adhesives. Gums. Resins.

BS 1133:Section 14:1985 Packaging code. Adhesive closing and sealing tape.

BS 1133:Section 16:1987 Packaging code. Adhesives for packaging.

BS 1203:1979 Specification for synthetic resin adhesives (phenolic and aminoplastic) for plywood.

BS 1204:1993 Specification for Type MR Phenolic and Aminoplastic Synthetic Resin Adhesives for Wood

BS 3046:1981 Specification for adhesives for hanging flexible wallcoverings.

BS 3544:1962 Methods of test for polyvinyl acetate adhesives for wood.

BS 3887:1991 Specification for pressure sensitive adhesive closing and sealing tapes.

BS 3924:1978 Specification for pressure-sensitive adhesive tapes for electrical insulating purposes.

BS 4071:1966 Specification for polyvinyl acetate (PVA) emulsion adhesives for wood.

BS 4129:1990 Specification for welding primers and weld-through sealants, adhesives and waxes for resistance welding of sheet steel.

BS 4584:Part 15:1978 Specification for metal-clad base materials for printed circuits. Adhesive coated polymeric films PETP-F-15 and PI-F-15.

BS 4781:1990 Specification for pressure-sensitive adhesive plastics labels for permanent use.

BS 4841:Part 3:1994 Rigid polyurethane (PUR) and polyisocyanurate (PIR) foam for building applications. Specification for two types of laminated board (roofboards) with auto-adhesively bonded reinforcing facings for use as roofboard thermal insulation for built-up roofs.

BS 5131 Methods of test for footwear and footwear materials. Adhesives:

- **Section 1.1:1991** Resistance of adhesive joints to heat (creep test).
- **Section 1.2:1991** Resistance of adhesive joints to peeling.
- **Section 1.3:1991** Preparation of test assemblies using adhesives (other than hot melt adhesives) for heat resistance (creep) and peel tests.
- **Section 1.4:1979** Heat activation life of adhesives.
- **Section 1.6:1979** Recommended environmental storage conditions for adhesive joints prior to heat resistance or peeling tests.
- **Section 1.7:1991** Preparation of test assemblies using hot melt adhesives for heat resistance (creep) and peel tests.
- **Section 1.8:1981** Rate of bond strength development in shear of hot melt adhesives for lasting.
- **Section 1.9:1985** Measurement of green strength of adhesive joints.

BS 5350: Methods of test for adhesives:

- **Part 0:1987** General introduction.
- **Part A1:1976** Adherend preparation.

BS 5350 (continued)

Part A2:1990 Guide to the selection of adherend materials.

Part B1:1978 Determination of density.

Part B2:1976 Determination of solids content.

Part B4:1993; ISO 10364:1993 Determination of pot life.

Part B5:1976 Determination of gelation time.

Part B8:1990 Determination of viscosity.

Part B9:1984 Determination of resistance to sagging (flow after application).

Part C1:1986 Determination of cleavage strength of adhesive bonds.

Part C4:1986 Determination of impact resistance of adhesive bonds (Warning: withdrawn document).

Part C5:1990 Determination of bond strength in longitudinal shear.

Part C6:1990 Determination of bond strength in direct tension in sandwich panel.

Part C7:1990 Determination of creep and resistance to sustained application of force.

Part C9:1990 Floating roller peel test.

Part C12:1994; ISO 11339:1993 180° peel test for flexible-to-flexible bonded assemblies (Tepee test).

Part C13:1990 Climbing drum peel test.

Part C14:1979 90° peel test for a rigid-to-rigid assembly.

Part C15:1990 Determination of bond strength in compressive shear.

Part D4:1976 Determination of staining potential.

Part E1:1976 Guide to statistical analysis.

Part E2:1979 Guide to sampling.

Part F1:1983 Performance tests for flooring adhesives.

Part G1:1987 Determination of torque strength of anaerobic adhesives on threaded fasteners.

Part G2:1987 Determination of static shear strength of anaerobic adhesives.

Part G3:1987 Determination of ability of anaerobic adhesives to set on metal surfaces.

Part H1:1981 Determination of heat stability of hot melt adhesives in the. application equipment.

Part H2:1982 Determination of low temperature flexibility or cold crack temperature.

Part H3:1984 Determination of heat resistance of hot-melt adhesives.

Part H4:1984 Determination of maximum open time of hot-melt adhesives (oven method).

BS 5442 Adhesives for construction:

Part 1:1989 Classification of adhesives for use with flooring materials.

Part 2:1989 Classification of adhesives for use with interior wall and ceiling coverings (excluding decorative flexible material in roll form).

Part 3:1979 Classification of adhesives for use with wood.

BS 5609:1986 Specification for printed pressure-sensitive, adhesive-coated labels for marine use, including requirements for label base material.

BS 5980:1980 Specification for adhesives for use with ceramic tiles and mosaics.

BS 6138:1989 Glossary of terms used in the adhesives industry.

BS 7116:1990 Specification for double sided pressure sensitive adhesive tapes.

BS EN 204:1991 Classification of non-structural adhesives for joining of wood and derived timber products.

BS EN 205:1991 Test methods for wood adhesives for non-structural applications. Determination of tensile shear strength of lap joints.

BS EN 301:1992 Adhesives, phenolic and aminoplastic, for load-bearing timber structures: classification an performance requirements.

BS EN 302 Adhesives for load-bearing timber structures: test methods:

Part 1:1992 Determination of bond strength in longitudinal tensile shear.

Part 2:1992 Determination of resistance to delamination (Laboratory method).

Part 3:1992 Determination of the effect of acid damage to wood fibres by temperature and humidity cycling on the transverse tensile strength.

Part 4:1992 Determination of the effects of wood shrinkage on the shear strength.

BS EN 582:1994 Thermal spraying. Determination of tensile adhesive strength.

BS EN 2243 Test methods for structural adhesives:

Part 2:1991 Peel metal-metal.

Part 3:1992 Test methods. Peeling test metal-honeycomb core.

Part 4:1991 Metal-honeycomb core flatwise tensile test.

Part 5:1992 Ageing tests.

BS EN 26922:1993; ISO 6922:1987 Adhesives. Determination of tensile strength of butt joints.

BS EN 28510 Adhesives. Peel test for a flexible-bonded to-rigid test specimen assembly:

Part 1:1993; ISO 8510-1:1990 90° Peel.

Part 2:1993; ISO 8510-2:1990 180° Peel.

BS EN 29142:1993; ISO 9142:1990 Adhesives. Guide to the selection of standard laboratory ageing conditions for testing bonded joints.

BS EN 29653:1994; ISO 9653:1991 Adhesives. Test method for shear impact strength of adhesive bonds.

DIN Standards

Deutsche Institut für Normung - Germany

DIN 933 M1.6 to M52 hexagon head screws threaded up to the head; product grades A and B. **Now: ISO 4017**.

DIN 934 Hexagon nuts with metric coarse and fine pitch thread; product grades A and B. **Now: ISO 4032**.

DIN 50017 Atmospheres and there technical application; Condensation water test atmospheres.

DIN 51755 Testing of mineral oils and other combustible liquids; determination of flash point by the closed tester according to Abel-Pensky, method for determination of the flash point below +5°C up to approximately -30°C.

DIN 51757 Testing of petroleum and related materials; determination of density.

DIN 52451 Testing of sealing materials for buildings; determination of change in volume after temperature load; pycnometer method.

DIN 52612 Testing of thermal insulating materials; determination of thermal conductivity by means of the guarded hot plate apparatus; conversion of the measured values for building applications.

DIN 53269 Adhesives for coverings for floors; walls and ceilings; dispersion adhesives for back fittings of textile floorings; requirements; test methods.

DIN 53282 Testing of adhesives for metals and adhesively bonded metal joints; T-peel test.

DIN 53283 Testing of adhesives for metals and adhesively bonded metal joints; determination of the shear strength of single lap joints in shear by tension loading.

DIN 53286 Testing of adhesives for metals and adhesively bonded metal joints; conditions for testing at different temperatures.

DIN 53289 Testing of adhesives for metals and adhesively bonded metal joints; floating roller peel test.

E DIN 53452 Testing of plastics; bending test.

E DIN 53454 Testing of plastics; compression test.

E DIN 53455 Testing of plastics; tensile test.

DIN 53456 Testing of plastics; indentation hardness test.

DIN 53457 Testing of plastics; determination of the elastic modulus by tensile, compression and bend testing.

DIN 53479 Testing of plastics and elastomers; determination of density.

DIN 53481 Now: DIN VDE 0303 Part 2 Testing of electrical insulating materials; determination of the electrical breakdown voltage and electric strength.

DIN 53482 Now: DIN VDE 0303 Part 3 Methods of test for materials for electrical purposes; measuring of the electrical resistance of non-metallic materials.

DIN 53483 Testing of insulting materials; Determination of dielectric properties; measuring cells for liquids for frequencies up to 100 MHz.

DIN 53495 Testing of plastics; determination of water absorption.

DIN 53504 Testing of vulcanised rubber; determination of tensile strength at break, tensile strength, elongation at break and stress values in a tensile test.

DIN 53505 Testing of rubber, elastomers and plastics; Shore hardness testing A and D.

DIN 53515 Testing of rubber and elastomer and plastic foils; Tear propagation test using the Graves angle specimen with incision.

DIN 54452 Testing of adhesives for metals and adhesively bonded metal joints; compression shear test.

DIN 54455 Testing of adhesives for metals and adhesively bonded metal joints; torsional shear test.

DIN 68131 Castors for revolving chairs and seats.

EN Standards

Euro Norme - Europe

EN 204 Classification of Non-structural Adhesives for Joining of Wood and Derived Timber Products.

EN 205 Test Methods for Wood Adhesives for Non-structural Applications - Determination of Tensile Shear Strength of Lap Joints.

EN 301:1992 Adhesives, phenolic and aminoplastic, for load-bearing timber structures: classification an performance requirements.

EN 302 Adhesives for load-bearing timber structures: test methods:

- **Part 1:1992** Determination of bond strength in longitudinal tensile shear.
- **Part 2:1992** Determination of resistance to delamination (Laboratory method).
- **Part 3:1992** Determination of the effect of acid damage to wood fibres by temperature and humidity cycling on the transverse tensile strength.
- **Part 4:1992** Determination of the effects of wood shrinkage on the shear strength.

EN 542:1995 Adhesives. Determination of density. (Norme Française T76-090).

EN 543:1995 Adhesives. Determination of apparent density of powder and granule adhesives. (Norme Française T76-091).

EN 582:1994 Thermal spraying. Determination of tensile adhesive strength.

EN 827:1995 Adhesives. Determination of conventional solids content and constant mass solids content. (Norme Française T76-101).

EN 924:1995 Adhesives. Solvent-borne and solvent-free adhesives. Determination of flash point, (Norme Française T76-093).

EN 1464:1995 Adhesives. Determination of peel resistance of high-strength adhesive bonds. Floating roller method. (Norme Française T76-112).

EN 1465:1995 Adhesives. Determination of tensile lap-shear strength of rigid-to-rigid bonded assemblies. (Norme Française T76-107).

EN 2243 Aerospace Series - Structural Adhesives - Test Methods:

- **Part 2:1991** Peel metal-metal.
- **Part 3:1992** Peeling test metal - honeycomb core.
- **Part 4:1991** Metal-honeycomb core flatwise tensile test.
- **Part 5:1992** Ageing tests.

EN 2374 Aerospace Series - Glass Fibre Reinforced Mouldings and Sandwich Composites - Production of Test Panels.

EN 2497 Aerospace Series - Dry Abrasive Blasting of Titanium and Titanium Alloys.

EN 26922:1993; ISO 6922:1987 Adhesives. Determination of tensile strength of butt joints.

EN 28510 Adhesives. Peel test for a flexible-bonded to-rigid test specimen assembly:

- **Part 1:1993; ISO 8510-1:1990** 90° Peel.
- **Part 2:1993; ISO 8510-2:1990** 180° Peel.

EN 29142:1993; ISO 9142:1990 Adhesives. Guide to the selection of standard laboratory ageing conditions for testing bonded joints.

EN 29653:1994; ISO 9653:1991 Adhesives. Test method for shear impact strength of adhesive bonds.

EN 60454 Specifications for Pressure Sensitive Adhesive Tapes for Electrical Purposes:

- **Part 1** General Requirements.
- **Part 2** Methods of Test.

prEN 2667:Part 6 Test Method for the Determination of Water Absorption of Structural Foam Film Adhesives.

prEN 2757 Test Method for Determining the Drying and Ignition Residues of Adhesive Primers.

ISO Standards

International Standards Organisation - Switzerland

ISO 4017:1988 M1.6 to M52 hexagon head screws threaded up to the head; product grades A and B. **Was: DIN 933**.

ISO 4032:1986 Hexagon nuts with metric coarse and fine pitch thread; product grades A and B. **Was: DIN 934**.

ISO 4578:1990 Ed.2 Adhesives - Determination of peel resistance of high-strength adhesive bonds - Floating roller method.

ISO 4587:1995 Ed.2 Adhesives - Determination of tensile lap-shear strength of rigid-to-rigid bonded assemblies.

ISO 4588:1989 Ed.1 Adhesives - Preparation of metal surfaces for adhesive bonding.

ISO 6237:1987 Ed.1 Adhesives - Wood-to-wood adhesive bonds - Determination of shear strength by tensile loading.

ISO 6238:1987 Ed.1 Adhesives - Wood-to-wood adhesive bonds - Determination of shear strength by compression loading.

ISO 6922:1987 Ed.1; BS EN 26922:1993 Adhesives - Determination of tensile strength of butt joints.

ISO 7387:1983 Ed.1 Adhesives with solvents for assembly of uPVC pipe elements - Characterisation.
Part I: Basic test methods.

ISO 8510:1990 Ed.1; BS EN 28510-1 & -2:1993 Adhesives - Peel test for a flexible-bonded-to-rigid test specimen assembly.
Part 1: 90 degree peel.
Part 2: 180 degree peel.

ISO 9142:1990 Ed.1; BS EN 29142:1993 Adhesives - Guide to the selection of standard laboratory ageing conditions for testing bonded joints.

ISO 9653:1991 Ed.1; BS EN 29653:1994 Adhesives - Test method for shear impact strength of adhesive bonds.

ISO 9664:1993 Ed.1 Adhesives - Test methods for fatigue properties of structural adhesives in tensile shear (Norme Française T76-111).

ISO 9665:1993 Ed.1; BS 647:1993 Adhesives - Animal glues - Methods for sampling and testing.

ISO 10123:1990 Ed.1 Adhesives - Determination of shear strength of anaerobic adhesives using pin-and-collar specimens.

ISO 10354:1992 Ed.1 Adhesives - Characterisation of durability of structural-adhesive-bonded assemblies - Wedge rupture test.

ISO 10363-1992 Ed.1 Hot-melt adhesives. - Determination of thermal stability. (Norme Française T76-120).

ISO 10364:1993 Ed.1; BS 5350:Part B4:1993 Adhesives - Determination of working life (pot life) of multi-component adhesives.

ISO 10365.1992 Ed.1 Adhesives - Designation of main failure patterns (Norme Française T76-130).

ISO 10964:1993 Ed.1 Adhesives - Determination of torque strength of anaerobic adhesives on threaded fasteners.

ISO 11003:1993 Ed.1 Adhesives - Determination of shear behaviour of structural bonds.
Part 1: Torsion test method using butt-bonded hollow cylinders.
Part 2: Thick-adherend tensile-test method.

ISO 11339:1993 Ed.1; BS 5350:Part C12:1994 Adhesives - 180 degree peel test for flexible-to-flexible bonded assemblies (T-peel test).

ISO 11343:1993 Ed.1 Adhesives - Determination of dynamic resistance to cleavage of high strength adhesive bonds under impact conditions - Wedge impact method.

ISO 13445:1995 Ed.1 Adhesives - Determination of shear strength of adhesive bonds between rigid substrates by the block shear method.

NF Standards

Norme Française (AFNOR) - France

NF B51-032:1981 Laminated bonded wood - Test of compression shear.

F01-811:1989 Railway rolling stock. Jointing sticking.

L17-500:1991 Aerospace series. Structural adhesives. Classification and qualification.

L17-511:1990 Aerospace series. Structural adhesives. Measurement of the creeping of a bonded joint.

P85-600:1994 Jointing products. Adhesives for structural sealant glazing systems. Specifications.

NF T76-001:1981 Adhesives. Vocabulary.

NF T76-011:1980 Adhesives. Classification of the main adhesives.

NF T76-090, NF EN 542:1995 Adhesives. Determination of density. (European Standard EN 542).

NF T76-091, NF EN 543:1995 Adhesives. Determination of apparent density of powder and granule adhesives. (European Standard EN 543).

NF T76-093, NF EN 924:1995 Adhesives. Solvent-borne and solvent-free adhesives. Determination of flash point, (European Standard EN 924).

NF T76-101, NF EN 827:1995 Adhesives. Determination of conventional solids content and constant mass solids content. (European Standard EN 827).

NF T76-102:1982 Adhesives. Measurement of the Brookfield viscosity of adhesives in the liquid state.

NF T76-103:1983 Adhesives. Determination of pH adhesives, their base products and similar preparation.

NF T76-104:1983 Adhesives. Methods of sampling adhesives and raw materials used for adhesives.

NF T76-105:1976 Adhesives. Determination of the white point temperature.

NF T76-106:1981 Adhesives. Determination of the conventional melting point of adhesives hot-melt. ("Ball and Ring" method).

NF T76-107, NF EN 1465:1995 Adhesives. Determination of tensile lap-shear strength of rigid-to-rigid bonded assemblies. (European Standard EN 1465).

T76-108:1979 Adhesives. Compressive shear strength of adhesives for parallel bonding of wood.

NF T76-110:1981 Adhesives. Determination of ash content.

NF T76-111, NF EN ISO 9664:1995 Adhesives. Test methods for fatigue properties of structural adhesives in tensile shear. (European Standard EN ISO 9664).

NF T76-112, NF EN 1464: 1995 Adhesives. Determination of peel resistance of high-strength adhesive bonds. Floating roller method. (European Standard EN 1464).

NF T76-113:1983 Water adhesives. Aminoplastic. Determination of conventional gel time at different temperatures.

T76-114:1983 Structural adhesives. Cleavage test using wedge.

NF T76-116:1987 Adhesives. Reactive adhesives with organic binders. Determination of conventional reactivity.

NF T76-117:1984 Water based adhesives. Resorcinol formaldehyde, resorcinol-phenol formaldehyde. Conventional determination of gel setting time at 23 degrees C.

NF T76-120, NF EN ISO 10363:1995 Hot-melt adhesives. Determination of thermal stability. (European Standard EN ISO 10363).

NF T76-121:1984 Structural adhesives. Determination of shear strength bonds axis-rings system.

NF T76-122:1985 Adhesives. Characterisation of structural adhesives. Performance test under permanent shear stress.

NF T76-124:1987 Structural adhesives. Specification. Determination of functional torques on bonded threaded components.

NF T76-125:1986 Adhesives. Conventional determination of setting strength.

NF T76-126:1987 Adhesives for flexible floor coverings. Peel resistance test at an angle of 90 degrees. (Floating Roller method).

NF T76-130, NF EN ISO 10365:1995 Adhesives. Designation of main failure patterns. (European Standard EN ISO 10365).

NF T76-131:1989 Adhesives for plastic flexible floor coverings. List of characterisation test methods.

NF T76-132:1989 Adhesives for plastic flexible floor coverings. PVC material for peel test.

NF T76-141:1988 Structural adhesives. Shear test for determining the stress/strain curve of an adhesive in a glued assembly. Test method.

NF T76-142:1988 Structural adhesives. Preparation method of structural adhesive plates for characterisation test pieces.

T76-143:1990 Structural adhesives. Characterisation of a surface finish. Measurement of the adherence by three-point bending.

NF T76-201:1982 Adhesives. Preparation of metallic substrates for test pieces intended for laboratory tests.

NF T76-300:1983 Adhesives. List of characterisation test methods.

NF T76-303:1983 Organic solvent adhesives. Lists of characterisation methods.

SPECIFICATIONS & STANDARDS BY INDUSTRY SECTOR AND USES

Terminology

ASTM D 907-91b Terminology of Adhesives.

BS 6138:1989 Glossary of terms used in the adhesives industry.

NF T76-001:1981 Adhesives. Vocabulary.

Aerospace

This industry sector has, quite understandably, placed very exacting demands on the materials used in the construction of civil aircraft, military aircraft and space structures. This includes the adhesives used to assemble parts and to repair structures. Adhesive bonding technology has been subjected to rigorous research, development, testing and stress analysis. There are several types of the standards and specifications applied:

- **Aircraft constructors**: Developed within a particular company, e.g. Airbus, Boeing, etc. often with reference to internationally accepted standards; although modifications are commonplace. Aircraft companies have worked hard to develop reliable standards and test procedures. The results of which have also been incorporated into other nationally or internationally recognised standards.

- **Space structure constructors**: Often assessed to aircraft standards; since the materials used are similar (many aircraft companies also build space structures). Other tests are needed to cover the different conditions encountered between space and earth. For example, out- and off-gassing (volatile substances liberated in the vacuum of space) are of great concern since the substances may affect crew members or condense on nearby equipment. Both ESA and NASA space agencies have comprehensive documentation relating to all aspects of the selection and safe, reliable use of materials for spacecraft and associated equipment.

- **Military Specifications:** There are a considerable number of specifications for adhesives and similar products used in military equipment, e.g. Def. Standards (UK); or MIL- and MMM (USA). Some of these have been adopted for wider use.

As a consequence, adhesives for aerospace applications can be quoted to international test method standards, as meeting a particular company specification, and/or a military-derived specification. Adhesive manufacturers often state whether a product meets (or exceeds) the requirements of a stated specification.

AECMA

Association Européen des Constructeurs de Materiel Aerospatiale

Examples of relevant standards include:

***pr* EN 2000** QA Requirements for Manufacture and Procurement of EN-aerospace Standard Products

***pr* EN 2310** Non-metallic Materials: requirements and methods for Inflammability for Qualification

Note: *pr* indicates provisional.

Green papers are available for:

EN 2243-01 Structural adhesives Single-lap shear test method.

EN 2243-02 Peel (metal-to-metal) Test Method.

EN 2243-03 Peeling Test: Metal-to-Honeycomb Test Method.

EN 2243-04 Metal-to-Honeycomb Core Flatwise Tensile Test Method.

EN 2243-05 Ageing Tests.

Space Agencies

Examples of those applicable to adhesives are:

ESA PSS-03-210 Adhesive Bonding Handbook for Advanced Structural Materials

ESA PSS-01-70 Material, Mechanical-Part and Process Selection and Quality Control for ESA Space Systems.

ESA PSS-01-700 The Technical Reporting and Approval Procedure for Materials, Mechanical-Parts and Processes.

ESA PSS-01-701 Data for the Selection of Space Materials.

ESA PSS-01-702 A Thermal Vacuum Test for the Screening of Space Materials.

ESA PSS-01-704 A Thermal Cycling Test for the Screening of Space Materials and Processes.

ESA PSS-01-705 The Detection of Organic Contamination of Surfaces by Infrared Spectroscopy.

ESA PSS-01-706 The Particle and UV Radiation Testing of Space Materials.

ESA PSS-01-711 Product Assurance Requirements for Micro- VCM Apparatus and Associated Equipment.

ESA PSS-01-721 Flammability testing for the screening of space materials.

ESA PSS-01-722 The control of limited-life materials.

ESA PSS-01-729 The Determination of Offgassing Products from Materials and Assembled Articles to be used in a Manned Space Vehicle Crew Compartment.

Note: See also **ESA PSS-01-20** and related specifications for contamination and cleanliness control requirements.

NASA NHB 8060.1 Flammability, Odour and Offgassing Requirements and Test Procedures for Materials in Environments that Support Combustion.

Aircraft manufacturers

Two examples of standard procedures used for pre-bonding preparation of metals quoted by many adhesive suppliers and widely used are those developed by the Boeing Airplane Company (BAC).

BAC 5555 Phosphoric Acid Anodising of Aluminium Alloys for Structural Bonding.

BAC 5890 Anodising of Titanium for Adhesive Bonding.

US Federal specifications

List reproduced courtesy of Aerospheres (UK)

A-A-272 Caulking Compounds.

A-A-373 Glazing Compound.

DOD-C-24176 Cement Epoxy, Metal Repair.

HH-C-536 Caulking Compound for Gasket Connections.

MIL- R-46082 Retaining Compounds, Single Component.

MIL-A-101 Adhesive, water-resistant.

MIL-A-1154 Adhesive-Bonds Vulcanised Synthetic Rubber Parts.

MIL-A-13883 Adhesive, Synthetic Rubber.

MIL-A-14042 Adhesive Epoxy.

MIL-A-21016 Linoleum and Tile Adhesive.

MIL-A-21366 Adhesive Plastic Table Top to Aluminium.

MIL-A-22010 Adhesive, Solvent Type Polyvinylchloride.

MIL-A-22397 Phenol and Resorcinol Resin Base.

MIL-A-22895 Adhesive, Metal Identification Plate.

MIL-A-24179 Adhesive, Flexible.

MIL-A-25463 Adhesive, Metallic Structure Sandwich Construction.

MIL-A-3167 Adhesives-for Plastic Inhibitors.

MIL-A-3316 Fire-resistant Adhesive.

MIL-A-374 Adhesive, Paste for Demolition Charges.

MIL-A-3920 Adhesive, Optical, Thermosetting.

MIL-A-3941 Water-resistant Label Adhesive.

MIL-A-43316 Adhesive, Patching.

MIL-A-43365 Adhesive, Repair of Radome, Air.

MIL-A-46050 Adhesive, Special Rapid Room Temperature.

MIL-A-46051 Adhesive, Room Temperature and Intermediate Temperature.

MIL-A-46091 Adhesive, Brake Lining to Metal.

MIL-A-46106 Adhesive, Sealant-Silicone, General Purpose.

MIL-A-46146 Adhesive, Sealants, Silicone, RTV, Non corrosive.

MIL-A-4833 Adhesive, Cellulose Nitrate Base.

MIL-A-5090 Adhesive, Heat Resistant Metal to Metal.

MIL-A-5092 Light Stress Bonding Only.

MIL-A-52194 Adhesive, Epoxy.

MIL-A-5540 Polychloroprene Adhesive.

MIL-A-60091 Adhesive for Bonding Demolition Charges.

MIL-A-81236 Adhesive, Epoxy Resins.

MIL-A-82484 Adhesive and Sealing Compounds.

MIL-A-8576 Adhesive, Acrylic Monomer Base.

MIL-A-9117 Synthetic Elastomeric Sealant.

MIL-C-12850 Natural Liquid Rubber Cement.

MIL-C-14064 Grinding Disk Cement.

MIL-C-15705 Caulking Compound.

MIL-C-18255 Caulking Compound with Synthetic Rubber Base.

MIL-C-18969 Caulking Compound-Watertight Exterior Hull Seams of Vessels.

MIL-C-23092 Cement, Natural Rubber.

MIL-C-27315 Coating Systems, Elastomeric.

MIL-C-27725 Coating, Corrosion Prevent. Air. Fuel Tanks

MIL-C-5539 Natural Rubber Cement.

MIL-C-83019 Protective Coating, Integral Fuel Tank.

MIL-C-83231 Coatings, Polyurethane, Rain, Erosion Resistant.

MIL-C-8514 Resin-Acid Metal Pretreatment Compound for Aircraft.

MIL-C-897 Cement Rubber.

MIL-D-17951 Sealing Compound for Use with Deck Covering.

MIL-G-413 Marine Glue.

MIL-P-20628 Putty, Sealing.

MIL-P-23236 Paint Coating, Ship, Fuel, Salt Tanks.

MIL-R-17882 Epoxy Resin for Pipe Repair.

MIL-S-11030 Non curing, Polysulphide Base Sealing Compound.

MIL-S-11031 Adhesive, Curing Compound for Bonding Metal to Metal.

MIL-S-12158 Metal to Metal Flange Joint Sealer.

MIL-S-12935 Synthetic Resin Lumber Knot Sealing Compound.

MIL-S-15204 High Temperature Sealing Compound.

MIL-S-17377 Compound Composed of Plastic or Resinous Binders.

MIL-S-20541 Liquid Rubber Cement.

MIL-S-22473 Anaerobic Single Component Sealing Compound.

MIL-S-23498 Sealing Compound, Bearing-Preservation.

MIL-S-24340 Sealing, Ship Deck Polyurethane.

MIL-S-3927 Sealing Compound, Thread.

MIL-S-4383 Sealing Compound-Synthetic Rubber Base.

MIL-S-45180 Thread and Gasket Sealing Compounds.

MIL-S-46163 Sealing, Lubricating and Wicking Compounds Thread-Locking.

MIL-S-7502 High Adhesion Sealant.

MIL-S-7916 Thread and Gasket Sealing Compound.

MIL-S-81732 Sealing Compound, Electrical.

MIL-S-81733 Sealing, Coating, Corrosion Inhibitive.

MIL-S-83315 Sealing, Aluminium Structure.

MIL-S-83318 Sealant, Quick Repair Integral Fuel Tank.

MIL-S-83430 Sealing Compound, Integral Fuel Tanks and Fuel Cell Cavities.

MIL-S-8516 Sealing Compound-Protects Electrical Components.

MIL-S-8784 Sealing Compound for Aircraft Fuel Tank.

MIL-S-8802 High Temperature Sealant.

MIL-T-5542 Thread Compound, Antiseize and Sealing.

MIL-T-83483 Thread Compound, Antiseize, Molybdenum Disulphide Petrolatum.

MMM-A-001058 Adhesive, Rubber Base Pressurised Dispensers.

MMM-A-001993 Adhesive, Epoxy, Flexible, Filled.

MMM-A-100 Animal Glue.

MMM-A-105 Adhesive and Sealing Compounds.

MMM-A-110 Fast Drying Paste.

MMM-A-115 Asphalt Tile Cement.

MMM-A-121 Adhesive, Binding.

MMM-A-122 High Strength Adhesive.

MMM-A-125 Casein Glue in Powder Form.

MMM-A-132 Adhesive, Heat Resistant.

MMM-A-134 Adhesive, Epoxy Resin, Metal to Metal Structural Bonding.

MMM-A-137 Strong Bonding Linoleum Cement.

MMM-A-138 Adhesive, Metal to Wood.

MMM-A-139 Adhesive, Natural or Synthetic.

MMM-A-150 Paste for Fibrous Acoustical Materials.

MMM-A-1617 Adhesive, Rubber Base, General Purpose.

MMM-A-1754 Adhesive and Sealing Compound, Epoxy, Metal Filled.

MMM-A-177 Adhesive-Paste.

MMM-A-178 Adhesive, Paper Label.

MMM-A-179 Adhesive, Paper Label.

MMM-A-180 Thermoplastic Synthetic Resin.

MMM-A-181 Phenol, Melamine or Resorcinol Resin Base.

MMM-A-182 Adhesive, Innertube Repair.

MMM-A-187 Epoxy Resin Paste.

MMM-A-188 Thermosetting Urea-Resin.

MMM-A-189 Synthetic Rubber for Hot and Cold Bonding.

MMM-A-193 Adhesive, Vinyl Acetate Resin Emulsion.

MMM-A-1931 Adhesive, Epoxy, Silver Filled, Conductive.

MMM-A-250 Water-resistant Rubber Base Liquid Adhesive.

MMM-A-260 Water-resistant Liquid Adhesive.

MMM-B-00350 Binder Adhesive, Epoxy Resin-Flexible.

SS-S-1996 Sealer, Water and Weather Resistant.

TT-C-00598 Caulking Compound.

TT-C-1796 Caulking Compounds, Metal Seam and Wood Seam.

TT-F-320 Filler for Cracks in Wood, Metal, Concrete and Cement.

TT-F-322 Metal Surface Dent Filler.

TT-F-336 Filler for Wood.

TT-F-340 Plastic Wood Filler.

TT-P-1536 Plumbing Fixture Setting Compound.

TT-P-781 Putty and Elastic Compound.

TT-P-791 Wood-Sash Glazing Putty.

TT-S-00230 Single Component Synthetic Rubber Base Joint Sealant.

TT-S-0227 Sealing Compound.

TT-S-1732 Sealing Compound , Pipe Joint and Thread.

VV-S-190 Solid Form Sealing Compound for Overseas Shipments.

Construction/Building

(See also: Wood and Wood-products)

ASTM C 557-73 (1985) Specification for Adhesives for Fastening Gypsum Wallboard to Wood Framing.

ASTM D 3498 Specification for Adhesives for Field-Gluing Plywood to Lumber Framing for Floor Systems (reference D-3632, Practice for Accelerated Ageing of Adhesive Joints by the Oxygen-Pressure Method and D-3931, Test Method for Determining Strength of Gap Filling Adhesive Bonds in Shear by Compression Loading).

ASTM D 3930 Specification for Adhesives for Wood-Based Materials for Construction of Manufactured Homes.

BS 5442 Adhesives for construction:

BS 5442 Part 1:1989 Classification of adhesives for use with flooring materials.

BS 5442 Part 2:1989 Classification of adhesives for use with interior wall and ceiling coverings (excluding decorative flexible material in roll form).

BS 5980:1980 Specification for adhesives for use with ceramic tiles and mosaics.

Metal Bonding

Many of these standards are applied in various industry sectors. (See also: Aerospace).

ASTM D 1002-72 (1983) Test Method for Strength Properties of Adhesives in Shear by Tension Loading (Metal-to-Metal).

ASTM D 1062-79 (1983) Test Method for Cleavage Strength of Metal-to Metal Adhesive Bonds.

ASTM D 1184-69 (1986) Test Method for Flexural Strength of Adhesive Bonded Laminated Assemblies.

ASTM D 1193 Test Methods for Resistance of Adhesives to Cyclic Laboratory Ageing Conditions.

ASTM D 1780 Practice for Conducting Creep Tests of Metal to-Metal Adhesives.

ASTM D 1781 Method for Climbing Drum Peel Test for Adhesives.

ASTM D 1828 Practice for Atmospheric Exposure of Adhesive-Bonded Joints and Structures.

ASTM D 1879 Practice for Exposure of Adhesive Specimens to High-Energy Radiation.

ASTM D 1976 Test Method for Peel Resistance of Adhesives (T-Peel Test).

ASTM D 2293 Test Method for Creep Properties of Adhesives in Shear by Compression Loading (Metal-to-Metal).

ASTM D 2294 Test Method for Creep Properties of Adhesives in Shear by Tension Loading (Metal-to-Metal).

ASTM D 2295 Test Method for Strength Properties of Adhesives in Shear by Tension Loading at Elevated Temperatures (Metal-to-Metal).

ASTM D 2557 Test Method for Strength Properties of Adhesives in Shear by Tension Loading in the Temperature Range from -267.9 to -55°C (-450 to 67°F).

ASTM D 2918 Practice for Determining Durability of Adhesive Joints Stressed in Peel.

ASTM D 2919 Test Method for Determining Durability of Adhesive Joints Stressed in Shear by Tension Loading.

ASTM D 3165 Test Method for Strength Properties of Adhesives in Shear by Tension Loading of Laminated Assemblies.

ASTM D 3166 Test Method for Fatigue Properties of Adhesives in Shear by Tension Loading (Metal/Metal).

ASTM D 3167 Test Method for Floating Roller Peel Resistance of Adhesives.

ASTM D 3433 Practice for Fracture Strength in Cleavage of Adhesives in Bonded Joints.

ASTM D 3528 Test Method for Strength Properties of Double Lap Shear Adhesive Joints by Tension Loading.

ASTM D 3762 Test Method for Adhesive-Bonded Surface Durability of Aluminium (Wedge Test).

ASTM D 3983 Test Method for Measuring Strength and Shear Modulus of Non-rigid Adhesives by the Thick Adherend Tensile Lap Specimen.

ASTM D 4027 Test Method for Measuring Shear Properties of Structural Adhesives by the Modified-Rail Test.

ASTM D 4501 Test Method for Shear Strength of Adhesive Bonds Between Rigid Substrates by the Block-Shear Method.

ASTM D 4562 Test Method for Shear Strength of Adhesives Using Pin-and Collar Specimen.

ASTM E 0229 Test Method for Shear Strength and Shear Modulus of Structural Adhesives.

BS EN 2243 Test methods for structural adhesives:

BS EN 2243 Part 2:1991 Peel metal-metal.

BS EN 2243 Part 3:1992 Test methods. Peeling test metal-honeycomb core.

BS 5350 Methods of test for adhesives:

BS 5350 Part 0:1987 General introduction.

BS 5350 Part C1:1986 Determination of cleavage strength of adhesive bonds.

BS 5350 Part C4:1986 Determination of impact resistance of adhesive bonds (Warning: withdrawn document).

BS 5350 Part C5:1990 Determination of bond strength in longitudinal shear.

BS 5350 Part C6:1990 Determination of bond strength in direct tension in sandwich panel.

BS 5350 Part C7:1990 Determination of creep and resistance to sustained application of force.

BS 5350 Part C9:1990 Floating roller peel test.

BS 5350 Part C13:1990 Climbing drum peel test.

BS 5350 Part C14:1979 90° peel test for a rigid-to-rigid assembly.

BS EN 28510 Part 1:1993; ISO 8510-1:1990 90° Peel.

BS EN 28510 Part 2:1993; ISO 8510-2:1990 180° Peel.

DIN 53282 Testing of adhesives for metals and adhesively bonded metal joints; T-peel test.

DIN 53283 Testing of adhesives for metals and adhesively bonded metal joints; determination of the shear strength of single lap joints in shear by tension loading.

DIN 53286 Testing of adhesives for metals and adhesively bonded metal joints; conditions for testing at different temperatures.

DIN 53289 Testing of adhesives for metals and adhesively bonded metal joints; floating roller peel test.

DIN 54452 Testing of adhesives for metals and adhesively bonded metal joints; compression shear test.

DIN 54455 Testing of adhesives for metals and adhesively bonded metal joints; torsional shear test.

EN 1464:1995 Adhesives. Determination of peel resistance of high-strength adhesive bonds. Floating roller method. (Norme Française T76-112).

EN 1465:1995 Adhesives. Determination of tensile lap-shear strength of rigid-to-rigid bonded assemblies. (Norme Française T76-107).

EN 2243 Aerospace Series - Structural Adhesives - Test Methods:

EN 2243 Part 2:1991 Peel metal-metal.

EN 2243 Part 4:1991 Metal-honeycomb core flatwise tensile test.

EN 26922:1993; ISO 6922:1987 Adhesives. Determination of tensile strength of butt joints.

EN 28510 Adhesives. Peel test for a flexible-bonded to-rigid test specimen assembly:

Part 1:1993; ISO 8510-1:1990 90° Peel.

Part 2:1993; ISO 8510-2:1990 180° Peel.

EN 29653:1994; ISO 9653:1991 Adhesives. Test method for shear impact strength of adhesive bonds.

ISO 4578:1990 Ed.2 Adhesives - Determination of peel resistance of high-strength adhesive bonds - Floating roller method.

ISO 4587:1995 Ed.2 Adhesives - Determination of tensile lap-shear strength of rigid-to-rigid bonded assemblies.

ISO 6922:1987 Ed.1; BS EN 26922:1993 Adhesives - Determination of tensile strength of butt joints.

ISO 8510:1990 Ed.1; BS EN 28510-1 & -2:1993 Adhesives - Peel test for a flexible-bonded-to-rigid test specimen assembly.

Part 1: 90 degree peel.

Part 2: 180 degree peel.

ISO 9653:1991 Ed.1; BS EN 29653:1994 Adhesives - Test method for shear impact strength of adhesive bonds.

ISO 9664:1993 Ed.1 Adhesives - Test methods for fatigue properties of structural adhesives in tensile shear (Norme Française T76-111).

ISO 10354:1992 Ed.1 Adhesives - Characterisation of durability of structural-adhesive-bonded assemblies - Wedge rupture test.

ISO 11003:1993 Ed.1 Adhesives - Determination of shear behaviour of structural bonds.

Part 1: Torsion test method using butt-bonded hollow cylinders.

Part 2: Thick-adherend tensile-test method.

ISO 11339:1993 Ed.1; BS 5350:Part C12:1994 Adhesives - 180 degree peel test for flexible-to-flexible bonded assemblies (T-peel test).

ISO 11343:1993 Ed.1 Adhesives - Determination of dynamic resistance to cleavage of high strength adhesive bonds under impact conditions - Wedge impact method.

ISO 13445:1995 Ed.1 Adhesives - Determination of shear strength of adhesive bonds between rigid substrates by the block shear method.

NF T76-107, NF EN 1465:1995 Adhesives. Determination of tensile lap-shear strength of rigid-to-rigid bonded assemblies. (European Standard EN 1465).

NF T76-111, NF EN ISO 9664:1995 Adhesives. Test methods for fatigue properties of structural adhesives in tensile shear. (European Standard EN ISO 9664).

NF T76-112, NF EN 1464: 1995 Adhesives. Determination of peel resistance of high-strength adhesive bonds. Floating roller method. (European Standard EN 1464).

T76-114:1983 Structural adhesives. Cleavage test using wedge.

NF T76-121:1984 Structural adhesives. Determination of shear strength bonds axis-rings system.

NF T76-122:1985 Adhesives. Characterisation of structural adhesives. Performance test under permanent shear stress.

NF T76-141:1988 Structural adhesives. Shear test for determining the stress/strain curve of an adhesive in a glued assembly. Test method.

Plastic Bonding

In addition to those listed here, test specifications for metal bonding are often used. Care is required when applying standards to some fibre-reinforced composites; especially peel tests as the adhered may fracture rather than deform.

ASTM D 3163-73 (1984) Test Method for Determining the Strength of Adhesively Bonded Rigid Plastic Lap-Shear Joints in Shear by Tension Loading.

ASTM D 3164-73 (1984) Test Method for Determining the Strength of Adhesively Bonded Plastic Lap-Shear Sandwich Joints in Shear by Tension Loading.

ASTM D 3807-79 (1984) Test Method for Strength Properties of Adhesives in Cleavage Peel by Tension Loading (Engineering Plastics-to-Engineering Plastics).

ASTM D 3929-80 (1984) Practice for Evaluating the Stress Cracking of Plastics by Adhesives Using the Bent-Beam Method.

E DIN 53452 Testing of plastics; bending test.

E DIN 53454 Testing of plastics; compression test.

E DIN 53455 Testing of plastics; tensile test.

DIN 53456 Testing of plastics; indentation hardness test.

DIN 53457 Testing of plastics; determination of the elastic modulus by tensile, compression and bend testing.

DIN 53505 Testing of rubber, elastomers and plastics; Shore hardness testing A and D.

DIN 53515 Testing of rubber and elastomer and plastic foils; Tear propagation test using the Graves angle specimen with incision.

Shoemaking

ASTM D 2558-69 (1984) Test Method for Evaluating Peel Strength of Shoe Sole-Attaching Adhesives (Intent to Withdraw).

BS 5131 Methods of test for footwear and footwear materials. Adhesives:

- **Section 1.1:1991** Resistance of adhesive joints to heat (creep test).
- **Section 1.2:1991** Resistance of adhesive joints to peeling.
- **Section 1.3:1991** Preparation of test assemblies using adhesives (other than hot melt adhesives) for heat resistance (creep) and peel tests.
- **Section 1.4:1979** Heat activation life of adhesives.
- **Section 1.6:1979** Recommended environmental storage conditions for adhesive joints prior to heat resistance or peeling tests.
- **Section 1.7:1991** Preparation of test assemblies using hot melt adhesives for heat resistance (creep) and peel tests.
- **Section 1.8:1981** Rate of bond strength development in shear of hot melt adhesives for lasting.
- **Section 1.9:1985** Measurement of green strength of adhesive joints.

Wood and Wood-products

ASTM D 0905 Test Method for Strength Properties of Adhesive Bonds in Shear by Compression Loading.

ASTM D 0906-82 (1987) Test Method for Strength Properties of Adhesives in Plywood Type Construction in Shear by Tension Loading.

ASTM D 1101-89 Test Methods for Integrity of Glue Joints in Structural Laminated Wood Products for Exterior Use.

ASTM D 2339-82 (1987) Test Method for Strength Properties of Adhesives in Two-Ply Wood Construction in Shear by Tension Loading.

ASTM D 2559-84 (1990) Specification for Adhesives for Structural Laminated Wood Products for Use Under Exterior (Wet Use) Exposure Conditions.

ASTM D 3024-84 Specification for Protein-Base Adhesives for Structural Laminated Wood Products for Use Under Interior (Dry Use) Exposure Conditions (Discontinued 1991).

ASTM D 3110-90 Specification for Adhesives Used in Nonstructural Glued Lumber Products.

ASTM D 3434-90 Practice for Multiple-Cycle Accelerated Ageing Test (Automatic Boil Test) for Exterior Wet Use Wood Adhesives.

ASTM D 3498-91 Specification for Adhesives for Field-Gluing Plywood to Lumber Framing for Floor Systems.

ASTM D 3535-90 Test Method for Resistance to Deformation Under Static Loading for Structural Wood Laminating Adhesives Used Under Exterior (Wet Use) Exposure Conditions.

ASTM D 3930-90a Specification for Adhesives for Wood-Based Materials for Construction of Manufactured Homes.

ASTM D 4299 Test Methods for Effect of Bacterial Contamination on Permanence of Adhesive Preparations and Adhesive Films.

ASTM D 4300 Test Methods for Effect of Mould Contamination on Permanence of Adhesive Preparations and Adhesive Films.

ASTM D 4317 Specification for Polyvinyl Acetate-Based Emulsion Adhesives.

ASTM D 4502 Test Method for Heat and Moisture Resistance of Wood-Adhesive Joints.

ASTM D 4680 Test Method for Creep and Time to Failure of Adhesives in Static Shear by Compression Loading (Wood-to-Wood).

ASTM D 4688-90 Test Methods for Evaluating Structural Adhesives for Fingerjointing Lumber.

ASTM D 4689 Specification for Adhesives, Casein-Type.

ASTM D 4690 Specification for Urea-Formaldehyde Resin Adhesives.

BS 1203:1979 Specification for synthetic resin adhesives (phenolic and aminoplastic) for plywood.

BS 1204:1993 Specification for type MR phenolic and aminoplastic synthetic resin adhesives for wood.

BS 3544:1962 Methods of test for polyvinyl acetate adhesives for wood.

BS EN 204:1991 Classification of non-structural adhesives for joining of wood and derived timber products.

BS 4071:1966 Specification for polyvinyl acetate (PVA) emulsion adhesives for wood.

BS EN 205:1991 Test methods for wood adhesives for non-structural applications. Determination of tensile shear strength of lap joints.

BS EN 301:1992 Adhesives, phenolic and aminoplastic, for load-bearing timber structures: classification an performance requirements.

BS EN 302 Adhesives for load-bearing timber structures: test methods:

- **Part 1:1992** Determination of bond strength in longitudinal tensile shear.
- **Part 2:1992** Determination of resistance to delamination (Laboratory method).

BS EN 302 (continued)

Part 3:1992 Determination of the effect of acid damage to wood fibres by temperature and humidity cycling on the transverse tensile strength.

Part 4:1992 Determination of the effects of wood shrinkage on the shear strength.

EN 204 Classification of Non-structural Adhesives for Joining of Wood and Derived Timber Products.

EN 205 Test Methods for Wood Adhesives for Non-structural Applications - Determination of Tensile Shear Strength of Lap Joints.

EN 301:1992 Adhesives, phenolic and aminoplastic, for load-bearing timber structures: classification an performance requirements.

EN 302 Adhesives for load-bearing timber structures: test methods:

Part 1:1992 Determination of bond strength in longitudinal tensile shear.

Part 2:1992 Determination of resistance to delamination (Laboratory method).

Part 3:1992 Determination of the effect of acid damage to wood fibres by temperature and humidity cycling on the transverse tensile strength.

Part 4:1992 Determination of the effects of wood shrinkage on the shear strength.

ISO 6237:1987 Ed.1 Adhesives - Wood-to-wood adhesive bonds - Determination of shear strength by tensile loading.

ISO 6238:1987 Ed.1 Adhesives - Wood-to-wood adhesive bonds - Determination of shear strength by compression loading.

NF B51-032:1981 Laminated bonded wood - Test of compression shear.

T76-108:1979 Adhesives. Compressive shear strength of adhesives for parallel bonding of wood.

Specifications - Types of Adhesives

ASTM D 1580-60 (1984) Specification for Liquid Adhesives for Automatic Machine Labelling of Glass Bottles (Intent to Withdraw).

ASTM D 1779-65 (1983) Specification for Adhesive for Acoustical Materials.

ASTM D 1874-62 (1986) Specification for Water- or Solvent-Soluble Liquid Adhesives for Automatic Machine Scaling of Top Flaps of Fiberboard Shipping Cases.

ASTM D 2851-86 Specification for Liquid Optical Adhesive.

ASTM D 4317-88 Specification for Polyvinyl Acetate-Based Emulsion Adhesives.

ASTM D 4689-90 Specification for Adhesive, Casein-Type.

ASTM D 4690-90 Specification for Urea-Formaldehyde Resin Adhesives.

ASTM D 4800-87 Guide for Classifying and Specifying Adhesives.

BS 1000[665]:1979 Universal Decimal Classification. Second English full edition. Oils. Fats. Waxes. Adhesives. Gums. Resins.

BS 1133:Section 14:1985 Packaging code. Adhesive closing and sealing tape.

BS 1133:Section 16:1987 Packaging code. Adhesives for packaging.

BS 1203:1979 Specification for synthetic resin adhesives (phenolic and aminoplastic) for plywood.

BS 1204:1993 Specification for type MR phenolic and aminoplastic synthetic resin adhesives for wood.

BS 3046:1981 Specification for adhesives for hanging flexible wallcoverings.

BS 3887:1991 Specification for pressure sensitive adhesive closing and sealing tapes.

BS 3924:1978 Specification for pressure-sensitive adhesive tapes for electrical insulating purposes.

BS 4071:1966 Specification for polyvinyl acetate (PVA) emulsion adhesives for wood.

BS 4129:1990 Specification for welding primers and weld-through sealants, adhesives and waxes for resistance welding of sheet steel.

BS 4781:1990 Specification for pressure-sensitive adhesive plastics labels for permanent use.

BS 5442 Part 3:1979 Classification of adhesives for use with wood.

BS 5980:1980 Specification for adhesives for use with ceramic tiles and mosaics.

DIN 53269 Adhesives for coverings for floors; walls and ceilings; dispersion adhesives for back fittings of textile floorings; requirements; test methods.

ISO 7387:1983 Ed.1 Adhesives with solvents for assembly of uPVC pipe elements - Characterisation

Part I: Basic test methods.

NF T76-011:1980 Adhesives. Classification of the main adhesives.

NF T76-131:1989 Adhesives for plastic flexible floor coverings. List of characterisation test methods.

NF T76-132:1989 Adhesives for plastic flexible floor coverings. PVC material for peel test.

NF T76-300:1983 Adhesives. List of characterisation test methods.

NF T76-303:1983 Organic solvent adhesives. Lists of characterisation methods.

P85-600:1994 Jointing products. Adhesives for structural sealant glazing systems. Specifications.

Surface Preparation

(See also: Aircraft Manufacturers)

ASTM D 2093-84 Practice for Preparation of Surfaces of Plastics Prior to Adhesive Bonding.

ASTM D 2651-90 Practice for Preparation of Metal Surfaces for Adhesive Bonding.

ASTM D 2674-72 (1984) Methods of Analysis of Sulfochromate Etch Solution Used in Surface Preparation of Aluminium.

ASTM D 3933-80 Practice for Preparation of Aluminium Surfaces for Structural Adhesives Bonding (Phosphoric Acid Anodising).

BS 5350 Part A1:1976 Adherend preparation.

EN 2497 Aerospace Series - Dry Abrasive Blasting of Titanium and Titanium Alloys.

ISO 4588:1989 Ed.1 Adhesives - Preparation of metal surfaces for adhesive bonding.

NF T76-142:1988 Structural adhesives. Preparation method of structural adhesive plates for characterisation test pieces.

T76-143:1990 Structural adhesives. Characterisation of a surface finish. Measurement of the adherence by three-point bending.

NF T76-201:1982 Adhesives. Preparation of metallic substrates for test pieces intended for laboratory tests.

Processing and Usage

ASTM D 1337-91 Test Method for Storage Life of Adhesives by Consistency and Bond Strength.

ASTM D 1338-91 Test Method for Working Life of Liquid or Paste Adhesives by Consistency and Bond Strength.

ASTM D 4497-89 Test Method for Determining the Open Time of Hot Melt Adhesives (Manual Method).

BS 5350 Part B4:1993; ISO 10364:1993 Determination of pot life.

BS 5350 Part B5:1976 Determination of gelation time.

BS 5350 Part G3:1987 Determination of ability of anaerobic adhesives to set on metal surfaces.

BS 5350 Part H4:1984 Determination of maximum open time of hot-melt adhesives (oven method).

ISO 10364:1993 Ed.1; BS 5350:Part B4:1993 Adhesives - Determination of working life (pot life) of multi-component adhesives.

NF T76-125:1986 Adhesives. Conventional determination of setting strength.

Chemical Properties

ASTM D 1488-86 Test Method for Amylaceous Matter in Adhesives.

ASTM D 1489-87 Test Method for Non-volatile Content of Aqueous Adhesives.

ASTM D 1490-82 (1987) Test Method for Non-volatile Content of Urea-Formaldehyde Resin Solutions.

ASTM D 1579-86 Test Method for Filler Content of Phenol, Resorcinol, and Melamine Adhesives.

ASTM D 1582-86 Test Method for Non-volatile Content of Phenol, Resorcinol, and Melamine Adhesives.

ASTM D 1583-86 Test Method for Hydrogen Ion Concentration of Dry Adhesive Films.

ASTM D 1994-91 Test Method for Determination of Acid Numbers of Hot Melt Adhesives.

ASTM D 4339-89 Test Method for Determination of the Odour of Adhesives.

ASTM D 4426-84 Test Method for Determination of Percent Non-volatile Content of Liquid Phenolic Resins Used for Wood Laminating.

ASTM D 4500-89 Test Method for Determining Grit, Lumps, or Undissolved Matter in Water-Borne Adhesives.

ASTM D 5040-90 Test Methods for Ash Content of Adhesives.

BS 5350 Part B2:1976 Determination of solids content

EN 827:1995 Adhesives. Determination of conventional solids content and constant mass solids content. (Norme Française T76-101).

NF T76-101, NF EN 827:1995 Adhesives. Determination of conventional solids content and constant mass solids content. (European Standard EN 827).

NF T76-103:1983 Adhesives. Determination of pH adhesives, their base products and similar preparation.

NF T76-116:1987 Adhesives. Reactive adhesives with organic binders. Determination of conventional reactivity.

Electrical Properties

ASTM D 1304-69 (1983) Methods of Testing Adhesives Relative to Their Use as Electrical Insulation.

ASTM D 2739-90 Test Method for Volume Resistivity of Conductive Adhesives.

DIN 53481 Now: DIN VDE 0303 Part 2 Testing of electrical insulating materials; determination of the electrical breakdown voltage and electric strength.

DIN 53482 Now: DIN VDE 0303 Part 3 Methods of test for materials for electrical purposes; measuring of the electrical resistance of non-metallic materials.

DIN 53483 Testing of insulting materials; Determination of dielectric properties; measuring cells for liquids for frequencies up to 100 MHz.

Physical Properties

ASTM D 570 *relating to* Moisture Absorption.

ASTM D 696 *relating to* Coefficient of Thermal Expansion.

ASTM D 898-90 Test Method for Applied Weight per Unit Area of Dried Adhesive Solids.

ASTM D 899-89 Test Method for Applied Weight per Unit Area of Liquid Adhesive.

ASTM D 1084-88 Test Methods for Viscosity of Adhesives.

ASTM D 1875-90 Test Method for Density of Adhesives in Fluid Form.

ASTM D 2183-69 (1982) Test Method for Flow Properties of Adhesives (Discontinued 1991).

ASTM D 2556-91 Test Method for Apparent Viscosity of Adhesives Having Shear-Rate-Dependent Flow Properties.

ASTM D 3418 *relating to* Glass Transition Temperature.

ASTM D 4498-85 (1989) Test Method for Heat-Fail Temperature in Shear of Hot Melt Adhesives.

ASTM D 4499-89 Test Method for Heat Stability of Hot-Melt Adhesives.

BS 5350 Part B1:1978 Determination of density.

BS 5350 Part B8:1990 Determination of viscosity

BS 5350 Part B9:1984 Determination of resistance to sagging (flow after application).

BS 5350 Part H3:1984 Determination of heat resistance of hot-melt adhesives.

DIN 51755 Testing of mineral oils and other combustible liquids; determination of flash point by the closed tester according to Abel-Pensky, method for determination of the flash point below +5°C up to approximately -30°C.

DIN 51757 Testing of petroleum and related materials; determination of density.

DIN 52451 Testing of sealing materials for buildings; determination of change in volume after temperature load; pycnometer method.

DIN 52612 Testing of thermal insulating materials; determination of thermal conductivity by means of the guarded hot plate apparatus; conversion of the measured values for building applications.

DIN 53479 Testing of plastics and elastomers; Determination of density.

DIN 53495 Testing of plastics; determination of water absorption.

EN 542:1995 Adhesives. Determination of density. (Norme Française T76-090).

EN 543:1995 Adhesives. Determination of apparent density of powder and granule adhesives. (Norme Française T76-091).

EN 924:1995 Adhesives. Solvent-borne and solvent-free adhesives. Determination of flash point, (Norme Française T76-093).

prEN 2667:Part 6 Test Method for the Determination of Water Absorption of Structural Foam Film Adhesives.

prEN 2757 Test Method for Determining the Drying and Ignition Residues of Adhesive Primers.

ISO 10363-1992 Ed.1 Hot-melt adhesives. - Determination of thermal stability. (Norme Française T76-120).

NF T76-090, NF EN 542:1995 Adhesives. Determination of density. (European Standard EN 542).

NF T76-091, NF EN 543:1995 Adhesives. Determination of apparent density of powder and granule adhesives. (European Standard EN 543).

NF T76-093, NF EN 924:1995 Adhesives. Solvent-borne and solvent-free adhesives. Determination of flash point, (European Standard EN 924).

NF T76-102:1982 Adhesives. Measurement of the Brookfield viscosity of adhesives in the liquid state.

NF T76-105:1976 Adhesives. Determination of the white point temperature.

NF T76-106:1981 Adhesives. Determination of the conventional melting point of adhesives hot-melt. ("Ball and Ring" method).

NF T76-110:1981 Adhesives. Determination of ash content.

NF T76-113:1983 Water adhesives. Aminoplastic. Determination of conventional gel time at different temperatures.

NF T76-117:1984 Water based adhesives. Resorcinol formaldehyde, resorcinol-phenol formaldehyde. Conventional determination of gel setting time at 23 degrees C.

NF T76-120, NF EN ISO 10363:1995 Hot-melt adhesives. Determination of thermal stability. (European Standard EN ISO 10363).

Mechanical Properties

ASTM D 0229 Test method for shear strength and shear modulus of structural adhesives.

ASTM D 997-78 (1983) Test Method for Tensile Properties of Adhesive Bonds.

ASTM D 905-8911 Test Method for Strength Properties of Adhesive Bonds in Shear by Compression Loading.

ASTM D 950-82 (1987) Test Method for Impact Strength of Adhesive Bonds.

ASTM D 1002-72 (1983) Test Method for Strength Properties of Adhesives in Shear by Tension Loading (Metal-to-Metal).

ASTM D 1062-79 (1983) Test Method for Cleavage Strength of Metal-to-Metal Adhesive Bonds.

ASTM D 1144-89 Practice for Determining Strength Development of Adhesive Bonds.

ASTM D 1184-69 (1986) Test Method for Flexural Strength of Adhesive Bonded Laminated Assemblies.

ASTM D 1344-7811 Method of Testing Cross-Lap Specimens for Tensile Properties of Adhesives (Discontinued 1987).

ASTM D 1780-72 (1983) Practice for Conducting Creep Tests of Metal-to-Metal Adhesives.

ASTM D 1781-76 (1986) Method for Climbing Drum Peel Test for Adhesives.

ASTM D 1876-72 (1983) Test Method for Peel Resistance of Adhesives (T-Peel Test).

ASTM D 1995-91 Test Method for Multi-Modal Strength of Autoadhesives (Contact Adhesives).

ASTM D 2095-72 (1983) Test Method for Tensile Strength of Adhesives by Means of Bar and Rod Specimens.

ASTM D 2293-69 (1980) Test Method for Creep Properties of Adhesives in Shear by Compression Loading (Metal-to-Metal).

ASTM D 2294-69 (1980) Test Method for Creep Properties of Adhesives in Shear by Tension Loading (Metal-to-Metal).

ASTM D 2295-72 (1983) Test Method for Strength Properties of Adhesives in Shear by Tension Loading at Elevated Temperatures (Metal-to-Metal).

ASTM D 2557-72 (1983) Test Method for Strength Properties of Adhesives in Shear by Tension Loading in the Temperature Range from -267.8 to -55°C (-450 to -67°F).

ASTM D 2918-71 (1987) Practice for Determining Durability of Adhesive Joints Stressed in Peel.

ASTM D 2919-90 Test Method for Determining Durability of Adhesive Joints Stressed in Shear by Tension Loading.

ASTM D 2979-88 Test Method for Pressure-Sensitive Tack of Adhesives Using an Inverted Probe Machine.

ASTM D 3111-88 Test Method for Flexibility Determination of Hot-Melt Adhesives by Mandrel Bend Test Method.

ASTM D 3121-89 Test Method for Tack of Pressure-Sensitive Adhesives by Rolling Ball.

ASTM D 3163-73 (1984) Test Method for Determining the Strength of Adhesively Bonded Rigid Plastic Lap-Shear Joints in Shear by Tension Loading.

ASTM D 3164-73 (1984) Test Method for Determining the Strength of Adhesively Bonded Plastic Lap-Shear Sandwich Joints in Shear by Tension Loading.

ASTM D 3165-73 (1979) Test Method for Strength Properties of Adhesives in Shear by Tension Loading of Laminated Assemblies.

ASTM D 3166-73 (1979) Test Method for Fatigue Properties of Adhesives in Shear by Tension Loading (Metal/Metal).

ASTM D 3167-76 (1986) Test Method for Floating Roller Peel Resistance of Adhesives.

ASTM D 3433-75 (1985) Practice for Fracture Strength in Cleavage of Adhesives in Bonded Joints.

ASTM D 3528-76 (1981) Test Method for Strength Properties of Double Lap Shear Adhesive Joints by Tension Loading.

ASTM D 3658-90 Test Method for Determining the Torque Strength of Ultraviolet (UV) Light-Cured Glass/Metal Adhesive Joints.

ASTM D 3762-79 (1988) Test Method for Adhesive-Bonded Surface Durability of Aluminium (Wedge Test).

ASTM D 3807-79 (1984) Test Method for Strength Properties of Adhesives in Cleavage Peel by Tension Loading (Engineering Plastics-to-Engineering Plastics).

ASTM D 3808-79 (1984) Practice for Qualitative Determination of Adhesion of Adhesives to Substrates by Spot Adhesion Test Method.

ASTM D 3929-80 (1984) Practice for Evaluating the Stress Cracking of Plastics by Adhesives Using the Bent-Beam Method.

ASTM D 3931-90 Test Method for Determining Strength of Gap-Filling Adhesive Bonds in Shear by Compression Loading.

ASTM D 3983-81 (1991) Test Method for Measuring Strength and Shear Modulus of Non-rigid Adhesives by the Thick Adherend Tensile Lap Specimen.

ASTM D 4027-81 (1991) Test Method for Measuring Shear Properties of Structural Adhesives by the Modified-Rail Test.

ASTM D 4338-90 Test Method for Flexibility Determination of Supported Adhesive Films by Mandrel Bend Test Method.

ASTM D 4501-85 Test Method for Shear Strength of Adhesive Bonds Between Rigid Substrates by the Block-Shear Method.

ASTM D 4562-90 Test Method for Shear Strength of Adhesives Using Pin-and-Collar Specimen.

ASTM D 4680-87 Test Method for Creep and Time to Failure of Adhesives in Static Shear by Compression Loading (Wood-to-Wood).

ASTM D 4896-89 Guide for Use of Adhesive-Bonded Single Lap-Joint Specimen Test Results.

ASTM D 5041-90 Test Method for Fracture Strength in Cleavage of Adhesives in Bonded Joints.

ASTM E 229-70 (1981) Test Method for Shear Strength and Shear Modulus of Structural Adhesives.

BS EN 582:1994 Thermal spraying. Determination of tensile adhesive strength.

BS EN 2243 Test methods for structural adhesives:

BS EN 2243 Part 2:1991 Peel metal-metal.

BS EN 2243 Part 3:1992 Test methods. Peeling test metal-honeycomb core.

BS 5350 Methods of test for adhesives:

BS 5350 Part 0:1987 General introduction.

BS 5350 Part C1:1986 Determination of cleavage strength of adhesive bonds.

BS 5350 Part C4:1986 Determination of impact resistance of adhesive bonds (Warning: withdrawn document).

BS 5350 Part C5:1990 Determination of bond strength in longitudinal shear.

BS 5350 Part C6:1990 Determination of bond strength in direct tension in sandwich panel.

BS 5350 Part C7:1990 Determination of creep and resistance to sustained application of force

BS 5350 Part C9:1990 Floating roller peel test.

BS 5350 Part C12:1994; ISO 11339:1993 180° peel test for flexible-to-flexible bonded assemblies (Tepee test).

BS 5350 Part C13:1990 Climbing drum peel test.

BS 5350 Part C14:1979 90° peel test for a rigid-to-rigid assembly.

BS 5350 Part C15:1990 Determination of bond strength in compressive shear.

BS 5350 Part F1:1983 Performance tests for flooring adhesives.

BS 5350 Part G1:1987 Determination of torque strength of anaerobic adhesives on threaded fasteners.

BS 5350 Part G2:1987 Determination of static shear strength of anaerobic adhesives.

BS 5350 Part H1:1981 Determination of heat stability of hot melt adhesives in the application equipment.

BS 5350 Part H2:1982 Determination of low temperature flexibility or cold crack temperature.

BS EN 26922:1993; ISO 6922:1987 Adhesives. Determination of tensile strength of butt joints.

BS EN 28510 Adhesives. Peel test for a flexible-bonded to-rigid test specimen assembly:

BS EN 28510 Part 1:1993; ISO 8510-1:1990 90° Peel.

BS EN 28510 Part 2:1993; ISO 8510-2:1990 180° Peel.

BS EN 29653:1994; ISO 9653:1991 Adhesives. Test method for shear impact strength of adhesive bonds

DIN 53282 Testing of adhesives for metals and adhesively bonded metal joints; T-peel test.

DIN 53283 Testing of adhesives for metals and adhesively bonded metal joints; determination of the shear strength of single lap joints in shear by tension loading.

DIN 53286 Testing of adhesives for metals and adhesively bonded metal joints; conditions for testing at different temperatures.

DIN 53289 Testing of adhesives for metals and adhesively bonded metal joints; floating roller peel test.

E DIN 53452 Testing of plastics; bending test.

E DIN 53454 Testing of plastics; compression test.

E DIN 53455 Testing of plastics; tensile test.

DIN 53456 Testing of plastics; indentation hardness test.

DIN 53457 Testing of plastics; determination of the elastic modulus by tensile, compression and bend testing.

DIN 53504 Testing of vulcanised rubber; determination of tensile strength at break, tensile strength, elongation at break and stress values in a tensile test.

DIN 53505 Testing of rubber, elastomers and plastics; Shore hardness testing A and D.

DIN 53515 Testing of rubber and elastomer and plastic foils; Tear propagation test using the Graves angle specimen with incision.

DIN 54452 Testing of adhesives for metals and adhesively bonded metal joints; compression shear test.

DIN 54455 Testing of adhesives for metals and adhesively bonded metal joints; torsional shear test.

EN 582:1994 Thermal spraying. Determination of tensile adhesive strength.

EN 1464:1995 Adhesives. Determination of peel resistance of high-strength adhesive bonds. Floating roller method. (Norme Française T76-112).

EN 1465:1995 Adhesives. Determination of tensile lap-shear strength of rigid-to-rigid bonded assemblies. (Norme Française T76-107).

EN 2243 Aerospace Series - Structural Adhesives - Test Methods.

EN 2243 Part 2:1991 Peel metal-metal.

EN 2243 Part 4:1991 Metal-honeycomb core flatwise tensile test.

EN 26922:1993; ISO 6922:1987 Adhesives. Determination of tensile strength of butt joints.

EN 28510 Adhesives. Peel test for a flexible-bonded to-rigid test specimen assembly:
Part 1:1993; ISO 8510-1:1990 90° Peel.
Part 2:1993; ISO 8510-2:1990 180° Peel.

EN 29653:1994; ISO 9653:1991 Adhesives. Test method for shear impact strength of adhesive bonds.

ISO 4578:1990 Ed.2 Adhesives - Determination of peel resistance of high-strength adhesive bonds - Floating roller method.

ISO 4587:1995 Ed.2 Adhesives - Determination of tensile lap-shear strength of rigid-to-rigid bonded assemblies

ISO 6922:1987 Ed.1; BS EN 26922:1993 Adhesives - Determination of tensile strength of butt joints.

ISO 8510:1990 Ed.1; BS EN 28510-1 & -2:1993 Adhesives - Peel test for a flexible-bonded-to-rigid test specimen assembly.
Part 1: 90 degree peel.
Part 2: 180 degree peel.

ISO 9653:1991 Ed.1; BS EN 29653:1994 Adhesives - Test method for shear impact strength of adhesive bonds.

ISO 9664:1993 Ed.1 Adhesives - Test methods for fatigue properties of structural adhesives in tensile shear (Norme Française T76-111).

ISO 10123:1990 Ed.1 Adhesives - Determination of shear strength of anaerobic adhesives using pin-and-collar specimens.

ISO 10354:1992 Ed.1 Adhesives - Characterisation of durability of structural-adhesive-bonded assemblies - Wedge rupture test.

ISO 10964:1993 Ed.1 Adhesives - Determination of torque strength of anaerobic adhesives on threaded fasteners.

ISO 11003:1993 Ed.1 Adhesives - Determination of shear behaviour of structural bonds.
Part 1: Torsion test method using butt-bonded hollow cylinders.
Part 2: Thick-adherend tensile-test method.

ISO 11339:1993 Ed.1; BS 5350:Part C12:1994 Adhesives - 180 degree peel test for flexible-to-flexible bonded assemblies (T-peel test).

ISO 11343:1993 Ed.1 Adhesives - Determination of dynamic resistance to cleavage of high strength adhesive bonds under impact conditions - Wedge impact method.

ISO 13445:1995 Ed.1 Adhesives - Determination of shear strength of adhesive bonds between rigid substrates by the block shear method.

NF T76-107, NF EN 1465:1995 Adhesives. Determination of tensile lap-shear strength of rigid-to-rigid bonded assemblies. (European Standard EN 1465).

NF T76-111, NF EN ISO 9664:1995 Adhesives. Test methods for fatigue properties of structural adhesives in tensile shear. (European Standard EN ISO 9664).

NF T76-112, NF EN 1464: 1995 Adhesives. Determination of peel resistance of high-strength adhesive bonds. Floating roller method. (European Standard EN 1464).

T76-114:1983 Structural adhesives. Cleavage test using wedge.

NF T76-121:1984 Structural adhesives. Determination of shear strength bonds axis-rings system.

NF T76-122:1985 Adhesives. Characterisation of structural adhesives. Performance test under permanent shear stress.

NF T76-124:1987 Structural adhesives. Specification. Determination of functional torques on bonded threaded components.

NF T76-126:1987 Adhesives for flexible floor coverings. Peel resistance test at an angle of 90 degrees. (Floating Roller method).

NF T76-141:1988 Structural adhesives. Shear test for determining the stress/strain curve of an adhesive in a glued assembly. Test method.

Environmental Properties

ASTM D 896-90 Test Method for Resistance of Adhesive Bonds to Chemical Reagents.

ASTM D 903-49 (1983) Test Method for Peel or Stripping Strength of Adhesive Bonds.

ASTM D 904-57 (1981) Practice for Exposure of Adhesive Specimens to Artificial (Carbon-Arc Type) and Natural Light.

ASTM D 1151-90 Test Method for Effect of Moisture and Temperature on Adhesive Bonds.

ASTM D 1183-70 (1987) Test Methods for Resistance of Adhesives to Cyclic Laboratory Ageing Conditions.

ASTM D 1382-64 (1987) Test Method for Susceptibility of Dry Adhesive Films to Attack by Roaches.

ASTM D 1383-64 (1987) Test Method for Susceptibility of Dry Adhesive Films to Attack by Laboratory Rats.

ASTM D 1828-70 (1987) Practice for Atmospheric Exposure of Adhesive-Bonded Joints and Structures.

ASTM D 1879-70 (1987) Practice for Exposure of Adhesive Specimens to High-Energy Radiation.

ASTM D 1916-88 Test Method for Penetration of Adhesives.

ASTM D 3310-90 Test Method for Determining Corrosivity of Adhesive Materials.

ASTM D 3482-90 Test Method for Determining Electrolytic Corrosion of Copper by Adhesives.

ASTM D 3632-77 (1990) Practice for Accelerated Ageing of Adhesive Joints by the Oxygen-Pressure Method.

ASTM D 4299-83 Test Methods for Effect of Bacterial Contamination on Permanence of Adhesive Preparations and Adhesive Films (Discontinued 1990. Replaced by D 4300 and D 4783.).

ASTM D 4300-88 Test Methods for Ability of Adhesive Films to Support or Resist the Growth of Fungi.

ASTM D 4502-85 (1990) Test Method for Heat and Moisture Resistance of Wood-Adhesive Joints.

ASTM D 4783-90 Test Methods for Resistance of Adhesive Preparations in Container to Attack by Bacteria, Yeast. and Fungi.

BS EN 2243 Part 5:1992 Ageing tests.

BS EN 29142:1993; ISO 9142:1990 Adhesives. Guide to the selection of standard laboratory ageing conditions for testing bonded joints.

DIN 50017 Atmospheres and their technical application; Condensation water test atmospheres.

EN 2243 Part 5:1992 Ageing tests.

EN 29142:1993; ISO 9142:1990 Adhesives. Guide to the selection of standard laboratory ageing conditions for testing bonded joints.

ISO 9142:1990 Ed.1; BS EN 29142:1993 Adhesives - Guide to the selection of standard laboratory ageing conditions for testing bonded joints.

Miscellaneous

ASTM D 816-82 (1988) Test Methods for Testing Rubber Cements.

ASTM D 1146-88 Test Method for Blocking Point of Potentially Adhesive Layers.

ASTM D 1581-60 (1984) Test Method for Bonding Permanency of Water- or Solvent-Soluble Liquid Adhesives for Labelling Glass Bottles (Intent to Withdraw).

ASTM D 1584-60 (1984) Test Method for Water Absorptiveness of Paper Labels (Intent to Withdraw).

ASTM D 1713-65 (1986) Test Method for Bonding Permanency of Water- or Solvent-Soluble Liquid Adhesives for Automatic Machine Sealing Top Flaps of Fiberboard Specimens (Intent to Withdraw).

ASTM D 1714-65 (1983) Test Method for Water Absorptiveness of Fiberboard Specimens for Adhesives (Intent to Withdraw).

ASTM D 2094-69 (1980) Practice for Preparation of Bar and Rod Specimens for Adhesion Tests.

ASTM D 3932-80 Practice for the Control of the Application of Structural Fasteners When Attached by Hot Melt Adhesives.

ASTM D 5113-90 Test Method for Determining Adhesive Attack on Rigid Cellular Polystyrene Foam.

BS 647:1993; ISO 9665:1993 Adhesives. Methods for sampling and testing animal glues.

BS 4584:Part 15:1978 Specification for metal-clad base materials for printed circuits. Adhesive coated polymeric films PETP-F-15 and PI-F-15.

BS 4841:Part 3:1994 Rigid polyurethane (PUR) and polyisocyanurate (PIR) foam for building applications. Specification for two types of laminated board (roofboards) with auto-adhesively bonded reinforcing facings for use as roofboard thermal insulation for built-up roofs.

BS 5350 Part A2:1990 Guide to the selection of adherend materials.

BS 5350 Part D4:1976 Determination of staining potential.

BS 5350 Part E1:1976 Guide to statistical analysis.

BS 5350 Part E2:1979 Guide to sampling.

DIN 933 M1.6 to M52 hexagon head screws threaded up to the head; product grades A and B. **Now: ISO 4017**.

DIN 934 Hexagon nuts with metric coarse and fine pitch thread; product grades A and B. **Now: ISO 4032**.

DIN 68131 Castors for revolving chairs and seats.

ISO 4017:1988 M1.6 to M52 hexagon head screws threaded up to the head; product grades A and B. **Was: DIN 933**

ISO 4032:1986 Hexagon nuts with metric coarse and fine pitch thread; product grades A and B. **Was: DIN 934**.

ISO 9665:1993 Ed.1; BS 647:1993 Adhesives - Animal glues - Methods for sampling and testing.

ISO 10365.1992 Ed.1 Adhesives - Designation of main failure patterns (Norme Française T76-130).

F01-811:1989 Railway rolling stock. Jointing sticking.

NF T76-104:1983 Adhesives. Methods of sampling adhesives and raw materials used for adhesives.

NF T76-130, NF EN ISO 10365:1995 Adhesives. Designation of main failure patterns. (European Standard EN ISO 10365).

Pressure sensitive

3J 12:1991 Specification for pressure-sensitive adhesive identification tape.

4J 10:1991 Specification for pressure-sensitive adhesive water resistant PVC tape.

4J 11:1991 Specification for pressure-sensitive adhesive paper masking tape.

BS 5609:1986 Specification for printed pressure-sensitive, adhesive-coated labels for marine use, including requirements for label base material.

BS 7116:1990 Specification for double sided pressure sensitive adhesive tapes.

EN 60454 Specifications for Pressure Sensitive Adhesive Tapes for Electrical Purposes:
Part 1 General Requirements.
Part 2 Methods of Test.

SOURCES

The information in this appendix has been compiled from that obtained from:

Aerospheres (UK), London. UK.

AFNOR Catalogue. Paris, France.
October 1995

ASTM
Annual Book of Standards - part 15.06 Adhesives.

British Standards Institute (BSI), Milton Keynes, UK.
July 1994.

DIN Standards Handbook.

ESA Publications Division
PO Box 299, 2200AG Noordwijk, NL.

ISO, Switzerland.
October 1995.

NASA Scientific and Technical Publications.
Sources for NASA Publications:
US Government Printing Office
Washington DC-20402 - USA
Tel: +1 202 783 3238

NTIS - National Technical Information Service
5285 Port Royal Road, Springfield
VA-22161 - USA
Tel: +1 703 487 4780 - Information

Also:
NASA documents - Public collections.
Federal Depository Library Program.
American Institute of Aeronautics and Astronautics.
British Lending Library.
ESA - European Space Agency.

Skeist, I. (Editor)
Handbook of Adhesives
Chapman & Hall, London (3rd Edition, 1990)
ISBN 0-442-28013-0

For details of Standards Organisations - see 'Useful Names and Addresses' directory.

Appendix B - GLOSSARY

Addition Polymerisation - chemical reaction to produce a polymer by the linking of monomer molecules without the production of any volatile substances. Chemical bonds are formed by the opening up of double or triple bonds in the (unsaturated) monomer.

Adherend - a material to be adhesively bonded.

Areal Weight - the weight per unit area of a sheet material (usually quoted for 1 square metre or 1 square foot).

Autoclave - a closed vessel for conducting adhesive bonding at increased temperature and pressure. Often used when co-curing adhesives and high-performance, fibre-reinforced composites.

Bondline Control - control of the thickness of the adhesive layer in a joint. Often by the inclusion in the adhesive of small glass spheres of a certain diameter, or a thin textile carrier material of known thickness.

Carrier - a thin textile material (either woven or non-woven) on which an adhesive is placed to form a film adhesive. Film adhesives are often offered with a choice of carrier type (material, e.g. nylon or glass, depending on the service temperature; form - matt or woven). The carrier aids bondline control and acts as a 'support' for the adhesive.

Coefficient of Thermal Expansion (CTE) - The change in length of a material, per unit length, produced by a 1 degree rise or fall in temperature. As this is not linear with temperature it is normally quoted as an average for a particular temperature range.

Cold Storage - Often a requirement for adhesives to meet a stated shelf life. Refrigerated (0-4°C) or freezer (-18°C) are common. Adhesives from cold-storage should be warmed gently to RT before opening the containers and using. Some adhesives are sensitive to frost, and should not be stored below ambient (RT) to avoid degradation. Manufacturers state storage requirements.

Cross-Linking - Chemical linking between adjacent polymer molecular chains. The creation of these bonds lead to the hardening or solidification of the polymer adhesive.

Co-cure - a manufacturing technique for producing adhesive bonds between composite adherends at the same time as consolidating and curing the composite, i.e. same process cycle for composite and adhesive.

Condensation (Elimination) Polymerisation - chemical reaction to produce a polymer by the linking of monomer molecules with the liberation of small volatile molecules. In many of these reactions, the substance given off is water vapour - hence the general term condensation.

Copolymer - a plastic formed by polymerising together two or more substances (monomers).

Cure - the chemical reaction process in which an adhesive obtains its final mechanical and physical properties.

Elongation - the permanent (plastic) deformation occurring during a tensile test. Expressed as a percentage of the original length of the test material.

EVA - Ethylenevinylacetate.

Exothermic - Denoting a chemical reaction that releases heat into its surroundings. Often occurs during cure of two-part adhesives.

FRP - Fibre reinforced plastic; composite materials.

Glass Transition Temperature (Tg) - the temperature at which an adhesive (or other material) undergoes a (normally) reversible change in behaviour. Thus called because materials tend to show a marked change in ductility at this temperature; below the Tg becoming glassy in their mechanical response. Other mechanical and physical characteristics may also be affected. Some environmental conditions, such as humidity, may alter the Tg of an adhesive.

Handling Strength - the time required for an adhesive to reach a certain state of cure (or set) such that the assembly can be moved without risk of damaging the joint. The adhesive is rarely fully cured at this point. Also an indication of production rate (press machines, quantity of jigs and fixtures, required, etc.).

Hardness - quoted for fully cured adhesives or polymers often using the Shore method (Hs). Two scales are commonly used for adhesives. Shore A for relatively soft (rubbery) materials, Shore D for relatively hard materials. The higher the number, the harder the material. Values are normally quoted at RT, although hardness at other temperatures may be stated.

Homopolymer - polymer formed from a single type of monomer molecule. cf. Copolymer.

Hot-Melt - see Thermoplastic.

Hygroscopic - Tending to absorb atmospheric moisture.

Lap Shear Strength (LSS) - A form of adhesive bond strength measured by pulling apart, in tension, two overlapped, bonded strips of material. One of the most commonly quoted mechanical properties of adhesives. Unfortunately the precise details of the adherends and conditions used for the test also have a great effect on the results obtained.

Melt Flow Index (MFI) - a measure of the viscosity of a molten thermoplastic (hot-melt adhesive). Given as the time taken for a quantity of the material to flow through a standard-sized orifice.

Mix Ratio - the proportion of a hardener or catalyst to be mixed with the resin part. These may be quoted by weight or by volume. (See also: No-mix).

Modified Adhesives - The inclusion of secondary compounds in an adhesive mix in order to promote certain of the chemical, physical or mechanical properties. e.g. the addition of rubbery compounds to epoxy resins to increase the flexibility and peel strength of the final bond.

Modulus - A material constant defined as the ratio between the applied stress and any resulting elastic deformation (reversible on removing the stress). There are several commonly used variants - depending on the loading method and the direction in which the deformation is measured relative to the direction of the applied stress. Young's modulus is that where both are co-axial.

No-mix - a two part adhesive in which the cure is initiated by applying the activator to one of the adherends. Avoids the need for measuring and mixing equipment, or errors resulting from poor measurements or mixing.

Nondestructive Testing (NDT, NDE, NDI, NDC) - Techniques used to examine the quality of adhesive bonds without causing damage to the item or bond. The terms NDT (Testing), NDE (Evaluation), NDI (inspection) and NDC (Characterisation) tend to be interchangeable. Nondestructive inspection systems may be manual or automated to some extent. Whilst effective for finding defects in the bond, no method currently exists which allows a reliable, quantitative measure of bond strength.

One-Part (One Pack) Adhesive - supplied as a single component, ready to use. These adhesives usually begin to cure with increasing temperature, irradiation or exposure to atmospheric moisture, and hence require special storage conditions.

Open Time - the period of time available for applying an adhesive and assembling the parts before being cured (heated and pressed, if required).

Outgassing/Offgassing - The giving-off of gases or vapours during cure or operation. Of particular interest for applications involving exposure to vacuum. Some adhesives are especially formulated to have a low volatile content for such uses; often denoted as 'Space Grades' to indicate their suitability for use in space vehicles. Depending on the test method, measurement of volatile substances is normally expressed as a percentage of Collectable Volatile Condensed Materials (CVCM or CVM) and Total Mass (Weight) Loss (TML, TWL).

Overlap - A simple adhesive joint, where the edge of one sheet overlaps with, and is bonded to, the edge of another.

PEEK - Polyetheretherketone thermoplastic.

Peel Strength - A form of bond strength, obtained by peeling apart the two adherends. There are a number of standard arrangements for carrying out testing of this quantity: T-Peel, Floating Roller Peel, Bell Peel, Honeycomb/Sandwich Peel, Climbing Drum Peel, etc. Unfortunately it is not possible to compare directly between the results from these various methods.

PES - Polyethersulfone thermoplastic.

PET - Polyethyleneterephthalate.

Phenolic Resin - Thermosetting resin with elevated temperature stability, e.g. phenol formaldehyde. May be blended with other polymers, or used as an adhesive. Water-proof versions used for wood bonding.

PI - Polyimide.

PMMA - Polymethylmethacrylate.

Polyamide - A polymer in which the structural units are linked by amide or thioamide groups; many polyamides are fibre-forming, e.g. Nylon. Thermoplastic hot-melt adhesives are often based on copolymers.

Polyester - Available as a thermoplastic or thermosetting resin. Thermosetting - used for the matrix phase of composite materials. Thermoplastic - base polymer for hot-melt adhesives; often a copolymer.

Polyimide - A polymer produced by reacting an aromatic dianhydride with an aromatic diamine; these are highly heat-resistant resins (> 315°C).

Polymer - A high molecular weight organic compound, natural or synthetic, whose structure can be represented by a repeated small unit (mer), e.g. polyethylene. Synthetic polymers are formed by addition or condensation polymerisation of monomers.

Polymerisation - The chemical reaction in which monomer molecules are linked together to form long-chain molecules (polymers).

Pot Life - Usually quoted for liquid or paste adhesives. The period after mixing during which an adhesive can be used. Manufacturer's quoted pot lives usually state the quantity of adhesive for which the time is valid because most two-part adhesives give out heat as they cure (exothermic). In larger quantities this heat builds up, greatly accelerating the cure - or, in extreme cases, causing chemical breakdown or burning of the mixture.

Primer - a substance, normally a liquid, which is applied to the adherends after a surface preparation treatment and before adhesive bonding. The primer protects the prepared surface from the environment and may also promote adhesion. Where advisable, appropriate primers are recommended by the adhesive manufacturers.

PS - Polysulfone; high temperature resistant thermoplastic.

PTFE - Polytetrafluoroethylene.

PU or PUR - Polyurethane.

PVA - Polyvinylacetate/acetal.

PVB - Polyvinylbutyral.

PVC - Polyvinylchloride.

PVdF - Polyvinylidene difluoride.

Reticulating Adhesive - Usually applied to unsupported film adhesives (i.e. with no textile carrier) where, during cure at higher temperatures, surface tension pulls the softened adhesive back to form fillets at corners. This property is useful when bonding face skins to honeycomb core materials.

Service Temperature - the maximum temperature at which the cured adhesive can be used continuously. At higher temperatures, adhesives progressively lose strength. For this reason test specifications normally define this as the temperature at which a certain level of Lap Shear Strength (LSS) is maintained, usually (and for the purposes of this book) this strength figure is 6.9 MPa (1000 psi). Service temperatures may also be quoted as a range. Most adhesives have a lower service temperature below 0°C. For joints in very cold applications (below -55°C) the choice of adhesive is very much reduced.

Set - see also Cure. Usually applied to thermoplastic or hot-melt adhesives. Often quoted as a time required for the adhesive to reach a handleable solid state.

SG - Specific gravity (Relative density). The density of a substance relative to that of water.

Shelf-Life - the period for which an adhesive may be stored before use. Manufactures provide guidance on the appropriate storage conditions. Some adhesives (notably one-part thermosetting adhesives) require refrigerated storage to prevent them from degrading.

SMC - Sheet moulding compound. A polymer material, often reinforced, widely used in the automotive industry.

Solid Content - the proportion (by weight) of the applied adhesive which remains after curing. If this is less than 100% then some component of the adhesive is lost (usually as a vapour) during curing.

Support - see Carrier.

Surface Preparation - a sequence of treatments during which the surfaces of the adherends are made suitable for bonding. Depending on the adherends and adhesives, this may range from simple cleaning or degreasing to abrasion or complex chemical treatments.

Syntactic - (highly ordered) For adhesives or materials this refers to foams made up of a hollow filler (often glass microspheres) with a resin binder. Syntactic materials are generally light and strong with a very regular structure.

Tack or Tack Time - Stickiness of a thermosetting film adhesive applied to an adherend. An important aid for placement and assembly. After a certain time, such adhesives begin to cure and loose their tack. Thermoplastic polymers are often not tacky, so the formulation is modified to include a sticky agent. Some adhesives quote a time to reach a tacky stage after they have been applied.

Tensile Strength (TS) - the maximum stress obtained (load per unit area) when pulling apart a material in pure tension.

Thermoplastic Adhesives - Thermoplastic polymers soften reversibly at increased temperatures, regaining their structure and properties on cooling. Adhesives based on these materials are normally supplied in a solid form (pellets, granules, sheets or rods) and require melting during the application and/or bonding process. The application temperature is often selected such that the material has a suitable viscosity to allow some flow in the joint during assembly. In some types of adhesives, this heating may also initiate a chemical change (melt-applied or reactive hot-melt adhesives). In general, thermoplastic, or hot-melt, adhesives are chemically stable under ambient conditions and do not require special storage conditions (cf. thermosetting).

Thermosetting Adhesive - Thermosetting polymers form a permanently cross-linked structure. They do not soften reversibly at increased temperatures, but tend to oxidise or break down chemically. Adhesives based on these materials require either moderate heating or catalysis using chemicals or radiation to start the curing (cross-linking) process. The heating required varies with the formulation of the adhesive. For example, many adhesives cure at room temperature (RT) whereas others cure unacceptably slowly without heating. In general, the higher the intended service temperature, the higher the cure temperature.

Thixotropic - adhesives whose viscosity becomes lower as they are disturbed (i.e. mixed) but thicken at rest to avoid slump or flow out of the joint region.

Tg - See: Glass Transition Temperature.

Tm - Melting temperature.

TM - Tensile or Young's modulus; also *E*.

Two-Part (Two Pack) Adhesive - two liquids/pastes, requiring mixing before use. Usually an adhesive (resin) and hardener (catalyst). Some two-parts are applied separately to the adherends, see **No-mix**.

Viscosity - A measure of the resistance of a fluid to flow when it is subjected to shear stress. Adhesives are available with a great range of viscosities. Some formulations include thickening agents to aid handling by reducing flow, sag or slump. Others have a very low viscosity to allow easy flow into crevices. Many formulations are thixotropic.

Working Life - Usually quoted for adhesives in a sheet or film form which begin to cure with increasing temperature. This is the equivalent of the pot-life of two-part adhesives.

Appendix C - MULTILINGUAL VOCABULARY

English	French	German	Italian	Spanish
1 Part	1 (mono)composant	Einkomponentige		
2 Part	2 (bi)composant	Zweikomponentige		
Abrasive	Abrasif	Schleifmittel	Abrasivo	Abrasivo
Accelerator	Durcisseur	Komponenten Harz		
Acid	Acide	Säure	Acido	Ácido
Acrylic	Acrylique	Acryl		
Adhere	Adhérer	Ankleben	Aderire	Adherirse
Adherend	Support	Untergründen, Klebflächen		
Adhesive	Colle, Adhésif	Klebstoffe	Adesivo, Colla	Adhesivo, Cola
Agent	Agent	Agent, Vertreter	Agente	Agente
Aligned	Aligné	Ausgerichtet	Allineato	Alineato
Alkali	Alcali	Alkali	Alcali	Álcali
Alloy	Alliage	Legierung	Lega	Aleación
Application	Mise en œuvre	Anwendung	Applicazione	Aplicación
Area	Superficie	Fläche	Superficie	Superficie
Areal Weight	Poids superficiel	Flachengewicht	Peso superficiale	
Asbestos	Amiante, Asbeste	Asbest		Amianto, Asbesto
Automatic	Automatique	Automatisch	Automatico	Automático
Beige	Beige	Beige	Beige	Crema, beis
Black	Noir(e)	Schwarz	Nero	Negro
Blue	Bleu(e)	Blau	Blu	Azul
Body-work (Automotive)	Carrosserie	Karosserie	Carrozzeria	Carrocería
Bond	Coller	Kleben	Incollare	Pegar
Bonded	Collage, Collé	Kleben, Verbinden		
Brown	Brun(e)	Braun	Bruno(a)	Pardo(a)
Cartridge	Cartouche	Kartuschen	Cartuccia	Cartucho
Ceramic	Céramique	Keramik	Ceramica	Cerámica
Chipboard	Agglomères, Panneau particules	Holzwollenplatten		
Chopped	Coupe	Zerhackt	Fibra tagliata	Cortar
Cleaning	Nettoyage	Saubermachen	Pulire	Limpiar
Cold	Froid	Kalt	Freddo	Frio
Colour (Color)	Couleur	Farbe	Colore	Color
Combination	Combinaison	Verbindung	Combinazione	Combinación
Composition	Composition	Zusammensetzung	Composizione	Composición
Compressive	Compressif	Drucke	Compressione	Compresión
Concrete	Béton	Beton	Calcestruzzo	Hormigón
Conductivity	Conductivité			
Contact Adhesive	Colle contact			
Contamination	Contamination	Ansteckung	Contaminazione	Contaminación

English	French	German	Italian	Spanish
Contents	Contenu	Inhalt	Contenuto	Contenido
Contraction	Contraction	Zusammenzeihung	Contrazione	Contracción
Copolymer	Copolymère			
Cork	Liège	Kork	Sughero	Corcho
Coverage	Surface couverte, Consommation			
Creep	Fluage	Kriechen	Scorrimento	
Cross Section	Coupe Transversale	Quarschnitt	Sezione Traversals	Sección Transversal
Cure	Polymérisation, Cuisson	Aushärten, Klebfuge	Cottura	
Cure Rate	Récticulation	Aushärten		
Cut	Couper	Schnitt	Tagliare	Cortar
Data Sheet	Fiche Technique	Merkblatt		
Density	Densité	Dichte	Densità	Densidad
Diameter	Diamètre	Durchmesser	Diametro	Diámetro
Direction	Direction	Richtung	Direzione	Dirección
Disperse	Disperser	Zerstreuen	Disperdere	Dispersar
Dispersion	Dispersion	Dispersion		Dispersión
Distribution	Distribution	Verteilung	Distribuzione	Distribución
Drop	Goutte	Tropfen	Goccia	Gota
Dry	Sèche, Sec	Trocken	Secco	Seco
Edge	Bord	Kante	Bordo	Borde
Electrical	Electrique	Elektrisch	Elettrico	Eléctrico
Elongation [to break]	Etirage, Allongement à la rupture	Reissdahnung, Bruchdehnung	Allungamento	
Emulsion	Émulsion	Emulsion	Emulsione	Emulsión
Etch	Décaper, Traité	Beiz		
Expansion	Dilatation	Ausdehnung	Espansione	Expansión
Fabric	Tissu	Gewebe	Tessuto	Tejido
Fatigue	Fatigue	Ermüdung	Fatica	Fatiga
Filler [Additive]	Matière de remplissage [Chargé]	[Füllstoffe]		
Film	Pellicule	Film	Strato sottile	Pelicula
Flash point	Point éclair	Flammpunkt		
Flexible	Souple	Flexibel	Flessibile	Flexible
Flexural	Flexure	Biege	Flessione	
Flow	Couler, Fluage	Fliessen	Scorrere	Correr
Foam	Écume	Schaum	Schiuma	Espuma
Full Cure	Prise définitif			
Gap	Trou, Vide	Lücke	Breccia	Vacio
Glass	Verre	Glas	Vetro	Vidrio
Green	Vert(e)	Grün	Verde	Verde
Grey	Gris(e)	Grau	Grigio(a)	Gris
Hard	Dur	Hart	Duro	Duro
Hardener	Durcisseur	Härter		
Hardness	Dureté	Härte	Durezza	Dureza
Honeycomb	Nid d'abeille	Waben	Favo	Panal
Hot	Chaud	Heiß	Caldo	Caliente
Hot-Melt	Thermofusible	Schmelzklebstoffe		
Humidity	Humidité	Feuchtigkeit	Umidità	Humedad
Impact	Impact, Choc	Aufprall, Stoss	Urto	Impacto
Initial Cure	Prise rapide, Prise superficielle			
Instructions (for use)	Mode d'emploi, Mise en Oeuvre	Anwendung	Istruzione	Modo de Empleo
Laminate	Stratifié	Schichstoffe, Lamellenplatten	Laminato	Laminado
Leather	Cuir	Leder	Cuoio, Pele	Cuero
Length	Longueur	Länge	Lunghezza	Largo
Liquid	Liquide	Flüssigkeit	Liquido	Liquido
Longitudinal	Longitudinal	Langsrichtung	Longitudinale	Longitudinal
Manufacturer	Fabricant	Hersteller	Fabbricante	Fabricante
Mass	Masse	Masse	Massa	Masa
Matrix	Matrice	Matrize	Matrice	Matriz
Melt	Fondre	Schmelzen	Sciogliere	Fundir
Metal	Métal	Metal	Metallo	Metal
Metallic	Métallique	Metallisc	Metallico(a)	Metálico(a)
Mix	Mélanger	Mischen	Mischiare	Mezclar
Modulus	Module	Modul	Modulo	

English	French	German	Italian	Spanish
Odour	Odeur	Geruch	Odore	Olor
Oil	Huile	Öl	Olio	Aceite
Opaque	Opaque	Undurchsichtig, Trüb	Opaco	Opaco
Packaging	Conditionnement	Leiferform, Gebinde	Imballaggio	Envoltura
Paste [Pasty]	Pâte [Pâteuse]	Paste, Masse [Pastös]	Impasto	Pasta, Masa
Peel	Écaillage, Pelage	Abblättern	Staccarsi	
Pink	Rose	Rosafarben	Rosa	Rosa
Plastic	Plastique	Plastik, Kunststoffe	Plastica	Plástico
Plywood	Contre-plaque	Sperrholz	(legno) compensato	Contrachapado
Polymerization	Polymérisation			
Porous	Poreux	Porös	Poroso	Poroso
Post-Cure	Cuisson (deuxième)			
Pot Life	Durée d'utilisation, Durée de vie	Topfzeit		
Precautions	Précautions	Vorsichtsmassnahmen	Precauzione	Precaución
Precursor	Précurseur	Ausgangsmaterial	Precursors	Precursor
Pressure	Pression	Druck	Pressione	Presión
Pretreatment	Prétraitement	Vorbehandlung	Pretrattamento	Pretratamiento
Primer	Primaire	Primer	Prima mano	Primera mano
Process	Procédé	Prozess/Verfahren	Procedimento	Procedimiento
Property	Propriété	Eigenschaft	Proprietà	Propiedad
Quality	Qualitie	Qualität	Calidad, Cualidad	Qualità
Rapid	Rapide	Schnell	Rapido	Rápido
Red	Rouge	Rot	Rosso(a)	Rojo(a)
Resin	Résine	Harz	Resina	Resina
Resist	Résiste, Tenue	Widerstehen, Beständigkeit	Resistenza	Resistencia
Resistivity	Résistivité	Spozfischer Widerstrand		
Rough	Rugueuse	Rauh	Ruvido	Áspero
Safety	Sécurité	Sicherheit	Sicurezza	Seguridad
Seal	Étanche	Abdichten	Sigillo	Sellar
Sealant	Mastic	Dichtungmasse		
Setting Time	Temps de Prise	Gelierzeit		
Shear	Cisaillement	Scher	Taglio	
Sheet	Feuille, Tôle (metal)	Blatt	Foglio	Hoja
Shelf Life	Durée de stockage	Lagerfähigkeit	Durata d'immagazzinoggio	
Shrink	Rétrécir, Retrait	Schrumpfen	Restringere	Encoger
Size	Appret	Appretur	Ensimaggio	
Skinning time	Formation de peau, Prise superficielle	Hautbildung		
Solution	Solution	Lösung	Soluzione	Solución
Solvent	Solvant	Lösend	Solvente	Solvente
Specific Gravity	Poids spécifique	Spezifisches Gewicht	Peso Specifico	
Specification	Spécification, Norme	Genau Angaben, Normen	Specificazione	Especificación
Sphere	Sphère	Sphäre	Sfera	Esfera
Splice	Epiasure	Spleissung	Intreccio	
Storage Conditions	Conservation	Lagerung		
Strength	Résistance	Stärke, Festigkeit	Resistenza, Forza	Fuerza
Supplier	Fournisseur	Lieferant	Fornitore	Abastecedor
Supported	Supporté			
Synthetic	Synthétique	Synthetisch	Sintético(a)	Sintetico
Tacky	Poisseux	Klebrig	Appiccicoso	Pegajoso
Technical data	Spécifications Technique	Technische Daten		
Temperature resistance	Résistance à la température	Temperaturbeständigkeit		
Tensile	Traction	Zug	Tensions	
Tension	Tension	Spannung	Tensione	Tensión
Test	Essais, Examen, Test	Test, Prüft	Prova	Prueba
Textile	Textile	Textil-, Web-	Tessile	Textil
Thermal	Thermique	Thermal	Termale	Termal
Thermoplastic	Thermoplastiques	Thermoplastische		
Thermoset	Thermodurcissable	Reaktionklebstoffe		
Thick	Épais	Dick	Spesso	Espeso
Thickness	Épaisseur	Dicke	Spessore	Espesor
Thin	Mince	Dünn	Sottile	Delgado
Thinners	Diluant	Verdünnungsmittel	Solvente	Solvente
Thixotropic	Thixotrope			
Transparent	Transparent	Durchsichtig	Trasparente	Transparente
Transverse	Transversal	Querrichtung	Trasversale	Transverso

English	French	German	Italian	Spanish
Unidirectional	Unidirectionnel	Unidirektionale	Unidirezionale	
Vacuum	Vide	Vakuum, Leere	Vuoto	Vacío
Vapour	Vapeur	Dunst	Vapore	Vapor
Viscosity	Viscosité	Viskosität		
Viscous	Visqueux	Zähflüssig	Vischioso	Viscoso
Volatile	Volatil	Fluchtig	Volatile	Volátil
Water Resistance	Tenue à l'eau	Wasserbestänigkeit		
Wetted	Mouiller	Naß Machen	Bagnare	Mojar
White	Blanc, Blanche	Weiß	Bianco(a)	Blanco(a)
Width	Largeur	Breite	Larghezza	Anchura
Wood	Bois	Holz	Legna	Leña
Wood products	Dérivés de bois	Holzprodukten		
Woven	Tisser	Weben	Tessere	Tejer
Yellow	Jaune	Gelb	Giallo(a)	Amarillo(a)
0 - zero, nought	zéro	null	zero	cero
1 - one	un(e)	ein	uno	uno
2 - two	deux	zwei	due	dos
3 - three	trois	drei	tre	tres
4 - four	quatre	vier	quatro	cuatro
5 - five	cinq	fünf	cinque	cinco
6 - six	six	sechs	sei	seis
7 - seven	sept	seiben	sette	siete
8 - eight	huit	acht	otto	ocho
9 - nine	neuf	neun	nove	nueve
10 - ten	dix	zehn	deici	diez
11 - eleven	onze	elf	undici	once
12 - twelve	douze	zwölf	dodici	doce
13 - thirteen	treize	dreizehn	tredici	trece
14 - fourteen	quatorze	vierzehn	quattordici	catorce
15 - fifteen	quinze	fünfzehn	quindici	quince
16 - sixteen	seize	sechzehn	sedici	dieciséis
17 - seventeen	dix-sept	seibzehn	diciassette	diecisiete
18 - eighteen	dix-huit	achtzehn	diciotto	dieciocho
19 - nineteen	dix-neuf	neunzehn	diciannove	diecinueve
20 - twenty	vingt	zwanzig	venti	veinte
21 - twenty-one	vingt-et-un	einundzwanzig	ventuno	veintiuno
30 - thirty	trente	dreißig	trenta	treinta
40 - forty	quarante	vierzig	quaranta	cuarenta
50 - fifty	cinquante	fünfzig	cinquanta	cincuenta
60 - sixty	soixante	sechzig	sessanta	sesenta
70 - seventy	soixante-dix	seibzig	settanta	setenta
80 - eighty	quatre-vingts	achtzig	ottanta	ochenta
90 - ninety	quatre-vingt-dix	neunzig	novanta	noventa
100 - hundred	cent	hundert	cento	cien
1000 - thousand	mille	tausend	mille	mil
1000000 - million	million	million	milione	un millón

Appendix D - CONVERSION FACTORS & UNITS

INTRODUCTION

The conversion factors listed here have been used throughout the book. Data originally supplied in a variety of metric and imperial units has been presented in a standardised set of units based on the *Système Internationale d'Unites* (S.I.), taking into account the differences between UK and US quantities. Multiple units and prefixes are also given.

Reference

D. R. Lide (Editor-in-Chief)
Handbook of Chemistry and Physics - A Ready-reference Book of Chemical & Physical Data
CRC Press Inc. (72nd Edition, 1991-1992)
ISBN 0-8493-0565-9

CONVERSION FACTORS

B ⇨ A Multiply by:	A	B	A ⇨ B Multiply by:
1.019716×10^{-5}	Atmosphere (At.)	Pascal (Pa, $N.m^{-2}$)	98066.5
0.980665	Bar	Atmosphere (At.)	1.019716
1×10^{-5}	Bar	Pascal (Pa, $N.m^{-2}$)	1×10^{5}
0.577790874	BTU/(hr.ft².°F)/ft	Watt/metre.°C ($W.m^{-1}.°C^{-1}$)	1.73073
6.9334664	BTU/(hr.ft².°F)/in	Watt/metre.°C ($W.m^{-1}.°C^{-1}$)	0.144228
0.0023884	Cal/sec.cm.°C	Watt/metre.°C ($W.m^{-1}.°C^{-1}$)	418.68
1000	Centipoise (cp)	Pascal second (Pa•s)	0.001
0.03531467	Cubic foot (cu.ft)	Litre (l)	28.316847
0.06102374	Cubic inch (cu.in)	Cubic centimetre (cc, cm^3, ml, millilitre)	16.387064
[B - 273.16]	Degrees Centigrade (°C)	Degrees Kelvin (K)	[A + 273.16]
[((B x 9) / 5) + 32]	Degrees Fahrenheit (°F)	Degrees centigrade (°C)	[((A - 32) x 5) / 9]
3.2808399	Foot (ft, ')	Meter (m)	0.3048
0.737561033	Foot pound (ft.lb)	Joule (J)	1.35582
0.737562	Foot pound (ft.lb)	Newton meter (Nm)	1.35582
29.037836	Foot pound per inch (ft.lb/in)	Newton metre/metre ($N.m.m^{-1}$)	0.034437828
0.3937008	Inch (in, ")	Centimetre (cm)	2.54
10.19716213	Kilogram centimetre (kg.cm)	Newton metre (N.m)	0.0980665
10.19716213	Kilogram per square centimetre (kg/cm^2)	Megapascal (MPa)	0.0980665
39.37008	Mil (thou, 0.001 inch)	Millimetre	0.0254
1	Newton per square millimetre ($N.mm^{-2}$)	Megapascal (MPa)	1
0.035273962	Ounce (oz)	Gram (g)	28.349523
0.8326742	Pint {UK}	Pint {USA Liq.}	1.200950
1.759754	Pint {UK}	Litre (l)	0.56826125
2.113376	Pint {USA Liq.}	Litre (l)	0.4731765
2.2046226	Pound (lb)	Kilogram (kg)	0.45359237
0.224809	Pound force (lb)	Newton (N)	4.44822
1.45038×10^{-4}	Pound force/square inch (psi)	Pascal (Pa, $N.m^{-2}$)	6894.76
0.06242796	Pounds per cubic foot (lb/cu.ft)	Kilograms per cubic meter ($kg.m^{-3}$)	16.01846
0.204816044	Pounds force per square foot ($lb.ft^{-2}$)	Kilogram per square metre ($kg.m^{-2}$)	4.88243
0.005710141	Pound force/linear inch (PLI, $lb.in^{-1}$)	Newton/metre ($N.m^{-1}$)	175.127
0.005710141	Pound force/linear inch (PLI, $lb.in^{-1}$)	Newton/millimetre ($N.mm^{-1}$)	0.175127
10.76391	Square foot (sq.ft)	Square meter (m^2)	0.09290304
1.550003	Square inch (sq.in)	Square centimetre (sq.cm, cm^2)	6.4516
0.145038	Thousand pound force/square inch (ksi)	Megapascal (MPa)	6.89476
0.06474881	Tons force per square inch (tsi)	Megapascal (MPa)	15.4443
1.093613298	Yard (yd)	Meter (m)	0.9144

MULTIPLE UNITS

Multiple	Prefix	Symbol	Multiple	Prefix	Symbol
10^{-1}	deci	d	10	deca	da
10^{-2}	centi	c	10^{2}	hecto	h
10^{-3}	milli	m	10^{3}	kilo	k
10^{-6}	micro	µ	10^{6}	mega	M
10^{-9}	nano	n	10^{9}	giga	G
10^{-12}	pico	p	10^{12}	tera	T
10^{-15}	femto	f	10^{15}	peta	P
10^{-18}	atto	a	10^{18}	exa	E

INDEX

INDEX OF PRODUCT CODES

The product code index lists alphabetically/numerically the product codes or names of all of the adhesives contained in the data sections of this book, showing the page number of their data sheet. The entries also give the manufacturer and type of each in the following format:

Product Code/Name Manufacturer (TYPE) Page Number

The type fields contain these abbreviations:

(ACR) Acrylic.
(ANA) Anaerobic.
(CYA) Cyanoacrylate.
(EPX) Epoxy.
(IMI) Imide - polyimide or bismaleimide.
(PHE) Phenolic.
(PUR) Polyurethane.
(SIL) Silicone.
(THM) Thermoplastic Hot Melt.
(Misc.) Miscellaneous.

For an index of suppliers and their product types, see page 415.

1

2

D

E

F

G

H

I

J

K

L

M

N

O

P

Q

R

S

T

U

V

X

Z

INDEX OF SUPPLIERS & PRODUCT TYPES

The suppliers & product types index lists alphabetically the names of all of the adhesives suppliers contained in this book, showing the page numbers of their data sheets for each type of adhesive.

3

A

B

C

D

E

F

G

H

I

K

L

M

N

O

P

Q

R

S

T

U

W

CONTRIBUTIONS FOR THE NEXT EDITION OF THE STRUCTURAL ADHESIVES DIRECTORY & DATA BOOK

To ensure that this publication remains up-to-date, we shall be revising its contents periodically. We welcome data on new products and changes to existing products or company information (address, telephone, fax, etc). Please complete this form and send it with new information. Please print clearly.

☐ **We have contributed information for this book. We wish to notify you of:**

- ☐ Change of name or address or telephone numbers.
- ☐ New products available.
- ☐ Products no longer available or replaced by new products.
- ☐ Products listed now have changed properties.
- ☐ Error in data (please specify precisely the nature of the error).

Please send ***BY POST*** *full* ***TECHNICAL DATA SHEETS*** *on the properties and characteristics of appropriate structural adhesive products*

☐ **We have not previously provided data, but now wish to contribute to the next edition of this book. We manufacture structural adhesives which can be added into the following sections of this book** (*please tick all those relevant and send this with the data sheets*):

- ☐ Acrylic
- ☐ Anaerobic
- ☐ Bismaleimide
- ☐ Cyanoacrylate
- ☐ Epoxy
- ☐ Phenolic (inc. resorcinol-formaldehyde)
- ☐ Polyimide
- ☐ Polyurethane (inc. isocyanate)
- ☐ Silicone
- ☐ Thermoplastic (hot-melt)
- ☐ Miscellaneous (please state chemical formulation)

Please do not send data on 'pressure sensitive' adhesives or generic types not listed above.

If possible, also indicate **typical applications**, e.g. car bodies, aerospace, packaging, construction industry, etc., and include any **qualifications** or **approvals** which the product has, i.e. to industry, international or government specifications.

Please send ***BY POST*** *full* ***TECHNICAL DATA SHEETS*** *on the properties and characteristics of each appropriate structural adhesive products.*

☐ **We are suppliers/distributors for (please state your company details and manufacturers' names, and send a letter-heading or business card):**

------------------------------ ------------------------------ ------------------------------

------------------------------ ------------------------------ ------------------------------

------------------------------ ------------------------------ ------------------------------

☐ **COMPANY INFORMATION:** A list of updated contact addresses (with a contact name if possible), telephone, fax, telex, E-mail numbers for all subsidiaries/offices across the world.

☐ **PERMISSION** Please indicate your agreement to reproduce any data provided in the directory & data book.

☐ **ADVERTISING**: Please indicate your interest in placing an advertisement in this book. There is no obligation.

Your name: ______________________ Position in Company: ______________________

Please send this completed page with your contributions or comments **by post** (not by fax) to:

RJ Technical Consultants
Bois Moreau
F-17770 JUICQ
Charente Maritime
FRANCE

Tel/Fax: +33 46 95 35 10

All queries should be addressed to either Bob Hussey or Jo Wilson. Sorry, we cannot accept data sent by fax as numerical information can often be difficult to read with any confidence.

Thank you for your cooperation and contributions